COMPUTATIONAL ORGANIC CHEMISTRY

COMPUTATIONAL ORGANIC CHEMISTRY

Second Edition

Steven M. Bachrach
Department of Chemistry
Trinity University
San Antonio, TX

Copyright © 2014 by John Wiley & Sons, Inc. All rights reserved.

Published by John Wiley & Sons, Inc., Hoboken, New Jersey.
Published simultaneously in Canada.

No part of this publication may be reproduced, stored in a retrieval system, or transmitted in any form or by any means, electronic, mechanical, photocopying, recording, scanning, or otherwise, except as permitted under Section 107 or 108 of the 1976 United States Copyright Act, without either the prior written permission of the Publisher, or authorization through payment of the appropriate per-copy fee to the Copyright Clearance Center, Inc., 222 Rosewood Drive, Danvers, MA 01923, (978) 750-8400, fax (978) 750-4470, or on the web at www.copyright.com. Requests to the Publisher for permission should be addressed to the Permissions Department, John Wiley & Sons, Inc., 111 River Street, Hoboken, NJ 07030, (201) 748-6011, fax (201) 748-6008, or online at http://www.wiley.com/go/permission.

Limit of Liability/Disclaimer of Warranty: While the publisher and author have used their best efforts in preparing this book, they make no representations or warranties with respect to the accuracy or completeness of the contents of this book and specifically disclaim any implied warranties of merchantability or fitness for a particular purpose. No warranty may be created or extended by sales representatives or written sales materials. The advice and strategies contained herein may not be suitable for your situation. You should consult with a professional where appropriate. Neither the publisher nor author shall be liable for any loss of profit or any other commercial damages, including but not limited to special, incidental, consequential, or other damages.

For general information on our other products and services or for technical support, please contact our Customer Care Department within the United States at (800) 762-2974, outside the United States at (317) 572-3993 or fax (317) 572-4002.

Wiley also publishes its books in a variety of electronic formats. Some content that appears in print may not be available in electronic formats. For more information about Wiley products, visit our web site at www.wiley.com.

Library of Congress Cataloging-in-Publication Data:

Bachrach, Steven M., 1959-
  Computational organic chemistry / by Steven M. Bachrach, Department of Chemistry, Trinity University, San Antonio, TX. – Second edition.
     pages cm
  Includes bibliographical references and index.
  ISBN 978-1-118-29192-4 (cloth)
 1. Chemistry, Organic–Mathematics. 2. Chemistry, Organic–Mathematical models. I. Title.
  QD255.5.M35B33 2014
  547.001'51–dc23
                                                                              2013029960

Printed in the United States of America.

10 9 8 7 6 5 4 3 2 1

To Carmen and Dustin

CONTENTS

Preface		xv
Acknowledgments		xxi
1. Quantum Mechanics for Organic Chemistry		1
1.1	Approximations to the Schrödinger Equation—The Hartree–Fock Method	2
	1.1.1 Nonrelativistic Mechanics	2
	1.1.2 The Born–Oppenheimer Approximation	3
	1.1.3 The One-Electron Wavefunction and the Hartree–Fock Method	3
	1.1.4 Linear Combination of Atomic Orbitals (LCAO) Approximation	4
	1.1.5 Hartree–Fock–Roothaan Procedure	5
	1.1.6 Restricted Versus Unrestricted Wavefunctions	7
	1.1.7 The Variational Principle	7
	1.1.8 Basis Sets	8
	1.1.8.1 Basis Set Superposition Error	12
1.2	Electron Correlation—Post-Hartree–Fock Methods	13
	1.2.1 Configuration Interaction (CI)	14
	1.2.2 Size Consistency	16
	1.2.3 Perturbation Theory	16
	1.2.4 Coupled-Cluster Theory	17
	1.2.5 Multiconfiguration SCF (MCSCF) Theory and Complete Active Space SCF (CASSCF) Theory	18
	1.2.6 Composite Energy Methods	20
1.3	Density Functional Theory (DFT)	22
	1.3.1 The Exchange-Correlation Functionals: Climbing Jacob's Ladder	24
	1.3.1.1 Double Hybrid Functionals	26
	1.3.2 Dispersion-Corrected DFT	26
	1.3.3 Functional Selection	28

1.4	Computational Approaches to Solvation	28
	1.4.1 Microsolvation	28
	1.4.2 Implicit Solvent Models	29
	1.4.3 Hybrid Solvation Models	34
1.5	Hybrid QM/MM Methods	35
	1.5.1 Molecular Mechanics	36
	1.5.2 QM/MM Theory	38
	1.5.3 ONIOM	39
1.6	Potential Energy Surfaces	40
	1.6.1 Geometry Optimization	42
1.7	Population Analysis	45
	1.7.1 Orbital-Based Population Methods	46
	1.7.2 Topological Electron Density Analysis	47
1.8	Interview: Stefan Grimme	48
	References	51

2. Computed Spectral Properties and Structure Identification — 61

2.1	Computed Bond Lengths and Angles	61
2.2	IR Spectroscopy	62
2.3	Nuclear Magnetic Resonance	66
	2.3.1 General Considerations	68
	2.3.2 Scaling Chemical Shift Values	69
	2.3.3 Customized Density Functionals and Basis Sets	71
	2.3.4 Methods for Structure Prediction	73
	2.3.5 Statistical Approaches to Computed Chemical Shifts	74
	2.3.6 Computed Coupling Constants	76
	2.3.7 Case Studies	77
	2.3.7.1 Hexacyclinol	77
	2.3.7.2 Maitotoxin	79
	2.3.7.3 Vannusal B	80
	2.3.7.4 Conicasterol F	81
	2.3.7.5 1-Adamantyl Cation	81
2.4	Optical Rotation, Optical Rotatory Dispersion, Electronic Circular Dichroism, and Vibrational Circular Dichroism	82
	2.4.1 Case Studies	85
	2.4.1.1 Solvent Effect	85
	2.4.1.2 Chiral Solvent Imprinting	86

		2.4.1.3	*Plumericin and Prismatomerin*	87
		2.4.1.4	*2,3-Hexadiene*	88
		2.4.1.5	*Multilayered Paracyclophane*	89
		2.4.1.6	*Optical Activity of an Octaphyrin*	90
2.5	Interview: Jonathan Goodman			90
	References			93

3. Fundamentals of Organic Chemistry — 99

3.1	Bond Dissociation Enthalpy			99
	3.1.1	Case Study of BDE: Trends in the R–X BDE		102
3.2	Acidity			104
	3.2.1	Case Studies of Acidity		107
		3.2.1.1	*Carbon Acidity of Strained Hydrocarbons*	107
		3.2.1.2	*Origin of the Acidity of Carboxylic Acids*	113
		3.2.1.3	*Acidity of the Amino Acids*	116
3.3	Isomerism and Problems With DFT			119
	3.3.1	Conformational Isomerism		119
	3.3.2	Conformations of Amino Acids		121
	3.3.3	Alkane Isomerism and DFT Errors		123
		3.3.3.1	*Chemical Consequences of Dispersion*	131
3.4	Ring Strain Energy			132
	3.4.1	RSE of Cyclopropane (28) and Cylcobutane (29)		138
3.5	Aromaticity			144
	3.5.1	Aromatic Stabilization Energy (ASE)		145
	3.5.2	Nucleus-Independent Chemical Shift (NICS)		150
	3.5.3	Case Studies of Aromatic Compounds		155
		3.5.3.1	*[n]Annulenes*	155
		3.5.3.2	*The Mills–Nixon Effect*	166
		3.5.3.3	*Aromaticity Versus Strain*	171
	3.5.4	π–π Stacking		173
3.6	Interview: Professor Paul Von RaguéSchleyer			177
	References			180

4. Pericyclic Reactions — 197

4.1	The Diels–Alder Reaction		198
	4.1.1	The Concerted Reaction of 1,3-Butadiene with Ethylene	199
	4.1.2	The Nonconcerted Reaction of 1,3-Butadiene with Ethylene	207

	4.1.3	Kinetic Isotope Effects and the Nature of the Diels–Alder Transition State	209
	4.1.4	Transition State Distortion Energy	214
4.2	The Cope Rearrangement		215
	4.2.1	Theoretical Considerations	217
	4.2.2	Computational Results	219
	4.2.3	Chameleons and Centaurs	227
4.3	The Bergman Cyclization		233
	4.3.1	Theoretical Considerations	237
	4.3.2	Activation and Reaction Energies of the Parent Bergman Cyclization	239
	4.3.3	The cd Criteria and Cyclic Enediynes	244
	4.3.4	Myers–Saito and Schmittel Cyclization	249
4.4	Bispericyclic Reactions		256
4.5	Pseudopericyclic Reactions		260
4.6	Torquoselectivity		267
4.7	Interview: Professor Weston Thatcher Borden		278
	References		282

5. Diradicals and Carbenes — 297

5.1	Methylene		298
	5.1.1	Theoretical Considerations of Methylene	298
	5.1.2	The H–C–H Angle in Triplet Methylene	299
	5.1.3	The Methylene and Dichloromethylene Singlet–Triplet Energy Gap	300
5.2	Phenylnitrene and Phenylcarbene		304
	5.2.1	The Low Lying States of Phenylnitrene and Phenylcarbene	305
	5.2.2	Ring Expansion of Phenylnitrene and Phenylcarbene	312
	5.2.3	Substituent Effects on the Rearrangement of Phenylnitrene	317
5.3	Tetramethyleneethane		324
	5.3.1	Theoretical Considerations of Tetramethyleneethane	326
	5.3.2	Is TME a Ground-State Singlet or Triplet?	326
5.4	Oxyallyl Diradical		330
5.5	Benzynes		333
	5.5.1	Theoretical Considerations of Benzyne	333
	5.5.2	Relative Energies of the Benzynes	336
	5.5.3	Structure of m-Benzyne	341

		5.5.4	The Singlet–Triplet Gap and Reactivity of the Benzynes	345

	5.6	Tunneling of Carbenes	349
		5.6.1 Tunneling control	353
	5.7	Interview: Professor Henry "Fritz" Schaefer	355
	5.8	Interview: Professor Peter R. Schreiner	357
		References	360

6. Organic Reactions of Anions — 373

6.1	Substitution Reactions	373
	6.1.1 The Gas Phase S_N2 Reaction	374
	6.1.2 Effects of Solvent on S_N2 Reactions	385
6.2	Asymmetric Induction Via 1,2-Addition to Carbonyl Compounds	391
6.3	Asymmetric Organocatalysis of Aldol Reactions	404
	6.3.1 Mechanism of Amine-Catalyzed Intermolecular Aldol Reactions	409
	6.3.2 Mechanism of Proline-Catalyzed Intramolecular Aldol Reactions	417
	6.3.3 Comparison with the Mannich Reaction	421
	6.3.4 Catalysis of the Aldol Reaction in Water	426
	6.3.5 Another Organocatalysis Example—The Claisen Rearrangement	429
6.4	Interview: Professor Kendall N. Houk	432
	References	435

7. Solution-Phase Organic Chemistry — 445

7.1	Aqueous Diels–Alder Reactions	446
7.2	Glucose	452
	7.2.1 Models Compounds: Ethylene Glycol and Glycerol	453
	7.2.1.1 Ethylene Glycol	453
	7.2.1.2 Glycerol	458
	7.2.2 Solvation Studies of Glucose	460
7.3	Nucleic Acids	468
	7.3.1 Nucleic Acid Bases	469
	7.3.1.1 Cytosine	469
	7.3.1.2 Guanine	473
	7.3.1.3 Adenine	475
	7.3.1.4 Uracil and Thymine	477
	7.3.2 Base Pairs	479

7.4	Amino Acids		489
7.5	Interview: Professor Christopher J. Cramer		492
	References		496

8. Organic Reaction Dynamics 505

8.1	A Brief Introduction To Molecular Dynamics Trajectory Computations		508
	8.1.1	Integrating the Equations of Motion	508
	8.1.2	Selecting the PES	510
	8.1.3	Initial Conditions	511
8.2	Statistical Kinetic Theories		512
8.3	Examples of Organic Reactions With Non-Statistical Dynamics		514
	8.3.1	[1,3]-Sigmatropic Rearrangement of Bicyclo[3.2.0]hex-2-ene	514
	8.3.2	Life in the Caldera: Concerted versus Diradical Mechanisms	518
		8.3.2.1 Rearrangement of Vinylcyclopropane to Cyclopentene	*520*
		8.3.2.2 Bicyclo[3.1.0]hex-2-ene 20	*524*
		8.3.2.3 Cyclopropane Stereomutation	*526*
	8.3.3	Entrance into Intermediates from Above	530
		8.3.3.1 Deazetization of 2,3-Diazabicyclo[2.2.1]hept-2-ene 31	*530*
	8.3.4	Avoiding Local Minima	533
		8.3.4.1 Methyl Loss from Acetone Radical Cation	*533*
		8.3.4.2 Cope Rearrangement of 1,2,6-Heptatriene	*534*
		8.3.4.3 The S_N2 Reaction: $HO^- + CH_3F$	*536*
		8.3.4.4 Reaction of Fluoride with Methyl Hydroperoxide	*538*
	8.3.5	Bifurcating Surfaces: One TS, Two Products	539
		8.3.5.1 C_2-C_6 Enyne Allene Cyclization	*540*
		8.3.5.2 Cycloadditions Involving Ketenes	*543*
		8.3.5.3 Diels–Alder Reactions: Steps toward Predicting Dynamic Effects on Bifurcating Surfaces	*547*
	8.3.6	Stepwise Reaction on a Concerted Surface	550
		8.3.6.1 Rearrangement of Protonated Pinacolyl Alcohol	*550*
	8.3.7	Roaming Mechanism	551
	8.3.8	A Roundabout S_N2 reaction	553
	8.3.9	Hydroboration: Dynamical or Statistical?	554

	8.3.10	A Look at the Wolff Rearrangement	555
8.4	Conclusions		557
8.5	Interview: Professor Daniel Singleton		558
	References		561

9. Computational Approaches to Understanding Enzymes — 569

9.1	Models for Enzymatic Activity	569
9.2	Strategy for Computational Enzymology	573
	9.2.1 High Level QM/MM Computations of Enzymes	576
	9.2.2 Chorismate Mutase	578
	9.2.3 Catechol-*O*-Methyltransferase (COMT)	582
9.3	*De Novo* Design of Enzymes	586
	References	592

Index — 599

PREFACE

In 1929, Dirac famously proclaimed that

> The fundamental laws necessary for the mathematical treatment of a large part of physics and *the whole of chemistry* (emphasis added) are thus completely known, and the difficulty lies only in the fact that application of these laws leads to equations that are too complex to be solved.[1]

This book is a testament to just how difficult it is to adequately account for the properties and reactivities of real chemical systems using quantum mechanics (QM).

Though QM was born in the mid-1920s, it took many years before rigorous solutions for molecular systems appeared. Hylleras[2] and others[3,4] developed nearly exact solutions to the single-electron diatomic molecule in the 1930s and 1940s. Reasonable solutions for multielectron multiatom molecules did not appear until 1960, with Kolos'[5,6] computation of H_2 and Boys'[7] study of CH_2. The watershed year was perhaps 1970 with the publication by Bender and Schaefer[8] on the bent form of triplet CH_2 (a topic of Chapter 5) and the release by Pople's[9] group of *Gaussian-70*, which is the first full-featured quantum chemistry computer package that was to be used by a broad range of theorists and nontheorists alike. So, in this sense, computational quantum chemistry is really only some five decades old.

The application of QM to organic chemistry dates back to Hückel's π-electron model of the 1930s.[10–12] Approximate quantum mechanical treatments for organic molecules continued throughout the 1950s and 1960s. Application of *ab initio* approaches, such as Hartree–Fock theory, began in earnest in the 1970s and really flourished in the mid-1980s, with the development of computer codes that allowed for automated optimization of ground and transition states and incorporation of electron correlation using configuration interaction or perturbation techniques.

In 2006, I began writing the first edition of this book, acting on the notion that the field of computational organic chemistry was sufficiently mature to deserve a critical review of its successes and failures in treating organic chemistry problems. The book was published the next year and met with a fine reception.

As I anticipated, immediately upon publication of the book, it was out of date. Computational chemistry, like all science disciplines, is a constantly changing field. New studies are published, new theories are proposed, and old ideas are replaced with new interpretations. I attempted to address the need for the book to remain current in some manner by creating a complementary blog at http://www.comporgchem.com/blog. The blog posts describe the results of new papers and how these results touch on the themes presented in the monograph. Besides providing an avenue for me to continue to keep my readers posted on current developments, the blog allowed for feedback from the readers. On a few occasions, a blog post and the article described engendered quite a conversation!

Encouraged by the success of the book, Jonathan Rose of Wiley approached me about updating the book with a second edition. Drawing principally on the blog posts, I had written since 2007, I knew that the ground work for writing an updated version of the book had already been done. So I agreed, and what you have in your hands is my perspective of the accomplishments of computational organic chemistry through early 2013.

The structure of the book remains largely intact from the *first edition*, with a few important modifications. Throughout this book. I aim to demonstrate the major impact that computational methods have had upon the current understanding of organic chemistry. I present a survey of organic problems where computational chemistry has played a significant role in developing new theories or where it provided important supporting evidence of experimentally derived insights. I expand the scope to include computational enzymology to point interested readers toward how the principles of QM applied to organic reactions can be extended to biological system too. I also highlight some areas where computational methods have exhibited serious weaknesses.

Any such survey must involve judicious selecting and editing of materials to be presented and omitted. In order to reign in the scope of the book, I opted to feature only computations performed at the *ab initio* level. (Note that I consider density functional theory to be a member of this category.) This decision omits some very important work, certainly from a historical perspective if nothing else, performed using semiempirical methods. For example, Michael Dewar's influence on the development of theoretical underpinnings of organic chemistry[13] is certainly underplayed in this book since results from MOPAC and its decedents are largely not discussed. However, taking a view with an eye toward the future, the principle advantage of the semiempirical methods over *ab initio* methods is ever-diminishing. Semiempirical calculations are much faster than *ab initio* calculations and allow for much larger molecules to be treated. As computer hardware improves, as algorithms become more efficient, ab initio computations become more practical for ever-larger molecules, which is a trend that certainly has played out since the publication of the first edition of this book.

The book is designed for a broad spectrum of users: practitioners of computational chemistry who are interested in gaining a broad survey or an entrée into a new area of organic chemistry, synthetic and physical organic chemists who might be interested in running some computations of their own and would like to learn of success stories to emulate and pitfalls to avoid, and graduate students interested in just what can be accomplished by computational approaches to real chemical problems.

It is important to recognize that the reader does not have to be an expert in quantum chemistry to make use of this book. A familiarity with the general principles of quantum mechanics obtained in a typical undergraduate physical chemistry course will suffice. The first chapter of this book introduces all of the major theoretical concepts and definitions along with the acronyms that so plague our discipline. Sufficient mathematical rigor is presented to expose those who are interested to some of the subtleties of the methodologies. This chapter is not intended to be of sufficient detail for one to become expert in the theories. Rather it will allow the reader to become comfortable with the language and terminology at a level sufficient to understand the results of computations and understand the inherent shortcoming associated with particular methods that may pose potential problems. Upon completing Chapter 1, the reader should be able to follow with relative ease a computational paper in any of the leading journals. Readers with an interest in delving further into the theories and their mathematics are referred to three outstanding texts, *Essential of Computational Chemistry* by Cramer,[14] *Introduction to Computational Chemistry* by Jensen,[15] and *Modern Quantum Chemistry: Introduction to Advanced Electronic Structure Theory* by Szabo and Ostlund.[16] In a way, this book serves as the applied accompaniment to these books.

How is the *second edition* different from the *first edition*? Chapter 1 presents an overview of computational methods. In this *second edition*, I have combined the descriptions of solvent computations and molecular dynamics computations into this chapter. I have added a discussion of QM/molecular mechanics (MM) computations and the topology of potential energy surfaces. The discussion of density functional theory is more extensive, including discussion of double hybrids and dispersion corrections. Chapter 2 of the *second edition* is mostly entirely new. It includes case studies of computed spectra, especially computed NMR, used for structure determination. This is an area that has truly exploded in the last few years, with computed spectra becoming an important tool in the structural chemists' arsenal. Chapter 3 discusses some fundamental concepts of organic chemistry; for the concepts of bond dissociation energy, acidity, and aromaticity, I have included some new examples, such as π-stacking of aromatic rings. I also added a section on isomerism, which exposes some major problems with families of density functionals, including the most commonly used functional, B3LYP.

Chapter 4 presents pericyclic reactions. I have updated some of the examples from the last edition, but the main change is the addition of bispericyclic reactions, which is a topic that is important for the understanding of many of the examples of dynamic effects presented in Chapter 8. Chapter 5 deals with radicals and carbenes. This chapter contains one of the major additions to the book: a detailed presentation of tunneling in carbenes. The understanding that tunneling is occurring in some carbenes was made possible by quantum computations and this led directly to the brand new concept of tunneling control.

The chemistry of anions is the topic of Chapter 6. This chapter is an update from the material in the *first edition*, incorporating new examples, primarily in the area of organocatalysis. Chapter 7, presenting solvent effects, is also updated to include some new examples. The recognition of the role of dynamic effects, situations where standard transition state theory fails, is a major triumph of computational organic chemistry. Chapter 8 extends the scope of reactions that are subject to dynamic effects from that presented in the *first edition*. In addition, some new types of dynamic effects are discussed, including the roundabout pathway in an S_N2 reaction and the roaming mechanism.

A major addition to the *second edition* is Chapter 9, which discusses computational enzymology. This chapter extends the coverage of quantum chemistry to a sister of organic chemistry—biochemistry. Since computational biochemistry truly deserves its own entire book, this chapter presents a flavor of how computational quantum chemical techniques can be applied to biochemical systems. This chapter presents a few examples of how QM/MM has been applied to understand the nature of enzyme catalysis. This chapter concludes with a discussion of *de novo* design of enzymes, which is a research area that is just becoming feasible, and one that will surely continue to develop and excite a broad range of chemists for years to come.

Science is an inherently human endeavor, performed and consumed by humans. To reinforce the human element, I interviewed a number of preeminent computational chemists. I distilled these interviews into short set pieces, wherein each individual's philosophy of science and history of their involvements in the projects described in this book are put forth, largely in their own words. I interviewed six scientists for the *first edition*—Professors Wes Borden, Chris Cramer, Ken Houk, Henry "Fritz" Schaefer, Paul Schleyer, and Dan Singleton. I have reprinted these interviews in this *second edition*. There was a decided USA-centric focus to these interviews and so for the *second edition*, I have interviewed three European scientists: Professors Stefan Grimme, Jonathan Goodman, and Peter Schreiner. I am especially grateful to these nine people for their time they gave me and their gracious support of this project. Each interview ran well over an hour and was truly a fun experience for me! This group of nine scientists is only a small fraction of the chemists who have been and are active participants within our discipline, and my apologies in advance to all those whom I did not interview for this book.

A theme I probed in all of the interviews was the role of collaboration in developing new science. As I wrote this book, it became clear to me that many important breakthroughs and significant scientific advances occurred through collaboration, particularly between a computational chemist and an experimental chemist. Collaboration is an underlying theme throughout the book, and perhaps signals the major role that computational chemistry can play; in close interplay with experiment, computations can draw out important insights, help interpret results, and propose critical experiments to be carried out next.

I intend to continue to use the book's ancillary Web site www.comporgchem.com to deliver supporting information to the reader. Every cited article that is available in some electronic form is listed along with the direct link to that article. Please keep in mind that the reader will be responsible for gaining ultimate access to the articles by open access, subscription, or other payment option. The citations are listed on the Web site by chapter, in the same order they appear in the book. Almost all molecular geometries displayed in the book were produced using the *GaussView*[17] molecular visualization tool. This required obtaining the full three-dimensional structure, from the article, the supplementary material, or through my reoptimization of that structure. These coordinates are made available for reuse through the Web site. Furthermore, I intend to continue to post (www.comporgchem.com/blog) updates to the book on the blog, especially focusing on new articles that touch on or complement the topics covered in this book. I hope that readers will become a part of this community and not just read the posts but also add their own comments, leading to what I hope will be a useful and entertaining dialogue. I encourage you to voice your opinions and comments. I wish to thank particular members of the computational chemistry community who have commented on the blog posts; comments from Henry Rzepa, Stephen Wheeler, Eugene Kwan, and Jan Jensen helped inform my writing of this edition. I thank Jan for creating the *Computational Chemistry Highlights* (http://www.compchemhighlights.org/) blog, which is an overlay of the computational chemistry literature, and for incorporating my posts into this blog.

REFERENCES

1. Dirac, P. "Quantum mechanics of many-electron systems," *Proc. R. Soc. A* **1929**, *123*, 714–733.
2. Hylleras, E. A. "Über die Elektronenterme des Wasserstoffmoleküls," *Z. Physik* **1931**, 739–763.
3. Barber, W. G.; Hasse, H. R. "The two centre problem in wave mechanics," *Proc. Camb. Phil. Soc.* **1935**, *31*, 564–581.
4. Jaffé, G. "Zur theorie des wasserstoffmolekülions," *Z. Physik* **1934**, *87*, 535–544.
5. Kolos, W.; Roothaan, C. C. J. "Accurate electronic wave functions for the hydrogen molecule," *Rev. Mod. Phys.* **1960**, *32*, 219–232.
6. Kolos, W.; Wolniewicz, L. "Improved theoretical ground-state energy of the hydrogen molecule," *J. Chem. Phys.* **1968**, *49*, 404–410
7. Foster, J. M.; Boys, S. F. "Quantum variational calculations for a range of CH_2 configurations," *Rev. Mod. Phys.* **1960**, *32*, 305–307.
8. Bender, C. F.; Schaefer, H. F., III "New theoretical evidence for the nonlinearity of the triplet ground state of methylene," *J. Am. Chem. Soc.* **1970**, *92*, 4984–4985.
9. Hehre, W. J.; Lathan, W. A.; Ditchfield, R.; Newton, M. D.; Pople, J. A.; Quantum Chemistry Program Exchange, Program No. 237: **1970**.
10. Huckel, E. "Quantum-theoretical contributions to the benzene problem. I. The Electron configuration of benzene and related compounds," *Z. Physik* **1931**, *70*, 204–288.
11. Huckel, E. "Quantum theoretical contributions to the problem of aromatic and non-saturated compounds. III," *Z. Physik* **1932**, *76*, 628–648.
12. Huckel, E. "The theory of unsaturated and aromatic compounds," *Z. Elektrochem. Angew. Phys. Chem.* **1937**, *43*, 752–788.
13. Dewar, M. J. S. *A Semiempirical Life*; ACS Publications: Washington, DC, **1990**.
14. Cramer, C. J. *Essentials of Computational Chemistry: Theories and Models*; John Wiley & Sons: New York, **2002**.
15. Jensen, F. *Introduction to Computational Chemistry*; John Wiley & Sons: Chichester, England, **1999**.
16. Szabo, A.; Ostlund, N. S. *Modern Quantum Chemistry: Introduction to Advanced Electronic Structure Theory*; Dover: Mineola, NY, **1996**.
17. Dennington II, R.; Keith, T.; Millam, J.; Eppinnett, K.; Hovell, W. L.; Gilliland, R. *GaussView*; Semichem, Inc.: Shawnee Mission, KS, USA, **2003**.

ACKNOWLEDGMENTS

This book is the outcome of countless interactions with colleagues across the world, whether in person, on the phone, through Skype, or by email. These conversations directly or indirectly influenced my thinking and contributed in a meaningful way to this book, and especially this second edition. In particular I wish to thank these colleagues and friends, listed here in alphabetical order: John Baldwin, David Birney, Wes Borden, Chris Cramer, Dieter Cremer, Bill Doering, Tom Cundari, Cliff Dykstra, Jack Gilbert, Tom Gilbert, Jonathan Goodman, Stephen Gray, Stefan Grimme, Scott Gronert, Bill Hase, Ken Houk, Eric Jacobsen, Steven Kass, Elfi Kraka, Jan Martin, Nancy Mills, Mani Paranjothy, Henry Rzepa, Fritz Schaefer, Paul Schleyer, Peter Schreiner, Matt Siebert, Dan Singleton, Andrew Streitwieser, Dean Tantillo, Don Truhlar, Adam Urbach, Steven Wheeler, and Angela Wilson. I profoundly thank all of them for their contributions and assistance and encouragements. I want to particular acknowledge Henry Rzepa for his extraordinary enthusiasm for, and commenting on, my blog. The library staff at Trinity University, led by Diane Graves, was extremely helpful in providing access to the necessary literature.

The cover image was prepared by my sister Lisa Bachrach. The image is based on a molecular complex designed by Iwamoto and co-workers (*Angew. Chem. Int. Ed.*, **2011**, *50*, 8342–8344).

I wish to acknowledge Jonathan Rose at Wiley for his enthusiastic support for the second edition and all of the staff at Wiley for their production assistance.

Finally, I wish to thank my wife Carmen for all of her years of support, guidance, and love.

S. M. B.

CHAPTER 1

Quantum Mechanics for Organic Chemistry

Computational chemistry, as explored in this book, will be restricted to quantum mechanical descriptions of the molecules of interest. This should not be taken as a slight upon alternate approaches. Rather, the aim of this book is to demonstrate the power of high level quantum computations in offering insight toward understanding the nature of organic molecules—their structures, properties, and reactions—and to show their successes and point out the potential pitfalls. Furthermore, this book will address the applications of traditional *ab initio* and density functional theory (DFT) methods to organic chemistry, with little mention of semiempirical methods. Again, this is not to slight the very important contributions made from the application of complete neglect of differential overlap (CNDO) and its progenitors. However, with the ever-improving speed of computers and algorithms, ever-larger molecules are amenable to *ab initio* treatment, making the semiempirical and other approximate methods for treatment of the quantum mechanics (QM) of molecular systems simply less necessary. This book is therefore designed to encourage the broader use of the more exact treatments of the physics of organic molecules by demonstrating the range of molecules and reactions already successfully treated by quantum chemical computation. We will highlight some of the most important contributions that this discipline has presented to the broader chemical community toward understanding of organic chemistry.

We begin with a brief and mathematically light-handed treatment of the fundamentals of QM necessary to describe organic molecules. This presentation is meant to acquaint those unfamiliar with the field of computational chemistry with a general understanding of the major methods, concepts, and acronyms. Sufficient depth will be provided so that one can understand why certain methods work well while others may fail when applied to various chemical problems, allowing the casual reader to be able to understand most of any applied computational chemistry paper in the literature. Those seeking more depth and details, particularly more derivations and a fuller mathematical treatment, should consult any of the

---

Computational Organic Chemistry, Second Edition. Steven M. Bachrach.
© 2014 John Wiley & Sons, Inc. Published 2014 by John Wiley & Sons, Inc.

three outstanding texts: *Essentials of Computational Chemistry* by Cramer,[1] *Introduction to Computational Chemistry* by Jensen,[2] and *Modern Quantum Chemistry: Introduction to Advanced Electronic Structure Theory* by Szabo and Ostlund.[3]

Quantum chemistry requires the solution of the time-independent Schrödinger equation,

$$\hat{H}\Psi(R_1, R_2 \ldots R_N, r_1, r_2 \ldots r_n) = E\Psi(R_1, R_2 \ldots R_N, r_1, r_2 \ldots r_n) \quad (1.1)$$

where $\hat{H}$ is the Hamiltonian operator, $\Psi(R_1, R_2 \ldots R_N, r_1, r_2 \ldots r_n)$ is the wavefunction for all of the nuclei and electrons, and E is the energy associated with this wavefunction. The Hamiltonian contains all the operators that describe the kinetic and potential energies of the molecule at hand. The wavefunction is a function of the nuclear positions **R** and the electron positions **r**. For molecular systems of interest to organic chemists, the Schrödinger equation cannot be solved exactly and so a number of approximations are required to make the mathematics tractable.

1.1 APPROXIMATIONS TO THE SCHRÖDINGER EQUATION—THE HARTREE–FOCK METHOD

1.1.1 Nonrelativistic Mechanics

Dirac[4] achieved the combination of QM and relativity. Relativistic corrections are necessary when particles approach the speed of light. Electrons near heavy nuclei will achieve such velocities, and for these atoms, relativistic quantum treatments are necessary for accurate description of the electron density. However, for typical organic molecules, which contain only first- and second-row elements, a relativistic treatment is unnecessary. Solving the Dirac relativistic equation is much more difficult than for nonrelativistic computations. A common approximation is to utilize an effective field for the nuclei associated with heavy atoms, which corrects for the relativistic effect. This approximation is beyond the scope of this book, especially since it is unnecessary for the vast majority of organic chemistry.

The complete nonrelativistic Hamiltonian for a molecule consisting of n electrons and N nuclei is

$$\hat{H} = -\frac{h^2}{2}\sum_I^N \frac{\nabla_I^2}{m_I} - \frac{h^2}{2m_e}\sum_i^n \nabla_i^2 - \sum_I^N \sum_i^n \frac{Z_I e'^2}{r_{Ii}} + \sum_{I<J}^N \frac{Z_I Z_J e'^2}{r_{IJ}} + \sum_i^n \frac{e'^2}{r_{ij}} \quad (1.2)$$

where the lowercase letter indexes the electrons and the uppercase one indexes the nuclei, h is the Planck's constant, m_e is the electron mass, m_I is the mass of nucleus I, and r is the distance between the objects specified by the subscript. For simplicity, we define

$$e'^2 = \frac{e^2}{4\pi\varepsilon_0} \quad (1.3)$$

1.1.2 The Born–Oppenheimer Approximation

The total molecular wavefunction $\Psi(\mathbf{R},\mathbf{r})$ depends on both the positions of all of the nuclei and the positions of all of the electrons. Since electrons are much lighter than nuclei, and therefore move much more rapidly, electrons can essentially instantaneously respond to any changes in the relative positions of the nuclei. This allows for the separation of the nuclear variables from the electron variables,

$$\Psi(R_1, R_2 \ldots R_N, r_1, r_2 \ldots r_n) = \Phi(R_1, R_2 \ldots R_N)\psi(r_1, r_2 \ldots r_n) \quad (1.4)$$

This separation of the total wavefunction into an electronic wavefunction $\psi(\mathbf{r})$ and a nuclear wavefunction $\Phi(\mathbf{R})$ means that the positions of the nuclei can be fixed, leaving it only necessary to solve for the electronic part. This approximation was proposed by Born and Oppenheimer[5] and is valid for the vast majority of organic molecules.

The potential energy surface (PES) is created by determining the electronic energy of a molecule while varying the positions of its nuclei. It is important to recognize that the concept of the PES relies upon the validity of the Born–Oppenheimer approximation so that we can talk about transition states and local minima, which are critical points on the PES. Without it, we would have to resort to discussions of probability densities of the nuclear–electron wavefunction.

The Hamiltonian obtained after applying the Born–Oppenheimer approximation and neglecting relativity is

$$\hat{H} = -\frac{1}{2}\sum_i^n \nabla_i^2 - \sum_I^N \sum_i^n \frac{Z_I}{r_{Ii}} + \sum_{i<j}^n \frac{1}{r_{ij}} + V^{\text{nuc}} \quad (1.5)$$

where V^{nuc} is the nuclear–nuclear repulsion energy. Eq. (1.5) is expressed in atomic units, which is why it appears so uncluttered. It is this Hamiltonian that is utilized in computational organic chemistry. The next task is to solve the Schrödinger equation (1.1) with the Hamiltonian expressed in Eq. (1.5).

1.1.3 The One-Electron Wavefunction and the Hartree–Fock Method

The wavefunction $\psi(\mathbf{r})$ depends on the coordinates of *all* of the electrons in the molecule. Hartree[6] proposed the idea, reminiscent of the separation of variables used by Born and Oppenheimer, that the electronic wavefunction can be separated into a product of functions that depend only on one electron,

$$\psi(r_1, r_2 \ldots r_n) = \phi_1(r_1)\phi_2(r_2) \ldots \phi_n(r_n) \quad (1.6)$$

This wavefunction would solve the Schrödinger equation exactly if it weren't for the electron–electron repulsion term of the Hamiltonian in Eq. (1.5). Hartree next rewrote this term as an expression that describes the repulsion an electron feels from the average position of the other electrons. In other words, the exact electron–electron repulsion is replaced with an effective field V_i^{eff} produced by the average positions of the remaining electrons. With this assumption, the separable functions ϕ_i satisfy the Hartree equations

$$\left(-\frac{1}{2}\nabla_i^2 - \sum_I^N \frac{Z_I}{r_{Ii}} + V_i^{\text{eff}}\right)\phi_i = E_i\phi_i \quad (1.7)$$

(Note that Eq. (1.7) defines a set of equations, one for each electron.) Solving for the set of functions ϕ_i is nontrivial because V_i^{eff} itself depends on all of the functions ϕ_i. An iterative scheme is needed to solve the Hartree equations. First, a set of functions $(\phi_1, \phi_2, \ldots, \phi_n)$ is assumed. These are used to produce the set of effective potential operators V_i^{eff}, and the Hartree equations are solved to produce a set of improved functions ϕ_i. These new functions produce an updated effective potential, which in turn yields a new set of functions ϕ_i. This process is continued until the functions ϕ_i no longer change, resulting in a self-consistent field (SCF).

Replacing the full electron–electron repulsion term in the Hamiltonian with V_i^{eff} is a serious approximation. It neglects entirely the ability of the electrons to rapidly (essentially instantaneously) respond to the position of other electrons. In a later section, we address how one accounts for this instantaneous electron–electron repulsion.

Fock[7,8] recognized that the separable wavefunction employed by Hartree (Eq. (1.6)) does not satisfy the Pauli exclusion principle.[9] Instead, Fock suggested using the Slater determinant

$$\psi(r_1, r_2 \ldots r_n) = \frac{1}{\sqrt{n!}} \begin{vmatrix} \phi_1(e_1) & \phi_2(e_1) & \cdots & \phi_n(e_1) \\ \phi_1(e_2) & \phi_2(e_2) & \cdots & \phi_n(e_2) \\ \vdots & \vdots & & \vdots \\ \phi_1(e_n) & \phi_2(e_n) & \cdots & \phi_n(e_n) \end{vmatrix} = |\phi_1 \phi_2 \cdots \phi_n| \quad (1.8)$$

which is antisymmetric and satisfies the Pauli exclusion principle. Again, an effective potential is employed, and an iterative scheme provides the solution to the Hartree–Fock (HF) equations.

1.1.4 Linear Combination of Atomic Orbitals (LCAO) Approximation

The solutions to the HF model, ϕ_i, are known as the *molecular orbitals* (MOs). These orbitals generally span the entire molecule, just as the atomic orbitals (AOs) span the space about an atom. Since organic chemists consider the atomic properties of atoms (or collection of atoms as functional groups) to persist to some extent when embedded within a molecule, it seems reasonable to construct the MOs as an expansion of the AOs,

$$\phi_i = \sum_\mu^k c_{i\mu} \chi_\mu \quad (1.9)$$

where the index μ spans all of the AOs χ of every atom in the molecule (a total of k AOs), and $c_{i\mu}$ is the expansion coefficient of AO χ_μ in MO ϕ_i. Eq. (1.9) defines the linear combination of atomic orbital (LCAO) approximation.

1.1.5 Hartree–Fock–Roothaan Procedure

Combining the LCAO approximation for the MOs with the HF method led Roothaan[10] to develop a procedure to obtain the SCF solutions. We will discuss here only the simplest case where all MOs are doubly occupied with one electron that is spin up and one that is spin down, also known as a *closed-shell wavefunction*. The open-shell case is a simple extension of these ideas. The procedure rests upon transforming the set of equations listed in Eq. (1.7) into matrix form

$$\mathbf{FC} = \mathbf{SC}\boldsymbol{\varepsilon} \tag{1.10}$$

where $\mathbf{S}$ is the overlap matrix, $\mathbf{C}$ is the $k \times k$ matrix of the coefficients $c_{i\mu}$, and $\boldsymbol{\varepsilon}$ is the $k \times k$ matrix of the orbital energies. Each column of $\mathbf{C}$ is the expansion of ϕ_i in terms of the AOs χ_μ. The Fock matrix $\mathbf{F}$ is defined for the $\mu\nu$ element as

$$F_{\mu\nu} = \left\langle \nu|\hat{h}|\mu \right\rangle + \sum_n^{n/2} [(jj|\mu\nu) - (j\nu|j\mu)] \tag{1.11}$$

where $\hat{h}$ is the core-Hamiltonian, corresponding to the kinetic energy of the electron and the potential energy due to the electron–nuclear attraction, and the last two terms describe the Coulomb and exchange energies, respectively. It is also useful to define the density matrix (more properly, the first-order reduced density matrix)

$$\mathbf{D}_{\mu\nu} = 2 \sum_i^{n/2} c_{i\mu}^* c_{i\nu} \tag{1.12}$$

The expression in Eq. (1.12) is for a closed-shell wavefunction, but it can be defined for a more general wavefunction by analogy.

The matrix approach is advantageous because a simple algorithm can be established for solving Eq. (1.10). First, a matrix $\mathbf{X}$ is found which transforms the normalized AOs χ_μ into the orthonormal set χ'_μ

$$\chi'_\mu = \sum_\mu^k \mathbf{X} \chi_\mu \tag{1.13}$$

which is mathematically equivalent to

$$\mathbf{X}^\dagger \mathbf{S} \mathbf{X} = 1 \tag{1.14}$$

where $\mathbf{X}^\dagger$ is the adjoint of the matrix $\mathbf{X}$. The coefficient matrix $\mathbf{C}$ can be transformed into a new matrix $\mathbf{C}'$

$$\mathbf{C}' = \mathbf{X}^{-1}\mathbf{C} \tag{1.15}$$

Substituting $\mathbf{C} = \mathbf{X}\mathbf{C}'$ into Eq. (1.10) and multiplying by $\mathbf{X}^\dagger$ gives

$$\mathbf{X}^\dagger\mathbf{F}\mathbf{X}\mathbf{C}' = \mathbf{X}^\dagger\mathbf{S}\mathbf{X}\mathbf{C}'\mathbf{e} = \mathbf{C}'\mathbf{e} \tag{1.16}$$

By defining the transformed Fock matrix

$$\mathbf{F}' = \mathbf{X}^\dagger\mathbf{F}\mathbf{X} \tag{1.17}$$

we obtain the Roothaan expression

$$\mathbf{F}'\mathbf{C}' = \mathbf{C}'\mathbf{e} \tag{1.18}$$

The Hartree–Fock–Roothaan algorithm is implemented by the following steps.

(1) Specify the nuclear position, the type of nuclei, and the number of electrons.
(2) Choose a basis set. The basis set is the mathematical description of the AOs. Basis sets are described in Section 1.1.8.
(3) Calculate all of the integrals necessary to describe the core Hamiltonian, the Coulomb and exchange terms, and the overlap matrix.
(4) Diagonalize the overlap matrix $\mathbf{S}$ to obtain the transformation matrix $\mathbf{X}$.
(5) Make a guess at the coefficient matrix $\mathbf{C}$ and obtain the density matrix $\mathbf{D}$.
(6) Calculate the Fock matrix and then the transformed Fock matrix $\mathbf{F}'$.
(7) Diagonalize $\mathbf{F}'$ to obtain $\mathbf{C}'$ and ϵ.
(8) Obtain the new coefficient matrix with the expression $\mathbf{C} = \mathbf{X}\mathbf{C}'$ and the corresponding new density matrix.
(9) Decide if the procedure has converged. There are typically two criteria for convergence, one based on the energy and the other on the orbital coefficients. The energy convergence criterion is met when the difference in the energies of the last two iterations is less than some pre-set value. Convergence of the coefficients is obtained when the standard deviation of the density matrix elements in successive iterations is also below some pre-set value. If convergence has not been met, return to step 6 and repeat until the convergence criteria are satisfied.

One last point concerns the nature of the MOs that are produced in this procedure. These orbitals are such that the energy matrix ϵ will be diagonal, with the diagonal elements being interpreted as the MO energy. These MOs are referred to as the *canonical orbitals*. One must be aware that all that makes them unique is that these orbitals will produce the diagonal matrix ϵ. Any new set of orbitals ϕ_i' produced from the canonical set by a unitary transformation

$$\phi'_i = \sum_j^n \mathbf{U}_{ji}\phi_i \qquad (1.19)$$

will satisfy the HF equations and give the exact same energy and electron distribution as that with the canonical set. No one set of orbitals is really any better or worse than another, as long as the set of MOs satisfies Eq. (1.19).

1.1.6 Restricted Versus Unrestricted Wavefunctions

The preceding development of the HF theory assumed a closed-shell wavefunction. The wavefunction for an individual electron describes its spatial extent along with its spin. The electron can be either spin up (α) or spin down (β). For the closed-shell wavefunction, each pair of electrons shares the same spatial orbital but each has a different spin—one is up and the other is down. This type of wavefunction is also called a *(spin)-restricted wavefunction* since the paired electrons are restricted to the same spatial orbital, leading to the restricted Hartree–Fock (RHF) method.

This restriction is *not* demanded. It is a simple way to satisfy the Pauli exclusion principle,[9] but it is not the only means for doing so. In an unrestricted wavefunction, the spin-up electron and its spin-down partner do not have the same spatial description. The Hartree–Fock–Roothaan procedure is slightly modified to handle this case by creating a set of equations for the α electrons and another set for the β electrons, and then an algorithm similar to that described above is implemented.

The downside to the (spin)-unrestricted Hartree–Fock (UHF) method is that the unrestricted wavefunction usually will not be an eigenfunction of the $\widehat{S}^2$ operator. Since the Hamiltonian and $\widehat{S}^2$ operators commute, the true wavefunction must be an eigenfunction of both of these operators. The UHF wavefunction is typically contaminated with higher spin states; for singlet states, the most important contaminant is the triplet state. A procedure called *spin projection* can be used to remove much of this contamination. However, geometry optimization is difficult to perform with spin projection. Therefore, great care is needed when an unrestricted wavefunction is utilized, as it must be when the molecule of interest is inherently open shell, like in radicals.

1.1.7 The Variational Principle

The variational principle asserts that any wavefunction constructed as a linear combination of orthonormal functions will have its energy greater than or equal to the lowest energy (E_0) of the system. Thus,

$$\frac{\left\langle \Phi | \widehat{H} | \Phi \right\rangle}{\left\langle \Phi | \Phi \right\rangle} \geq E_0 \qquad (1.20)$$

if

$$\Phi = \sum_i c_i \phi_i \qquad (1.21)$$

If the set of functions ϕ_i is infinite, then the wavefunction will produce the lowest energy for that particular Hamiltonian. Unfortunately, expanding a wavefunction using an infinite set of functions is impractical. The variational principle saves the day by providing a simple way to judge the quality of various truncated expansions—the lower the energy, the better the wavefunction! The variational principle is *not* an approximation to treatment of the Schrödinger equation; rather, it provides a means for judging the effect of certain types of approximate treatments.

1.1.8 Basis Sets

In order to solve for the energy and wavefunction within the Hartree–Fock–Roothaan procedure, the AOs must be specified. If the set of AOs is infinite, then the variational principle tells us that we will obtain the lowest possible energy within the HF–SCF method. This is called the *HF limit*, E_{HF}. This is *not* the actual energy of the molecule; recall that the HF method neglects instantaneous electron–electron interactions, otherwise known as *electron correlation*.

Since an infinite set of AOs is impractical, a choice must be made on how to truncate the expansion. This choice of AOs defines the *basis set*.

A natural starting point is to use functions from the exact solution of the Schrödinger equation for the hydrogen atom. These orbitals have the form

$$c = N x^i y^j z^k e^{-z(r-\mathbf{R})} \qquad (1.22)$$

where $\mathbf{R}$ is the position vector of the nucleus upon which the function is centered and N is the normalization constant. Functions of this type are called *Slater-type orbitals* (STOs). The value of ζ for every STO for a given element is determined by minimizing the atomic energy with respect to ζ. These values are used for every atom of that element, regardless of the molecular environment.

At this point, it is worth shifting nomenclature and discussing the expansion in terms of basis functions instead of AOs. The construction of MOs in terms of some set of functions is entirely a mathematical "trick," and we choose to place these functions at a nucleus since that is the region of greatest electron density. We are not using "AOs" in the sense of a solution to the atomic Schrödinger equation, but just mathematical functions placed at nuclei for convenience. To make this more explicit, we will refer to the expansion of *basis functions* to form the MOs.

Conceptually, the STO basis is straightforward as it mimics the exact solution for the single electron atom. The exact orbitals for carbon, for example, are *not* hydrogenic orbitals, but are similar to the hydrogenic orbitals. Unfortunately, with STOs, many of the integrals that need to be evaluated to construct the Fock matrix can only be solved using an infinite series. Truncation of this infinite series results in errors, which can be significant.

APPROXIMATIONS TO THE SCHRÖDINGER EQUATION—THE HARTREE–FOCK METHOD

Following on a suggestion of Boys,[11] Pople decided to use a combination of Gaussian functions to mimic the STO. The advantage of the Gaussian-type orbital (GTO),

$$\chi = N x^i y^j z^k e^{-\alpha(r-\mathbf{R})^2} \tag{1.23}$$

is that with these functions, the integrals required to build the Fock matrix can be evaluated exactly. The trade-off is that GTOs do differ in shape from the STOs, particularly at the nucleus where the STO has a cusp while the GTO is continually differentiable (Figure 1.1). Therefore, multiple GTOs are necessary to adequately mimic each STO, increasing the computational size. Nonetheless, basis sets comprising GTOs are the ones that are most commonly used.

A number of factors define the basis set for a quantum chemical computation. First, how many basis functions should be used? The minimum basis set has one basis function for every formally occupied or partially occupied orbital in the atom. So, for example, the minimum basis set for carbon, with electron occupation $1s^2 2s^2 2p^2$, has two s-type functions and p_x, p_y, and p_z functions, for a total of five basis functions. This minimum basis set is referred to as a *single zeta* (SZ) basis set. The use of the term *zeta* here reflects that each basis function mimics a single STO, which is defined by its exponent, ζ.

The minimum basis set is usually inadequate, failing to allow the core electrons to get close enough to the nucleus and the valence electrons to delocalize. An obvious solution is to double the size of the basis set, creating a double zeta (DZ) basis. So for carbon, the DZ basis set has four s basis functions and two p basis functions (recognizing that the term p *basis functions* refers here to the full set—p_x, p_y, and p_z functions), for a total of 10 basis functions. Further improvement can be made by choosing a triple zeta (TZ) or even larger basis set.

Since most of chemistry focuses on the action of the valence electrons, Pople[12,13] developed the split-valence basis sets, SZ in the core and DZ in the valence region. A double-zeta split-valence basis set for carbon has three s basis

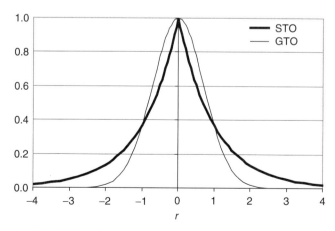

Figure 1.1 Plot of the radial component of Slater-type and Gaussian-type orbitals.

functions and two p basis functions for a total of nine functions, a triple-zeta split valence basis set has four s basis functions, and three p functions for a total of 13 functions, and so on.

For a vast majority of basis sets, including the split-valence sets, the basis functions are not made up of a single Gaussian function. Rather, a group of Gaussian functions are contracted together to form a single basis function. This is perhaps most easily understood with an explicit example: the popular split-valence 6-31G basis. The name specifies the contraction scheme employed in creating the basis set. The dash separates the core (on the left) from the valence (on the right). In this case, each core basis function is comprised of six Gaussian functions. The valence space is split into two basis functions, frequently referred to as the *inner* and *outer functions*. The inner basis function is composed of three contracted Gaussian functions, while each outer basis function is a single Gaussian function. Thus, for carbon, the core region is a single s basis function made up of six s-GTOs. The carbon valence space has two s and two p basis functions. The inner basis functions are made up of three Gaussians, and the outer basis functions are each composed of a single Gaussian function. Therefore, the carbon 6-31G basis set has nine basis functions made up of 22 Gaussian functions (Table 1.1).

Even large multizeta basis sets will not provide sufficient mathematical flexibility to adequately describe the electron distribution in molecules. An example of this deficiency is the inability to describe bent bonds of small rings. Extending the basis

TABLE 1.1 Composition of the Carbon 6-31G and 6-31+G(d) Basis Sets

	6-31G		6-31+G(d)	
	Basis functions	GTOs	Basis functions	GTOs
Core	s	6	s	6
Valence	s(inner)	3	s(inner)	3
	s(outer)	1	s(outer)	1
	p_x(inner)	3	p_x(inner)	3
	p_x(outer)	1	p_x(outer)	1
	p_y(inner)	3	p_y(inner)	3
	p_y(outer)	1	p_y(outer)	1
	p_z(inner)	3	p_z(inner)	3
	p_z(outer)	1	p_z(outer)	1
Diffuse			s(diffuse)	1
			p_y(diffuse)	1
			p_z(diffuse)	1
			p_z(diffuse)	1
Polarization			d_{xx}	1
			d_{yy}	1
			d_{zz}	1
			d_{xy}	1
			d_{xz}	1
			d_{yz}	1
Total		9	22	19	32

set by including a set of functions that mimic the AOs with angular momentum one greater than in the valence space greatly improves the basis flexibility. These added basis functions are called *polarization functions*. For carbon, adding polarization functions means adding a set of d GTOs while for hydrogen, polarization functions are a set of p functions. The designation of a polarized basis set is varied. One convention indicates the addition of polarization functions with the label "+P"; DZ+P indicates a DZ basis set with one set of polarization functions. For the split-valence sets, addition of a set of polarization functions to all atoms but hydrogen is designated by an asterisk, that is, 6-31G*, and adding the set of p functions to hydrogen as well is indicated by double asterisks, that is, 6-31G**. Since adding multiple sets of polarization functions has become broadly implemented, the use of asterisks has been deprecated in favor of explicit indication of the number of polarization functions within parentheses, that is, 6-311G(2df,2p) means that two sets of d functions and a set of f functions are added to nonhydrogen atoms and two sets of p functions are added to the hydrogen atoms.

For anions or molecules with many adjacent lone pairs, the basis set must be augmented with diffuse functions to allow the electron density to expand into a larger volume. For split-valence basis sets, this is designated by "+," as in 6-31+G(d). The diffuse functions added are a full set of additional functions of the same type as are present in the valence space. So, for carbon, the diffuse functions would be an added s basis function and a set of p basis functions. The composition of the 6-31+G(d) carbon basis set is detailed in Table 1.1.

The split-valence basis sets developed by Pople, though widely used, have additional limitations made for computational expediency that compromise the flexibility of the basis set. The correlation-consistent basis sets developed by Dunning[14–16] are popular alternatives. The split-valence basis sets were constructed by minimizing the energy of the atom at the HF level with respect to the contraction coefficients and exponents. The correlation-consistent basis sets were constructed to extract the maximum electron correlation energy for each atom. We will define the electron correlation energy in the next section. The correlation-consistent basis sets are designated as "cc-pVNZ," to be read as correlation-consistent polarized split-valence N-zeta, where N designates the degree to which the valence space is split. As N increases, the number of polarization functions also increases. So, for example, the cc-pVDZ basis set for carbon is DZ in the valence space and includes a single set of d functions, and the cc-pVTZ basis set is TZ in the valence space and has two sets of d functions and a set of f functions. The addition of diffuse functions to the correlation-consistent basis sets is designated with the prefix *aug-*, as in aug-cc-pVDZ. A set of even larger basis sets are the polarization consistent basis sets (called *pc-X*, where X is an integer) of Jensen,[17,18] and the *def2*-family developed the Ahlrichs[19] group. These modern basis sets are reviewed by Hill[20] and Jensen.[21]

Basis sets are built into the common computational chemistry programs. A valuable web-enabled database for retrieval of basis sets is available at the Molecular Science Computing Facility, Environmental and Molecular Sciences Laboratory "EMSL Gaussian Basis Set Order Form" (https://bse.pnl.gov/bse/portal).[22]

1.1.8.1 Basis Set Superposition Error Since in practice, basis sets must be of some limited size far short of the HF limit, their incompleteness can lead to a spurious result known as *basis set superposition error* (BSSE). This is readily grasped in the context of the binding of two molecules, A and B, to form the complex AB. The binding energy is evaluated as

$$E_{\text{binding}} = E_{AB}^{ab} - (E_A^a + E_B^b) \tag{1.24}$$

where a refers to the basis set on molecule A, b refers to the basis set on molecule B, and ab indicates the union of these two basis sets. Now in the supermolecule AB, the basis set a will be used to (1) describe the electrons on A, (2) describe, in part, the electrons involved in the binding of the two molecules, and (3) aid in describing the electrons of B. The same is true for the basis set b. The result is that the complex AB, by having a larger basis set than available to describe either A or B individually, is treated more completely and its energy will consequently be lowered, relative to the energy of A or B. The binding energy will therefore be larger (more negative) due to this superposition error.

The counterpoise method proposed by Boys and Bernardi[23] attempts to remove some of the effect of BSSE. The counterpoise correction is defined as

$$E_{CP} = E_{A*}^{ab} + E_{B*}^{ab} - (E_{A*}^a + E_{B*}^b) \tag{1.25}$$

The first term on the right-hand side is the energy of molecule A in its geometry of the complex (designated with the asterisk) computed with the basis set a and the basis functions of B placed at the position of the nuclei of B, but absent in the nuclei and electrons of B. These basis functions are called *ghost orbitals*. The second term is the energy of B in its geometry of the complex computed with its basis functions and the ghost orbitals of A. The last two terms correct for the geometric distortion of A and B from their isolated structure to the complex. The counterpoise-corrected binding energy is then

$$E_{\text{binding}}^{CP} = E_{\text{binding}} - E_{CP} \tag{1.26}$$

BSSE can, in principle, exist in any situation, including within a single molecule. There are two approaches toward removing this *intramolecular* BSSE. Asturiol et al.[24] propose an extension of the standard counterpoise correction: Divide the molecule into small fragments and apply the counterpoise correction to these fragments. For benzene, as an example, one can use C–H or (CH)$_2$ fragments.

Jensen[25]'s approach to remove *intramolecular* BSSE is to define the atomic counterpoise correction as

$$\Delta E_{ACP} = \sum E_A(\text{BasisSet}_A) - \sum E_A(\text{BasisSet}_{AS}) \tag{1.27}$$

where the sums run over all atoms in the molecule, and $E_A(\text{BasisSet}_A)$ is the energy of atom A using the basis set centered on atom A. The key definition is of the last term $E_A(\text{basisSet}_{AS})$; this is the energy of atom A using the basis set consisting of those functions centered on atom A and some subset of the basis functions centered on the other atoms in the molecule. The key assumption then is just how to select the subset of ghost functions to include in the calculation of the second term. For intramolecular corrections, Jensen suggests keeping only the orbitals on atoms at a certain bonded distance away from atom A. So, for example, ACP(4) would indicate that the energy correction is made using all orbitals on atoms that are four or more bonds away from atom A. Orbitals on atoms that are farther than some cut-off distance away from atom A may also be omitted.

Kruse and Grimme[26] proposed a correction for BSSE that relies on an empirical relationship based on the geometry of the molecule. They define energy terms on a per atom basis that reflects the difference between the energy of an atom computed with a particular basis set and the energy computed using a very large basis set. These atomic energies are scaled by an exponential decay based on the distances between atoms. This empirical correction, called *geometric counterpoise* (gCP), relies on four parameter; Kruse and Grimme report the values for a few combinations of method and basis set. The key advantage here is that this correction can be computed in a trivial amount of computer time, while the traditional CP corrections can be quite time consuming for large systems. They demonstrated that the B3LYP functional corrected for dispersion and gCP can provide quite excellent reaction energies and barriers.[27]

1.2 ELECTRON CORRELATION—POST-HARTREE–FOCK METHODS

The HF method ignores instantaneous electron–electron repulsion, also known as *electron correlation*. The electron correlation energy is defined as the difference between the *exact* energy and the energy at the HF limit

$$E_{\text{corr}} = E_{\text{exact}} - E_{\text{HF}} \tag{1.28}$$

How can we include electron correlation? Suppose the total electron wavefunction is composed of a linear combination of functions that depend on all n electrons

$$\Psi = \sum_i c_i \psi_i \tag{1.29}$$

We can then solve the Schrödinger equation with the *full* Hamiltonian (Eq. (1.5)) by varying the coefficients c_i so as to minimize the energy. If the summation is over an infinite set of these N-electron functions, ψ_i, we will obtain the exact energy. If, as is more practical, some finite set of functions is used, the variational principle tells us that the energy so computed will be above the exact energy.

The HF wavefunction *is* an N-electron function (itself composed of one-electron functions—the MOs). It seems reasonable to generate a set of functions from the HF wavefunction ψ_{HF}, sometimes called the *reference configuration*.

The HF wavefunction defines a single configuration of the n electrons. By removing electrons from the occupied MOs and placing them into the virtual (unoccupied) MOs, we can create new configurations, new N-electron functions. These new configurations can be indexed by how many electrons are relocated. Configurations produced by moving one electron from an occupied orbital to a virtual orbital are singly excited relative to the HF configuration and are called *singles* while those where two electrons are moved are called *doubles*, and so on. A simple designation for these excited configurations is to list the occupied MO(s), where the electrons are removed as a subscript and the virtual orbitals where the electrons are placed as the superscript. Thus, the generic designation of a singles configuration is ψ_i^a or ψ_S, a doubles configuration is ψ_{ij}^{ab} or ψ_D, and so on. Figure 1.2 shows an MO diagram for a representative HF configuration and examples of some singles and doubles configurations. These configurations are composed of spin-adapted Slater determinants, each of which is constructed from the arrangements of the electrons in the various, appropriate MOs.

1.2.1 Configuration Interaction (CI)

Using the definition of configurations, we can rewrite Eq. (1.29) as

$$\Psi_{CI} = c_0 \Psi_{HF} + \sum_i^{occ}\sum_a^{vir} c_i^a \psi_i^a + \sum_{i,j}^{occ}\sum_{a,b}^{vir} c_{ij}^{ab} \psi_{ij}^{ab} + \sum_{i,j,k}^{occ}\sum_{a,b,c}^{vir} c_{ijk}^{abc} \psi_{ijk}^{abc}$$

$$+ \sum_{i,j,k,l}^{occ}\sum_{a,b,c,d}^{vir} c_{ijkl}^{abcd} \psi_{ijkl}^{abcd} + \dots \quad (1.30)$$

In order to solve the Schrödinger equation, we need to construct the Hamiltonian matrix using the wavefunction of Eq. (1.30). Each Hamiltonian matrix element is

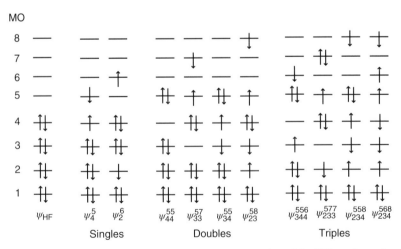

Figure 1.2 MO diagram indicating the electron occupancies of the HF configuration and representative examples of singles, doubles, and triples configurations.

the integral

$$\mathbf{H}_{xy} = \langle \psi_x | \hat{H} | \psi_y \rangle \quad (1.31)$$

where $\hat{H}$ is the full Hamiltonian operator (Eq. (1.5)) and ψ_x and ψ_y define some specific configuration. Diagonalization of this Hamiltonian then produces the solution—the set of coefficients that defines the configuration interaction (CI) wavefunction.[28] This is a rather daunting problem as the number of configurations is infinite in the exact solution, but still quite large for any truncated configuration set.

Fortunately, many of the matrix elements of the CI Hamiltonian are zero. Brillouin's theorem[29] states that the matrix element between the HF configuration and *any* singly excited configuration ψ_i^a is zero. The Condon–Slater rules provide the algorithm for computing any generic Hamiltonian matrix elements. One of these rules states that configurations that differ by three or more electron occupancies will be zero. In other words, suppose we have two configurations ψ_A and ψ_B defined as the Slater determinants $\psi_A = |\phi_1 \phi_2 \cdots \phi_{n-3} \phi_i \phi_j \phi_k|$ and $\psi_B = |\phi_1 \phi_2 \cdots \phi_{n-3} \phi_r \phi_s \phi_t|$, then

$$\langle \psi_A | \hat{H} | \psi_B \rangle = 0 \quad (1.32)$$

Therefore, the Hamiltonian matrix tends to be rather sparse, especially as the number of configurations included in the wavefunction increases.

Since the Hamiltonian is both spin- and symmetry-independent, the CI expansion only contains configurations that are of the spin and symmetry of interest. Even taking advantage of the spin, symmetry, and sparseness of the Hamiltonian matrix, we may nonetheless be left with a matrix of a size well beyond our computational resources.

Two approaches toward truncating the CI expansion to some manageable length are utilized. The first is to delete some subset of virtual MOs from being potentially occupied. Any configuration where any of the very highest energy MOs are occupied will be of very high energy and will likely contribute very little toward the description of the ground state. Similarly, we can freeze some MOs (usually those describing the core electrons) to be doubly occupied in all configurations of the CI expansion. Those configurations where the core electrons are promoted into a virtual orbital are likely to be very high in energy and unimportant.

The second approach is to truncate the expansion at some level of excitation. By Brillouin's theorem, the single excited configurations will not mix with the HF reference. By the Condon–Slater rules, this leaves the doubles configurations as the most important for including in the CI expansion. Thus, the smallest reasonable truncated CI wavefunction includes the reference and all doubles configurations (CID):

$$\psi_{CID} = c_0 \psi_{HF} + \sum_{i,j}^{occ} \sum_{a,b}^{vir} c_{ij}^{ab} \psi_{ij}^{ab} \quad (1.33)$$

The most widely employed CI method includes both the singles and doubles configurations (CISD):

$$\Psi_{CISD} = c_0\Psi_{HF} + \sum_i^{occ}\sum_a^{vir} c_i^a \psi_i^a + \sum_{i,j}^{occ}\sum_{a,b}^{vir} c_{ij}^{ab}\psi_{ij}^{ab} \quad (1.34)$$

where the singles configurations enter by their nonzero matrix elements with the doubles configurations. Higher order configurations can be incorporated, if desired.

1.2.2 Size Consistency

Suppose one was interested in the energy of two molecules separated far from each other. (This is not as silly as it might sound—it is the description of the reactants in the reaction A + B → C.) This energy could be computed by calculating the energy of the two molecules at some large separation, say 100 Å. An alternative approach is to calculate the energy of each molecule separately and then add their energies together. These two approaches should give the same energy. If the energies are identical, we call the computational method "size consistent."

While the HF method and the complete CI method (infinite basis set and all possible configurations) are size-consistent, a truncated CI is *not* size-consistent! A simple way to understand this is to examine the CID case for the H_2 dimer, with the two molecules far apart. The CID wavefunction for the H_2 molecule includes the double excitation configuration. So taking twice the energy of this monomer effectively includes the configuration where *all four* electrons have been excited. However, in the CID computation of the dimer, this configuration is not allowed; only doubles configurations are included—not this quadruple configuration. The Davidson[30] correction approximates the energy of the missing quadruple configurations as

$$E_Q = (1 - c_0)(E_{CISD} - E_{HF}) \quad (1.35)$$

1.2.3 Perturbation Theory

An alternative approach toward including electron correlation is provided by perturbation theory. Suppose we have an operator $\hat{O}$ that can be decomposed into two component operators

$$\hat{O} = \hat{O}^{(0)} + \hat{O}' \quad (1.36)$$

where the eigenvectors and eigenvalues of $\hat{O}^{(0)}$ are known. The operator $\hat{O}'$ defines a perturbation upon this known system to give the true operator. If the perturbation is small, then Rayleigh–Schrödinger perturbation theory provides an algorithm for finding the eigenvectors of the full operator as an expansion of the eigenvectors of $\hat{O}^{(0)}$. The solutions derive from a Taylor series, which can be truncated to whatever order is desired.

Møller and Plesset[31] developed the means for applying perturbation theory to molecular system. They divided the full Hamiltonian (Eq. (1.5)) into essentially the HF Hamiltonian, where the solution is known and a set of eigenvectors can be created (the configurations discussed above), and a perturbation component that is essentially the instantaneous electron–electron correlation. The HF wavefunction is correct through first-order Møller–Plesset (MP1) perturbation theory. The second-order correction (MP2) involves doubles configurations, as does MP3. The fourth-order correction (MP4) involves triples and quadruples. The terms involving the triples configuration are especially time consuming. MP4SDQ is fourth-order perturbation theory neglecting the triples contributions, an approximation that is appropriate when the highest occupied molecular orbital–lowest unoccupied molecular orbital (HOMO–LUMO) gap is large.

The major benefit of perturbation theory is that it is computationally more efficient than CI. MP theory, however, is not variational. This means that at any particular order, the energy may be above or below the actual energy. Furthermore, since the perturbation is really not particularly small, including higher order corrections are not guaranteed to converge the energy, and extrapolation from the energy determined at a small number of orders may be impossible. On the positive side, MP theory is size-consistent at any order.

Nonetheless, MP2 is quite a bit slower than HF theory. The resolution of the identity approximation (RI) makes MP2 nearly competitive with HF in terms of computational time. This approximation involves a simplification of the evaluation of the four-index integrals.[32,33]

Grimme[34–36] proposed an empirical variant of MP2 that generally provides improved energies. This is the spin-component-scaled MP2 (SCS-MP2) that scales the terms involving the electron pairs having the same spin (SS) differently than those with opposite spins (OS). The SCS-MP2 correlation correction is given as

$$E_C(SCS - MP2) = p_{OS} E_c^{OS}(MP2) + p_{SS} E_c^{SS}(MP2) \qquad (1.37)$$

where p_{OS} and p_{SS} are empirically fit terms, with best values of 6/5 and 1/3, respectively.

1.2.4 Coupled-Cluster Theory

Coupled-cluster theory, developed by Cizek,[37] describes the wavefunction as

$$\Psi = e^{\hat{T}} \psi_{HF} \qquad (1.38)$$

The operator $\hat{T}$ is an expansion of operators

$$\hat{T} = \hat{T}_1 + \hat{T}_2 + \cdots + \hat{T}_n \qquad (1.39)$$

where the $\hat{T}_i$ operator generates all of the configurations with i electron excitations. Since Brillouin's theorem states that singly excited configurations do not mix directly with the HF configuration, the $\hat{T}_2$ operator

$$\hat{T}_2 = \sum_{i,j}^{occ}\sum_{a,b}^{vir} t_{ij}^{ab}\psi_{ij}^{ab} \tag{1.40}$$

is the most important contributor to $\hat{T}$. If we approximate $\hat{T} = \hat{T}_2$, we have the CCD (coupled-cluster doubles) method, which can be written as the Taylor expansion:

$$\psi_{CCD} = e^{\hat{T}_2}\psi_{HF} = \left(1 + \hat{T}_2 + \frac{\hat{T}_2^2}{2!} + \frac{\hat{T}_2^3}{3!} + \ldots\right)\psi_{HF} \tag{1.41}$$

Because of the incorporation of the third and higher terms of Eq. (1.34), the CCD method *is* size consistent. Inclusion of the $\hat{T}_1$ operator is only slightly more computationally expensive than the CCD calculation and so the coupled-clusters CCSD (coupled-cluster singles and doubles) method is the typical coupled-cluster computation. Inclusion of the $\hat{T}_3$ operator is quite computationally demanding. An approximate treatment, where the effect of the triples contribution is incorporated in a perturbative treatment is the CCSD(T) method,[38] which has become the "gold standard" of computational chemistry—the method of providing the most accurate evaluation of the energy. CCSD(T) requires substantial computational resources and is therefore limited to relatively small molecules. Another downside to the CC methods is that they are not variational. A recent comparison of binding energy in a set of 24 systems that involve noncovalent interactions, an interaction that is very sensitive to the accounting of electron correlation, shows that errors in the bonding energy are less that 1.5 percent using the CCSD(T) method.[39] These errors are due to neglect of core correlation, relativity and higher order correlation terms (full treatment of triples and perturbative treatment of quadruples).

There are a few minor variations on the CC methods. The quadratic configuration interaction including singles and doubles (QCISD)[40] method is nearly equivalent to CCSD. Another variation on CCSD is to use the Brueckner orbitals. Brueckner orbitals are a set of MOs produced as a linear combination of the HF MOs such that all of the amplitudes of the singles configurations (t_i^a) are zero. This method is called *BD* and differs from CCSD method only in fifth order.[41] Inclusion of triples configurations in a perturbative way, BD(T), is frequently more stable (convergence of the wavefunction is often smoother) than in the CCSD(T) treatment.

1.2.5 Multiconfiguration SCF (MCSCF) Theory and Complete Active Space SCF (CASSCF) Theory

To motivate a discussion of a different sort of correlation problem, we examine how to compute the energy and properties of cyclobutadiene. An RHF calculation of rectangular D_{2h} cyclobutadiene **1** reveals four π MOs, as shown in Figure 1.3. The HF configuration for this molecule is

$$\psi_{HF} = |\cdots \pi_1^2 \pi_2^2| \tag{1.42}$$

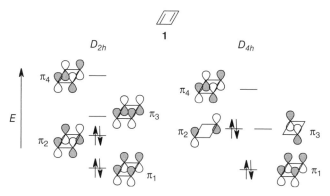

Figure 1.3 π-MO diagram of cyclobutadiene (**1**).

As long as the HOMO–LUMO energy gap (the difference in energy of π_2 and π_3) is large, this single configuration wavefunction is reasonable. However, as we distort cyclobutadiene more and more toward a D_{4h} geometry, the HOMO–LUMO gap grows smaller and smaller, until we reach the square planar structure where the gap is nil. Clearly, the wavefunction of Eq. (1.31) is inappropriate for D_{4h} cyclobutadiene, and also for geometries close to it because it does not contain any contribution from the degenerate configuration $|\cdots \pi_1^2 \pi_3^2|$.

Rather, a more suitable wavefunction for cyclobutadiene might be

$$\Psi = c_1 |\cdots \pi_1^2 \pi_2^2| + c_2 |\cdots \pi_1^2 \pi_3^2| \qquad (1.43)$$

This wavefunction appears to be a CI wavefunction with two configurations. Adding even more configurations would capture more of the *dynamic* electron correlation. The underlying assumption to the CI expansion is that the single-configuration reference, the HF wavefunction, is a reasonable description of the molecule. For cyclobutadiene, especially as it nears the D_{4h} geometry, the HF wavefunction does not capture the inherent multiconfigurational nature of the electron distribution. The MOs used to describe the first configuration of Eq. (1.43) are *not* the best for describing the second configuration. To capture this *nondynamic* correlation (often also called *static* correlation), we must determine the set of MOs that best describe *each* of the configurations of Eq. (1.43), giving us the wavefunction

$$\Psi_{\text{MCSCF}} = c_1 |\cdots \sigma_{11}^2 \pi_1^2 \pi_2^2| + c_2 |\cdots \sigma_{11'}^2 \pi_{1'}^2 \pi_{3'}^2| \qquad (1.44)$$

where the primed orbitals are different from the unprimed set. We have explicitly indicated the highest σ-orbital in the primed and unprimed sets to emphasize that all of the MOs are optimized within each configuration. In the multiconfiguration SCF (MCSCF)[42] method, the coefficient c_i of each configuration, along with the LCAO expansion of the MOs of each configuration, are solved for in an iterative, self-consistent way.

The question arises as to how to select the configurations for the MCSCF wavefunction. In the example of cyclobutadiene, one might wonder about also including the configurations where π_2 and π_3 are each singly occupied with net spin of zero,

$$\Psi_{MCSCF} = c_1|\cdots\sigma_{11}^2\pi_1^2\pi_2^2| + c_2|\cdots\sigma_{11'}^2\pi_{1'}^2\pi_{3'}^2| + c_3|\cdots\sigma_{11''}^2\pi_{1''}^2\pi_{2''}^1(\alpha)\pi_{3''}^1(\beta)| \quad (1.45)$$

Perhaps a more appropriate selection might also include configurations where the electrons from π_1 are excited into the higher lying π-orbitals. A goal of *ab initio* theory is to make as few approximations and as few arbitrary decisions as possible. In order to remove the possibility that an arbitrary selection of configurations might distort the result, the complete active space SCF (CASSCF)[43] procedure dictates that *all* configurations involving a set of MOs (the active space) and a given number of electrons comprise the set of configurations to be used in the MCSCF procedure. This set of configurations is indicated as CASSCF(n,m), where n is the number of electrons and m is the number of MOs of the active space (both occupied and virtual). So, an appropriate calculation for cyclobutadiene is CASSCF(4,4), where all four π-electrons are distributed in all possible arrangements among the four π-MOs.

Since MCSCF attempts to account for the nondynamic (static) correlation, really to correct for the inherent multiconfiguration nature of the electron distribution, how can one then also capture the dynamic correlation? The application of perturbation theory using the MCSCF wavefunction as the reference requires some choice as to the nonperturbed Hamiltonian reference. This had led to a number of variants of multireference perturbation theory. The most widely utilized is CASPT2N,[44] which is frequently referred to as *CASPT2* though this designation ignores other flavors developed by the same authors. Along with CCSD(T), CASPT2N is considered to be one of the more robust methods for obtaining the highest quality treatment of QM of molecules.

For molecules that require a multireference description, use of a single reference post-HF method can often fail since the dynamic correlation space is insufficient. Multireference post-HF methods are quite taxing in terms of computational resources and comprise a very active area of theoretical development.[45] A method that has shown some recent promise is multireference coupled cluster (MRCC) theory, and the implementation proposed by Mukherjee and coworkers[46,47] (often labeled as MkCC or MkMRCC) has garnered much interest.[48]

1.2.6 Composite Energy Methods

While rigorous quantum chemical methods are available, the best of them are exceptionally demanding in terms of computer performance (CPU time, memory, and hard disk needs). For all but the smallest molecules, these best methods are impractical.

How then to capture the effects of large basis sets and extensive accounting of electron correlation? The answer depends in part on what question one is seeking to answer—are we looking for accurate energies or structures or properties? Since all of these are affected by the choice of basis set and treatment of electron correlation, oftentimes to different degrees, which methods are used depends on what information we seek. As we will demonstrate in the following chapters, prediction of geometries is usually less demanding than obtaining accurate energies. We may then get by with relatively small basis sets and low orders of electron correlation treatment. Accurate energies are, however, quite sensitive to the computational method.

The composite methods were developed to provide an algorithm for obtaining accurate energies. They take the approach that the effect of larger basis sets, including the role of diffuse and polarization functions, and the effect of higher order treatment of electron correlation can be approximated as additive corrections to a lower level computation. One can thereby reproduce a huge computation, say a CCSD(T) calculation with the 6-311+G(3df,2p) basis set, by summing together the results of a series of much smaller calculations.

This first model chemistry, called *G1*,[49] was proposed by Pople[50] and Curtiss[51] in the late 1980s, but was soon replaced by the more accurate G2 and G3 model chemistries. The latest version is called *G4*.[52] The baseline calculation is to compute the energy at MP4 with the 6-31G(d) basis set using the geometry optimized at B3LYP/6-31G(2df,p). Corrections for various deficiencies are then made as additions to this baseline energy. The steps for carrying out the G4 calculation are listed in the following.

(1) Optimize the geometry at B3LYP/6-31G(2df,p) and compute the zero-point vibrational energy (ZPVE), using the computed frequencies scaled by 0.9854. Use this geometry for all subsequent single-point energy computations.
(2) Compute the baseline energy: E[MP4/6-31G(d)].
(3) Correct for diffuse functions: E[MP4/6-31+G(d)] – E[MP4/6-31G(d)].
(4) Correct for higher order polarization functions: E[MP4/6-31G(2df,p)] – E[MP4/6-31G(d)].
(5) Correct for better treatment of electron correlation: E[CCSD(T)/6-31G(d)] – E[MP4/6-31G(d)].
(6) Correct for larger basis sets and additivity assumptions: E[MP2(full)/G3LargeXP] – E[MP2/6-31G(2df,p)] – E[MP2/6-31+g(d)] + E[MP2/6-31G(d)].
(7) Correct for extrapolation to the HF limit: $(E[\text{HF/aug-cc-pV5Z}] - E[\text{HF/aug-cc-pVQZ}]e^{-1.63})/(1 - e^{-1.63})$.
(8) Apply a correction for spin-orbit coupling and an empirical correction for higher level corrections.
(9) Compute the G4 energy as $E[\text{G4}] = \text{ZPVE}(1) + (2) + (3) + (4) + (5) + (6) + (7) + (8)$.

Subsequently, the G4(MP2)[53] model was proposed, whose major advantage is avoiding a number of the larger MP4 computations and extrapolating to the HF limit using smaller basis sets. Using a large test set of some 454 energies from experiment, such as enthalpies of formation, ionization potentials, electron affinities, and proton affinities, the average error with G4 is 0.80 and 0.99 kcal mol^{-1} with G4(MP2). These are both much better than with any of the previous G*n* models.

There are other series of composite methods: the CBS-n models of Petersson,[54,55] the HEAT (high accuracy extrapolated ab initio thermochemistry) approach of Stanton,[56,57] the Wn models of Martin,[58–60] and the quite recently developed FPD (Feller, Peterson, Dixon) method.[61] All of these composite methods are conceptually similar, just varying in which quantum methods are used for the baseline and the corrections and what sets of compounds and what properties will be used in the ultimate fitting procedure.[62] Because of the fitting of the calculated energy to some experimental energy (often atomization energies), these composite methods have an element of semiempirical nature to them. The focal-point scheme developed by Allen and Schaefer[63] combines (1) the effect of basis set by extrapolating the energies from calculations with large basis sets (up to cc-pV6Z), (2) the effect of higher order correlation by extrapolation of energies from higher order MP (up to MP5) or CC (up to CCSDT), and (3) corrections for the assumed additivity of basis set and correlation effects. It produces extraordinary accuracy without resorting to any empirical corrections, but the size of the computations involved generally restricts application to molecules with less than 10 atoms.

1.3 DENSITY FUNCTIONAL THEORY (DFT)

The electronic wavefunction is dependent on $3n$ variables: the x, y, and z coordinates of each electron. As such, it is quite complicated and difficult to readily interpret. The total electron density $\rho(\mathbf{r})$ is dependent on just three variables: the x, y, and z positions in space. Since $\rho(\mathbf{r})$ is simpler than the wavefunction and is also observable, perhaps it might offer a more direct way to obtain the molecular energy.

The Hohenberg–Kohn[64] existence theorem proves just that: there exists a unique functional such that

$$E[r(\mathbf{r})] = E_{elec} \quad (1.46)$$

where E_{elec} is the exact electronic energy. Furthermore, they demonstrated that the electron density obeys the variational theorem. This means that given a specific electron density, its energy will be greater than or equal to the exact energy. These two theorems constitute the basis of density functional theory (DFT). The hope is that evaluation of Eq. (1.46) might be easier than traditional *ab initio* methods because of the simpler variable dependence.

Before proceeding with an explanation of how this translates into the ability to compute properties of a molecule, we need to define the term *functional*. A mathematical *function* is one that relates a scalar quantity to another scalar quantity, that is, $y = f(x)$. A mathematical *functional* relates a function to a scalar quantity and is denoted within brackets, that is, $y = F[f(x)]$. In Eq. (1.46), the function $\rho(\mathbf{r})$ depends on the spatial coordinates, and the energy depends on the values (is a functional) of $\rho(\mathbf{r})$.

In order to solve for the energy via the DFT method, Kohn and Sham[65] proposed that the functional has the form

$$E[\rho(\mathbf{r})] = T_{e'}[\rho(\mathbf{r})] + V_{ne}[\rho(\mathbf{r})] + V_{ee}[\rho(\mathbf{r})] + E_{xc}[\rho(\mathbf{r})] \quad (1.47)$$

where V_{ne}, the nuclear–electron attraction term, is

$$V_{ne}[\rho(\mathbf{r})] = \sum_I^N \int \frac{Z_I}{|\mathbf{r} - \mathbf{r}_I|} \rho(\mathbf{r}) d\mathbf{r} \quad (1.48)$$

and V_{ee}, the classical electron–electron repulsion term, is

$$V_{ee}[\rho(\mathbf{r})] = \frac{1}{2} \int \int \frac{\rho(\mathbf{r}_1)\rho(\mathbf{r}_2)}{|\mathbf{r}_1 - \mathbf{r}_2|} d\mathbf{r}_1 d\mathbf{r}_2 \quad (1.49)$$

The real key, though, is the definition of the first term of Eq. (1.47). Kohn and Sham defined it as the kinetic energy of *noninteracting* electrons whose density is the same as the density of the real electrons, the true interacting electrons. The last term is called the *exchange-correlation functional*, and is a catchall term to account for all other aspects of the true system.

The Kohn–Sham procedure is to solve for the orbitals that minimize the energy, which reduces to the set of pseudoeigenvalue equations

$$\hat{h}_i^{KS} \chi_i = E_i \chi_i \quad (1.50)$$

This is closely analogous to the Hartree equations (Eq. (1.7)). The Kohn–Sham orbitals are separable by definition (the electrons they describe are *noninteracting*) analogous to the HF MOs. Eq. (1.50) can, therefore, be solved using a similar set of steps as was done in the Hartree–Fock–Roothaan method.

So for a similar computational cost as the HF method, DFT produces the energy of a molecule that *includes the electron correlation*! This is the distinct advantage of DFT over the traditional *ab initio* methods discussed previously—it is much more computationally efficient in providing the correlation energy.

DFT is not without its own problems, however. While the Hohenberg–Kohn theorem proves the existence of a functional that relates the electron density to the energy, it offers no guidance as to the form of that functional. The real problem is the exchange-correlation term of Eq. (1.47). There is no way of deriving this term, and so a series of different functionals have been proposed, leading to a lot of different DFT methods. A related problem with DFT is that if the chosen functional fails, there is no way to systematically correct its performance. Unlike with CI, where one can systematically improve the result by increasing the number and type of configurations employed in the wavefunction expansion, or with perturbation theory, where one can move to arbitrarily higher order corrections, if a given functional does not provide a suitable result, one must go back to square one and select a new functional. Paraphrasing Cramer's[1] description of the contrast between wavefunction theory (WFT) and DFT, HF and the various post-HF electron correlation methods provide an *exact* solution to an *approximate* theory, while DFT provides an *exact* theory with an *approximate* solution.

1.3.1 The Exchange-Correlation Functionals: Climbing Jacob's Ladder

We should begin this discussion by remarking that the exchange-correlation functional is poorly named as it is not the same as exchange and correlation defined within WFT. Rather, it is named this way as a weak analogy to those classic quantum mechanical terms. The exchange-correlation functional is generally written as a sum of two components, an exchange part and a correlation part. This is an assumption, an assumption that we have no way of knowing is true or not. These component functionals are usually written in terms of an energy density ε,

$$E_{XC}[\rho(\mathbf{r})] = E_X[\rho(\mathbf{r})] + E_C[\rho(\mathbf{r})] = \int \rho(\mathbf{r})\varepsilon_X[\rho(\mathbf{r})]d\mathbf{r} + \int \rho(\mathbf{r})\varepsilon_C[\rho(\mathbf{r})]d\mathbf{r} \quad (1.51)$$

Perdew[66,67] has described a hierarchy of approximate treatments of the exchange-correlation term as "Jacob's Ladder." The ladder is grounded in HF theory here on earth, and reaches to heaven, where the exact functional is found. Along the way are five rungs, each defining a set of assumptions made in creating an exchange-correlation expression.

The first rung comprises the local density approximation (LDA), which assumes that the value of ε_x could be determined from just the value of the density. A simple example of the LDA is Dirac's treatment of a uniform electron gas, which gives

$$\varepsilon_X^{LDA} = -C_X\rho^{1/3} \quad (1.52)$$

This can be extended to the local spin density approximation (LSDA) for those cases where the α and β densities are not equal. Slater's X_α method is a scaled form of Eq. (1.52), and often the terms "LSDA" and "Slater" are used interchangeably.

Local correlation functionals were developed by Vosko et al.,[68] which involve a number of terms and empirical scaling factors. The most popular versions are called *VWN* and *VWN5*. The combination of a local exchange and a local correlation energy density is the SVWN method.

In order to make improvements over the LSDA, one has to assume that the density is not uniform. The approach that has been taken is to develop functionals that are dependent on not only the electron density but also derivatives of the density. This constitutes the generalized gradient approximation (GGA) and is the second rung on Jacob's Ladder. The third rung, meta-GGA functional, includes a dependence on the Laplacian of the density ($\nabla^2\rho$) or on the orbital kinetic energy density (τ). The fourth row, the hyper-GGA or hybrid functionals, includes a dependence on exact (HF) exchange. Finally, the fifth row incorporates the unoccupied Kohn–Sham orbitals. This is most widely accomplished within the so-called *double hybrid functionals*.

Close inspection of functionals might lead to the eyes glazing over amid a sea of random acronyms in an alphabet soup. For full mathematical details, the interested reader is referred to the books by Cramer[1] or Jensen[2] or the monograph by Koch and Holhausen, *A Chemist's Guide to Density Functional Theory*.[69]

We will present here just a few of the more widely utilized functionals. The DFT method is denoted with an acronym that defines the exchange functional and the correlation functional in that order. For the exchange component, the most widely used is one proposed by Becke.[70] It introduces a correction term to LSDA that involves the density derivative. The letter "B" signifies its use as the exchange term. Of the many correlation functionals, the two most widely used are due to Lee, Yang, and Parr[71] (referred to as *LYP*) and by Perdew and Wang[72] (referred to as *PW91*). While the PW91 functional depends on the derivative of the density, the LYP functional depends on $\nabla^2 \rho$, so it is of the third rung. So the BPW91 designation indicates use of the Becke exchange functional with the Perdew–Wang (19)91 correlation functional. Another popular second-rung functional is PBE0 (sometimes referred to as *PBE1PBE*).[73,74]

The hybrid methods (fourth rung) combine the exchange-correlation functionals with some admixture of the HF exchange term and are among the most widely used functionals. The dominant DFT method for the 1990s and well into the 2000s is the hybrid B3LYP functional,[75,76] which includes Becke's exchange functional along with the LYP correlation functional:

$$E_{XC}^{B3LYP} = (1-a)E_X^{LSDA} + aE_X^{HF} + bE_X^B + (1-c)E_C^{LSDA} + cE_C^{LYP} \quad (1.53)$$

The three variables (a, b, and c) are the origin of the "3" in the acronym. Since these variables are fit to reproduce the experimental data, B3LYP (and all other hybrid methods) contains some degree of "semiempirical" nature.

A nonempirical third-rung functional is TPSS,[77] which includes a term based on the kinetic energy density of the occupied Kohn–Sham orbitals. The M06 family of hybrid meta functionals[78–81] also includes a kinetic energy density term and is highly parameterized to fit experimental data. The M06-2x variant, which adds a nonlocal HF exchange term in a fashion similar to Eq. (1.53) (and is thus a forth-rung functional), has become quite popular due to its performance[80] in situations that have been notoriously problematic for other DFT methods, such as noncovalent interactions including π–π stacking and transition metal–transition metal bonds.

An approach to address the proper r^{-1} decay of the exchange-correlation potential is the LC scheme, where LC indicates a long-range correction.[82] HF exchange is only applied to long-range electron correlation, with a functional defined as

$$E_{xc}^{LC-DFA} = E_x^{LR-HF} + E_x^{SR-DFA} + E_C^{DFA} \quad (1.54)$$

where DFA indicates some density functional approximation and LR and SR refer to long-range and short-range corrections. The ωB97 method employs the B97[83] functional and makes short- and long-range corrections to it.[84] A hybrid variant, ωB97-X,[84] includes a short-range HF exchange as

$$E_{xc}^{\text{LC-DFA}} = E_x^{\text{LR-HF}} + c_c E_x^{\text{SR-HF}} + E_x^{\text{SR-DFA}} + E_C^{\text{DFA}} \qquad (1.55)$$

1.3.1.1 Double Hybrid Functionals

Incorporation of the unoccupied Kohn–Sham orbitals to create fifth-rung methods has been difficult. The one approach that has seen some application to organic chemistry is the semiempirical double hybrid approach of Grimme.[85,86] The approach begins with a hybrid GGA functional of the form

$$E_{XC}^{\text{hybrid-GGA}} = a_1 E_X^{\text{GGA}} + a_2 E_C^{\text{GGA}} + (1 - a_1) E_X^{\text{HF}} \qquad (1.56)$$

where any semilocal exchange and correlation functionals can be used. A normal Kohn–Sham calculation is performed and the resulting orbitals are then used for a standard Moller–Plesset second-order perturbation theory computation:

$$E_C^{\text{PT2}} = \frac{1}{4} \sum_{ijab} \frac{|\langle ij|ab \rangle|^2}{(\varepsilon_i + \varepsilon_j - \varepsilon_a - \varepsilon_b)} \qquad (1.57)$$

where ij label occupied and ab unoccupied Kohn–Sham orbitals with corresponding energy ε. The total exchange-correlation term is then given as the sum

$$E_{XC}^{\text{double-hybrid}} = E_{XC}^{\text{hybrid-GGA}} + (1 - a_2) E_C^{\text{PT2}} \qquad (1.58)$$

The B2-PLYP double hybrid, constructed using Becke88 exchange and the LYP correlation functionals, and the mPW2-PLYP, which uses the modified Perdew–Wang exchange term, both have two fit parameters a_1 and a_2. (Thus the naming convention with the exchange term listed first, the "2" indicating the two parameters, the "P" indicating the perturbation correction, and the exchange term given last.) As a test of the performance of the double hybrids, the mean absolute deviation for the G3/05 test set, which includes a large number of experimental thermochemical data, is 2.5 kcal mol^{-1} for B2-PLYP and 2.1 kcal mol^{-1} for mPW2-PLYP. These deviations are significantly better than for standard DFT methods like B3LYP and TPSS, though twice as large as that of G3. Martin[87] has proposed modified parameters to provide better performance for activation barriers. One major drawback to the double hybrid method is that since a standard MP2 computation must be performed, much of the computational expediency of DFT is lost.

1.3.2 Dispersion-Corrected DFT

London dispersion interactions are, in general, omitted from traditional density functionals. These long-range attractions are an essential component of wide swaths of chemistry and physics, including systems such as DNA, protein folding, nanoarchitectures, and molecular recognition. We will discuss in detail in later chapters a few situations where proper accounting of dispersion is essential, such as π-stacking and hydrogen bonding.

Though a number of approaches have been made for incorporating dispersion into DFT,[88] the most widely utilized method is to append a semiclassical correction. This approach, called *DFT-D*, was developed by Grimme.[89] A dispersion term is added to the Kohn–Sham DFT energy (Eq. (1.47)) of the form

$$E_{\text{disp}} = \sum_{AB} \sum_{n=6,8,10...} s_n \frac{C_n^{AB}}{R_{AB}^n} f_{\text{damp}}(R_{AB}) \quad (1.59)$$

This builds off of the long-range Lennard–Jones-type potential. The expansion is in the inverse of the distance between atom pairs to the sixth power, eighth power, and so on. Each of these terms has a coefficient C_n^{AB} that depends on the particular atom pair AB. Each term is scaled by a coefficient s_n and a damping term f_{damp}. The coefficients C and s must be defined for each density functional method, that is, a set for B3LYP, another set for PBE0, and so on. Addition of the dispersion correction is denoted by adding "-D" to the name, such as B3LYP-D. The first two versions of this correction, termed now *-D1*[89] and *-D2*[90] include only the $n = 6$ term. The latest version, indicated by the suffix "-D3" includes two terms, for $n = 6$ and $n = 8$.[91] The second term aids in describing the mid-range dispersion along with the long-range dispersion generally thought to follow the R^{-6} relation.

One can even combine this damped atom–atom dispersion correction to the LC hybrid functionals. The ωB97X functional has been modified to include dispersion correction to create the ωB97X-D functional.[92] Performance of this functional is significantly better than dispersion-corrected standard functionals, like B3LYP-D, and improves performance in the G3/05 thermodynamic data set by 0.6 kcal mol^{-1} over ωB97-X.

A few different alternatives have been proposed for the damping term. In a series of tests on systems particularly sensitive to dispersion, the different damping functions have similar effects.[93] The damping function of Becke and Johnson[94] leads to a dispersion energy term (for a particular order n)

$$E_{\text{disp}} = -\frac{1}{2} \sum_{AB} \frac{C_n^{AB}}{R_{AB}^n + \text{const}} \quad (1.60)$$

Use of this damping term within the D3 correction "-D3(BJ)" is the method now promoted by Grimme. Mean deviations for a benchmark data set that involve weak interactions, chemical reactions, and conformations are reduced by at least 1 kcal mol^{-1} with the inclusion of the D3 corrections for a range of functionals.[88,91] Even the performance of the double hybrid methods can be improved by about 0.5 kcal mol^{-1} with the D3 correction.[95] In a study of molecular geometry in weakly interacting systems, the B3LYP-D3 model produced excellent results.[96] Since the added computational time for including the dispersion correction is relatively small, this option should be routinely taken.

Kozuch[97] and Martin have proposed combining the double hybrid method with the SCS-MP2 treatment along with a dispersion correction. This so-called *DSD-DFT* method (dispersion corrected, spin-component-scaled double hybrid) has the exchange-correlation term

$$E_{XC}^{DSD-DFT} = c_X E_X^{DFT} + (1 - c_X)E_X^{HF} + c_C E_C^{DFT} + c_O E_O^{MP2} + c_S E_S^{MP2} + s_6 E_D \quad (1.61)$$

where c_X indicates the contribution of DFT exchange, c_C is the contribution of DFT correlation, c_O and c_S indicate the contribution of the opposite- and same-spin MP2 terms, and s_6 indicates the contribution of the dispersion term. Their best performing DSD-DFT functional, benchmarked against a number of standards including kinetic and thermochemical databases, incorporates the PBEP86 functional.[98] This approach is likely to garner much attention in the future.

1.3.3 Functional Selection

The plethora of functionals is overwhelming to the beginner, and even gives pause to the long-time practitioner. Just how does one select the right functional for the problem at hand? Hopefully, some guidance can be found in the many examples to be described in the later chapters. Of particular interest will be Section 3.3.3 where we present a series of critical studies of many widely used functionals (including the once nearly ubiquitous B3LYP functional) that show serious failures for seemingly very simple organic molecules.

Unfortunately, there is no single functional that appears to work best for all chemical systems. One should be guided by functional performance for analogous compounds or reactions that have been previously examined. It does seem clear that some accounting for dispersion must be included in the functional for optimal performance, but this still leaves a wide selection of functionals.

The list of functionals that have been developed and utilized is quite long and continues to grow. Instead of providing here a comprehensive list, the interested user is directed to examine the latest density functional popularity poll organized by Swart et al.[99] Functionals are voted on by computational chemists and the results lead to a *Primera divisió* and *Segona divisió*. Among the most favored functionals in the 2012 poll are B2PLYP, B3LYP, B3LYP*, B3LYP-D, B3PW91, B97-D, BLYP, BP86, CAM-B3LYP, and M06-2X.

1.4 COMPUTATIONAL APPROACHES TO SOLVATION

1.4.1 Microsolvation

Imagine the environment about a solute in a water solution. The solute is in direct contact with a small number of water molecules, interacting with them through perhaps, hydrogen bonds, electrostatic, and van der Waals interactions. This constitutes the first solvation sphere. The water molecules in the first sphere are in direct contact with neighboring water molecules, primarily through a hydrogen-bonding network. These water molecules in the second solvation sphere are weakly interacting with the solute, but in turn are hydrogen bonded to a further larger layer of water molecules. This goes on and on, creating ever larger spheres (shells) of water molecules.

Now imagine computing the energy of this collection of solute and water molecules. The sheer number of molecules is enormous, making a quantum computation intractable. Even if one could compute the energy of this large ensemble, it would be for a *single* configuration. Many, many configurations would have to be sampled in order to create a Boltzmann distribution and an appropriate free energy of the system.

Clearly, one must truncate the number of solvation shells to limit the number of water molecules to some reasonable value. But just how many water molecules are necessary to obtain bulk liquid water? Certainly a calculation of the solute and just the first solvation shell does not capture the effect of bulk water. Without the next solvation shell, the water molecules in the first shell do not have these neighboring water molecules to interact with via hydrogen bonding. Instead, the water molecules might seek out additional favorable interactions with the solute or be forced to have some dangling O–H bonds and lone pairs.

Nonetheless, this technique of including a small number of explicit solvent molecules within the full quantum mechanical treatment can effectively treat the most important interactions between the first solvent shell and the solute. In the case of water, these interactions can involve fairly strong hydrogen bonds, interactions that can be poorly treated with the continuum method described below. Computations with a few solvent molecules (typically no more than four or five molecules) are termed *microsolvation*. Generally, full geometry optimizations are performed, allowing the solute and all solvent molecules to fully relax. With every additional solvent molecule, the configurational space increases and so care must be taken to properly sample the space in order to locate all reasonable optimized geometries. Given that solvent–solute and solvent–solvent interactions can be weak, the potential energies surfaces are often flat, leading to frequent difficulties in locating relative minimum energy structures.

The microsolvation computations[100–103] are excellent models of, for example, small hydrated clusters that can be observed in the gas phase.[104,105] Still under intense computational study is how informative microsolvation computations are for understanding true solution phase chemistry.

1.4.2 Implicit Solvent Models

Suppose that we were to average out the effects of all of the solvent molecules, effectively integrating over the coordinates describing the solvent molecules.[106] This would dramatically simplify the description of the solvent molecules, and thereby simplify the computation of the energy of the solute–solvent system. This is the general principle behind the implicit solvent models. The solvent is described by a single term, its dielectric constant, and we just need to treat the interaction of the solute with this field.

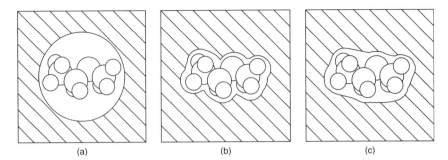

Figure 1.4 Schematic of a solute embedded within a dielectric medium with (a) a spherical cavity (b) a cavity formed from interlocking atomic spheres of radii equal to the van der Waals radii, and (c) a solvent accessible surface.

The implicit solvation model begins by creating a cavity inside the polarizable medium to hold the solute molecule. There is an energy cost for cavitation, ΔG_{cav}, due to solvent–solvent interactions that are removed. The solute is then placed into the cavity (see Figure 1.4), resulting in electrostatic and nonelectrostatic interactions between the solute and the polarizable medium. The solute induces a polarization of the dielectric medium and the medium induces a polarization upon the solute. This electrostatic contribution, ΔG_{elec}, takes into account these attractive terms along with the energetic cost of polarizing the solute and solvent. The main nonelectrostatic contribution to the solvation energy, $\Delta G_{nonelec}$, is dispersion, but other factors may come into play such as the nonelectrostatic component of hydrogen bonding. The total solvation energy is then a sum of these three energetic contributions:

$$\Delta G_{solvation} = \Delta G_{elec} + \Delta G_{cav} + \Delta G_{nonelec} \quad (1.62)$$

The Poisson equation (Eq. (1.63)) describes the electrostatics of a dielectric medium with an embedded charged species:

$$4\pi\varepsilon\sigma(r_s) = (\varepsilon - 1)\mathbf{F}(r_s) \quad (1.63)$$

where $\mathbf{F}(r)$ and $\sigma(r)$ are the electric field perpendicular to the cavity and the charge density at point r on the surface of the cavity. If the solute carries a nonzero charge, the Poisson–Boltzmann equation applies. Once this equation is solved, the Hamiltonian is defined as

$$\mathbf{H} = \mathbf{H}_0 + \mathbf{V}_s \quad (1.64)$$

where the potential $\mathbf{V}_s$ is

$$\mathbf{V}_s(\mathbf{r}) = \int \frac{\sigma(r_s)}{|\mathbf{r} - r_s|} dr_s \quad (1.65)$$

Note that the solute charge distribution (from its wavefunction) enters into the definition of the potential $\mathbf{V}_s$, and so this equation must be solved iteratively, thus giving rise to the term *self-consistent reaction field* (*SCRF*).

The shape of the cavity affects the difficulty in solving the Poisson equation. If the cavity is a sphere, then an analytical solution is obtained. If the solute is an ion with charge q, we get the Born equation

$$\Delta G_{\text{elec}} = -\left(1 - \frac{1}{\varepsilon}\right)\frac{q}{2a} \tag{1.66}$$

where a is the radius of the spherical cavity. This spherical cavity SCRF model is also referred to as the *Onsager model*. If the solute has a fixed dipole moment μ, we have the Kirkwood–Onsager model where

$$\Delta G_{\text{elec}} = -\left(\frac{\varepsilon - 1}{2\varepsilon + 1}\right)\frac{\mu^2}{a^2} \tag{1.67}$$

The Kirwood–Onsager model can be readily extended to using higher order multipole moments.

As shown in Figure 1.4a, most molecules are not spherical, and so use of a spherical cavity large enough to enclose the whole molecule may mean that solvent will be quite far away from the molecule in some directions. The Kirkwood–Westheimer model uses an ellipsoidal cavity, which comes closer to mimicking a true molecular surface.

A more appropriate cavity can be created by representing each atom of the solute molecule as a sphere with a particular radius that reflects the atom size. This is often the van der Waals radius, but can be a scaled van der Waals radius, or even radii that are empirically determined to fit solvation energies. The union of these spheres makes up the cavity. Figure 1.4b displays the van der Waals surface about a solute embedded within a dielectric. A cavity constructed in this manner might still have deficiencies when one considers that the cavity should reflect the closest contact between the solute and solvent. These surfaces might contain valleys and pockets that reach inwards toward the molecular interior but are too small to accommodate a real solvent molecule. The *solvent accessible surface* is obtained by effectively rolling a ball of the size of the solvent molecule along the van der Waals surface of the solute. The solvent accessible surface cavity is shown in Figure 1.4c.

In the *polarized continuum method* (PCM), the Poisson equation is solved by using the apparent charge distribution on the cavity surface.[107] The cavity is formed by interlocking spheres centered at the nuclei. The spheres are then described as a set of spherical triangles, creating a tessellated surface.[108] The virtual charges assigned to the centers of these triangles are based on the iterative solution of the Poisson equation using the boundary element method, creating the solvent reaction field.

An alternative to solving the Poisson equation is an approximation based on the Born equation (Eq. (1.66)). The generalized Born (GB) equation gives the electronic energy

$$\Delta G_{\text{elec}} = -\frac{1}{2}\left(1 - \frac{1}{\varepsilon}\right)\sum_{k,k'}^{\text{atoms}} q_k q_{k'} f_{kk'} \tag{1.68}$$

where $f_{kk'}'$ is the effective Coulomb integral and has the form

$$f_{kk'} = (r_{kk'}^2 + \alpha_k \alpha_{k'} e^{r_{kk'}^2/d})^{-1/2} \quad (1.69)$$

The effective Born radius of atom k, α_k, can be computed from the Poisson equation or by a pairwise screening procedure. The other key component is the way the charges on each atom are assessed. Any of the methods for obtaining populations as discussed later in this chapter may be used; often, the Mulliken charge is used. Cramer and Truhlar[109,110] have advocated using charges derived from the Lowdin population, which are then scaled to better reproduce the dipole moment. The key advantage of this method is that the actual computation of Eq. (1.68) is analytical, while the solution of the Poisson equation (using the interlocking spheres cavity) is numerical.

The remaining portion of the solvation energy, energies attributable to cavitation, dispersion, and other short-range forces, is usually treated in an empirical fashion. All of these effects are usually grouped together, in a term called ΔG_{CDS}, and computed as

$$\Delta G_{CDS} = \sum_k^{atoms} A_k \sigma_k \quad (1.70)$$

In this equation, A_k is the exposed surface area (perhaps the exposed solvent accessible surface area) of atom k, and σ_k is the "surface tension" of the atom k. This surface tension is really just an empirical value that provides the best fit to some experimental value(s); usually, these surface tensions are fit to the experimental solvation energy less the electrostatic contribution. Since the available experimental solvation energies comprise a limited set, particularly for nonaqueous solvents, this fitting procedure can have a large error.

To tackle organic solvents, the *solvation model* (SMx) of Cramer and Truhlar[111] defines the surface tensions as

$$\sigma_k = \sum_j \Gamma_j n_k^{\Gamma_j} \quad (1.71)$$

Here, the values $n_k^{\Gamma_j}$ are fit to a number of *solvent* properties Γ_j, including surface tension, index of refraction, and hydrogen bonding acidity and basicity. For water, however, SMx employs atomic surface tensions that are fit to the experimental solvation energies of a large number of neutral solutes.

Many different implicit solvation models have been proposed and modified over the course of the past 10–15 years. While all of quantum chemistry is plagued by a proliferation of acronyms, the situation concerning the names of these implicit solvation models is excruciatingly obtuse. Most of the popular methods have undergone many variations and updates, leading to subtle name changes—sometimes no name changes at all. Models with identical names can mean different things in different computer programs, sometimes even with different versions of the same computer program! The 1996 review by Cramer and Truhlar[112] provides a good accounting of the various different methods and acronyms, though it is now somewhat outdated. They provide a simple four-component classification scheme of the different implicit solvent models, namely

- How are the electrostatics treated (usually by the Poisson equation or the GB approximation)?
- How is the cavity defined? Most methods now use some variation on the interlocking spheres approach, but the size of these spheres is very much method dependent.
- How is the charge distribution of the solute determined? Typically, this is through some quantum mechanical population method, with perhaps a scaling procedure.
- How are nonelectrostatic effects treated?

With this in mind, we now summarize the most widely used implicit solvent models. The polarized continuum model (PCM) utilizes a cavity of interlocking sphere with the size typically 20 percent larger than the van der Waals radius. The electrostatics are treated using the Poisson equation and nonelectrostatic effects are evaluated using a form of Eq. (1.70). The method has undergone quite a few improvements and variations over the years since its first[107] publication, spearheaded by the groups of Tomasi,[108,113–115] Cossi and Barone,[116,117] and Orozco and Luque.[118,119] The most current implementation, called *IEF-PCM*[114] for the integral equation formalism, improves the performance of the code, especially in obtaining gradients. A variant is to define the cavity based on the electron density. IPCM defines the cavity based on a fixed value of the electron density from a gas-phase calculation, while SCIPCM adjusts the cavity in a self-consistent way.[120] A formalism for computing a smooth solvation potential, developed by York and Karplus[121] in 1999 but not implemented within a major quantum chemical code until nearly a decade later, significantly improves the computational performance of PCM. In particular, this formalism dramatically improves geometry optimization of a solute in solution, especially for transition states. Mennucci[122] has authored an excellent review of PCM and the various implementations.

The SMx approach by Cramer and Truhlar has undergone a series of improvements since its first implementation, SM1, in 1991.[110,111,123,124] The current version is SM12.[125] SMx utilizes the GB procedure with scaled atomic charges for treating the electrostatics using an interlocking sphere cavity. The nonelectrostatic effects are handled using surface tensions fit to solvation energies for water and to a variety of solvent properties (Eq. (1.71)) for organic solvents.

The conductor-like screening model (COSMO) approach replaces the dielectric medium with a conducting medium (basically a medium that effectively has an infinite dielectric constant).[126,127] Interlocking spheres are used to generate the cavity. The conductor-like screening has been implemented as a PCM version, called *CPCM*.[128,129]

We conclude this section by briefly noting some recent comparisons of the implicit solvation models. Table 1.2 compares the errors in solvation free

TABLE 1.2 Mean Unsigned Error (kcal mol^{-1}) for the Solvation Free Energies Computed with Different Methods[a]

Method	Aqueous Neutral[b]	Organic Neutrals[c]	Ions[d]
SM8[e]	0.55	0.61	4.31
SM5.43R[e]	4.87	5.99	9.73
IEF-PCM/UAHF[f]	1.18	3.94	8.15
C-PCM/GAMESS[g]	1.57	2.78	8.39
PB/Jaguar[h]	0.86	2.28	6.72
3PM[i]	2.65	1.49	8.60

[a]Ref. 130.
[b]274 data points.
[c]666 data points spread among 16 solvents.
[d]332 data points spread among acetonitrile, water, DMSO, and methanol.
[e]Using mPW1PW/6-31G(d).
[f]Using mPW1PW/6-31G(d) and the UAHF atomic radii in *Gaussian*.
[g]Using B3LYP/6-31G(d) and conductor-PCM in *GAMESS*.
[h]Using B3LYP/6-31G(d) and the PB method in *Jaguar*.
[i]Using average experimental values.

energies computed using the SM8 model against some other popular continuum methods.[130] The SM8 and SM5.43R computations were performed with at mPW1PW/6-31G(d), the DFT method for which SM8 was parameterized, though it should be noted that performance with other functionals and basis sets is essentially as good. The IEF-PCM computations were performed at B3LYP/6-31G(d) and *Gaussian*-03. The C-PCM computations were done at B3LYP/6-31G(d) with *GAMESS*, and the Poisson–Boltzmann computations were carried out with the same method with the *Jaguar* suite. Clearly, SM8 provides much better results for all three test sets: neutral solutes in aqueous or organic solvents and ions in acetonitrile, DMSO, methanol, or water. The errors are quite large in particular for the ions. The simple-minded model 3PM, a "three-parameter model," sets the solvation energies of all the neutral solute in water to the average experimental value (-2.99 kcal mol^{-1}), and similarly for the neutral solutes in organic solvents (-5.38 kcal mol^{-1}), and for ions (-65.0 kcal mol^{-1}). This "three-parameter model" outperforms *most of the continuum methods*!

1.4.3 Hybrid Solvation Models

The two methods described above have complementary strengths and weaknesses. The implicit solvation models treat the bulk, long-range effect of solvation but may underestimate local effects within the first solvation shell, especially if hydrogen bonding can occur between solute and solvent. Microsolvation explicitly addresses these local effects but neglects long-range solvation effects. Perhaps a combination of the two approaches might offer a treatment that combines the best of both methods.

This hybrid solvation model surrounds the solute with a small number of explicit solvent molecules, and then embeds this cluster into the implicit dielectric field. Local effects are addressed by the full quantum mechanical treatment of the interaction between the solute and the few explicit solvent molecules. Long-range effects are included through the interaction of the cluster with the dielectric field. A decision is still needed as to how many explicit solvent molecules should be included within the cluster, recognizing that each additional solvent molecule increases the size of the calculation and expands the configurational space that must be explored.

Nonetheless, this hybrid solvent model has been utilized in a number of problems. The example we describe here is for the computation of acid dissociation constants in water. Cramer and Truhlar[131] computed the free energy of dissociation for 57 species, mostly organic compounds using the SM6 implicit solvation model. The correlation between the computed free energy of dissociation and the experimental pK_a is poor, only 0.76. Inspection of the predicted free energies showed a few particular outliers, some of which are listed in Table 1.3. They then included a single explicit water molecule for the description of the conjugate bases when either (1) the anion contained three or fewer atoms or (2) when the anion contained one or more oxygen atoms that carry a more negative charge than does the oxygen in water. There were 20 such molecules in their test suite of 57 compounds, and their free energy of dissociation was recomputed using SM6 of the mono-hydrated anion, some of which are listed in Table 1.3. The resulting correlation between these computed energies and the pK_a values is much improved with $r^2 = 0.86$.

1.5 HYBRID QM/MM METHODS

While computers continue to become ever faster and quantum mechanical algorithms become ever more efficient, the bottom line remains that many chemically

TABLE 1.3 Calculated (SM6) Dissociation Free Energies (kcal mol^{-1}) and Experimental pK_a[a]

Compound	ΔG Implicit Solvent	ΔG Implicit Solvent + One Explicit Water	pK_a Experimental
2,2,2-Trifluoroethanol	24.4	19.7	12.4
2-Methoxyethanol	29.7	23.6	14.8
Benzyl alcohol	32.0	25.7	15.4
1,2-Ethanediol	21.6	15.7	15.4
Methanol	27.8	21.8	15.5
Allyl alcohol	28.6	22.2	15.5
Water	9.7	17.3	15.7
Ethanol	30.2	24.0	15.9
1-Propanol	30.0	22.7	16.1
2-Propanol	31.8	26.0	17.1
t-Butyl alcohol	34.6	27.5	19.2

[a]Ref. 131.

interesting systems, like enzymes, are simply too large to treat in any realistic amount of time. Molecular mechanics (MM) is very efficient but has a serious drawback in its inability to properly treat many important aspects of chemistry, especially bond making and bond breaking. Perhaps there is a way to combine the best of both worlds: MM for the majority of a system and QM for the portion that needs high level, accurate treatment?

The hybrid QM/MM method does just that.[132–134] A system is partitioned into two spaces (Figure 1.5). The first region contains the part of the molecule or system that can be addressed using solely an MM approach. The second (and smaller) region contains the part of the molecule or system that will be treated with some quantum mechanical approach. Critical to this approach is defining a method to treat the interaction between the two regions. We begin by first describing MM in very general terms, and then follow with the description of the major approaches toward constructing a QM/MM theory. The last section will describe a popular implementation of QM/MM theory that provides greater flexibility in the treatment of the larger (MM) subregion.

1.5.1 Molecular Mechanics

The quantum mechanical methods described so far all properly treat the electrons as quantum particles. A vastly simpler approach toward obtaining molecular structures and energies is to treat atoms as classical particles. The potential energy is then just a function of the nuclear coordinates. MM, also referred to as *force field* methods, involves specifying the various functions used to relate nuclear positions to energy and fitting these functions to experimental data. It is a highly empirical approach, dependent on the choice of reference data, the functional form, and selection of parameters.

The typical MM[1,2] approach is to divide the total energy into components related to classical concepts:

$$E_{MM} = E_{stretch} + E_{bend} + E_{torsion} + E_{vdW} + E_{el} + E_{cross} \qquad (1.72)$$

Each of these terms will be briefly discussed in turn.

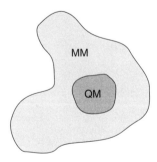

Figure 1.5 Schematic of the partitioning of a system into a QM (primary) subspace and an MM (secondary) subspace.

The first term E_{stretch} is the energy associated with stretching a bond. Typically, one assumes a harmonic potential for this stretching, making an analogy between a chemical bond and a classical spring. This term can be written as

$$E_{\text{stretch}} = \sum_{\text{bonds}} k_{AB}(r_{AB} - r_{AB,\text{eq}})^2 \tag{1.73}$$

where r_{AB} is the distance between atoms A and B, $r_{AB,\text{eq}}$ is the average or equilibrium bond distance for atoms A and B, and k_{AB} is, by analogy, the force constant for the AB bond. The parameters $r_{AB,\text{eq}}$ and k_{AB} must fit to experimental data. One should keep in mind that parameters are usually needed not just for every atom pair AB but also for every hybridization of each atom. So, parameters are needed for the C_{sp^3}–C_{sp^3} bond, the C_{sp^2}–C_{sp^3} bond, the C_{sp^2}–C_{sp^2} bond, and so on. Some MM methods adopt cubic and higher order terms into Eq. (1.73) and others use a Morse potential.

The second term is the energy associated with bending about the angle formed of atoms ABC. This too is generally regarded as behaving like a spring:

$$E_{\text{bend}} = \sum_{\text{bond angles}} k_{ABC}(\theta_{ABC} - \theta_{ABC,\text{eq}})^2 \tag{1.74}$$

where θ_{ABC} is the angle of A–B–C, $\theta_{ABC,\text{eq}}$ is the normal value of that angle, and k_{ABC} is the force constant. Again parameters will be needed for different atomic hydrizations.

The third term E_{torsion} provides the energy associated with torsional rotations. It is generally written as

$$E_{\text{torsion}} = \sum_{\text{torsions}} \sum_n V_n \cos(n\omega) \tag{1.75}$$

Usually the torsional expression includes terms through $n = 3$, providing potentials that are periodic in 360° ($n = 1$), 180° ($n = 2$), and 120° ($n = 3$).

The E_{vdW} term mimics the van der Waals interactions between two nonbonded atoms. A typical representation is the Lennard–Jones potential so that

$$E_{\text{vdW}} = \sum_{\substack{\text{nonbonded} \\ \text{atoms}}} \frac{c_6}{r^6} - \frac{c_{12}}{r^{12}} \tag{1.76}$$

The electrostatic energy term E_{el} accounts for electrostatic interactions between nonbonded atoms. This is usually treated by assigning point charges to atoms, defined by some empirical fitting scheme. The electrostatic energy is

$$E_{\text{el}} = \sum_{\substack{\text{nonbonded} \\ \text{atoms}}} \frac{q_A q_B}{r_{AB}} \tag{1.77}$$

The last term of Eq. (1.72) accounts from coupling (cross terms) between the above factors. So, one might include the coupling that can exist, for example, between two bond stretches or between a bond stretch and an angle bend.

The many different MM methods differ in the exact form of these terms, which terms are included, and the parameterization of the many constants involved in each term. Some of the more common force fields are MM2,[135] AMBER,[136] OPLS,[137] and CHARMM.[138]

1.5.2 QM/MM Theory

The basic assumption of the hybrid QM/MM methods is that a system can be divided into two regions, as in Figure 1.5. (See references 139–142 for a more detailed discussion of QM/MM theory.) The smaller region contains the portion of the system that must be treated quantum mechanically. This might be, for example, the active site of enzyme including the substrate, or the solute in a solution, or a guest molecule within a host. This region is often labeled QM or the primary subspace. The remainder of the system will be treated with MM. This region is often labeled as MM or the secondary subspace.

The energy of the system divided in this way is

$$E_{QM/MM} = E_{QM} + E_{MM} + E_{int} \qquad (1.78)$$

where E_{QM} is the energy of the primary subspace determined using a QM method, E_{MM} is the energy of the secondary subspace evaluated using MM, and E_{int} is the energy of the interaction between the two subspaces. How one handles the interaction energy is key to the effectiveness of the QM/MM approach.

There are two general approaches to the interaction energy.[143] Mechanical embedding treats the interaction exclusively through MM. In other words, an MM computation is performed on the entire system, both the primary and secondary spaces, and this gives the sum of the last two terms of Eq. (1.78). The alternative approach is called *electrostatic embedding*. Here, the primary subspace is computed with a QM method in the presence of the electrostatic field of the secondary subspace. The effective QM Hamiltonian is modified typically with a term describing point-charge distributions on all of the atoms in the secondary space.

The advantages and disadvantages of each method were nicely summarized by Lin and Truhlar.[141] Mechanical embedding is simple and cheap to compute, though MM parameters for all of the atoms in the primary subspace may not be available (a problem that arises more often with metals in enzyme active sites). Further, and more seriously, mechanical embedding completely omits any electrostatic perturbation of the primary subspace by the secondary subspace. On the other hand, electronic embedding accounts for this perturbation of the primary subspace by the secondary subspace but at significantly greater computational expense. In addition, one must make a fundamental arbitrary choice on how to construct the charge distribution of the secondary subspace.

In addition to considering the evaluation of the interaction energy, one must consider just where to place the boundary between the two subspaces. In some situations, the boundary can be in regions between molecules, passing through areas of low electron density. An example of this circumstance might be a study of solvation, where the solute molecule is the QM subspace and all of the solvent molecules are in the MM subspace. The coupling between these two spaces is weak and should be well accounted for by the embedding schemes discussed above.

However, the boundary may have to slice between chemically bonded atoms. This might occur, for example, when defining the QM space as the active center of a protein and the MM space contains the remaining portions of the molecule. In such a situation, the boundary would cleave a bond, leaving behind in the QM space an atom with an unfilled valency, that is, if the boundary cuts through a C_{sp^3}–C_{sp^3} bond, one would be left with a carbon radical.

Though there are a number of different approaches toward handling this sliced bond problem, involving localized fixed orbitals at the boundary,[144–146] the method most broadly used is to cap off these dangling bonds with an atom. The typical choice is to use a hydrogen as this "link atom" mainly selected for computational efficiency, though ideally one should choose a link atom with an electronegativity as similar to the replaced atom as possible. The link atom is usually positioned on the line connecting the two atoms at the boundary that have been severed. (These can be called *Q1* and *M1* residing in the QM and MM spaces, respectively.) Instead of placing the link atom at a fixed distance away from Q1, say the typical Q1–H distance, Morokuma[147] has advocated for scaling the position based on the Q1–M1 distance. The bond energies of the sliced bonds are computed with MM alone, via Eq. (1.73).

Since the link atom is very close to atoms in the MM region, the point charges assigned to the MM atoms that are nearby may unduly polarize the density in the QM space. Different approaches toward reducing the size of the charges on MM atoms near link atoms have been used,[133,148] including assigning charges of zero to atom M1 (and also to atoms neighboring M1) or scaling the charges of these nearby MM atoms.

1.5.3 ONIOM

The ONIOM approach (our Own N-layer Integrated molecular Orbital molecular Mechanics) developed by the Morokuma group takes an alternative approach to hybrid QM/MM.[140,147,149] The distinct advantages of the ONIOM approach are that the low level treatment does not have to be MM and one can chose an arbitrary number of subspaces, not just two.

In the ONIOM approach, the energy is written as

$$E_{\text{ONIOM}} = E_{\text{real,low}} + E_{\text{model,high}} - E_{\text{model,low}} \quad (1.79)$$

Here, the entire system is referred to as the *real* space, and the *model* space is the region of the molecule that requires greater computational care. The computational methods are called *low*, indicating some relative inexpensive method (like MM), and *high*, indicating some more exacting computational method (like QM). One can view Eq. (1.79) as correcting the low level energy of the entire system by adding the energy of the model region computed at a higher level. In order to not double count the model region, you subtract off the energy of that region computed at the low level. In terms of relating ONIOM back to QM/MM, the interaction energy is contained within the $E_{\text{real,low}}$ term and implies a mechanical embedding approach.

This approach can be readily extended to any number of layers. A three-layer ONIOM computation (Figure 1.6) divides the entire system into a real space (the whole system, S_0) and an intermediate space (S_1), which wholly encloses an interior space (S_2). The energy is then

$$E_{\text{ONIOM}} = E_{S_0,M_0} + E_{S_1,M_1} - E_{S_1,M_0} + E_{S_2,M_2} - E_{S_2,M_1} \qquad (1.80)$$

Link atoms are used in ONIOM to handle the unfulfilled valences caused by slicing bonds to create the different regions. The original formulation of ONIOM handles the interaction energy using mechanical embedding; the MM computation of the real (entire) system is by definition mechanical embedding. However, ONIOM has now been modified to include electronic embedding, including reducing the MM charges for atoms within three bonds of the boundary to zero.[150]

It is important to recognize that the low level computation, and the low and intermediate level computations in a three-layer ONIOM, does not have to be MM. Any computational method can be used for any layer. For example, a three-layer ONIOM might use a semiempirical computation for the real system (S_0), a DFT computation for the intermediate layer (S_1), and a CASSCF computation for the high layer (S_2).

1.6 POTENTIAL ENERGY SURFACES

Most computations begin with a search for chemically interesting structures. Typically, this means locating the structures of ground states and transition states and

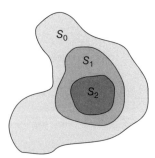

Figure 1.6 Schematic of a three-layer ONIOM approach. A low level method (M_0) is applied to the entire system S_0, an intermediate level method (M_1) is applied to region S_1, and a high level method (M_2) applied to region S_2.

perhaps the path that takes reactants through the transition state and on to product. In other words, one needs to characterize the PES, which relates energy to the atomic coordinates.

In general, interesting structures on the PES are *critical points*, points where the gradient of the energy vanishes. Critical points are characterized by the eigenvalues of the Hessian matrix evaluated at the point of interest. The matrix elements of the Hessian are defined as

$$H_{ij} = \frac{\partial^2 E}{\partial q_i \partial q_j} \quad (1.81)$$

where q_i is an atomic coordinate (say, for example, the y-coordinate of the seventh atom). The eigenvalues correspond to the curvatures associated with the normal vibrational modes. If all of the eigenvalues of the vibrational modes are positive, then the critical point is a local energy minimum. If there is one and only one negative eigenvalue, then the critical point is a transition state.

A PES for a generic reaction is shown in Figure 1.7. The reactant sits in an energy well on the left side. Motion to the right leads to an increase in energy until the transition state is reached. Continued motion to the right past the transition state leads to the product on the left.

At the transition state, the negative eigenvalue of the Hessian matrix corresponds with the eigenvector that is downhill in energy. This is commonly referred to as the *reaction coordinate*. Tracing out the steepest descent from the transition state, with the initial direction given by the eigenvector with the negative eigenvalue, gives the *minimum energy path* (MEP). If this is performed using mass-weighted coordinates, the path is called the *intrinsic reaction coordinate* (IRC).[151]

In most circumstances, the MEP or IRC will connect reactant to TS to product, but other topologies are possible. For example, Figure 1.8 represents a PES that will

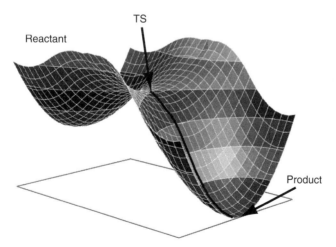

Figure 1.7 Diagramatic PES for a general reaction. The heavy line traces the reaction path connecting the TS to product.

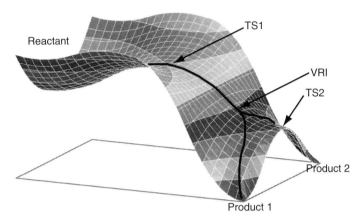

Figure 1.8 A PES with a bifurcation that takes place after **TS1** at the valley ridge inflection (**VRI**). The reaction path splits at the VRI leading in one direction to **Product 1** and in the other direction to **Product 2**.

be important in the discussion of reaction dynamics discussed in Chapter 8. Here, the reaction path connects the reactant to **TS1** and then moving to the right, it connects to a second transition state **TS2**. **TS2** is the transition state that interconverts **Product 1** and **Product 2**.

Let's consider what happens after the molecule passes over **TS1**. The molecule continues to follow the reaction path downhill until it reaches the *valley ridge inflection* (**VRI**) point. At this point, the gradient in one direction perpendicular to the reaction path becomes zero, and after this point, the PES actually falls downhill faster in the direction off of the reaction path than continuing on the reaction path to **TS2**. It is important to note that the VRI is not a critical point; the gradients are not all zero at a VRI point. So at the VRI, the pathway diverges off of the ridge that leads to **TS2**, in one direction toward **Product 1** and in a second direction toward **Product 2**. What is so unusual about this type of surface is that the reaction paths that lead from the reactant over **TS1** then proceed onward to two products (**Product 1** and **Product 2**), without crossing any more transition states. In other words, this surface has a TS that leads to two products!

There are other types of critical points on PESs; however, they do not typically have chemical relevance. For example, Figure 1.9 displays a PES with a hilltop, a critical point with two negative eigenvalues of the Hessian matrix. Stanton and McIver[152] have demonstrated that all hilltops can be avoided, meaning that a path from reactant to product can be found that skirts around the hilltop, passing through a lower energy transition state.

1.6.1 Geometry Optimization

The first step in performing a quantum chemical calculation is to select an appropriate method from the ones discussed above. We will discuss the relative merits and demerits of the methods in the remaining chapters of this book. For now, we assume

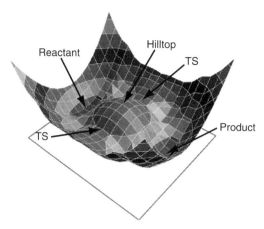

Figure 1.9 A PES with a hilltop. Paths connect reactant to product avoiding the hilltop, passing through one of the two TSs.

that we can choose a method that will be suitable for the task at hand. The nomenclature for designating the method is "method/basis set," such as MP2/6-31+G(d), which means that the energy is computed using the MP2 theory with the 6-31+G(d) basis set.

Next, we need to choose the geometry of the molecule. It is antithetical to the idea of *ab initio* methods to arbitrarily choose a geometry; rather, more consistent is to find the best geometry predicted by the QM itself. In other words, we should optimize the geometry of the molecule such that a minimum energy structure is found.

There are many, many methods for optimizing the value of a function and detailed discussion of these techniques is inappropriate here.[153,154] The general procedure is to start with a guess of the molecular geometry and then systematically change the positions of the atoms in such a way that the energy decreases, continuing to vary the positions until the minimum energy is achieved. So how does one decide how to alter the atomic positions; that is, should a particular bond be lengthened or shortened? If the derivative of the energy with respect to that bond distance is positive, it means that the energy will *increase* with an increase in the bond separation. Computation of all of the energy gradients with respect to the positions of the nuclei will offer guidance then in which directions to move the atoms. But how far should the atoms be moved; that is, how much should the bond distance be decreased? The second derivatives of the energy with respect to the atomic coordinates provide the curvature of the surface, which can be used to determine just how far each coordinate needs to be adjusted.

Efficient geometry optimization, therefore, typically requires the first and second derivatives of the energies with respect to the atomic coordinates. Computation of these derivatives is always more time consuming than the evaluation of the energy itself. Further, analytical expression of the first and second energy derivatives is not available for some methods. The lack of these derivatives may be a deciding factor as to which method might be appropriate for geometry optimization. An economical procedure is to evaluate the first derivatives and then make an educated guess at the second derivatives, which can be updated numerically as each new geometry is evaluated.

The optimization procedure followed in many computational chemistry programs is as follows:

(1) Make an initial guess of the geometry of the molecule.
(2) Compute the energy and gradients of this structure. Obtain the Hessian matrix as a guess or by analytical or numerical computation.
(3) Decide if the geometry meets the optimization criteria. If so, we are done.
(4) If the optimization criteria not met, use the gradients and Hessian matrix to suggest a new molecular geometry. Repeat Step 2, with the added option of obtaining the new Hessian matrix by numerical updating of the old one.

What are the criteria for determining if a structure has been optimized? A local energy minimum will have all of its gradients equal to zero. Driving a real-world quantum chemical computation all the way until every gradient vanishes will involve a huge number of iterations with very little energy change in many of the last steps. Typical practice is to set a small but nonzero value as the maximum acceptable gradient.

Testing of the gradient alone is not sufficient for defining a local energy minimum. Computing the full and accurate Hessian matrix can confirm the nature of the critical point, be it a local minimum (with only positive eigenvalues of the Hessian matrix), transition state (one negative eigenvalue), or some other higher order saddle point.

Optimization to local energy minima are usually straightforward, with tried-and-true optimization techniques such as quasi-Newton methods well established. Optimization to a transition state can be more difficult as the traditional optimization techniques work best within a quadratic region near the optimized structure. One therefore needs to have a very good initial guess of the TS geometry for these methods to converge. In addition, since the optimization must proceed uphill direction in one direction (the reaction coordinate), knowledge of that direction often necessitates computation of the Hessian matrix.

There are two widely used methods to aid in finding an initial geometry in the correct vicinity of the TS. The linear synchronous transit (LST) method locates the highest energy geometry along a line connecting reactant and product. The quadratic synchronous transit (QST) approach attempts to find the maximum energy on a curved path connecting reactant and product.[154,155] After locating these geometries, a quasi-Newton method (or some related optimization technique) is employed to fine-tune the geometry.

The Hessian matrix is useful in others ways, too. The square root of the mass-weighted Hessian eigenvalue is proportional to the vibrational frequency ω_i.

Within the harmonic oscillator approximation, the ZPVE is obtained as

$$\text{ZPVE} = \sum_i^{\text{vibrations}} \frac{\hbar\omega_i}{2} \quad (1.82)$$

The eigenvector associated with the diagonal mass-weighted Hessian defines the atomic motion associated with that particular frequency. The vibrational frequencies can also be used to compute the entropy of the molecule and ultimately the Gibbs free energy.[1]

The molecular geometry is less sensitive to computational method than is its energy. Since geometry optimization can be computationally time consuming, often a molecular structure is optimized using a smaller, lower level method and then the energy is computed with a more accurate higher level method. For example, one might optimize the geometry at the HF/6-31G(d) level and then compute the energy of *that* geometry using the CCSD(T)/cc-pVTZ method. This computation is designated "CCSD(T)/cc-pVTZ//HF/6-31G(d)" with the double slashes separating the method used for the single-point energy calculation (on the left-hand side) from the method used to optimize the geometry (on the right-hand side).

1.7 POPULATION ANALYSIS

We next take on the task of analyzing the wavefunction and electron density. All of the wavefunctions described in this chapter are represented as very long lists of coefficients. Making sense of these coefficients is nigh impossible, not just because there are so many coefficients or not just because these coefficients multiply Gaussian functions that have distinct spatial distributions, but fundamentally because the wavefunction itself has no physical interpretation. Rather, the square of the wavefunction at a point is the probability of locating an electron at that position. It is therefore more sensible to examine the electron density $\rho(\mathbf{r})$. Plots of the electron density reveal a rather featureless distribution; molecular electron density looks very much like a sum of spherical densities corresponding to the atoms in the molecule. The classical notions of organic chemistry, like a build-up of density associated with a chemical bond, or a lone pair, or a π-cloud are not readily apparent—as seen in isoelectronic surfaces of ammonia **2** and benzene **3** in Figure 1.10.

The notion of transferable atoms and functional groups pervades organic chemistry; a methyl group has some inherent, common characteristics whether the methyl group is in hexane, toluene, or methyl acetate. One of these characteristics is, perhaps, the charge carried by an atom (or a group of atoms) within a molecule. If we can determine the number of electrons associated with an atom in a molecule, which we call the *gross atomic population* $N(k)$, then the charge carried by the atom (q_k) is its atomic number Z_k less its population

$$q_k = Z_k - N(k) \quad (1.83)$$

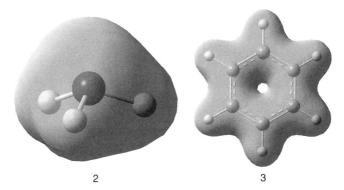

Figure 1.10 Isoelectronic surface of the total electron density of ammonia **2** and benzene **3**. Note the lack of lone pairs or a π-cloud. (*See insert for color representation of this figure.*)

Since there is no operator that produces the "atomic population," it is not an observable and so the procedure for computing $N(k)$ is arbitrary. There are two classes of methods for computing the atomic population: those based on the orbital population and those based on a spatial distribution.[156]

1.7.1 Orbital-Based Population Methods

Of the orbital-based methods, the earliest remains the most widely used method: that developed by Mulliken[157] and called the *Mulliken population*. The total number of electrons in a molecule N must equal the integral of $\rho(\mathbf{r})$ over all spaces. For simplicity, we will examine the case of HF wavefunction; then this integral can be expressed as

$$n = \int \psi_{HF}\psi_{HF}\,d\mathbf{r} = \sum_i^{MOs} N(i) \sum_r^{AOs} c_{ir}^2 + 2\sum_i^{MOs} N(i) \sum_{r>s}^{AOs} c_{ir}c_{is}S_{rs} \quad (1.84)$$

where $N(i)$ is the number of electrons in MO ϕ_i and S_{rs} is the overlap integral of AOs χ_r and χ_s. Mulliken then collected all terms of Eq. (1.84) for a given atom k to define the *net atom population* $n(k)$

$$n(k) = \sum_i^{MOs} N(i) \sum_{r_k} c_{ir_k}^2 \quad (1.85)$$

and the *overlap population* $N(k,l)$

$$N(k,l) = \sum_i^{MOs} N(i) \sum_{r_k,s_l} c_{ir_k}c_{is_l}S_{r_ks_l} \quad (1.86)$$

The net atomic population neglects the electrons associated with the overlap between two atoms. Mulliken arbitrarily divided the overlap population equally between the two atoms, producing the *gross atomic population*

$$N(k) = \sum_i^{MOs} N(i) \sum_{r_k} c_{ir_k} \left(c_{ir_k} + \sum_{s_l \neq k} c_{is_l} S_{r_k s_l} \right) \quad (1.87)$$

The Mulliken population is easy to compute and understand. All electrons that occupy an orbital centered on atom k "belong" to that atom. However, Mulliken populations suffer from many problems. If a basis set is not balanced, the population will reflect this imbalance. Orbital populations can be negative or greater than zero. This deficiency can be removed[158] by using orthogonal basis functions (the Löwdin orbitals[159]). But perhaps most serious is that the Mulliken procedure totally neglects the spatial aspect of the AOs (basis functions). Some basis functions can be quite diffuse, and electrons in these orbitals might in fact be closer to a neighboring atom than to the nuclei upon which the function is centered. Nonetheless, the Mulliken procedure assigns these electrons back to the atom upon which the AO is centered. The natural population analysis (NPA) of Weinhold[160] creates a set of AOs that have maximal occupancy, effectively trying to create the set of *local* AOs. NPA charges, while trivially more expensive to compute, suffer fewer of the problems that plague Mulliken analysis. Natural bond orbitals (NBOs), related to the orbitals that diagonalize the one-electron density matrix, provide an alternative description of chemical bonding that is quite consistent with traditional methods, like Lewis dot structures and Pauling's hybridization and valence bond models.[161,162]

1.7.2 Topological Electron Density Analysis

The alternative approach is to count the number of electrons in an atom's "space." The question is how to define the volume an individual atom occupies within a molecule. The topological electron density analysis (sometimes referred to as *atoms-in-molecules* or AIM) developed by Bader[163] uses the electron density itself to partition molecular space into atomic volumes.

The molecular electron density is composed of overlapping radially decreasing spheres of density. One can think of each nucleus as being the location of the "mountain peak" in the electron density. Between two neighboring atoms, there will then be a "valley" separating the two "mountains." The "pass" through the valley defines the boundary between the "mountains."

To do this in a more rigorous way, the local maxima and minima of the electron density are defined as critical points, the positions where

$$\nabla \rho(\mathbf{r}) = \left(\frac{\partial}{\partial x} + \frac{\partial}{\partial y} + \frac{\partial}{\partial z} \right) \rho(\mathbf{r}) = 0 \quad (1.88)$$

The type of electron density critical point is defined by the matrix **L**,

$$\mathbf{L}_{ij} = \frac{\partial^2 \rho(\mathbf{r})}{\partial r_i \partial r_j} \quad (1.89)$$

where r_i is the x, y, or z coordinates. Each critical point is then classified by the *rank*, the number of nonzero eigenvalues of **L**, and the *signature*, the number of positive eigenvalues less the number of negative eigenvalues. The nuclei are $(3,-3)$ critical points, where the density is at a local maximum in all three directions. The bond critical point $(3,-1)$ is a minimum along the path between two bonded atoms, and a maximum in the directions perpendicular to this path.

A gradient path follows the increasing electron density toward a local maximum. The collection of all such paths that terminate at the bond critical point forms a curtain, which is the surface that separates the two neighboring atoms from each other. If we locate all of these surfaces (known as *zero-flux surfaces*) about a given atom, it defines the atomic basin Ω_k, a unique volume that contains a single nucleus. All gradient paths that originate within this basin terminate at the atomic nucleus. We can integrate the electron density within the atomic basin to obtain the electron population of the atom

$$N(k) = \int_{\Omega_k} \rho(\mathbf{r}) d\mathbf{r} \quad (1.90)$$

The bond critical point is the origin of two special gradient paths; each one traces the ridge of maximum electron density from the bond critical point to one of the two neighboring nuclei. The union of these two gradient paths is the *bond path*, which usually connects atoms that are joined by a chemical bond.

The inherent value of the topological method is that these atomic basins are defined *by* the electron density distribution of the molecule.[164] No arbitrary assumptions are required. The atomic basins are quantum mechanically well-defined spaces, individually satisfying the virial theorem. Properties of an atom defined by its atomic basin can be obtained by integration of the appropriate operator within the atomic basin. The molecular property is then simply the sum of the atomic properties.

1.8 INTERVIEW: STEFAN GRIMME

Interviewed August 30, 2012

"I am looking for solutions that work in practice," says Stefan Grimme, Universität Professor at Universität Bonn. "I want to have methods that are as black box as possible, as robust as possible, that can be used for large systems. Large chemical systems are always my focus. I am not always looking for high accuracy methods; a little bit of empiricism is ok. But this kind of empiricism must be physically well-founded." These principles have guided Grimme's pursuit of new quantum chemical methods.

Grimme's three major contributions to quantum chemistry all derive from a desire to deal with large systems, where standard quantum mechanical approaches, be they wavefunction or density functional methods, are computationally intractable. His work on dispersion corrections came about through work on surface chemistry, such as for adsorption, where dispersion is inherent. Grimme was not the first to suggest a dispersion correction; early attempts date back to Ahlrich's corrected HF in the 1970s. After a few years of working with various approaches, Grimme found success with the "-D" correction, first published in 2004. He has twice since developed improvements, the latest being the "-D3" correction. He indicates that "my name is connected with it because my corrections are the best around, but others had the original idea. I just developed them further and improved them."

In contrast, the SCS-MP2 idea came to Grimme suddenly, and he had it all worked out in about 30 min. "I thought 'MP2 is a nice method; it can be used for large systems (which is again my point), but is too inaccurate for some reactions.' I saw that there are physically these different spin components of the correlation, and no one had ever looked at their relations to questions of accuracy." The computer program he was using was able to print out the values of these components, so he could immediately test out his hypothesis. "I want to put my ideas into practice very quickly! I don't want to spend my time on tiny improvements. I normally will invest a week or a month on a new idea, but not a year," he explains. Within a couple of days, he knew the method was "a great idea." He observed huge improvement in a few simple test cases. "Writing the paper took many months," he notes "but from the idea to the first results only one day!"

Grimme's third major methodological development is the notion of double hybrid density functionals. This development came when he recognized that the principle of perturbation theory, expressed through MP2, could be applied to DFT. He characterizes this realization as "not a big step."

While he recognizes that SCS-MP2 is perhaps his most important individual contribution, he asserts that the dispersion correction work is much more significant. "There are many problems that *cannot* be solved without dispersion. We can live without double hybrids or SCS-MP2, but proper accounting for dispersion is critical within DFT."

Grimme expressed some surprise that adoption of his dispersion corrections took so long. Today he is seeing exponential growth in the use of dispersion corrections. "This is really quite normal," he realizes. "People need time to recognize new ideas. I was just too naïve."

Getting quantum chemical methods into some of the standard programs is very important to any methods developer. "It is necessary to have your methods in some code that is noncommercial, like *Orca* by Frank Neese. Everybody can use it. But having it in *Gaussian* is fantastic. When you have something in *Gaussian* connected to your name, you are totally on the safe side."

As we discussed the general state of large-molecule computational methodologies, Grimme noted that he, along with many others, have repeatedly pointed out the problems with the B3LYP functional. Nonetheless, B3LYP remains a very widely used functional, a phenomenon that is often discussed in his group. He claims "I would not in general say that people should not use B3LYP. Rather they should understand it. When you properly correct B3LYP for its over-repulsive behavior, dispersion correct it, then it's not bad. It's not the best functional. It has the nice property that it is the best tested functional. It works in very many cases. This is a very important property, it's robustness in a sense. But this only holds when you properly dispersion-correct it."

Grimme concurs that CCSD(T) remains the gold standard method. "It is our go-to method. If you have more-or-less a single reference situation, as in most chemical reactions, then CCSD(T) is OK. In cases where this approximation breaks down, we need then to choose on a case-by-case basis. The situation is no longer black box. There is much room for development for this situation – to make these more difficult cases black box."

The Grimme group has evolved a standard protocol for addressing new problems. "This is important when you have so many different projects that you don't get confused, that the students don't get confused," he notes. "We use TPSS with dispersion correction for geometry optimization – its fast in *Orca*. For the energies, we employ a hybrid or double hybrid functional. To correct for solvation we use COSMO-RS. Most problematic is the computation of the costly frequencies, where we often run into numerical problems for large systems."

At the beginning of his career, Grimme was in close contact with experimentalists. This began with his examination of electronic spectroscopy of cyclophanes as he sought ways to compute its CD spectrum. Grimme's work on large systems was largely inspired by the supramolecular chemist Fritz Vögtle and the organic chemist Gerhard Erker. "Usually experimentalists come to me with their problems. I could have applied standard methods but they inspired me to think of new ways to answer their problems accurately and efficiently."

Occasionally, he has approached an experimentalist with a computationally inspired problem. In working with Bader's Atoms-In-Molecules theory, Grimme was interested in whether bond paths really correlate with bonds. Grimme proposed to Erker to synthesize 4,5-dideuterated phenanthrene "and we got spectroscopists to measure this!"

When asked about problems and bottlenecks facing computational chemist today, Grimme commented "In my experience with larger systems, the problem is conformational space: the degrees of freedom, the sheer number of isomers and conformers. What quantum chemists often fail to recognize is that it's not only the electrons, it's the nuclei as well! We can narrow down the error in the electronic energy to 1 kcal mol^{-1}, but the error in the ZPE can easily reach 3 kcal mol^{-1}, which is not uncommon for a big system. With all those degrees of freedom, the errors can mount up. There is room for improvement here in dealing with large conformational space. It's not just computing power, there is factorial growth. And there is also the entropy term. The harmonic oscillator approximation is simply too crude. But moving beyond the harmonic approximation is just impossible now. Its unavoidable today, but we know that this approach is seriously deficient. We are starting to think about this more. If you want to do accurate thermochemistry for a big system you are faced with this problem."

Finally, he noted a real dearth of solid computational methods to deal with the electronic structure of transition metal compounds. "For main groups, we have appropriate methods," he says. "But for metallic clusters, this is really hard. Computations of metal surfaces at high accuracy are impossible today. This is a case where the physicists are leading the chemists." These problems will no doubt keep Grimme inspired for years to come!

REFERENCES

1. Cramer, C. J. *Essentials of Computational Chemistry: Theories and Models*; John Wiley & Sons: New York, **2002**.
2. Jensen, F. *Introduction to Computational Chemistry*; John Wiley & Sons: Chichester, England, **1999**.
3. Szabo, A.; Ostlund, N. S. *Modern Quantum Chemistry: Introduction to Advanced Electronic Structure Theory*; Dover: Mineola, NY, **1996**.
4. Dirac, P. A. M. "The quantum theory of the electron," *Proc. R. Soc. London Ser. A* **1928**, *117*, 610–624.
5. Born, M.; Oppenheimer, R. "Zur quantentheorie der molekeln," *Ann. Phys.* **1927**, *84*, 457–484.
6. Hartree, D. R. "The wave mechanics of an atom with a non-Coulomb central field. I. Theory and methods," *Proc. Cambridge Philos. Soc.* **1928**, *24*, 89–110.
7. Fock, V. "Näherungsmethode zur lösung des quantenmechanischen mehrkörperproblems," *Z. Phys.* **1930**, *61*, 126–148.
8. Fock, V. "'Selfconsistent field' mit austausch für natrium," *Z. Phys.* **1930**, *62*, 795–805.
9. Pauli, W. "Über den zusammenhang des abschlusses der elektronengruppen im atom mit der komplexstruktur der spektren," *Z. Phys.* **1925**, *31*, 765–783.
10. Roothaan, C. C. J. "New developments in molecular orbital theory," *Rev. Mod. Phys.* **1951**, *23*, 69–89.
11. Boys, S. F. "Electronic wave functions. I. A general method of calculation for the stationary states of any molecular system," *Proc. R. Soc. Lond. Ser. A* **1950**, *A200*, 542–554.
12. Ditchfield, R.; Hehre, W. J.; Pople, J. A. "Self-consistent molecular-orbital methods. IX. An extended gaussian-type basis for molecular-orbital studies of organic molecules," *J. Chem. Phys.* **1971**, *54*, 724–728.
13. Hehre, W. J.; Ditchfield, R.; Pople, J. A. "Self-consistent molecular orbital methods. XII. Further extensions of Gaussian-type basis sets for use in molecular orbital studies of organic molecules," *J. Chem. Phys.* **1972**, *56*, 2257–2261.

14. Dunning, T. H., Jr. "Gaussian basis sets for Use in correlated molecular calculations. I. The atoms boron through neon and hydrogen," *J. Chem. Phys.* **1989**, *90*, 1007–1023.
15. Woon, D. E.; Dunning, T. H., Jr. "Gaussian basis sets for Use in correlated molecular calculations. III. The atoms aluminum through argon," *J. Chem. Phys.* **1993**, *98*, 1358–1371.
16. Woon, D. E.; Dunning, T. H., Jr. "Gaussian basis sets for Use in correlated molecular calculations. V. Core-valence basis sets for boron through neon," *J. Chem. Phys.* **1995**, *103*, 4572–4585.
17. Jensen, F. "Polarization consistent basis sets: principles," *J. Chem. Phys.* **2001**, *115*, 9113–9125.
18. Jensen, F. "Polarization consistent basis sets. II. Estimating the Kohn–Sham basis set limit," *J. Chem. Phys.* **2002**, *116*, 7372–7379.
19. Weigend, F.; Ahlrichs, R. "Balanced basis sets of split valence, triple zeta valence and quadruple zeta valence quality for H to Rn: design and assessment of accuracy," *Phys. Chem. Chem. Phys.* **2005**, *7*, 3297–3305.
20. Hill, J. G. "Gaussian basis sets for molecular applications," *Int. J. Quantum Chem.* **2013**, *113*, 21–34.
21. Jensen, F. "Atomic orbital basis sets," *WIREs Comput. Mol. Sci.*, **2012**, *3*, 273–295.
22. Molecular Science Computing Facility, E. a. M. S. L. "EMSL Gaussian basis set order form," 2005, URL: https://bse.pnl.gov/bse/portal
23. Boys, S. F.; Bernardi, F. "The calculation of small molecular interactions by the differences of separate total energies. Some procedures with reduced errors," *Mol. Phys.* **1970**, *19*, 553–566.
24. Asturiol, D.; Duran, M.; Salvador, P. "Intramolecular basis set superposition error effects on the planarity of benzene and other aromatic molecules: a solution to the problem," *J. Chem. Phys.* **2008**, *128*, 144108.
25. Jensen, F. "An atomic counterpoise method for estimating inter- and intramolecular basis set superposition errors," *J. Chem. Theory Comput.* **2010**, *6*, 100–106.
26. Kruse, H.; Grimme, S. "A geometrical correction for the inter- and intra-molecular basis set superposition error in Hartree-Fock and density functional theory calculations for large systems," *J. Chem. Phys.* **2012**, *136*, 154101–154116.
27. Kruse, H.; Goerigk, L.; Grimme, S. "Why the standard B3LYP/6-31G* model chemistry should not be used in DFT calculations of molecular thermochemistry: understanding and correcting the problem," *J. Org. Chem.* **2012**, *77*, 10824–10834.
28. Sherrill, C. D.; Schaefer, H. F., III. "The configuration interaction method: advances in highly correlated approaches," *Adv. Quantum Chem.* **1999**, *34*, 143–269.
29. Brillouin, L. "Les champs self-consistents de Hartree et de Fock," *Actualitiés Sci. Ind.* **1934**, *71*, 159.
30. Langhoff, S. R.; Davidson, E. R. "Configuration interaction calculations on the nitrogen molecule," *Int. J. Quantum Chem.* **1974**, *8*, 61–72.
31. Møller, C.; Plesset, M. S. "Note on an approximation treatment for many-electron systems," *Phys. Rev.* **1934**, *48*, 618–622.
32. Vahtras, O.; Almlof, J.; Feyereisen, M. W. "Integral approximations for LCAO-SCF calculations," *Chem. Phys. Lett.* **1993**, *213*, 514–518.

33. Bernholdt, D. E.; Harrison, R. J. "Large-scale correlated electronic structure calculations: the RI-MP2 method on parallel computers," *Chem. Phys. Lett.* **1996**, *250*, 477–484.
34. Grimme, S. "Improved second-order Møller-Plesset perturbation theory by separate scaling of parallel- and antiparallel-spin pair correlation energies," *J. Chem. Phys.* **2003**, *118*, 9095–9102.
35. Schwabe, T.; Grimme, S. "Theoretical thermodynamics for large molecules: walking the thin line between accuracy and computational cost," *Acc. Chem. Res.* **2008**, *41*, 569–579.
36. Grimme, S.; Goerigk, L.; Fink, R. F. "Spin-component-scaled electron correlation methods," *WIREs Comput. Mol. Sci.* **2012**, *2*, 886–906.
37. Cizek, J. "On the correlation problem in atomic and molecular systems. Calculation of wavefunction components in Ursell-type expansion using quantum-field theoretical methods," *J. Chem. Phys.* **1966**, *45*, 4256–4266.
38. Raghavachari, K.; Trucks, G. W.; Pople, J. A.; Head-Gordon, M. "A fifth-order perturbation comparison of electron correlation theories," *Chem. Phys. Lett.* **1989**, *157*, 479–483.
39. Řezáč, J.; Hobza, P. "Describing noncovalent interactions beyond the common approximations: how accurate is the "gold standard," CCSD(T) at the complete basis set limit?," *J. Chem. Theor. Comput.* **2013**, *9*, 2151–2155.
40. Pople, J. A.; Head-Gordon, M.; Raghavachari, K. "Quadratic configuration interaction. A general technique for determining electron correlation energies," *J. Chem. Phys.* **1987**, *87*, 5968–5975.
41. Handy, N. C.; Pople, J. A.; Head-Gordon, M.; Raghavachari, K.; Trucks, G. W. "Size-consistent Brueckner theory limited to double substitutions," *Chem. Phys. Lett.* **1989**, *164*, 185–192.
42. Shepard, R. "The multiconfiguration self-consistent field method," *Adv. Chem. Phys.* **1987**, *69*, 63–200.
43. Roos, B. "The complete active space self-consistent field method and its applications in electronic structure calculations," *Adv. Chem. Phys.* **1987**, *69*, 399–445.
44. Andersson, K.; Malmqvist, P.-Å.; Roos, B. O. "Second-order perturbation theory with a complete active space self-consistent field reference function," *J. Chem. Phys.* **1992**, *96*, 1218–1226.
45. Lyakh, D. I.; Musiał, M.; Lotrich, V. F.; Bartlett, R. J. "Multireference nature of chemistry: the coupled-cluster view," *Chem. Rev.* **2011**, *112*, 182–243.
46. Mahapatra, U. S.; Datta, B.; Mukherjee, D. "A size-consistent state-specific multireference coupled cluster theory: Formal developments and molecular applications," *J. Chem. Phys.* **1999**, *110*, 6171–6188.
47. Chattopadhyay, S.; Mahapatra, U. S.; Mukherjee, D. "Property calculations using perturbed orbitals via state-specific multireference coupled-cluster and perturbation theories," *J. Chem. Phys.* **1999**, *111*, 3820–3831.
48. Evangelista, F. A.; Allen, W. D.; Schaefer III, H. F. "Coupling term derivation and general implementation of state-specific multireference coupled cluster theories," *J. Chem. Phys.* **2007**, *127*, 024102–024117.
49. Pople, J. A.; Head-Gordon, M.; Fox, D. J.; Raghavachari, K.; Curtiss, L. A. "Gaussian-1 theory: a general procedure for prediction of molecular energies," *J. Chem. Phys.* **1989**, *90*, 5622–5629.

50. Curtiss, L. A.; Raghavachari, K.; Trucks, G. W.; Pople, J. A. "Gaussian-2 theory for molecular energies of first- and second-row compounds," *J. Chem. Phys.* **1991**, *94*, 7221–7230.
51. Curtiss, L. A.; Raghavachari, K.; Redfern, P. C.; Rassolov, V.; Pople, J. A. "Gaussian-3 (G3) theory for molecules containing first and second-row atoms," *J. Chem. Phys.* **1998**, *109*, 7764–7776.
52. Curtiss, L. A.; Redfern, P. C.; Raghavachari, K. "Gaussian-4 theory," *J. Chem. Phys.* **2007**, *126*, 084108.
53. Curtiss, L. A.; Redfern, P. C.; Raghavachari, K. "Gaussian-4 theory using reduced order perturbation theory," *J. Chem. Phys.* **2007**, *127*, 124105.
54. Montgomery, J. A., Jr.; Frisch, M. J.; Ochterski, J. W.; Petersson, G. A. "A complete basis Set model chemistry. VI. Use of density functional geometries and frequencies," *J. Chem. Phys.* **1999**, *110*, 2822–2827.
55. Montgomery, J. A., Jr.; Frisch, M. J.; Ochterski, J. W.; Petersson, G. A. "A complete basis set model chemistry. VII. Use of the minimum population localization method," *J. Chem. Phys.* **2000**, *112*, 6532–6542.
56. Harding, M. E.; Vazquez, J.; Ruscic, B.; Wilson, A. K.; Gauss, J.; Stanton, J. F. "High-accuracy extrapolated ab initio thermochemistry. III. Additional improvements and overview," *J. Chem. Phys.* **2008**, *128*, 114111–114115.
57. Tajti, A.; Szalay, P. G.; Csaszar, A. G.; Kallay, M.; Gauss, J.; Valeev, E. F.; Flowers, B. A.; Vazquez, J.; Stanton, J. F. "HEAT: high accuracy extrapolated ab initio thermochemistry," *J. Chem. Phys.* **2004**, *121*, 11599–11613.
58. Martin, J. M. L.; de Oliveira, G. "Towards standard methods for benchmark quality ab initio thermochemistry—W1 and W2 theory," *J. Chem. Phys.* **1999**, *111*, 1843–1856.
59. Karton, A.; Rabinovich, E.; Martin, J. M. L.; Ruscic, B. "W4 theory for computational thermochemistry: in pursuit of confident sub-kJ/mol predictions," *J. Chem. Phys.* **2006**, *125*, 144108–144117.
60. Boese, A. D.; Oren, M.; Atasoylu, O.; Martin, J. M. L.; Kallay, M.; Gauss, J. "W3 theory: robust computational thermochemistry in the kJ/mol accuracy range," *J. Chem. Phys.* **2004**, *120*, 4129–4141.
61. Dixon, D. A.; Feller, D.; Peterson, K. A. "A practical guide to reliable first principles computational thermochemistry predictions across the periodic table," In *Annual Reports in Computational Chemistry*; Wheeler, R. A., Ed.; Elsevier: **2012**; Vol. 8, p 1–28.
62. Martin, J. M. L. "Computational thermochemistry: a brief overview of quantum mechanical approaches," *Ann. Rep. Comput. Chem.* **2005**, *1*, 31–43.
63. Császár, A. G.; Allen, W. D.; Schaefer, H. F., III "In pursuit of the ab initio limit for conformational energy prototypes," *J. Chem. Phys.* **1998**, *108*, 9751–9764.
64. Hohenberg, P.; Kohn, W. "Inhomogeneous electron gas," *Phys. Rev.* **1964**, *136*, B864–B871.
65. Kohn, W.; Sham, L. J. "Self-consistent equations including exchange and correlation effects," *Phys. Rev.* **1965**, *140*, A1133–A1138.
66. Perdew, J. P.; Schmidt, K. "Jacob's ladder of density functional approximations for the exchange-correlation energy," In *Density Functional Theory and its Application to Materials*; Doren, V. V., Alsenoy, C. V., Geerlings, P., Eds.; AIP-Press: Melville, **2001**, p 1–20.

67. Perdew, J. P.; Ruzsinszky, A.; Tao, J.; Staroverov, V. N.; Scuseria, G. E.; Csonka, G. I. "Prescription for the design and selection of density functional approximations: more constraint satisfaction with fewer fits," *J. Chem. Phys.* **2005**, *123*, 062201.
68. Vosko, S. H.; Wilk, L.; Nusair, M. "Accurate spin-dependent electron liquid correlation energies for local spin density calculations: a critical analysis," *Can. J. Phys.* **1980**, *58*, 1200–1211.
69. Koch, W.; Holthaisen, M. C. *A Chemist's Guide to Density Functional Theory*; Wiley-VCH: Weinheim, Germany, **2000**.
70. Becke, A. D. "Density-functional exchange-energy approximation with correct asymptotic behavior," *Phys. Rev. A* **1988**, *38*, 3098–3100.
71. Lee, C.; Yang, W.; Parr, R. G. "Development of the Colle-Salvetti correlation-energy formula into a functional of the electron density," *Phys. Rev. B* **1988**, *37*, 785–789.
72. Perdew, J. P.; Wang, Y. "Accurate and simple analytic representation of the electron-gas correlation energy," *Phys. Rev. B* **1992**, *45*, 13244–13249.
73. Perdew, J. P.; Burke, K.; Ernzerhof, M. "Generalized gradient approximation made simple," *Phys. Rev. Lett.* **1996**, *77*, 3865–3868 (errata **1997**, *3878*, 1396).
74. Adamo, C.; Barone, V. "Toward reliable density functional methods without adjustable parameters: the PBE0 model," *J. Chem. Phys.* **1999**, *110*, 6158–6170.
75. Becke, A. D. "Density-functional thermochemistry. III. The role of exact exchange," *J. Chem. Phys.* **1993**, *98*, 5648–5652.
76. Stephens, P. J.; Devlin, F. J.; Chabalowski, C. F.; Frisch, M. J. "Ab initio calculation of vibrational absorption and circular dichroism spectra using density functional force fields," *J. Phys. Chem.* **1994**, *98*, 11623–11627.
77. Tao, J.; Perdew, J. P.; Staroverov, V. N.; Scuseria, G. E. "Climbing the density functional ladder: nonempirical metaÂ-generalized gradient approximation designed for molecules and solids," *Phys. Rev. Lett.* **2003**, *91*, 146401.
78. Zhao, Y.; Truhlar, D. G. "Hybrid meta density functional theory methods for thermochemistry, thermochemical kinetics, and noncovalent interactions: the MPW1B95 and MPWB1K models and comparative assessments for hydrogen bonding and van der Waals interactions," *J. Phys. Chem. A* **2004**, *108*, 6908–6918.
79. Zhao, Y.; Truhlar, D. G. "Design of density functionals that Are broadly accurate for thermochemistry, thermochemical kinetics, and nonbonded interactions," *J. Phys. Chem. A* **2005**, *109*, 5656–5667.
80. Zhao, Y.; Schultz, N. E.; Truhlar, D. G. "Design of density functionals by combining the method of constraint satisfaction with parametrization for thermochemistry, thermochemical kinetics, and noncovalent interactions," *J. Chem. Theory Comput.* **2006**, *2*, 364–382.
81. Zhao, Y.; Truhlar, D. "The M06 suite of density functionals for main group thermochemistry, thermochemical kinetics, noncovalent interactions, excited states, and transition elements: two new functionals and systematic testing of four M06-class functionals and 12 other functionals," *Theor. Chem. Acc.* **2008**, *120*, 215–241.
82. Iikura, H.; Tsuneda, T.; Yanai, T.; Hirao, K. "A long-range correction scheme for generalized-gradient-approximation exchange functionals," *J. Chem. Phys.* **2001**, *115*, 3540–3544.
83. Becke, A. D. "Density-functional thermochemistry. V. Systematic optimization of exchange-correlation functionals," *J. Chem. Phys.* **1997**, *107*, 8554–8560.

84. Chai, J.-D.; Head-Gordon, M. "Systematic optimization of long-range corrected hybrid density functionals," *J. Chem. Phys.* **2008**, *128*, 084106–084115.
85. Grimme, S. "Semiempirical hybrid density functional with perturbative second-order correlation," *J. Chem. Phys.* **2006**, *124*, 034108–034116.
86. Schwabe, T.; Grimme, S. "Towards chemical accuracy for the thermodynamics of large molecules: new hybrid density functionals including non-local correlation effects," *Phys. Chem. Chem. Phys.* **2006**, *8*, 4398–4401.
87. Tarnopolsky, A.; Karton, A.; Sertchook, R.; Vuzman, D.; Martin, J. M. L. "Double-hybrid functionals for thermochemical kinetics," *J. Phys. Chem. A* **2007**, *112*, 3–8.
88. Grimme, S. "Density functional theory with London dispersion corrections," *WIREs Comput. Mol. Sci.* **2011**, *1*, 211–228.
89. Grimme, S. "Accurate description of van der Waals complexes by density functional theory including empirical corrections," *J. Comput. Chem.* **2004**, *25*, 1463–1473.
90. Grimme, S. "Semiempirical GGA-type density functional constructed with a long-range dispersion correction," *J. Comput. Chem.* **2006**, *27*, 1787–1799.
91. Grimme, S.; Antony, J.; Ehrlich, S.; Krieg, H. "A consistent and accurate ab initio parametrization of density functional dispersion correction (DFT-D) for the 94 elements H-Pu," *J. Chem. Phys.* **2010**, *132*, 154104–154119.
92. Chai, J.-D.; Head-Gordon, M. "Long-range corrected hybrid density functionals with damped atom-atom dispersion corrections," *Phys. Chem. Chem. Phys.* **2008**, *10*, 6615–6620.
93. Grimme, S.; Ehrlich, S.; Goerigk, L. "Effect of the damping function in dispersion corrected density functional theory," *J. Comput. Chem.* **2011**, *32*, 1456–1465.
94. Becke, A. D.; Johnson, E. R. "Exchange-hole dipole moment and the dispersion interaction," *J. Chem. Phys.* **2005**, *122*, 154104–154105.
95. Schwabe, T.; Grimme, S. "Double-hybrid density functionals with long-range dispersion corrections: higher accuracy and extended applicability," *Phys. Chem. Chem. Phys.* **2007**, *9*, 3397–3406.
96. Hujo, W.; Grimme, S. "Performance of non-local and atom-pairwise dispersion corrections to DFT for structural parameters of molecules with noncovalent interactions," *J. Chem. Theor. Comput.* **2013**, *9*, 308–315.
97. Kozuch, S.; Gruzman, D.; Martin, J. M. L. "DSD-BLYP: a general purpose double hybrid density functional including spin component scaling and dispersion correction," *J. Phys. Chem. C* **2010**, *114*, 20801–20808.
98. Kozuch, S.; Martin, J. M. L. "DSD-PBEP86: in search of the best double-hybrid DFT with spin-component scaled MP2 and dispersion corrections," *Phys. Chem. Chem. Phys.* **2011**, *13*, 20104–20107.
99. Swart, M. "Popularity poll density functionals 2012 (DFT2012)," **2012**, URL: http://www.marcelswart.eu/
100. Bickelhaupt, F. M.; Baerends, E. J.; Nibbering, N. M. M. "The effect of microsolvation on E_2 and S_N2 reactions: theoretical study of the model system $F^- + C_2H_5F + nHF$," *Chemistry Eur. J.* **1996**, *2*, 196–207.
101. Okuno, Y. "Theoretical examination of solvent reorganization and nonequilibrium solvation effects in microhydrated reactions," *J. Am. Chem. Soc.* **2000**, *122*, 2925–2933.

102. Raugei, S.; Cardini, G.; Schettino, V. "Microsolvation effect on chemical reactivity: the case of the Cl$^-$ + CH$_3$Br S$_N$2 reaction," *J. Chem. Phys.* **2001**, *114*, 4089–4098.

103. Re, S.; Morokuma, K. "ONIOM study of chemical reaction in microslvation clusters: (H$_2$O)$_n$CH$_3$Cl + OH$^-$(H$_2$O)$_m$ (n+m=1 and 2)," *J. Phys. Chem. A* **2001**, *105*, 7185–7197.

104. Takashima, K.; Riveros, J. M. "Gas-phase solvated negative ions," *Mass Spectrom. Rev.* **1998**, *17*, 409–430.

105. Laerdahl, J. K. "Gas phase nucleophilic substitution," *Int. J. Mass Spectrom.* **2002**, *214*, 277–314.

106. Roux, B.; Simonson, T. "Implicit solvent models," *Biophys. Chem.* **1999**, *78*, 1–20.

107. Miertus, S.; Scrocco, E.; Tomasi, J. "Electrostatic interaction of a solute with a continuum. A direct utilization of ab initio molecular potentials for the prevision of solvent effects," *Chem. Phys.* **1981**, *55*, 117–129.

108. Cossi, M.; Barone, V.; Cammi, R.; Tomasi, J. "Ab initio study of solvated molecules: a new implementation of the polarizable continuum model," *Chem. Phys. Lett.* **1996**, *255*, 327–335.

109. Thompson, J. D.; Cramer, C. J.; Truhlar, D. G. "Parameterization of charge model 3 for AM1, PM3, BLYP, and B3LYP," *J. Comput. Chem.* **2003**, *24*, 1291–1304.

110. Kelly, C. P.; Cramer, C. J.; Truhlar, D. G. "SM6: a density functional theory continuum solvation model for calculating aqueous solvation free energies of neutrals, ions, and solute-water clusters," *J. Chem. Theory Comput.* **2005**, *1*, 1133–1152.

111. Li, J.; Zhu, T.; Hawkins, G. D.; Winget, P.; Liotard, D. A.; Cramer, C. J.; Truhlar, D. G. "Extension of the platform of applicability of the SM5.42R universal solvation model," *Theor. Chem. Acc.* **1999**, *103*, 9–63.

112. Cramer, C. J.; Truhlar, D. G. In *Solvent Effects and Chemical Reactivity*; Tapia, O., Bertran, J., Eds.; Kluwer: Dordrecht, **1996**, p 1–80.

113. Tomasi, J.; Persico, M. "Molecular interactions in solution: an overview of methods based on continuous distributions of the solvent," *Chem. Rev.* **1994**, *94*, 2027–2094.

114. Cancè, E.; Mennucci, B.; Tomasi, J. "A new integral equation formalism for the polarizable continuum model: theoretical background and applications to isotropic and anisotropic dielectrics," *J. Chem. Phys.* **1997**, *107*, 3032–3041.

115. Barone, V.; Cossi, M.; Tomasi, J. "Geometry optimization of molecular structures in solution by the polarizable continuum model," *J. Comput. Chem.* **1998**, *19*, 404–417.

116. Cossi, M.; Rega, N.; Scalmani, G.; Barone, V. "Polarizable dielectric model of solvation with inclusion of charge penetration effects," *J. Chem. Phys.* **2001**, *115*, 5691–5701.

117. Cossi, M.; Scalmani, G.; Rega, N.; Barone, V. "New developments in the polarizable continuum model for quantum mechanical and classical calculations on molecules in solution," *J. Chem. Phys.* **2002**, *117*, 43–54.

118. Orozco, M.; Luque, F. J. "Theoretical methods for the description of the solvent effect in biomolecular systems," *Chem. Rev.* **2000**, *100*, 4187–4226.

119. Curutchet, C.; Orozco, M.; Luque, J. F. "Solvation in octanol: parametrization of the continuum MST model," *J. Comput. Chem.* **2001**, *22*, 1180–1193.

120. Foresman, J. B.; Keith, T. A.; Wiberg, K. B.; Snoonian, J.; Frisch, M. J. "Solvent effects. 5. Influence of cavity shape, truncation of electrostatics, and electron correlation on ab initio reaction field calculations," *J. Phys. Chem.* **1996**, *100*, 16098–16104.

121. York, D. M.; Karplus, M. "A smooth solvation potential based on the conductor-like screening model," *J. Phys. Chem. A* **1999**, *103*, 11060–11079.
122. Mennucci, B. "Polarizable continuum model," *WIREs Comput. Mol. Sci.* **2012**, *2*, 386–404.
123. Cramer, C. J.; Truhlar, D. G. "General parameterized SCF model for free energies of solvation in aqueous solution," *J. Am. Chem. Soc.* **1991**, *113*, 8305–8311.
124. Hawkins, G. D.; Cramer, C. J.; Truhlar, D. G. "Universal quantum mechanical model for solvation free energies based on gas-phase geometries," *J. Phys. Chem. B* **1998**, *102*, 3257–3271.
125. Marenich, A. V.; Cramer, C. J.; Truhlar, D. G. "Generalized born solvation model SM12," *J. Chem. Theor. Comput.* **2013**, *9*, 609–620.
126. Klamt, A.; Schüürmann, G. "COSMO: a new approach to dielectric screening in solvents with explicit expressions for the screening energy and its gradient," *J. Chem. Soc., Perkin Trans. 2* **1993**, 799–805.
127. Klamt, A.; Jonas, V.; Burger, T.; Lohrenz, J. C. W. "Refinement and parametrization of COSMO-RS," *J. Phys. Chem. A* **1998**, *102*, 5074–5085.
128. Barone, V.; Cossi, M. "Quantum calculation of molecular energies and energy gradients in solution by a conductor solvent model," *J. Phys. Chem. A* **1998**, *102*, 1995–2001.
129. Cossi, M.; Rega, N.; Scalmani, G.; Barone, V. "Energies, structures, and electronic properties of molecules in solution with the C-PCM solvation model," *J. Comput. Chem.* **2003**, *24*, 669–681.
130. Marenich, A. V.; Olson, R. M.; Kelly, C. P.; Cramer, C. J.; Truhlar, D. G. "Self-consistent reaction field model for aqueous and nonaqueous solutions based on accurate polarized partial charges," *J. Chem. Theor. Comput.* **2007**, *3*, 2011–2033.
131. Kelly, C. P.; Cramer, C. J.; Truhlar, D. G. "Adding explicit solvent molecules to continuum solvent calculations for the calculation of aqueous acid dissociation constants," *J. Phys. Chem. A* **2006**, *110*, 2493–2499.
132. Warshel, A.; Levitt, M. "Theoretical studies of enzymic reactions: dielectric, electrostatic and steric stabilization of the carbonium ion in the reaction of lysozyme," *J. Mol. Biol.* **1976**, *103*, 227–249.
133. Singh, U. C.; Kollman, P. A. "A combined ab initio quantum mechanical and molecular mechanical method for carrying out simulations on complex molecular systems: applications to the $CH_3Cl + Cl^-$ exchange reaction and gas phase protonation of polyethers," *J. Comput. Chem.* **1986**, *7*, 718–730.
134. Field, M. J.; Bash, P. A.; Karplus, M. "A combined quantum mechanical and molecular mechanical potential for molecular dynamics simulations," *J. Comput. Chem.* **1990**, *11*, 700–733.
135. Allinger, N. L. "Conformational analysis. 130. MM2. A hydrocarbon force field utilizing V1 and V2 torsional terms," *J. Am. Chem. Soc.* **1977**, *99*, 8127–8134.
136. Cornell, W. D.; Cieplak, P.; Bayly, C. I.; Gould, I. R.; Merz, K. M.; Ferguson, D. M.; Spellmeyer, D. C.; Fox, T.; Caldwell, J. W.; Kollman, P. A. "A second generation force field for the simulation of proteins, nucleic acids, and organic molecules," *J. Am. Chem. Soc.* **1995**, *117*, 5179–5197.
137. Damm, W.; Frontera, A.; Tirado–Rives, J.; Jorgensen, W. L. "OPLS all-atom force field for carbohydrates," *J. Comput. Chem.* **1997**, *18*, 1955–1970.

138. Brooks, B. R.; Bruccoleri, R. E.; Olafson, B. D.; States, D. J.; Swaminathan, S.; Karplus, M. "CHARMM: a program for macromolecular energy, minimization, and dynamics calculations," *J. Comput. Chem.* **1983**, *4*, 187–217.
139. Gao, J.; Truhlar, D. G. "Quantum mechanical methods for enzyme kinetics," *Ann. Rev. Phys. Chem.* **2002**, *53*, 467–505.
140. Vreven, T.; Morokuma, K. "Hybrid methods: ONIOM(QM:MM) and QM/MM," *Ann. Rep. Comput. Chem.* **2006**, *2*, 35–51.
141. Lin, H.; Truhlar, D. "QM/MM: what have we learned, where are we, and where do we go from here?," *Theor. Chem. Acc.* **2007**, *117*, 185–199.
142. Senn, H. M.; Thiel, W. "QM/MM methods for biomolecular systems," *Angew. Chem, Int. Ed.* **2009**, *48*, 1198–1229.
143. Bakowies, D.; Thiel, W. "Hybrid models for combined quantum mechanical and molecular mechanical approaches," *J. Phys. Chem.* **1996**, *100*, 10580–10594.
144. Théry, V.; Rinaldi, D.; Rivail, J.-L.; Maigret, B.; Ferenczy, G. G. "Quantum mechanical computations on very large molecular systems: the local self-consistent field method," *J. Comput. Chem.* **1994**, *15*, 269–282.
145. Assfeld, X.; Rivail, J.-L. "Quantum chemical computations on parts of large molecules: the ab initio local self consistent field method," *Chem. Phys. Lett.* **1996**, *263*, 100–106.
146. Gao, J.; Amara, P.; Alhambra, C.; Field, M. J. "A generalized hybrid orbital (GHO) method for the treatment of boundary atoms in combined QM/MM calculations," *J. Phys. Chem. A* **1998**, *102*, 4714–4721.
147. Dapprich, S.; Komiromi, I.; Byun, K. S.; Morokuma, K.; Frisch, M. J. "A new ONIOM implementation in Gaussian98. Part I. The calculation of energies, gradients, vibrational frequencies and electric field derivatives," *J. Mol. Struct. (THEOCHEM)* **1999**, *461–462*, 1–21.
148. Waszkowycz, B.; Hillier, I. H.; Gensmantel, N.; Payling, D. W. "Combined quantum mechanical-molecular mechanical study of catalysis by the enzyme phospholipase A2: an investigation of the potential energy surface for amide hydrolysis," *J. Chem. Soc., Perkin Trans. 2* **1991**, 2025–2032.
149. Chung, L. W.; Hirao, H.; Li, X.; Morokuma, K. "The ONIOM method: its foundation and applications to metalloenzymes and photobiology," *WIREs Comput. Mol. Sci.* **2012**, *2*, 327–350.
150. Vreven, T.; Byun, K. S.; Komaromi, I.; Dapprich, S.; Montgomery, J. A.; Morokuma, K.; Frisch, M. J. "Combining quantum mechanics methods with molecular mechanics methods in ONIOM," *J. Chem. Theory Comput.* **2006**, *2*, 815–826.
151. Gonzalez, C.; Schlegel, H. B. "Reaction path following in mass-weighted internal coordinates," *J. Phys. Chem.* **1990**, *94*, 5523–5527.
152. Stanton, R. E.; McIver, J. W. "Group theoretical selection rules for the transition states of chemical reactions," *J. Am. Chem. Soc.* **1975**, *97*, 3632–3646.
153. Schlegel, H. B. "Optimization of equilibrium geometries and transition structures," *J. Comput. Chem.* **1982**, *3*, 214–218.
154. Schlegel, H. B. "Exploring potential energy surfaces for chemical reactions: an overview of some practical methods," *J. Comput. Chem.* **2003**, *24*, 1514–1527.
155. Peng, C.; Schlegel, H. B. "Combining synchronous transit and quasi-Newton methods to find transition states," *Isr. J. Chem.* **1993**, *33*, 449–454.

156. Bachrach, S. M. In *Reviews in Computational Chemistry*; Lipkowitz, K. B., Boyd, D. B., Eds.; Wiley-VCH: New York, **1994**; Vol. 5, p 171–228.
157. Mulliken, R. S. "Electronic population analysis on LCAO-MO molecular wave functions. I," *J. Chem. Phys.* **1955**, *23*, 1833–1840.
158. Cusachs, L. C.; Politzer, P. "On the problem of defining the charge on an atom in a molecule," *Chem. Phys. Lett.* **1968**, *1*, 529–531.
159. Löwdin, P.-O. "On the orthogonality problem," *Adv. Quantum Chem.* **1970**, *5*, 185–199.
160. Reed, A. E.; Weinstock, R. B.; Weinhold, F. "Natural population analysis," *J. Chem. Phys.* **1985**, *83*, 735–746.
161. Weinhold, F. "Natural bond orbital analysis: a critical overview of relationships to alternative bonding perspectives," *J. Comput. Chem.* **2012**, *33*, 2363–2379.
162. Glendening, E. D.; Landis, C. R.; Weinhold, F. "Natural bond orbital methods," *WIREs Comput. Mol. Sci.* **2012**, *2*, 1–42.
163. Bader, R. F. W. *Atoms in Molecules: A Quantum Theory*; Clarendon Press: Oxford, UK, **1990**.
164. Bader, R.; Matta, C. "Atoms in molecules as non-overlapping, bounded, space-filling open quantum systems," *Found. Chem.* **2012**, 1–24.

CHAPTER 2

Computed Spectral Properties and Structure Identification

Once the wavefunction is in hand, all observable properties can, at least in principle, be computed. This can include all varieties of spectral properties, which are particularly valuable in ascertaining molecular structure. The full theoretical and computational means for computing spectral properties are mathematically involved and beyond the scope of this chapter. Instead, we will focus on the use of quantum chemical computations to help identify chemical structure. The purpose of this chapter is to inspire *the routine use of computed spectra to aid in structural identification*.

As discussed in Chapter 1, the full three-dimensional structure of a compound can be optimized with almost any of the quantum computational techniques. Since most quantum chemical computations are still performed on a single molecule in the gas phase, these computed structures can be most readily compared to gas-phase experimental structures. In the following chapters, we will present a number of case studies where computed and experimental geometries are compared. To get a sense of the quality of computed geometries, a few selected cases are discussed next.

2.1 COMPUTED BOND LENGTHS AND ANGLES

Schaefer and Allinger[1] examined a series of small organic compounds with a number of different computational methods. Table 2.1 compares the C–C and C–H bond lengths (r_e) in some very small compounds, and finds the coupled-cluster singles and doubles (CCSD/TZ2+P) results in excellent agreement with the limited experimental values. Importantly, the bond lengths predicted by the much more computationally efficient B3LYP/6-31G* are very similar to the CCSD results. For a group of 10 small hydrocarbons, including some alkenes and alkynes, the B3LYP/6-31G* C–C and C–H distances are on an average too small (compared with r_g values) by only 0.0031 and 0.016 Å, respectively. Similar excellent performance is also seen for a group of 19 organic compounds containing oxygen or nitrogen.

Computational Organic Chemistry, Second Edition. Steven M. Bachrach.
© 2014 John Wiley & Sons, Inc. Published 2014 by John Wiley & Sons, Inc.

TABLE 2.1 Comparison of Computed and Experiment Bond Lengths (r_e, Å)[a]

Cmpd	Bond	MP2 6-31G**	MP2 TZ2P+f	B3LYP 6-31G*	CCSD TZ2P+f	Expt
Ethane	C–C	1.523	1.523	1.530	1.532	
	C–H	1.089	1.087	1.095	1.090	
Ethene	C=C	1.335	1.331	1.332	1.330	
Ethyne	C≡C	1.219	1.210	1.205	1.202	1.203
	C–H	1.062	1.061	1.067	1.062	1.062
Formaldehyde	C=O	1.221	1.211	1.207	1.203	1.203
	C–H	1.104	1.099	1.110	1.100	1.099

[a]Ref. 1.

TABLE 2.2 Mean Absolute Error in Computed Versus Experimental Equilibrium Bond Distances (Å) for 19 Small Molecules[a]

| | cc-pVxZ | | | aug-cc-pVxZ | | |
	D	T	Q	D	T	Q
HF	0.0180	0.0249	0.0259	0.0188	0.0249	0.0259
MP2	0.0134	0.0055	0.0051	0.0149	0.0051	0.0048
CCSD	0.0116	0.0057	0.0072	0.0113	0.0058	0.0069
CCSD(T)	0.0168	0.0020	0.0013	0.0173	0.0019	0.0010

[a]Ref. 2.

In a study of 19 small molecules, including both inorganics and organics (like formaldehyde, ethane, and ethyne), Bak et al.[2] compared the computed and experimental equilibrium bond distances and angles. Geometries were optimized at Hartree–Fock (HF), MP2, CCSD, and CCSD(T) using a series of correlation consistent basis sets. A sampling of mean absolute errors in the bond distances are given in Table 2.2. Error of less than 0.005 Å can be had with MP2/cc-pVTZ, but to get to errors below 0.002 Å, CCSD(T) with at least cc-pVTZ is necessary. Mean absolute errors in bond angles are less than a degree for MP2/cc-pVTZ, CCSD/cc-pVTZ, and CCSD(T)/cc-pVTZ.

Oftentimes one can use a quite resource-efficient method to obtain excellent geometries. Sequential ethynyl[3] or cyano[4] substitution of benzene leads to a systematic increase in the mean bond length within the ring. As seen in Table 2.3, quite excellent reproduction of this trend, and the bond distances themselves, can be had with even B3LYP/6-31G*.

2.2 IR SPECTROSCOPY

Vibrational frequencies, used to predict IR and Raman spectra, are computed from the Hessian matrix, assuming a harmonic oscillator approximation. Errors in the

TABLE 2.3 Mean Distances (Å) of the Ring C–C Bond[a]

	 benzene	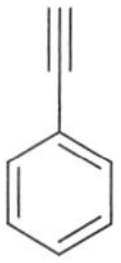 phenylacetylene	 1,4-diethynylbenzene	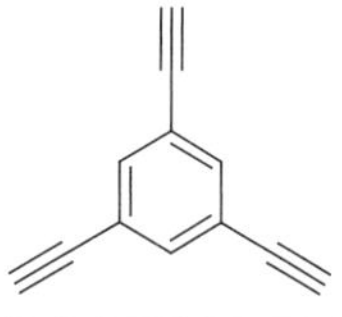 1,3,5-triethynylbenzene
MP2/6-31G*	1.397	1.399	1.401	1.403
B3LYP/6-31G*	1.397	1.399	1.402	1.404
Expt	1.398	1.401	1.402	1.404

	 benzene	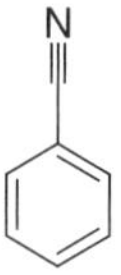 benzonitrile	 1,4-dicyanobenzene	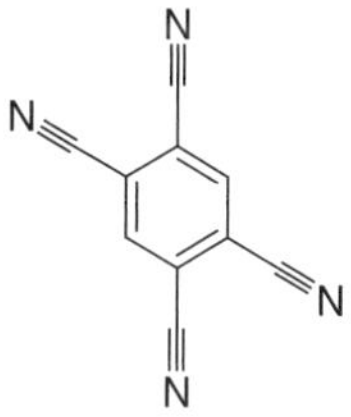 1,2,4,5-tetracyanobenzene
MP2/cc-pVDZ	1.393	1.395	1.396	1.400
B3LYP/6-31G*	1.397	1.398	1.400	1.405
Expt	1.398	1.399	1.400	1.405

[a] Refs 3, 4.

predicted frequencies can be attributed then to (1) the use of an incomplete basis set, (2) incomplete treatment of electron correlation, and (3) the anharmonicity of the potential energy surface. The first two can be assessed by examining a series of computations with different basis sets and treatments of electron correlation, looking for an asymptotic trend. In terms of treating the anharmonicity, recently developed techniques demonstrate how one can directly compute the anharmonic vibrational frequencies.[5] Nonetheless, computation of the anharmonic frequencies are significantly more computationally intensive than computation of the harmonic frequencies.

Due to the harmonic approximation, most methods will overestimate the vibrational frequencies. Listed in Table 2.4 are the mean absolute deviations of the vibrational frequencies for a set of 32 simple molecules with different computational methods. A clear trend is that as the method improves in accounting for electron correlation, the predicted vibrational frequencies are in better accord with experiment.

Another view of the dependence of the vibrational frequencies upon computational method is given in Table 2.5, where the computed vibrational frequencies of formaldehyde and ethyne are compared with experimental values. Again, as the basis set is improved and as the accounting for electron correlation becomes more complete, the computed vibrational frequencies become more in accord with

TABLE 2.4 Mean Absolute Deviation (MAD) of the Vibrational Frequencies (cm^{-1}) for 32 Molecules[a]

Method	MAD
HF	144
MP2	99
CCSD(T)	31
BPW	69
BLYP	59
B3LYP	31
*m*PW1PW	39

[a]Computed using the 6-311G(d,p) basis set. See Ref. 6.

experiment. The density functional theory (DFT) methods provide about as good vibrational frequencies as MP2 or CISD.

While the computed vibrational frequencies are in error, they appear to be systematically in error. Pople[10] proposed scaling the values of the vibrational frequencies to improve their overall agreement with experiment. The problem with scaling the frequencies is that a unique *scaling factor* must be determined for every different computational level, meaning a scaling factor has to be determined for every combination of computational method and basis set. Radom[11] has established the scaling factors for a number of computational levels, including HF, MP2, and DFT with various basis sets by fitting the frequencies from 122 molecules. Scaling factors for additional methods have been suggested by Schlegel[12], and the Truhlar group[13] has developed a large set of scaling factors for various combinations of methods and basis sets and maintains a Web page of these values.[14] It is also important to recognize that the vibrational frequencies enter into the calculation of the zero-point vibrational energy (ZPVE), and a different scaling factor is required to produce the appropriately scaled ZPVEs. Careful readers may have noted a scaling factor of 0.9854 applied to the ZPVE in step 9 of the G4 composite method (Section 1.2.6).

Inclusion of anharmonicity improves the agreement between experimental and computed vibrational frequencies. However, Jacobsen and co-workers[15] have noted that this improvement may not be worth the substantial increase in computational cost. In examining a set of 176 molecules that include 2738 vibrational modes, they find that for HF, MP2, B3LYP, and PBE0 with the 6-31G(d) or 6-31+G(d,p) basis sets, the root mean square deviation (RMSD) error in the vibrational frequencies obtained from scaled harmonic frequencies are just as good as the scaled anharmonic frequencies. They suggest that computing the anharmonic correction is only valuable when one has a very high quality potential energy surface, which requires much larger basis sets.

A common use of computed vibrational frequencies is to ascertain the identity of an unknown structure by comparison with experimental IR spectra. A few recent examples of the positive identification intermediates will suffice here. In the attempt to prepare benzocyclobutenylidene (**1**), an unknown compound was detected. By comparing the experimental IR spectrum with the computed IR

TABLE 2.5 Vibrational Frequencies (cm^{-1}) and Mean Absolute Deviation (MAD) from Experiment[a,b]

	Formaldehyde						
Method	$\omega_1(a_1)$	$\omega_2(a_1)$	$\omega_3(a_1)$	$\omega_4(b_1)$	$\omega_5(b_2)$	$\omega_6(b_2)$	MAD
HF/6-311G(d,p)	3081	1999	1648	1337	3147	1364	134
HF/cc-pVDZ	3109	2013	1637	1325	3183	1360	142
HF/cc-pVTZ	3084	1999	1652	1337	3153	1370	137
HF/aug-cc-pVTZ	3087	1992	1648	1335	3155	1367	135
B3LYP/6-31G(d)	2916	1849	1563	1199	2967	1280	37
B3LYP/aug-cc-pVDZ	2892	1803	1513	1194	2960	1245	37
B3LYP/aug-cc-pVTZ	2885	1813	1530	1198	2940	1263	38
M06-2x/6-31G(d)	2990	1907	1570	1215	3055	1291	45
M06-2x/aug-cc-pVDZ	2960	1865	1526	1214	3036	1260	35
M06-2x/aug-cc-pVTZ	2959	1875	1541	1220	3026	1276	35
ωB97X-D/6-31G(d)	2948	1884	1569	1213	3007	1289	36
ωB97X-D/aug-cc-pVDZ	2916	1844	1522	1209	2990	1256	34
ωB97X-D/aug-cc-pVTZ	2912	1854	1535	1211	2974	1271	36
MP2/6-311G(d,p)	2964	1777	1567	1211	3030	1291	19
MP2/aug-cc-pVTZ	2973	1753	1540	1198	3048	1267	22
CISD/6-311G(d,p)	3007	1879	1595	1243	3071	1316	56
CISD/aug-cc-pVTZ	3021	1875	1580	1245	3090	1305	57
QCISD/6-311G(d,p)	2945	1913	1566	1207	3003	1290	39
QCISD/aug-cc-pVTZ	2955	1803	1545	1204	3021	1275	21
CCSD/aug-cc-pVTZ	2961	1817	1549	1206	3030	1279	24
CCSD(T)/aug-cc-pVTZ	2932	1765	1530	1181	3000	1262	18
Expt[c]	2978	1778	1529	1191	2997	1299	

	Ethyne					
Method	$\omega_1(\sigma_g)$	$\omega_2(\sigma_g)$	$\omega_3(\sigma_u)$	$\omega_4(\pi_g)$	$\omega_5(\pi_u)$	MAD
HF/6-311G(d,p)	3676	2215	3562	815	875	171
HF/cc-pVDZ	3689	2224	3577	784	866	170
HF/cc-pVTZ	3674	2213	3556	807	868	166
HF/aug-cc-pVTZ	3674	2210	3554	810	869	166
B3LYP/6-31G(d)	3542	2088	3441	535	775	54
B3LYP/aug-cc-pVDZ	3523	2062	3419	557	727	35
B3LYP/aug-cc-pVTZ	3517	2068	3412	666	769	30
M06-2x/6-31G(d)	3554	2126	3445	643	805	57
M06-2x/aug-cc-pVDZ	3527	2099	3412	622	763	29
M06-2x/aug-cc-pVTZ	3528	2107	3414	715	797	55
ωB97X-D/6-31G(d)	3556	2110	3450	596	769	50
ωB97X-D/aug-cc-pVDZ	3539	2909	3429	590	732	38
ωB97X-D/aug-cc-pVTZ	3526	2089	3417	672	758	35

TABLE 2.5 (*Continued*)

Method	\multicolumn{5}{c}{Ethyne}	
	$\omega_1(\sigma_g)$	$\omega_2(\sigma_g)$	$\omega_3(\sigma_u)$	$\omega_4(\pi_g)$	$\omega_5(\pi_u)$	MAD
MP2/6-311G(d,p)	3550	1970	3460	562	770	45
MP2/aug-cc-pVTZ	3534	1968	3432	601	754	25
CISD/6-311G(d,p)	3587	2088	3484	642	804	63
CISD/aug-cc-pVTZ	3583	2090	3467	682	798	68
QCISD/6-311G(d,p)	3536	2023	3438	577	774	31
QCISD/aug-cc-pVTZ	3527	2035	3416	630	768	17
CCSD/aug-cc-pVTZ	3529	2044	3417	634	770	21
CCSD(T)/aug-cc-pVTZ	3503	1995	3394	593	748	15
Expt[d]	3495	2008	3415	624	747	

[a]Ref. 7.
[b]All DFT computations performed by the author.
[c]Ref. 8.
[d]Ref. 9.

spectra of a number of different proposed intermediates **2–5**, the cycloalkyne **3** was verified as the first intermediate detected. **3** then rearranges to **4** under further photolysis, and the structure of **4** was confirmed by comparison of its computed and experimental IR spectra.[16] McMahon[17] has used computed IR spectra to aid in confirming a conformational change upon annealing of triplet 1,3-diphenylpropynylidene **6**. In the third example, the carbene **7** and the strained allene **8**, which can be interconverted by irradiation at 302 nm, were identified by the comparison of their experimental and computed IR spectra.[18] The computed IR spectra were particularly helpful in identifying the stereochemistry of **7**.

The last example concerns the structures of the low energy conformations of glycine in the gas phase. Allen et al.[19] applied the focal point method to determine the structures of the two lowest energy conformations **9a** and **9b** (Figure 2.1). Careful examination of the microwave and electron diffraction data along with the focal point computations indicate that **9b** is nonplanar with a small barrier (about $20 \, \text{cm}^{-1}$) interconverting the two mirror image conformations through the planar structure. CCSD(T)/CBS computations predict that **9a** is $0.64 \, \text{kcal mol}^{-1}$ lower in energy than **9b**, with **9c** another $0.60 \, \text{kcal mol}^{-1}$ higher still. Raman spectroscopy of jet-cooled glycine identified six peaks below $500 \, \text{cm}^{-1}$.[20] All of these peaks could be matched up with vibrations from the three lowest energy conformations by comparison to the computed frequencies, see Table 2.6.

2.3 NUCLEAR MAGNETIC RESONANCE

Nuclear magnetic resonance (NMR) spectroscopy probes the energy required to flip a nuclear spin in the presence of a magnetic field. Computation of this effect

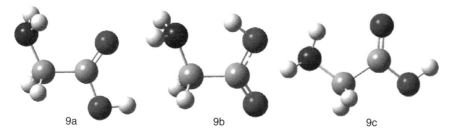

Figure 2.1 B3LYP/cc-pVTZ optimized structures of the three lowest energy conformers of glycine **9**.

requires, among other terms, derivatives of the kinetic energy of the electrons. This necessitates a definition of the origin of the coordinate system, called the *gauge origin*. The magnetic properties are independent of the gauge origin, but this is only true when an exact wavefunction is utilized. Since this is not a practical option, a

TABLE 2.6 Comparison of Computed and Experimental Frequencies (cm^{-1}) of Glycine[a]

Conformer	MP2[b]	B3LYP[b]	Expt
9a	211	213	203.7
	257	258	250
	467	463	458
9b	321	310	302.9
9c	193	174	171.1
	283	280	274.8

[a] Ref. 20.
[b] With the aug-cc-pVTZ basis set.

choice of gauge origin is necessary. The two commonly used methods are the *individual gauge for localized orbitals* (IGLOs)[21] and *gauge-including atomic orbitals* (GIAOs).[22,23] While there are differences in these two methods, implementations of these methods in current computer programs are particularly robust and both methods can provide good results.

Given that the chemical shifts for any nucleus can be computed, what are the appropriate methods to employ to obtain accurate values? We next describe a number of approaches toward computing chemical shifts and coupling constants, particularly probing for adequate treatments of the quantum mechanics (QM) and basis sets. Additionally, we examine some procedures for using computed NMR spectra to assist in identifying chemical structure. (Tantillo[24] has written an excellent review of computation approaches to chemical shifts.) We will end this section with a few case studies where computation played an important role in determining the chemical structure.

2.3.1 General Considerations

An obvious first concern is what would be a reasonable quantum mechanical approach (method and basis set) for computing NMR chemical shifts. We will consider this in the next section.

The second concern regards geometry. Since the NMR experiment examines molecules on a long time frame, multiple conformations of a molecule might need to be considered. Instead of simply optimizing to the lowest energy structure, a set of low energy conformations must be obtained. Given the Boltzmann population distribution,

$$\frac{P_i}{P_j} = e^{(G_i - G_j)/k_B T} \quad (2.1)$$

in general, all conformations with 2–3 kcal mol^{-1} of the lowest energy structure need to be identified. Then, the chemical shifts must be computed for each of these low energy conformations and averaged by their Boltzmann population as

$$\delta = \sum_i \frac{\delta_i e^{G_i/RT}}{\sum_j \delta_j e^{G_j/RT}} \quad (2.2)$$

where i and j index the set of low energy conformations.

Since a vast majority of NMR experiments are performed in solution, incorporation of solvent within the computation is a reasonable expectation. One might anticipate inclusion of solvent using a continuum method as described in Section 1.4.2. In a consistent manner, the molecular geometry should be optimized with the solvent field, and the chemical shifts computed with this geometry and with the solvent field. As will be demonstrated below, optimization in the solvent field turns out to oftentimes be unnecessary and the gas-phase geometry will suffice.

No systematic study of the effect of different solvation models has been performed. A few reports have compared specific cases such as the study of cationic and anionic alanines[25], which shows a significant improvement in the chemical shift prediction using polarized continuum method (PCM) or better still a hybrid solvation approach (Section 1.4.3). However, the linear scaling correction discussed below can often account for the systematic solvent effect and so sometimes one can get away without any solvent computation at all.

Chemical shifts are typically reported relative to some reference compound.

$$\delta_i = \sigma_{\text{ref}} - \sigma_i \quad (2.3)$$

where σ_i is the isotropic shielding constant of nucleus i in the compound of interest, and σ_{ref} is the isotropic shielding constant of a nucleus in a reference compound, often tetramethylsilane (TMS). This subtraction can correct for some systematic errors, but some errors might be related to different chemical environments. Again, the linear scaling approach described in the next section can correct for some of this type of error.

An alternative approach is to make use of multiple reference compounds. The idea here is that errors in computing the electron density near, for example, an sp^2 carbon may be different than the errors in the density near an sp^3 carbon, and both may have errors quite different than the errors in the density near the methyl carbon of TMS. Instead of choosing a single reference (TMS) to use in Eq. (2.3), Sarotti and Pellegrinet[26] suggest using multiple reference compounds to deal with different kinds of chemical environments. They propose using methanol as the reference for sp^3 carbon atoms and using benzene as the reference for sp^2 and sp carbons. The ^{13}C chemical shifts for a set of 50 organic compounds representing a broad array of functional groups were computed at mPW1PW91/6-311+G(2d,p)//B3LYP/6-31G(d) using this multireference model. The average RMSD in the ^{13}C chemical shifts from experiment is 4.6 ppm, less than half that found when TMS is used as the sole reference.

2.3.2 Scaling Chemical Shift Values

Rablen[27] examined the proton NMR shifts of 80 organic molecules using three different DFT functionals (B3LYP, B3P86, and B3PW91) and three different basis sets (6-311++G(d,p), 6-311++G(2df,p), and 6-311++G(3df,2p). While the correlation between the experimental and computed chemical shifts was quite reasonable with all the methods (correlation coefficients typically better than 0.993), there were systematic differences. In analogy with the scaling of vibrational frequencies, Rablen suggested scaling the computed isotropic shielding constants to match the experimental chemical shifts. A least square fit of the computed and experimental values provided the linear scaling factor (the slope of the line) along with the intercept, which ideally should be the isotropic shielding constant for TMS. With these scaled values, Rablen proposed two computational models: a high level model based on the computed shift at GIAO/B3LYP/6-311++G(2df,p)//B3LYP/6-31+G(d) and a more economical model based on the computed shift at GIAO/B3LYP/6-311++G(d,p)//B3LYP/6-31+G(d). The root mean square error (RMSE) is less than 0.15 ppm for both models.

Suggesting that the relatively high error found in Rablen's study comes from the broad range of chemical structures used in the test sample, Pulay[28] examined two separate sets of closely related molecules: a set of 14 aromatic molecules and a set of eight cyclic amide[29] molecules. Again using a linear scaling procedure, the predicted B3LYP/6-311+G(d,p) proton chemical shifts have a RMSE of only 0.04 ppm for the aromatic test suite. This same computational level did well for the amides (RMSE = 0.10 ppm), but the best agreement with the experimental values in D_2O was with the HF/6-311G(d,p) values (RMSE = 0.08 ppm).

Based on these studies, two different approaches are routinely taken toward computing NMR chemical shifts. The first is to determine a scaling factor for the particular problem at hand. This will mean choosing a particular computational method, that is, a particular density functional and a basis set. Next, a set of molecules related to the molecule of interest are selected. The computed chemical shifts of these reference compounds are compared to their experimental values to determine the linear scaling factors. This linear scaling is then applied to the computed chemical shifts of the molecule of interest.

The downside to this approach is that a new scaling factor must be determined for every new study. The second approach is to take a more general view, that a single scaling factor might be appropriate for a broad set of compounds. One will still need a unique scaling factor for every computational method, but perhaps it is possible to also settle on a generic computational method.

Jain et al.[30] examined the dependence of the agreement between experimental and computed ^1H chemical shifts on choice of the density functional, basis set, and inclusion of solvent (chloroform modeled using PCM). They employed the same 80-compound test set in the study presented above. Geometries were optimized at B3LYP/6-31G*. The linear scaled computed chemical shifts, as expected, have RMSDs from the experiment that are typically 0.05–0.15 pm better than the unscaled values.

Table 2.7 presents some of their results. We will discuss the WP04 results in Section 2.3.3. Inspection of the B3LYP results reveals a few very important trends. First, overall predictive power is quite good using the geometries optimized in the

TABLE 2.7 RMS Error (ppm) of the Computed Linearly Scaled Chemical Shifts for ^{1}H

Functional	Basis Set	Solvent	RMS
B3LYP	6-31G(d,p)	No	0.165
	6-31G(d,p)	Yes	0.129
	cc-pVDZ	No	0.173
	cc-pVDZ	Yes	0.132
	aug-cc-pVDZ	No	0.133
	aug-cc-pVDZ	Yes	0.106
	cc-pVTZ	No	0.143
	cc-pVTZ	Yes	0.118
	aug-cc-pVTZ	No	0.140
	aug-cc-pVTZ	Yes	0.117
WP04	6-31G(d,p)	No	0.140
	6-31G(d,p)	Yes	0.119
	cc-pVDZ	No	0.146
	cc-pVDZ	Yes	0.115
	aug-cc-pVDZ	No	0.112
	aug-cc-pVDZ	Yes	0.103
	cc-pVTZ	No	0.124
	cc-pVTZ	Yes	0.113
	aug-cc-pVTZ	No	0.122
	aug-cc-pVTZ	Yes	0.112

gas phase. Second, inclusion of the solvent in all cases improves performance. Given the dramatic improvements in the computational performance of PCM (see Section 1.4), computation of the chemical shifts should routinely be carried out with PCM. Third, the RMSD decreases with increasing size of the basis set. However, really quite reasonable results can be had with quite affordable basis sets; PCM/B3LYP/6-31G(d,p) has an RMS of only 0.129 ppm. This is the recommended method by these authors for those unwilling or unable to use the WP04 functional.

Lodewyk et al.[24] have extended the approach of Jain et al. to include ^{13}C chemical shifts. The RMSE results for the ^{13}C chemical shifts of their recommended computational models are listed in Table 2.8. Errors are larger than for proton chemical shifts, but quite reasonable results can again be had with modest computational effort. Again, it is important to note that optimization of the geometry in solution is unnecessary. Tantillo[31] maintains a Web site compiling the recommended methods and scaling factors for ^{1}H and ^{13}C chemical shift prediction.

2.3.3 Customized Density Functionals and Basis Sets

An alternative approach to computing chemical shifts is to create a density functional designed expressly for the purpose of computing chemical shifts. Wiitala et al.[32] have done this with the WP04 and WC04 functionals. They define the

TABLE 2.8 RMS Error (ppm) of the Computed Linearly Scaled Chemical Shifts for ^{13}C

Method	RMS
PCM/B3LYP/6-311+G(2d,p)//B3LYP/6-31+G(d,p)	2.24
PCM/B3LYP/6-311+G(2d,p)//B3LYP/6-311+G(2d,p)	2.12
PCM/mPW1PW91/6-311+G(2d,p)//B3LYP/6-31+G(d,p)	2.06
PCM/mPW1PW91/6-311+G(2d,p)//B3LYP/6-311+G(2d,p)	1.91
PCM/PBE0/6-311+G(2d,p)//B3LYP/6-31+G(d,p)	2.08
PCM/PBE0/6-311+G(2d,p)//B3LYP/6-311+G(2d,p)	1.93

exchange-correlation functional as

$$E_{XC} = P_2 E_X^{HF} + P_3 E_X^B + P_4 E_X^{LSDA} + P_5 E_C^{LYP} + P_6 E_C^{LSDA} \qquad (2.4)$$

where the Ps are parameters to be fit. This is the same functional form as for B3LYP (see Eq. (1.53)) but with different values of the parameters P. A total of 43 different molecules were used for this fitting procedure. Chemical shifts were computed for proton and carbon, and the parameters P were adjusted (between 0 and 1) to minimize the difference between the predicted and the experimental chemical shifts. The values of the parameters are substantially different among WC04, WP04, and B3LYP. The RMSD between the computed and experimental chemical shifts for a set of 43 molecules is listed in Table 2.7. As can be seen, the WP04 and WC04 outperform the other methods, but this advantage is largely eliminated when linear scaled shifts are used (Table 2.9).

The performance of WP04 with the test set of 80 compounds used by Jain et al. is, however, much better.[30] In fact, they recommend the use of linear scaled chemical shifts computed with WP04. Proton chemical shifts computed at PCM/WP04/6-31G(d,p) have an RMSE of only 0.140 ppm; considering the very small basis set and the fact that geometries are optimized for the gas phase, this is an incredibly computationally efficient method. Slightly better performance can be had with the PCM/WP04/aug-cc-pVDZ, though these computations are much more time consuming.

TABLE 2.9 RMS Deviations in the ^{13}C and ^{1}H Chemical Shifts^a

	Unscaled		Linear Scaled	
	^{13}C	^{1}H	^{13}C	^{1}H
WC04	3.8	0.20	3.8	0.14
WP04	7.6	0.13	3.5	0.10
B3LYP	7.7	0.19	3.0	0.10
PBE1	6.9	0.22	2.8	0.11
mPW1PW91	7.0	0.21	2.8	0.11

^{a}All calculations were performed at PCM(chloroform)/*method*/6-311+G(2d,p)//PCM(chloroform)/B3LYP/6-31G(d).

In a similar vein, one might consider optimizing a basis set for use in computing NMR chemical shifts. Jensen[33] has done this, modifying his *pc-X* family (see Section 1.1.8) by adding a set of tight p functions and decontracting the valence p functions. The resulting family of basis sets, called *pcS-X*, performs better than comparably sized basis sets in predicting chemical shifts for hydrogen and first row atoms.

2.3.4 Methods for Structure Prediction

Computed NMR chemical shifts have been used numerous times to help identify chemical structure. This might be to identify the correct structure from a number of possible constitutional isomers or the proper structure from a selection of possible diastereomers. How one judges the correct selection is the topic of this section.

A common procedure, initially used by Barone and Bifulco,[34,35] makes use of the correlation coefficient and mean absolute errors, exemplified by the following two examples. A variety of NMR experiments along with mass spectrometry (MS) enabled Vera et al.[36] to determine the structure of aplysqualenol A **10**, with some doubts as to the stereochemistry at C_{15} and C_{22}. The structures of the four diastereomers due to variation in the stereochemistry at these positions were optimized HF/3-21G(2d,p) and then ^{13}C chemical shifts were computed at mPW91PW91/6-31G(2d,p). These shifts were obtained by linearly scaling them to fit the experimental values as closely as possible. Comparison of the computational values for the four isomers with the experimental chemical shifts revealed that mean absolute error for the (15R,22R) isomer was 2.9 ppm, the least among the four diastereomers. Further, its correlation coefficient r^2 relating the computed-to-experimental values was the largest of the four isomers, thus suggesting this particular stereochemistry for **10**.

Fattorusso and co-workers[37] identified a component of wormwood called artarborol. Correlation spectroscopy (COSY) and rotating frame nuclear overhauser effect spectroscopy (ROESY) experiments allowed for deduction of four possible diastereomeric structures of artarborol, **2–5**. Low energy conformers of **11–14** were obtained through a molecular mechanics (MM) search. These conformers were screened to identify those having a dihedral angle of around 90° for the C-8 and C-9 protons due to a low coupling constant between these protons. Only conformers of **11** and **13** satisfied this criterion. Next, five low energy conformers, two of **11** and three of **13**, were optimized at mPW1PW91/6-31G(d,p). The ^{13}C chemical shifts were computed and averaged according Eq. (2.2). These shifts were then compared with the experimental values. The correlation factor for the computed shifts for **2** ($r^2 = 0.9997$) was much better than that of **13** ($r^2 = 0.9713$). The average deviation of the chemical shifts (after being corrected using the fitting procedure from the above correlation) was only 0.8 ppm for **2** but 2.0 ppm for **13**. They conclude that the structure of artarborol is **11**.

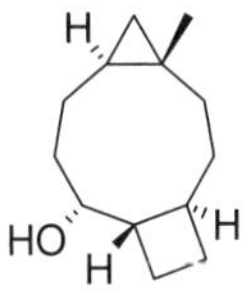

11 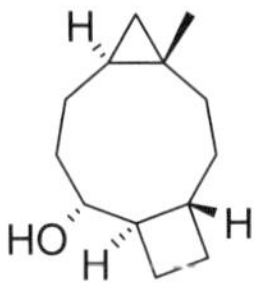   **12** 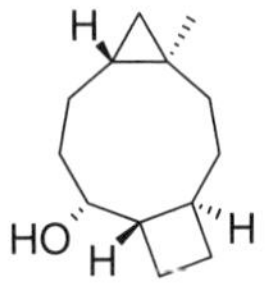   **13** 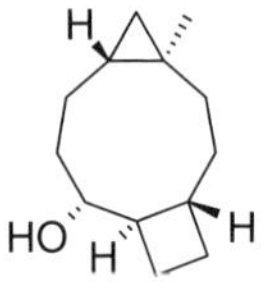   **14**

2.3.5 Statistical Approaches to Computed Chemical Shifts

An issue with the use of the correlation coefficient is deciding just how large a value is needed to indicate a good match. More difficult still is to meaningfully discriminate between correlation coefficients for two different isomers that are close in value. The same concern applies to the mean average deviation. Some statistical measure of the goodness of the fit would be very useful in discriminating the options. Goodman has proposed two schemes that fit this request.

Goodman[38] suggests that in attempting to identify the structure of diastereomers, it is not the values of the chemical shifts that matter. Rather, the *differences* in the chemical shifts of pairs of diastereomers are critical in identifying which one is which. Since the difference in computed chemical shifts of diastereomers will likely mean a cancellation of systematic errors,[39,40] these differences are likely to be more accurate than the chemical shifts themselves.

Using Goodman's notation,[38] suppose you have experimental NMR data on diastereomers **A** and **B** and the computed NMR shifts for structures **a** and **b**. The key is deciding whether **A** correlates with **a** or **b** and the same for **B**. Goodman proposed three variants on how to compare the chemical shift differences, and we will discuss here the simplest version, called *CP1*. The difference in the chemical shift of nucleus *I* between the two diastereomers is

$$\Delta \exp_I = \delta_{A_I} - \delta_{B_I} \tag{2.5}$$

and similarly the differences in the computed chemical shifts are

$$\Delta \mathrm{cal}_I = \delta_{a_I} - \delta_{b_I} \tag{2.6}$$

CP1 is then defined as

$$\mathrm{CP1} = \frac{\sum_I \Delta \exp_I \Delta \mathrm{cal}_I}{\sum_I (\Delta \exp_I)^2} \tag{2.7}$$

Goodman shows in a number of examples that CP1 and its variants provide measures of when a computed structure's chemical shifts agree with the experimental values. Using Bayes' theorem, a confidence level can also be assigned to these CP values.

In a subsequent paper, Goodman provided an extension for the situation where you have a single experimental NMR spectrum and you are trying to determine which of a number of diasteromeric structures best accounts for this spectrum.[41] Not only does this prescription provide a means for identifying the best structure, it also provides a confidence level. The method, called *DP4*, works as follows. With linear scaled computed chemical shifts for all reasonable diastereomers in hand, the error in each chemical shift is computed (i.e., for each atom *I* and for each potential structure *A*, $e_I^A = \delta_I^A - \delta_I^{\exp}$). Determine the probability of this error for each nucleus of a particular structure using the Student's *t*-test (with mean, standard deviation, and degrees of freedom as found using their database of over 1700 ^{13}C and over 1700 ^{1}H chemical shifts). Lastly, the DP4 probability is computed as the product of these probabilities divided by the sum of the product of the probabilities over all possible diastereomers. A Java applet is available to compute DP4, given the experimental and computed chemical shifts.[42]

Smith and Goodman[41] demonstrate that for a broad range of natural products, the DP4 method does an outstanding job in identifying the correct diastereomer, and an even better job of not misidentifying a wrong structure to the spectrum. Performance is markedly better than using the correlation coefficient or mean absolute error.

It should also be noted that Goodman and Smith used a rather simple computational approach in testing out their DP4 method. A MM conformational search of every diastereomer was first performed to identify low energy conformers. All conformations within 10 kJ mol^{-1} of the global minimum were selected, and their ^{13}C and ^{1}H NMR chemical shifts were computed at B3LYP/6-31G(d,p). These shifts were then Boltzmann averaged. Note that geometries were not reoptimized at the DFT level and that solvent was omitted, and yet the resultant chemical shifts were successful in discriminating diastereomers for a set of 36 natural products!

An example of the application of the DP4 procedure is in the determination of the structure of nobilisitine A, whose originally proposed structure was **15**.[43] Banwell and co-workers[44] synthesized the enantiomer of **15**, but its NMR did not correspond to that of nobilisitine A. Lodewyk and Tantillo[45] examined **15** and seven diastereomers, all of which have a *cis* fusion between the saturated five and six-member rings. Low energy conformations were computed for each of these diastereomers at B3LYP/6-31+G(d,p). NMR shielding constants were then computed in solvent (using a continuum approach) at mPW1PW91/6-311+G(2d,p). A Boltzmann weighting and linear scaling was applied to the chemical shifts. The resultant-computed NMR shifts for the eight diasteomers showed deviations from the experiment, which could not definitively identify the structure. However, DP4 analysis indicated that **16** is the correct structure with a probability of 99.8 percent.

2.3.6 Computed Coupling Constants

Bally and Rablen[30] followed up their important study of the appropriate basis sets and density functional needed to compute NMR chemical shifts with an examination of procedures for computing proton–proton coupling constants.[46] They performed a comparison of 165 experimental with computed proton–proton coupling constants from 66 small, rigid molecules. They tested a variety of basis sets and functionals, along with questioning whether all four components that lead to nuclear–nuclear spin coupling constants are required, or if just the Fermi contact term would suffice.

A number of findings are relevant here. First, following on the work of Deng et al.[47] the hydrogen basis set must be expanded in the inner regions. This is done by uncontracting the core functions and adding a very tight s-function to standard basis sets. Bally and Rablen append the term *u+1s[H]* to designate this basis set. Second, using a variety of basis sets (6-31G(d), 6-31G(d,p), 6-311G(d,p), cc-pVDZ, and cc-pVTZ), all with the uncontracted core functions, leads to very small differences (<0.10 Hz) in the ^{1}H–^{1}H coupling constants. Third, computing the coupling constants with only the Fermi contact term leads to an error of about 0.1–0.2 Hz, an inconsequential difference given the tremendous computation cost savings. Lastly, a large number of density functionals provide coupling constants with RMSDs from experiments of less than 1 Hz; a representative set of results are shown in Table 2.10.

Their recommended method, one that is computationally very efficient, provides excellent agreement with the experimental coupling constants. The steps are as follows:

(1) Optimize the geometry at B3LYP/6-31G(d).
(2) Calculate only the proton–proton Fermi contact terms at B3LYP/6-31G(d,p) u+1s[H].
(3) Scale the Fermi contact terms by 0.9155 to obtain the proton–proton coupling constants.

This methodology provides coupling constants with a mean error of 0.51 Hz, and when applied to a probe set of 61 coupling constants from 37 different molecules (including a few that require a number of conformers and thus a Boltzmann-weighted averaging of the coupling constants), the mean error is only 0.56 Hz.

TABLE 2.10 RMS Deviation (Hz) from Experiment of Computed Proton–Proton Coupling Constants

Method	$J_{\text{rms}}{}^{a}$
B3LYP	0.51
M06-2x	0.69
O3LYP	0.70
B3PW91	0.78
ωB97X-D	0.78
WP04	0.86
PBE0	0.90
mPW1PW91	0.97
TPSSh	1.23
M06	3.73

^{a}Linear scaled coupling constants evaluated using the Fermi contact term only, averaged over the five different basis sets.

Bally and Rablen supply a set of scripts to automate the computation of the coupling constants according to this prescription using *GAUSSIAN-09*; these scripts are available in the supporting materials of the paper and also on the Cheshire Web site.[31]

Williamson[48] devised an experimental method to obtain $^{1}J_{\text{CC}}$ and $^{3}J_{\text{CC}}$ coupling constants, and in an interesting extension, he also computed these coupling constants. Their initial test was on strychnine. The coupling constants were computed in two steps. First, the Fermi contact contribution was computed at B3LYP/6-31+G(d,p) with an uncontracted core and additional tight polarization functions, as suggested by Deng et al.[47] Second, the remaining terms (spin-dipolar, paramagnetic spin-orbit, and diamagnetic spin-orbit coupling) were computed with the 6-31+G(d,p) set without modifications. The two computed terms were added to give the final estimate. The mean absolute deviation values for the computed and experimental $^{1}J_{\text{CC}}$ and $^{3}J_{\text{CC}}$ coupling constants are 1.0 and 0.4 Hz, respectively, both well within the experimental error.

2.3.7 Case Studies

Detailed below are three case studies where computed chemical shifts helped to assign (or dramatically could have helped assign) chemical structure. Further examples can be found in the review articles of Tantillo[24] and Bifulco.[49]

2.3.7.1 Hexacyclinol

Hexacyclinol was isolated from *Panus rudis*, a type of mushroom. Based on spectroscopic studies, Grafe proposed **17** as its structure.[50] La Clair[51] claimed to have synthesized a substance with this structure in 2006. This article became a *cause célèbre* in the blogosphere,[52] with serious doubts cast upon the veracity of the author's claims.

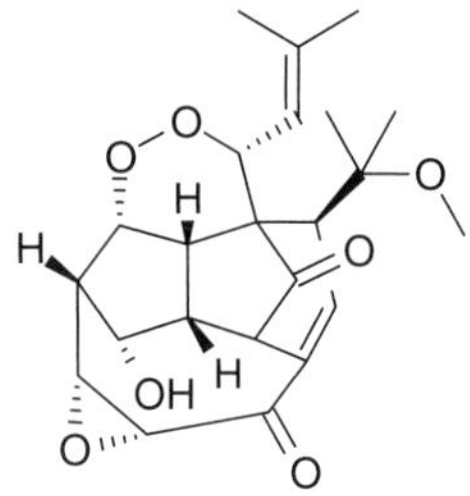

Rychnovsky[53] doubted that the molecule actually possessed the unusual structure of **17**. He proposed to compute the NMR shifts based on the optimized structure of **17** and compare them with the experimental values. Given the very large size of hexacyclinol, the computational approach would have to be rather limited. Therefore, whatever (small) method was to be employed would have to be tested for adequate predictive performance with known compounds. Rychnovsky selected the three diterpenes elisapterosin B **18**, elisabethin A **19**, and maoecrystal V **20** to benchmark his computations. His computational approach was to first utilize a Monte Carlo search with the Merck molecular force field (MMFF) force field to identify the lowest energy conformer. This conformer was then optimized at HF/3-21G and the chemical shifts were computed using this geometry with GIAO/mPW1PW91/6-31G(d,p).

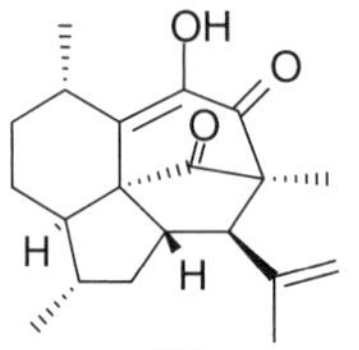

The computed ^{13}C chemical shifts for these test compounds were then plotted against the experimental values and a linear fit was determined to correct the computed values. The average ^{13}C chemical shift difference between computation and experiment was less than 2 ppm, and no difference exceeded 5 ppm. Structure **17** was then optimized, its ^{13}C chemical shifts computed and then corrected using the fitting procedure developed for the three test compounds. These computed chemical shifts were in very poor agreement with the experimental values; the average deviation was 6.8 ppm and five shifts differ by more than 10 ppm. Rychnovsky concluded that this poor agreement discredits the proposed structure **17**.

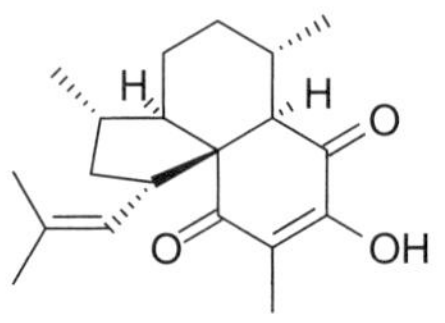

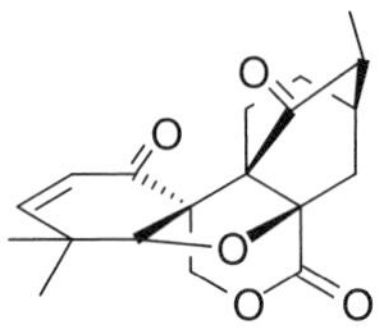

Rychnovsky proposed that hexacyclinol is in fact the by-product from workup of the natural product panepophenanthrin, also obtained from *P. rudis*, resulting in structure **21**. Optimization of **21** led to two low energy conformers. The second lowest conformer has a predicted ^{13}C NMR spectrum in very close agreement with experiment, with an average chemical shift deviation of 1.8 ppm and a maximum difference of 5.8 ppm. These differences are consistent with those found in the diterpenes test set.

21

Structure **21** was subsequently synthesized by Porco,[54] its X-ray structure obtained, and its NMR spectra was found to be completely consistent with the original hexacyclinol compound reported by Grafe. In addition, Bagno[55] recomputed the ^{1}H and ^{13}C chemical shifts of **17** and **21** at B97-2/cc-pVTZ/B3LYP/6-31G(d,p). The mean absolute error between the experimental and computed structures is half that for **21** than for the originally proposed structure **17**. The computed coupling constants also are in much better agreement with **21**.

The impact of the hexacyclinol story, particularly of Rychnovsky's paper, is hard to minimize. This study dramatically demonstrated the power of computational chemistry to aid in structure determination. The fact that the study was undertaken by a synthetic organic chemist spurred many others to utilize this tool as an adjunct in their efforts to identify structure, including the next two case studies.

2.3.7.2 Maitotoxin Nicolaou and Frederick[56] examined the somewhat controversial structure of the marine natural product maitotxin. For the sake of brevity, the structure of maitotoxin, one of the largest nonprotein molecules known, is not reproduced here; the interested reader should check out its Wikipedia entry. The structure of maitotoxin has been extensively studied, but in 2006, Gallimore and Spencer[57] questioned the stereochemistry of the J/K ring juncture. A fragment of maitotoxin that has the previously proposed stetreochemistry is **22**. Gallimore and Spencer argued for a reversed stereochemistry at this juncture (**23**), one that would be more consistent with the biochemical synthesis of the maitotoxin. Nicolaou noted that reversing this stereochemistry would lead to other stereochemical changes in order for the structure to be consistent with the NMR spectrum. Their alternative is given as **24**.

Nicolaou and Frederick computed ^{13}C NMR of the three proposed fragments **22–24** at B3LYP/6-31G. They benchmarked this method against brevetoxin B, where the average error is 1.24 ppm, but they provide no error analysis—particularly no linear regression so that scaled chemical shift data might be employed. The best agreement between the computed and experimental chemical shifts is for **22**, with average difference of 2.01 ppm. The differences are 2.85 ppm for **23** and 2.42 ppm for **24**. These computations support the original structure of maitotoxin.

2.3.7.3 Vannusal B
Vannusal B, isolated from *Euplotes vannus*, was originally proposed to have the structure **25a**.[58] Subsequent synthetic and NMR studies have settled on **25b** as its actual structure.[59]

Saelli et al. computed the ^{13}C chemical shifts of **25a**, **25b**, and six other diastereomers at M06/pcS-2//B3LYP/6-31g(d,p).[60] The computed chemical shifts of **25a** poorly correlate with the experimental chemical shifts of vannusal B, with a low correlation coefficient of 0.9580 and a maximum error of 16.2 ppm. They point out that had DFT computations been performed in 1992, structure **25a** might never have been proposed! On the other hand, the correlation between the computed chemical shifts of **25b** with the experimental values is excellent ($r^2 = 0.9948$), with a maximum error of 3.0 ppm. Comparison of computed and experimental H–H coupling constants of model compounds of the "northeast" section of the molecule verified the correct structure and it is **25b**.

2.3.7.4 Conicasterol F

Butts and Bifulco[61] were interested in the structure of conicasterol F (**26**). Standard NMR analysis led to two possible diasteromers **26a** and **26b**. The interesting aspect of their approach toward discriminating between these two options was to compare not only the computed and experimental ^{13}C chemical shifts, but also to compare the distances between protons, determined from the optimized structure and the rotating-frame nuclear overhauser effect (ROE).

26a   26b

The optimized geometries (MPW1PW91/6-31G) and ^{13}C chemical shifts were determined for both diastereomers. Comparison of 15 distances between protons determined by the ROE experiment and by computation led to a mean absolute error of 7.8 percent for **26a** and 3.0 percent for **26b**, suggesting that the latter is the correct structure. Similar comparison was then made between the experimental chemical shifts of 12 of the carbon atoms with the computed values of the two isomers. The mean absolute error in the chemical shifts of **26a** is 3.7 ppm, but only 0.8 ppm for **26b**. Both methods give the same conclusion: conicasterol F has structure **26b**.

2.3.7.5 1-Adamantyl Cation

What is required in order to compute *very* accurate NMR chemical shifts? Harding et al. take on the interesting spectrum of 1-adamantyl cation to try to discern the important factors in computing its ^{13}C and ^{1}H chemical shifts.[62] The chemical shifts of 1-adamantyl cation were computed at B3LYP/def2-QZVPP and MP2/qz2p//MP2/cc-pVTZ. The RMSE (compared to experiment) for the carbon chemical shifts is large: 12.76 for B3LYP and 6.69 for MP2. The proton shifts are predicted much more accurately with an RMSE of 0.27 and 0.19 ppm, respectively.

A better geometry does improve the agreement. The RMSE of the ^{13}C chemical shifts, computed at MP2/tzp with the HF, MP2, and CCSD(T) geometries, are 9.55, 5.62, and 5.06 ppm, respectively. Unfortunately, the computed chemical shifts at CCSD(T)/qz2p//CCSD(T)/cc-pVTZ are still in error; the RMS is 4.78 ppm for the carbon shifts and 0.26 ppm for the proton shifts. Including a correction for the zero-point vibrational effects and adjusting to a temperature of 193 K to match the experiment does reduce the error. The remaining error is attributed to basis set incompleteness in the NMR computation, a low level treatment of the zero-point vibrational effects, neglect of the solvent, and use of a reference in the experiment that was not dissolved in the same media as the adamantyl cation.

So, we can offer no definitive answer to our opening question, other than that extreme caution must be used when attempting to compute highly accurate chemical shifts!

2.4 OPTICAL ROTATION, OPTICAL ROTATORY DISPERSION, ELECTRONIC CIRCULAR DICHROISM, AND VIBRATIONAL CIRCULAR DICHROISM

Optical rotation (OR), optical rotatory dispersion (ORD), electronic circular dichroism (ECD), and vibrational circular dichroism (VCD) provide spectral information unique to enantiomers, allowing for the determination of absolute configuration. Recent theoretical developments in DFT, using time-dependent density functional theory (TD-DFT), provide the means for computing OR, ORD, and ECD.[63,64] Similar theoretical development with coupled-cluster theory, making use of linear response theory, allows for computation of these properties as well.[65,66] Again, the technical details of these computations are beyond the scope of this book. Interested readers are referred to the review article by Crawford[67] as a starting point.

While HF fails to adequately predict OR, a study of eight related alkenes and ketones at the B3LYP/6-31G* level demonstrated excellent agreement between the calculated and experimental ORs (reported as $[\alpha]_D$, with units understood throughout this discussion as deg [dm g/cm^3]$^{-1}$, see Table 2.11) and the ORD spectra.[68] A subsequent, more comprehensive study on a set of 65 molecules (including alkanes, alkenes, ketones, cyclic ethers, and amines) was carried out by Frisch.[69] Overall, the agreement between the experimental $[\alpha]_D$ values and those computed at B3LYP/aug-cc-pVDZ//B3LYP/6-31G* is reasonable; the RMSD for the entire set is 28.9. An RMSE this large, however, implies that molecules with small rotations might actually be computed with the wrong *sign*, the key feature needed to discriminate the absolute configuration of enantiomers. In fact, Frisch identified eight molecules in his test set where the computed $[\alpha]_D$ is of the wrong sign (Table 2.12). Frisch concludes, contrary to the authors of the earlier study, that determination of absolute configuration is not always "simple and reliable." Kongsted[70] also warns that vibrational contributions to the OR can be very important, especially for molecules that have conformational flexibility. In this case, he advocates using the "effective geometry," the geometry that minimizes the electronic plus ZPVE.

Crawford and Stephens[71] have compared the performance of B3LYP with CCSD for computing ORs for a set of 13 small, rigid alkanes, alkenes, and ketones. Geometries were optimized at B3LYP/6-31G*. ORs were computed at

TABLE 2.11 Comparison of Experimental and Calculated Optical Rotation[a] for Ketones and Alkenes[b]

	$[\alpha]_D$(Expt)	$[\alpha]_D$(B3LYP)		$[\alpha]_D$(Expt)	$[\alpha]_D$(B3LYP)
	−180	−251		−40	−121
	−44	−85		−68	−99
	+59	+23		−36	−50
	+7	+13		−15	+27

[a] $[\alpha]_D$ in deg·[dm·g/cm^3]$^{-1}$.
[b] Computed at B3LYP/6-31G*, see Ref. 68.

TABLE 2.12 Compounds for Which Calculated Optical Rotation[a] Disagree in Sign with the Experimental Value[b]

	$[\alpha]_D$(Expt)	$[\alpha]_D$(Comp)		$[\alpha]_D$(Expt)	$[\alpha]_D$(Comp)
	−15.9	3.6		−78.4	13.1
	6.6	−11.3		14.4	−9.2
	29.8	−10.1		−59.9	20.0

[a] $[\alpha]_D$ in deg·[dm g/cm^3]$^{-1}$.
[b] Computed at B3LYP/aug-cc-pVDZ//B3LYP/6-31G*, see Ref. 69.

B3LYP/aug-cc-pVDZ but for the CCSD computations, aug-cc-pVDZ was used for carbon and oxygen, while cc-pVDZ was used for hydrogen. For the entire set of compounds, the mean absolute deviations between the computed and experimental $[\alpha]_D$ values are 31.7 for B3LYP and 56.5 for CCSD. This includes a very large error of the CCSD prediction for (1S,4S)-norbornenone (exp: −1146, B3LYP: −1216, CCSD: −740). Omitting this compound gives mean absolute deviations for both methods of about 27. Sources of errors include (1) use of a relatively small basis set, though examples with larger basis sets do not seem to be much improved, (2) lack of triple or higher excitations, (3) neglect of vibration and temperature effects, and (4) comparison of computed gas-phase values compared to experimental solution phase values.

In a benchmark study of 45 compounds (including three organometallic compounds), the effect of long-range corrections on ORs was tested.[72] Neither the CAM-B3LYP functional[73] (which is a range-separated density functional) nor LC-PBE0 preformed any better than B3LYP. In addition, the LPol-ds basis set,[74] a customized basis set proposed for OR computations, offers no general improvement over the aug-cc-pVDZ basis set used in the other benchmark studies mention above.

Grimme[75] extended the double hybrid density functional formalism to be able to compute ECD. For a test suite of six molecules, performance of B2PLYP is notably better than for B3LYP.

Crawford[67] highlights in his review article three particular studies that demonstrate the varied results one might expect to attain with computed optical activities. In particular, he notes molecules where the CC approach performs much better than DFT, where DFT performs better than CC, but for the wrong reason, and a case where DFT appears to perform better than CC.

(P)-(+)-[4]-Triangulane **27**, synthesized by de Meijere,[76,77] exhibits amazing optical activity: varying from 193 deg/[dm (g/ml)] at 589 nm to 648 deg/[dm (g/ml)] at 365 nm. This is remarkable given its lack of a long wavelength chromophore or chiral atom. B3LYP and CCSD computations of the OR of **27** were done at five different wavelengths.[78] B3LYP overestimates the value of $[\alpha]$ by about 15 percent, but CCSD underestimates the experimental values by less than 2 percent across the wavelengths. It is readily apparent that CCSD outperforms B3LYP. The poor performance of the DFT method is linked to the first electronic excitation energy that is too small (7.21 eV with CCSD vs 6.24 eV with B3LYP).

27

The ORD of (S)-methyloxirane shows a change of sign: −8.39 at 633 nm and +7.39 at 355 nm.[79] CCSD predicts a reasonable value at 633 nm but gets the wrong sign at the shorter wavelength.[66] On the other hand, B3LYP does predict the sign change. However, this seemingly correct result is due to (once again) underestimation of the excitation energy. The correct answer is obtained here with B3LYP but for the wrong reason!

Lastly, $[\alpha]_D$ for (1S,4S)-norbornenone is -1146 deg/[dm (g/ml)]. B3LYP does a very reasonable job in predicting a value of -1216. However, CCSD grossly underestimates this value at -740.[65,71] Though B3LYP again underestimates the excitation energy, it appears to get the energy and rotational strength near the liquid-phase values. This is a case where B3LYP outperforms CCSD.

The upshot here is that a standard method for computing ORs, ORD, ECD, and VCD is still a work in progress. We next present a few case studies where computation has played a role in structure determination. The interested reader should use these examples for guidance in performing his/her own computations.

2.4.1 Case Studies

2.4.1.1 Solvent Effect The method most widely utilized for computing optical activities is B3LYP/aug-cc-pVDZ, as advocated by the Stephens and Gaussian groups.[64,69,71] Can one use a smaller, more computationally efficient basis set? Is the neglect of solvent justified? To address these points, Rosini[80] examined five related cyclohexene oxides with known absolute configuration and ORs. Low energy conformations (within 2 kcal mol^{-1} of the minimum) were optimized at in the gas-phase B3LYP/6-31G(d) and B3LYP/aug-cc-pVDZ. Geometries were also reoptimized using PCM to simulate the solvent environment of the experiments. Both the gas- and solution-phase $[\alpha]_D$ values were computed with both basis sets and Boltzmann averaged. We discuss the results for two of these compounds, (+)-chaloxone **28** and (+)-epoxydon **29**.

There are five low energy conformations of **28**. The computed Boltzmann-weighted ORs with both basis sets, in the gas and solution phases are listed in Table 2.13. While the gas-phase B3LYP/6-31G(d) average value is far off the experimental value, it does predict the correct sign, and since all of the five conformers give rise to a positive rotation, any error in their relative energies will not affect the overall sign of the rotation. The computed gas-phase value with the larger basis set using geometries optimized with that basis set is in better agreement with experiment. However, the solution-phase values are acceptable, indicating the potential importance of including solvent effects.

TABLE 2.13 Boltzmann-Weighted Values of $[\alpha]_D$ for 28 and 29

	Gas			Solution		
Cmpd	6-31G(d)// 6-31G(d)	aug-cc-pVDZ// 6-31G(d)	aug-cc-pVDZ// aug-cc-pVDZ	6-31G(d)// 6-31G(d)	aug-cc-pVDZ// 6-31G(d)	Expt
28	+378	+372	+333	+318	+322	+271
29	−16	+11	+57	+4	+32	+93

This is reinforced in the study of **29**. Again, five low energy conformers were optimized, and the Boltzmann-weighted $[\alpha]_D$ computed for gas and solution phases are listed in Table 2.13. In this case, the small basis set performs very poorly. The gas-phase B3LYP/6-31G(d) value of $[\alpha]_D$ is −16, predicting the wrong sign, let alone the wrong magnitude. Things improve with the larger basis set, which predicts a value of +57. Since the lowest energy conformer is levorotatory and the other four are dextrorotatory, the computed relative energies are key to getting the correct prediction. This is made even more poignant with the solution results, where both computations predict a positive rotation. In fact, if the PCM/B3LYP/aug-cc-pVDZ geometries are used, the predicted rotation is +61. The bottom line is that 6-31G(d) provides marginal results, but the larger aug-cc-pVDZ is much better suited for computing ORs. Furthermore, inclusion of solvent effects within computation of ORs and related properties are likely to be necessary.

2.4.1.2 Chiral Solvent Imprinting Beratan and Wipf[81] have examined the role of solvent organization about a chiral molecule in producing optical activity. They generated 1000 configurations of benzene arrayed about either (*S*)- or (*R*)-methyloxirane via a Monte Carlo simulation. Each configuration was then stripped off every benzene molecule greater than 0.5 nm from the center-of-mass of methyloxirane, leaving usually 8–10 solvent molecules. The OR was then computed at four wavelengths using TDDFT at BP86/SVP. (The authors note that though the Gaussian group[64,69,71] recommends B3LYP/aug-ccpVDZ, using the nonhybrid density functional BP86 allows the use of resolution-of-the-identity techniques that make the computations about six orders of magnitude faster—of critical importance given the size of the clusters and the sheer number of them!) OR is then obtained by averaging over the ensemble.

The computed ORs disagree with the experiment by about 50 percent in magnitude but have the correct sign across the four different wavelengths. Use of the COSMO (implicit solvent) model provides the wrong sign at short wavelengths, justifying the need of this microsolvation approach. Most intriguing is that the computed optical activity of the solvent molecules in the configuration about the solute, but without including the central methyloxirane molecule, is nearly identical to that of the whole cluster! In other words, the optical activity is due to the dissymmetric distribution of the solvent molecules about the chiral molecule, not the chiral molecule itself! It is the imprint of the chiral molecule on the solvent ordering that accounts for nearly all of the optical activity. This provides considerable guidance as to why computation of the optical activity of methyloxirane, described in the previous section, is so fraught with difficulties.

In fact, a recent approach delineates just how much effort is needed to obtain good results for the OR of methyloxirane. Cappelli and Barone[82] have developed a QM/MM procedure where methyloxirane is treated with DFT (B3LYP/aug-cc-pVDZ). Then, 2000 arrangements of water about methyloxirane were obtained from an MD simulation. For each of these configurations, a supermolecule containing methyloxirane and all water molecules within 16 Å was identified. The waters of the supermolecule were treated with a polarized force field. This supermolecule is embedded into bulk water employing a conductor-polarizable continuum model (C-PCM). Lastly, inclusion of vibrational effects and averaging over the 2000 configurations gives a predicted OR at 589 nm that is of the correct sign (which is not accomplished with a gas-phase or simple PCM computation) and is within 10 percent of the correct value.

2.4.1.3 Plumericin and Prismatomerin Isolation and characterization of plumericin led Albers-Schönberg and Schmid[83] to assign it structure **30**. Subsequent analysis of its ECD spectrum and comparison with computed semiempirical spectrum suggested that the actual structure is the enantiomer of **30**.[84]

Stephens reexamined this compound, obtaining its experimental VCD spectrum. The four lowest energy conformers of **30** were optimized at B3LYP/6-31G*, B3LYP/TZ2P, and B3PW91/TZ2P; their relative energies are very similar with each method, and the Boltzmann population is dominated by just two conformers. The VCD spectrum was computed at B3PW91/TZ2p for the two lowest energy conformers and averaged together. This resulting VCD-computed spectrum is in very close agreement with the experimental VCD spectrum, arguing for the original structural assignment. Further confirmation was supplied by computing the ECD spectrum at this same level and comparing it back to the earlier experimental ECD spectrum; their agreement is also excellent.

Stephens[85] next examined the related iridoid prismatomerin **31**. The OR of plumericin **30** is $[\alpha]_D = +204$. The structurally related prismatomerin **31** has $[\alpha]_D = -136$, suggesting that the core polycyclic portion may have opposite absolute configuration. Stephens prepared the acetate of **31** and experimentally determined its VCD spectrum. Using the same computational procedure as described above, 16 low energy conformers of 21 were identified. The VCD spectrum for the 12 lowest energy conformers was then computed at B3PW91/TZ2p and summed together using their Boltzmann weights. The computed spectrum for the enantiomer with the same absolute configuration as **30** matches the experimental spectrum. Thus, **30** and **31** have the same absolute configuration. Stephens concludes with the warning that optical activity of analogous compounds can be quite different and is not suitable for obtaining configuration information. Rather, VCD is a much more suitable test, especially when experimental and computed spectra are compared.

2.4.1.4 2,3-Hexadiene
Wiberg et al.[86] produced a tour-de-force experimental and computational studies of the ORD of 2,3-hexadiene **32**. This seemingly normal compound offers some confounding problems!

The compound 2,3-hexadiene exists in three conformations, as shown in Figure 2.2. The cis conformer is the lowest energy form, but the other two are only 0.3 kcal mol^{-1} higher in energy (computed at G3), meaning that all three will be present to a significant extent at 0 °C. The OR for each conformer was determined using B3LYP/aug-cc-pVDZ and CCSD/aug-ccpVDZ. (We use here just the velocity gauge results.) While there is some disagreement in the values determined by the two methods, what is most interesting is the large dependence of $[\alpha]_D$ on the conformation (see Table 2.14).

The ORD spectrum of **32** was taken for neat liquid and in the gas phase. The computed and experimental ORs are listed in Table 2.15. Two interesting points can be made from this data. First, the optical activity of **32** is strongly affected

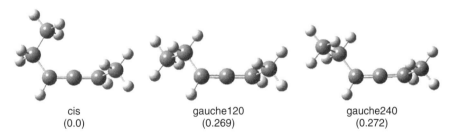

cis
(0.0)

gauche120
(0.269)

gauche240
(0.272)

Figure 2.2 Conformations and relative energies (G3) of 2,3-hexadiene **32**.

TABLE 2.14 Calculated $[\alpha]_D$ for **32**

	cis	gauche120	gauche240	Average[a]
B3LYP/aug-cc-pVDZ	205.2	415.9	−179.8	156.8
CCSD/aug-cc-pVDZ	208.5	376.7	−120.6	163.8

[a]Boltzmann-weighted average based on the G3 energies listed in Figure 2.2.

TABLE 2.15 Boltzmann-Weighted Computed and Experimental Optical Rotations of 32

	Computed		Experiment	
nm	B3LYP	CCSD	Liquid	Gas
633	134.7	140.6		122
589	156.8	163.8	86.5	
546	183.8	203.6	102.0	
365	409.7	492.5	243.3	
355	427.5	489.3		511

by phase. Second, the computed ORs, especially the CCSD values, are in fairly good agreement with the gas-phase experimental values.

A hypothesis to account for the large difference in the gas- and liquid-phase ORDs for **32** is that the conformational distribution changes with the phase. The gas- and liquid-phase ORDs of 2,3-pentadiene, also examined in this study, show the same strong phase dependence, even though this compound exists as only one conformer. A Monte Carlo simulation of gas- and liquid-phase 2,3-pentadienes was performed to assess potential differences in conformational distributions in the two phases. Though the range of dihedral angle distributions spans about 60°, the population distribution is nearly identical in the gas and liquid phases. Therefore, conformation distribution cannot explain the difference in the gas and liquid ORDs.

The authors also tested for the vibrational dependence on the OR. While there is a small correction due to vibrations, it is not enough to account for the differences due to the phase. The origin of this phase effect remains unexplained.

2.4.1.5 Multilayered Paracyclophane

The three-layer paracylophane **33** was synthesized and the two stereoisomers separated. The question facing Kobayashi and colleagues[87] was "is the (+)-isomer of *R* or *S* stereochemistry?" Comparison of the computed and experimental ORs resolved the issue.

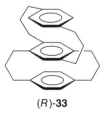

(*R*)-**33**

They located seven conformations of (*R*)-**33** at B3LYP/TZVP and computed the $[\alpha]_D$ values for each conformer. The lowest energy conformer has $[\alpha]_D = -171.7$ deg/[dm (g/ml)], and all conformers have ORs that are negative, ranging from

−124.4 to −221.8 deg/[dm (g/ml)]. These values are consistent with the experimental observation of the (−)-**33** isomer, whose $[\alpha]_D = -123$ deg/[dm (g/ml)]. The computed CD spectrum of the seven (R)-**33** conformations are similar to the experimental spectra of (−)-**33**. Thus, the two enantiomers are R-(−)-**33** and S-(+)-**33**.

2.4.1.6 Optical Activity of an Octaphyrin
The octaphyrin **34** has been prepared and its crystal structure and ECD spectra reported.[88] The X-ray structure identified the compound as having the *M,M* helical structure. The OR, however, could not be determined.

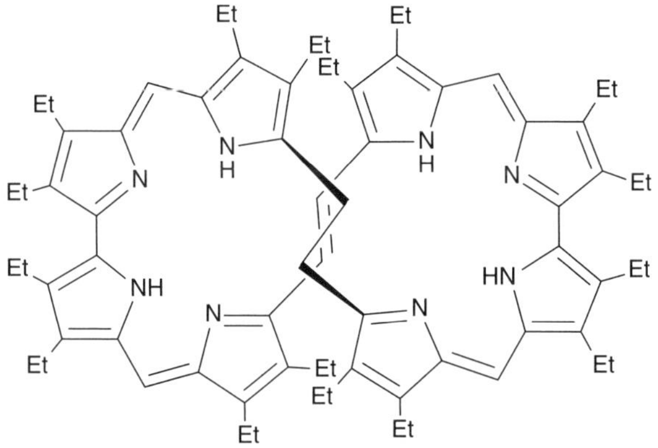

34

Rzepa[89] reported the computed ECD spectrum and optical activity of **34** and some related compounds. These computed spectra were obtained using TD-DFT/B3LYP/6-31G(d) method with the C-PCM treatment of the dichloromethane solvent. The computed ECD spectrum matches nicely with the experimental one, except that the *signs at 570 and 620 nm are opposite*. Rzepa suggests that either the compound is really of *P,P* configuration or the authors of experimental work have erroneously switched their assignments.

The computed value of $[\alpha]_D$ of **34** is about −4000 deg/[dm (g/ml)]. However, what is truly fantastic is the magnitude of the optical activity of the dication of **34** produced by loss of two electrons. This dication should be aromatic and it is predicted to have $[\alpha]_{1000} = -31{,}597$ deg/[dm (g/ml)]!

2.5 INTERVIEW: JONATHAN GOODMAN

Interviewed August 28, 2012

Jonathan Goodman is Reader in Chemistry at the University of Cambridge University and Clare College. He became interested in computational chemistry during his graduate studies with Professor Ian Paterson. He was exploring the boron-mediated aldol condensation and wondered if computations might be able to assist in rationalizing the selectivity they were observing in the experiments. Goodman had the good fortune of meeting Scott Kahn who was a post-doctoral associate at that time with Dudley Williams in a neighboring laboratory. Kahn collaborated with Goodman and taught him how to perform quantum computations. Despite the fact that they were limited by computational resources to examining the reaction of the BH_2-mediated enolate of ethanol with formaldehyde in the gas phase, a reaction decidedly different from their experiments, the computations revealed a key new insight: the enol borinates did not have to be flat, maximizing conjugation, but could be twisted! Goodman remarked that "we weren't expecting the calculations to tell us the answer; we were hoping the calculations would give us an insight to how to think about these systems." Subsequent increases in computer power allowed them to go to a seven heavy-atom system that permitted a "huge amount of additional complexity." This early success led to a post-doctoral stint with Clark Still at Columbia where Goodman was able to continue to explore the interplay between computations and experiments.

In establishing his own independent research program back at Cambridge, Goodman aimed to find problems that were accessible to both experiment and computations. Early on he developed an empirical rule: "If it's easy to do the experiment, it's next to impossible to do the computations and vice versa." In his early studies, this meant the avoidance of transition metals and reactions with strong solvent effects. His rule is not so strictly true today; the continued improvement in computational hardware and algorithmic advances, especially DFT, allows one to judiciously include transition metals and solvent effects. However, Goodman suggests that this rule remains " the reason why synthetic chemistry often don't do computations today: unless you have some experience with computations, it's very difficult to tell a computation that will be done in 10 min from one that will take a thousand years." Some significant skill remains in choosing an experiment where computations are tractable. He notes, for example, that polymers remain a singular challenge.

Goodman considers his work in NMR analysis to be a significant scientific achievement. This project was motivated by his experimental studies, in particular, answering the question "how do you know what you have made?" Goodman feels that there is a "certain prejudice that this ought to be easy; it is straightforward to calculate the NMR spectrum, and get reasonable agreement." But he notes that problems crop up when isomers, especially diastereomers, are possible. "You have a set of computed NMR for the set of isomers and all are different from each other—and different from the experiment! So which is which? How sure can you be?" Working with his doctoral student Steven Smith, Goodman developed statistical methods for not just assigning a structure but also providing a confidence measure. These techniques, especially DP4, have allowed for uncovering misassigned spectra and structural identification of natural products.

Goodman is generally pleased with the acceptance and uptake of the DP4 procedure, but recognizes that progress still is needed in making the method more accessible, especially to synthetic chemists. He is focused on making the entire process from computing the spectra to assigning the structure easier to accomplish. Goodman is especially eager to find a way to incorporate this process within a laboratory notebook.

Goodman's choice of merging cheminformatics with computations was driven by his recognition of the closing gap between experiments and calculations. Goodman says "there has always been a tension between the experimental chemist and the computational chemist: the computational guy thinks the experimentalist should do the experiment better; the experimentalist thinks the theoretician is out of touch with reality and should do something useful. That's where informatics comes in. It's not just that you have a number, say a diastereomer ratio of 9 : 1. The experiments and calculations will be different, not just because they are measuring different things. Whether the different things the experiment and calculation are telling you can be put together to tell you more than you have individually; that's informatics—getting a united whole." Interestingly, Goodman has essentially trained himself in informatics!

Goodman is quite modest regarding his contribution toward understanding the role of the enzyme in catalysis, with Luis Simón, then a post doc in his group. He naïvely thought that to catalyze a reaction, one should simply lower the transition state, but he notes that he "needed to read Pauling more closely!" His operating picture is now that "if a catalyst is brilliant at stabilizing the TS but even better at stabilizing the ground state, then it will not be a good catalyst—nature realizes this. Enzymes do not stabilize the TS as well as they maximally could because they need to avoid stabilizing the ground state too." While he feels that this assessment seems "pretty obvious now," he notes that his papers in this area met with significant resistance from referees, they were rejected many times, and that the publication process took more than a couple of years.

Goodman expresses serious dismay regarding progress in the area of data mining. He despairs that so much data is controlled by publishers and copyright holders and this prevents our abilities to effectively mine this resource. "We could know all sorts of stuff, but it's just out of reach," he says. "The task is technically straightforward with enormous potential but we can't do it because of license restrictions and that's frustrating. I need to collaborate with lawyers as much as with chemists!" In collaboration with his Cambridge colleague Peter Murray-Rust, they developed an experimental data checker for the Royal Society of Chemistry. He wistfully speaks about getting access to a large literature backfile and "checking to see what data is good and what data is not."

Goodman remains firm in his commitment to a high goal for computations: "If we want to understand a reaction we need to be able to draw a picture on a fume cupboard that an experimentalist will understand."

REFERENCES

1. Ma, B.; Lii, J.-H.; Schaefer, H. F.; Allinger, N. L. "Systematic comparison of experimental, quantum mechanical, and molecular mechanical bond lengths for organic molecules," *J. Phys. Chem.* **1996**, *100*, 8763–8769.
2. Bak, K. L.; Gauss, J.; Jorgensen, P.; Olsen, J.; Helgaker, T.; Stanton, J. F. "The accurate determination of molecular equilibrium structures," *J. Chem. Phys.* **2001**, *114*, 6548–6556.
3. Campanelli, A. R.; Arcadi, A.; Domenicano, A.; Ramondo, F.; Hargittai, I. "Molecular structure and benzene ring deformation of three ethynylbenzenes from gas-phase electron diffraction and quantum chemical calculations," *J. Phys. Chem. A* **2006**, *110*, 2045–2052.
4. Campanelli, A. R.; Domenicano, A.; Ramondo, F.; Hargittai, I. "Molecular structure and benzene ring deformation of three cyanobenzenes from gas-phase electron diffraction and quantum chemical calculations," *J. Phys. Chem. A* **2008**, *112*, 10998–11008.
5. Barone, V. "Anharmonic vibrational properties by a fully automated second-order perturbative approach," *J. Chem. Phys.* **2005**, *122*, 014108.
6. Adamo, C.; Barone, V. "Exchange functionals with improved long-range behavior and adiabatic connection methods without adjustable parameters: the mPW and mPW1PW models," *J. Chem. Phys.* **1998**, *108*, 664–675.
7. Galabov, B.; Yamaguchi, Y.; Remington, R. B.; Schaefer, H. F. "High level ab initio quantum mechanical predictions of infrared intensities," *J. Phys. Chem. A* **2002**, *106*, 819–832.
8. Wohar, M. M.; Jagodzinski, P. W. "Infrared spectra of H_2CO, $H_2{}^{13}CO$, D_2CO, and $D_2{}^{13}CO$ and anomalous values in vibrational force fields," *J. Mol. Spectrosc.* **1991**, *148*, 13–19.
9. Strey, G.; Mills, I. M. "Anharmonic force field of acetylene," *J. Mol. Spectrosc.* **1976**, *59*, 103–115.
10. Pople, J. A.; Schlegel, H. B.; Krishnan, R.; Defrees, D. J.; Binkley, J. S.; Frisch, M. J.; Whiteside, R. A.; Hout, R. F.; Hehre, W. J. "Molecular orbital studies of vibrational frequencies," *Int. J. Quantum Chem., Quantum Chem. Symp.* **1981**, *15*, 269–278.
11. Scott, A. P.; Radom, L. "Harmonic vibrational frequencies: an evaluation of Hartree-Fock, Moller-Plesset, quadratic configuration interaction, density functional theory, and semiempirical scale factors," *J. Phys. Chem.* **1996**, *100*, 16502–16513.
12. Halls, M. D.; Velkovski, J.; Schlegel, H. B. "Harmonic frequency scaling factors for Hartree-Fock, S-VWN, B-LYP, B3-LYP, B3-PW91 and MP2 with the Sadlej pVTZ electric property basis set," *Theor. Chem. Acc.* **2001**, *105*, 413–421.
13. Alecu, I. M.; Zheng, J.; Zhao, Y.; Truhlar, D. G. "Computational thermochemistry: scale factor databases and scale factors for vibrational frequencies obtained from electronic model chemistries," *J. Chem. Theor. Comput.* **2010**, *6*, 2872–2887.
14. Zheng, J.; Alecu, I. M.; Lynch, B. J.; Zhao, Y.; Truhlar, D. G. "Database of frequency scaling factors for electronic structure methods," **2003**, URL: http://comp.chem.umn.edu/freqscale/index.html
15. Jacobsen, R. L.; Johnson, R. D.; Irikura, K. K.; Kacker, R. N. "Anharmonic vibrational frequency calculations are not worthwhile for small basis sets," *J. Chem. Theor. Comput.* **2013**, *9*, 951–954.

16. Nicolaides, A.; Matsushita, T.; Yonezawa, K.; Sawai, S.; Tomioka, H.; Stracener, L. L.; Hodges, J. A.; McMahon, R. J. "The elusive benzocyclobutenylidene: a combined computational and experimental attempt," *J. Am. Chem. Soc.* **2001**, *123*, 2870–2876.
17. DePinto, J. T.; DeProphetis, W. A.; Menke, J. L.; McMahon, R. J. "Triplet 1,3-diphenylpropynylidene (Ph-C-C-C-Ph)," *J. Am. Chem. Soc.* **2007**, *129*, 2308–2315.
18. Nikitina, A. F.; Sheridan, R. S. "Geometry and aromaticity in highly strained heterocyclic allenes: characterization of a 2,3-didehydro-2H-thiopyran," *Org. Lett.* **2005**, *7*, 4467–4470.
19. Kasalová, V.; Allen, W. D.; Schaefer III, H. F.; Czinki, E.; Császár, A. G. "Molecular structures of the two most stable conformers of free glycine," *J. Comput. Chem.* **2007**, *28*, 1373–1383.
20. Balabin, R. M. "Conformational equilibrium in glycine: experimental jet-cooled Raman spectrum," *J. Phys. Chem. Lett.* **2009**, *1*, 20–23.
21. Schindler, M.; Kutzelnigg, W. "Theory of magnetic susceptibilities and NMR chemical shifts in terms of localized quantities. II. Application to some simple molecules," *J. Chem. Phys.* **1982**, *76*, 1919–1933.
22. London, F. "Quantum theory of interatomic currents in aromatic compounds," *J. Phys. Radium* **1937**, *8*, 397–409.
23. Ditchfield, R. "Self-consistent perturbation theory of diamagnetism. I. A gage-invariant LCAO(linear combination of atomic orbitals) method for NMR chemical shifts," *Mol. Phys.* **1974**, *27*, 789–807.
24. Lodewyk, M. W.; Siebert, M. R.; Tantillo, D. J. "Computational prediction of ^{1}H and ^{13}C chemical shifts: a useful tool for natural product, mechanistic, and synthetic organic chemistry," *Chem. Rev.* **2012**, *112*, 1839–1862.
25. Dračínský, M.; Bouř, P. "Computational analysis of solvent effects in NMR spectroscopy," *J. Chem. Theor. Comput.* **2009**, *6*, 288–299.
26. Sarotti, A. M.; Pellegrinet, S. C. "A multi-standard approach for GIAO ^{13}C NMR calculations," *J. Org. Chem.* **2009**, *74*, 7254–7260.
27. Rablen, P. R.; Pearlman, S. A.; Finkbiner, J. "A comparison of density functional methods for the estimation of proton chemical shifts with chemical accuracy," *J. Phys. Chem. A* **1999**, *103*, 7357–7363.
28. Wang, B.; Fleischer, U.; Hinton, J. F.; Pulay, P. "Accurate prediction of proton chemical shifts. I. Substituted aromatic hydrocarbons," *J. Comput. Chem.* **2001**, *22*, 1887–1895.
29. Wang, B.; Hinton, J. F.; Pulay, P. "Accurate prediction of proton chemical shifts. II. Peptide analogues," *J. Comput. Chem.* **2002**, *23*, 492–497.
30. Jain, R.; Bally, T.; Rablen, P. R. "Calculating accurate proton chemical shifts of organic molecules with density functional methods and modest basis sets," *J. Org. Chem.* **2009**, *74*, 4017–4023.
31. Tantillo, D. J. "Chemical shift repository for computed NMR scaling factors, with coupling constants added too," **2012**, URL: http://www.cheshirenmr.info/index.htm
32. Wiitala, K. W.; Hoye, T. R.; Cramer, C. J. "Hybrid density functional methods empirically optimized for the computation of ^{13}C and ^{1}H chemical shifts in chloroform solution," *J. Chem. Theory Comput.* **2006**, *2*, 1085–1092.
33. Jensen, F. "Basis Set convergence of nuclear magnetic shielding constants calculated by density functional methods," *J. Chem. Theory Comput.* **2008**, *4*, 719–727.

34. Barone, G.; Gomez-Paloma, L.; Duca, D.; Silvestri, A.; Riccio, R.; Bifulco, G. "Structure validation of natural products by quantum-mechanical GIAO calculations of ^{13}C NMR chemical shifts," *Chem.-Eur. J.* **2002**, *8*, 3233–3239.
35. Barone, G.; Duca, D.; Silvestri, A.; Gomez-Paloma, L.; Riccio, R.; Bifulco, G. "Determination of the relative stereochemistry of flexible organic compounds by ab initio methods: conformational analysis and Boltzmann-averaged GIAO ^{13}C NMR chemical shifts," *Chem.-Eur. J.* **2002**, *8*, 3240–3245.
36. Vera, B.; Rodríguez, A. D.; Avilés, E.; Ishikawa, Y. "Aplysqualenols A and B: squalene-derived polyethers with antitumoral and antiviral activity from the Caribbean Sea slug Aplysia dactylomela," *Eur. J. Org. Chem.* **2009**, *2009*, 5327–5336.
37. Fattorusso, C.; Stendardo, E.; Appendino, G.; Fattorusso, E.; Luciano, P.; Romano, A.; Taglialatela-Scafati, O. "Artarborol, a nor-caryophyllane sesquiterpene alcohol from artemisia arborescens. Stereostructure assignment through concurrence of NMR data and computational analysis," *Org. Lett.* **2007**, *9*, 2377–2380.
38. Smith, S. G.; Goodman, J. M. "Assigning the stereochemistry of pairs of diastereoisomers using GIAO NMR shift calculation," *J. Org. Chem.* **2009**, *74*, 4597–4607.
39. Belostotskii, A. M. "Calculated chemical shifts as a fine tool of conformational analysis: an unambiguous solution for haouamine alkaloids," *J. Org. Chem.* **2008**, *73*, 5723–5731.
40. Poza, J. J.; Jiménez, C.; Rodríguez, J. "J-based analysis and DFT–NMR assignments of natural complex molecules: application to 3β,7-dihydroxy-5,6-epoxycholestanes," *Eur. J. Org. Chem.* **2008**, *2008*, 3960–3969.
41. Smith, S. G.; Goodman, J. M. "Assigning stereochemistry to single diastereoisomers by GIAO NMR calculation: the DP4 probability," *J. Am. Chem. Soc.* **2010**, *132*, 12946–12959.
42. Goodman, J. M.; Smith, S. G. "Assignment of stereochemistry and structure using NMR and DP4" **2010**, URL: http://www-jmg.ch.cam.ac.uk/tools/nmr/DP4/
43. Evidente, A.; Abou-Donia, A. H.; Darwish, F. A.; Amer, M. E.; Kassem, F. F.; Hammoda, H. A.m. Motta, A. "Nobilisitine a and B, two masanane-type alkaloids from Clivia nobilis," *Phytochemistry*, **1999**, *51*, 1151–1155.
44. Schwartz, B. D.; Jones, M. T.; Banwell, M. G.; Cade, I. A. "Synthesis of the enantiomer of the structure assigned to the natural product nobilisitine A," *Org. Lett.* **2010**, *12*, 5210–5213.
45. Lodewyk, M. W.; Tantillo, D. J. "Prediction of the structure of nobilisitine a using computed NMR chemical shifts," *J. Nat. Prod.* **2011**, *74*, 1339–1343.
46. Bally, T.; Rablen, P. R. "Quantum-chemical simulation of ^{1}H NMR spectra. 2. Comparison of DFT-based procedures for computing proton-proton coupling constants in organic molecules," *J. Org. Chem.* **2011**, *76*, 4818–4830.
47. Deng, W.; Cheeseman, J. R.; Frisch, M. J. "Calculation of nuclear spin-spin coupling constants of molecules with first and second row atoms in study of basis set dependence," *J. Chem. Theor. Comput.* **2006**, *2*, 1028–1037.
48. Williamson, R. T.; Buevich, A. V.; Martin, G. E. "Experimental and theoretical investigation of 1JCC and nJCC coupling constants in strychnine," *Org. Letters* **2012**, *14*, 5098–5101.
49. Di Micco, S.; Chini, M. G.; Riccio, R.; Bifulco, G. "Quantum mechanical calculation of NMR parameters in the stereostructural determination of natural products," *Eur. J. Org. Chem.* **2010**, *2010*, 1411–1434.

50. Schlegel, B.; Hartl, A.; Dahse, H.-M.; Gollmick, F. A.; Grafe, U.; Dorfelt, H.; Kappes, B. "Hexacyclinol, a new antiproliferative metabolite of *Panus rudis* HKI 0254," *J. Antibiot.* **2002**, *55*, 814–817.

51. La Clair, J. J. "Total syntheses of hexacyclinol, 5-*epi*-hexacyclinol, and desoxohexacyclinol unveil an antimalarial prodrug motif," *Angew. Chem. Int. Ed.* **2006**, *45*, 2769–2773.

52. Halford, B. "Hexacyclinol debate heats up," *Chem. Eng. News*, **2006**, *84 (31, July 28)*, 11.

53. Rychnovsky, S. D. "Predicting NMR spectra by computational methods: structure revision of hexacyclinol," *Org. Lett.* **2006**, *8*, 2895–2898.

54. Porco, J. A. J.; Shun Su, S.; Lei, X.; Bardhan, S.; Rychnovsky, S. D. "Total synthesis and structure assignment of (+)-hexacyclinol," *Angew. Chem. Int. Ed.* **2006**, *45*, 5790–5792.

55. Saielli, G.; Bagno, A. "Can two molecules have the same NMR spectrum? Hexacyclinol revisited," *Org. Lett.* **2009**, *11*, 1409–1412.

56. Nicolaou, K. C.; Frederick, M. O. "On the structure of maitotoxin," *Angew. Chem. Int. Ed.* **2007**, *46*, 5278–5282.

57. Gallimore, A. R.; Spencer, J. B. "Stereochemical uniformity in marine polyether ladders – implications for the biosynthesis and structure of maitotoxin," *Angew. Chem. Int. Ed.* **2006**, *45*, 4406–4413.

58. Guella, G.; Dini, F.; Pietra, F. "Metabolites with a novel C30 backbone from marine ciliates," *Angew. Chem. Int. Ed.* **1999**, *38*, 1134–1136.

59. Nicolaou, K. C.; Ortiz, A.; Zhang, H.; Dagneau, P.; Lanver, A.; Jennings, M. P.; Arseniyadis, S.; Faraoni, R.; Lizos, D. E. "Total synthesis and structural revision of vannusals a and B: synthesis of the originally assigned structure of vannusal B," *J. Am. Chem. Soc.* **2010**, *132*, 7138–7152.

60. Saielli, G.; Nicolaou, K. C.; Ortiz, A.; Zhang, H.; Bagno, A. "Addressing the stereochemistry of complex organic molecules by density functional theory-NMR: vannusal B in retrospective," *J. Am. Chem. Soc.* **2011**, *133*, 6072–6077.

61. Chini, M. G.; Jones, C. R.; Zampella, A.; D'Auria, M. V.; Renga, B.; Fiorucci, S.; Butts, C. P.; Bifulco, G. "Quantitative NMR-derived interproton distances combined with quantum mechanical calculations of ^{13}C chemical shifts in the stereochemical determination of conicasterol F, a nuclear receptor ligand from *theonella swinhoei*," *J. Org. Chem.* **2012**, *77*, 1489–1496.

62. Harding, M. E.; Gauss, J.; Schleyer, P. v. R. "Why benchmark-quality computations are needed to reproduce 1-adamantyl cation NMR chemical shifts accurately," *J. Phys. Chem. A*, **2011**, *115*, 2340–2344.

63. Cheeseman, J. R.; Frisch, M. J.; Devlin, F. J.; Stephens, P. J. "Hartree-Fock and density functional theory ab initio calculation of optical rotation using GIAOs: basis set dependence," *J. Phys. Chem. A*, **2000**, *104*, 1039–1046.

64. Stephens, P. J.; Devlin, F. J.; Cheeseman, J. R.; Frisch, M. J. "Calculation of optical rotation using density functional theory," *J. Phys. Chem. A*, **2001**, *105*, 5356–5371.

65. Ruud, K.; Stephens, P. J.; Devlin, F. J.; Taylor, P. R.; Cheeseman, J. R.; Frisch, M. J. "Coupled-cluster calculations of optical rotation," *Chem. Phys. Lett.* **2003**, *373*, 606–614.

66. Tam, M. C.; Russ, N. J.; Crawford, T. D. "Coupled cluster calculations of optical rotatory dispersion of (*S*)-methyloxirane," *J. Chem. Phys.* **2004**, *121*, 3550–3557.

67. Crawford, T. D.; Tam, M. C.; Abrams, M. L. "The current state of ab initio calculations of optical rotation and electronic circular dichroism spectra," *J. Phys. Chem. A*, **2007**, *111*, 12057–12068.
68. Giorgio, E.; Viglione, R. G.; Zanasi, R.; Rosini, C. "Ab initio calculation of optical rotatory dispersion (ORD) curves: a simple and reliable approach to the assignment of the molecular absolute configuration," *J. Am. Chem. Soc.* **2004**, *126*, 12968–12976.
69. Stephens, P. J.; McCann, D. M.; Cheeseman, J. R.; Frisch, M. J. "Determination of absolute configurations of chiral molecules using ab initio time-dependent density functional theory calculations of optical rotation: how reliable are absolute configurations obtained for molecules with small rotations?," *Chirality* **2005**, *17*, S52–S64.
70. Kongsted, J.; Pedersen, T. B.; Jensen, L.; Hansen, A. E.; Mikkelsen, K. V. "Coupled cluster and density functional theory studies of the vibrational contribution to the optical rotation of (S)-propylene oxide," *J. Am. Chem. Soc.* **2006**, *128*, 976–982.
71. Crawford, T. D.; Stephens, P. J. "Comparison of time-dependent density-functional theory and coupled cluster theory for the calculation of the optical rotations of chiral molecules," *J. Phys. Chem. A*, **2008**, *112*, 1339–1345.
72. Srebro, M.; Govind, N.; de Jong, W. A.; Autschbach, J. "Optical rotation calculated with time-dependent density functional theory: the OR45 benchmark," *J. Phys. Chem. A*, *115*, 10930–10949.
73. Yanai, T.; Tew, D. P.; Handy, N. C. "A new hybrid exchange-correlation functional using the Coulomb-attenuating method (CAM-B3LYP)," *Chem. Phys. Lett.* **2004**, *393*, 51–57.
74. Baranowska, A.; Łączkowski, K. Z.; Sadlej, A. J. "Model studies of the optical rotation, and theoretical determination of its sign for β-pinene and trans-pinane," *J. Comput. Chem.*, *31*, 1176–1181.
75. Goerigk, L.; Grimme, S. "Calculation of electronic circular dichroism spectra with time-dependent double-hybrid density functional theory," *J. Phys. Chem. A* **2008**, *113*, 767–776.
76. de Meijere, A.; Khlebnikov, A. F.; Kostikov, R. R.; Kozhushkov, S. I.; Schreiner, P. R.; Wittkopp, A.; Yufit, D. S. "The first enantiomerically pure triangulane (M)-trispiro[2.0.0.2.1.1]nonane is a σ-[4]heliccne," *Angew. Chem. Int. Ed.* **1999**, *38*, 3474–3477.
77. de Meijere, A.; Khlebnikov, A. F.; Kozhushkov, S. I.; Kostikov, R. R.; Schreiner, P. R.; Wittkopp, A.; Rinderspacher, C.; Menzel, H.; Yufit, D. S.; Howard, J. A. K. "The first enantiomerically pure [n]triangulanes and analogues: σ-[n]helicenes with remarkable features," *Chem. Eur. J.* **2002**, *8*, 828–842.
78. Crawford, T. D.; Owens, L. S.; Tam, M. C.; Schreiner, P. R.; Koch, H. "Ab initio calculation of optical rotation in (P)-(+)-[4]triangulane," *J. Am. Chem. Soc.* **2005**, *127*, 1368–1369.
79. Wilson, S. M.; Wiberg, K. B.; Cheeseman, J. R.; Frisch, M. J.; Vaccaro, P. H. "Nonresonant optical activity of isolated organic molecules," *J. Phys. Chem. A*, **2005**, *109*, 11752–11764.
80. Mennucci, B.; Claps, M.; Evidente, A.; Rosini, C. "Absolute configuration of natural cyclohexene oxides by time dependent density functional theory calculation of the optical rotation: the absolute configuration of (-)-sphaeropsidone and (-)-episphaeropsidone revised," *J. Org. Chem.* **2007**, *72*, 6680–6691.

81. Mukhopadhyay, P.; Zuber, G.; Wipf, P.; Beratan, D. N. "Contribution of a solute's chiral solvent imprint to optical rotation," *Angew. Chem. Int. Ed.* **2007**, *46*, 6450–6452.
82. Lipparini, F.; Egidi, F.; Cappelli, C.; Barone, V. "The optical rotation of methyloxirane in aqueous solution: a never ending story?," *J. Chem. Theor. Comput.* **2013**, *9*, 1880–1884.
83. Albers-Schonberg, G.; Schmid, H. "The structure of plumericin and related compounds," *Chimia*, **1960**, *14*, 127–128.
84. Elsässer, B.; Krohn, K.; Nadeem Akhtar, M.; Flörke, U.; Kouam, S. F.; Kuigoua, M. G.; Ngadjui, B. T.; Abegaz, B. M.; Antus, S.; Kurtán, T. "Revision of the absolute configuration of plumericin and isoplumericin from plumeria rubra," *Chem. Biodiversity*, **2005**, *2*, 799–808.
85. Stephens, P. J.; Pan, J. J.; Krohn, K. "Determination of the absolute configurations of pharmacological natural products via density functional theory calculations of vibrational circular dichroism: the bew cytotoxic iridoid prismatomerin," *J. Org. Chem.* **2007**, *72*, 7641–7649.
86. Wiberg, K. B.; Wang, Y. G.; Wilson, S. M.; Vaccaro, P. H.; Jorgensen, W. L.; Crawford, T. D.; Abrams, M. L.; Cheeseman, J. R.; Luderer, M. "Optical rotatory dispersion of 2,3-hexadiene and 2,3-pentadiene," *J. Phys. Chem. A*, **2008** 112, 2415–2422.
87. Muranaka, A.; Shibahara, M.; Watanabe, M.; Matsumoto, T.; Shinmyozu, T.; Kobayashi, N. "Optical resolution, absolute configuration, and chiroptical properties of three-layered [3.3]paracyclophane(1)," *J. Org. Chem.* **2008**, *73*, 9125–9128.
88. Werner, A.; Michels, M.; Zander, L.; Lex, J.; Vogel, E. "'Figure eight' cyclooctapyrroles: enantiomeric separation and determination of the absolute configuration of a binuclear metal complex," *Angew. Chem. Int. Ed.* **1999**, *38*, 3650–3653.
89. Rzepa, H. S. "The chiro-optical properties of a lemniscular octaphyrin," *Org. Lett.* **2009**, *11*, 3088–3091.

CHAPTER 3

Fundamentals of Organic Chemistry

When students confront organic chemistry for the first time, the discipline appears overwhelming, seemingly a huge collection of unrelated facts: reactants, reagents, solvents, temperature, and products. The beauty and power of organic chemistry is that patterns and rules can be discerned to help organize and generalize these facts. Organic chemists think in terms of functional group transformations and reaction mechanisms to bring order to the field, allowing chemists to design complex syntheses of new materials with great success.

This chapter presents five powerful organizing concepts of organic chemistry. Since reactions involve making and breaking bonds, we begin the chapter by discussing the two ways bonds can be cleaved. Homolytic cleavage, creating two radicals, provides the classic definition of bond energy. The alternative is heterolytic cleavage that creates a cation/anion pair. The heterolytic cleavage of a C–H bond defines the acid strength of a molecule. Next, we discuss isomerism, which is the phenomenon by which molecules have identical molecular formulae but differ in some important structural way. This is an entrée into the importance of dispersion and medium- and long-range correlation and the problems many density functionals have in properly accounting for these effects.

We then turn to two general concepts that define the stability (or lack thereof) of cyclic molecules. The first is the ring strain energy (RSE) that tends to destabilize molecules possessing small rings with distorted bond angles. This chapter concludes with a discussion of aromaticity, a crucially important organizational concept that dates back to the 1880s, yet still challenges careful definition.

The emphasis in this and the subsequent chapters will be on the role computational chemistry has played in helping to bring definition and understanding to organic problems.

3.1 BOND DISSOCIATION ENTHALPY

Chemistry is fundamentally about making and breaking bonds. Thus, in order to understand a reaction at its most basic level, one must account for how much energy

---

Computational Organic Chemistry, Second Edition. Steven M. Bachrach.
© 2014 John Wiley & Sons, Inc. Published 2014 by John Wiley & Sons, Inc.

is required to cleave the appropriate bonds and how much energy will be released when the new bonds are formed. The energy associated with a chemical bond, called the bond dissociation enthalpy (BDE), is measured as the energy released in the reaction

$$A-B \rightarrow A\cdot + B\cdot \qquad \text{(Reaction 3.1)}$$

This is a homolytic cleavage, resulting in two radicals. Computation of the BDE requires simply computing the energies of each of the three species, the molecule AB and the two radicals A· and B·.

While this computation might seem straightforward at first, the open-shell radicals are inherently different than their closed-shell counterpart: the unpaired electron does not "have" as much correlation energy as the other electrons in the radical or in the closed-shell molecule because it does not have a partner electron sharing the same space. Some computational methods, therefore, will not treat the open-shell and closed-shell species on equivalent footings. A simple manifestation of this is that a spin-restricted wavefunction is quite suitable for the closed-shell parent, but a spin-unrestricted wavefunction is likely to be required for computing the energies of the open-shell radicals. Representative computed BDEs for a number of simple bond cleavages using a variety of different methodologies are compared with experimental values in Table 3.1. In all of these cases, the necessary radicals are computed with the spin-unrestricted wavefunction.

TABLE 3.1 Bond Dissociation Enthalpy (ΔH_{298}, kcal mol^{-1}) of Some Simple Organic Molecules

	CH_3CH_2-H	CH_2CH-H	$HCC-H$	CH_3O-H	$CH_3CH_2-CH_3$	CH_3-CH_3
HF/6-31G*	78.03	87.72	98.85	64.90	62.44	60.85
HF/6-31+G*	74.37	81.52	99.66	65.89	57.78	58.84
HF/aug-cc-pVTZ	77.41	87.79	101.62	69.85	59.21	57.16
B3LYP/6-31G*	100.11	109.12	131.14	94.80	85.91	89.05
B3LYP/6-31+G*	99.07	108.63	131.49	95.98	82.86	85.78
B3LYP/6-311++G(d,p)[a]	99.0	108.2	134.5	99.3		84.4
B3LYP/aug-cc-pVTZ	98.45	108.26	134.07	99.61	81.34	84.42
MP2/6-31G*	92.87	107.85	136.74	93.94	90.84	90.95
MP2/6-31+G*	91.83	107.41	136.93	95.88	88.65	88.42
MP2/aug-cc-pVTZ	102.31	117.51	147.89	111.01	92.78	92.37
G2MP2	102.68	112.24	135.18	104.40	90.76	91.05
G3[a]	102.0	110.2	133.4	105.0		88.3
CBS-Q[a]	101.6	110.3	133.2	105.4		90.0
Expt[b]	101.1	110.7	133.32	104.6	89.0	90.1

[a]Ref. 1.
[b]Ref. 2.

The Hartree–Fock (HF) method seriously underestimates the BDE for all of the examples in Table 3.1. This results from the inherent lesser electron correlation in the radicals than in the parent species, and electron correlation is completely neglected in the HF method. Inclusion of some electron correlation, as in the MP2 method, dramatically improves the estimated BDEs. In a study of the BDE of the C–H bond in 19 small molecules, Radom[3] found that the mean absolute deviation (MAD) at the MP2/6-311+G(2df,p)//MP2/6-31G(d) level is 3.6 kcal mol^{-1}. This error is generally larger than that associated with the experimental method, and so DPEs obtained using MP2 may be problematic.

On the other hand, the composite methods very accurately predict BDEs. The MAD for the molecules listed in Table 3.1 using the G2MP2 method is only 1.3 kcal mol^{-1}. The G3 and CBS-Q methods were evaluated for their ability to predict the BDE for a broad selection of molecules, including hydrocarbons, alkyl halides, alcohols, ethers, and amines.[1] The MAD for the G3 values (165 values) is 1.91 kcal mol^{-1} and that for CBS-Q (161 values) is 1.89 kcal mol^{-1}. Radom determined the C–H BDEs for a set of 22 small organic molecules. The G3(MP2) method performed well, with its MAD of 0.9 kcal mol^{-1}, and CBS did even better, with its MAD of only 0.6 kcal mol^{-1}.[3] Clearly, extensive treatment of electron correlation is required to compute reasonable BDEs, but the computational demands of the composite methods make them prohibitive for any but the smallest molecules. Since density functional theory (DFT) incorporates electron correlation at a fraction of the computational cost of more traditional *ab initio* methods, it is potentially an attractive procedure. The MAD values of the BDEs predicted by the B3LYP computations for the molecules of Table 3.1 are 3–4 kcal mol^{-1}, somewhat better than the MP2 methods. For the set of 165 BDEs used in Feng's study,[1] the MAD values of the B3LYP values using the 6-31G(d), 6-311++G(d,p), and 6-311++D(3df,2p) basis sets were 3.66, 4.63, and 4.23 kcal mol^{-1}, respectively. For a set of 23 C–H BDEs, Radom[3] found that the MAD of the UB3LYP/6-311+G(3df,2p)//UB3LYP/6-31G(d) values was 2.7 kcal mol^{-1}, but the estimates were improved using the *restricted* DFT method: the MAD was then only 1.3 kcal mol^{-1}. While these deviations are a bit larger than experimental errors, the DFT values of the BDEs are systematically lower than the experimental values. In fact, the DFT values correlate quite well with experimental or G3 predicted BDEs.[4] Somewhat disappointing is the performance of double hybrid methods, whose mean deviation from experiment of the C–H BDE for a set of 23 small methyl-substituted organic compounds is 4–5 kcal mol^{-1}.[5] For this set of compounds, the RMPWB1K functional performed best.

However, more detailed evaluations of the performance of DFT methods have revealed serious deficiencies. In a study of 57 different bond cleavages, including representative examples of C–H, C–C, O–H, C–N, C–O, C–S, and C–X bonds, both the B3LYP/6-311+G(d,p) and B3PW91/6-311+G(d,p) very poorly predicted the BDEs; their MAD values were 5.7 and 4.4 kcal mol^{-1}, respectively, with some individual bond energies in error by 14 kcal mol^{-1}.[6] On the other hand, the MPW1P86/6-311+G(d,p) BDEs are in satisfactory agreement with experiment (MAD = 1.7 kcal mol^{-1}). Gilbert[7] investigated the C–C BDE

TABLE 3.2 Experimental and Predicted C–C BDE (kcal mol^{-1})^a

	Expt^b	G3MP2	MP2^c	B3LYP^c
CH_3-CH_3	90.2 ± 0.2	85.9	86.6	81.5
$CH_3CH_2-CH_3$	88.5 ± 0.5	85.9	87.3	79.1
$(CH_3)_2CH-CH_3$	88.2 ± 0.9	85.7	87.8	76.5
$CH_3CH_2-CH_2CH_3$	86.8 ± 0.6	86.0	88.1	76.7
$(CH_3)_3CH-CH_3$	86.9 ± 0.7	85.6	88.2	73.7
$(CH_3)_2CH-CH_2CH_3$	86.1 ± 0.9	85.4	88.1	73.0
$(CH_3)_3C-CH_2CH_3$	84.5 ± 0.9	84.7	87.9	69.2
$(CH_3)_2CH-CH(CH_3)_2$	84.5 ± 1.1	84.1	87.3	68.2
$(CH_3)_3C-CH(CH_3)_2$	81.5 ± 1.1	82.6	86.3	63.2
$(CH_3)_3C-C(CH_3)_3$	77.1 ± 1.0	79.8	84.1	56.0
MAD^d		1.7	2.8	13.7

^{a}Ref. 7.
^{b}Ref. 9.
^{c}MP2 and B3LYP were performed with the 6-311++G(d,p) basis set.
^{d}Mean absolute deviation.

for a number of branched hydrocarbons using the G3MP2, MP2/6-311++G(d,p), and B3LYP/6-311++G(d,p) methods (Table 3.2). Consistent with the other composite methods, G3MP2 quite accurately predicts the BDEs, with the MAD of 1.7 kcal mol^{-1}. The MAD is greater with the MP2 values, but still marginally acceptable at 2.8 kcal mol^{-1}. However, B3LYP completely fails to adequately predict BDEs, especially for the larger molecules where the errors are on the order of 20 kcal mol^{-1}! This error results from its systematic underestimation of the heats of formation of linear alkanes; Curtiss et al.[8] estimated an error of nearly 0.7 kcal mol^{-1} per bond with B3LYP. So, while B3LYP might be acceptable for small molecules, its error increases with the size of the molecule, becoming unacceptable even with molecules as small as propane. Further discussion of the problems inherent within DFT for computing bond dissociation energies and related effects will be presented in Section 3.3.

Undeniably, the composite methods are best suited for prediction of bond dissociation enthalpies. However, these methods may be impractical for many molecules of interest. In these cases, extreme caution must be used with any other computational method. The best situation would be to test the method of choice with a series of related molecules, comparing these computed BDEs with either experimental or high level computed values to obtain some sort of correlation that would justify the use of the more moderate computational method.

3.1.1 Case Study of BDE: Trends in the R–X BDE

The standard thought on alkyl radical stability is that it increases with α-substitution, that is, Me < Et < i-Pr < t-Bu, a trend based on the BDEs of simple alkanes. However, it has also been long recognized that the trend in R–X BDE is dependent on the nature of the X group.[10] These differences are clearly

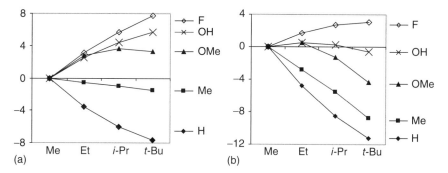

Figure 3.1 Trends in relative C–X BDE (0 K, kcal mol^{-1}) computed at (a) G3 and (b) B3LYP/6-311+(3df,2p).[11] Adapted with permission from Izgorodina, E. I.; Coote, M. L.; Radom, L. *J. Phys. Chem. A.* **2005**, *109*, 7558–7566. Copyright 2005 American Chemical Society.

presented in the plot shown in Figure 3.1a, where BDEs relative to Me–X are plotted against increasing α-substitution of the R–X molecule. The BDE of the C–H bond decreases with increasing α-substitution. The same is also true for the C–C bond, though the size of the effect is diminished. However, the exact *opposite* trend is observed for the C–F and C–OH bonds, and for the C–OMe series, the BDE actually turns over.

As expected, the G3 composite method accurately reproduces the experimental BDEs with a MAD of only 0.7 kcal mol^{-1}.[11] The B3LYP values are much inferior: the MAD is 7 kcal mol^{-1}, well outside the error limits of the experiments. Even more discouraging is that the relation between α-substitution and C–X BDE for the different X groups is not reproduced (Figure 3.1b). While the trend of decreasing C–H and C–C BDEs with increasing α-substitution is predicted by the B3LYP values, the α-effect is magnified. For the C–F BDE, B3LYP again predicts the correct trend that the bonds become stronger with increasing α-substitution, but the size of the effect is diminished. For the C–OMe bonds, B3LYP predicts a turnover too early with α-substitution. The C–OH trend, which is for increasing bond energies with α-substitution, is incorrectly predicted at B3LYP to have a turnover. B3LYP underestimates BDEs, and the error becomes worse with increasing size of the molecule, consistent with the data in Table 3.2. This is yet another cautionary tale concerning the use of B3LYP to predict bond dissociation enthalpies (and see Section 3.3).

As to the origins of the trends expressed in Figure 3.1a, Coote et al.[12] noted that the R–X bond lengths increase in the order Me < Et < *i*-Pr < *t*-Bu regardless of the group X. Thus, the distance increases independent of the bond strength; only sterics seemingly affect the bond length. So, for example, with X=F and OH, increasing the α-substitution from Me to *t*-But leads to longer and stronger bonds.

The key to understanding these trends is to examine how the charge on the X group varies with α-substitution. The NBO charges are listed in Table 3.3 and two distinct trends are apparent. Not unexpectedly, as X becomes more electronegative,

TABLE 3.3 NBO Charge on X of R–X^a

R	X=CH$_3$	X=OCH$_3$	X=OH	X=F
Me	0.000	−0.298	−0.282	−0.387
Et	0.008	−0.303	−0.285	−0.393
i-Pr	0.018	−0.309	−0.289	−0.400
t-Bu	0.028	−0.317	−0.294	−0.407

^{a}Computed at MP2, see Ref. 12.

the negative charge on X becomes greater. For any alkyl group R, the polarity of the C–X bond increases as C–C < C–O < C–F. More interestingly, as the R group becomes more branched, then the charge on X becomes more negative. In other words, α-substitution increases the electron-donating capability of the R group. The upshot is that as α-substitution increases, the C–X bond becomes more polar. Coote et al. argue that the R–X bond can be considered within valence-bond theory as a resonance hybrid of three structures: R:X ↔ R$^+$ X$^-$ ↔ R$^-$ X$^+$. As X becomes more electronegative, the R$^+$X$^-$ structure becomes more important. Also, as the degree of α-substitution increases, the R$^+$X$^-$ structure becomes more important.

This ionic contribution adds strength to the R–X bond, especially so when X is very electronegative or if R is highly branched. When X=F, the ionic character is very important and increasing α-substitution enhances this polarity, leading to the C–F strength increasing with α-substitution. When X=H or CH$_3$, the ionic contribution is minimal and so the stability of the alkyl radical dominates the trend in BDEs. When X=O, the two trends work in opposite directions, with alkyl stability more important with minimal alkyl branching and ionic stabilization important with higher alkyl branching. This argument confirms the one put forth by Zavitsas[13] based on Pauling electronegativity equation.

3.2 ACIDITY

As an alternative to what was discussed in the previous section, bond cleavage can occur heterolytically, with the two bonding electrons accompanying one of the atoms of the former bond, creating an anion and cation pair. The most important example of the heterolytic cleavages is for A–H bonds, where cleavage results in the anion A$^-$ and a proton (Reaction 3.2). This process is better known as defining the chemistry of an acid, and the energy required to perform this cleavage determines the acidity of the A–H species.

$$A - H \rightarrow A^- + H^+ \quad \quad \text{(Reaction 3.2)}$$

Since computational chemistry typically models the gas phase, Reaction 3.2 defines the gas-phase acidity of the Brønsted acid AH. This can also be termed the deprotonation energy (DPE). Gas-phase acidities have been well studied by

TABLE 3.4 Deprotonation Enthalpies[a] for Acetic Acid, Acetone, and Propene

Method	Acetic Acid	Acetone	Propene
HF/6-31G	354.71	383.91	412.66
HF/6-31G(d)	359.67	390.90	415.72
HF/6-31+G(d)	348.90	377.74	398.63
HF/6-31+G(d,p)	353.02	378.46	398.81
HF/6-311+G(2d,p)	353.69	378.90	398.97
MP2/6-31G	360.60	385.05	412.35
MP2/6-31G(d)	356.72	386.43	412.02
MP2/6-31+G(d)	339.91	367.41	389.50
MP2/6-31+G(d,p)	346.15	370.88	392.64
MP2/6-311+G(2d,p)	345.16	367.65	388.34
MP4/6-31G[b]	354.68	386.09	414.87
MP4/6-31G(d)[b]	358.16	388.39	413.72
MP4/6-31+G(d)[b]	341.11	369.37	391.99
MP4/6-31+G(d,p)[b]	347.53	373.10	395.50
MP4/6-311+G(2d,p)[b]	346.42	370.15	391.45
B3LYP/6-31G	354.41	381.82	407.81
B3LYP/6-31G(d)	359.18	385.85	408.82
B3LYP/6-31+G(d)	342.28	368.01	388.45
B3LYP/6-31+G(d,p)	345.66	368.43	388.35
B3LYP/6-311+G(2d,p)	346.10	368.05	387.90
G2MP2	346.90	369.81	390.45
Expt[c]	348.7	368.8	391.1

[a] Enthalpies computed using the unscaled vibrational frequencies.
[b] Computed at MP4SDTQ//MP2 with the vibrational frequencies from the MP2 calculation.
[c] Ref. 15.

experiment,[14,15] often as the reverse of Reaction 3.2, defining in this case the proton affinity of the Brønsted base A⁻.

Computed acidities have been reported since the dawn of the age of computational chemistry. Instead of taking a historical walk through this development, we present the highlights of the methodological developments necessary to accurately compute DPEs motivated by the results presented in Table 3.4. This table presents the DPEs of acetic acid, acetone, and propene computed with the HF, MP2, MP4, B3LYP, and G2MP2 methods, using a variety of basis sets. For all cases, the geometry of the acid and its conjugate base was completely optimized at the level under consideration, except for the MP4 energies, which were computed using the MP2 geometries. The computed enthalpy of deprotonation is reported to allow for the best comparison with the experimental enthalpies of reaction.

A very striking dependency on the makeup of the basis set is readily apparent from the data of Table 3.4. While the DPEs increase by 1–6 kcal mol^{-1} with the addition of polarization functions, the effect of added diffuse functions is much larger. Regardless of the computational method, diffuse functions reduce the DPEs by some 10–20 kcal mol^{-1}! The role of diffuse functions is to properly describe the tail behavior of orbitals, especially those in electron-rich systems where electron–electron repulsion leads to a more extensive distribution than usual. Kollmar[16] first recognized that diffuse functions dramatically aid in determining the proton affinity of methyl anion. However, it was a communication by Schleyer[17] that really highlighted their critical necessity in describing anions in order to obtain reasonable DPEs and popularized their wide use.

Even with diffuse functions, expanded valence space, and multiple polarization functions, the DPEs predicted by the HF method are too high, from 5 to 10 kcal mol^{-1}. Inclusion of electron correlation is needed to improve the computational results. In fact, DPEs obtained using MP2/6-31+G(d) are frequently within 3 kcal mol^{-1} of experiment,[18] although that is not the case for acetic acid. For some carbon acids, higher orders of perturbation were not helpful.[18] For the examples listed in Table 3.4, the MP4 DPEs are all larger than the values obtained at MP2, and the MP4/6-311+G(2d,p) DPEs are within 1.4 kcal mol^{-1} of the experimental values. Clearly, quite reasonable DPEs can be obtained at the MP2 level, with slight improvement achieved by moving to MP4.

DFT offers partial treatment of electron correlation at drastically reduced computational cost relative to other post-HF methods. How well does DFT account for acidity? DPEs computed using the popular B3LYP method are listed in Table 3.4. As with the other methods, diffuse functions are essential for reasonable estimation of DPEs using B3LYP. With a set of diffuse functions, even the smallest basis set tried (6-31+G(d)) produced quite satisfactory results; the DPEs of acetone and propene are predicted within 2.5 kcal mol^{-1} of the experimental values, but that of acetic acid is off by 6.5 kcal mol^{-1}. For a set of 45 acids, the average absolute error for the B3LYP/6-31+G(d)-predicted DPEs is 4.7 kcal mol^{-1}.[19] Improvement of the basis set leads to smaller errors; with the B3LYP/6-311+G(2d,p) method, even the DPE of acetic acid is predicted with an error less than 2.5 kcal mol^{-1}. For the set of 45 acids, the average error is only 2.1 kcal mol^{-1} with the B3LYP/6-311++G(3df,2pd) level.

Merrill and Kass[20] have examined a serious of alternative functionals for predicting the DPEs of a set of 35 anions. They find that not all functionals are appropriate for predicting DPEs. In particular, Slater's local exchange function, whether coupled with a local (S-VWN) or nonlocal (S-LYP) correlation functional, fails to adequately predict DPEs, having average unsigned errors of 10–20 kcal mol^{-1}. The BVWN and BLYP functionals perform somewhat better, with average errors of about 4 kcal mol^{-1} when aug-cc-pVTZ or similar basis sets are employed. However, the B3PW91 functional, with diffuse-augmented, polarized double-, or triple-zeta basis sets having average errors of 2.5 kcal mol^{-1} or less. Thus, very reasonable estimates of acidities can be made with fairly inexpensive methods, as long as a hybrid functional (B3LYP, B3PW91, etc.) is employed.

Since larger basis sets and increased treatment of correlation improve the prediction of acidities, the composite methods potentially offer particular value. In the paper describing the CBS-4 and CBS-Q methods, Petersson[21] notes the excellent agreement with experiment these two methods give for the DPEs of benzene and ethylene. Wiberg[22] examined 16 small acids using the CBS-4, CBS-Q, G2MP2, and

TABLE 3.5 Gas-Phase DPEs for Some Carbon Acids[a]

Compound	MP2[b]	B3LYP[c]	G2MP2	Expt[d]
Ethyne	379.3	376.7	377.4	379.5
Cyclopentadiene	348.0	353.7	354.0	353.9
Ethane	420.9	418.9	420.5	420.1
Cyclopropane	414.0	413.1	413.1	412
Toluene	384.4	380.9	382.9	380.8
Cyclohexane	413.5	411.9	414.5	418.3
Average unsigned error	2.9	2.0	1.6	

[a] ΔH (298 K) in kilocalorie per mole. Ref. 23.
[b] MP2/6-311++G(d,p).
[c] B3LYP/6-311++G(d,p).
[d] Ref. 15.

G2 methods. All four of these methods give excellent results, with average unsigned errors ranging from 1.8 kcal mol^{-1} with CBS-4 to 1.1 kcal mol^{-1} with G2.

Since we will discuss next a number of interesting carbon acidity cases, it is worth concluding here with a look at the relative performance of some reasonable methods in predicting the DPEs of simple hydrocarbons. Table 3.5 presents the gas-phase DPEs for six compounds using methods that should be adequate, based on the discussions above. Excluding the DPE of cyclohexane, G2MP2, MP2/6-311++G(d,p), and B3LYP/6-311++G(d,p) all produce acceptable results, with B3LYP doing a very fine job (average error of only 1.1 kcal mol^{-1}).[23] Given the quality of the predictions made by these computational methods, the very large disagreement between the computed and experimental DPEs of cyclohexane calls into question the experimental result.

3.2.1 Case Studies of Acidity

3.2.1.1 Carbon Acidity of Strained Hydrocarbons

The difference in the formal hybridization at carbon accounts for the relative acidities of alkanes versus alkenes versus akynes. The greater the s-character of the carbon of the C–H bond of interest, the more acidic is that proton. Of interest here is extension of this idea to strained compounds, which may employ greater or lesser s-character to accommodate the unusual geometries. Will these compounds have acidities that reflect their hybridization? Are strained hydrocarbons more acidic than their unstrained analogs?

The general answer to these questions is yes. In 1989, DePuy and coworkers[24] measured the gas-phase DPEs of a number of small alkanes using the reaction of hydroxide with alkyltrimethylsilanes in a flowing afterglow-selected ion flow tube. Their experimental DPEs for some simple alkanes are listed in Table 3.6. The DPE of cyclopropane, 411.5 kcal mol^{-1}, is much lower than that of a typical acyclic alkane (415–420 kcal mol^{-1}). Cyclobutane is less acidic than cyclopropane, reflecting the diminished s-character of its C–H bonds. Also listed in Table 3.6 are

TABLE 3.6 DPEs (ΔH, kcal mol^{-1}) of Some Simple Alkanes

Alkane	Expt[a]	MP2[b]
Ethane	420.1	420.5
Propane (C_1–H)	415.6	415.6
(C_2–H)	419.4	417.9
Cyclopropane	411.5	413.7
Cyclobutane	417.4	417.2

[a]Ref. 24 with error bars of ± 2 kcal mol^{-1}.
[b]MP2/6-31+G*.

the computed DPEs using MP2/6 31+G*, and again very good agreement is seen between the computed and experimental data.

Over the past 15 years, theoreticians and experimentalists have examined more exotic strained hydrocarbons. Noting a previous study indicating that cubane exchanged hydrogen faster than cyclopropane,[25] Ritchie and Bachrach[26] computed the DPE of cubane and cyclopropane. They first obtained the energy difference between the neutral species and its conjugate base at MP2/6-31+G//HF/6-31+G. They then corrected for the absence of polarization function by adding the difference in the DPE using the HF/6-31+G* and HF/6-31+G energies and corrected for the difference in zero-point vibrational energies using vibrational frequencies computed at AM1. Their estimated value for cyclopropane was 413.1 kcal mol^{-1}, within the error bars of the experimental value. Their estimate for the DPE of cubane was 406.3 kcal mol^{-1}, indicating that cubane was in fact much more acidic than cyclopropane. Eaton and Kass[27] subsequently measured the gas-phase DPE of cubane to be 404 ± 3 kcal mol^{-1}. This is in excellent agreement with the computational prediction. Eaton and Kass also obtained a value of 404.7 kcal mol^{-1} for the DPE of cubane at a slightly better computational level (MP2/6-31+G(d)//MP2/6-31+G(d)+ZPE). Rayne and Forest[28] have evaluated the DPE with some composite methods, obtaining values consistent with experiment and the previous computations, namely 406.0 (G4MP2) and 406.9 (G4) kcal mol^{-1}.

Kass and Prinzbach[29] examined the acidity of the next larger hydrocarbon analog of the platonic solids, dodecahedrane **1**. The MP2/6-31+G(d)//HF/6-31+G(d) computed DPE of **1** is 406.7 kcal mol^{-1}. This level of theory overestimates the DPEs of related hydrocarbons by about 1.6 kcal mol^{-1}, leading to a corrected estimate of 405.1 kcal mol^{-1} for the DPE of **1**, similar to that of cubane. (The DPE of **1** is 405.4 kcal mol^{-1} when computed with B3LYP/6-31+G(d).) In the FTMS experiments, **1** was deprotonated by amide to produce the dodecahedryl anion **1cb**, which then deprotonates ethylamine, dimethylamine, and water. This allows an experimental assignment of 402 ± 2 kcal mol^{-1} for the DPE of **1**, somewhat below the computational estimate. Kass and Prinzbach argued that the increased acidity of **1** relative to acyclic simple hydrocarbons (see Table 3.6) is that the anionic charge is spread over a number of carbon atoms in **1cb**. The Mulliken charge for the deprotonated carbon is only −0.38 e, significantly less than the charge at the deprotonated carbon of the cubyl anion (−0.86 e).

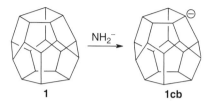

3,3-dimethylcyclopropene (**2**) is another highly strained species. Since the ^{13}C–H coupling constant, a measure of the s-character of the C–H bond, for **2** is similar to that of acetylene, their acidities and C–H bond strengths are likely to be similar. Kass computed the DPE of **2** to be 381.9 (B3LYP/6-31+G(d)) and 383.6 (G3) kcal mol^{-1}.[30] Gas-phase experiments using both bracketing techniques via flowing afterglow spectroscopy and kinetic exchange between **2** and methoxide in an FTMS give $\Delta H_{acid} = 382.7 \pm 1.3$ kcal mol^{-1}, indicating excellent agreement between experiment and computation. It is also very similar to the acidity of acetylene, $\Delta H_{acid} = 376.9$ kcal mol^{-1}, as expected.

Next, Kass determined the electron affinity of **2r** by electron transfer from **2cb** to a set of reference compounds. The experimental value of 37.6 ± 3.5 kcal mol^{-1} is in close agreement with the B3LYP/6-31+G(d) and G3 values of 35.1 and 37.4 kcal mol^{-1}, respectively. Using the thermodynamic cycle shown in Scheme 3.1, Kass was able to estimate the bond dissociation energy of **2** as 106.7 kcal mol^{-1}. This is again in excellent accord with the computed values: 103.3 (B3LYP) and 107.2 (G3) kcal mol^{-1}. This bond dissociation energy is much more similar to that of methane (104.9 kcal mol^{-1}) than acetylene (132.8 kcal mol^{-1}). It appears that the radical **2r** is more stable than expected.

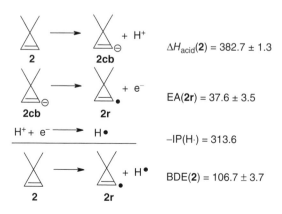

Scheme 3.1

2cb

Charges
1: −0.64
2: 0.09
3: −0.25

2r

Spins
1: 0.59
2: 0.07
3: 0.34

Scheme 3.2

Kass argued that **2r** is stabilized by an interaction between the radical orbital on C_1 and $\sigma(C_2-C_3)$ orbital. This allows the radical to delocalize, especially onto C_3. As listed in Scheme 3.2, the spin densities at C_1 and C_3 are quite similar, 0.59 and 0.34, respectively. A further manifestation of this delocalization is the short C_1-C_3 distance (1.493 Å) in **2r** compared to that in **2** (1.512 Å). This orbital interaction is negligible in the anion **2cb** because it is a filled–filled interaction. The charge density is much more localized onto C_1 for **2a** than is the spin density for **2r**. Consequently, **2r** is more stable than expected, leading **2** to have a smaller BDE than anticipated. **2** thus has the "acidity of an acetylene but the bond energy of methane."[30]

The last example is the acidity of the ethynyl-expanded cubane **3**. Inspired by the synthesis of the methoxy-substituted buta-1,3-diynediyl expanded cubane **4**,[31] Bachrach[32] examined some of the properties of its smaller homolog **3**. Since the s-character of the C−H bonds of **3** should be large and its conjugate base **3cb** can potentially delocalize the anion into three neighboring alkynyl units, **3** might be quite acidic.

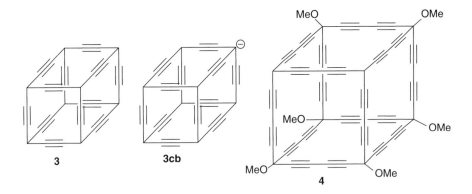

Merrill and Kass[20] demonstrated that MP2/6-31+G(d) DPEs are in excellent agreement with experimental values due to fortuitous cancellations of errors. However, computation of **3**, and (especially), **3cb** were prohibitively expensive with this method. B3LYP/6-31+G(d) is a much more tractable method; however, it

TABLE 3.7 Free Energy (kcal mol^{-1}) for Deprotonation

Compound	B3LYP/6-31+G(d)	MP2/6-31+G(d)	Expt
Cubane	399.8	397.6	396.5[a]
Propyne **5**	371.6	377.1	
2-Butyne **6**	377.8	382.5	381.5[b]
Pent-1,4-diyne **7**	346.8	353.7	
3-Ethynylpenta-1,4-diyne **8**	325.9	334.1	
3	316.1	(325.2)[c]	
		(327.6)[d]	

[a]Ref. 27.
[b]Ref. 34.
[c]Estimated using Eq. (3.1).
[d]Estimated using Eq. (3.2).[33]

underestimates the value of the DPE. Bachrach therefore computed the free energy for deprotonation of a number of compounds related to **3**, namely cubane and **5–8**, with both the MP2 and B3LYP methods (Table 3.7). These values correlate well, providing the relationship in Eq. (3.1) for estimating the MP2 DPE from the B3LYP value. The estimated free energy of deprotonation of **3** is extremely small—only 325.2 kcal mol^{-1}. In a follow-up study, Bachrach[33] developed a more refined method for estimating the MP2 value of the DPE: Eq. (3.2), where N_{3MR}, N_{4MR}, and N_{trip} are the number of three-member rings, four-member rings, and triple bonds, respectively. This estimate for the free energy of deprotonation of **3** with Eq. (3.2) is 327.6 kcal mol^{-1}. Compounds with similar acidities include methyl- and fluorobenzoic acids and lysine!

$$\Delta G(\text{MP2}) = 0.936 \Delta G(\text{B3LYP}) + 29.3 \tag{3.1}$$

$$\Delta G(\text{MP2}) = 0.897 \Delta G(\text{B3LYP}) - 2.031 N_{3MR} - 0.509 N_{4MR} + 0.062 N_{trip} + 42.28 \tag{3.2}$$

What makes **3** such a strong acid? The propargyl proton is acidic because the resulting anion can be delocalized into the π-bond of the adjacent triple bond, giving rise to resonance structures **9A** and **9B**. This latter resonance structure is likely to predominate since the anion resides in a formal sp^2 orbital rather than the formal sp^3 orbital of **9B**. The participation of **9B** requires that the $C_t\equiv C_t-R$ angle be bent, reflecting the allenic character of the anion. The structures of **5cb**, **7cb**, **8cb**, and **3cb** are drawn in Figure 3.2. The $C_t\equiv C_t-H$ angles in **5cb**, **7cb**, and **8cb** are 121.9°, 141.5°, and 153.9°, respectively. The angle is wider with the increasing number of triple bonds reflecting delocalization into each triple bond, resulting in less negative charge at each terminal carbon and less sp^2 character. This is also evident in the decreasing terminal C–C bond lengths in the series **5cb**, **7cb**, and **8cb**, suggesting lessened allenic character to *each* triple bond. The participation of the allenic resonance structure is seen in **3cb** by the strong kink along each edge that terminates at the anion carbon: the $C_t\equiv C_t-C$ angle is 136.6°.

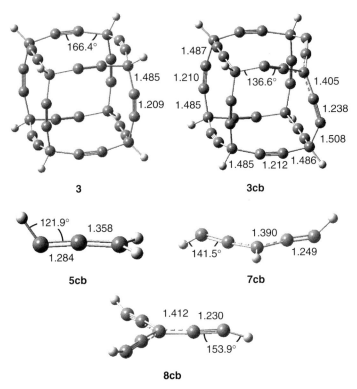

Figure 3.2 B3LYP/6-31+G* optimized geometries.[32]

The addition of more triple bond fragments that can conjugate with the anionic center allows for even greater stabilization of the anion, leading to the reduction in DPE in the series **5** (377.1 kcal mol^{-1}) to **7** (353.7 kcal mol^{-1}) to **8** (334.1 kcal mol^{-1}). Since the anion of **3cb** can also interact with three triple bonds, it is reasonable to expect that the DPE of **3** should be similar to that of **8**.

However, **3** is about 9 kcal mol^{-1} more acidic than **8**. The additional factor contributing to the enhanced acidity of **3** is strain relief upon deprotonation. The RSE (see Section 3.4) is 48 kcal mol^{-1}, but that of **3cb** is only 29.2 kcal mol^{-1}. This energy difference in strain energies perfectly accounts for the difference in DPEs of **3** and **8**.

Predicted to be even more acidic is the hydrocarbon **10**, the diethynyl-expanded cubane. The free energy of deprotonation of **10** is 309.4 kcal mol^{-1} using Eq. (3.2).[33] The DPE of **10** is so small due to the resonance stabilization of the conjugate base. The formal anion can be delocalized into the three adjacent diethynyl groups, giving rise to seven different resonance structures. This is an amazingly small DPE for a pure hydrocarbon. Organic molecules with comparable acidities include trinitrotoluene (TNT) and 2,3-dinitrophenol.

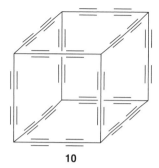

10

3.2.1.2 Origin of the Acidity of Carboxylic Acids

Why are carboxylic acids so acidic? This question makes sense only when the acidity of carboxylic acids is compared with some reference compound(s). The usual comparison is between carboxylic acids and alcohols, since an O–H bond is cleaved when both act as acids. The standard textbook explanation for the difference in acidity between alcohols and carboxylic acids resides in the stability of the resulting anions. So, for example, when methanol loses a proton, the negative charge in the resulting anion (methoxide) is localized largely on oxygen. However, when formic acid loses a proton, its conjugate base (**11**) can be written as two equivalent resonance structures, indicating delocalization of charge onto both oxygen atoms. This delocalization reduces the electron–electron repulsion. Therefore, **11** is more stable than methoxide, resulting in the enhanced acidity of formic acid over methanol.

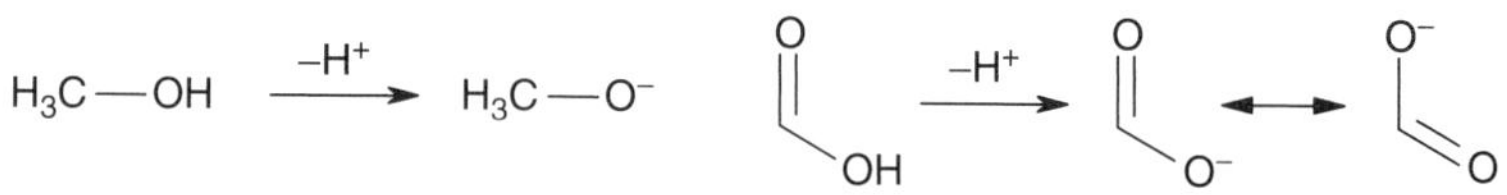

$$H_3C-OH \xrightarrow{-H^+} H_3C-O^- \qquad \underset{OH}{\overset{O}{\|}}\!\!\!\diagdown \xrightarrow{-H^+} \underset{O^-}{\overset{O}{\|}}\!\!\!\diagdown \longleftrightarrow \underset{O}{\overset{O^-}{\|}}\!\!\!\diagdown$$

11

In 1984, Siggel and Thomas[35] first presented the case that resonance plays a minor role in explaining the acidity of carboxylic acids. They separated the deprotonation process into two components. The first process is the loss of the proton from the acid. This energy can be computed as the negative of the electrostatic potential $(-V)$ at the site of the acidic proton within the neutral acid. The second step is the reorganization of the electrons and the nuclei to produce the ground state of the anion. The energy associated with this step, which is the relaxation energy R, Siggel and Thomas associated with the resonance stabilization. Siggel and Thomas reported the difference in the potential and relaxation energies for acetic acid and 2-propanol at HF/3-21+G; the values are $\Delta V = -39.7$ kcal mol^{-1}

and $\Delta R = 10.6$ kcal mol^{-1}. They conclude that the enhanced acidity is largely due to not to resonance but rather due to the less negative potential at the hydrogen in the acid than in the alcohol. The carbonyl group inductively polarized the O–H bond, making the hydrogen more positively charged than the hydrogen of an alcohol. Schleyer[36] challenged this argument on the basis that the relaxation energy R does correspond to the concept of resonance energy typically used by organic chemists. He rightfully pointed out that the reference used by Siggel and Thomas, the hypothetical anion formed by the loss of a proton with no relaxation of the electron density nor nuclear geometry, is not the appropriate reference. The typical model for resonance energy is instead the energy difference between the anion and a hypothetical anion, where all of the electron pairs are localized within *one* Lewis structure.

This, however, was not the end of the challenge to the orthodox model for the acidity of carboxylic acids. Siggel et al.[37] examined the changes in charge distribution upon deprotonation of formic acid and methanol (Table 3.8). While the absolute magnitude of the charges obtained by the topological method and natural population analysis (NPA) differ dramatically, the changes in atomic charge upon deprotonation predicted by the two methods are similar. The oxygen of the hydroxy group gains about 0.2 e after deprotonation of either formic acid or ethanol. The resonance model suggests that this oxygen in formic acid should gain fewer electrons than that of ethanol, since charge should be delocalized to the other oxygen. However, the carbonyl oxygen is quite negatively charged and so it can accept only a little additional charge upon deprotonation of formic acid. Since extensive charge redistribution is not occurring upon deprotonation of formic acid, Siggel et al. argue that the enhanced acidity is due to inductive polarization in formic acid. The carbonyl group withdraws charge from the proton, reducing the potential at hydrogen, making it easier to remove as a bare proton.

While a number of authors[38–41] disputed these results and conclusions, two papers published in the early 2000s have added considerable support to the notion of the importance of the inductive effect in explaining the acidity of carboxylic acids.

Rablen[42] employed carefully constructed isodesmic reactions to assess the relative contribution of resonance (ΔE_{res}) and electrostatics (ΔE_{el}) in enhancing the acidity of carboxylic acids. In order to assess the enhanced acidity of acetic acid,

TABLE 3.8 Atomic Charges^a of Formic Acid, Ethanol, and Their Conjugate Bases

Compound	Topological			NPA		
	C=O	C=O	O–H	C=O	C=O	O–H
HCO$_2$H	−1.39	+1.97	−1.30	−0.69	+0.81	−0.79
HCO$_2^-$	−1.50	+2.11	−1.50	−0.90	+0.81	−0.90
		C$_1$	O		C$_1$	O
CH$_3$CH$_2$OH		+0.67	−1.25		−0.04	−0.82
CH$_3$CH$_2$O$^-$		+1.06	−1.49		+0.05	−1.05

^{a}Computed at HF/6-31+G*, see Ref. 37.

Rablen chose *tert*-butanol as the reference system, arguing that C–C bonds are less polar than C–H bonds and thereby eliminates polarization effects that may be inherent in smaller alcohol reference molecules, like methanol or ethanol. The proton exchange shown in Reaction 3.3 defines the relative acidities of acetic acid versus *tert*-butanol. Using the CBS-Q method, acetic acid is 27.9 kcal mol^{-1} more acidic than *tert*-butanol.

$$\Delta E = -27.9 \text{ kcal mol}^{-1}$$

(Reaction 3.3)

Rablen next partitioned this enhanced acidity into a resonance and electrostatic contribution, as shown in Eq. (3.3). The factor of 2 for the electrostatic component reflects the contribution of two additional C–O bonds (one σ and one π) in acetic acid relative to *tert*-butanol.

$$\Delta E_{\text{acid enhancement}} = \Delta E_{\text{res}} + 2\Delta E_{\text{el}} \qquad (3.3)$$

With two unknowns in Eq. (3.3), another equation is needed in order to solve for the values of these contributions. Rablen suggested that acetic acid could act as a base and experience the same type of resonance stabilization as afforded by an acetate ion. Reaction 3.4 presents the relative *basicity* of acetic acid versus acetone. At CBS-Q, acetic acid is 6.2 kcal mol^{-1} less basic than acetone. This energy difference can again be partitioned into an electrostatic and resonance contribution, but in this case resonance enhances the basicity, while electrostatic interactions diminish it by destabilizing the resulting conjugate acid (Eq. (3.4)). Solving for the resonance and electrostatic contributions using Eqs (3.3) and (3.4) yields $\Delta E_{\text{el}} = -11.4$ kcal mol^{-1} per bond and $\Delta E_{\text{res}} = 5.2$ kcal mol^{-1}. Therefore, resonance accounts for only 20 percent of the enhanced acidity of acetic acid, the remainder coming from electrostatic effects.

(Reaction 3.4)

$$\Delta E_{\text{base enhancement}} = \Delta E_{\text{res}} - \Delta E_{\text{el}} \qquad (3.4)$$

An alternative approach toward quantifying the contributions of resonance and electrostatic/inductive effects was offered by Holt and Karty.[43] They assessed the resonance contribution toward the enhanced acidity of carboxylic acids by comparing the DPEs of a series of conjugated hydroxyaldehydes in their planar (**12**) versus perpendicular (**13**) conformations. This perpendicular reference should have the same inductive effect as the planar conformation but should lack the resonance effect. The DPEs for these two sets of compounds with $n = 1-3$ are listed in

TABLE 3.9 B3LYP/6-31+G* DPEs (kcal mol^{-1}) of 12–14[a]

n	12	13	14
1	321.5	334.1	351.9
2	314.8	327.1	339.4
3	310.9	322.0	331.3

[a]Ref. 43.

Table 3.9. A plot of the difference in these DPEs with n is linear, and extrapolating back to $n = 0$ gives a value of 13.5 kcal mol^{-1}. This is an estimate of the resonance energy contribution to the acidity enhancement of formic acid.

To assess the inductive effect, Holt and Karty employed the reference series **14**. The DPEs of these compounds are listed in Table 3.9. The terminal methyl group should make no inductive contribution. The resonance energy of the molecules in the series **14** should be equal to their analog in the **13** series. Therefore, the difference in DPEs between **13** and **14** should reflect solely the inductive effect of the carbonyl group. This difference decreases with increasing n, reflecting the diminishing effect of induction with increasing distance between the groups. Using a exponential function of n to fit these DPE differences, Holt and Karty extrapolate a value of 24.2 kcal mol^{-1} when $n = 0$. This provides the estimate for the inductive/electrostatic contribution to the acidity enhancement of formic acid. Combined with their estimate for the resonance energy above, this estimates that the DPE of formic acid is 37.7 kcal mol^{-1} less than that of methanol. The B3LYP/6-31+G* computed difference for these two compounds is actually 39.0 kcal mol^{-1}, showing good internal consistency. With the inductive contribution estimated to be almost twice the size of the resonance contribution, Holt and Karty's results are in complete agreement with Rablen's results. These results, coupled with the charge distribution results of Siggel et al. provide strong evidence that the classic textbook explanation that resonance stabilization of the resulting anion accounts for why carboxylic acids are acidic is misleading, if not erroneous. The inductive effect is the dominant player.

3.2.1.3 Acidity of the Amino Acids

The critical role that amino acids play in biochemistry demands special attention from physical organic chemists. Of particular interest is the acidity of these compounds. Perhaps surprising is that a systematic examination of the gas phase acidities of most of the amino acids was not reported until 1992.[44] Poutsma and Kass[45] combined experimental and computational studies of all 20 common naturally occurring amino acids. The

TABLE 3.10 Deprotonation Energies (kJ mol⁻¹) of the Amino Acids[a]

	Expt	Comp
GLY	1434 ± 9	1434
PRO	1431 ± 9	1430
VAL	1431 ± 8	1430
ALA	1430 ± 8	1432
ILE	1423 ± 8	1426
TRP	1421 ± 9	1422
LEU	1419 ± 10	1428
PHE	1418 ± 18	1417
LYS	1416 ± 7	1415
TYR	1413 ± 11	1419
MET	1407 ± 9	1412
CYS	1395 ± 9	1396
SER	1391 ± 22	1392
THR	1388 ± 10	1397
ASN	1385 ± 9	1384
GLN	1385 ± 11	1378
ARG	1381 ± 9	1387
HIS	1375 ± 8	1374
GLU	1348 ± 2	1349
ASP	1345 ± 14	1345

[a]Ref. 45.

experiments were performed using a quadrupole ion trap with an electrospray ionization source and determining the kinetic acidity. The computations were performed at B3LYP/6-311++G**//B3LYP/6-31+G*, following an MM search to identify low lying conformations. The computed acidities were obtained relative to acetic acid, that is, $R-CH_2COOH + OAc^- \rightarrow R-CH_2COO^- + HOAc$, to remove some systematic errors in the prediction of acidity values. Table 3.10 presents the experimental and computational deprotonation enthalpies of the amino acids.

The agreement between the experimental and computational deprotonation enthalpies is quite good. The correlation coefficient of a plot of the two sets of values is 0.9799, but all the computed values lie within the error bars of each experiment. While an ordered list by DPE of the 20 amino acids are not quite the same using the experimental and computed values, there are really very small disagreements.

Embedded within this list of DPE of the amino acids are two unexpected results concerning what is the most acidic proton. The first case is cysteine, where one might ask which is more acidic, the carboxylate proton or the thiol proton? Kass and Poutsma[46] note that, in general, the gas-phase acidity of carboxylic acids is greater than thiols; the DPE of propanoic acid ($CH_3CH_2CO_2H$) is 347.7 kcal mol⁻¹ at G3B3 (347.2 expt [15]), about 6 kcal mol⁻¹ less than that of ethanethiol (CH_2CH_2SH: 355.0 at G3B3 and 354.2 expt [15]). They optimized the structure of cysteine and its conjugate base at B3LYP/aug-cc-pVDZ. We will discuss cysteine conformers

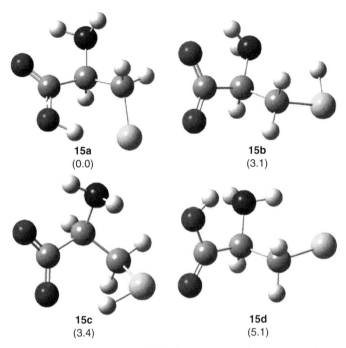

Figure 3.3 B3LYP/aug-cc-pVDZ optimized structures of the conjugate base of cysteine **15**. Relative energies (kcal mol^{-1}) computed at G3B3.

in Section 3.3.2. The four lowest energy conformers of the conjugate base **15** are shown in Figure 3.3, along relative energies computed at G3B3.

The lowest energy structure of deprotonated cysteine **15a** comes about by loss of the thiol proton! The lowest energy cysteine conjugate base from loss of the carboxylate proton, **15b**, is 3.1 kcal mol^{-1} higher in energy. Apparently, the hydrogen bonding network in **15a** is quite favorable, especially important is the hydrogen bond from the hydroxyl proton to the sulfur anion, which is able to make up for the inherent favorability of a carboxylate over a thiolate anion. The G3B3-computed DPE of cysteine is 333.3 kcal mol^{-1} (for removal of the thiol proton). Kass and Poutsma[46] determined the DPE of cysteine using both kinetic and thermodynamic methods. The kinetic method gives a value of 332.9 ± 3.3 kcal mol^{-1}, while the thermodynamic method gives 334.4 ± 3.3 kcal mol^{-1}. These are in fine agreement with the computed value.

Another unconventional deprotonation is seen with tyrosine. Gas-phase experiments by Kass indicate that deprotonation of tyrosine leads to a 70 : 30 mixture of the phenoxide **16** to carboxylate **17** anions.[47] Kass optimized the structures of tyrosine and its two possible conjugate bases **16** and **17** at B3LYP/aug-cc-pVDZ, and computed the two DPEs at G3B3. **17** is predicted to be 0.2 kcal mol^{-1} lower in energy than **16** at B3LYP and slightly more stable at G3B3 (0.5 kcal mol^{-1}). However, both computational methods fail to predict the DPEs of acetic acid and phenol in unequal ways. For example, the DPE of acetic acid is predicted to have a DPE of 0.1 kcal mol^{-1} less than experiment, but the computed value of the DPE of phenol is 1.2 kcal mol^{-1} higher than the experiment. When the DPEs are corrected for this error, the phenolic proton is predicted to be 0.4 kcal mol^{-1} more acidic than the carboxylate proton at B3LYP and 0.9 kcal mol^{-1} more acidic at G3B3. These computations are in accord with the experimental observation of a surplus of **16** over **17**. Both the cysteine and tyrosine acidities point out the importance of solvation; in solution, the carboxylate proton is lost exclusively for each compound.

3.3 ISOMERISM AND PROBLEMS WITH DFT

3.3.1 Conformational Isomerism

Conformational isomerism of alkanes is fundamental toward understanding organic structures. As such, it has been subject to many computational studies. The simplest alkane expressing conformational isomers is butane, which has two stable conformers, trans and gauche. Table 3.11 presents some of the more recent computational results regarding the relative energies of these conformers and rotational transition states that interconvert them. The composite methods (focal point and W1h-val) predict an energy difference between the trans and gauche conformers in very close agreement with experiment. The agreement between the experimental and computed rotational barriers between the trans and gauche conformers is excellent, but the computations predict a much rotational higher barrier through the cis conformer than does experiment. Given the performance of the focal point method, some doubt must be cast upon the experimental value of this barrier.

Martin examined the conformational energies of butane, pentane, and hexane with a variety of methods.[49] Table 3.12 compares the root mean square deviation of the conformational energies of hexane relative to those at W1h-val, a high end composite method.[53,54] The traditionally used functionals (B3LYP, PBE, etc.) overestimate conformer energies, while the M06 family underestimates the interaction energies that occur in GG-type conformers. The "-D" dispersion correction tends to overcorrect and leads to wrong energy ordering of conformers. However, the double hybrid functionals (B2-PLYP and B2K-PLYP) with the dispersion correction provide quite nice agreement with the benchmarks. Somewhat worrisome is that all of the density functionals examined have issues in geometry prediction, particularly in the backbone dihedral angles. So, for example, B3LYP misses the τ_1 dihedral angle in the GG pentane conformer by 5° and M06-2x misses the τ_2 angle in the pentane TG conformer by 2.4°.

TABLE 3.11 Relative Energies (kcal mol^{-1}) of the Conformers of Butane

Method	trans	gauche	t–g	cis
HF/6-311G**[a]	0.0	0.95	3.65	6.31
MP2/6-311G**[a]	0.0	0.51	3.40	5.92
MP2/cc-pVTZ[b]	0.0	0.561		
B3LYP/pc-2[b]	0.0	0.898		
M06-2x/pc-2[b]	0.0	0.522		
B2-PLYP/pc-2[b]	0.0	0.986		
CCSD(T)/cc-pVTZ//MP2/6-311G(2df,p)[a]	0.0	0.59	3.31	5.48
CCSD(T)/cc-pvQZ//MP2/cc-pVTZ[b]	0.0	0.606		
Focal point[c]	0.0	0.621	3.307	5.497
W1h-val[b]	0.0	0.598		
Expt[d]	0.0	0.67	3.62	3.95
Expt[e]	0.0	0.660		

[a]Ref. 48.
[b]Ref. 49.
[c]Ref. 50.
[d]Ref. 51.
[e]Ref. 52.

TABLE 3.12 Root Mean Square Deviation of Hexane Conformational Energies[a]

	B3LYP	M06-2x	PBE0	B2K-PLYP	B2-PLYP
No dispersion	0.90	0.31	0.69	0.38	0.61
With dispersion	0.90	0.31	0.52	0.13	0.34

[a]Ref. 49.

It should be noted that intramolecular BSSE is conformer dependent. Gauche conformers are stabilized relative to trans conformers by BSSE; atoms and their atomic orbitals are nearer to each other in the gauche conformation and can thereby better utilize neighboring basis functions in a superposition fashion than can be done in the trans conformation. This can be seen in the conformational study of the small alkanes by Balabin,[55] who corrected for intramolecular BBSE using the approach of Asturiol et al.[56] (see Section 1.1.8.1). Listed in Table 3.13 are the energy differences between the gauche and trans conformers of butane and the all-gauche and all-trans conformers of hexane with and without correction for intramolecular BSSE. Even with the large cc-pVTZ basis set, there is a noticeable BSSE correction of 0.044 kcal mol^{-1} with both MP2 and CCSD(T). For hexane, the energy gap is increased even more by inclusion of BSSE; the energy gap increases by about 0.4 kcal mol^{-1} (nearly 50 percent) with inclusion of BSSE. In fact, intramolecular BSSE increases linearly with the size of the molecule, such that the BSSE correction can swamp out the inherent stability of the all-trans conformer entirely when the chain contains 10 or more carbon atoms.

TABLE 3.13 Energy Difference (kcal mol⁻¹) Between the G and T Conformers of Butane and the GGG and TTT Conformers of Hexane[a]

Method	6-311G(d,p)	6-311++G(d,p)	cc-pVDZ	cc-pVTZ
		Butane		
MP2	0.519	0.546	0.649	0.571
	0.628	*0.569*	*0.829*	*0.614*
	0.109	**0.024**	**0.111**	**0.044**
CCSD(T)	0.540	0.575	0.686	0.657
	0.643	*0.593*	*0.766*	*0.701*
	0.103	**0.018**	**0.080**	**0.044**
		Hexane		
MP2	0.719	0.696		
	1.400	*1.098*		
	0.681	**0.402**		
CCSD(T)	0.958	0.990		
	1.620	*1.375*		
	0.662	**0.385**		

[a] The conformer energy difference is given in regular text, the energy difference including intramolecular BSSE is given in italics, and the net effect of intramolecular BSSE (the difference between the above two values) is given in bold.

3.3.2 Conformations of Amino Acids

Alanine **18** is the smallest chiral common amino acid. It has been the subject of many computational studies but we focus here on three recent reports. Császár[57] identified 13 conformations of alanine and optimized the structures at HF/6-31G**, HF/6-311++G**, B3LYP/6-311++G**, and MP2/6-311++G**. The relative energies for some of the low lying conformations, computed with the latter three methods, are reported in Table 3.14. The two lowest energy conformations correspond with the two conformations identified in an earlier microwave experiment.[58]

The Alonso group has developed a laser ablation molecular beam Fourier transform microwave spectroscopy experiment that allows for the study of large biomolecules that previously were difficult, if not impossible, to study in the gas phase. The instrument produces high resolution spectra that allow for resolving quadrupole splitting, thereby providing additional means for identifying the structure. These studies invariably involve comparison of the microwave data with computational results, where the symbiotic relationship of experiment and computation allows for clear and direct structural and conformational identification. In their study of alanine, the high resolution experiment allowed for definitive identification of the two low energy conformations, by comparison of the experimental and computed rotational constants and the ^{14}N quadrupole coupling constants. The MP4 energies of the low energy conformer are listed in Table 3.14.

TABLE 3.14 Relative Energies of Alanine Conformers

	MP4[a]	Focal Point[b]	HF[c]	B3LYP[c]	MP2[c]
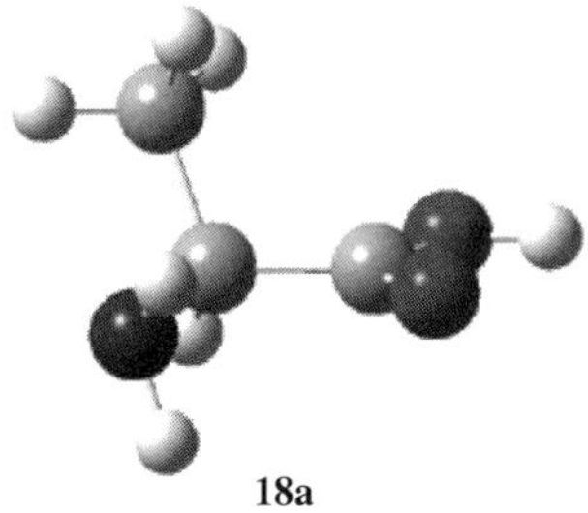 18a	0.0 (0.0)	0.0	0.0	0.0	0.0
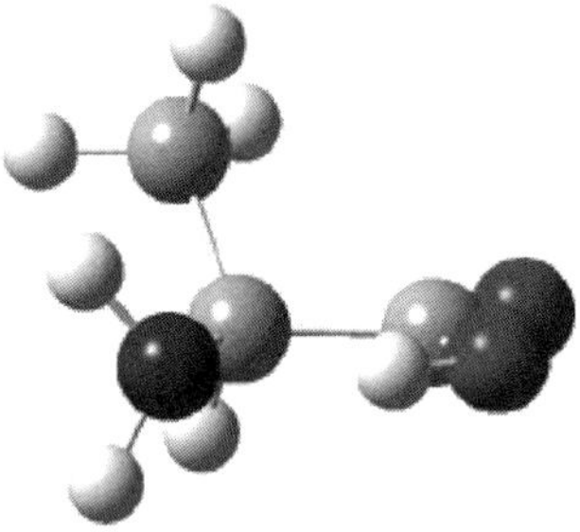 18b	0.70 (1.13)	0.50	2.42	0.03	0.15
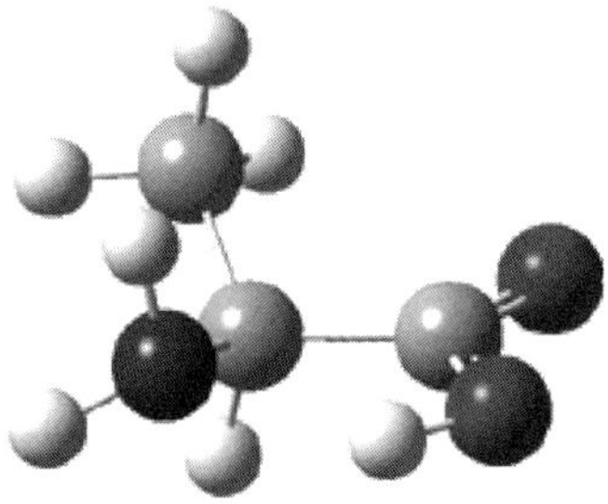 18c	0.85 (1.08)		2.48	0.01	0.47
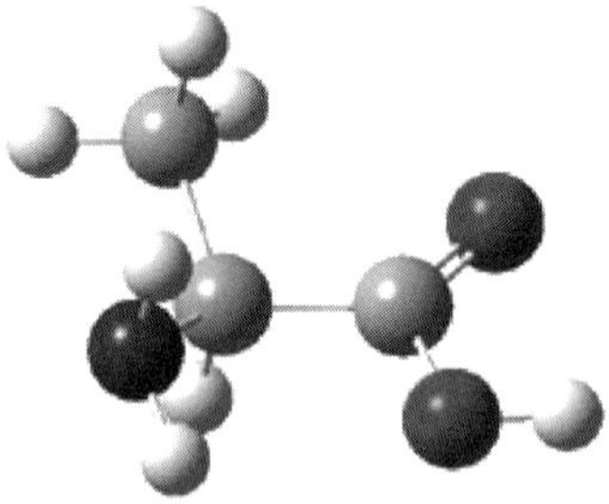 18d	1.03 (1.07)		1.20	1.07	0.96

[a] MP4/6-311++G(d,p)//MP2/6-311++G(d,p), Ref. 60, relative free energy within parentheses.
[b] Ref. 59.
[c] With the 6-311++G** basis set, Ref. 57.

Császár and Allen[59] applied the focal point method to the structure and energies of alanine conformers. The HF/CBS estimate places **18b** below **18a**, but this is reversed at MP2. With the correction for CCSD and CCSD(T), and core electron energies, the energy gap is only 0.10 kcal mol^{-1}, favoring **18a**. Zero-point vibrational energy (ZPVE) favors **18a** by 0.40 kcal mol^{-1}, for a prediction that **18a** is 0.05 kcal mol^{-1} lower in energy than **18b**. It is interesting that most of this energy difference arises from differences in their ZPVE. The authors also discuss the structures of these two conformers, obtained through a combination of theoretical treatment and revisiting the experimental measurements.

Cysteine **19** possesses a more complex conformational space, thanks to the additional thiol group, along with the possibility of hydrogen bond-like behavior involving the thiol. In particular, the thiol proton can potentially interact with either oxygen's or nitrogen's lone pair. Gronert[61] identified 42 different conformers of cysteine at the MP2/6-31+G* level. Dobrowolski et al. located 51 conformers at B3LYP/aug-cc-pVDZ.[62]

In 2008, Alonso and coworkers[63] reported a combined experimental and computational study of the conformations of cysteine. They located 11 low lying conformers at MP4/6-311++G(d,p)//MP2/6-311++G(d,p), and the five lowest energy structures are shown in Table 3.15. Using laser ablation molecular beam Fourier transform microwave spectroscopy, they identified six conformers present in the gas phase. Comparing the computed rotational constants and ^{14}N nuclear quadrupole coupling tensor components with the experiment, they were able to decidedly match up all six experimental conformers with computed structures. Of the five low energy conformers listed in Table 3.15, four of them were identified in the experiment.

Császár and Allen[64] applied the focal point method to the question of the cysteine conformers. They performed a broad conformation search, initially examining over 11,000 structures. These reduced to 71 unique conformations identified at MP2/cc-pvTZ. The lowest 11 energy structures were further optimized at MP2(FC)/aug-cc-pV(T+d)Z. The relative energies, computed with the focal point method for the five conformers found by Alonso, are listed in Table 3.15. The agreement between the focal point method energies and those obtained at MP4 by Alonso is quite good; both methods identify the same two lowest energy conformers. Further, the geometries of the conformers identified by Alonso are very similar to the corresponding geometries found at this higher computational level, and the rotational constants and ^{14}N nuclear quadrupole coupling tensor components agree as well.

3.3.3 Alkane Isomerism and DFT Errors

Alkanes are so fundamental to organic chemistry and comprise such a simple class of compounds that it is absolutely essential that any computational method worth employing must treat alkanes accurately. To the surprise of many practicing organic chemists, the B3LYP functional, which had by the turn of the twenty-first century become the *de facto* method for computing organic systems, turns out to fail in describing some seemingly simple properties of alkanes. This failure is not just

TABLE 3.15 Relative Energies of Cysteine Conformers

Conformer	MP4[a]	Focal Point[b]	B3LYP[c]
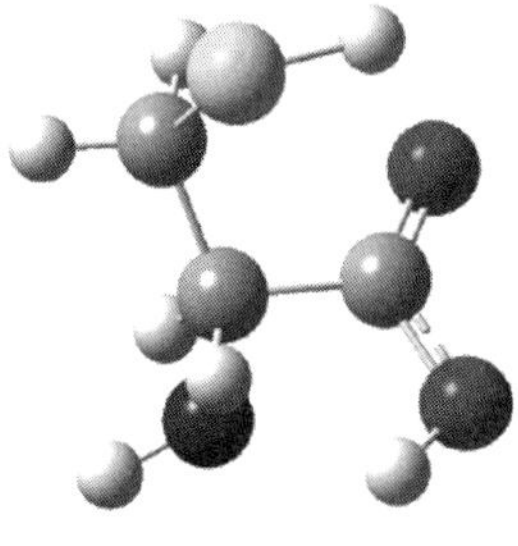 19a	0.0	0.0	0.0
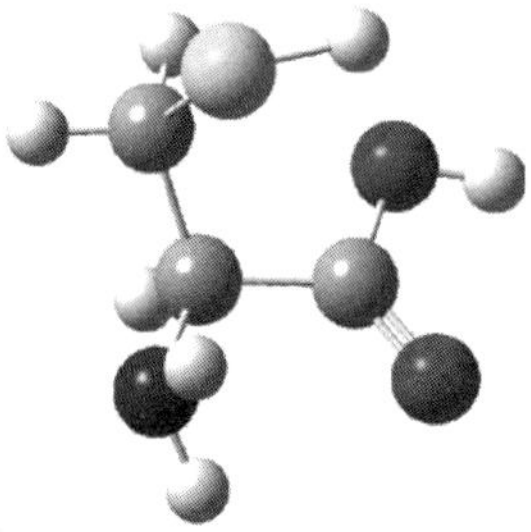 19b	0.93	1.14	0.59
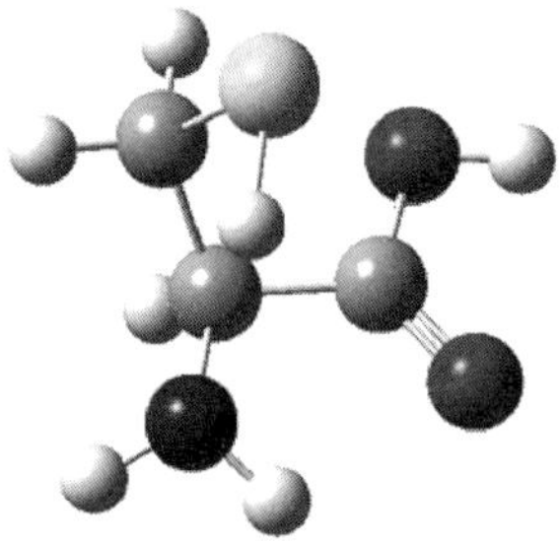 19c	1.21	1.42	
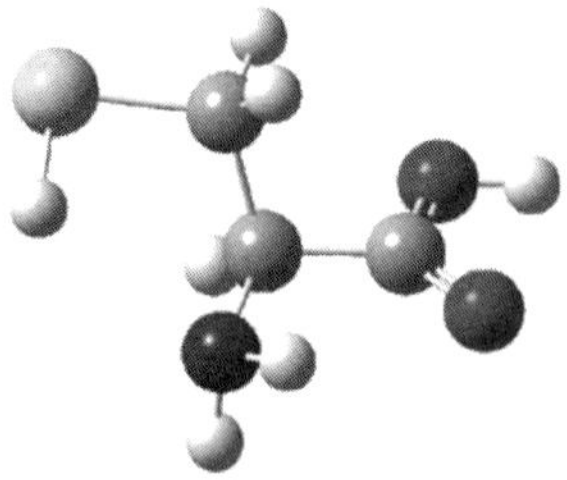 19d	1.29	1.39	0.24

(*continued*)

TABLE 3.15 (*Continued*)

Conformer	MP4[a]	Focal Point[b]	B3LYP[c]
19e	1.51	2.06	0.83

[a] Relative energies at MP4/6-311++G(d,p)//MP2/6-311++G(d,p), Ref. 63.
[b] Focal point, Ref. 64.
[c] Relative free energies at B3LYP/aug-cc-pVTZ, Ref. 62.

of the B3LYP functional, but many widely used functionals are also plagued with these same faults.

Hints of these failures began with the increasing error in predicted bond dissociation energy with the size of the compound. This work by Curtiss et al.[8] and Check and Gilbert[7] was discussed in Section 3.1 and pointed toward some systemic problem of B3LYP.

In 2006, three different groups pointed out significant problems with B3LYP and many other density functionals in computing the relative energies of structural isomers of alkanes. Grimme[65] reported on the relative energies of two C_8H_{18} isomers, 2,2,3,3-tetramethylbutane, and *n*-octane. The relative energy computed with a variety of different methods is listed in Table 3.16. The branched isomer is more stable than the linear isomer, yet all of the DFT methods Grimme examined, including the double hybrid B2PLYP functional, predict the opposite relative energy!

Schreiner also compared the energy of alkane isomers.[68] The three lowest energy isomers of $C_{12}H_{12}$ that he computed are **20–22**. These three relatively ordinary hydrocarbons are predicted to have very different relative energies depending on the computational method (Table 3.17). CCSD(T) predicts that **21** is about 15 kcal mol^{-1} less stable than **1** and that **3** lies another 10 kcal mol^{-1} higher in energy. MP2 exaggerates the separation by a few kcal mol^{-1}. HF predicts that **20** and **21** are degenerate. The large HF component within B3LYP leads to this functional's poor performance.

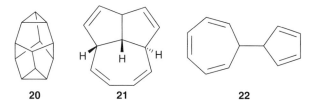

TABLE 3.16 Energy (kcal mol^{-1}) of 2,2,3,3-Tetramethylbutane Relative to Octane

Method	ΔE
Expt[a]	1.9 ± 0.5
MP2[h,c]	4.6
SCS-MP2[h,c]	1.4
B2PLYP[h,c]	−3.5
PBE[h,c]	−5.5
B3LYP[h,c]	−8.4
BLYP[h,c]	−9.9
HF[h,c]	−11.5
M05-2x[b,d]	2.0
M05-2x[d,e]	1.4

[a]Ref. 66.
[b]Using the cQZV3P basis set and MP2/TZV(d,p) optimized geometries.
[c]Ref. 65.
[d]Computed at M05-2x/6-311+G(2df,2p).
[e]Ref. 67.

TABLE 3.17 Energies (kcal mol^{-1}) of 21 and 22 Relative to 20

Method	21	22
CCSD(T)/cc-pVDZ//MP2(fc)/aug-cc-pVDZ[a]	14.3	25.0
CCSD(T)/cc-pVDZ//B3LYP/6-31+G(d)[a]	14.9	25.0
MP2(fc)/aug-cc-pVDZ[a]	21.6	29.1
HF/6-311+(d)[a]	0.1	6.1
B3LYP/6-31G(d)[a]	4.5	7.2
B3LYP/aug-cc-pvDZ[a]	0.4	3.1
B3PW91/aug-cc-pVDZ[a]	16.8	23.5
KMLYP/6-311+G(d,p)[a]	28.4	41.7
M05-2X/6-311+G(d,p)[b]	16.9	25.4
M05-2X/6-311+G(2df,2p)[b]	14.0	21.4

[a]Ref. 68.
[b]Ref. 67.

The last of the three papers, by the Schleyer group,[69] was a study of the bond separation reaction for linear alkanes (Reaction 3.5):

$$CH_3(CH_2)_m CH_3 + mCH_4 \rightarrow (m+1)\, CH_3CH_3 \quad \text{(Reaction 3.5)}$$

The energies for the alkanes propane through decane are plotted in Figure 3.4. As expected, G3 closely mimics the experimental values, but all of the DFT methods examined, including B3LYP perform poorly, and the agreement with experiment gets worse as the molecules get larger.

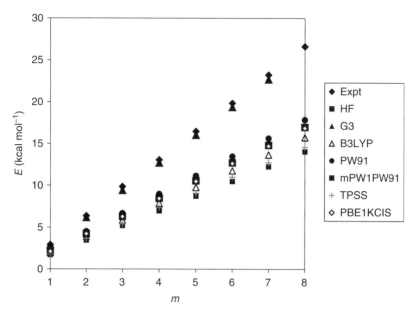

Figure 3.4 Energy (kcal mol^{-1}) of Reaction 3.5 for alkanes $CH_3(CH_2)_mCH_3$ obtained with different computational methods. All computations (except G3) were performed with the 6-311+G(d,p) basis set.

Schreiner and Schleyer[70] expanded on these last two papers by computing the bond separation energy for 72 hydrocarbons whose experimental heats of formation are known. These compounds include alkenes, alkynes, rings, and cage compounds and aromatics. The MAD values of the computed bond separation energies using a variety of density functionals are listed in Table 3.18. Of note is the large MAD value for B3LYP and the MAD of many other functionals are of 4–5 kcal mol^{-1}.

TABLE 3.18 Mean Absolute Deviation (kcal mol^{-1}) of the Computed Bond Separation Energy Compared with Experiment[a]

Method	MAD
B1B95	4.93
PBE	4.14
MPW1B95	4.09
B3LYP	3.99
PW91	3.88
B3PW91	3.64
MPWB1K	3.34
MPW1PW91	3.23
M05-2X	2.13

[a]Ref. 70. All computations were performed using the 6-311+G(d,p) basis set.

TABLE 3.19 Mean Absolute Deviation (kcal mol^{-1}) of 10 Simple Cycloaddition Reactions[a]

Method	MAD
B3LYP[b]	11.7
MPW1K[b]	5.1
MP2[c]	3.3
mPW1PW91[b]	2.0
M05-2x[d]	1.5
M06-2x[d]	1.3
SCS-MP2[c]	1.2

[a]Ref. 71.
[b]Calculated with the 6-311+G(2df,2p) basis set with the 6-31+G(d,p) geometry.
[c]Calculated with the cc-pVTZ basis set with the B3LYP/6-31+G(d,p) geometry.
[d]Calculated with the 6-311+G(2df,2p) basis set with the B3LYP/6-31+G(d,p) geometry.

Houk[71] noted that many density functionals fail in predicting the reaction energy of simple cycloadditions, such as Reactions 3.6 and 3.7. The MAD values between the reaction energy at CSBS-QB3 and a number of functionals are listed in Table 3.19. The error, which is astonishingly 12 kcal mol^{-1} with B3LYP, is ascribed to a "startling overestimation of the energy of conversion of π bonds into σ bonds."

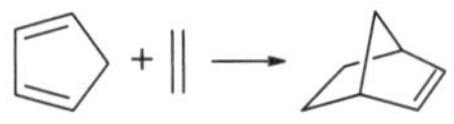

(Reaction 3.6)

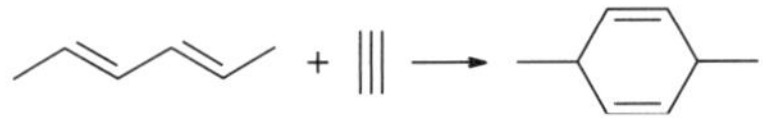

(Reaction 3.7)

There are two serious questions to ask at this point: (1) why do B3LYP and other traditional density functionals fail to describe relative energies of hydrocarbon isomers?; and (2) are there any functionals that reduce this error and might prove to be versatile?

Grimme[65] suggested in his article that the failure of B3LYP is due to inadequate treatment of medium range electron correlation. This was disputed by Brittain and coworkers[72] and by Song and coworkers.[73]

Brittain and coworkers[72] determined the energies of the methyl exchange reactions (Reactions 3.8a–d) with R = methyl, ethyl, *iso*-propyl, and *tert*-butyl. Energies determined with a number of density functionals, MP2 and G3(MP2)-RAD were compared with values at CCSD(T)/cc-pVTZ. The two wavefunction methods perform quite well, but all of the functionals (B3LYP, B3P86, KMLYP, MPW91B95, PBE, TPSS, etc.) show large errors, especially with the *t*-butyl cases. Computing the energies using the HF densities, that is, the kinetic, potential, and Coulomb energies are the exact HF values, leads to no improvement, so that only the DFT exchange and correlation terms can be the culprit. Subsequent removal of the correlation energy from the PBE functional (and also for BLYP and PW91), implicates exchange as the ultimate culprit.

$$R-Me + Me-H \rightarrow R-H + Me-Me \quad \text{(Reaction 3.8a)}$$

$$R-Me + \cdot Me \rightarrow R\cdot + Me-Me \quad \text{(Reaction 3.8b)}$$

$$R-Me + Me^- \rightarrow R^- + Me-Me \quad \text{(Reaction 3.8c)}$$

$$R-Me + Me^+ \rightarrow R^+ + Me-Me \quad \text{(Reaction 3.8d)}$$

Song and coworkers[73] evaluated the role of long range corrections to the energies of Reaction 3.5. They applied a long range correction, as defined in Eq. (1.53), and also a local response dispersion (LRD) method.[74] The MAD from experimental values for a number of methods are listed in Table 3.20. The significant improvement in the reaction energies with the incorporation of LC, and particularly with both LC and LRD corrections, has led these authors to conclude that the lack of long range exchange is the main culprit for the generic failure of so many density functionals.

Grimme, however, maintains that medium range correlation is the problem.[75] Krieg and Grimme[76] carefully detailed the problems in using experimental data as the reference in evaluating bond separation reactions like Reaction 3.5. First, bond separation reactions require many equivalents of methane and ethane, often over 20 methane molecules are employed. Any error in the experimental heats of formation of these compounds will be additive, and recent experimental errors are ± 0.07 and ± 0.10 kcal mol^{-1} for methane and ethane. Second, large alkanes will exist in a number of conformations, but most computations (including the studies of Song[73] and Brittain[72]) employ only the lowest energy conformer, typically the extended all s-trans conformer. Neglect of the conformational enthalpy was estimated for hexane to be 0.68 kcal mol^{-1}, and this would only be larger for longer hydrocarbons. Since methane and ethane have no conformers, this error is not canceled in Reaction 3.5. Third, ZPVE and rotational energy associated with conformational isomers is again not treated and does not cancel. Lastly, anharmonicity is often not accounted for

TABLE 3.20 mmm Mean Absolute Deviation (MAD, kcal mol^{-1}) for Energies of Reaction 3.5^a

Method	MAD
LC-PBE+LRD	0.17
ωB97XD	2.00
LC-PBE	2.24
ωB97X	2.91
M06-2x	3.34
B3LYP	5.97

^{a}Ref. 73. Deviation from experimental values for propane through decane.

and it too will not cancel in Reaction 3.5. Krieg and Grimme argue that comparison to high level computations is much less problematic and much easier to put into practice.

Deciding to compare the bond separation energies of Reaction 3.1 to CCSD(T)/CBS energies, Grimme[75] systematically look at a series of methods. HF systematically underestimates the energies, and since HF includes exact treatment of exchange and no treatment of Coulomb correlation, the error is clearly with the latter. MP2 overcorrects; in other words it "over-correlates" but the SCS-MP2 method adjusts for this and its agreement with the CCSD(T)/CBS energies is excellent (in error by less than 0.5 kcal mol^{-1}). The energies of Reaction 3.5 were then evaluated with four density functionals that differ in their degree of inclusion of Fock exchange: PBE and BLYP include no exchange, PBE0 (which has 25 percent Fock exchange) and BHLYP (which has 50 percent Fock exchange). The two pairs, PBE with PBE0 and BLYP with BHLYP, each gives similar poor energies, indicating that exchange is not the cause of the errors. Inclusion of the "-D3" dispersion correction does improve the error with all four functionals. In fact the errors for the four DFT-D3 methods are nearly identical, further supporting the notion that exchange is not the culprit. While inclusion of the "-D3" term, which includes long range and medium range dispersion, does improve matters, errors still persist. The missing component is medium range correlation, which can be addressed using a double hybrid functional. The errors in Reaction 3.5 when B2PLYP-D3 or B2GPLYP-D3 are employed are half of that with a DFT-D3 method, and less than 2 kcal mol^{-1} for any given bond separation reaction involving an alkane up to $C_{12}H_{26}$. It seems clear that medium range correlation is the missing piece of most functionals.

Are there functionals that address this problem and properly account for medium range correlation? Zhao and Truhlar[67] responded very quickly after the initial flurry of papers were published that pointed out these problems by noting that the highly parameterized M05-2x[77] functional appears to fit the bill. As indicated in Table 3.16, M05-2x predicts that the branched alkane 2,2,3,3-tetrmethylbutane is more stable than the linear alkane *n*-octane. The energy of Reaction 3.5 for hexane and octane is quite nicely predicted by M05-2x, and lastly, it also correctly predicts the relative energies of **20–22** (Table 3.17). Brittain et al. noted that M05-2x addressed the concerns posed by Reaction 3.8.[72]

The successor functional M06-2x[78] also appears to address these problems. In comparison of the DFT energies to CCSD(T) energies for Reaction 3.8a, while B3LYP has a mean average error of 1.75 kcal mol^{-1}, the error with M06-2x is only 0.09 kcal mol^{-1}.[79] In a set of isomerization energies of C_4 through C_8 alkanes, the mean average error for B3LYP was 3.15 kcal mol^{-1} but only 0.34 kcal mol^{-1} for M06-2x.[79] Similarly, for a set of 34 hydrocarbon isomerization energies, the mean average errors using MO5-2x and M06-2x were 1.34 and 1.17 kcal mol^{-1}, respectively.[80] Exactly how these functionals address the medium range correlation problem is difficult to discern.

Additionally, the ωB97X-D[81] functional shows promise. For example, its mean absolute error for Reaction 3.8a is 0.48 kcal mol^{-1} and for the isomerization energy of the C_4 through C_8 alkanes is 0.83 kcal mol^{-1}.[79] Nonetheless, as of 2012, there is no single density functional that serves as a panacea for all computational organic chemistry problems. It remains essential that proper benchmarking of functionals appropriate to the problem under study be routinely performed.

3.3.3.1 Chemical Consequences of Dispersion While dispersion has long been recognized as a fundamental interaction in chemistry, recognition of its importance in organic chemistry is ripe for a renaissance. Dispersion is likely to play an important role in large molecules, where the surface areas can be significant. Computations of very large molecules have only become feasible over the past decade with the advent of high performance computer clusters and DFT. The development of dispersion corrections, like "-D3," and functionals that inherently account for dispersion, like M06-2x, allows us to appropriately account for dispersion. We detail here two examples of organic compounds whose very structures are determined by dispersion.

Hexaphenylethane **23** is unstable and dissociates into two trityl radicals.[82] Similarly, the *para-t*-butyl derivative **24** is also unstable.[83] The standard explanation for the instabilities of these two crowded compounds is steric congestion. That argument, however, fails to account for the stability (a melting point of 214 °C!) of the seemingly even more crowded all-*meta-t*-butyl derivative **25**.[84] The C–C distance in the ethyl fragment is very long (1.67 Å), which is perhaps an indication of its steric congestion.

23: $R_1 = R_2 = H$
24: $R_1 = t$-But, $R_2 = H$
25: $R_1 = H$, $R_2 = t$-But

Grimme and Schreiner[85] employed DFT to ascertain the origin of the stability of **25**. TPSS/TZV(2d,2p) computations on **23–25** indicate that separation into their two radical fragments is very exoergonic. However, when the "-D3" dispersion correction is included, **23** and **24** remain unstable relative to their diradical fragments, but **25** is stable by 13.7 kcal mol^{-1}. In fact, when the dispersion correction is left off of the *t*-butyl groups but remains on for the rest of the molecule, **25** becomes unstable. Thus, it is *dispersion* brought on by the intertwining of the 12 *t*-butyl groups that makes **25** stable to dissociation. The surface areas of **23** and **24** are smaller than that of **25** and so they exhibit less attractive dispersion and cannot overcome the congestion.

This suggests that very highly congested molecules may be held together by attractive dispersion forces, leading to potentially very long bonds. Such a molecule is **26**, which was prepared by Schreiner and coworkers[86] in 2011. The X-ray structure of **26** displays a long C–C bond of 1.647 Å between the two diamantanyl fragments. Computations of **26** were performed with four density functionals, all of which predict a long C–C bond; M06-2x/6-31G(d,p) predicts a distance of 1.648 Å. The bond dissociation energy of 43.9 kcal mol^{-1} is predicted for **26** at B3LYP/6-31G(d,p), a value seemingly consistent with the long CC bond. However, B3LYP does not account for dispersion. The H⋯H distances between the two diamantanyl groups range from 1.94 to 2.28 Å, suggesting that there could be appreciable dispersion stabilization. The BDE of **26** computed with B3LYP+D, B97D, or M06-2x (all of which account for dispersion to some extent) ranges from 65 to 71 kcal mol^{-1}. Dispersion increases the BDE by more than 20 kcal mol^{-1}. **26** is a stable molecule with a strong, yet long C–C bond, and a good deal of the strength results from the energy associated with dispersion interactions across the entire molecule. Other molecules possessing dispersion-stabilized long bonds have been examined by Schreiner[87] as well, like **27**.

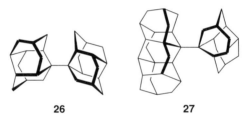

26   27

3.4 RING STRAIN ENERGY

Among the many interesting properties of cyclic compounds, none has captured the attention of physical and synthetic organic chemists as has their strain energy. Rings, especially small rings, require angles about the atoms that deviate from their normal value in acyclic species. The angular deviation results in a high energy molecule, a ring strain. This RSE can be exploited to drive reactions, oftentimes reactions that are difficult in the absence of the ring. For example, nucleophilic substitution of ethers is uncommon, while substitution reactions of epoxides are widely utilized.[88]

We begin by addressing the methods for quantifying the strain energy. A cyclic compound is strained because it has greater energy than an unstrained analog. The key concept in evaluating RSE is, therefore, finding a suitable reference system.

What are the ring strain energies of the simple hydrocarbon rings, cyclopropane (**28**) and cyclobutane (**29**)? These rings are made up of methylene groups,

CH_2 fragments, just like the interior carbons in straight-chain, unstrained alkanes. The difference in energy between, say, *n*-hexane and *n*-pentane can be attributed to the energy of the extra, unstrained methylene group; we call this energy increment due to the CH_2 group $E(CH_2)$. Using experimental gas-phase heats of formation,[89] $E(CH_2) = \Delta H_f(n\text{-hexane}) - \Delta H_f(n\text{-pentane}) = (-39.94 - -35.11)$ kcal mol^{-1} = -4.83 kcal mol^{-1}. If cyclopropane was unstrained, its energy would be three times this increment: $3 \times E(CH_2) = 3(-4.83) = -14.49$ kcal mol^{-1}. Similarly, the unstrained reference for cyclobutane has an energy of $4 \times E(CH_2) = -19.32$ kcal mol^{-1}. The energy difference between the actual heats of formation of **28** or **29** and their respective reference energies gives their RSE.

$$\text{RSE}(\mathbf{28}) = \Delta H_f(\mathbf{28}) - 3 \times E(CH_2) = 12.74 - (-14.49) = 27.23 \text{ kcal mol}^{-1}$$

$$\text{RSE}(\mathbf{29}) = \Delta H_f(\mathbf{29}) - 4 \times E(CH_2) = 6.79 - (-19.32) = 26.11 \text{ kcal mol}^{-1}$$

Benson[90,91] built upon this idea to create a procedure to determine conventional ring strain energy (CRSE). Benson devised group equivalents (GEs), the average energy of chemical groups based on regression analysis of many organic compounds. These *GEs* when summed together for a given molecule provide the energy of this normal unstrained species. The difference in energy between the actual heat of formation of the cyclic compound and the sum of the appropriate GE is the conventional RSE. The CRSEs for cyclopropane and cyclobutane are

$$\text{CRSE}(\mathbf{28}) = \Delta H_f(\mathbf{28}) - 3 \times \text{GE}(C - (C)_2(H)_2)$$
$$= 12.74 - 3 \times (-4.95) = 27.69 \text{ kcal mol}^{-1}$$
$$\text{CRSE}(\mathbf{29}) = \Delta H_f(\mathbf{29}) - 4 \times \text{GE}(C - (C)_2(H)_2)$$
$$= 6.79 - 4 \times (-4.95) = 26.59 \text{ kcal mol}^{-1}$$

This approach is not directly amenable to computational chemistry since most calculations do not provide heats of formation. While some attempts to create "ab initio" equivalents have been published,[92–95] their real failure is that equivalents are not transferable between methods and basis sets, necessitating huge sets of such "ab initio" equivalents.

The alternative approach is to choose molecule(s) that can serve as reference(s) for the strained species. The energy of a balanced chemical reaction where the strained molecule is the reactant and the product is the unstrained reference directly provides the RSE. One simply needs to compute the energies of all species in the appropriate chemical equation, determine the reaction energy, and the RSE is at hand. This method works only if the sole difference between the reactants and the products is the presence of the ring on the reactant side and its absence on the product side.

The choice of molecule(s) to serve as the unstrained reference is not unique, nor is the chemical reaction that is ultimately used to compute the RSE. Three methods for generating an appropriate chemical reaction have found utility: the isodesmic reaction, the homodesmotic reaction, and the group equivalent reaction. The RSE is then equal to the negative of the energy of these reactions. For this discussion, we will determine the RSE of cyclobutane (**29**), cyclobutene (**30**), and oxetane (**31**).

The isodesmic reaction conserves the number of bonds of a given formal type.[96] This is accomplished by choosing reference molecules that contain two heavy atoms, preserving the formal bond between them. The four C–C bonds of cyclobutane are preserved in the reference as four ethane molecules. Cyclobutene, having one C=C and three C–C bonds, needs a molecule of ethene and three molecules of ethane as reference. The two C–O bond and two C–C bonds of oxetane translate into two molecules of ethanol and two molecules of ethane. Balancing the reaction to conserve all other bonds leads to the isodesmic Reactions 3.9i–3.11i.

☐ + 4 CH$_4$ ⟶ 4 ── (Reaction 3.9i)

☐ + 4 CH$_4$ ⟶ = + 3 ── (Reaction 3.10i)

☐ + 4 CH$_4$ + H$_2$O ⟶ 2 ── OH + 2 ── (Reaction 3.11i)

While the isodesmic reaction is simple to define and compute, the reactants and products differ by more than just the ring. For example, in Reaction 3.9i, the 8 secondary C–H bonds of cyclobutane and the 16 C–H bonds of methane must correlate with the 24 primary C–H bonds of ethane. In Reaction 3.10i, the two C_{sp^2}–C_{sp^3} bonds of cyclobutene correlate with the C_{sp^3}–C_{sp^3} bond in ethane. These bonds are not of equivalent energy, and so these reaction energies contain contributions from these differences along with the RSE.

The homodesmotic reaction[97,98] attempts to remove these contributions by conserving (1) the number of carbon (and other heavy) atoms and their state of hybridization and (2) the number of carbon (and other heavy) atoms bearing zero, one, two, and three H atoms. This effectively preserves the bond types, that is, C_{sp^3}–C_{sp^3}, C_{sp^2}–C_{sp^3}, and so on. Reference molecules usually contain three heavy atoms. The homodesmotic reactions used to obtain the RSEs of **29–31** are Reactions 3.9h–3.11h, respectively.

☐ + 4 ── ⟶ 4 ⋀ (Reaction 3.9h)

☐ + = + 3 ── ⟶ 2 ⋀ + 2 ⋀ (Reaction 3.10h)

☐ + 4 ── ⟶ ⋀O⋀ + 3 ⋀ (Reaction 3.11h)

While the homodesmotic reaction is a notable improvement over the isodesmic scheme, it too suffers from inequivalences between reactant and product other than the presence/absence of the ring. In Reaction 3.11h, for example, oxetane has two carbons classified by Benson as C–(C)(O)(H)$_2$ whose GE = −8.5 kcal mol^{-1}. These carbons correspond to C–(C)$_2$(H)$_2$ groups in the product (the methylene of propane), but its GE is −4.95 kcal mol^{-1}. The energy of Reaction 3.11h reflects, therefore, not just the RSE but also the difference in the group equivalent energies.

The group equivalent reaction[99] was developed to eliminate this error. The algorithm for constructing the GE reaction is simple. Every equivalent group in the cyclic molecule must be paired with an equivalent group in a short acyclic molecule in the product. This will typically require molecules having three heavy atoms, though occasionally larger molecules are needed. Oxetane contains the following equivalent groups: C–(C)$_2$(H)$_2$, 2 C–(C)(O)(H)$_2$, and O–(C)$_2$. The shortest acyclic molecules that contain these groups are propane, ethanol, and dimethyl ether, respectively. These molecules constitute the reference set. The reaction is then balanced by adding small acyclic molecules, usually containing two heavy atoms, selected to balance the equivalent groups in the products. The group equivalent reactions for **29–31** are given as Reactions 3.9g–3.11g. The group equivalent reaction effectively conserves the next-nearest neighbors.

☐ + 4 —— → 4 ∧   (Reaction 3.9g)

☐ + = + — → 2 ∨∨   (Reaction 3.10g)

☐O + 2 —— +2 ——OH → ∧O∧ + ∧∧ + 2 ∧∧OH   (Reaction 3.11g)

It is worth noting that Wheeler et al.[100] have delineated a hierarchy of reaction schema that define various levels of conserved components. This includes a more rigorous definition of isodesmic and homodesmotic reactions, along with providing higher order reactions that conserve larger fragments of molecules.

The estimates of the RSE for small cyclic alkanes, cyclo-olefins, and cyclic ethers using all three reactions are listed in Table 3.21. The failure of the isodesmic reaction is readily apparent by examining the RSE of the cyclic alkanes. For the rings having five or more carbon atoms, the isodesmic method predicts that there is *no* strain energy. Rather, it predicts that these cyclic species are more stable than their acyclic counterparts. The homodesmotic and GE reactions, which are identical for the cyclic alkanes, provide RSEs in excellent agreement with CRSE estimates.

For the cycloalkenes, the isodesmic reaction again fails miserably to predict RSE. The homodesmotic reaction systematically overestimates the RSE by about 0.5 kcal mol^{-1} relative to CRSE. The RSEs from the GE reaction are in excellent agreement with CRSE.

The RSEs of the cyclic ethers demonstrate the usefulness of the GE scheme. Figure 3.5 plots the RSE against ring size for the cyclic ethers computed with the different schemes. As before, the isodesmic reaction fails by estimating positive

TABLE 3.21 RSE (kcal mol^{-1}) Evaluated Using the Isodesmic, Homodesmotic, and Group Equivalent Reactions[a]

Method	△ 28	□ 29	⬠	⬡	⬢	⬣
Isodesmic	19.48	15.77	−7.03	−16.01	−12.50	−11.76
Homodesmotic/	27.72	26.77	6.72	0.48	6.74	10.23
Group equiv.	*28.03*	*26.34*	*6.54*	*0.51*	*6.69*	*11.70*
CRSE[b]	27.6 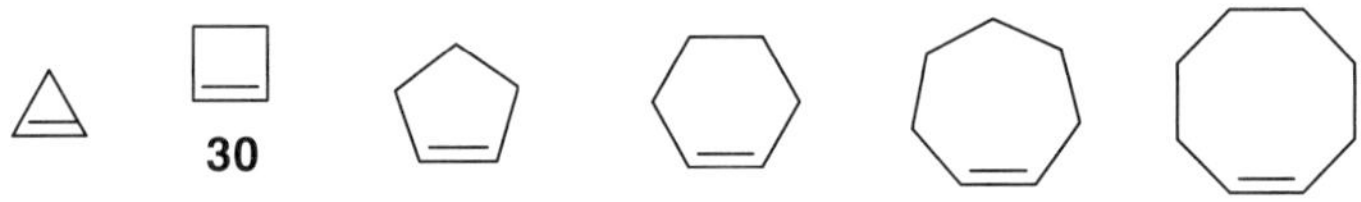	26.2	6.3	0	6.4	9.9

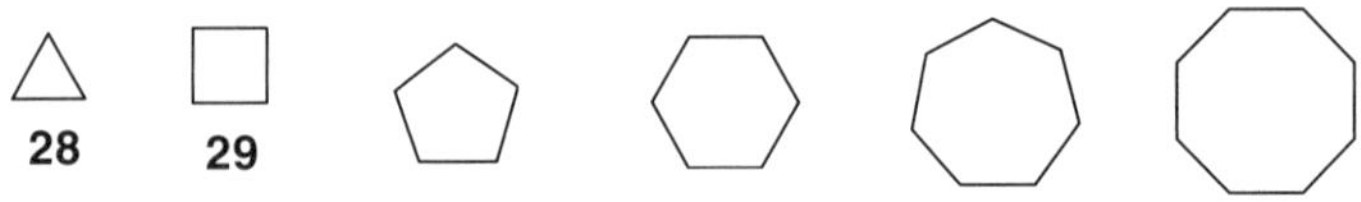

Method	△=	□= 30	⬠=	⬡=	⬢=	⬣=
Isodesmic	40.39	13.86	−13.24	−20.29	−19.05	−21.06
Homodesmotic	54.18	30.40	6.05	1.74	5.74	6.48
Group equiv.	53.54	29.92	5.57	1.27	5.26	6.00
	55.01	*31.20*	*6.09*	*1.39*	*5.78*	
CRSE[b]	53.7	29.8	5.9	1.4	5.4	6.0

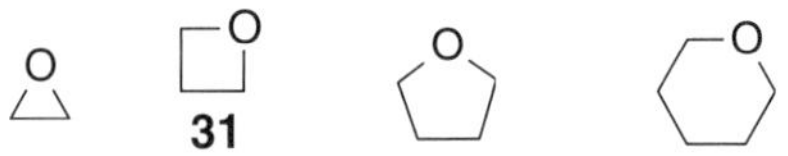

Method	△O	□O 31	⬠O	⬡O
Isodesmic	10.42	6.00	−16.54	−23.66
Homodesmotic	21.39	19.72	−0.07	−4.45
Group equiv.[c]	27.51	25.84	6.05	1.67
	27.11	*25.59*	*6.25*	*1.47*
Group equiv.[d]	27.65	25.98	6.19	1.82
	27.52	*26.00*	*6.65*	*1.87*
CRSE[b]	27.6	26.4	6.7	2.2

[a]Values in normal text are derived from experimental heats of formation.[89] Values in italics are from G2MP2-computed energies.
[b]Refs 90, 91.
[c]Evaluated using the reaction.

$$\text{O}\triangleright(CH_2)_{n-2} + 2\,CH_3OH + n{-}2\,CH_3CH_3 \longrightarrow CH_3OCH_3 + n{-}3\,CH_3CH_2CH_3 + 2\,CH_3CH_2OH$$

[d] Evaluated using the reaction.

$$\text{O}\triangleright(CH_2)_{n-2} + n{-}1\,CH_3CH_3 \longrightarrow CH_3CH_2OCH_2CH_3 + n{-}3\,CH_3CH_2CH_3$$

reaction energies for the larger rings. The homodesmic reaction fails here in the same way. However, the GE reaction closely mimics the CRSE.

The group equivalent algorithm does not specify a unique reaction, but only a specification that groups be conserved. The last two entries of Table 3.21 demonstrate two different group equivalent reactions that can be used to estimate the RSE

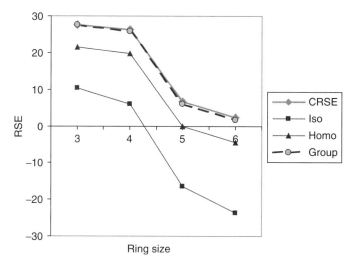

Figure 3.5 Comparison of RSE versus ring size for the cyclic ethers evaluated using CRSE, isodesmic, homodesmotic, and group equivalent reactions.

of the cyclic ethers. The results are very similar, indicating that larger reference molecules are not needed to obtain consistent results.

The last issue presented by Table 3.21 is the ability to compute RSE. The composite method G2MP2, which closely mimics the results of the more rigorous G2 method, but at a fraction of the computational cost, was used to compute the energies of the species present in the group equivalent reactions for the cycloalkanes, cycloalkenes, and cyclic ethers. The agreement between the RSE predicted with the G2MP2 energies and the experimental heats of formation is outstanding. Figure 3.6a displays the correlation between the experimental and G2MP2 RSE for the entire set of compounds listed in Table 3.21. This plot clearly demonstrates that reaction energies for GE reactions can be very accurately computed.

Gordon[101] has shown that RSE computed using HF/6-31G(d)//HF/3-21G* and MP2/6-31G(d)//HF/3-21G* energies and the homodesmotic equation are in good agreement for the smallest cycloalkanes, differing from each other by no more than 1.5 kcal mol^{-1}. The RSE values for all of the cyclic species shown in Table 3.21 have been computed at G2MP2, HF/6-31G(d), and B3LYP/6-31+G(d) using the group equivalent reaction. These values have then been compared with the RSE obtained using the group equivalent reaction and experimental heats of formation in the plots shown in Figure 3.6. All three show an excellent linear relationship, indicating that correlation effects cancel in the group equivalent reaction. The best fit is with the composite G2MP2 method, but certainly satisfactory results can be obtained with the relatively cheap HF/6-31G(d) method. On the other hand, the computed values of the RSEs of trioxirane (**32**) and triaziridine (**33**) are dependent upon the method.[102] The RSE values of **32** at HF/6-31G*, MP2/6-31G*, and G2 are 38.8, 29.2, and 33.3 kcal mol^{-1}, respectively, while those of **33** are 31.9, 30.1, and 23.4 kcal mol^{-1}, respectively. The computational difficulty with these molecules

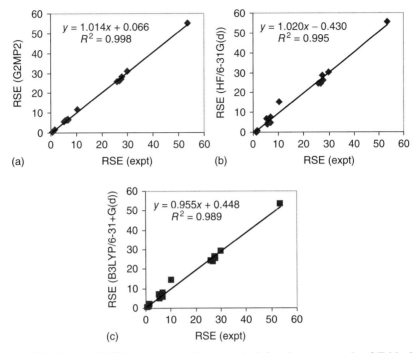

Figure 3.6 Plots of RSE (group equivalent reaction) for the compounds of Table 3.21 determined using experimental heats of formation versus those determined using (a) G2MP2 energies, (b) HF/6-31G(d) energies, or (c) B3LYP/6-31+G(d) energies.

is their many adjacent atoms with lone pairs that require correlated methods for reasonable treatment of the electron distribution.

32 (O₃ triangle) **33** (HN—NH with NH above)

3.4.1 RSE of Cyclopropane (28) and Cylcobutane (29)

A long vexing problem for physical organic chemists is why the ring strain energies of cyclopropane (27.6 kcal mol^{-1}) and cyclobutane (26.2 kcal mol^{-1}) are so similar. If one just considers the C–C–C angles in these two compounds, the distortion of the normal sp^3 bond angle from 109.5° to 60° in cyclopropane seems as if it should entail much greater strain than the much smaller distortion (to ~90°) necessary in cyclobutane. Their similar RSEs can be interpreted as (1) some effect is stabilizing cyclopropane or (2) some effect is destabilizing cyclobutane.[103] Most work has emphasized the former—cyclopropane is more stable than might be expected based solely on its ring angles.

In order to rationalize the RSE of cyclopropane and cyclobutane, it is necessary to consider all effects that may stabilize or destabilize them. This process was most completely spelled out by Cremer and Gauss,[104] whose arguments are largely presented next.

The simplest effect, and earliest to be identified, is the angular deviation about the carbon atoms, called *Baeyer strain*.[105,106] If the bending about the carbon is considered to be like a spring, then the Baeyer strain, $\Delta E(B)$, is proportional to the square of the angle deviation, $\Delta \alpha$.

$$\Delta E(B) = \frac{nk_\alpha}{2}(\Delta \alpha)^2 \qquad (3.5)$$

where k_α is the CCC bending force constant. Choosing the bending force constant of propane (1.071 mdyn Å rad^{-2})[107] gives the Bayer strain of cyclopropane and cyclobutane as 173 and 36 kcal mol^{-1}, respectively. The former value is astonishingly large, suggesting that the simple definitions utilized here are inappropriate. Nevertheless, it does reinforce the notion that cyclopropane appears to be much more stable than it ought to be!

The first assumption is that the bond angle is 60°. Certainly, this is the angle formed by the internuclear line segments. However, Walsh[108] argued for bent bonds in strained rings and so the bonding electron density may not necessarily follow the internuclear path. The topological method developed by Bader[109] rigorously defines the maximum electron density path between two bonding atoms, called the *bond path*. The angle formed between two bond paths originating from the nucleus defines a more reasonable bond angle (called β), reflecting how the electron density deviates from that in a normal unstrained molecule. The values of β in cyclopropane and cyclobutane are 78.8° and 96.5°, respectively, leading to Baeyer strain energies of 66.4 kcal mol^{-1} for **28** and 15.9 kcal mol^{-1} for **29**. The second assumption concerns the value of k_α. Cremer and Gauss[104] argue for a smaller angle force constant since 1,3 carbon–carbon interactions are absent in **28**, but are present in propane. Usage of their best guess for the force constant ($k^* = 0.583$ mdyn Å rad^{-2}) and the angle β leads to the Bayer strain in cyclopropane of 41.3 and 13.0 kcal mol^{-1} in cyclobutane. Cremer[103] later revised the value of the Baeyer strain for cyclopropane, raising it to 46.3 kcal mol^{-1} with the inclusion of anharmonicity effects.

A Newman projection down the C–C bond of cyclopropane makes readily apparent the next type of strain to be considered; the C–H bonds are all eclipsing, giving rise to what is called *Pitzer strain*, $\Delta E(P)$.[110] Planar cyclobutane has all of its C–H bonds eclipsing as well, but the Pitzer strain can be diminished somewhat when cyclobutane adopts a puckered conformation, leading in fact to its ground-state form. Since the rotational barrier in ethane is nearly 3 kcal mol^{-1}, a first estimate at the Pitzer strain due to a pair of eclipsing C–H bonds is 1 kcal mol^{-1}. This leads to $\Delta E(P)$ of 6 kcal mol^{-1} for **28** and 8 kcal mol^{-1} for **29**. This needs to be tempered by the fact that C–C–H angle is wider in **28** (118.1°) and **29** (114.6°) than in ethane (111.7°). Substituting the appropriate value for this angle and computing the energy difference between staggered and eclipsed ethanes reduces $\Delta E(P)$ by about 2 kcal mol^{-1} for both **28** and **29**. The value for **29** is further reduced due to its puckering, which changes the H–C–C–H dihedral angle from 0.0° to 25.5°. This reduces the Pitzer strain on the order of cos(3*25.5°) or 2 kcal mol^{-1} for **29**. Therefore, the final corrected values of $\Delta E(P)$ are 4.0 and 3.9 kcal mol^{-1} for **28** and **29**, respectively.

Dunitz and Shomaker[111] offered another type of interaction that plays a role in determining RSE: nonbonding 1,3-CC repulsions, $\Delta E(DS)$. Cyclopropane does not suffer from this repulsion; however, both cyclobutane and propane are affected. The 1,3-CC repulsion is stronger in cyclobutane than propane because of the shorter distances between the affected carbon atoms as well as the fact that there are two such repulsions in **29** but only one in propane. No truly satisfactory way of estimating the strength of this effect has been offered. Bauld[112] has estimated its magnitude by taking the difference between the computed CNDO/2 energy with and without the Fock matrix elements corresponding to the 1,3-CC interactions. Cremer and Gauss[104] argued that these resulting values for $\Delta E(DS)$ are too large. They proposed a scheme of scaling these values and simultaneously fitting them and the force constant necessary for the Baeyer strain. Their estimated values of $\Delta E(DS)$ are 12.5 and 3.2 kcal mol^{-1} for cyclobutane and propane, respectively. Keep in mind that since cyclopropane does not have any 1,3-CC interactions, its value of $\Delta E(DS)$ is exactly zero.

Coulson and Moffitt[113] were the first to suggest that the strain energies of small rings might be offset by changes in the hybridization at carbon, resulting in stronger C–H bonds. For example, the C–C bonds of cyclopropane must have high p-character, thereby enhancing the s-character of the bonds to hydrogen. Since the s-character of the hybrid orbitals of carbon used to bond to hydrogen is larger in **28** than **29**, the RSE of the former will be reduced in **28** relative to **29**. The key is how to evaluate the energy associated with the C–H bonds.

The experimental bond dissociation energies of cyclopropane[114] and propane[115] are 106.3 ± 0.3 and 98.6 ± 0.4 kcal mol^{-1}, respectively. Since there are six C–H bonds in cyclopropane, each of which is 7.7 kcal mol^{-1} stronger than the analogous bond in propane, one might conclude that cyclopropane is stabilized by 46.2 kcal mol^{-1} due to its strong C–H bonds. This analysis is flawed by neglecting the fact that the RSE of the cyclopropyl radical is *not* equivalent to that in cyclopropane.

In order to quantify the stabilizing effect of rehybridization at carbon in small rings, two separate computational tacks have been followed. The first approach attempts to determine the energy of a bond in a single molecule using its electron density, while the second approach makes use of isodesmic reaction energies.

Cremer and Gauss[104] estimated the C–C and C–H bond energies based on the topological method.[109] First, they obtained the quantity $N(A,B)$ by integrating the density over the surface between the two bonded atoms A and B.

$$N(A, B) = \mathbf{R}(A, B) \oint_{AB} \rho(\mathbf{r}) n_A(\mathbf{r}) dS(\mathbf{r}) \qquad (3.6)$$

The bond energy was then calculated by multiplying $N(A,B)$ by a scaling factor $\alpha(A,B)$ and dividing by the distance along the bond path between atoms A and B,

TABLE 3.22 Bond Energies (kcal mol^{-1}) Computed Using Topological Approaches for 28, 29, and Propane

Molecule	Bond	BE(A,B)^a	BE(A,B)^b
28	C–C	71.0	73.2
	C–H	106.6	105.8
29	C–C	73.9	79.1
	C–H$_{ax}$	105.9	104.3
	C–H$_{eq}$	105.9	
Propane	C–C	81.9	87.0
	sec-C–H	105.5	104.3
	C–H$_i$	105.2	
	C–H$_p$	104.8	

^{a}Computed using Eq. (3.6), see Ref. 104.
^{b}Computed using Eq. (3.7), see Ref. 116.

R_b. Thus, the bond energy BE(A,B) is given by

$$\text{BE}(A, B) = \frac{\alpha(A, B)N(A, B)}{R_b} \quad (3.7)$$

The quantities α(C,H) and α(C,C) were fit to the atomization energies of methane and ethane. The resulting bond energies for cyclopropane, cyclobutane, and propane computed using HF/6-31G** densities are listed in Table 3.22. Since the C–C bonds in **28** are 10.9 kcal mol^{-1} weaker than in propane, its strain energy is about 33 kcal mol^{-1}. However, since its C–H bonds are 1.1 kcal mol^{-1} stronger than in propane, the C–H bonds stabilize cyclopropane by about 6 kcal mol^{-1}. On the other hand, the C–H bonds of **29** are only slightly stronger (0.4 kcal mol^{-1}) than in propane, leading to a stabilization of about 3 kcal mol^{-1}, 50 percent that of **28**.

Grimme[117] suggested an alternate method for evaluating the bond energy again using topological electron density properties, but using only properties that can be evaluated at the bond critical point (BCP), thereby simplifying the computations. He proposed the relationship shown in Eq. (3.8) for evaluating bond energies.

$$\text{BE}(A, B) = \frac{c_1^{AB} E(\mathbf{r}_{BCP(AB)})}{(c_2^{AB} + \rho(\mathbf{r}_{BCP(AB)}))} - c_3^{AB}(R_b - R_e) \quad (3.8)$$

Schleyer[116] refit the parameters for this equation with a larger set of test molecules computed with better methods than Grimme had done. The values for the bond energies of **28, 29**, and propane using Schleyer's parameters for Eq. (3.8) are presented in Table 3.22. Using cyclohexane as the reference for the "strainless" C–H bond, Schleyer estimated that the C–H bond in cyclopropane is 1.95 kcal mol^{-1} stronger, leading to a stabilization energy of 11.7 kcal mol^{-1}.

The C–H bond in cyclobutane is only 0.46 kcal mol^{-1} stronger than that in cyclohexane, giving a stabilization of only 3.7 kcal mol^{-1}.

Borden,[118] employed Reaction 3.12 to estimate the bond energy in; it is exothermic, $\Delta H = -10.0$ at G2MP2. Borden claims that this demonstrates the unusually strong C–H bonds of cyclopropane, a claim also supported in similar work by Bach.[119,120] Reaction 3.12 can be recast in group equivalent reaction form as Reaction 3.13. Its energy is very close to the isodesmic counterpart: -9.4 kcal mol^{-1}. Nonetheless, both of these reactions assume that the RSE values of the parent three-member ring and its radical are equivalent. The RSE of cyclopropane is 28.0 kcal mol^{-1} at G2MP2 (Table 3.21), while that of cyclopropyl radical, computed using Reaction 3.14, is 37.4 kcal mol^{-1}. It is not obvious how one can partition these RSEs in order to deduce how much of their difference is due to differential C–H bond energies. At best, one can cautiously conclude that the C–H bond is stronger in cyclopropane than in propane (or cyclobutane), leading to some reduction in its RSE. The Schleyer estimates of the stabilization due to rehybridization for **28** and **29**, namely 11.7 and 3.7 kcal mol^{-1}, respectively, appear to be suitable compromise values.

$$\triangle\text{C}^{\cdot} + \triangle \longrightarrow \triangle + \text{C}^{\cdot} \qquad \text{(Reaction 3.12)}$$

$$\triangle\text{C}^{\cdot} + \text{/\textbackslash/\textbackslash} \longrightarrow \triangle + \text{/\textbackslash C}^{\cdot}\text{\textbackslash} \qquad \text{(Reaction 3.13)}$$

$$\triangle\text{C}^{\cdot} + \text{—} + 2 \text{—C}^{\cdot} \longrightarrow \text{C}^{\cdot} + 2 \underset{\text{C}^{\cdot}}{\diagdown} \qquad \text{(Reaction 3.14)}$$

We now assess the state of all of these contributions. In Table 3.23, we list the contributions attributed to Bayer strain, Pitzer strain, Dunitz–Shomaker strain, and rehybridization that lead to strengthened C–H bonds. We list two values for this

TABLE 3.23 Decomposition of the RSE of **28** and **29**[a]

	28	29
$\Delta E(B)$	46.3	13.0
$\Delta E(P)$	4.0	3.9
$\Delta E(DS)$	0.0	12.0
ΔE(CH strengthening)	-6.4^b $(-11.7)^c$	-2.8^b $(-3.7)^c$
Subtotal	43.9 (38.6)	26.1 (25.2)
RSE	27.6	26.2
$\Delta E(\sigma)$	-16.3 (-11.0)	~ 0

[a]Ref. 103.
[b]Ref. 104.
[c]Ref. 117.

last quantity, those provided by Cremer and the revised values of Schleyer. Summing these contributions leads to a value of 25–26 kcal mol^{-1} for cyclobutane. This is nearly identical to the CRSE of cyclobutane, thereby accounting for its strain in total. On the other hand, the sum of these contributions is 38.6 kcal mol^{-1} (43.9 kcal mol^{-1} using the smaller value for the effect of CH strengthening) for cyclopropane, which is 11–16 kcal mol^{-1} *larger* than its RSE. In other words, cyclopropane is about 11 kcal mol^{-1} *more stable* than what these effects predict.

Dewar suggested one further type of stabilization for cyclopropane.[121,122] In analogy with benzene, which is stabilized by aromaticity provided by six π-electrons, Dewar argued for σ-aromaticity provided by the six σ-electrons of the C–C bonds of cyclopropane. Dewar's arguments lay with simple MO theory and energetic analysis.

Cremer later provided firmer footing for the concept of σ-aromaticity. First, he noted that the electron density in the interior of the cyclobutane ring is much less than in the internuclear region. The value of the electron density at the C–C BCP is about 1.68 e Å^{-3}, but only 0.554 e Å^{-3} at the ring critical point. On the other hand, the density at the ring critical point of cyclopropane is 1.379 e Å^{-3}, or about 82 percent of the density at the BCP.[123] Cremer argued that this is the evidence of surface delocalization of electron density, a manifestation of σ-aromaticity. Second, using the arguments that are summarized in Table 3.23, Cremer[103,104] suggested that the difference between the subtotal (ΔE(B) + ΔE(P) + ΔE(DS) + ΔE(CH strengthening) and the RSE is the stabilization energy associated with σ-aromaticity, $\Delta E(\sigma)$. Cyclobutane, which exhibits little to no surface delocalization and whose subtotal of the energetic consequences of the standard components of ring strain fully accounts for its RSE, does not possess any σ-aromaticity. For cyclopropane, however, the value of $\Delta E(\sigma)$ is substantial, ranging from -11 to -16 kcal mol^{-1}. Dewar[122] and Cremer[104,123] ascribe this energetic stabilization to σ-aromaticity.

The rationale for σ-aromaticity lies largely in the energetic partitioning scheme of the RSE displayed in Table 3.23. The value of each component relies quite heavily on the underlying assumptions. The component with least certainty is the strengthening of the C–H bonds in small rings. No clear, unambiguous way exits to define this bond energy, and so the magnitude of σ-aromaticity in three-membered rings remains poorly defined.

Schleyer has offered support for σ-aromaticity based on magnetic properties.[124] The nucleus-independent chemical shift (NICS) value at the center of the cyclopropane ring is -42.8 ppm. This very negative value results from the combination of C–C (-24.3 ppm) and C–H (-18.0 ppm) shielding. In contrast, the NICS values at the center of cyclopentane and cyclohexane are -5.2 ppm ($+6.0$ due to the C–C bonds) and -1.8 ppm ($+6.6$ ppm due to the C–C bonds), respectively. Schleyer argued that the negative NICS value for **28** indicates diatropic magnetic effects, just as in the aromatic prototype benzene, while cyclopentane and cyclohexane exhibit no unusual magnetic shielding.

Furthermore, the NICS value for cyclobutane is 2.6 ppm, with a very large, positive C–C contribution of $+15.2$ ppm. This paratropic magnetic effect is opposite to that of cyclopropane and is suggestive of perhaps σ-*antiaromaticity* in cyclobutane.

Schleyer offers no energetic evaluation of the destabilization afforded cyclobutane by this antiaromatic contribution, but if it were present, it would offer another factor toward understanding the close RSEs of **28** and **29**.

Fowler et al.[125] have computed the ring current in cyclopropane and cyclobutane. The plot of the current density of cyclopropane indicates a large diatropic current circling the outside of the ring. In the interior of the ring is a smaller paratropic current. This is consistent with σ-aromaticity. In contrast, the ring current map for cyclobutane shows a strong paratropic current in the inside of the ring with a weaker diatropic current on the outside of the ring, consistent with weak σ-antiaromaticity.

It should be noted, however, that the concept of σ-aromaticity is fairly limited in its applicability—to just three-member rings—certainly in contrast to the notion of aromaticity. For example, aromaticity can be used to group compounds with differing ring sizes (benzene, naphthalene, cyclopentadienyl anion, tropylium cation, etc.), but no one is suggesting that cyclopentane, which has 10 σ-electrons, exhibits σ-aromaticity.

3.5 AROMATICITY

Aromaticity is among the most important organizational concepts within organic chemistry, dating back into the 1800s.[126,127] Aromaticity is plagued, however, by being rather ill-defined.[128] In fact, there is no single definition of aromaticity, but rather a collection of properties associated with aromatic molecules. These properties include the following.

- Aromatic compounds are stable relative to some nonaromatic reference.
- Aromatic compounds display magnetic properties related to a ring current. These include unusual chemical shifts, especially of protons, large magnetic anisotropies, and large diamagnetic susceptibility exaltations.
- Aromatic compounds tend to be planar and have equivalent C–C bond lengths, even though formal Lewis structures indicate alternating single and double bonds about the rings.
- Even though their Lewis dot structures show C=C bonds, aromatic compounds prefer substitution reactions to addition reactions.

Debate has raged during the past century over exactly what constitutes an aromatic compound and what properties can best be used to judge relative aromaticity.[129–131] A survey of the literature might bring one to despair. Stabilization energy can be rather small for some "aromatic" compounds, for example, only 5.3 kcal mol^{-1} for pyrrole.[132] Some "aromatic" compounds are nonplanar, like corranulene,[133,134] while others express bond alternation (defined as the difference in the lengths of the C–C bonds about the ring, ΔCC), like [30]-annulene.[135] Some "aromatic" compounds will readily undergo addition reactions: phenanthene and anthracene add bromine. Perhaps there really is only one truly aromatic compound: benzene![136] More recent efforts have been toward developing a linear regression among the various aromaticity measures that might produce a single equation that can be used to rank-order aromatic compounds.[137–141] These efforts are challenged by whether the component measures are truly orthogonal or not.

Before addressing a number of critical questions of evaluating properties associated with aromatic compounds using computational chemistry, a word of caution concerning geometries and methodologies is needed. One might imagine that of all the properties of benzene, getting the structure to be planar would seem to be perhaps most fundamental, and likely a trivial result of any geometry optimization. Unfortunately, as demonstrated by Schleyer and Schaefer,[142] this turns out not to be true. Any one-electron wavefunction method, like HF and DFT, predicts a planar benzene structure. Many widely used correlated methods predict that planar benzene has one or more *imaginary* frequencies, indicating that a nonplanar structure is more stable than the planar one. For example, MP2/6-311++G(d,p) and CISD/6-311++G(d,p) both predict one imaginary frequency for planar benzene: $936i$ and $247i$ cm^{-1}, respectively. Even computations of planar benzene at MP2 with very large sp [C(8s7p)/H(7s)] or spd [C(8s7p6d)/H(7s6p)] basis sets have at least one imaginary frequency. The culprit is improperly large correlation energy between the s and p electrons, resulting from an intramolecular basis set incompleteness error that can be removed only with careful balance of the spd basis functions with higher angular momentum functions. The Dunning correlation consistent basis sets and atomic natural orbital (ANO) basis sets minimize these errors, and properly predict that benzene is planar using any of the major electron correlation methods. Application of intramolecular basis set superposition corrections, as defined in Section 1.1.8, removes the imaginary frequency corresponding to an out-of-plane distortion for benzene with all combinations of either MP2 or CISD and the 6-31+G*, 6-311G, or 6-311++G basis sets.[56]

We will not concern ourselves here with the attempt to discern a "true" definition of aromaticity, nor with the largely unproductive[143] desire to rank-order relative aromaticity. Rather, we will focus on how computational chemistry has been used to help define two of the criteria of aromaticity. We will discuss how to compute the stabilization energy of an aromatic species and then how we can compute magnetic properties that can be related to aromaticity. We will then conclude with a discussion of how these measures have been employed to investigate the nature of annulenes and the Mills–Nixon effect.

3.5.1 Aromatic Stabilization Energy (ASE)

Aromatic compounds are unusually stable. The classic demonstration of this stability is with the heat of hydrogenation of benzene, $\Delta H_{rxn} = -49.2$ kcal mol^{-1} (Reaction 3.15) and the heat of hydrogenation of cyclohexene, $\Delta H_{rxn} = -28.3$ kcal mol^{-1} (Reaction 3.16). If we can consider benzene as being composed of three normal C–C double bonds, than we might expect the heat of hydrogenation of benzene to be three times that of Reaction 3.15 or -84.9 kcal mol^{-1}. The difference between this estimated heat of hydrogenation of benzene and its actual value,

35.7 kcal mol^{-1}, is due to the stability of this aromatic ring.

$$\text{benzene} + 3\,H_2 \longrightarrow \text{cyclohexane} \quad \text{(Reaction 3.15)}$$

$$\text{cyclohexene} + H_2 \longrightarrow \text{cyclohexane} \quad \text{(Reaction 3.16)}$$

As with the evaluation of RSE discussed previously, evaluating the stabilization afforded by aromaticity requires comparison with some sort of nonaromatic reference. In the following discussion, all computed energies were done at the G2MP2 level, which is the same method that we used in the previous section to obtain accurate RSEs.

Computational evaluation of the stability of aromatic compounds really began with Roberts'[144] report of Hückel delocalization energy for a series of aromatic hydrocarbons. The delocalization energy tends to increase with the size of the compound, and some compounds predicted to have large delocalization energies turn out to be unstable. The error here is not so much with the computational method, which is of course very rudimentary, but rather with the reference compound.[145] In the Hückel approach, the π-energy of the aromatic compound is compared with the π-energy of the appropriate number of ethylene molecules, that is, for benzene, the reference is three ethylene molecules.

One can recast the use of ethylene as a reference into an isodesmic reaction (Reaction 3.17), where its reaction energy should indicate the stabilization due to aromaticity for benzene. One can use the experimental ΔH_f or the computed energies to obtain the overall energy of this reaction. The resulting estimate for the stabilization energy (~65 kcal mol^{-1}) is large due to additional differences between reactants and products besides just aromaticity. Isodesmic Reactions 3.18 and 3.19 might improve matters. These reactions are less exothermic.

$$3\,CH_2{=}CH_2 + 3\,CH_3{-}CH_3 \longrightarrow \text{benzene} + 6\,CH_4$$
$\Delta H_{rxn}(\text{expt}) = -64.2$ kcal mol^{-1}
$\Delta H_{rxn}(\text{G2MP2}) = -63.9$ kcal mol^{-1} (Reaction 3.17)

$$3\,CH_2{=}CH_2 + \text{cyclohexane} \longrightarrow \text{benzene} + 6\,CH_3{-}CH_3$$
$\Delta H_{rxn}(\text{expt}) = -48.5$ kcal mol^{-1}
$\Delta H_{rxn}(\text{G2MP2}) = -47.5$ kcal mol^{-1} (Reaction 3.18)

$$3\,\text{cyclohexene} \longrightarrow \text{benzene} + 2\,\text{cyclohexane}$$
$\Delta H_{rxn}(\text{expt}) = -35.7$ kcal mol^{-1}
$\Delta H_{rxn}(\text{G2MP2}) = -35.8$ kcal mol^{-1} (Reaction 3.19)

Recalling our discussion of isodesmic reaction in the previous section, it is clear that Reactions 3.17–3.19 contain energetic consequences for other effects besides aromaticity, including changes in hybridization. In particular, delocalization effects are not conserved. It is important to distinguish "delocalization" effects from "resonance" effects from "aromatic" effects. The first refers to stabilization when double bonds come into conjugation, and the second occurs when multiple resonance structures (Kekule structures for aromatic compounds) are needed. Aromatic compounds benefit from both of these stabilizing effects. The third category is necessary to distinguish the stability of, say, the allyl anion from that of benzene—both require multiple resonance structures, but allyl anion, unlike benzene, is not aromatic. Aromatic stabilization energy (ASE) might be argued as arising from resonance in a cyclic species.

The Pauling–Wheland[146,147] definition of "resonance energy" captures all three of these electronic effects. It is the difference in the energy of the actual molecule and the energy of its single most important contributing resonance structure. Mo and Schleyer[148] have employed a block-localized wavefunction to compute the energy of the virtual 1,3,5-cyclohexatriene molecule that corresponds to one of the standard Kekule resonance structures of benzene. The resonance energy of benzene, taken as the energy difference between the virtual 1,3,5-cyclohexatriene and benzene, is 57.5 kcal mol^{-1}.

Dewar proposed the first reasonable definition of ASE.[149] He defined the reference energy as

$$E_{\text{ref}} = n'E' + n''E'' + n_{\text{CH}}E_{\text{CH}} \tag{3.9}$$

where E', E'', and E_{CH} are the average energies of the formal single and double C–C bonds and C–H bonds, respectively, in acyclic polyenes. The Dewar resonance energy (DRE) of benzene is then computed as DRE(benzene) = ΔH_f(benzene) − ($3E' + 3E'' + 6E_{\text{CH}}$). This reference includes the effect of delocalization. In other words, the reference incorporates the fact the C_{sp^2}–C_{sp^2} bond is stronger than a typical C–C bond and that the double bond in a conjugated system is not equivalent to a double bond in a typical alkene.

This idea can be recast into a homodesmotic reaction that conserves the number of C_{sp^2}–C_{sp^2} bonds by choosing 1,3-butadiene as the reference, Reaction 3.20. The energy of this reaction is −21.5 kcal mol^{-1}, significantly less than the isodesmic Reactions 3.17–3.19. One might question the use of *trans*-1,3-butadiene as a suitable reference, since the stereochemistry about the formal C–C single bonds of benzene is cis.[150] Reaction 3.21 employs *cis*-1,3-butadiene as the reference, and gives the ASE for benzene as −30.8 kcal mol^{-1}. An alternative is to use 1,3-cyclohexadiene as the reference, which constrains the diene to the s-cis conformation, leading to Reaction 3.22.[150] The close agreement of the ASE values from Reactions 3.21 and 3.22 suggest that there is little ring strain in benzene, not an unexpected result.

ΔH_{rxn}(expt) = −21.5 kcal mol^{-1}

ΔH_{rxn}(G2MP2) = −22.0 kcal mol^{-1} (Reaction 3.20)

ΔH_{rxn}(G2MP2) = −30.8 kcal mol^{-1} (Reaction 3.21)

3 [cyclohexene] + [benzene] → [cyclohexadiene] + 3 [cyclohexane] $\Delta H_{rxn}(\text{expt}) = -30.5$ kcal mol^{-1}
$\Delta H_{rxn}(\text{G2MP2}) = -30.4$ kcal mol^{-1}

(Reaction 3.22)

An example of the importance of trying to separate out the effect of ring strain from aromaticity is in the evaluation of ASE for furan. The simple homodesmotic Reaction 3.23 predicts that its ASE is -7.1 kcal mol^{-1} (G2MP2). Reaction 3.24, proposed by Cyranski and Schleyer,[132] is a homodesmotic reaction that attempts to balance the RSE in the products and the reactants; it predicts that the ASE of furan is -12.2 kcal mol^{-1}. The difference in these two estimates, about 5 kcal mol^{-1}, is a reasonable estimate of the RSE of furan, based on the RSEs of tetrahydrofuran and cyclopentadiene (see Table 3.21).

[butadiene] + [divinyl ether] → [furan] + 2 [ethene] $\Delta H_{rxn}(\text{G2MP2}) = -7.1$ kcal mol^{-1}

(Reaction 3.23)

[cyclopentadiene] + 2 [2,3-dihydrofuran] + [cyclopentane] →
[furan] + 2 [2,5-dihydrofuran] + [tetrahydrofuran] $\Delta H_{rxn}(\text{G2MP2}) = -12.2$ kcal mol^{-1}

(Reaction 3.24)

One might question whether 1,3-butadiene is sufficiently long enough to capture the conjugation of benzene. The conjugation length in benzene, one might argue, is six carbons. Perhaps a better reference would be (3*E*)-1,3,5-hexadiene, which also has a conjugation length of six carbons. Hess and Schaad[151] proposed the "hyperhomodesmotic" Reaction 3.25 to determine the ASE of benzene. At G2MP2, it predicts a value of -19.4 kcal mol^{-1}. This is about 2 kcal mol^{-1} lower than the value predicted with Reaction 3.20. Therefore, the larger conjugation length might reduce the ASE values from Reaction 3.21 and 3.22 to about -28 kcal mol^{-1}. Since the formal double bonds in benzene are cis, a wiser reference is likely to be (3*Z*)-1,3,5-hexatriene, Reaction 3.26.[143] This reaction predicts the ASE of benzene as -24.3 kcal mol^{-1}. This value is probably a bit too low since the reference conformation is s-trans, rather than the s-cis of benzene (see the difference in ASE from Reactions 3.20 and 3.21). Unfortunately, the bis-s-cis conformation of (3*Z*)-1,3,5-hexatriene is nonplanar and likely suffers from steric interactions between its interior terminal hydrogen atoms.

3 [butadiene] → [benzene] + 3 [ethene] $\Delta H_{rxn}(\text{G2MP2}) = -19.4$ kcal mol^{-1} (Reaction 3.25)

3 [(3Z)-hexatriene] → [benzene] + 3 [ethene] $\Delta H_{rxn}(\text{G2MP2}) = -24.3$ kcal mol^{-1}

(Reaction 3.26)

Schleyer proposed one last alternative method for gauging ASE. Noting that many of these better methods (especially those analogous to Reaction 3.24) require computation of many compounds, he developed the "isomerization stabilization energy" (ISE) method, particularly useful for strained aromatic systems.[152] ISE measures the energy realized when an isomeric compound converts into its aromatic analog. Benzene itself cannot be analyzed by the ISE method, however, toluene can, and the ASE values of toluene and benzene are expected to be quite similar. The conversions of two different isomers into toluene provide the ISE for toluene (Reactions 3.27a and 3.27b). Both of these reactions do not conserve s-cis/s-trans diene conformations. Reaction 3.28 can be added once to Reaction 3.27a and twice to Reaction 3.27b to give the corrected ISE values of −32.0 and −28.9 kcal mol⁻¹, respectively.

ΔH_{rxn}(G2MP2) = −31.5 kcal mol⁻¹ (Reaction 3.27a)

ΔH_{rxn}(G2MP2) = −27.9 kcal mol⁻¹ (Reaction 3.27b)

ΔH_{rxn}(G2MP2) = −0.5 kcal mol⁻¹ (Reaction 3.28)

An interesting application of the ISE method is toward assessing the ASE of corannulene **34**. The energy of homodesmotic Reaction 3.29 is quite different whether one chooses s-*cis* or s-*trans* butadiene: 67.5 kcal mol⁻¹ with the former and 14.8 kcal mol⁻¹ with the latter. Exactly how one balances the total number of cis/trans relationships is problematic, but worse still is that Reaction 3.29 does not remove the effect of strain and nonplanarity of **34**. Dobrowolski et al.[134] proposed a series of reactions based on ISE, such as Reaction 3.30. Using B3LYP/6-311G** energies with ZPVE, the energy of Reaction 3.30 is 46.7 kcal mol⁻¹. Using all of the reaction variations, the mean value is 44.7 kcal mol⁻¹, with a standard deviation of only 1.2 kcal mol⁻¹. It is clear that corannulene has a rather substantial ASE.

(Reaction 3.29)

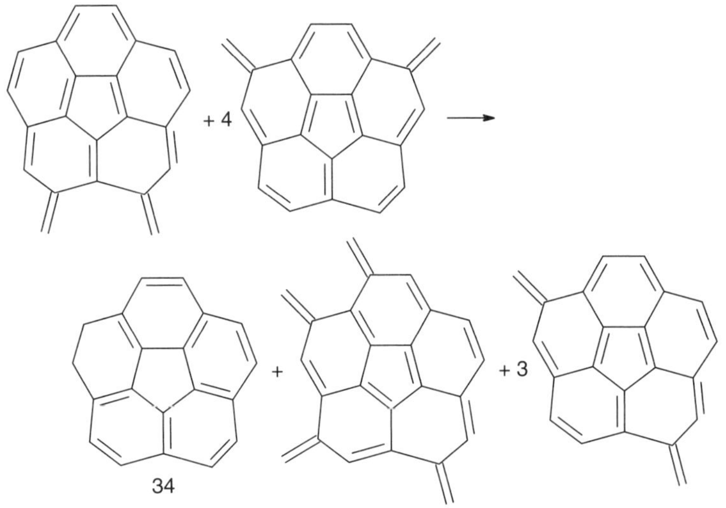

(Reaction 3.30)

While the different reaction schemes presented here show some variation, the methods that most carefully attempt to conserve all chemical properties other than "aromaticity" do come to an accord. These methods, typified in Reactions 3.18, 3.22, and 3.23, suggest the ASE of benzene is about -28 kcal mol^{-1}. Application of these techniques to other aromatics should provide quite reasonable estimates of their stabilization energies, and these will be utilized in upcoming sections of this book. It should, however, be noted that *all* of these procedures for estimating ASE contain arbitrary references—as Schleyer put it: "no choice is free from objection."[153]

3.5.2 Nucleus-Independent Chemical Shift (NICS)

Magnetic properties are often pointed to as being very distinctive for aromatic compounds.[154] Until fairly recently, the two most widely utilized magnetic properties for identifying and ranking aromatics were diamagnetic susceptibility exaltation (Λ^m) and ^{1}H NMR chemical shifts. Diamagnetic susceptibility exaltation is the increase of the molar susceptibility above that computed as a sum of atomic and bond increments. Dauben[155,156] popularized the notion that the value of the diamagnetic susceptibility exaltation can indicate aromatic character, that is, compounds with $\Lambda^m > 0$ are aromatic, those with $\Lambda^m \approx 0$ are nonaromatic, and those with $\Lambda^m < 0$ are antiaromatic. While Schleyer[126] has advocated that "magnetic susceptibility exaltation is the only unique applicable criterion" for aromaticity, it is not without its critics. Primary concerns include (1) that the value of Λ^m is dependent on a nonunique reference system to garner the increments and (2) Λ^m is dependent on the square of the area of the ring, making direct comparison of Λ^m values for different compounds problematic.

^{1}H NMR chemical shifts for aromatic systems generally appear in the region from 7 to 9 ppm, a downfield shift of about 2 ppm from typical alkenes. These are

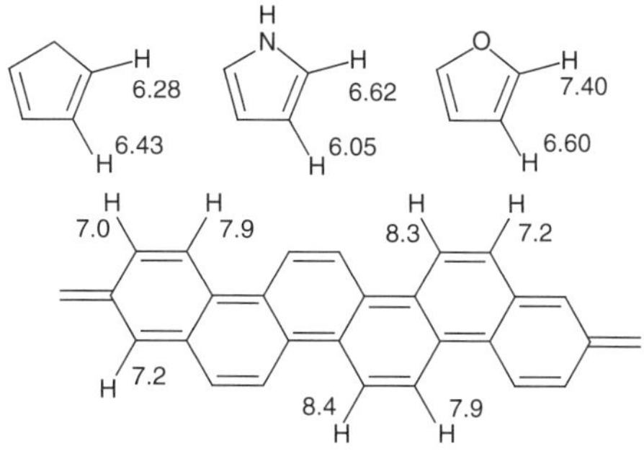

Figure 3.7 ^{1}H NMR chemical shifts (δ^1H, in parts per million) for cyclopentadiene, pyrrole, furan, and **35**, the last computed at PW91/IGLO-III//B3LYP/6-311+G**.[159]

for protons on the outside of a ring. The effect is even more dramatic for protons in the interior of a ring, such as in [18]annulene where the outside protons have their chemical shift at 9.17 ppm but the inner protons appear at −2.96 ppm.[157,158] Establishment of a relative aromaticity scale based on chemical shifts has not been forthcoming. Furthermore, Schaefer and Schleyer have argued that downfield chemical shifts do not necessarily indicate aromatic behavior.[159] They cite two prime examples. First, the protons of cyclopentadiene, pyrrole, and furan (the latter two are aromatic while the first is nonaromatic) have comparable chemical shifts (see Figure 3.7). Second, they computed the ^{1}H NMR shifts for **35**, shown also in Figure 3.7. Its proton chemical shifts range from 7.0 to 8.4 ppm, all well within the aromatic range; however, **35** is not an aromatic compound. It does not satisfy the $4n + 2$ rule and it displays significant bond alternation of 0.08 Å. Lazzeretti[160] criticized this work, arguing that the unusual downfield shifts in **35** do result from diatropic π-ring currents. Nonetheless, Lazzeretti agrees that downfield proton shifts are "not reliable aromaticity indicators."

In 1996, Schleyer[161] proposed a new magnetic measure of aromaticity, one that he hoped would avoid some of the pitfalls facing other aromaticity measures, namely dependence on some arbitrary reference and dependence on the ring size. Schleyer advocated the use of the absolute magnetic shielding computed at the geometric center of the ring. In order to comply with NMR chemical shift convention, the sign of this computed value is reversed to give the "NICS." When evaluated at the geometric center of the ring, they are called *NICS(0)*. Representative NICS(0) values for some aromatic, nonaromatic, and antiaromatic compounds are listed in Table 3.24.

It must first be remarked that there is *no* way to compare these computed NICS values with an experimental measurement, since there is *no* nucleus (typically) at the center of aromatic rings. The values in Table 3.24 demonstrate some small basis set dependence. Schleyer[161] recommended the use of diffuse functions for the evaluation of NICS. NICS is a local measure, a magnetic property at a single point. There are concerns over using such a local property to evaluate the global nature of

TABLE 3.24 NICS(0) (in ppm) for a Series of Hydrocarbons

Compound		NICS(0)^a	NICS(0)^b
Cyclohexane		−2.2	
Adamantane		−1.1	
Benzene		−9.7	−11.5
Naphthalene		−9.9	−11.4
Anthracene	(outer ring)	−8.2	−9.6
	(central ring)	−13.3	−14.3
Azulene	(five-ring)	−19.7	
	(seven-ring)	−7.0	
		−7.6	
		−14.3	−19.4
	(six-ring)		−12.5
	(five-ring)		−19.5
	(six-ring)		−12.4
	(five-ring)		−16.4
Cyclobutadiene		27.6	
Cyclooctatetraene	(planar)	30.1	
	(six-ring)	−2.5	
	(four-ring)	22.5	
	(six-ring)	−5.1	
	(four-ring)	19.0	
	(six-ring)	−8.6	
	(five-ring)	2.9	
			−4.2
	(six-ring)		−11.3
	(five-ring)		−1.4
	(six-ring)		−10.7
	(five-ring)		0.5

^{a}Computed at GIAO/6-31+G*//B3LYP/6-31G*.[161]
^{b}Computed at GIAO/6-31G*//B3LYP/6-31G*.[162]

a molecule, such as whether it is aromatic or not.[154] This is particularly troubling when the molecule has multiple rings.

Nevertheless, the NICS values appear to readily classify standard molecules into three discrete categories. Aromatic molecules possess NICS values that are negative. The values at the center of the six-member rings of benzene and naphthalene and anthracene are −9.7 and −9.9, respectively. Charged aromatic molecules also have negative NICS values; the values for cyclopentadienyl anion and tropylium cation are −14.3 and −7.6 ppm, respectively. Nonaromatic compounds like cyclohexane and adamantane have NICS values near zero. Lastly, antiaromatic molecules such as cyclopentadiene and planar D_{4h} cyclooctatetrane have NICS values that are positive, 27.6 and 30.1 ppm, respectively.

Schleyer[161] demonstrated the utility of NICS as an aromaticity metric by comparing NICS values with the values of ASE for a series of five-member ring systems. The correlation is quite acceptable, suggesting that NICS might be useful for rank-ordering relative aromatic character. An example of NICS used in this way was in the study of a series of dimethyldihydropyrene derivatives. NICS and other aromaticity measures were found to correlate.[163]

About a year later, Schleyer[153] advocated evaluating NICS at a point above the ring center. Dissecting the total NICS value into its σ- and π-contributions indicated a strong paramagnetic σ-contribution, one that rapidly falls off with distance.[164] For benzene, the π-contribution, which is diamagnetic, also falls off with distance above the ring (see Table 3.25). However, the total NICS value reaches a maximum at a value of 1 Å above the ring center. For the antiaromatic compound cyclobutadiene, the σ-contribution to NICS again falls off sharply with distance above the ring. However, the π-contribution, which is diamagnetic in cyclobutadiene, is negligible at the ring center and reaches a maximum at about 1 Å above the ring center. Thus, the positive NICS for cyclobutadiene when evaluated at its ring center is entirely due to σ-effects. In order to more clearly measure the magnetic consequence of ring current, which is associated with aromaticity, evaluating NICS at a distance of 1 Å above the ring center might prove more reliable. The choice of where to evaluate NICS is clearly arbitrary, and so NICS fails in reaching the goal to remove all arbitrary components of aromaticity measure. Stanger[165] has proposed using NICS evaluated along the line perpendicular to the ring center, calling this a NICS scan, a simplification of the 2D grid approach suggested by Klod and Kleinpeter[166] and by Schleyer.[164] Decomposing this into in-plane and

TABLE 3.25 Dissected NICS (ppm) for Benzene and Cyclobutadiene[a]

Compound	x^b	NICS(Total)	σ	π	CH	Core
Benzene	0	−8.8	+13.8	−20.7	−1.3	−0.6
	1	−10.6	+2.3	−9.6	−3.0	−0.3
Cyclobutadiene	0	+20.8	+23.2	−0.2	−1.6	−0.4
	1	+12.7	+1.5	+14.1	−2.8	0.0

[a]Evaluated at PW91/IGLO-III//B3LYP/6-311+G** from Ref. 164.
[b]Distance in angstroms above the ring center.

out-of-plane curves creates sets of plots that appear to differentiate aromatic from nonaromatic compounds. The in-plane curves are similar for aromatic and antiaromatic species, but the out-of-plane curves are dramatically different: antiaromatic compounds have large positive values that decline with increasing distance above the ring, while for aromatic compounds, the out-of-plane curve is negative near the ring center with a minimum at some small distance above the ring.

Pople's[167] original description of the induced magnetic field, which points opposite the applied magnetic field in the center of the ring, gives rise to the downfield chemical shifts of the protons of benzene. Lazzeretti[168] evaluated the out-of-plane component of the magnetic shielding tensor ($\sigma_\parallel$) on a 2D grid to demonstrate the shielding and deshielding zones of π-systems, particularly aromatic systems. These plots clearly demonstrate the "cones" one typically finds in introductory organic texts to explain NMR effects. Lazzeretti[160] further reinforced this notion by pointing out that aromaticity should relate solely to this magnetic effect in the direction perpendicular to the ring plane. These comments inspired Schleyer to propose another NICS variation.[169] Both NICS(0) and NICS(1) make use of the total isotropic shielding, the average of the three diagonal elements (xx, yy, and zz) of the shielding tensor. The component of the shielding tensor that is perpendicular to the aromatic ring (usually taken as zz) is consistent with Pople's argument of the induced magnetic field that arises from the aromatic character of the molecule. When evaluated at the ring center, this is called $NICS(0)_{zz}$. Perhaps even closer to the Pople model is to evaluate this perpendicular tensor component using only the π MOs, giving rise to $NICS(0)_{\pi zz}$. In comparing these new NICS indices with the NICS(0) and NICS(1) for a set of 75 mono- and polyheterocyclic five-member rings, Schleyer finds the best correlation with the ASE when using the $NICS(0)_{\pi zz}$ (the r^2 correlation coefficient is 0.980). This is better than the correlation with either NICS(0) ($r^2 = 0.946$) or NICS(1) ($r^2 = 0.935$). A significant benefit of this measure is its much broader range of than for the isotropic NICS(0).

NICS methods (NICS(0) and NICS(1)) have found widespread use in evaluating relative aromatic and antiaromatic characters for a variety of systems—aromatic transition states,[170–173] antiaromatic dications,[174–176] and aromatic bowls[177–179]

An interesting application of NICS is in characterizing planar cyclooctatetraenes. Cyclooctatetraene is tub-shaped to avoid the unfavorable anitaromaticity of a planar $4n$ π-electron system. Cyclooctatetraene is stabilized in the tub geometry by double hyperconjugation: $\sigma(CC) \rightarrow \pi^*(CC)$ and $\sigma(CH) \rightarrow \pi^*(CC)$.[180] Nonetheless, two examples of (nearly) planar cyclooctatetraene have been synthesized: **36** and **37**. Since **36** displays large bond alternation ($\Delta R = 0.169$ Å), one might expect both paratropic and diatropic ring currents to be attenuated. However, the NICS value of **36** is 10.6 ppm, indicative of considerable antiaromatic character, though this NICS value is much reduced from that in planar cyclooctatetraene constrained to the ring geometry of **36** (22.1 ppm).[181] Compound **37** is nearly planar in both the X-ray and computed (B3LYP/6-31G(d,p)) structures.[182] Its NICS(0) value is 17.4 ppm, indicative of antiaromatic character. So, both **36** and **37** are antiaromatic yet isolable!

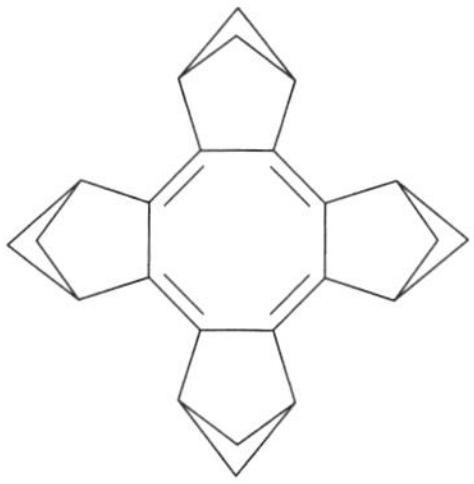

                    36 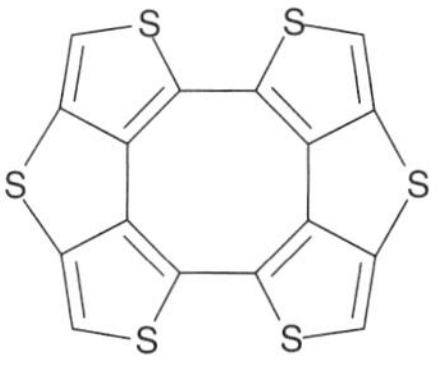   37

3.5.3 Case Studies of Aromatic Compounds

3.5.3.1 [n]Annulenes   The Hückel rule—that compounds containing $4n + 2$ π-electrons are stable and those containing $4n$ π-electrons are unstable—inspired the study of the monocyclic hydrocarbons consisting of alternating single and double C–C bonds, known as the *annulenes*. The prototype annulene, [6]annulene, better known as *benzene*, is the paradigm of aromatic compounds. Its counterpart is [4]annulene, cyclobutadiene, the archetype of antiaromatic compounds. We take up now the question of the larger annulenes. Are the [$4n + 2$]annulenes aromatic? What is the balance between energetic stabilization due to delocalization and destabilization from ring strain? Are the [$4n$]annulenes antiaromatic? Computational chemists have weighed in greatly on these issues, inspiring the synthesis of some very interesting hydrocarbons.

We begin with what should be the next-larger aromatic annulene after benzene, namely [10]annulene. The first characterization of [10]annulene was reported by Masamune and coworkers[183] in 1971. They identified two different isomers. The first isomer, type I, exhibits a single ^{1}H NMR signal ($\delta = 4.43$ ppm) and a single ^{13}C NMR signal ($\delta = 130.4$ ppm) at temperatures from $-40\,°C$ to $-160\,°C$. Based on this NMR data, Masamune argued that type I has the all cis configuration, and though the planar form is likely to be unstable due to ring strain, equivalent nonplanar boat-like forms (**38**) would be able to interconvert through low energy transition states. The NMR spectra of other isomer, type II, exhibits strong temperature dependence. At $-40\,°C$, both the ^{1}H and ^{13}C NMR show a single peak. At $-100\,°C$, however, the proton spectrum has two signals and the ^{13}C spectrum has five signals. Masamune argued that type II must have a single trans double bond and that migration of the trans double bond about the ring gives rise to the temperature dependence of its NMR spectra. Masamune suggested the "twist" conformation (**39**) for type II.

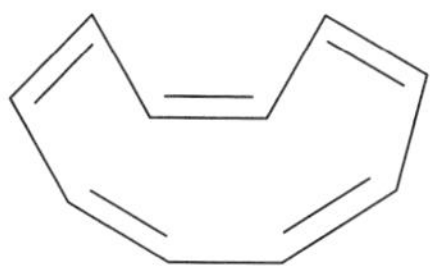

                    38 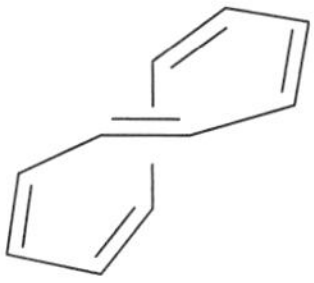   39

The first ab initio study of [10]annulene was the work of Schaefer and coworkers.[184] Eight different isomers of [10]annulene were optimized at MP2/DZP//HF/DZP. They first ruled out the two planar isomers, **40** having D_{10h} symmetry and **41** having D_{5h} symmetry, since both have multiple imaginary frequencies (3 and 2, respectively). The boat-like conformation suggested by Masamune (**38**) has one imaginary frequency. The two lowest energy local minima have the twist conformation (**39**) or a naphthalene-like conformation (**42**). An azulene-like isomer (**43**) was located from a large MM3 stochastic conformational search. The relative energies of these isomers are listed in Table 3.26. Based on these results, Schaefer argued that Masamune's type II could be the twist isomer **39**, but that type I could not be the boat form **38** since it is a transition state.

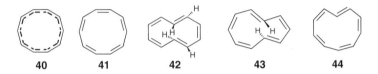

40   41   42   43   44

In a follow-up study, Schaefer and Schleyer discovered another isomer of [10]annulene, this one in a heart-like conformation (**44**). Masamune's original suggestion was that **44** would be the transition state interconverting mirror images of the twist isomer. Surprisingly, **44** is lower in energy than the twist isomer **39** at both the MP2 and B3LYP levels (Table 3.26). There is little bond alternation in **44** ($\Delta(C-C) = 0.050$ Å), its magnetic susceptibility exaltation is very large (−66.9 ppm cgs), and its predicted ^{1}H NMR signals are −6.2 ppm for the inward

TABLE 3.26 Relative Energies (kcal mol^{-1}) of [10]Annulene Isomers

Isomer	Rel. E^a	Rel. E^b	Rel. E^c	Rel. E^d	Rel. E^e	Rel. E^f	Rel. E^g	Rel. E^h	Rel. E^i
39	0.0	0.0	0.0	0.0	0.0	0.0	0.0	0.0	0.0
42	0.62	0.53			2.04	1.74	1.40	1.15	1.7
43	5.11	−4.15			8.61			4.60	6.3
38	6.02	6.81			5.66			4.42	5.4
40	13.79								
41	22.04								
44		−7.06	−6.99	+3.52	5.99	6.29	4.24	2.05	5.0

^{a}Calculated at MP2/DZP//HF/DZP with ZPE at HF/DZP, Ref. 184.
^{b}Calculated at MP2/TZ2P//MP2/DZd, Refs 185, 186.
^{c}Calculated at B3LYP/TZP, Ref. 185.
^{d}Calculated at CCSD(T)/cc-pVDZ//B3LYP/DZd, Ref. 187.
^{e}Calculated at CCSD(T)/DZd//MP2/DZd, Ref. 186.
^{f}Calculated at CCSD(T)/DZd, Ref. 186.
^{g}Calculated at CCSD(T)/TZ2P//CCSD(T)/DZd, Ref. 186.
^{h}Calculated at KMLYP/6-311+G**, Ref. 188.
^{i}Calculated at CCSDT(T)/cc-pvDZ//CCSD/6-31G, Ref. 189.

proton and 7.9–8.6 ppm for the outward protons. All of these values are consistent with **44** having significant aromatic character.

Unfortunately, the situation became much more complicated. Schaefer and Luthi noted that **44** would give rise to six ^{1}H NMR signals, not the five noted in the experiment. In addition, single-point computations at CCSD(T) reverse the relative energies of the twist and heart isomers (Table 3.26), suggesting that the heart isomer cannot be described by just a single configuration wavefunction.[187]

In 1999, Schaefer and Stanton[186] reinvestigated the isomers of [10]annulene by optimizing the structures of **39**, **42**, and **44** at CCSD(T)/DZd. These are shown in Figure 3.8, along with the structures, optimized at other levels, for the remaining isomers. Isomers **39** and **42** display large bond alternation, but the C–C distances in **44** vary by only 0.046 Å. Their relative energies are listed in Table 3.26. The two lowest energy isomers at CCSD(T)/DZd//MP2/DZd are the twist isomer **39** and the naphthalene isomer **42**. The heart-like isomer **44**, which as noted before was very stable at the MP2 and B3LYP levels, is much higher than these two isomers.

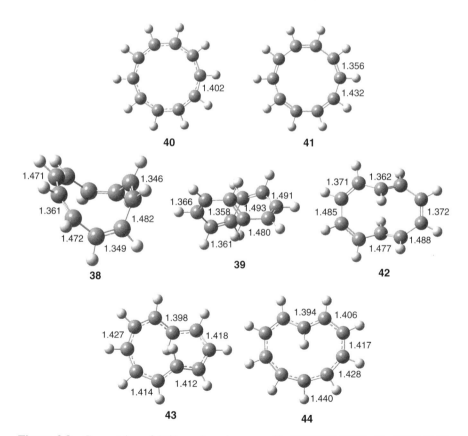

Figure 3.8 Geometries of [10]annulene isomers; **40**: B3LYP/6-31G*; **41**: HF/6-31G*; **38**: B3LYP/DZd; **39**: CCSD(T)/DZd; **42**: CCSD(T)/DZd; **43**: B3LYP/DZd; and **44**: CCSD(T)/DZd.

TABLE 3.27 Computed[a] ^{13}C Chemical Shifts for 39, 42, and 44 Compared with the Experimental Values for Type II

Conformer	δ(ppm)
39	128.0 130.7 130.9 131.2 131.9
42	126.1 128.3 128.6 133.2 144.8
44	125.6 126.1 127.1 129.8 130.3 136.2
Type II (expt)	128.4 131.4 131.6 132.3 132.5

[a]Chemical shifts estimated by the additivity equation: σ = σ(MP2/TZP) − σ(CCSD(T)/DZP) − σ(MP2/DZP), Ref.190.

Since the geometries of these isomers were found with what may be inappropriate wavefunctions, reoptimization with the CCSD(T) method is warranted. However, the relative ordering of these three isomers is unchanged upon this reoptimization, and so these best energy computations indicate that the two most favorable conformations of [10]annulene are **39**, which agrees with Masamune's type II, and **42**.

Stanton[190] confirmed the assignment of type II as the twist isomer by computing the ^{1}H NMR shifts of **39**, **42**, and **44**, shown in Table 3.27. The heart isomer (**44**) gives rise to six signals, inconsistent with the five observed in the experiment. The differences between the experimental chemical shifts and the computed values for isomer **39** are less than 1.1 ppm but 13.1 ppm for **42**. The relative spacings between the peaks also match much better with **39**. This still leaves open for question the exact nature of Masamune's type I isomer.

Castro and Karney[189] examined the mechanism for the possible interconversion of [10]annulene isomers. Masamune's original explanation of the NMR of type I and type II forms of [10]annulene involves interconverting structures, so location of the appropriate transition states might help account for the NMR spectra.

The transition state that connects **39** to **44** is 10.1 kcal mol^{-1} above **39** (Figure 3.9a) at CCSDT(T)/cc-pvDZ//CCSD/6-31G. This is consistent with the variable temperature NMR corresponding to Masamune's type II isomer, and his interpretation of this structure, with the proviso that **39** is not the transition state but an intermediate. The psuedorotational barrier of **38** is essential negligible; it lies in a very shallow well with little energy difference between C_2 and C_s forms. Therefore, **38** is consistent with an NMR spectrum that shows a single carbon peak despite increasing the temperature—Masamune's type I isomer.

Isomer **39** isomerizes via a bond shift. This barrier is computed to be 3.9 kcal mol^{-1} (Figure 2.1b). This small barrier implies that at higher temperature, it would show three ^{13}C NMR signals, inconsistent with either of Masamune observed [10]annulene structures. Isomer **43** can undergo a conformational change through a barrier of only 2.7 kcal mol^{-1} (Figure 3.9b). This process creates five unique carbon atoms. Bond shifting of **43** has a barrier of only 0.6 kcal mol^{-1}, so when combined with the conformational process, all 10 carbon atoms are equivalent. This could be consistent with the type I NMR spectrum. Castro and Karney, however, identified a transition state that connects **42** to **43**, with

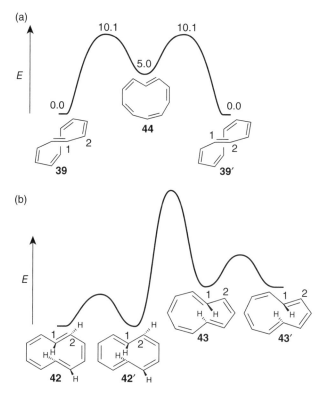

Figure 3.9 Potential energy surfaces (CCSDT(T)/cc-pvDZ//CCSD/6-31G) for (a) the bond shift of **39** through **44** and (b) the bond shift of **42**, its conformational change into **43** and its conformational change to **43'**.[189]

a barrier of 12.6 kcal mol^{-1} relative to **43**. At higher temperature, **43** would convert to **42** and so the NMR would show three signals. Since this does not occur, they discount **42** or **43** as the type I structure. The net result is that in fact Masamune was correct after all: the type I compound is **38** and the type II compound is **39**.

The energetics of the structures of 1,2-didehydro[10]annulene differs markedly from that of [10]annulene.[191] The lowest energy conformation has one trans double bond and has the heart-like conformation (**45**). The next lowest energy isomer, lying 17.0 kcal mol^{-1} above **45** at CCSD(T)//cc-pvDZ//B3LYP/6-31G*, is the all cis form, with C_{2v} symmetry (**46**). The removal of the two hydrogens from [10]annulene relieves some angle strain, allowing **45** to be the most stable isomer. Both express little bond alternation. **46** is a planar structure, while **45** is nearly so. The predicted ^{1}H NMR shifts are from 7.8 to 8.4 ppm for the external hydrogens and −1.5 ppm for the internal hydrogen of **45**, while chemical shifts of the external hydrogen atoms of **46** are from 8.3 to 8.8 ppm. Thus, these two isomers both possess aromatic character.

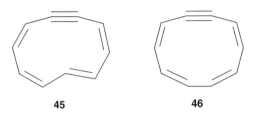

45 **46**

Controversy surrounds the nature of [14]annulene and [18]annulene as well. As the annulenes become larger, a more localized structure is expected since the ASE per carbon is predicted to decrease with increasing size.[135,192] Larger annulenes should thus become more polyene-like.[193] This turnover from delocalized, aromatic structures to localized polyene structures has been suggested to occur with [18]annulene.[135]

The X-ray crystal structures of both [14]annulene and [18]annulene have been interpreted as evidence of aromatic character. [14]annulene is nonplanar due to steric interactions between the internal hydrogen atoms, but the bond alternation is only 0.057 Å, about the same as in naphthalene.[194] The two X-ray structures of [18]annulene support a D_{6h} geometry with C–C bond distances differing by 0.037 Å at 80 K[195] and 0.020 Å at 111 K.[196]

B3LYP/6-311G* and MP2/6-31G* optimizations of [14]annulene both converged to a C_s structure, with $\Delta(CC)$ = 0.014 and 0.013 Å, respectively.[197] Optimization of the geometry of [18]annulene gave a D_{6h} structure with ΔCC = 0.017 Å at B3LYP/6-311G* and 0.016 Å at MP2/6-31G*. Baldridge and Siegel[198] also optimized the structure of [18]annulene with three different DFT methods, finding a D_{2h} minimum with each. At B3PW91/DZ(2df,2p), $\Delta(CC)$ = 0.018 Å. All of these calculations suggest a delocalized, and aromatic structure for [18]annulene, consistent with the x-ray experiments.[195,196] The only major concern was the very poor agreement between the computed ^{1}H NMR shifts of [18] annulene (9.7 and −15.3 ppm) and the experimental[157,158] values (9.17 and −2.96 ppm). The authors suggest that "(t)he discrepancy between experiment and theory may arise ... because the ring current is overestimated at these computational levels."

The Schaefer and Schleyer studies on [10]annulene certainly support this suggestion. DFT and MP2 provided unsatisfactory descriptions of the different isomers of [10]annulene, artificially favoring delocalized structures, such as **44**. Noting that since various DFT methods overestimate the degree of delocalization and that HF underestimates it, Schaefer and Schleyer proposed using DFT methods with large HF components to study [14]- and [18]-annulene.[188] Using the KMLYP/6-311+G** and BHLYP/6-311+G** methods, they found that D_2 and C_{2h} forms of [14]annulene are transition states. The lowest energy isomer of [14]annulene is **47**, having C_s symmetry (Figure 3.10). For [18]annulene, the D_{6h}, D_{3h}, and D_3 structures have one imaginary frequency, and the lowest energy structure, **48**, has C_2 symmetry.

Justification for these low symmetry isomers comes from their computed ^{1}H NMR spectra. The chemical shifts for **47** range from 0.0 to −0.4 ppm for its inner

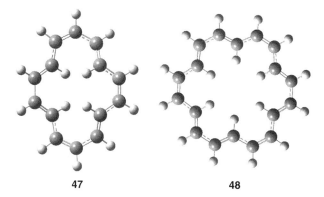

Figure 3.10 BHLYP/6-311+G** optimized geometries of **47** and **48**.[188]

hydrogens and 7.4 to 8.2 ppm for its outer hydrogens, in fine agreement with the experimental values of 0.0 and 7.6 ppm.[199] The X-ray prediction of [14]annulene is for a C_{2h} structure. Computed chemical shifts for this geometry are −6.1 ppm for the inner protons and 8.7–9.4 ppm for the outer protons, completely different than the experimental values. The BHLYP- and KMLYP-predicted chemical shifts for **48** are −2.3 to −2.9 ppm for the inner protons and 8.9 to 9.4 ppm for the outer protons, in excellent agreement with the experimental values[158] of −2.96 and 9.17 ppm. Again the computed chemical shifts (about −11 and 11.5 ppm) for the D_{6h} structure, which is predicted by the X-ray experiments, are in total disagreement with the experimental values. Schaefer and Schleyer suggest that the X-ray structures are incorrect, and that lower symmetry forms are found for these two annulenes. They also note that both [14]- and [18]annulene express aromatic character in these NMR shifts. Nevertheless, the bond alternation is each is appreciable: $\Delta(CC) = 0.101$ Å for **47** and 0.100 Å for **48**. Thus, they conclude, "conjugated $4n + 2$ π-electron systems may have appreciable bond alternation and still be delocalized and aromatic."

We turn our attention to the [$4n$]annulenes. Heilbronner[200] suggested the idea that a conjugated, cyclic system might be stabilized, that is, aromatic,[201] if there was a twist along the π orbitals, creating a Möbius topology. Schleyer[202] investigated the possibility of a Möbius [12]-, [16]-, and [20]annulenes using B3LYP/6-31G*. The lowest energy [12]annulene has Hückel topology, but **49**, which has a Möbius topology, lies only 4.4 kcal mol^{-1} higher in energy. Its bond alternation is rather small, $\Delta(CC) = 0.070$ Å, (Figure 3.11) and its NICS(0) value is −14.6 ppm, both characteristic of an aromatic compound. An isomer of [16]annulene, **50**, expresses even stronger aromatic behavior: its $\Delta(CC)$ is very small, only 0.019 Å, and its NICS(0) = −14.5 ppm. However, **50** is 15.8 kcal mol^{-1} higher in energy than the most favorable Hückel [16]annulene, and it is even 10 kcal mol^{-1} higher than the lowest energy Möbius [16]annulene, though this conformer shows little aromatic character. Lastly, the Möbius [20]annulene, **51**, lies 6.2 kcal mol^{-1} above the best Hückel form, but it does show some aromatic character, with a large negative NICS (−10.2 ppm) and slight bond alternation ($\Delta(CC) = 0.049$ Å).

Figure 3.11 B3LYP/6-31G* optimized structures of the Möbius annulenes **49–51**.[202]

Herges performed a Monte Carlo search of the [16]annulene isomers, identifying 153 structures that were then optimized at B3LYP/6-31G*.[203] The three lowest energy isomers are of Hückel topology. **50** is predicted to be some 10 kcal mol^{-1} above the lowest energy Möbius isomer, which is 5 kcal mol^{-1} above the lowest energy Hückel structure. A principal component analysis of these [16]annulene isomers using a set of typical aromaticity metrics, including NICS, separates the annulenes into two groups: the set of Hückel isomers and the set of Möbius isomers. However, the relative energy of the [16]annulenes does not correlate with any of the aromaticity measures.

These calculations inspired the synthesis of a derivative of [16]annulene with a Möbius topology, **52**, by Herges.[203,204] A total of five isomers were isolated, and crystal structures of three of them were obtained. Of these three, two have Möbius topology and one is of Hückel topology. For the Möbius isomer of C_2 symmetry, the bond alternation in the polyene bridge is 0.120 Å in the X-ray structure and 0.095 Å in the B3LYP/6-31G* structure. Herges computed an ASE, using a modification of Schleyer's ISE method,[205] of 4.04 kcal mol^{-1}, concluding that it is "moderately aromatic."

Subsequent computations by Schleyer and coworkers[206] have cast serious doubts on the aromatic nature of **52**. They reoptimized the geometry of **52** at B3LYP/6-311+G** and its geometry is compared with that of the X-ray structure in Figure 3.12. They first noted that the bond alternation within the bridge is larger than the alternation found in long chain polyenes. Further, they pointed out that the bond alternation should be examined for the entire 16-carbon path and not just the

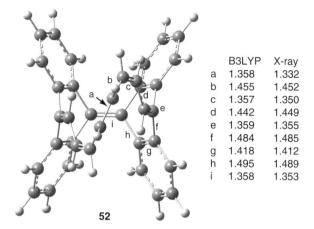

Figure 3.12 Comparison of the B3LYP/6-311+G**[206] and experimental[204] distances of **52**.

bridge. The degree of alternation is even larger for the entire path than just within the bridge, 0.135 Å at B3LYP and 0.157 Å in the X-ray structure. The computed NICS(0) value is only −3.4 ppm, much below that found in the parent **50**. The ISE value of **52** is nearly identical to the value obtained for an acyclic analog of the bridge alone, and the ISE value when corrected for s-cis/s-trans relations is only 0.6 kcal mol⁻¹. These authors conclude that **52** is not aromatic. They attribute the lack of aromatic character to large deviations from planarity about the double bonds and the tendency of the C–C bonds within benzene rings to not delocalize into fused rings. The [16]annulene core of **52** requires the formal participation of two bonds that are also shared with a benzene ring. However, the electrons in these bonds preferentially "localize" to the benzene ring and do not fully participate in creating a delocalized 16-member ring.[207] Herges has countered that a principal component analysis of 25 isomers of **52** divides them into a set of Hückel isomers and a set of Möbius, and the latter reflect aromatic properties. Herges claims that the benzoannelation in **52** may "reduce its aromaticity but not to zero."[203]

It seems clear that *definitive* evidence of a synthesized aromatic Möbius annulene has not been produced. We note in passing that a few examples of aromatic Möbius hexaporphyrins have been reported.[208–211] Castro and Karney[212] have suggested that an aromatic Möbius annulene has been prepared, but as a transition state. The optimized transition state **54** for the "twist-coupled bond shifting" process then connects the tri-*trans*-[12]annulene **53** with the di-*trans*-[12]annulene **55**. This transition state, shown in Figure 3.13, is 18.0 kcal mol⁻¹ above **53** at CCSD(T)//cc-pvDZ//BH&HLYP/6-311+G**, which corresponds well with the activation barrier for the conversion of **53** to **55** observed by Oth[199] as the rate-limiting step in producing **56**. **54** has aromatic properties—its NICS value is −13.9 ppm and its bond alternation is only 0.032 Å—supporting the notion of a Möbius annulene transition state.

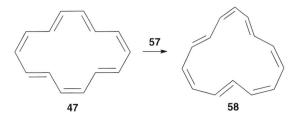

Castro and Karney conjectured that the bond shift that takes **47** into **58** should proceed through a Möbius antiaromatic transition state.[213] They located a Möbius TS **57**, shown in Figure 3.13. The value of NICS(0) is +19.0 and the computed chemical shifts of the inner hydrogens are 26.4 and 26.7 ppm, all indicative of its Möbius antiaromatic nature.

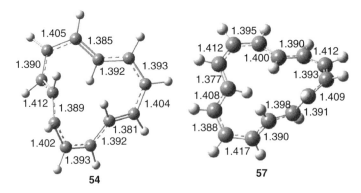

We next turn to the possibility of a cationic Möbius annulene. Schleyer had noted in the solvolysis of **59** the production of **61** with mondeuteration at all positions.[214] This implicated the [9]annulene cation **60** as an intermediate, which Yakali[215] suggested might have Möbius topology. Schleyer subsequently identified the Möbius structure of **60** at B3LYP/6-311+G** as being aromatic, based on its very negative NICS(0) value of −13.4 ppm.[216] Herges and Grimme[217] find that the Möbius form of **60** (all proteo-form) is 1 kcal mol^{-1} more stable than the Hückel form at B3LYP/cc-pVTZ but this reverses with the KMLYP

Figure 3.13 BH and HLYP/6-311+G** optimized structures of **54** and **57**.[212,213]

functional. A definitive CCSD(T)/CBS//CCSD(T)cc-pVTZ computation including solvation corrections indicates that they differ by only 0.04 kcal mol^{-1}. Laser flash photolysis of proteo-**59** suggests, however, that only the Hückel structure is formed, and that its short lifetime is due to rapid electrocyclic ring closure.

In a follow-up study, Herges[218] has examined the larger annulene cations, specifically [13]-, [17]-, and [21]-annulenes. The Möbius form of [13]-annulene cation (**62m**) is predicted to be 11.0 kcal mol^{-1} lower in energy that the Hückel (**62h**) form at B3LYP/6-311+G**. The Möbius cation **62m** is likely aromatic, having NICS(0) = −8.95. Electrocyclic ring closure of **62m** requires passing through a barrier of at least 20 kcal mol^{-1}, suggesting that **62m** is a realistic target for preparation and characterization (Figure 3.14). The energy difference between the Möbius and Hückel structures of the larger annulenes is very dependent on computational method, but in all cases the difference is small. Thus, Herges concludes that the [13]-annulene cation should be the target of synthetic effort toward identification of a Möbius annulene. Additionally, Herges finds that the [14]annulene dication in its Möbius form is predicted to be lower in energy than its Hückel isomer, by 7.8 kcal mol^{-1} at B2PLYP/6-311+G** and 5.4 kcal mol^{-1} at CCSD(T)/cc-pVTZ.[219] The NICS value of the Hückel [14]annulene dication is −15.6 ppm, indicating it is aromatic. This is another potential synthetic target.

The common Möbius strip has a single half twist. Ribbons can be even more twisted and so higher order annulenes might be possible, expressing potential aromatic or antiaromatic properties.[220] Rzepa et al.[221] have found examples of these higher order twisted annulenes, such as $C_{14}H_{14}$ (**63**) with one full twist,[222] $C_{16}H_{16}^{2-}$

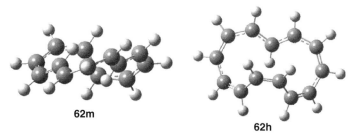

Figure 3.14 B3LYP/6-311+G** optimized structures of the [13]annulene cation **62m** and **62h**.[219]

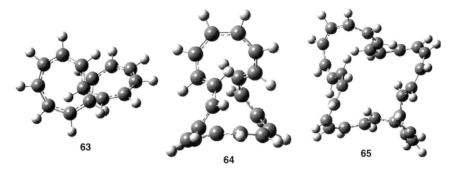

Figure 3.15 Optimized structures of higher order twisted annulenes **63–65**.[221]

(**64**) with three half twists, and $C_{20}H_{20}^{2+}$ (**65**) with four half twists (Figure 3.15). These higher order twisted annulenes are aromatic, as determined by a variety of measures. For example, all express negative NICS values, all have positive diamagnetic exaltations, and all have positive ASE. The full twisted [14]annulene **63** has been implicated in the rearrangement pathway of **65** to **47**.[223] Herges[224] has prepared a substituted [36]annulene, and four different isomers were isolated, two of which have a full twist. Though not isolated in experiment, a computed low lying conformer has three half-twists.

3.5.3.2 The Mills–Nixon Effect

The Mills–Nixon effect is bond localization within a benzene ring due to annelation of one or more small rings. Mills and Nixon actually were interested in isolating benzene in one of its two Kekule tautomeric structures.[225] While this two-well model for benzene was thoroughly discredited by the Pauling resonance model, the idea of bond localization within a formal benzene ring remains of both theoretical and experimental interests.[226] The search for compounds demonstrating the Mills–Nixon effect is the subject of this section.

The idea behind bond localization in annelated benzenes is exemplified by compound **66**, tricyclopropabenzene. Of the two standard Kekule structures of **66**, one places the double bonds within the three-member ring (endo) and the other places the double bonds outside (exo) of the small ring. Since cyclopropene is much more strained than cyclopropane, the avoidance of strain energy suggests that the double bonds might be localized into the exo positions. Compounds **66** and **67** are the prime test subjects for this idea of strain-induced bond localization (SIBL).[227]

Both **66** and **67** have been examined with a variety of computational techniques, from HF[228,229] to CASSCF[230] to DFT.[227,231] The X-ray crystal structure[232] of **67** has been determined. Bond localization can be measured as the difference in the length of the $R(\text{C}-\text{C}_{\text{endo}})$ and $R(\text{C}-\text{C}_{\text{exo}})$ bonds, $\Delta R = R_{\text{endo}} - R_{\text{exo}}$. The computed and experimental values of these distances and ΔR are listed in Table 3.28. Bond localization is not observed in any of the calculations of **66** and **67**, and the crystal structure of **67** also indicates only a very small difference in the bond lengths about the central ring. Clearly, the Mills–Nixon effect is *not* operational in these two compounds.

Was the idea of SIBL then just wishful thinking? In order to assess the effect of angle strain alone upon a benzene ring, Stanger[228] optimized the geometry of benzene with fixed C–C–H angles. Since rehybridization at hydrogen is minimal and hydrogen does not participate in the π-system, Stanger argued that relief of angle strain should be the only driving force behind any bond localization in this model. At HF/3-21G, he found that ΔR increased from 0.0 Å in normal benzene to 0.2427 Å when the C–C–H angle is 90°. Stanger found that there was a correlation between the distortion angle α, where $\alpha = (120° - \theta_{\text{C}-\text{C}-\text{H}})$, and ΔR.

$$\Delta R = A \sin^2 \alpha + B \qquad (3.10)$$

While the DFT computations indicate a smaller bond localization that predicted at HF/3-21G, distinct bond localization *does* result from C–C–H angular distortion from the normal 120° value.

So why is there little bond localization in **66** and **67**, especially since each suffers from large angular distortions ($\theta_{\text{C}-\text{C}-\text{C}}$ equals 62.6° in **66** and 93.2° in **66**)? Using the DFT values, Eq. (3.10) predicts that ΔR should be 0.143 Å for **67**, significantly greater than its actual value of 0.025 Å. Stanger suggested that the bonds in the three- and four-member rings are highly bent, thereby relieving some of the strain at the carbon atoms common to both the benzene and small ring. There is both experimental and computational support for this notion of bent bonds in the annelated benzenes. The electron density maps available from the X-ray crystal structures of **67–69** show maxima that are shifted outside of the four-member rings. While the quality of the electron density map[232] of **67** does not allow for a measurement of size of this shift, these maxima could be located for **68** and **69**.[236] The bond angles measured along the maximum electron density paths are 106° in **68** and 107° in **69**, much larger than their angles measured from the positions of the nuclei (about 93.5° in both **68** and **69**). The topological electron density analysis defines bond angles as the union of two bond paths at the central atom. Yanez[233] has computed this topological angle for **66** at HF/6-31G* and MP2/6-31G*, finding values of 85.0° and 81.7°, respectively. Again, these electron-density-defined angles are much larger than the angles defined by the positions of the nuclei (63.0° and 62.8°, respectively).

TABLE 3.28 Length (Å) of the Endo and Exo Bonds of the Central Benzene Ring and Their Difference (ΔR) in 66–67 and 70–73

Method	R_{endo}	R_{exo}	ΔR
66			
HF/6-31G(d)[a]	1.359	1.356	0.003
MP2/6-31G*[b]	1.377	1.367	0.010
B3LYP/6-31G*[c]	1.384	1.361	0.023
B3LYP/aug-cc-pVDZ[d]	1.390	1.363	0.027
B3LYP/6-311+G(d,p)	1.384	1.356	0.027
CASSCF(6,6)/6-31G(d)[e]	1.371	1.361	0.010
67			
HF/3-21G[f]	1.408	1.361	0.047
HF/6-31G(d)[a]	1.391	1.378	0.013
HF/6-311G**[g]	1.391	1.377	0.014
MP2/6-31G*[g]	1.406	1.389	0.017
B3LYP/6-31G*[c]	1.410	1.386	0.024
B3LYP/aug-cc-pVDZ[d]	1.412	1.389	0.023
B3LYP/6-311+G(d,p)	1.408	1.383	0.025
CASSCF(6,6)/6-31G(d)[e]	1.402	1.386	0.016
Expt[g]	1.401	1.383	0.018
	1.413	1.390	0.023
70			
HF/6-31G(d)[a]	1.440	1.344	0.096
MP2/6-31G(d)[h]	1.430	1.373	0.057
B3LYP/aug-cc-pVDZ	1.443	1.366	0.077
CASSCF(6,6)/6-31G(d)[e]	1.451	1.353	0.098
Expt[i]	1.438	1.349	0.089
71			
HF/3-21G[e]	1.523	1.309	0.214
HF/6-31G(d)[a]	1.500	1.317	0.183
B3LYP/6-31G*[c]	1.515	1.338	0.177
B3LYP/aug-cc-pVDZ[d]	1.516	1.342	0.174
B3LYP/6-311+G(d,p)	1.514	1.335	0.178
72			
HF/6-31G(d)[a]	1.481	1.329	0.152
CASSCF(6,6)/6-31G(d)[e]	1.451	1.353	0.098
B3LYP/6-311+G**	1.489	1.346	0.143
Expt[h]	1.495	1.335	0.160

[a] Ref. 229.
[b] Ref. 233.
[c] Ref. 227.
[d] Ref. 231.
[e] Ref. 230.
[f] Ref. 228.
[g] Ref. 232.
[h] Ref. 234.
[i] Ref. 235.

The lack of significant bond alternation in **66** and **67** appears to be attributable to bent bonds relieving some of the strain. Baldridge and Siegel[229] offered up **70** as a strained annelated benzene whose bridgehead carbon should be reluctant to participate in a bent bond to the aromatic ring. Their HF/6-31G* optimized structure of **70** has endo and exo bond lengths of 1.440 and 1.344 Å, respectively, and appreciable bond alternation of almost 0.1 Å. Subsequent calculations that include electron correlation (Table 3.28) show similar alternation. This prediction was confirmed with the synthesis and X-ray structure determination of **70** by Bürgi and coworkers;[235] its endo and exo bond lengths are 1.438 and 1.349 Å, respectively. **70** clearly demonstrates that the bond alternation is properly attributed to SIBL.

With the distinct bond alternation of **70**, might its aromaticity be reduced? The NICS values at the center of the six-member ring of benzene, **68**, and **70** computed at B3LYP/6-31G* are −9.65, −10.13, and −7.34 ppm, respectively.[237] The NICS values for **66** and **67** at B3LYP/aug-cc-pVDZ are −10.7 and −9.0 ppm, respectively. These NICS values indicate that **66** and **67**, with little bond alternation, are certainly aromatic. While the NICS value of **70** is somewhat reduced from that of benzene, it is nevertheless fairly negative and suggests appreciable aromatic character.

A second category of benzannelated compounds that might express the Mills–Nixon effect are exemplified by **71** and **72**. Vollhardt's 1986 synthesis of starphenylene (**72**) revitalized interest in the Mills–Nixon effect.[234] The X-ray crystal structure of **72** indicates a long endo distance of 1.495 Å and a short exo distance of 1.335 Å—strong evidence of the Mills–Nixon effect. Computations have reproduced this large bond alternation in **72** (Table 3.28).

71 is a smaller analog of starphenylene. The computed structure of **66** also shows bond alternation; in fact, it is larger than in any of the other examples (Table 3.28).

The DFT computations suggest that ΔR is about 0.17 Å, comparable to the difference in length between normal C–C single and double bonds.

Vollhardt[234] and Baldridge and Siegel[229] have attributed the bond localization in **71** and **72** to avoidance of the cyclobutadiene structure. Proposing that the annelated molecule will distort to maximize the number of $4n + 2$ π-electron rings and minimize the number of $4n$ π-electron rings, Baldridge and Siegel examined the series of compounds displayed in Scheme 3.3. These compounds are divided into three categories based on whether the annelation is $4n + 2$ (aromatic), neutral if it does not participate in the π-system, or $4n$ (antiaromatic). As discussed before, there is a little bond alternation in the neutral examples **66** and **67**. Since the nitrogen p-orbital contributes two electrons, the three-member rings of **74** are formally antiaromatic, just like the four-member rings of **71**. For both of these compounds, ΔR is quite large—the endo bond is long to minimize the π-contribution from the bond shared by the two rings. This reduces the antiaromatic character of the annelated ring. On the other hand, the p-orbital of boron is empty, making the three-member rings of **73** potentially aromatic. The four-member rings of **75** are similarly potentially aromatic. The vales of ΔR for **73** and **75** are very negative (a reversed Mills–Nixon effect). The endo bond is quite short in these compounds. By building up the π-character of the bond shared by the two rings, both rings can express aromatic character. The fact that bond localization is greater in **71** than in **72** can be understood in terms of the avoidance of the $4n$ ring. In **71**, the four-member ring will always have full participation of the two π-electrons from its double bond. This forces the doubles bonds of the six-member ring to localize completely to the exo positions to minimize any antiaromatic character. However, in **73**, the outside phenyl rings do not fully contribute 2 π-electrons to the four-member ring due to delocalization, allowing the central ring to delocalize some π-density into the endo positions.

Borden,[238] on the contrary, argued that the bond alternation is a consequence of the interaction of π MOs of the annelating groups with the π* MOs of benzene. Stanger[128,227] has argued that the localization in the molecules of Scheme 3.3 can be solely attributed to SIBL by properly accounting for the angle formed by the bond paths via Eq. (3.10).

The NICS(0) value for **71** is +1.3 ppm.[231] The NICS(1) value for the central six-member ring of **72** is −1.1 ppm.[239] These NICS measures indicate that the central six-member ring of both compounds is not aromatic. This is consistent with the extreme bond localization in these molecules. The experimental heat of hydrogenation of **72** is −71.7 kcal mol^{-1}.[240] When this is corrected for the strain in the resulting cyclohexyl ring, the heat of hydrogenation of **72** is nearly identical to that of three times the heat of hydrogenation of cyclohexene. Vollhardt[240] concluded from these studies that the central ring of **72** behaves as if it were cyclohexatriene and not benzene.

Though the explanation first offered by Mills and Nixon to account for bond alternation in annelated species is discredited, bond alternation is certainly produced by angle strain. Somewhat more controversial is the still lingering arguments that π-effects cause bond alternation. Perhaps the answer to the question posed by

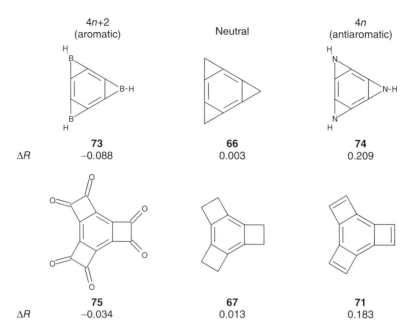

Scheme 3.3 Bond alternation of aromatic neutral and antiaromatic benzanneleated species computed at HF/6-31G(d).[229]

Stanger[228] in the title of his 1991 article "Is the Mills Nixon effect real?" is both "no" and "yes"—no, bond alternation does not result from a freezing out of a particular Kekule structure, but yes, bond alternation is most certainly observed in a number of "aromatic" compounds.

3.5.3.3 Aromaticity Versus Strain The Mills–Nixon effect discussed in the previous section is mostly a strain effect: angles about an aromatic carbon are forced to be nonstandard (120°) resulting in bond alternation. Strain can be introduced in other ways to an aromatic system,[241] such as by tying opposite ends of an aromatic fragment together with a short linker, forcing these ends to move out of plane. This type of strain has been nicely introduced by attaching a bridge to the 2 and 7 positions of pyrene (**76**).[242] By adjusting the length of the bridge, the pyrene unit can be bent out of plane, introducing a strain and a potential loss of aromaticity.

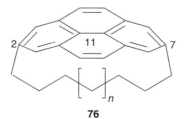

Cyranski et al.[243] examined the series of pyrenophanes where the tethering chain has 6–12 carbon atoms. The nonplanar distortion is measured as the bend angle α, defined as the angle made between the outside ring plane and the horizon, or more simply, the angle $C_{11}-C_2-C_7$. Relative ASE is computed using Reaction 3.31, which cleverly avoids the complication of knowing (1) what is the ASE of pyrene itself and (2) what is the strain energy in these compounds.

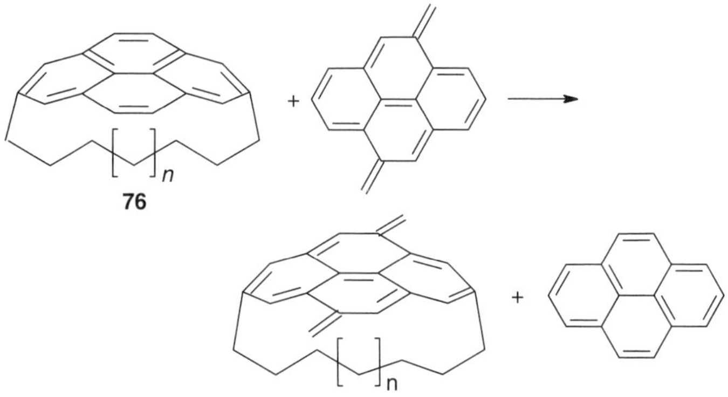

(Reaction 3.31)

The results of B3LYP/6-311G** computations for this series of pyrenophanes is given in Table 3.29. The bending angle smoothly increases with decreasing length of the tether. The ASE decreases in the same manner. The ASE correlates quite well with the bending angle, as does the relative magnetic susceptibility exaltation. The NICS(1) values become less negative with decreasing tether length.

All of these trends are consistent with reduced aromaticity with increased out-of-plane distortion of the pyrene framework. What may be surprising is the relatively small loss of aromaticity in this sequence. Even though the bend angle is as large as almost 40°, the loss of relative ASE is only 16 kcal mol^{-1}, only about a quarter of the ASE of pyrene itself. Apparently, aromatic systems are fairly robust.

TABLE 3.29 Computed Values for the Pyrenophanes 76[a]

n	α^b	ΔASE^c	Rel. Λ^d	NICS(1)^e
6	39.7	−15.8	18.8	−7.8, −4.1
7	32.7	−12.1	17.5	−8.7, −4.5
8	26.5	−10.6	14.3	−9.6, −5.2
9	21.3	−7.5	11.3	−10.6, −5.5
10	15.9	−6.2	9.5	−11.3, −6.2
11	11.0	−3.4	7.0	−12.0, −6.4
12	7.2	−3.1	6.3	−12.6, −7.0
Pyrene	0.0	0.0	0.0	−13.9, −7.8

[a]Ref. 243.
[b]In degrees.
[c]In kilocalorie per mole, from Reaction 3.27.
[d]In centimeter–gram–second parts per million.
[e]In parts per million, for the outer and inner rings.

3.5.4 π–π Stacking

The stacking of aromatics compounds, one planar molecule lying atop another, is a frequently observed structural motif, most famously present as the stacked nucleotide base pairs in DNA. This stacked arrangement is also important in molecular recognition problems including host–guest chemistry and self-assembled superstructures. This motif goes by the name of "π–π stacking" or "π–π interactions" and is generally associated with stacked aromatic compounds.

Computational studies on the π–π stacking problem began with the simplest model: the benzene dimer. While not the only possible configurations, the benzene dimer structures most studied are the sandwiches **77s** (D_{6h}) and **77s'** (D_{6d}), the parallel displaced **77pd**, and the T-shaped configurations **77t** and **77t'**.

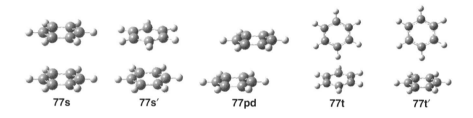

It was recognized early on that the binding energy of the benzene dimer is quite sensitive to computational method. The potential energy surface of **77** is completely repulsive at the HF level, as well as with any DFT that does not include a dispersion correction, such as B3LYP.[244] Furthermore, the weakly bound benzene dimer is susceptible to basis set superposition error and counterpoise corrections are likely to be necessary.[245]

There are only a few gas-phase experiments of the benzene dimer. Some experiments[246–248] point to a T-shape dimer while others[249,250] implicate a sandwich or more likely a displaced structure. Hole-burning experiments find evidence for multiple different structures.[251] The experimental estimates of the binding energy of **77** are 1.6 ± 0.2 and 2.4 ± 0.4 kcal mol^{-1}.[252]

A summary of the computational results for the binding energy of the benzene dimer **77** is listed in Table 3.30. Most of these computations were performed by holding the geometry of the benzene monomer fixed and varying the separation between the two rings, along with including a counterpoise correction. (An estimate of the size of the counterpoise correction is 3–4 kcal mol^{-1} at MP2/aug-cc-pVDZ and 1.5 kcal mol^{-1} at MP2/aug-cc-pVTZ.[253]) Consensus among these computations is that the sandwich structure **77s** is the least stable, and in fact, it is not a local energy minimum. Concerning the other two structures, all which can be safely stated is that the energies of **77t** and **77pd** are very close, that MP2 overestimates the binding energy, and that dispersion-corrected DFT provides quite reasonable energy estimates at a fraction of the computational cost.

The standard model of the nature of the π–π stacking, developed by Hunter and Saunders[258] and promoted by Cozzi et al.[259] partitions the interaction into a

TABLE 3.30 Binding Energy (kcal mol^{-1}) of the Benzene Dimer 77

Method	77t	77pd	77s
MP2/aug-cc-pVDZ[a]	2.96	3.94	2.56
MP2/aug-cc-pVQZ[b]	3.54	4.95	3.64
SCS-MP2/QZV3P[c]	2.49	2.97	
CCSD(T)/aug-cc-pVDZ[a]	2.17	2.00	1.21
est. CCSD(T)/CBS[b]	2.39	2.74	1.99
est. CBS CCSD(T)[d]	2.74	2.78	1.81
est. CBS(Δha(DT)Z) CCSD(T)[e]	5.0	3.5	3.9
KT1/TZ2P[f]			1.58
PW91/TZ2P[f]			0.45
BLYP-D/TZVP[g]	2.80	2.88	1.84
B2PLYP-D/QZV3P[c]	2.82		2.62

[a] Ref. 245, only the s-orbitals are used for hydrogen.
[b] Ref. 253.
[c] Ref. 254.
[d] Ref. 255.
[e] Ref. 256.
[f] Ref. 244.
[g] Ref. 257.

number of components. Dispersion accounts for the majority of the stabilization energy, and many subsequent high level computations corroborate the major role of dispersion. However, it is electrostatics that determines the geometry, and this involves the interaction of the π-electrons of one monomer with the σ-system of the other monomer. When substituents are added to one ring, the standard model predicts that electron withdrawing groups (EWGs) will remove some π-density, thereby reducing the electron repulsion between the rings and leading to a larger binding energy. Electron donating groups (EDGs) will have just the opposite effect. If a substituent is placed on both rings, Cozzi predicts that when EWGs are on both rings, the binding will be strongest, while EDGs on each ring will lead to the weakest bonding. Hunter and Saunders emphasize more of the σ-effect and suggest that the strongest binding will be when one ring has an EWG and the other ring has an EDG.

The Sherill group[260,261] has reported a number of computational studies of substituted dimers. In their studies of a benzene-substituted benzene dimer, they noted that for the sandwich structure, all substituents (OH, Me, F, and CN) lead to stronger binding than that in the parent benzene dimer. Using a different computational method and including the additional substituents NH_2 and NO_2, Hobza and Kim[262] also found that all substituents increased the binding energy relative to the benzene dimer, regardless of whether the dimer was in the sandwich, T-shape, or displaced configuration. Similar results were obtained by Grimme[263] using DFT-D. As an interesting aside, Lewis[264] notes that the relative increase in binding energy when a single substituent is added to one of the rings in the benzene dimer are comparable when computed at MP2/6-311G** including the counterpoise correction and the much more computationally efficient DFT M05-2x/6-311G** *without* counterpoise correction.

A number of computational studies have looked at the benzene dimer with multiple substituents, such as **78** and **79** and **80**.[26] All of them show an additivity effect (i.e., increased binding) with each new substituent, regardless of whether the substituent is an EWG or an EDG, and regardless of whether the substitution is on one ring or on both rings. This additivity is not seen in the T-shape configuration.[265]

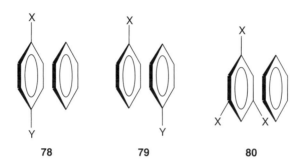

While Sherrill had suggested that the binding energy of the substituted benzene dimer complexes is due to local interactions related to the substituent,[265,266] Houk and Wheeler[267] greatly amplified this notion. For a set of 24 dimers with one substituent on one ring, they noted that the binding energy relative to **77s** correlates with σ_m, as long as the benzene dimer itself is removed from the set. A model of these 24 dimers was created by replacing Ph–X with H–X, with the substituent placed in the same position as in the actual substituted dimer complex. The relative binding energies of the HX model complexes correlates well ($r = 0.91$) with the relative binding energies of the actual substituted benzene dimers. This model removes the second π-system entirely, yet the substituent binding effect is retained. They conclude that the binding in benzene dimers is driven by an electrostatic attraction between the quadrupole of benzene and the local dipole induced by the substituent. Wheeler[268] subsequently extended the model system to one of HX with propene. The relative binding energy of this model, which removes entirely any aromatic component, also correlates with the bonding energy of the substituted benzene dimer.

The relation between the relative binding energy of substituted benzene dimers and σ_m has been questioned. Sherill[269] considered not just monosubstituted benzenes, but also 1,3,5-trisubstituted benzenes (**80**) and hexasubstituted benzenes, with the substituents Me, F, OH, NH_2, CH_2OH, and CN. No correlation is found between the relative binding energies and the sum of the σ_m values. Lewis[264] employed 11 different substituents and examined mono-, di-, tri-, and tetrasubstituted benzenes in a complex with unsubstituted benzene, for a total of 66 different dimers. Again, no correlation with the relative binding energies is

TABLE 3.31 B2PLYP-D/QZV3P Interaction Energy (kcal mol^{-1})

	Number of Rings			
	1	2	3	4
Saturated	3.09	5.92	7.72	10.48
Aromatic	2.62	6.81	11.46	16.33

observed with the sum of the σ_m values, though a correlation is seen with the sum of the absolute values of σ_m. Lewis points out that Houk and Wheeler[267] missed the curvature in the plot due to too few examples with EDGs. He also points out that there is no reasonable rationale for using σ_m values in the first place. Finally, he notes that while dispersion is the greatest factor of the four factors deciding the binding energy, the dispersion, exchange, and induction components sum to approximately the same value for all of these dimers, implicating electrostatics as the deciding factor.

Is there anything special about π–π stacking at all? Grimme[254] compared stacked benzene analogs (benzene, naphthalene, anthracene, and tetracene) with their stacked saturated analogs. The B2PLYP-D interaction energies of these dimers are listed in Table 3.31. The interaction energy of the saturated and aromatic dimers with one or two rings is similar, suggesting that for small systems, there is *no special π–π stacking*. However, for the larger rings, the aromatic dimers have significantly greater interaction energy than their saturated analogs, which comes about by a much greater dispersion component in the former. Grimme concludes that special noncovalent π–π stacking interaction *does manifest in larger aromatic systems*.

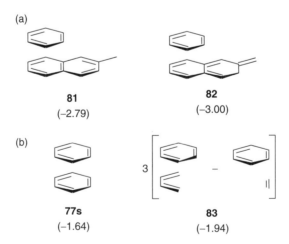

Figure 3.16 Models of nonaromatic analogs of the benzene dimer. The value within parentheses is the binding energy in kilocalorie per mole computed at (a) SCS-MP2/TZVPP and (b) CCSD(T)/AVTZ.

Wheeler[270] took a different approach toward assessing the role of aromatics in π–π stacking. He made use of two nonaromatic models of benzene (Figure 3.16). With the first model, he compared the binding energy in **81** with that of **82**. **82** contains a nonaromatic analog of 2-methylnaphthalene (present in **81**), which conserves the number of π-electrons. The binding energy is 0.21 kcal mol^{-1} *greater* in the nonaromatic model.

The second model (see Figure 3.16) decomposes benzene into three equivalents of 1,3-butadiene less three equivalents of ethylene. This model **83** places the smaller fragments in the position of the second benzene in **77s**. The binding energy of this nonaromatic model **83** is −0.30 kcal mol^{-1} *greater* than the binding energy in **77s**. Wheeler concludes that aromaticity is *not* a requirement for π-stacking; in fact, conjugated polyenes might have a greater stacking attraction than do the aromatics.

Where does this all leave us concerning "π-stacking"? Iverson[271] argues that the term is loaded with context that no longer has any value. He suggests that "π-stacking" and "π–π interactions" are terms that are misleading and should be avoided. The computational results discussed here certainly point toward that confusion.

3.6 INTERVIEW: PROFESSOR PAUL VON RAGUÉ SCHLEYER

Interviewed March 28, 2006

Professor Paul Schleyer is the Graham Perdue Professor of Chemistry at the University of Georgia. Prior to that, he was a professor at the University at Erlangen (codirector of the Organic Institute) and the founding director of its Computer Chemistry Center. Schleyer began his academic career at Princeton University.

Professor Schleyer's involvement in computational chemistry dates back to the 1960s, when his group was performing MM and semiempirical computations as an adjunct to his predominantly experimental research program. This situation dramatically changed when Professor John Pople invited Schleyer to visit Carnegie-Mellon University in 1969 as the NSF Center of Excellence Lecturer. From discussions with Dr. Pople, it became clear to Schleyer that "ab initio methods could look at controversial subjects like the nonclassical carbocations. I became hooked on it!" The collaboration between Pople and Schleyer that originated from that visit lasted well over 20 years, and covered such topics as substituent effects, unusual structures that Schleyer terms "rule-breaking," and organolithium chemistry. This collaboration started while Schleyer was at Princeton but continued after his move to Erlangen, where Pople came to visit many times. The collaboration was certainly of peers. "It would be unfair to say that the ideas came from me, but it's clear that the projects we worked on would not have been chosen by Pople. Pople added a great deal of insight and he would advise me on what was computationally possible," Schleyer recalls of this fruitful relationship.

Schleyer quickly became enamored with the power of ab initio computations to tackle interesting organic problems. His enthusiasm for computational chemistry eventually led to his decision to move to Erlangen—they offered unlimited (24/7) computer time, while Princeton's counteroffer was just 2 h of computer time per week. He left Erlangen in 1998 due to enforced retirement. However, his adjunct status at the University of Georgia allowed for a smooth transition back to the United States, where he now enjoys a very productive collaborative relationship with Professor Fritz Schaefer.

Perhaps the problem that best represents how Schleyer exploits the power of ab initio computational chemistry is the question of how to define and measure aromaticity. Schleyer's interest in the concept of aromaticity spans his entire career. He was drawn to this problem because of the pervasive nature of aromaticity across organic chemistry. Schleyer describes his motivation: "Aromaticity is a central theme of organic chemistry. It is re-examined by each generation of chemists. Changing technology permits that re-examination to occur." His direct involvement came about by Kutzelnigg's development of a computer code to calculate chemical shifts. Schleyer began the use of this program in the 1980s and applied it first to structural problems. His group "discovered in this manner many experimental structures that were incorrect."

To assess aromaticity, Schleyer first computed the lithium chemical shifts in complexes formed between lithium cation and the hydrocarbon of interest. The lithium cation would typically reside above the aromatic ring and its chemical shift would be affected by the magnetic field of the ring. While this met with some success, Schleyer was frustrated by the fact that lithium was often not positioned especially near the ring, let alone in the center of the ring. This led to the development of NICS, where the virtual chemical shift can be computed at any point in space. Schleyer advocated using the geometric center of the ring, then later a point 1 Å above the ring center.

Over time, Schleyer came to refine the use of NICS, advocating an examination of NICS values on a grid of points. His most recent paper posits using just the component of the chemical shift tensor perpendicular to the ring evaluated at the center of the ring. This evolution reflects Schleyer's continuing pursuit of a simple measure of aromaticity. "Our endeavor from the beginning was to select one NICS point that we could say characterizes the compound," Schleyer says. "The problem is that chemists want a number which they can associate with a phenomenon rather than a picture. The problem with NICS was that it was not soundly based conceptually from the beginning because cyclic electron delocalization-induced ring current was not expressed solely perpendicular to the ring. It's only *that* component which is related to aromaticity."

The majority of our discussion revolved around the definition of aromaticity. Schleyer argues that "aromaticity can be defined perfectly well. It is the manifestation of cyclic electron delocalization which is expressed in various ways. The problem with aromaticity comes in its quantitative definition. How big is the aromaticity of a particular molecule? We can answer this using some properties. One of my objectives is to see whether these various quantities are related to one another. That, I think, is still an open question."

Schleyer further detailed this thought, "The difficulty in writing about aromaticity is that it is encrusted by two centuries of tradition, which you cannot avoid. You have to stress the interplay of the phenomena. Energetic properties are most important, but you need to keep in mind that aromaticity is only 5 percent of the total energy. But if you want to get as close to the phenomenon as possible, then one has to go to the property most closely related, which is magnetic properties." This is why he focuses upon the use of NICS as an aromaticity measure. He is quite confident in his new NICS measure employing the perpendicular component of the chemical shift tensor. "This new criteria is very satisfactory," he says. "Most people who propose alternative measures do not do the careful step of evaluating them against some basic standard. We evaluate against aromatic stabilization energies."

Schleyer notes that his evaluation of the ASE of benzene is larger than many other estimates. This results from the fact that, in his opinion, "all traditional equations for its determination use tainted molecules. Cyclohexene is tainted by hyperconjugation of about 10 kcal mol^{-1}. Even cyclohexane is very tainted, in this case by 1,3-interactions." An analogous complaint can be made about the methods Schleyer himself employs: NICS is evaluated at some arbitrary point or arbitrary set of points, the block-diagonalized "cyclohexatriene" molecule is a *gedanken* molecule. When pressed on what then to use as a reference that is not 'tainted', Schleyer made this trenchant comment: "What we are trying to measure is virtual. Aromaticity, like almost all concepts in organic chemistry, is virtual. They're not measurable. You can't measure atomic charges within a molecule. Hyperconjugation, electronegativity, everything is in this sort of virtual category. Chemists live in a virtual world. But science moves to higher degrees of refinement." Despite its inherent 'virtual' nature, "Aromaticity has this 200 year history. Chemists are interested in the unusual stability and reactivity associated with aromatic molecules. The term survives, and remains an enormously fruitful area of research."

His interest in the annulenes is a natural extension of the quest for understanding aromaticity. Schleyer was particularly drawn to [18]-annulene because it can express the same D_{6h} symmetry as does benzene. His computed chemical shifts for the D_{6h} structure differed significantly from the experimental values, indicating that the structure was clearly wrong. "It was an amazing computational exercise," Schleyer mused, "because practically every level you used to optimize the geometry gave a different structure. MP2 overshot the aromaticity, HF and B3LYP undershot it. Empirically, we had to find a level that worked. This was not very intellectually satisfying but was a pragmatic solution." Schleyer expected a lot of flak from crystallographers about this result, but in fact none occurred. He hopes that the X-ray structure will be redone at some point.

Reflecting on the progress of computational chemistry, Schleyer recalls that "physical organic chemists were actually antagonistic toward computational chemistry at the beginning. One of my friends said that he thought I had gone mad. In addition, most theoreticians disdained me as a black-box user." In those early years as a computational chemist, Schleyer felt disenfranchised from the physical organic chemistry community. Only slowly has he felt accepted back into this camp. "Physical organic chemists have adopted computational chemistry; perhaps, I hope to think, due to my example demonstrating what can be done. If you can show people that you can compute chemical properties, like chemical shifts to an accuracy that is useful, computed structures that are better than experiment, then they get the word sooner or later that maybe you'd better do some calculations." In fact, Schleyer considers this to be his greatest contribution to science—demonstrating by his own example the importance of computational chemistry toward solving relevant chemical problems. He cites his role in helping to establish the *Journal of Computational Chemistry* in both giving name to the discipline and stature to its practitioners.

Schleyer looks to the future of computational chemistry residing in the breadth of the periodic table. "Computational work has concentrated on one element, namely carbon," Schleyer says. "The rest of the periodic table is waiting to be explored." On the other hand, he is dismayed by the state of research at universities. In his opinion, "the function of universities is to do pure research, *not* to do applied research. Pure research will not be carried out at any other location." Schleyer sums up his position this way—"Pure research is like putting money in the bank. Applied research is taking the money out." According to this motto, Schleyer's account is very much in the black.

REFERENCES

1. Feng, Y.; Liu, L.; Wang, J.-T.; Huang, H.; Guo, Q.-X. "Assessment of experimental bond dissociation energies using composite ab initio methods and evaluation of the performances of density functional methods in the calculation of bond dissociation energies," *J. Chem. Inf. Comput. Sci.* **2003**, *43*, 2005–2013.

2. Blanksby, S. J.; Ellison, G. B. "Bond dissociation energies of organic molecules," *Acc. Chem. Res.* **2003**, *36*, 255–263.

3. Henry, D. J.; Parkinson, C. J.; Mayer, P. M.; Radom, L. "Bond dissociation energies and radical stabilization energies associated with substituted methyl radicals," *J. Phys. Chem. A* **2001**, *105*, 6750–6756.

4. Feng, Y.; Liu, L.; Wang, J.-T.; Zhao, S.-W.; Guo, Q.-X. "Homolytic C–H and N–H bond dissociation energies of strained organic compounds," *J. Org. Chem.* **2004**, *69*, 3129–3138.

5. Menon, A. S.; Wood, G. P. F.; Moran, D.; Radom, L. "Bond dissociation energies and radical stabilization energies: an assessment of contemporary theoretical procedures," *J. Phys. Chem. A* **2007**, *111*, 13638–13644.

6. Yao, X.-Q.; Hou, X.-J.; Jiao, H.; Xiang, H.-W.; Li, Y.-W. "Accurate calculations of bond dissociation enthalpies with density functional methods," *J. Phys. Chem. A* **2003**, *107*, 9991–9996.

7. Check, C. E.; Gilbert, T. M. "Progressive systematic underestimation of reaction energies by the B3LYP model as the number of C–C bonds increases: why organic chemists should use multiple DFT models for calculations involving polycarbon hydrocarbons," *J. Org. Chem.* **2005**, *70*, 9828–9834.

8. Redfern, P. C.; Zapol, P.; Curtiss, L. A.; Raghavachari, K. "Assessment of Gaussian-3 and density functional theories for enthalpies of formation of C_1-C_{16} alkanes," *J. Phys. Chem. A* **2000**, *104*, 5850–5854.

9. Luo, Y.-R. *Handbook of Bond Dissociation Energies in Organic Compounds*; CRC Press: New York, **2002**.

10. Rüchardt, C. "Relations between structure and reactivity in free-radical chemistry," *Angew. Chem. Int. Ed. Engl.* **1970**, *9*, 830–843.

11. Izgorodina, E. I.; Coote, M. L.; Radom, L. "Trends in R–X bond dissociation energies (R = Me, Et, *i*-Pr, *t*-Bu; X = H, CH_3, OCH_3, OH, F): a surprising shortcoming of density functional theory," *J. Phys. Chem. A* **2005**, *109*, 7558–7566.

12. Coote, M. L.; Pross, A.; Radom, L. "Variable trends in R–X bond dissociation energies (R = Me, Et, *i*-Pr, *t*-Bu)," *Org. Lett.* **2003**, *5*, 4689–4692.

13. Matsunaga, N.; Rogers, D. W.; Zavitsas, A. A. "Pauling's electronegativity equation and a new corollary accurately predict bond dissociation enthalpies and enhance current understanding of the nature of the chemical bond," *J. Org. Chem.* **2003**, *68*, 3158–3172.

14. Lias, S. G.; Bartmess, J. E.; Holmes, J. L.; Levin, R. D.; Mallard, W. G. "Gas-phase ion and neutral thermochemistry," *J. Phys. Chem. Ref. Data* **1988**, Suppl. 17.

15. Linstrom, P. J.; Mallard, W. G. "NIST chemistry webBook, NIST standard reference database number 69," 2012, URL: http://webbook.nist.gov

16. Kollmar, H. "The stability of alkyl anions. A molecular orbital theoretical study," *J. Am. Chem. Soc.* **1978**, *100*, 2665–2669.

17. Chandrasekhar, J.; Andrade, J. G.; Schleyer, P. v. R. "Efficient and accurate calculation of anion proton affinities," *J. Am. Chem. Soc.* **1981**, *103*, 5609–5612.

18. Saunders, W. H., Jr. "Ab initio and semi-empirical investigation of gas-phase carbon acidity," *J. Phys. Org. Chem.* **1994**, *7*, 268–271.

19. Burk, P.; Koppel, I. A.; Koppel, I.; Leito, I.; Travnikova, O. "Critical test of performance of B3LYP functional for prediction of gas-phase acidities and basicities," *Chem. Phys. Lett.* **2000**, *323*, 482–489.

20. Merrill, G. N.; Kass, S. R. "Calculated gas-phase acidities using density functional theory: is it reliable?," *J. Phys. Chem.* **1996**, *100*, 17465–17471.

21. Ochterski, J. W.; Petersson, G. A.; Montgomery, J. A., Jr. "A complete basis set model chemistry. V. Extensions to six or more heavy atoms," *J. Chem. Phys.* **1996**, *104*, 2598–2619.

22. Ochterski, J. W.; Petersson, G. A.; Wiberg, K. B. "A comparison of model chemistries," *J. Am. Chem. Soc.* **1995**, *117*, 11299–11308.

23. Topol, I. A.; Tawa, G. J.; Caldwell, R. A.; Eissenstat, M. A.; Burt, S. K. "Acidity of organic molecules in the gas phase and in aqueous solvent," *J. Phys. Chem. A* **2000**, *104*, 9619–9624.

24. DePuy, C. H.; Gronert, S.; Barlow, S. E.; Bierbaum, V. M.; Damrauer, R. "The gas-phase acidities of the alkanes," *J. Am. Chem. Soc.* **1989**, *111*, 1968–1973.
25. Luh, T.-Y.; Stock, L. M. "Kinetic acidity of cubane," *J. Am. Chem. Soc.* **1974**, *96*, 3712–3713.
26. Ritchie, J. P.; Bachrach, S. M. "Comparison of the calculated acidity of cubane with that of other strained and unstrained hydrocarbons," *J. Am. Chem. Soc.* **1990**, *112*, 6514–6517.
27. Hare, M.; Emrick, T.; Eaton, P. E.; Kass, S. R. "Cubyl anion formation and an experimental determination of the acidity and C–H bond dissociation energy of cubane," *J. Am. Chem. Soc.* **1997**, *119*, 237–238.
28. Rayne, S.; Forest, K. "Gas-phase enthalpies of formation, acidities, and strain energies of the [m,n]polyprismanes ($m \geq 2$; $n = 3-8$; $m \times n \leq 16$): a CBS-Q//B3, G4MP2, and G4 theoretical study," *Theor. Chem. Acc.* **2010**, *127*, 697–709.
29. Broadus, K. M.; Kass, S. R.; Osswald, T.; Prinzbach, H. "Dodecahedryl anion formation and an experimental determination of the acidity and C–H bond dissociation energy of dodecahedrane," *J. Am. Chem. Soc.* **2000**, *122*, 10964–10968.
30. Fattahi, A.; McCarthy, R. E.; Ahmad, M. R.; Kass, S. R. "Why does cyclopropene have the acidity of an acetylene but the bond energy of methane?," *J. Am. Chem. Soc.* **2003**, *125*, 11746–11750.
31. Manini, P.; Amrein, W.; Gramlich, V.; Diederich, F. "Expanded cubane: synthesis of a cage compound with a C56 core by acetylenic scaffolding and gas-phase transformations into fullerenes," *Angew. Chem. Int. Ed.* **2002**, 4339–4343.
32. Bachrach, S. M. "Structure, deprotonation energy, and cation affinity of an ethynyl-expanded cubane," *J. Phys. Chem. A* **2003**, *107*, 4957–4961.
33. Bachrach, S. M.; Demoin, D. W. "Computational studies of ethynyl- and diethynyl-expanded tetrahedranes, prismanes, cubanes, and adamantanes," *J. Org. Chem.* **2006**, *71*, 5105–5116.
34. de Visser, S. P.; van der Horst, E.; de Koning, L. J.; van der Hart, W. J.; Nibbering, N. M. M. "Characterization of isomeric $C_4H_5^-$ anions in the gas phase; theory and experiment," *J. Mass Spectrom.* **1999**, *34*, 303–310.
35. Siggel, M. R.; Thomas, T. D. "Why are organic acids stronger acids than organic alcohols?," *J. Am. Chem. Soc.* **1986**, *108*, 4360–4363.
36. Burk, P.; Schleyer, P. v. R. "Why are carboxylic acids stronger acids than alcohols? The electrostatic theory of Siggel–Thomas revisited," *J. Mol. Struct. (THEOCHEM)* **2000**, *505*, 161–167.
37. Siggel, M. R. F.; Streitwieser, A. J.; Thomas, T. D. "The role of resonance and inductive effects in the acidity of carboxylic acids," *J. Am. Chem. Soc.* **1988**, *110*, 8022–8028.
38. Exner, O. "Why are carboxylic acids and phenols stronger acids than alcohols?," *J. Org. Chem.* **1988**, *53*, 1810–1812.
39. Dewar, M. J. S.; Krull, K. L. "Acidity of carboxylic acids: due to delocalization or induction?," *J. Chem. Soc., Chem. Commun.* **1990**, 333–334.
40. Perrin, C. L. "Atomic size dependence of Bader electron populations: significance for questions of resonance stabilization," *J. Am. Chem. Soc.* **1991**, *113*, 2865–2868.
41. Hiberty, P. C.; Byrman, C. P. "Role of π-electron delocalization in the enhanced acidity of carboxylic acids and enols relative to alcohols," *J. Am. Chem. Soc.* **1995**, *117*, 9875–9880.

42. Rablen, P. R. "Is the acetate anion stabilized by resonance or electrostatics? A systematic structural comparison," *J. Am. Chem. Soc.* **2000**, *122*, 357–368.
43. Holt, J.; Karty, J. M. "Origin of the acidity enhancement of formic acid over methanol: resonance versus inductive effects," *J. Am. Chem. Soc.* **2003**, *125*, 2797–2803.
44. O'Hair, R. A. J.; Bowie, J. H.; Gronert, S. "Gas phase acidities of the α- amino acids," *Int. J. Mass Spectrom. Ion Processes* **1992**, *117*, 23–36.
45. Jones, C. M.; Bernier, M.; Carson, E.; Colyer, K. E.; Metz, R.; Pawlow, A.; Wischow, E. D.; Webb, I.; Andriole, E. J.; Poutsma, J. C. "Gas-phase acidities of the 20 protein amino acids," *Int. J. Mass Spectrom.* **2007**, *267*, 54–62.
46. Tian, Z.; Pawlow, A.; Poutsma, J. C.; Kass, S. R. "Are carboxyl groups the most acidic sites in amino acids? Gas-phase acidity, H/D exchange experiments, and computations on cysteine and its conjugate base," *J. Am. Chem. Soc.* **2007**, *129*, 5403–5407.
47. Tian, Z.; Wang, X.-B.; Wang, L.-S.; Kass, S. R. "Are carboxyl groups the most acidic sites in amino acids? Gas-phase acidities, photoelectron spectra, and computations on tyrosine, p-hydroxybenzoic acid, and their conjugate bases," *J. Am. Chem. Soc.* **2009**, *131*, 1174–1181.
48. Smith, G. D.; Jaffe, R. L. "Quantum chemistry study of conformational energies and rotational energy barriers in n-alkanes," *J. Phys. Chem.* **1996**, *100*, 18718–18724.
49. Gruzman, D.; Karton, A.; Martin, J. M. L. "Performance of Ab initio and density functional methods for conformational equilibria of C_nH2_{n+2} alkane isomers (n = 4–8)," *J. Phys. Chem. A* **2009**, *113*, 11974–11983.
50. Allinger, N. L.; Fermann, J. T.; Allen, W. D.; Schaefer Iii, H. F. "The torsional conformations of butane: definitive energetics from ab initio methods," *J. Chem. Phys.* **1997**, *106*, 5143–5150.
51. Herrebout, W. A.; van der Veken, B. J.; Wang, A.; Durig, J. R. "Enthalpy difference between conformers of n-butane and the potential function governing conformational interchange," *J. Phys. Chem.* **1995**, *99*, 578–585.
52. Balabin, R. M. "Enthalpy difference between conformations of normal alkanes: Raman spectroscopy study of *n*-pentane and *n*-butane," *J. Phys. Chem. A* **2009**, *113*, 1012–1019.
53. Martin, J. M. L.; de Oliveira, G. "Towards standard methods for benchmark quality ab initio thermochemistry—W1 and W2 theory," *J. Chem. Phys.* **1999**, *111*, 1843–1856.
54. Parthiban, S.; Martin, J. M. L. "Assessment of W1 and W2 theories for the computation of electron affinities, ionization potentials, heats of formation, and proton affinities," *J. Chem. Phys.* **2001**, *114*, 6014–6029.
55. Balabin, R. M. "Enthalpy difference between conformations of normal alkanes: effects of basis set and chain length on intramolecular basis set superposition error," *Mol. Phys.* **2011**, *109*, 943–953.
56. Asturiol, D.; Duran, M.; Salvador, P. "Intramolecular basis set superposition error effects on the planarity of benzene and other aromatic molecules: a solution to the problem," *J. Chem. Phys.* **2008**, *128*, 144108.
57. Császár, A. G. "Conformers of gaseous α-alanine," *J. Phys. Chem.* **1996**, *100*, 3541–3551.
58. Godfrey, P. D.; Firth, S.; Hatherley, L. D.; Brown, R. D.; Pierlot, A. P. "Millimeter-wave spectroscopy of biomolecules: alanine," *J. Am. Chem. Soc.* **1993**, *115*, 9687–9691.

59. Jaeger, H. M.; Schaefer, H. F.; Demaison, J.; Császár, A. G.; Allen, W. D. "Lowest-lying conformers of alanine: pushing theory to ascertain precise energetics and semiexperimental R_e structures," *J. Chem. Theory Comput.* **2010**, *6*, 3066–3078.
60. Blanco, S.; Lesarri, A.; López, J. C.; Alonso, J. L. "The gas-phase structure of alanine," *J. Am. Chem. Soc.* **2004**, *126*, 11675–11683.
61. Gronert, S.; O'Hair, R. A. J. "Ab initio studies of amino acid conformations. 1. The conformers of alanine, serine, and cysteine," *J. Am. Chem. Soc.* **1995**, *117*, 2071–2081.
62. Dobrowolski, J. C.; Rode, J. E.; Sadlej, J. "Cysteine conformations revisited," *J. Mol. Struct. (THEOCHEM)* **2007**, *810*, 129–134.
63. Sanz, M. E.; Blanco, S.; López, J. C.; Alonso, J. L. "Rotational probes of six conformers of neutral cysteine," *Angew. Chem. Int. Ed.* **2008**, *47*, 6216–6220.
64. Wilke, J. J.; Lind, M. C.; Schaefer, H. F.; Császár, A. G.; Allen, W. D. "Conformers of gaseous cysteine," *J. Chem. Theor. Comput.* **2009**, *5*, 1511–1523.
65. Grimme, S. "Seemingly simple stereoelectronic effects in alkane isomers and the implications for Kohn-Sham density functional theory," *Angew. Chem. Int. Ed.* **2006**, *45*, 4460–4464.
66. NIST. "NIST chemistry webBook," **2005**, URL: http://webbook.nist.gov/
67. Zhao, Y.; Truhlar, D. G. "A density functional that accounts for medium-range correlation energies in organic chemistry," *Org. Lett.* **2006**, *8*, 5753–5755.
68. Schreiner, P. R.; Fokin, A. A.; Pascal, R. A.; DeMeijere, a. "Many density functional theory approaches fail to give reliable large hydrocarbon isomer energy differences," *Org. Lett.* **2006**, *8*, 3635–3638.
69. Wodrich, M. D.; Corminboeuf, C.; Schleyer, P. v. R. "Systematic errors in computed alkane energies using B3LYP and other popular DFT functionals," *Org. Lett.* **2006**, *8*, 3631–3634.
70. Wodrich, M. D.; Corminboeuf, C.; Schreiner, P. R.; Fokin, A. A.; Schleyer, P. v. R. "How accurate are DFT treatments of organic energies?," *Org. Lett.* **2007**, *9*, 1851–1854.
71. Pieniazek, S. N.; Clemente, F. R.; Houk, K. N. "Sources of error in DFT computations of C–C bond formation thermochemistries: π→σ transformations and error cancellation by DFT methods," *Angew. Chem. Int. Ed.* **2008**, *47*, 7746–7749.
72. Brittain, D. R. B.; Lin, C. Y.; Gilbert, A. T. B.; Izgorodina, E. I.; Gill, P. M. W.; Coote, M. L. "The role of exchange in systematic DFT errors for some organic reactions," *Phys. Chem. Chem. Phys.* **2009**, *11*, 1138–1142.
73. Song, J.-W.; Tsuneda, T.; Sato, T.; Hirao, K. "Calculations of alkane energies using long-range corrected DFT combined with intramolecular van der Waals correlation," *Org. Lett.* **2010**, *12*, 1440–1443.
74. Sato, T.; Nakai, H. "Density functional method including weak interactions: dispersion coefficients based on the local response approximation," *J. Chem. Phys.* **2009**, *131*, 224104–224112.
75. Grimme, S. "n-Alkane isodesmic reaction energy errors in density functional theory are due to electron correlation effects," *Org. Lett.* **2010**, *12*, 4670–4673.
76. Krieg, H.; Grimme, S. "Thermochemical benchmarking of hydrocarbon bond separation reaction energies: Jacob's ladder is not reversed!," *Mol. Phys.* **2010**, *108*, 2655–2666.

77. Zhao, Y.; Schultz, N. E.; Truhlar, D. G. "Design of density functionals by combining the method of constraint satisfaction with parametrization for thermochemistry, thermochemical kinetics, and noncovalent interactions," *J. Chem. Theory Comput.* **2006**, *2*, 364–382.

78. Zhao, Y.; Truhlar, D. "The M06 suite of density functionals for main group thermochemistry, thermochemical kinetics, noncovalent interactions, excited states, and transition elements: two new functionals and systematic testing of four M06-class functionals and 12 other functionals," *Theor. Chem. Acc.* **2008**, *120*, 215–241.

79. Mardirossian, N.; Parkhill, J. A.; Head-Gordon, M. "Benchmark results for empirical post-GGA functionals: difficult exchange problems and independent tests," *Phys. Chem. Chem. Phys.* **2011**, *13*, 19325–19337.

80. Song, J.-W.; Tsuneda, T.; Sato, T.; Hirao, K. "An examination of density functional theories on isomerization energy calculations of organic molecules," *Theor. Chem. Acc.* **2011**, *130*, 851–857.

81. Chai, J.-D.; Head-Gordon, M. "Systematic optimization of long-range corrected hybrid density functionals," *J. Chem. Phys.* **2008**, *128*, 084106–084115.

82. McBride, J. M. "The hexaphenylethane riddle," *Tetrahedron* **1974**, *30*, 2009–2022.

83. Selwood, P. W.; Dobres, R. M. "The diamagnetic correction for free radicals," *J. Am. Chem. Soc.* **1950**, *72*, 3860–3863.

84. Kahr, B.; Van Engen, D.; Mislow, K. "Length of the ethane bond in hexaphenylethane and its derivatives," *J. Am. Chem. Soc.* **1986**, *108*, 8305–8307.

85. Grimme, S.; Schreiner, P. R. "Steric crowding can stabilize a labile molecule: solving the hexaphenylethane riddle," *Angew. Chem. Int. Ed.* **2011**, *50*, 12639–12642.

86. Schreiner, P. R.; Chernish, L. V.; Gunchenko, P. A.; Tikhonchuk, E. Y.; Hausmann, H.; Serafin, M.; Schlecht, S.; Dahl, J. E. P.; Carlson, R. M. K.; Fokin, A. A. "Overcoming lability of extremely long alkane carbon-carbon bonds through dispersion forces," *Nature* **2011**, *477*, 308–311.

87. Fokin, A. A.; Chernish, L. V.; Gunchenko, P. A.; Tikhonchuk, E. Y.; Hausmann, H.; Serafin, M.; Dahl, J. E. P.; Carlson, R. M. K.; Schreiner, P. R. "Stable alkanes containing very long carbon–carbon bonds," *J. Am. Chem. Soc.* **2012**, *134*, 13641–13650.

88. Smith, M. B.; March, J. *March's Advanced Organic Chemistry: Reactions, Mechanisms, and Structure*; Wiley: New York, **2001**.

89. Pedley, J. B.; Naylor, R. D.; Kirby, S. P. *Thermochemical Data of Organic Compounds*; 2nd ed.; Chapman and Hall: London, **1986**.

90. Benson, S. W.; Cruickshank, F. R.; Golden, D. M.; Haugen, G. R.; O'Neal, H. E.; Rodgers, A. S.; Shaw, R.; Walsh, R. "Additivity rules for the estimation of thermochemical properties," *Chem. Rev.* **1969**, *69*, 279–324.

91. Benson, S. W. *Thermochemical Kinetics: Methods for the Estimation of Thermochemical Data and Rate Parameters*; 2nd ed.; Wiley: New York, **1976**.

92. Wiberg, K. B. "Group equivalents for converting ab initio energies to enthalpies of formation," *J. Comp. Chem.* **1984**, *5*, 197–199.

93. Ibrahim, M. R.; Schleyer, P. v. R. "Atom equivalents for relating ab initio energies to enthalpies of formation," *J. Comp. Chem.* **1985**, *6*, 157–167.

94. Cioslowski, J.; Liu, G.; Piskorz, P. "Computationally inexpensive theoretical thermochemistry," *J. Phys. Chem. A* **1998**, *102*, 9890–9900.

95. Guthrie, J. P. "Heats of formation from DFT calculations: an examination of several parameterizations," *J. Phys. Chem. A* **2001**, *105*, 9196–9202.
96. Hehre, W. J.; Ditchfield, R.; Radom, L.; Pople, J. A. "Molecular orbital theory of the electronic structure of organic compounds. V. Molecular theory of bond separation," *J. Am. Chem. Soc.* **1970**, *92*, 4796–4801.
97. George, P.; Trachtman, M.; Bock, C. W.; Brett, A. M. "An alternative approach to the problem of assessing destabilization energies (strain energies) in cyclic hydrocarbons," *Tetrahedron* **1976**, *32*, 317–323.
98. George, P.; Trachtman, M.; Brett, A. M.; Bock, C. W. "Comparison of various isodesmic and homodesmotic reaction heats with values derived from published ab initio molecular orbital calculations," *J. Chem. Soc., Perkin Trans. 2* **1977**, 1036–1047.
99. Bachrach, S. M. "The group equivalent reaction: an improved method for determining ring strain energy," *J. Chem. Ed.* **1990**, *67*, 907–908.
100. Wheeler, S. E.; Houk, K. N.; Schleyer, P. v. R.; Allen, W. D. "A hierarchy of homodesmotic reactions for thermochemistry," *J. Am. Chem. Soc.* **2009**, *131*, 2547–2560.
101. Boatz, J. A.; Gordon, M. S.; Hilderbrandt, R. L. "Structure and bonding in cycloalkanes and monosilacycloalkanes," *J. Am. Chem. Soc.* **1988**, *110*, 352–358.
102. Alcamí, M.; Mó, O.; Yáñez, M. "G2 ab initio calculations on three-membered rings: role of hydrogen atoms," *J. Comp. Chem.* **1998**, *19*, 1072–1086.
103. Cremer, D. "Pros and cons of σ-aromaticity," *Tetrahedron* **1988**, *44*, 7427–7454.
104. Cremer, D.; Gauss, J. "Theoretical determination of molecular structure and conformation. 20. Reevaluation of the strain energies of cyclopropane and cyclobutane – CC and CH bond energies, 1,3 interactions, and σ-aromaticity," *J. Am. Chem. Soc.* **1986**, *108*, 7467–7477.
105. Baeyer, A. v. "Über polyacetylenverbindungen," *Chem. Ber.* **1885**, *18*, 2269–2281.
106. Huisgen, R. "Adolf von Baeyer's scientific achievements – a legacy," *Angew. Chem. Int. Ed. Engl.* **1986**, *25*, 297–311.
107. Snyder, R. G.; Schachtschneider, J. H. "A valence force field for saturated hydrocarbons," *Spectrochim. Acta* **1965**, *21*, 169–195.
108. Walsh, A. D. "Structures of ethylene oxide, cyclopropane, and related molecules," *Trans. Faraday Soc.* **1949**, *45*, 179–190.
109. Bader, R. F. W. *Atoms in Molecules – A Quantum Theory*; Oxford University Press: Oxford, **1990**.
110. Pitzer, K. S. "Strain energies of cyclic hydrocarbons," *Science* **1945**, *101*, 672.
111. Dunitz, J. D.; Schomaker, V. "The molecular structure of cyclobutane," *J. Chem. Phys.* **1952**, *20*, 1703–1707.
112. Bauld, N. L.; Cessac, J.; Holloway, R. L. "1,3(Nonbonded) carbon/carbon interactions. The common cause of ring strain, puckering, and inward methylene rocking in cyclobutane and of vertical nonclassical stabilization, pyramidalization, puckering, and outward methylene rocking in the cyclobutyl cation," *J. Am. Chem. Soc.* **1977**, *99*, 8140–8144.
113. Coulson, C. A.; Moffitt, W. E. "The properties of certain strained hydrocarbons," *Phil. Mag.* **1949**, *40*, 1–35.

114. Baghal-Vayjooee, M. H.; Benson, S. W. "Kinetics and thermochemistry of the reaction atomic chlorine + cyclopropane .dblarw. hydrochloric acid + cyclopropyl. Heat of formation of the cyclopropyl radical," *J. Am. Chem. Soc.* **1979**, *101*, 2838–2840.

115. Seakins, P. W.; Pilling, M. J.; Niiranen, J. T.; Gutman, D.; Krasnoperov, L. N. "Kinetics and thermochemistry of R + HBr .dblarw. RH + Br reactions: determinations of the heat of formation of C_2H_5, i-C_3H_7, sec-C_4H_9 and t-C_4H_9," *J. Phys. Chem.* **1992**, *96*, 9847–9855.

116. Exner, K.; Schleyer, P. v. R. "Theoretical bond energies: a critical evaluation," *J. Phys. Chem. A* **2001**, *105*, 3407–3416.

117. Grimme, S. "Theoretical bond and strain energies of molecules derived from properties of the charge density at bond critical points," *J. Am. Chem. Soc.* **1996**, *118*, 1529–1534.

118. Johnson, W. T. G.; Borden, W. T. "Why are methylenecyclopropane and 1-methylcylopropene more "strained" than methylcyclopropane?," *J. Am. Chem. Soc.* **1997**, *119*, 5930–5933.

119. Bach, R. D.; Dmitrenko, O. "The effect of substitutents on the strain energies of small ring compounds," *J. Org. Chem.* **2002**, *67*, 2588–2599.

120. Bach, R. D.; Dmitrenko, O. "Strain energy of small ring hydrocarbons. Influence of C–H bond dissociation energies," *J. Am. Chem. Soc.* **2004**, *126*, 4444–4452.

121. Dewar, M. J. S. "σ-Conjugation and σ-aromaticity," *Bull. Soc. Chim. Belg.* **1979**, *88*, 957–967.

122. Dewar, M. J. S. "Chemical implications of σ conjugation," *J. Am. Chem. Soc.* **1984**, *106*, 669–682.

123. Kraka, E.; Cremer, D. "Theoretical determination of molecular structure and conformation. 15. Three-membered rings: bent bonds, ring strain, and surface delocalization," *J. Am. Chem. Soc.* **1985**, *107*, 3800–3810.

124. Moran, D.; Manoharan, M.; Heine, T.; Schleyer, P. v. R. "σ-Antiaromaticity in cyclobutane, cubane, and other molecules with saturated four-membered rings," *Org. Lett.* **2003**, *5*, 23–26.

125. Fowler, P. W.; Baker, J.; Mark Lillington, M. "The ring current in cyclopropane" *Theor. Chem. Acta* **2007**, *118*, 123–127.

126. Schleyer, P. v. R.; Jiao, H. "What is aromaticity?," *Pure. Appl. Chem.* **1996**, *68*, 209–218.

127. Krygowski, T. M.; Cyrañski, M. K.; Czarnocki, Z.; Häfelinger, G.; Katritzky, A. R. "Aromaticity: a theoretical concept of immense practical importance," *Tetrahedron* **2000**, *56*, 1783–1796.

128. Stanger, A. "What is ... aromaticity: a critique of the concept of aromaticity – can it really be defined?," *Chem. Commun.* **2009**, 1939–1947.

129. Minkin, V. I.; Glukhovtsev, M. N.; Simkin, B. Y. *Aromaticity and Antiaromaticity: Electronic and Structural Aspects*; John Wiley & Sons: New York, **1994**.

130. Schleyer, P. v. R. "Aromaticity," *Chem. Rev.* **2001**, *101*, 1115–1566.

131. Cyranski, M. K. "Energetic aspects of cyclic Pi-electron delocalization: evaluation of the methods of estimating aromatic stabilization energies," *Chem. Rev.* **2005**, *105*, 3773–3811.

132. Cyranski, M. K.; Schleyer, P. v. R.; Krygowski, T. M.; Jiao, H.; Hohlneicher, G. "Facts and artifacts about aromatic stability estimation," *Tetrahedron* **2003**, *59*, 1657–1665.

133. Hedberg, L.; Hedberg, K.; Cheng, P.-C.; Scott, L. T. "Gas-phase molecular structure of corannulene, C20H10. An electron-diffraction study augmented by ab initio and normal coordinate calculations," *J. Phys. Chem. A* **2000**, *104*, 7689–7694.

134. Dobrowolski, M. A.; Ciesielski, A.; Cyranski, M. K. "On the aromatic stabilization of corannulene and coronene," *Phys. Chem. Chem. Phys.* **2011**, *13*, 20557–20563.

135. Choi, C. H.; Kertesz, M. "Bond length alternation and aromaticity in large annulenes," *J. Chem. Phys.* **1998**, *108*, 6681–6688.

136. *Aromaticity, Pseudo-Aromaticity, Anti-Aromaticity, Proceedings of an International Symposium*; Bergmann, E. D.; Pullman, B., Eds.; Israel Academy of Sciences and Humanities: Jerusalem, **1971**; Vol. 33 see the following exchange: E. Heilbronner: "Now could you point out a molecule, except benzene, which classifies as 'aromatic'?" B. Binsch: "Benzene is a perfect example!" E. Heilbronner: "Name a second one." B. Binsch: "It is, of course, a question of degree".

137. Katritzky, A. R.; Barczynski, P.; Musumarra, G.; Pisano, D.; Szafran, M. "Aromaticity as a quantitative concept. 1. A statistical demonstration of the orthogonality of classical and magnetic aromaticity in five- and six-membered heterocycles," *J. Am. Chem. Soc.* **1989**, *111*, 7–15.

138. Jug, K.; Koester, A. M. "Aromaticity as a multi-dimensional phenomenon," *J. Phys. Org. Chem.* **1991**, *4*, 163–169.

139. Schleyer, P. v. R.; Freeman, P. K.; Jiao, H.; Goldfuss, B. "Aromaticity and antiaromaticity in five-membered C4H4X ring systems: classical and magnetic concepts may not be orthogonal," *Angew. Chem. Int. Ed. Engl.* **1995**, *34*, 337–340.

140. Katritzky, A. R.; Karelson, M.; Sild, S.; Krygowski, T. M.; Jug, K. "Aromaticity as a quantitative concept. 7. Aromaticity reaffirmed as a multidimensional characteristic," *J. Org. Chem.* **1998**, *63*, 5228–5231.

141. Cyranski, M. K.; Krygowski, T. M.; Katritzky, A. R.; Schleyer, P. v. R. "To what extent can aromaticity be defined uniquely?," *J. Org. Chem.* **2002**, *67*, 1333–1338.

142. Moran, D.; Simmonett, A. C.; Leach, F. E.; Allen, W. D.; Schleyer, P. v. R.; Schaefer, H. F. III, "Popular theoretical methods predict benzene and arenes to be nonplanar," *J. Am. Chem. Soc.* **2006**, *128*, 9342–9343.

143. Baldridge, K. K.; Siegel, J. S. "Stabilization of benzene versus oligoacetylenes: not another scale for aromaticity," *J. Phys. Org. Chem.* **2004**, *17*, 740–742.

144. Roberts, J. D.; Streitwieser, A. J.; Regan, C. M. "Small-ring compounds. X. Molecular orbital calculations of properties of some small-ring hydrocarbons and free radicals," *J. Am. Chem. Soc.* **1952**, *74*, 4579–4582.

145. Schaad, L. J.; Hess, B. A., Jr. "Dewar resonance energy," *Chem. Rev.* **2001**, *101*, 1465–1476.

146. Pauling, L. *The Nature of the Chemical Bond*; Cornell University Press: Ithaca, NY, **1960**.

147. Wheland, G. W. *The Theory of Resonance*; John Wiley: New York, **1944**.

148. Mo, Y.; Schleyer, P. v. R. "An energetic measure of aromaticity and antiaromaticity based on the pauling-wheland resonance energies," *Chem. Eur. J.* **2006**, *12*, 2009–2020.

149. Dewar, M. J. S.; De Llano, C. "Ground states of conjugated molecules. XI. Improved treatment of hydrocarbons," *J. Am. Chem. Soc.* **1969**, *91*, 789–795.

150. Schleyer, P. v. R.; Manoharan, M.; Jiao, H.; Stahl, F. "The acenes: is there a relationship between aromatic stabilization and reactivity?," *Org. Lett.* **2001**, *3*, 3643–3646.
151. Hess, B. A., Jr.; Schaad, L. J. "Ab initio calculation of resonance energies. Benzene and cyclobutadiene," *J. Am. Chem. Soc.* **1983**, *105*, 7500–7505.
152. Schleyer, P. v. R.; Puhlhofer, F. "Recommendations for the evaluation of aromatic stabilization energies," *Org. Lett.* **2002**, *4*, 2873–2876.
153. Schleyer, P. v. R.; Jiao, H.; Hommes, N. J. R. v. E.; Malkin, V. G.; Malkina, O. "An evaluation of the aromaticity of inorganic rings: refined evidence from magnetic properties," *J. Am. Chem. Soc.* **1997**, *119*, 12669–12670.
154. Gomes, J. A. N. F.; Mallion, R. B. "Aromaticity and ring currents," *Chem. Rev.* **2001**, *101*, 1349–1384.
155. Dauben, H. J., Jr.; Wilson, J. D.; Laity, J. L. "Diamagnetic susceptibility exaltation in hydrocarbons," *J. Am. Chem. Soc.* **1969**, *91*, 1991–1998.
156. Dauben, H. J.; Wilson, J. D.; Laity, J. L. "Diamagnetic susceptibility exaltations as a criterion of aromaticity," In *Nonbenzenoid Aromatics*; Snyder, J. P., Ed.; Academic Press: New York, **1971**; Vol. 2, p 167–206.
157. Jackman, L. M.; Sondheimer, F.; Amiel, Y.; Ben-Efraim, D. A.; Gaoni, Y.; Wolovsky, R.; Bothner-By, A. A. "The nuclear magnetic resonance spectroscopy of a series of annulenes and dehydro-annulenes," *J. Am. Chem. Soc.* **1962**, *84*, 4307–4312.
158. Stevenson, C. D.; Kurth, T. L. "Isotopic perturbations in aromatic character and new closely related conformers found in [16]- and [18]annulene," *J. Am. Chem. Soc.* **2000**, *122*, 722–723.
159. Wannere, C. S.; Corminboeuf, C.; Allen, W. D.; Schaefer, H. F., III; Schleyer, P. v. R. "Downfield proton chemical shifts are not reliable aromaticity indicators," *Org. Lett.* **2005**, *7*, 1457–1460.
160. Faglioni, F.; Ligabue, A.; Pelloni, S.; Soncini, A.; Viglione, R. G.; Ferraro, M. B.; Zanasi, R.; Lazzeretti, P. "Why downfield proton chemical shifts are not reliable aromaticity indicators," *Org. Lett.* **2005**, *7*, 3457–3460.
161. Schleyer, P. v. R.; Maerker, C.; Dransfeld, A.; Jiao, H.; Hommes, N. J. R. v. E. "Nucleus-independent chemical shifts: a simple and efficient aromaticity probe," *J. Am. Chem. Soc.* **1996**, *118*, 6317–6318.
162. Jiao, H.; Schleyer, P. v. R.; Mo, Y.; McAllister, M. A.; Tidwell, T. T. "Magnetic evidence for the aromaticity and antiaromaticity of charged fluorenyl, indenyl, and cyclopentadienyl systems," *J. Am. Chem. Soc.* **1997**, *119*, 7075–7083.
163. Williams, R. V.; Armantrout, J. R.; Twamley, B.; Mitchell, R. H.; Ward, T. R.; Bandyopadhyay, S. "A theoretical and experimental scale of aromaticity. The first nucleus-independent chemical shifts (NICS) study of the dimethyldihydropyrene nucleus," *J. Am. Chem. Soc.* **2002**, *124*, 13495–13505.
164. Schleyer, P. v. R.; Manoharan, M.; Wang, Z.-X.; Kiran, B.; Jiao, H.; Puchta, R.; van Eikema Hommes, N. J. R. "Dissected nucleus-independent chemical shift analysis of -aromaticity and antiaromaticity," *Org. Lett.* **2001**, *3*, 2465–2468.
165. Stanger, A. "Nucleus-independent chemical shifts (NICS): distance dependence and revised criteria for aromaticity and antiaromaticity," *J. Org. Chem.* **2006**, *71*, 883–893.
166. Klod, S.; Kleinpeter, E. "Ab initio calculation of the anisotropy effect of multiple bonds and the ring current effect of arenes—application in conformational and configurational analysis," *Chem. Soc., Perkin Trans. 2* **2001**, 1893–1898.

167. Pople, J. A. "Proton magnetic resonance of hydrocarbons," *J. Chem. Phys.* **1956**, *24*, 1111.
168. Viglione, R. G.; Zanasi, R.; Lazzeretti, P. "Are ring currents still useful to rationalize the benzene proton magnetic shielding?," *Org. Lett.* **2004**, *6*, 2265–2267.
169. Fallah-Bagher-Shaidaei, H.; Wannere, C. S.; Corminboeuf, C.; Puchta, R.; Schleyer, P. v. R. "Which NICS aromaticity index for planar π rings is best?," *Org. Lett.* **2006**, *8*, 863–866.
170. Herges, R.; Jiao, H.; Schleyer, P. v. R. "Magnetic properties of aromatic transition states: the Diels-Alder reactions," *Angew. Chem. Int. Ed. Engl.* **1994**, *33*, 1376–1378.
171. Jiao, H.; Schleyer, P. v. R. "The Cope rearrangement transition structure is not diradicaloid, but is it aromatic?," *Angew. Chem. Int. Ed. Engl.* **1995**, *34*, 334–337.
172. Cabaleiro-Lago, E. M.; Rodriguez-Otero, J.; Varela-Varela, S. M.; Pena-Gallego, A.; Hermida-Ramon, J. M. "Are electrocyclization reactions of (3Z)-1,3,5-hexatrienone and nitrogen derivatives pseudopericyclic? A DFT study," *J. Org. Chem.* **2005**, *70*, 3921–3928.
173. Martín-Santamaría, S.; Lavan, B.; Rzepa, H. S. "Hückel and Möbius aromaticity and trimerous transition state behaviour in the pericyclic reactions of [10], [14], [16] and [18]annulenes," *J. Chem. Soc., Perkin Trans.* **2000**, *2*, 1415–1417.
174. Levy, A.; Rakowitz, A.; Mills, N. S. "Dications of fluorenylidenes. The effect of substituent electronegativity and position on the antiaromaticity of substituted tetrabenzo[5.5]fulvalene dications," *J. Org. Chem.* **2003**, *68*, 3990–3998.
175. Mills, N. S.; Levy, A.; Plummer, B. F. "Antiaromaticity in fluorenylidene dications. Experimental and theoretical evidence for the relationship between the HOMO/LUMO gap and antiaromaticity," *J. Org. Chem.* **2004**, *69*, 6623–6633.
176. Piekarski, A. M.; Mills, N. S.; Yousef, A. "Dianion and dication of tetrabenzo[5.7]fulvalene. Greater antiaromaticity than aromaticity in comparable systems," *J. Am. Chem. Soc.* **2008**, *130*, 14883–14890.
177. Dinadayalane, T. C.; Deepa, S.; Reddy, A. S.; Sastry, G. N. "Density functional theory study on the effect of substitution and ring annelation to the rim of corannulene," *J. Org. Chem.* **2004**, *69*, 8111–8114.
178. Schulman, J. M.; Disch, R. L. "Properties of phenylene-based hydrocarbon bowls and archimedene," *J. Phys. Chem. A* **2005**, *109*, 6947–6952.
179. Kavitha, K.; Manoharan, M.; Venuvanalingam, P. "1,3-Dipolar reactions involving corannulene: How does its rim and spoke addition vary?," *J. Org. Chem.* **2005**, *70*, 2528–2536.
180. Wu, J. I.; Fernández, I.; Mo, Y.; Schleyer, P. v. R. "Why cyclooctatetraene is highly stabilized: the importance of "two-way" (double) hyperconjugation," *J. Chem. Theor. Comput.* **2012**, *8*, 1280–1287.
181. Nishinaga, T.; Uto, T.; Inoue, R.; Matsuura, A.; Treitel, N.; Rabinovitz, M.; Komatsu, K. "Antiaromaticity and reactivity of a planar cyclooctatetraene fully annelated with bicyclo[2.1.1]hexane units," *Chem. Eur. J.* **2008**, *14*, 2067–2074.
182. Ohmae, T.; Nishinaga, T.; Wu, M.; Iyoda, M. "Cyclic tetrathiophenes planarized by silicon and sulfur bridges bearing antiaromatic cyclooctatetraene core: syntheses, structures, and properties," *J. Am. Chem. Soc.* **2009**, *132*, 1066–1074.
183. Masamune, S.; Hojo, K.; Hojo, K.; Bigam, G.; Rabenstein, D. L. "Geometry of [10]annulenes," *J. Am. Chem. Soc.* **1971**, *93*, 4966–4968.

184. Xie, Y.; Schaefer, H. F., III; Liang, G.; Bowen, J. P. "[10]Annulene: the wealth of energetically low-lying structural isomers of the same $(CH)_{10}$ connectivity," *J. Am. Chem. Soc.* **1994**, *116*, 1442–1449.

185. Sulzbach, H. M.; Schleyer, P. v. R.; Jiao, H.; Xie, Y.; Schaefer, H. F. III, "A [10]annulene isomer may be aromatic, after all!," *J. Am. Chem. Soc.* **1995**, *117*, 1369–1373.

186. King, R. A.; Crawford, T. D.; Stanton, J. F.; Schaefer, H. F., III "Conformations of [10]annulene: more bad news for density functional theory and second-order perturbation theory," *J. Am. Chem. Soc.* **1999**, *121*, 10788–10793.

187. Sulzbach, H. M.; Schaefer, H. F., III; Klopper, W.; Luthi, H.-P. "Exploring the boundary between aromatic and olefinic character: bad news for second-order perturbation theory and density functional schemes," *J. Am. Chem. Soc.* **1996**, *118*, 3519–3520.

188. Wannere, C. S.; Sattelmeyer, K. W.; Schaefer, H. F. III; Schleyer, P. v. R. "Aromaticity: The alternating CC bond length structures of [14]-, [18]-, and [22]annulene," *Angew. Chem. Int. Ed.* **2004**, *43*, 4200–4206.

189. Castro, C.; Karney, W. L.; McShane, C. M.; Pemberton, R. P. "[10]Annulene: bond shifting and conformational mechanisms for automerization," *J. Org. Chem.* **2006**, *71*, 3001–3006.

190. Price, D. R.; Stanton, J. F. "Computational study of [10]annulene NMR spectra," *Org. Lett.* **2002**, *4*, 2809–2811.

191. Navarro-Vázquez, A.; Schreiner, P. R. "1,2-Didehydro[10]annulenes: structures, aromaticity, and cyclizations," *J. Am. Chem. Soc.* **2005**, *127*, 8150–8159.

192. Wannere, C. S.; Schleyer, P. v. R. "How aromatic Are large (4n + 2) annulenes?," *Org. Lett.* **2003**, *5*, 865–868.

193. Longuet-Higgins, H. C.; Salem, L. "Alternation of bond lengths in long conjugated chain molecules," *Proc. Roy. Soc. London* **1959**, *A251*, 172–185.

194. Chiang, C. C.; Paul, I. C. "Crystal and molecular structure of [14]annulene," *J. Am. Chem. Soc.* **1972**, *94*, 4741–4743.

195. Bregman, J.; Hirshfeld, F. L.; Rabinovich, D.; Schmidt, G. M. J. "The crystal structure of [18]annulene. I. X-ray study," *Acta Cryst.* **1965**, *19*, 227–234.

196. Gorter, S.; Rutten-Keulemans, E.; Krever, M.; Romers, C.; Cruickshank, D. W. J. "[18]-annulene, $C_{18}H_{18}$, structure, disorder and Hueckel's 4n + 2 rule," *Acta Crystallogr. B* **1995**, *51*, 1036–1045.

197. Choi, C. H.; Kertesz, M.; Karpfen, A. "Do localized structures of [14]- and [18]annulenes exist?," *J. Am. Chem. Soc.* **1997**, *119*, 11994–11995.

198. Baldridge, K. K.; Siegel, J. S. "Ab initio density functional vs Hartree Fock predictions for the structure of [18]annulene: evidence for bond localization and diminished ring currents in bicycloannelated [18]annulenes," *Angew. Chem. Int. Ed. Engl.* **1997**, *36*, 745–748.

199. Oth, J. F. M. "Conformational mobility and fast bond shift in the annulenes," *Pure Appl. Chem.* **1971**, *25*, 573–622.

200. Heilbronner, E. "Hückel molecular orbitals of Möbius-type conformations of annulenes," *Tetrahedron Lett.* **1964**, *5*, 1923–1928.

201. Rzepa, H. S. "Möbius aromaticity and delocalization," *Chem. Rev.* **2005**, *105*, 3697–3715.

202. Castro, C.; Isborn, C. M.; Karney, W. L.; Mauksch, M.; Schleyer, P. v. R. "Aromaticity with a twist: Möbius [4n]annulenes," *Org. Lett.* **2002**, *4*, 3431–3434.

203. Ajami, D.; Hess, K.; Köhler, F.; Näther, C.; Oeckler, O.; Simon, A.; Yamamoto, C.; Okamoto, Y.; Herges, R. "Synthesis and properties of the first Möbius annulenes," *Chem. Eur. J.* **2006**, *12*, 5434–5445.

204. Ajami, D.; Oeckler, O.; Simon, A.; Herges, R. "Synthesis of a Möbius aromatic hydrocarbon," *Nature* **2003**, *426*, 819–821.

205. Wannere, C. S.; Moran, D.; Allinger, N. L.; Hess, B. A., Jr.; Schaad, L. J.; Schleyer, P. v. R. "On the stability of large [4n]annulenes," *Org. Lett.* **2003**, *5*, 2983–2986.

206. Castro, C.; Chen, Z.; Wannere, C. S.; Jiao, H.; Karney, W. L.; Mauksch, M.; Puchta, R.; Hommes, N. J. R. v. E.; Schleyer, P. v. R. "Investigation of a putative Möbius aromatic hydrocarbon. The effect of benzannelation on Möbius [4n]annulene aromaticity," *J. Am. Chem. Soc.* **2005**, *127*, 2425–2432.

207. Clar, E. *The Aromatic Sextet*; Wiley: London, **1972**.

208. Shimizu, S.; Aratani, N.; Osuka, A. "*Meso*-trifluoromethyl-substituted expanded porphyrins," *Chem. Eur. J.* **2006**, *12*, 4909–4918.

209. Rzepa, H. S. "Lemniscular hexaphyrins as examples of aromatic and antiaromatic double-twist Möbius molecules," *Org. Lett.* **2008**, *10*, 949–952.

210. Tanaka, Y.; Saito, S.; Mori, S.; Aratani, N.; Shinokubo, H.; Shibata, N.; Higuchi, Y.; Yoon, Z. S.; Kim, K. S.; Noh, S. B.; Park, J. K.; Kim, D.; Osuka, A. "Metalation of expanded porphyrins: a chemical trigger used to produce molecular twisting and Möbius aromaticity," *Angew. Chem. Int. Ed.* **2008**, *47*, 681–684.

211. Tokuji, S.; Shin, J.-Y.; Kim, K. S.; Lim, J. M.; Youfu, K.; Saito, S.; Kim, D.; Osuka, A. "Facile formation of a benzopyrane-fused [28]hexaphyrin that exhibits distinct Möbius aromaticity," *J. Am. Chem. Soc.* **2009**, *131*, 7240–7241.

212. Castro, C.; Karney, W. L.; Valencia, M. A.; Vu, C. M. H.; Pemberton, R. P. "Möbius aromaticity in [12]annulene: cis-trans isomerization via twist-coupled bond shifting," *J. Am. Chem. Soc.* **2005**, *127*, 9704–9705.

213. Moll, J. F.; Pemberton, R. P.; Gutierrez, M. G.; Castro, C.; Karney, W. L. "Configuration change in [14]annulene requires Möbius antiaromatic bond shifting," *J. Am. Chem. Soc.* **2006**, *129*, 274–275.

214. Schleyer, P. v. R.; Barborak, J. C.; Su, T. M.; Boche, G.; Schneider, G. "Thermal bicyclo[6.1.0]Nonatrienyl chloride-dihydroindenyl chloride rearrangement," *J. Am. Chem. Soc.* **1971**, *93*, 279–281.

215. Yakali, E. "Genesis and bond relocation of the cyclononatetraenyl cation and related compounds," Dissertation, Syracuse University, **1973**.

216. Mauksch, M.; Gogonea, V.; Jiao, H.; Schleyer, P. v. R. "Monocyclic $(CH)_9^+$—a heilbronner Möbius aromatic system revealed," *Angew. Chem. Int. Ed.* **1998**, *37*, 2395–2397.

217. Bucher, G.; Grimme, S.; Huenerbein, R.; Auer, A. A.; Mucke, E.; Köhler, F.; Siegwarth, J.; Herges, R. "Is the [9]annulene cation a Möbius annulene?," *Angew. Chem. Int. Ed.* **2009**, *48*, 9971–9974.

218. Mucke, E.-K.; Kohler, F.; Herges, R. "The [13]annulene cation is a stable Möbius annulene cation," *Org. Lett.* **2010**, *12*, 1708–1711.

219. Mucke, E.-K.; Schönborn, B.; Köhler, F.; Herges, R. "Stability and aromaticity of charged Möbius[4n]annulenes," *J. Org. Chem.* **2010**, *76*, 35–41.

220. Fowler, P. W.; Rzepa, H. S. "Aromaticity rules for cycles with arbitrary numbers of half-twists," *Phys. Chem. Chem. Phys.* **2006**, *8*, 1775–1777.

221. Wannere, C. S.; Rzepa, H. S.; Rinderspacher, B. C.; Paul, A.; Allan, C. S. M.; Schaefer III, H. F.; Schleyer, P. v. R. "The geometry and electronic topology of higher-order charged Mobius annulenes," *J. Phys. Chem. A* **2009**, *113*, 11619–11629.
222. Rzepa, H. S. "A double-twist Möbius-aromatic conformation of [14]annulene," *Org. Lett.* **2005**, *7*, 4637–4639.
223. Okoronkwo, T.; Nguyen, P. T.; Castro, C.; Karney, W. L. "[14]Annulene: cis/trans isomerization via two-twist and nondegenerate planar bond shifting and Möbius conformational minima," *Org. Lett.* **2010**, *12*, 972–975.
224. Mohebbi, A.; Mucke, E. K.; Schaller, G.; Köhler, F.; Sönnichsen, F.; Ernst, L.; Näther, C.; Herges, R. "Singly and doubly twisted [36]annulenes: synthesis and calculations," *Chem. Eur. J.* **2010**, *16*, 7767–7772.
225. Mills, W. H.; Nixon, I. G. "Stereochemical influences on aromatic substitution. Substitution derivatives of 5-hydroxyhydrindene," *J. Chem. Soc.* **1930**, 2510–2524.
226. Siegel, J. S. "Mills–Nixon effect: wherefore art thou?," *Angew. Chem. Int. Ed. Engl.* **1994**, *33*, 1721–1723.
227. Stanger, A. "Strain-induced bond localization. The heteroatom case," *J. Am. Chem. Soc.* **1998**, *120*, 12034–12040.
228. Stanger, A. "Is the Mills-Nixon effect real?," *J. Am. Chem. Soc.* **1991**, *113*, 8277–8280.
229. Baldridge, K. K.; Siegel, J. S. "Bond alternation in triannelated benzenes: dissection of cyclic π from Mills-Nixon effects," *J. Am. Chem. Soc.* **1992**, *114*, 9583–9587.
230. Sakai, S. "Theoretical study on the aromaticity of benzenes annelated to small rings," *J. Phys. Chem. A* **2002**, *106*, 11526–11532.
231. Bachrach, S. M. "Aromaticity of annulated benzene, pyridine and phosphabenzene," *J. Organomet. Chem.* **2002**, *643–644*, 39–46.
232. Boese, R.; Bläser, D.; Billups, W. E.; Haley, M. M.; Maulitz, A. H.; Mohler, D. L.; Vollhardt, K. P. C. "The effect of fusion of angular strained rings on benzene: crystal structures of 1,2-dihydrocyclobuta[a]cyclopropa[c]-, 1,2,3,4-tetrahydrodicyclobuta[a,c]-, 1,2,3,4-tetrahydrodicyclobuta[a,c]cyclopropa[e]-, and 1,2,3,4,5,6-hexahydrotricyclobuta[a,c,e]benzene," *Angew. Chem. Int. Ed. Engl.* **1994**, *33*, 313–317.
233. Mo, O.; Yanez, M.; Eckert-Maksic, M.; Maksic, Z. B. "Bent bonds in benzocyclopropenes and their fluorinated derivatives," *J. Org. Chem.* **1995**, *60*, 1638–1646.
234. Diercks, R.; Vollhardt, K. P. C. "Tris(benzocyclobutadieno)benzene, the triangular [4]phenylene with a completely bond-fixed cyclohexatriene ring: cobalt-catalyzed synthesis from hexaethynylbenzene and thermal ring opening to 1,2:5,6:9,10-tribenzo-3,4,7,8,11,12-hexadehydro[12]annulene," *J. Am. Chem. Soc.* **1986**, *108*, 3150–3152.
235. Bürgi, H.-B.; Baldridge, K. K.; Hardcastle, K.; Frank, N. L.; Gantzel, P.; Siegel, J. S.; Ziller, J. "X-ray diffraction evidence for a cyclohexatriene motif in the molecular structure of tris(bicyclo[2.1.1]hexeno)benzene: bond alternation after the refutation of the Mills-Nixon theory," *Angew. Chem. Int. Ed. Engl.* **1995**, *34*, 1454–1456.
236. Boese, R.; Bläser, D. "Structures and deformation electron densities of 1,2-dihydrocyclobutabenzene and 1,2,4,5-tetrahydrodicyclobuta[a,d]benzene," *Angew. Chem. Int. Ed. Engl.* **1988**, *27*, 304–305.
237. Alkorta, I.; Elguero, J. "Can aromaticity be described with a single parameter? Benzene vs. cyclohexatriene," *New J. Chem.* **1999**, *23*, 951–954.

238. Bao, X.; Hrovat, D.; Borden, W. "The effects of orbital interactions on the geometries of some annelated benzenes," *Theor. Chem. Acc.* **2011**, *130*, 261–268.
239. Schulman, J. M.; Disch, R. L.; Jiao, H.; Schleyer, P. v. R. "Chemical shifts of the [N]phenylenes and related compounds," *J. Phys. Chem. A* **1998**, *102*, 8051–8055.
240. Beckhaus, H.-D.; Faust, R.; Matzger, A. J.; Mohler, D. L.; Rogers, D. W.; Ruchardt, C.; Sawhney, A. K.; Verevkin, S. P.; Vollhardt, K. P. C.; Wolff, S. "The heat of hydrogenation of (a) cyclohexatriene," *J. Am. Chem. Soc.* **2000**, *122*, 7819–7820.
241. Hopf, H. *Classics in Hydrocarbon Chemistry: Syntheses, Concepts, Perspectives*; Wiley-VCH: Weinheim, Germany, **2000**.
242. Bodwell, G. J.; Fleming, J. J.; Mannion, M. R.; Miller, D. O. "Nonplanar aromatic compounds. 3. A proposed new strategy for the synthesis of buckybowls. Synthesis, structure and reactions of [7]-, [8]- and [9](2,7)pyrenophanes," *J. Org. Chem.* **2000**, *65*, 5360–5370.
243. Dobrowolski, M. A.; Cyranski, M. K.; Merner, B. L.; Bodwell, G. J.; Wu, J. I.; Schleyer, P. v. R. "Interplay of π-electron delocalization and strain in [n](2,7)pyrenophanes," *J. Org. Chem.* **2008**, *73*, 8001–8009.
244. Swart, M.; van der Wijst, T.; Guerra, C. F.; Bickelhaupt, F. M. "π-π stacking tackled with density functional theory," **2007**, *13*, 1245–1257.
245. Hobza, P.; Selzle, H. L.; Schlag, E. W. "Potential energy surface for the benzene dimer. Results of ab Initio CCSD(T) calculations show two nearly isoenergetic structures: T-shaped and parallel-displaced," *J. Phys. Chem.* **1996**, *100*, 18790–18794.
246. Steed, J. M.; Dixon, T. A.; Klemperer, W. "Molecular beam studies of benzene dimer, hexafluorobenzene dimer, and benzene–hexafluorobenzene," *J. Chem. Phys.* **1979**, *70*, 4940–4946.
247. Arunan, E.; Gutowsky, H. S. "The rotational spectrum, structure and dynamics of a benzene dimer," *J. Chem. Phys.* **1993**, *98*, 4294–4296.
248. Felker, P. M.; Maxton, P. M.; Schaeffer, M. W. "Nonlinear Raman studies of weakly bound complexes and clusters in molecular beams," *Chem. Rev.* **1994**, *94*, 1787–1805.
249. Bornsen, K. O.; Selzle, H. L.; Schlag, E. W. "Spectra of isotopically mixed benzene dimers: details on the interaction in the vdW bond," *J. Chem. Phys.* **1986**, *85*, 1726–1732.
250. Law, K.; Schauer, M.; Bernstein, E. R. "Dimers of aromatic molecules: (benzene)$_2$, (toluene)$_2$, and benzene–toluene," *J. Chem. Phys.* **1984**, *81*, 4871–4882.
251. Scherzer, W.; Kraetzschmar, O.; Selzle, H. L.; Schlag, E. W. "Structural isomers of the benzene dimer from mass selective hole-burning spectroscopy," *Z. Naturforsch. A* **1992**, *47*, 1248–1252.
252. Grover, J. R.; Walters, E. A.; Hui, E. T. "Dissociation energies of the benzene dimer and dimer cation," *J. Phys. Chem.* **1987**, *91*, 3233–3237.
253. Sinnokrot, M. O.; Valeev, E. F.; Sherrill, C. D. "Estimates of the ab initio limit for π–π interactions: the benzene dimer," *J. Am. Chem. Soc.* **2002**, *124*, 10887–10893.
254. Grimme, S. "Do special noncovalent π–π stacking interactions really exist?," *Angew. Chem. Int. Ed.* **2008**, *47*, 3430–3434.
255. Sinnokrot, M. O.; Sherrill, C. D. "Highly accurate coupled cluster potential energy curves for the benzene dimer: sandwich, T-shaped, and parallel-displaced configurations," *J. Phys. Chem. A* **2004**, *108*, 10200–10207.

256. Sherrill, C. D.; Takatani, T.; Hohenstein, E. G. "An assessment of theoretical methods for nonbonded interactions: comparison to complete basis set limit coupled-cluster potential energy curves for the benzene dimer, the methane dimer, benzene−methane, and benzene−H2S," *J. Phys. Chem. A* **2009**, *113*, 10146–10159.

257. Pitoňák, M.; Neogrády, P.; Řezáč, J.; Jurečka, P.; Urban, M.; Hobza, P. "Benzene dimer: high-level wave function and density functional theory calculations," *J. Chem. Theor. Comput.* **2008**, *4*, 1829–1834.

258. Hunter, C. A.; Sanders, J. K. M. "The nature of π-π interactions," *J. Am. Chem. Soc.* **1990**, *112*, 5525–5534.

259. Cozzi, F.; Cinquini, M.; Annunziata, R.; Dwyer, T.; Siegel, J. S. "Polar/π interactions between stacked aryls in 1,8-diarylnaphthalenes," *J. Am. Chem. Soc.* **1992**, *114*, 5729–5733.

260. Sinnokrot, M. O.; Sherrill, C. D. "Unexpected substituent effects in face-to-face π-stacking interactions," *J. Phys. Chem. A* **2003**, *107*, 8377–8379.

261. Sinnokrot, M. O.; Sherrill, C. D. "Substituent effects in $\pi-\pi$ interactions: sandwich and T-shaped configurations," *J. Am. Chem. Soc.* **2004**, *126*, 7690–7697.

262. Lee, E. C.; Kim, D.; Jurečka, P.; Tarakeshwar, P.; Hobza, P.; Kim, K. S. "Understanding of assembly phenomena by aromatic−aromatic interactions: benzene dimer and the substituted systems," *J. Phys. Chem. A* **2007**, *111*, 3446–3457.

263. Grimme, S.; Antony, J.; Schwabe, T.; Muck-Lichtenfeld, C. "Density functional theory with dispersion corrections for supramolecular structures, aggregates, and complexes of (bio)organic molecules," *Org. Biomol. Chem.* **2007**, *5*, 741–758.

264. Watt, M.; Hardebeck, L. K. E.; Kirkpatrick, C. C.; Lewis, M. "Face-to-face arene−arene binding energies: dominated by dispersion but predicted by electrostatic and dispersion/polarizability substituent constants," *J. Am. Chem. Soc.* **2011**, *133*, 3854–3862.

265. Ringer, A. L.; Sinnokrot, M. O.; Lively, R. P.; Sherrill, C. D. "The effect of multiple substituents on sandwich and T-shaped $\pi-\pi$ interactions," *Chem. Eur. J.* **2006**, *12*, 3821–3828.

266. Arnstein, S. A.; Sherrill, C. D. "Substituent effects in parallel-displaced $\pi-\pi$ interactions," *Phys. Chem. Chem. Phys.* **2008**, *10*, 2646–2655.

267. Wheeler, S. E.; Houk, K. N. "Substituent effects in the benzene dimer are due to direct interactions of the substituents with the unsubstituted benzene," *J. Am. Chem. Soc.* **2008**, *130*, 10854–10855.

268. Wheeler, S. E. "Local nature of substituent effects in stacking interactions," *J. Am. Chem. Soc.* **2011**, *133*, 10262–10274.

269. Ringer, A. L.; Sherrill, C. D. "Substituent effects in sandwich configurations of multiply substituted benzene dimers are not solely governed by electrostatic control," *J. Am. Chem. Soc.* **2009**, *131*, 4574–4575.

270. Bloom, J. W. G.; Wheeler, S. E. "Taking the aromaticity out of aromatic interactions," *Angew. Chem. Int. Ed.* **2011**, *50*, 7847–7849.

271. Martinez, C. R.; Iverson, B. L. "Rethinking the term "pi-stacking"," *Chem. Sci.* **2012**, *3*, 2191–2201.

CHAPTER 4

Pericyclic Reactions

Into the mid-1960s, a large swath of organic chemistry was a sea of confusing reactivity. Seemingly related or analogous reagents gave unrelated products. Reactions would often give outstanding regio- and stereoselectivities, but with no apparent rhyme or reason. Some examples are shown in Scheme 4.1. Alkenes do not react under thermal conditions to give cyclobutanes, yet an alkene and diene will add to give cyclohexenes—the famous Diels–Alder reaction. Placing a donor substituent on the diene and an acceptor substituent on the dienophile accelerates the Diels–Alder reaction, and the resulting major product is the more congested 3,4-substituted cyclohexene rather than 3,5-substituted cyclohexene. *cis*-2,3-dimethylcyclobutene opens upon heating to give *cis*,*trans*-2,4-hexadiene, but *cis*-5,6-dimethyl-1,3-cyclohexadiene opens to give *trans*,*trans*-2,4,6-octatriene. [1,3]-Hydrogen migrations are very rare, but [1,5]-hydrogen migrations are facile.

Perhaps the greatest achievement of theoretical organic chemistry was the development of systematic rules that organized all of these (and more) disparate reactions. Beginning in 1965, Woodward and Hoffmann[1] developed the concept of conservation of orbital symmetry for pericyclic reactions, reactions in which "concerted reorganization of bonding occurs through a cyclic array of continuously bonded atoms." Fukui[2,3] developed an alternate viewpoint focusing on the interaction of frontier molecular orbitals. Zimmerman[4] extended these ideas with the concept of Hückel and Möbius orbitals. Together, these ideas consolidated electrocyclizations, cycloadditions, sigmatropic rearrangements, chelotropic reactions, and group addition reactions into one conceptual framework. Detailed discussion of these organizing concepts for pericyclic reactions (particularly the ideas of orbital correlation diagrams, frontier molecular orbital theory, and the aromatic transition state) is outside the scope of this book. The interested reader is referred to a number of fine introductory texts.[5–7]

Given the great success of the conservation of orbital symmetry to understand pericyclic reactions, it should come as no surprise that computational chemistry has been widely applied to this area.[8–10] Instead of surveying this broad literature, we focus our discussion on a few reactions where either computational chemistry has served to broaden our insight into pericyclic reactions or where these studies

---

Computational Organic Chemistry, Second Edition. Steven M. Bachrach
© 2014 John Wiley & Sons, Inc. Published 2014 by John Wiley & Sons, Inc.

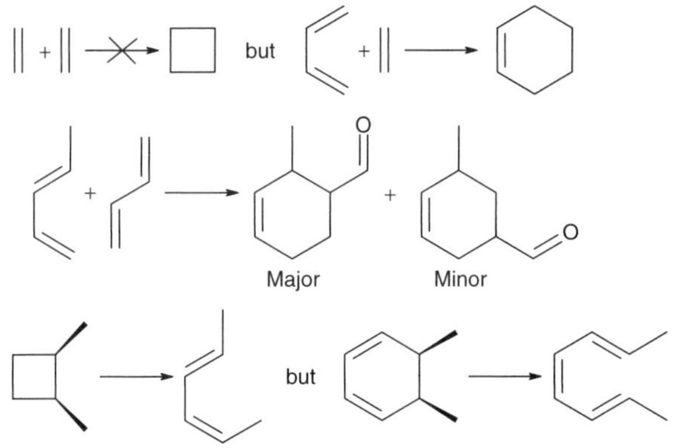

Scheme 4.1

have helped discern the limitations of computational methods. We begin using the Diels–Alder reaction and Cope rearrangements to help define which computational methods are appropriate for studying pericyclic reactions. These studies have served as paradigms for the selection of computational methods to a broad range of organic chemistry. Next, we discuss the role computational chemistry played in determining the mechanism of the Bergman cyclization and defining the category of bispericyclic and psuedopericyclic reactions. We conclude with a discussion of torquoselectivity, an extension of the orbital symmetry rules developed largely from the results of computational chemistry.

4.1 THE DIELS–ALDER REACTION

The Diels–Alder reaction is among the most important of the pericyclic reactions. By simultaneously creating a ring, often with great regioselectivity, and setting up to four stereocenters, the Diels–Alder reaction finds widespread application in synthetic chemistry. As the first pericyclic reaction typically presented in introductory organic chemistry textbooks, the Diels–Alder reaction has virtually become the paradigm for pericyclic reactions. It is therefore not surprising that the Diels–Alder reaction has been the subject of considerable attention by the computational chemistry community. Perhaps more surprising is that some controversy has also been aroused concerning the nature of the reaction mechanism. Focusing on the parent Diels–Alder reaction of butadiene with ethylene to give cyclohexene (Reaction 4.1), we discuss in this section (1) how the concerted activation barrier depends on the computational method, (2) the computational dependence of the reaction energy, (3) the competition between the concerted and stepwise paths, and (4) the use of kinetic isotope effects (KIEs) to distinguish these two possible mechanisms.

(Reaction 4.1)

4.1.1 The Concerted Reaction of 1,3-Butadiene with Ethylene

The reaction of 1,3-butadiene with ethylene to give cyclohexene is an allowed reaction if the ethylene fragment approaches the butadiene fragment from one face, preserving a plane of symmetry as indicated in **4**. The Hückel-type arrangement of the p-orbitals shown in **4** involves six electrons and so should express some aromatic stabilization.

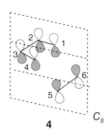

4

We first take up the issue of computing the activation barrier for this parent Diels–Alder reaction. Rowley and Steiner[11] experimentally determined an activation barrier of 27.5 ± 0.5 kcal mol^{-1} for Reaction 4.1. Using thermal corrections computed at B3LYP/6-31G*, Houk and coworkers[12] estimate that $\Delta H^{\ddagger}(0\ \mathrm{K}) = 23.3 \pm 2$ kcal mol^{-1}. A large number of computational estimates of the activation energy have been reported, many of which are listed in Table 4.1.

A number of important trends can be drawn from Table 4.1, which are trends that have influenced how computational chemists approach related (and sometimes even largely unrelated) problems. Hartree–Fock (HF) self-consistent field (SCF) computations vastly overestimate the barrier, predicting a barrier twice as large as experiment.[13–15] The omission of any electron correlation more seriously affects the transition state, where partial bonds require correlation for proper description, than the ground-state reactants. Inclusion of nondynamical correlation is also insufficient to describe this reaction; complete active space self-consistent field (CASSCF) computations also overestimate the barrier by some 20 kcal mol^{-1}.[16–20]

Clearly, dynamical correlation must be included. The first study to obtain a reasonable value for the barrier was the work of Bach,[14] which employed perturbation theory. MP2 lowers the barrier substantially from the HF result; unfortunately, it predicts a barrier that is too low, in the range 16–20 kcal mol^{-1}. MP3 raises the barrier but predicts now too high a value (27–28 kcal mol^{-1}). Finally, MP4 provides very good agreement with experiment, but triples configurations must be included to get this best fit (E_a values of 22.4–22.8 kcal mol^{-1}). Very similar trends were obtained by Jorgensen[21] in his examination of the Diels–Alder reaction of cyclopentadiene with ethylene and related dienophiles.

Perturbation corrections also dramatically improve upon the CASSCF results. MRMP2[18] and CASPT2[12] predict an activation barrier within a few kcal mol^{-1} of the experimental value. Lischka[20] used the multireference-averaged quadratic coupled cluster (MR-AQCC) method to obtain a barrier of 23.7 kcal mol^{-1}, and

TABLE 4.1 Computed Activation Energy[a] (kcal mol^{-1}) and R_{16} Distance (Å) for the Concerted Pathway of Reaction 4.1

Method	R_{16}	E_a
HF/3-21G//HF/3-21G[b]	2.210	35.9
HF/6-31G*//HF/6-31G*[c]	2.201	45.0
HF/6-31G**//HF/6-31G**[d]	2.273	45.9
CASSCF(6,6)/3-21G// CASSCF(6,6)/3-21G[e]	2.217	37.3
CASSCF(6,6)/6-31G*//B3LYP/6-31G*[f]		43.8
MRMP2-CAS(6,6)/6-31G*//B3LYP/6-31G*[g]		28.6
CASSCF(6,6)/6-31G*// CASSCF(6,6)/6-31G*[e]	2.223	43.8
CASSCF(6,6)/6-31G(d,p)//CASSCF(6,6)/6-31(d,p)[h]	2.221	44.5
CASSCF(6,6)-MP2/6-31G(d,p)//CASSCF(6,6)/6-31(d,p)[h]		39.4
CASSCF(6,6)/6-311+G(d,p)//CASSCF(6,6)/6-31(d,p)[h]		37.0
CASSCF(6,6)-MP2/6-311+G(d,p)//CASSCF(6,6)/6-31(d,p)[h]		40.9
CASPT2/6-31G*//CASSCF(6,6)/6-31G*[i]		25.0
CASSCF(6,6)/6-31G*//CASSCF(6,6)/6-31G*[j]	2.223	43.6
MR-AQCC/6-31G*//MR-AQCC/6-31G*[j]	2.240	25.7
MR-AQCC/6-31G**//MR-AQCC/6-31G**[j]	2.236	25.3
MR-AQCC/6-311G**//MR-AQCC/6-311G**[j]	2.241	24.2
MR-AQCC/6-311G(2d,p)//MR-AQCC/6-311G(2d,p)[j]	2.240	23.7
MP2/6-31G*//HF/6-31G*[c]		16.6
MP2/6-31G*//MP2/6-31G*[k]	2.285	18.5
MP2/6-31G*//B3LYP/6-31G*[g]		20.4
MP2/6-31G**//MP2/6-31G**[d]	2.244	15.9
MP2/6-311G(d,p)//B3LYP/6-31G[l]		17.5
MP3/6-31G*//HF/6-31G*[c]		26.9
MP3/6-311G(d,p)//B3LYP/6-31G[l]		28.2
MP4(SDQ)/6-31G*//HF/6-31G*[c]		29.0
MP4(SDTQ)/6-31G*//HF/6-31G*[c]		21.9
MP4(SDTQ)/6-31G*//MP2/6-31G*[k]		22.4
MP4(SDTQ)/6-311G(d,p)//B3LYP/6-31G[l]		22.8
BLYP/6-31G*//BLYP/6-31G*[m]	2.286	23.1
B3LYP/6-31G*//B3LYP/6-31G*[m,n]	2.272	24.9
B3LYP/6-31G**//B3LYP/6-31G**[o]	2.268	22.4
B3LYP/6-31+G**//B3LYP/6-31+G**[i]		27.2
B3LYP/6-311+G(2d,p)//B3LYP/6-311+G(2d,p)[i]		26.2
BPW91/6-31G*//BPW91/6-31G*[i]		19.9
MPW1K/6-31+G**//MPW1K/6-31+G*[i]		24.4
KMLYP/6-31G*//KMLYP/6-31G*[i]		21.1
KMLYP/6-311G//KMLYP/6-311G[i]		22.4
OLYP/6-31G(d)//OLYP/6-31G(d)[p]		26.7
OLYP/6-311+G(2d,p)//OLYP/6-311+G(2d,p)[p]		30.1
OLYP/6-311G(2df,2pd)// OLYP/6-311G(2df,2pd)[p]		29.2
O3LYP/6-31G(d)//OLYP/6-31G(d)		26.8

TABLE 4.1 (*Continued*)

Method	R_{16}	E_a
O3LYP/6-311+G(2d,p)//OLYP/6-311+G(2d,p)[p]		30.1
O3LYP/6-311G(2df,2pd)// OLYP/6-311G(2df,2pd)[p]		29.3
M06-2x/6-31G(d)	2.269	20.4
M06-2x/cc-pVTZ	2.239	23.2
ωB97X-D/6-31G(d)	2.265	21.7
ωB97X-D/cc-pVTZ	2.236	25.1
B2PLYP/cc-pVDZ	2.265	24.0
QCISD(T)/6-31G*//CASSCF(6,6)/6-31G*[e]		25.5
QCISD(T)/6-31G*//B3LYP/6-31G*[f]		25.0
CCSD/6-311G(d,p)//B3LYP/6-31G[l]		30.8
CCSD(T)/6-311G(d,p)//B3LYP/6-31G[l]		25.7
CCSD(T)/6-31G*//B3LYP/6-31G*[g]		27.6
G2MP2/6-311+G(3df,2p)[l]		24.6
G2MS/6-311+G(2df,2p)[l]		23.9
CBS-QB3[i]		22.9
Estimated[i]		23.3 ± 2

[a] Activation energy relative to s-*trans*-1,3-butadiene + ethylene. Most calculations include zero-point vibrational energy.
[b] Ref. 13.
[c] Ref. 14.
[d] Ref. 15.
[e] Ref. 16.
[f] Ref. 17.
[g] Ref. 18.
[h] Ref. 19.
[i] Ref. 12.
[j] Ref. 20.
[k] Ref. 28.
[l] Ref. 27.
[m] Ref. 23.
[n] Ref. 24.
[o] Ref. 25.
[p] Ref. 26.

correcting for complete interacting space predicts a barrier of 22.2 kcal mol^{-1} in excellent agreement with experiment. The MR-AQCC method is a configuration interaction (CI) expansion from a multiconfiguration reference, correcting for size extensivity using a couple cluster approach.[22]

All of the gradient-corrected density functional methods provide very fine estimates of the barrier for Reaction 4.1. In particular, B3LYP ($E_a = 23-27$ kcal mol^{-1}),[12,23–25] MPW1K[12] ($E_a = 24.4$ kcal mol^{-1}), and KMLYP[12] ($E_a = 21-22$ kcal mol^{-1}) produce excellent values. OLYP and O3LYP, however, overestimate the barrier by 4–8 kcal mol^{-1}.[26] The recently developed functionals that account for some long-range effects, ωB97X-D and M06-2x, provide quite reasonable estimates of the barrier, as does the double hybrid B2PLYP method.

There are only a few reports of CI computations of the Diels–Alder reaction. A QCISD(T)/6-31G* study using B3LYP/6-31G* geometries gives a fine value for the barrier (25.0 kcal mol^{-1}).[17] Coupled-cluster singles and doubles (CCSD) computations give a barrier that is too high, but inclusion of the triples configurations (similar to what was found with MP4) results in a reasonable barrier (26–28 kcal mol^{-1}).[27]

Finally, there are three reports of composite method studies of Reaction 4.1. Morokuma[27] has applied two variations on the G2 method, both of which employ smaller basis sets at various levels. These two variants give excellent values for the activation barrier: 24.6 and 23.9 kcal mol^{-1} with G2MP2 and G2MS, respectively. The CBS-QB3 method predicts a barrier of 22.9 kcal mol^{-1}.[12] These composite methods, developed and tested on grossly differing reactions, nevertheless provide excellent activation energies and are likely to be applicable to a broad range of organic reactions. Their computational expense will continue, however, to severely limit their practical utility.

We can also extract some trends concerning the effect the computational method has on the geometry of the transition state. Figure 4.1 presents the B3LYP/6-31G* optimized TS structure (**4**). All the methods (HF, MP, CASSCF, CI, and density functional theory (DFT)) predict a transition state with C_s symmetry, which demands synchronous formation of the two new σ-bonds. Here, s-*cis*-1,3-butadiene is the reactive conformer, even though it is less stable than s-*trans*-1,3-butadiene. Johnson[29] has located the transition state for the concerted reaction of s-*trans*-1,3-butadiene with ethylene to give *trans*-cyclohexene (Reaction 4.2) with a variety of methods. The B3LYP/6-31G* transition state **5** is shown in Figure 4.1. The *cis*-TS (**4**) is significantly lower in energy than the *trans*-TS (**5**) at all levels: anywhere from 16.1 kcal mol^{-1} at MP2/6-31G* to 19.7 kcal mol^{-1} at CASSCF(6,6)/3-21G. Therefore, it is safe to assume that the concerted reaction proceeds from the s-*cis* conformation of 1,3-butadiene.

(Reaction 4.2)

5

Kraka and Cremer[25] have carefully detailed the reaction pathway for this concerted process that maintains C_s symmetry. It is easiest to understand by tracing the path forward and backward from the C_s transition state. Proceeding backward from the transition state, the two fragments separate, maintaining C_s symmetry by restricting butadiene to a planar cis conformation. Eventually, a bifurcation point is reached, at an energy of 3.9 kcal mol^{-1} above isolated reactants. This point corresponds to where the pathway divides by breaking symmetry and allowing butadiene to rotate from the cis into the trans conformation. In the forward direction,

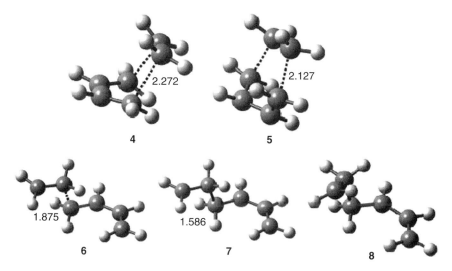

Figure 4.1 B3LYP/6-31G* optimized structures of **4–8**.

cyclohexene is formed in the chair (C_s) conformation. This is another bifurcation point; the forward pathway divides here to give two paths toward the twist (C_2) conformation, which lies 5.7 kcal mol^{-1} below the bifurcation point. Thus, the concerted C_s reaction path lies strictly between these two bifurcation points.

The most important geometrical parameter of the concerted TS is R_{16}, which is the distance of the forming σ-bond between the two fragments. This value is listed for some computations in Table 4.1. The most striking observation is that the range of the computed distance is small, only varying from 2.201 Å to 2.286 Å. The CASSCF and MR-AQCC optimizations cluster around 2.22–2.24 Å, while the DFT values cluster around 2.27–2.28 Å, showing little variation with basis set. On the other hand, at the HF level, R_{16} increases by 0.07 Å in going from 6-31G* to 6-31G**, but with MP2, the same basis set change results in a decrease in R_{16} by 0.04 Å.

This brings up the question of the role of the basis set. In terms of geometry, while there is some variation with the size of the basis set, all the methods with varying basis set size provide extremely similar TS geometries. With the most suitable computational methods (MP4(SDTQ), B3LYP, and MR-AQCC), the activation energy varies by less than 2 kcal mol^{-1} with differing basis sets. Therefore, quite reasonable results can be produced with the relatively small 6-31G* basis set with these QM methods.

One further consideration with regard to basis set choice is the degree of basis set superposition error (BSSE). In general, very little is understood of how BSSE might affect the computation of a transition state, its geometry, and energy. However, Kobko and Dannenberg[30] have surveyed exactly these effects for Reaction 4.1 with the HF, B3PW91, B3LYP, and MP2 methods. Some of their results are gathered in Table 4.2. As expected, the 3-21G basis set exhibits extremely large

TABLE 4.2 Comparison of Normal and Counterpoise-Corrected TSs for Reaction 4.1[a]

Method	$\Delta\Delta H^{\ddagger}$[b]	$\Delta\Delta R_{16}$[c]
HF/3-21G	10.02	−0.0147
HF/6-31G	3.89	−0.0077
HF/6-31G**	3.60	−0.0055
HF/6-311++G**	0.72	−0.0015
B3PW91/6-31G**	2.72	−0.0087
B3PW91/6-311++G**	0.69	−0.0033
B3LYP/6-311++G**	0.62	−0.0005
MP2/3-21G	13.16	−0.0495
MP2/6-31G**	7.07	−0.0269
MP2/6-311++G**	3.96	−0.0133

[a] Ref. 30.
[b] $\Delta\Delta H^{\ddagger} = \Delta H^{\ddagger}$(CP-corrected) $- \Delta H^{\ddagger}$(normal) in kilocalories per mole.
[c] $\Delta\Delta R_{16} = \Delta R_{16}$(CP-corrected) $- \Delta R_{16}$(normal) in angstroms.

BSSE and is unsuitable for obtaining reasonable barrier heights. Keeping in mind that the counterpoise correction overestimates BSSE, the error in the activation barrier is still large with the widely used 6-31G** basis set with all four computational methods. However, the 6-311++G** basis set, which is quite large, expresses minimal, and likely acceptable, BSSE. A similar situation applies with regard to the R_{16} error due to BSSE. Figure 4.2 provides a simple pictorial representation of the effect of BSSE on the potential energy surface (PES). BSSE is maximal in the product and nil for the reactants. Subtracting off this error results in a reaction that is now less exothermic. Application of the Hammond postulate suggests that the

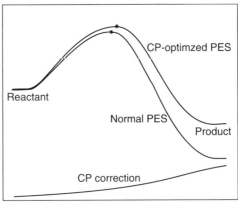

Figure 4.2 Comparison of the normal and counterpoise-corrected (CP) potential energy surfaces for the Diels–Alder reaction Reprinted with permission from Kobko, N.; Dannenberg, J. J. *J. Phys. Chem. A* **2001**, *105*, 1944–1950. Copyright 2001 American Chemical Society.

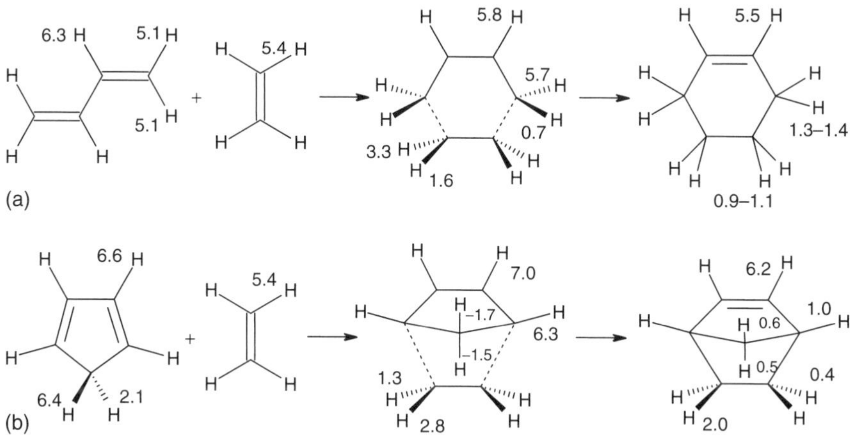

Figure 4.3 IGLO-calculated ^{1}H NMR shifts for the reaction of ethylene with (a) 1,3-butadiene or (b) cyclopentadiene.[28]

counterpoise-corrected (CP) TS will be later (i.e., a shorter R_{16}) than the normally computed TS. Interestingly, BSSE is much smaller for the DFT methods than MP2, supplying further reason for their use in computing activation energies of organic reactions.

Concerted pericyclic transition states have been argued as being favorable due to possessing some aromatic character. Calculation of the magnetic properties of the Diels–Alder transition state supports this notion. Shown in Figure 4.3 are the computed ^{1}H NMR shifts for the reactions of ethylene with butadiene or cyclopentadiene.[28] In particular, note the large upfield shifts of the methylene protons in the cyclopentadiene TS, indicative of a strong ring current. These TSs also express large, negative magnetic susceptibilities, indicative of aromaticity. Nucleus-independent chemical shift (NICS) values have also been computed for the TS of Reaction 4.1.[31] The value is dependent on the position inside the ring; however, all of the points examined in the interior region have large negative values (−21.4 to −27.2).

To summarize this section, for the Diels–Alder reaction, HF and CASSCF vastly overestimate the reaction barrier. Dynamical correlation is essential for the description of the Diels–Alder transition state. MP2 underestimates the barrier. MP4 and CI methods both provide very good results, but triples configurations must be included. The preferred method, when one combines both computational efficiency and accuracy, is clearly DFT. It is likely the strong performance of B3LYP with pericyclic reactions, typified by the Diels–Alder results described here, that propelled this method to be one of the most widely used among computational organic chemists.

While the performance of most of the post-HF methods and most density functionals is quite reasonable for predicting activation barriers, the same cannot be said about reaction energies. The energy of Reaction 4.1, computed at many levels, is

TABLE 4.3 Energy (kcal mol^{-1}) of Reaction 4.1

Method	ΔE_{rxn}	Error
HF/6-31+G(d,p)	−30.5	7.1
HF/6-311+G(2df,2p)	−27.5	−10.1
MP2/6-31+G(d,p)//B3LYP/6-31+G(d,p)[a]	−45.4	−7.8
MP2/6-311+G(2df,2p)//B3LYP/6-31+G(d,p)[a]	−43.3	−5.7
SCS-MP2/6-31+G(d,p)//B3LYP/6-31+G(d,p)[a]	−41.5	−3.9
SCS-MP2/6-311+G(2df,2p)//B3LYP/6-31+G(d,p)[a]	−39.6	−2.0
B3LYP/6-31+G(d,p)[a]	−31.3	6.6
B3LYP/6-311+G(2df,2p)[a]	−28.0	10.3
mPW1PW91/6-31+G(d,p)[a]	−40.3	−2.7
PBE0/6-31G(d)	−50.0	−12.4
PBE0/cc-pVTZ	−43.0	−5.4
M06-2x/6-31G(d)	−47.0	−9.4
M06-2x/cc-pVTZ	−39.6	−2.0
ωB97X-D/6-31G(d)	−48.6	−11.0
ωB97X-D/cc-pVTZ	−41.1	−3.5
B2PLYP/6-31G(d)	−40.0	−2.4
B2PLYP/cc-pVDZ	−44.7	−7.1
G3[a]	−37.8	−0.2
CBS-QB3[a]	−38.3	−0.7
Expt	−37.6 ± 0.5	

[a]Ref. 32.

listed in Table 4.3. The composite methods (G3 and CBS-QB3) do an excellent job in matching the experimental reaction energy, and this is also true for other simple Diels–Alder reactions.[32] SCS-MP2 performs somewhat less well. The modern functionals M06-2x and ωB97X-D are the best performers of the DFT. Somewhat disheartening is the very poor performance of most of the other functionals, including B3LYP and PBE0; for a series of Diels–Alder reactions, the mean average deviation is 8 kcal mol^{-1}. Houk[32] has discerned that the error is derived from the inability of many functionals to properly treat the π→σ transformation. Any reaction where the number of π- and σ-bonds are not conserved will pose this problem, and so proper care is needed when choosing a computational method.

Accurate accounting of dispersion may also be necessary in determining reasonable reaction energies. A nice example of this is the Diels–Alder reaction of fullerene C$_{60}$ with 1,3-butadiene, where the experimental estimate for the reaction enthalpy is −19.8 ± 2.2 kcal mol^{-1}.[33,34] B3LYP/6-31G(d) predicts the reaction energy as −6.6 kcal mol^{-1}, severely underestimating the exothermicity. However, with incorporation of Grimme's dispersion correction (B3LYP-D), the reaction energy is −23.3 kcal mol^{-1}, a considerably better estimate.[35]

4.1.2 The Nonconcerted Reaction of 1,3-Butadiene with Ethylene

An alternative mechanism for the Diels–Alder reaction involves a stepwise pathway (Reaction 4.3). Here, one end of the butadiene and ethylene fragments approach each other, passing through transition state **6**, to form a diradical intermediate **7**. In a second distinct chemical step, the second new σ-bond is formed through transition state **8** to give cyclohexene.

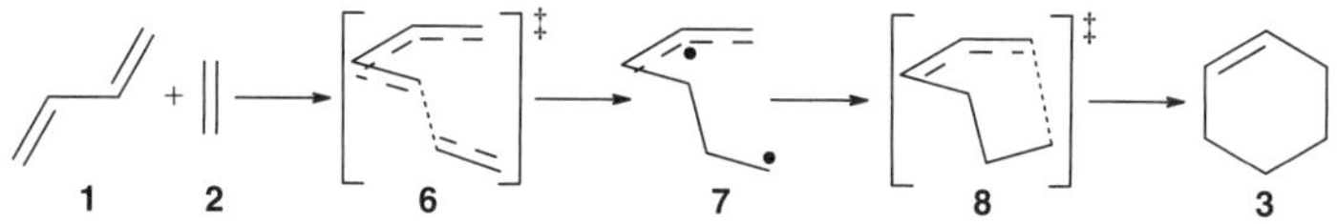

(Reaction 4.3)

Houk and coworkers[13] provided the experimental evidence in favor of the concerted mechanism. They reacted 1,1,4,4-tetradeuterio-1,3-butadiene with *cis*- or *trans*-1,2-dideuterioethene to give the tetradeuteriocyclohexenes (Scheme 4.2). The cyclohexenes were then reacted with *meta*-chloroperoxybenzoic acid (MCPBA) to make the epoxides, which were characterized by ^{1}H NMR. Less than 1 percent of the nonstereoconserved isomers were observed. If a stepwise reaction was occurring through a diradical intermediate (**7**), extensive scrambling would be expected. They conclude that the concerted activation energy must be at least 3.7 kcal mol^{-1} below that of the stepwise reaction. This "energy of concert" is in agreement with the estimate of 5.7 kcal mol^{-1} by Doering and Roth.[36,37]

This stepwise pathway has been examined numerous times with differing computational levels. There are three conformations of the diradical intermediate to consider (anti, gauche-in, and gauche-out) and corresponding transition states to create them. We discuss only the lowest energy pathway, the one invoking the anti conformation. Listed in Table 4.4 are the energies of the critical points along Reaction 4.3. The B3LYP/6-31G* optimized structures of **6**–**8** are drawn in Figure 4.1. Locating the second transition state (**8**) is more difficult than locating the first. **8** corresponds mostly to rotation about the C_2–C_3 bond, changing the

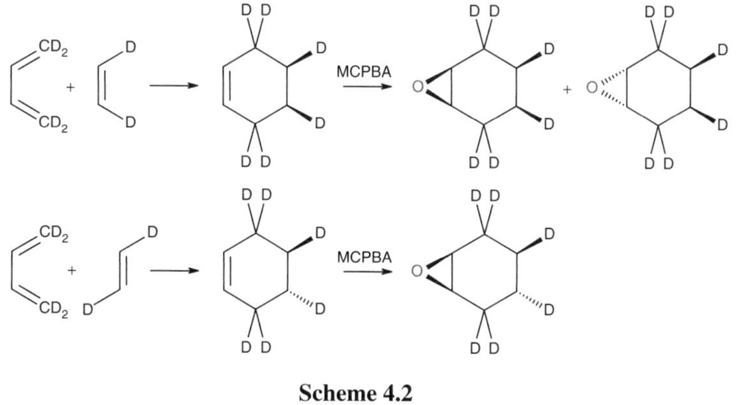

Scheme 4.2

TABLE 4.4 Computed Energies (kcal mol^{-1}) for the Entrance (6) and Exit (8) TSs and Intermediate (7) for the Stepwise Pathway of Reaction 4.3[a]

Method	6	7	8	4
HF/6-31G*//B3LYP/6-31G*[b]	21.3 (8.8)			46.6
MP2/6-31G*//B3LYP/6-31G*[b]	49.9 (42.0)			20.4
CASSCF(6,6)/3-21G//CASSCF(6,6)/3-21G[c]	43.1	41.1		37.3
CASSCF(6,6)/6-31G*//CASSCF(6,6)/6-31G*[c]	45.7	40.7		43.8
CASSCF(6,6)/6-31G*//B3LYP/6-31G*[d]		40.7		43.8
CASSCF(6,6)/6-31G*//B3LYP/6-31G*[b]	46.6 (42.0)			45.1
CASSCF(6,6)/6-31G(d,p)//CASSCF(6,6)/6-31G(d,p)[e]	44.6	40.4	43.5	44.5
CASSCF(6,6)-MP2/6-31G(d,p)//CASSCF(6,6)/6-31G(d,p)[e]	41.1	35.1	38.3	39.4
CASSCF(6,6)/6-311+G(d,p)//CASSCF(6,6)/6-31G(d,p)[e]	46.4	42.5	45.5	47.0
CASSCF(6,6)-MP2/6-311+G(d,p)//CASSCF(6,6)/6-31G(d,p)[e]	41.7	35.3	38.1	40.9
CASSCF(6,6)/6-31G*//CASSCF(6,6)/6-31G*[f]	45.6	40.7	43.9	43.6
MR-AQCC/6-31G*//CASSCF(6,6)/6-31G*[f]	32.0	29.7	32.6	25.7
MR-AQCC/6-31G**//CASSCF(6,6)/6-31G*[f]	31.8	29.5	32.4	25.3
UB3LYP/6-31G*//UB3LYP/6-31G*[g]	33.6 (28.2)	29.0 (27.6)		24.8
UQCISD(T)/6-31G*//CASSCF(6,6)/6-31G*[c]	39.2	30.3		29.4
RQCISD(T)/6-31G*//CASSCF(6,6)/6-31G*[c]	35.7	29.8		25.5
QCISD(T)/6-31G*//B3LYP/6-31G*[b]	40.7 (32.1)			27.6
CCSD(T)/6-31G*//B3LYP/6-31G*[b]	40.7 (32.0)			27.6

[a] Spin-corrected energies are presented in parentheses.
[b] Ref. 18.
[c] Ref. 16.
[d] Ref. 17.
[e] Ref. 19.
[f] Ref. 20.
[g] Ref. 24.

conformation from anti to gauche-in. Once C_1 and C_6 near each other, they collapse to form the σ-bond without further barrier.

The major question is which pathway is preferred: the concerted one passing through **4** or the stepwise one, whose rate-limiting step is through **6**. The HF method, which grossly overestimates the barrier of the concerted process, favors the stepwise pathway, indicating a very low barrier of only 8.8 kcal mol^{-1} after spin correction.[18] This estimate is substantially below the experimental value, and one should conclude that the HF method is entirely inappropriate for describing the Diels–Alder (and likely all other pericyclic reactions).

CASSCF favors the concerted TS over the stepwise TS, but by less than 2 kcal mol^{-1}. In all cases, the diradical intermediate **7** lies 4–5 kcal mol^{-1} below **4**.[16,18–20] As we will discuss in more detail in the section on the Cope rearrangement, CASSCF tends to overstabilize diradicals, favoring them over closed-shell species, and so these results here are likely spurious.

Perturbation theory, which largely corrected the errors of the HF and CASSCF methods for the concerted pathway, fails to properly estimate the energy of **6** relative to **4**. MP2 suggests that the stepwise barrier is 22 kcal mol^{-1} above the concerted TS,[18] while CASSCF-MP2 suggests a separation of less than 2 kcal mol^{-1}. However, the MR-AQCC method predicts that **6** lies about 6 kcal mol^{-1} above **4**,[20] in accord with experimental evidence. The CI methods also largely fail, predicting that the stepwise path is disfavored by about 10 kcal mol^{-1}.[16,18]

On the other hand, DFT performs extraordinarily well. Spin contamination is a problem with DFT methods such as B3LYP and so an unrestricted wavefunction must be used and spin correction applied. Using this approach, Houk found **4** to lie 3.4 kcal mol^{-1} below **6** and 2.8 kcal mol^{-1} below **7**. Coupled with the computed KIEs, discussed next, the results of computational studies indicate that the Diels–Alder reaction proceeds by the concerted mechanism.

4.1.3 Kinetic Isotope Effects and the Nature of the Diels–Alder Transition State

With all of the theoretical analysis pointing toward an allowed concerted pathway for the Diels–Alder reaction, it is perhaps unexpected that in the mid-1980s, a small controversy surrounded the mechanism. This controversy surfaced with Dewar's[38] proposal that all concerted reactions could not normally be synchronous, or in other words, when multiple bonds are made or broken in a single chemical step, the degree of bond making/breaking is not synchronous. Dewar's work was inspired by semiempirical calculations on the Diels–Alder cycloaddition and the Cope rearrangement that found decidedly asymmetric transition states, if not outright stepwise pathways.

All of the ab initio calculations that include electron correlation to some extent clearly favor the concerted pathway for Reaction 4.1. All of these computations also identified a transition state with C_s symmetry, indicating perfectly synchronous bond formation. One method for distinguishing a synchronous from an asynchronous transition state is by secondary kinetic isotope effects (KIEs). Isotopic substitution alters the frequencies for all vibrations in which that isotope is involved. This leads to a different vibrational partition function for each isotopically labeled species. Bigeleisen and Mayer[39] determined the ratio of partition functions for isotopically labeled species. Incorporating this into the Eyring transition state theory results in the ratio of rates for the isotopically labeled species (Eq. (4.1)).[40] Computation of the vibrational frequencies is thus sufficient to predict KIEs, as long as the vibrational frequencies can be computed accurately.

$$\frac{k_H}{k_D} = \frac{v_H^{\ddagger}}{v_D^{\ddagger}} \frac{\prod_{}^{3N-6} \frac{u_H}{u_D} \prod_{}^{3N-6} \frac{[1-\exp(-u_H)]}{[1-\exp(-u_D)]}}{\prod_{}^{3N^{\ddagger}-7} \frac{u_H^{\ddagger}}{u_D^{\ddagger}} \prod_{}^{3N^{\ddagger}-7} \frac{[1-\exp(-u_H^{\ddagger})]}{[1-\exp(-u_D^{\ddagger})]}} \frac{\exp\left(\sum^{3N-6} \frac{(u_H - u_D)}{2}\right)}{\exp\left(\sum^{3N^{\ddagger}-7} \frac{(u_H^{\ddagger} - u_D^{\ddagger})}{2}\right)} \quad (4.1)$$

Since the terminal carbons of both the diene and dienophile change the hybridization from sp^2 to sp^3, the corresponding C–H out-of-plane vibration frequency increases during the Diels–Alder reaction.[41] Therefore, deuterium or tritium substitution at the termini positions will potentially lead to differing reaction rates. An inverse 2°-KIE is expected for the forward reaction ($k_H/k_D < 1$), while the reverse reaction would experience a normal 2°-KIE ($k_H/k_D > 1$). Houk and Singleton have pioneered the use of both experimental and computational KIEs to determine the nature of pericyclic TSs, and we discuss here their studies of the Diels–Alder reaction.

Houk first examined the 2°-KIE for Reaction 4.1 using the MP2, CASSCF, and UHF methods to optimize the appropriate transition states and compute the necessary vibrational frequencies.[40] The computed KIE are compared with experimental results for analogous reactions in Table 4.5. While the agreement between experimental and computation is not exact, one important result can be readily extracted: the predicted KIEs for the concerted transition state are in much better agreement with experiment than those for the stepwise transition state. In particular, note that the stepwise KIEs are all normal and the concerted are all inverse, in agreement with the experimental inverse KIEs.

Employing NMR to quantify the amount of remaining labeled and unlabeled reactants after considerable reaction, Singleton[44] was able to determine very accurate KIEs using natural isotope abundance. The KIEs for the Diels–Alder reaction of isoprene and maleic anhydride (Reaction 4.4) are shown in Figure 4.4a. Computed B3LYP/6-31G* KIEs for Reaction 4.1 are presented for the concerted (Figure 4.4b) and stepwise (Figure 4.4c) pathways.[24] The most salient feature is that the KIEs for the stepwise reaction are completely inconsistent with the experimental values, especially in their prediction of normal KIE at the two termini positions, whereas only inverse KIEs are found. Shown in Figure 4.5d and e are the B3LYP/6-31G* KIEs for Reaction 4.4 passing through either the concerted endo or exo transition states.[45] The agreement between the computed and experimental KIEs is excellent, and is truly outstanding if one adjusts the value of the KIE for the reference methyl group from 1.00 (assumed in the original experimental work) to the values computed at B3LYP. These studies confirm that KIEs can be accurately computed and applied to a variety of reactions.[46–50]

TABLE 4.5 Comparison of Experimental and Computed[a] Kinetic Isotope Effects for Some Diels–Alder Reactions

Reactants	Expt	Concerted MP2/6-31G* (CASSCF/ 6-31G*)	Stepwise (6) UHF/6-31G* (CASSCF/ 6-31G*)
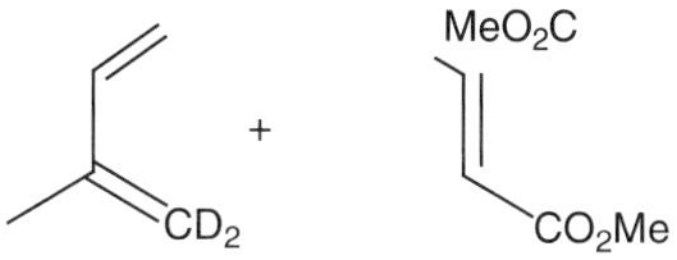	0.92[42]	0.93 (0.95)	1.02 (0.97)
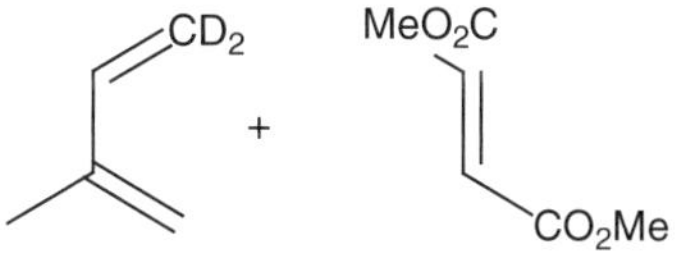	0.93[42]	0.93 (0.95)	1.02 (0.97)
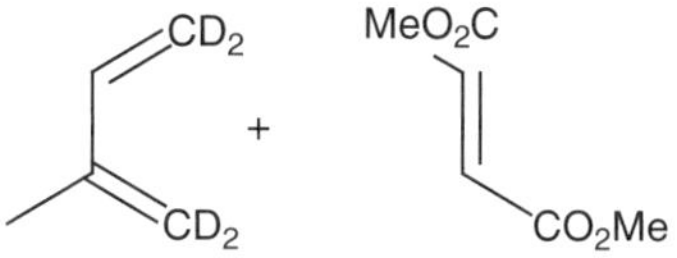	0.85[42]	0.87 (0.90)	1.06 (0.94)
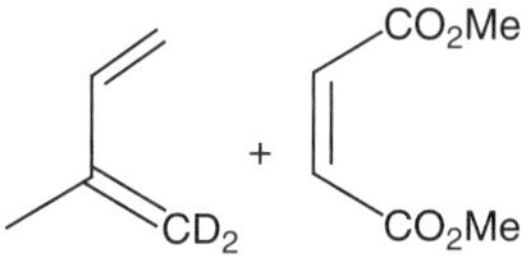	0.92[42]	0.93 (0.95)	1.02 (0.97)
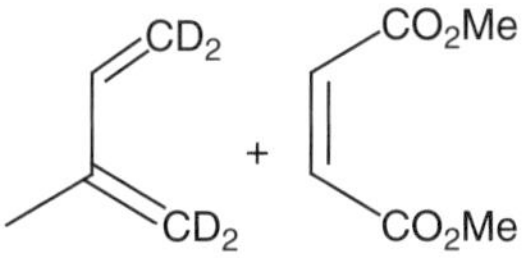	0.87[42]	0.87 (0.90)	1.06 (0.94)
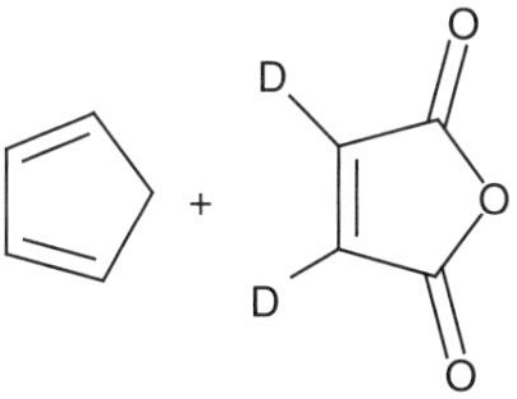	0.92[43]	0.89 (0.92)	1.08 (1.04)
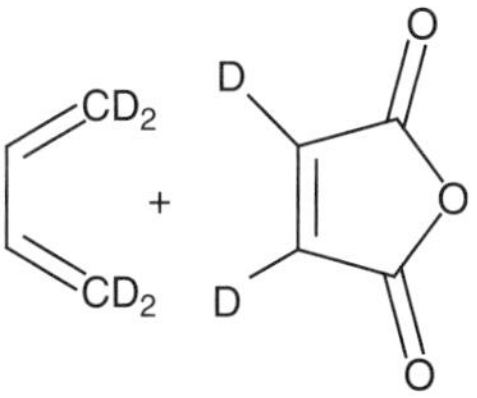	0.76[43]	0.81 (0.85)	1.07 (0.92)

[a] Theoretical KIEs based on the reaction of ethylene with 1,3-butadiene.[40]

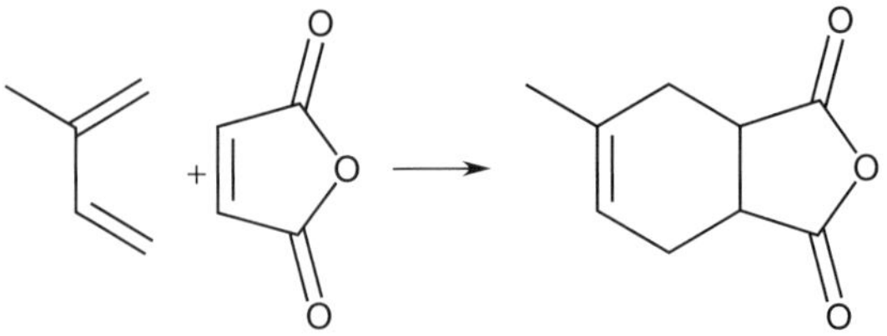

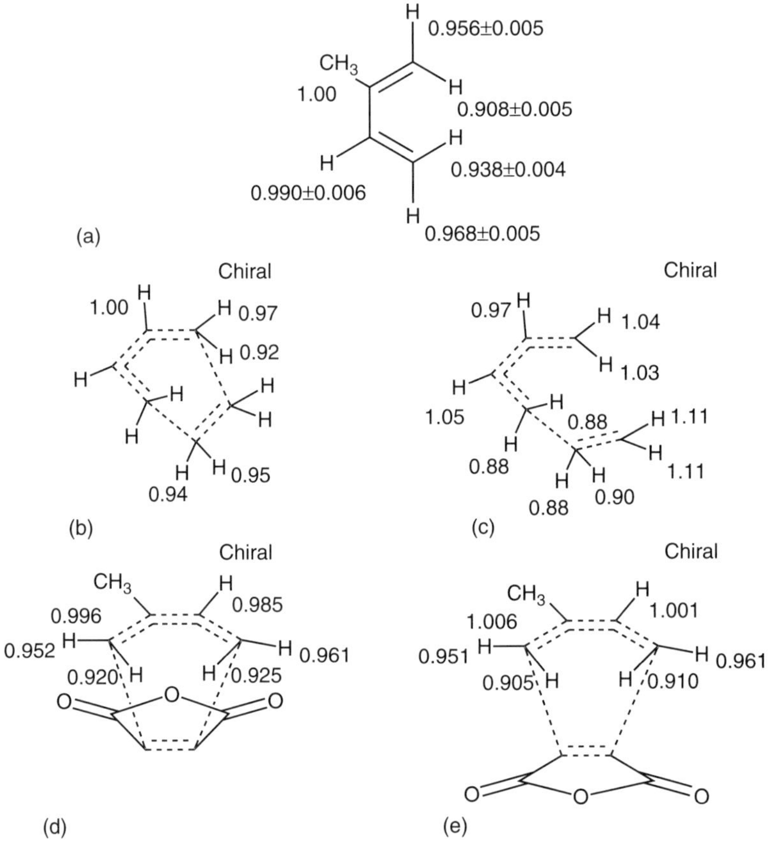

Figure 4.4 KIE for some Diels–Alder cycloadditions. (a) Experimental values for Reaction 4.4,[44] B3LYP/6-31G* computed values[24] for the (b) concerted and (c) stepwise pathways for the reaction of ethylene and 1,3-butadiene, and B3LYP/6-31G* computed values[45] for the (d) endo, and (e) exo pathways for Reaction 4.4.

These KIE studies indicate a concerted reaction, but what about its synchronicity? Using isoprene instead of 1,3-butadiene as the diene component breaks the symmetry of the cycloaddition. This symmetry break is seen in the transition states for the reaction of isoprene with maleic anhydride (Figure 4.5). This reaction undoubtedly proceeds through a concerted transition state, based on the KIE discussed above. The computed TSs show slight asymmetry: the forming σ-bonds

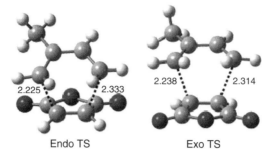

Figure 4.5 B3LYP/6-31G* structures for the endo and exo TSs for Reaction 4.4.[45]

differ by about 0.1 Å. The asynchronicity is small and might be understood as simply being the necessary result of the lack of symmetry of the reactants.

Similarly, Morokuma[27] located four transition states (**9–12**) for the Diels–Alder cyclization of acrylic acid with 2,4-pentadienoic acid. These transition states are shown in Figure 4.6. All show rather large asynchronous bond formation, with a difference in the new σ-bond distances of as much as 0.5 Å!

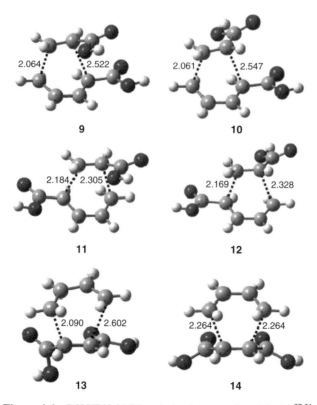

Figure 4.6 B3LYP/6-31G* optimized geometries of **9–14**.[27,51]

More interesting would be Diels–Alder cycloadditions between reactants that could potentially conserve C_s symmetry, and thereby pass through a synchronous transition state, but rather prefer an asynchronous route. Somewhat remarkably, there are examples of such cases. The transition state (**13**) for the cyclization of 1,3-butadiene with maleic acid is predicted to be highly asynchronous.[51] In fact, the two forming σ-bond distances differ by 0.512 Å. Imposing C_s symmetry during the optimization leads to a second-order saddle point (SOSP) **14** that is only 1.6 kcal mol^{-1} above the first-order saddle point **13**. The experimental KIEs for the reaction of isoprene with dimethyl maleate were obtained and compared with computed KIEs of analogous systems. A clear cut distinction between a pathway through a synchronous TS or through an asynchronous TSs cannot be made. In this case, KIEs are consistent with either path. Nonetheless, highly asynchronous Diels–Alder TS certainly appear feasible.

4.1.4 Transition State Distortion Energy

An ofttimes convenient model for analyzing reaction barriers is to dissect the barrier into a distortion energy (E_d) and an interaction energy (E_i).[52] The distortion energy, sometimes also referred to as the *deformation energy*[53] or *activation strain*,[54] is defined as the sum of the energy needed to distort each reactant molecules into its geometry in the transition state. The interaction energy is the energy released by allowing these distorted reactant molecules to come together in the actual transition state geometry. The activation barrier is then given as

$$E_a = E_d + E_i \qquad (4.2)$$

This approach has proven to be quite useful in a number of applications; two cycloaddition examples are examined here.

Ess and Houk[55,56] applied this model to 1,3-dipolar cycloadditions to ethene and ethyne. B3LYP/6-31G(d) and CBS-QB3 computations were carried out for the reactions of nine 1,3-dipoles shown in Scheme 4.3. The activation energies of these 18 reactions do not correlate with the reaction energies; thus, there is no correlation to the effect that the more exothermic is the reaction, the lower will be its activation barrier. Rather, the activation energies correlate extremely well with the distortion energy ($r^2 = 0.97$). Ess and Houk argue that the TS is achieved when the π-orbitals of the dipole align well with the π-orbitals of the dipolarophile. This

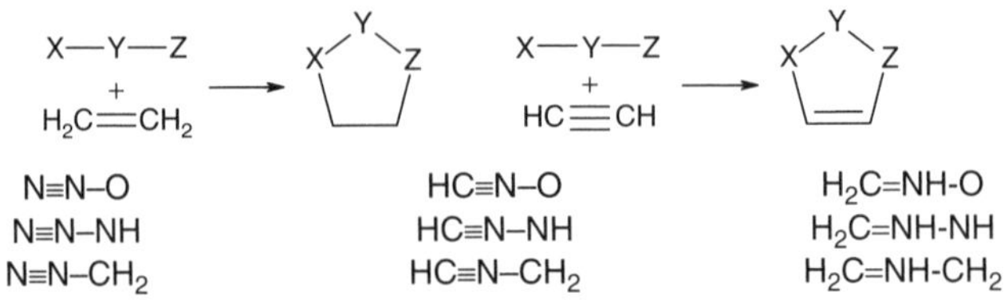

X—Y—Z + $H_2C\!=\!CH_2$ → [ring]   X—Y—Z + HC≡CH → [ring]   [ring]

N≡N–O
N≡N–NH
N≡N–CH$_2$

HC≡N–O
HC≡N–NH
HC≡N–CH$_2$

$H_2C=NH$–O
$H_2C=NH$–NH
$H_2C=NH$–CH$_2$

Scheme 4.3

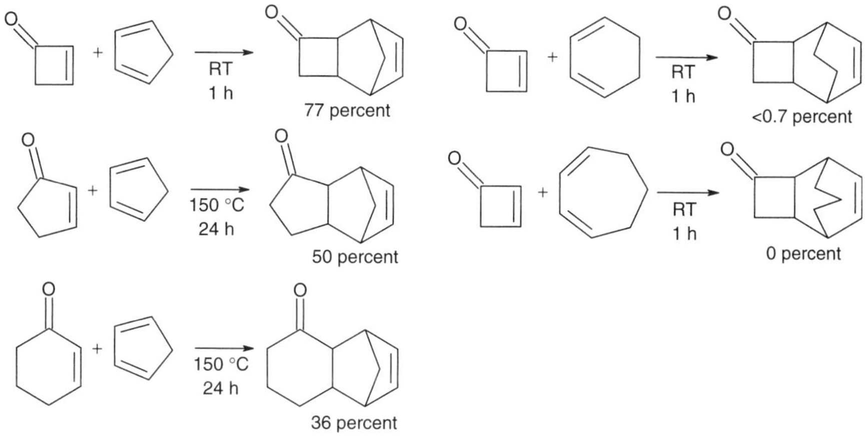

Scheme 4.4

requires distorting the dipole, principally by bending about the central atom. Those dipoles with stiff angles require greater distortion energy and therefore a higher barrier, while the dipoles with soft angles will distort easily, leading to small values of E_d and E_a.

Danishefsky[57] noted that cyclobutenone is a much more active dienophile than either cyclopentenone or cyclohexenone toward cyclopentadiene, but it is only weakly active toward cyclhexenone (Scheme 4.4). Paton et al.[58] examined the five reactions shown in Scheme 4.4, along with the Diels–Alder reaction of pent-3-en-2-one and cyclopropenone with cyclopentadiene, at M06-2x/6-31G(d), including both endo and exo orientations. A plot of the activation energy versus the reaction energy shows no correlation ($r^2 = 0.14$). Again, this implies that just because a reaction is more exothermic does *not* mean it will have a lower activation barrier. The argument that there is greater strain relief afforded by the reaction of the smaller four-membered ring of cyclobutanone than in the larger cyclohexanone is too simplistic and invalid here.

Rather, the plot of the activation energy versus the distortion energy shows a very strong correlation ($r^2 = 0.93$). The principal geometric change in the dienophile in going from the reactant to the transition state is the bending of the alkenyl hydrogens out-of-plane. This is easier to do in cyclobutanone than in cyclohexanone because of the greater s-character of the C–H bonds in the smaller ring. The key element is minimizing the energy needed to reach the transition state; the energy release afterwards has little bearing on the size of the barrier.

4.2 THE COPE REARRANGEMENT

The Cope rearrangement is perhaps the premier example of [3,3]-sigmatropic rearrangements. The simplest case is the degenerate rearrangement of 1,5-hexadiene

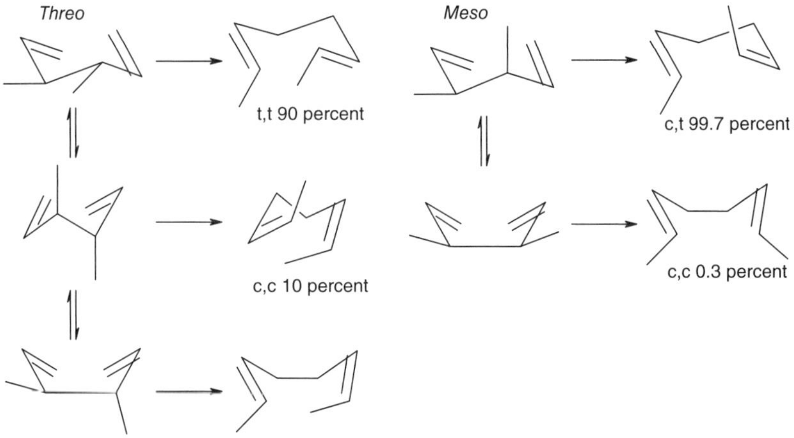

Scheme 4.5

15 shown in Reaction 4.5. Though first discovered in the 1940s, the mechanism of this reaction remained controversial into the 1990s.[59]

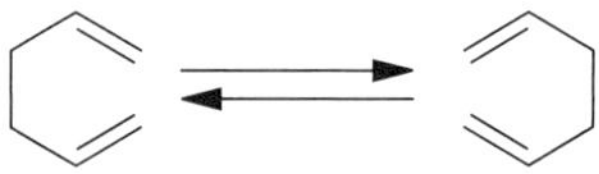

(Reaction 4.5)

15

The classic experiment by Doering and Roth[60] addressed the stereochemistry of the Cope rearrangement. Heating *threo-* or *meso*-3,4-dimethyl-1,5-hexadiene gives mixtures of octadienes that indicate a preference for the reaction to occur through a chair-like transition state (see Scheme 4.5). They estimated that the chair pathway was preferred over the boat by at least 5.7 kcal mol^{-1} in free energy, a figure later supported by Goldstein's[61] experiments with deuterated 1,5-hexadiene. Therefore, our discussion will focus primarily on the reaction in the chair conformation.

More contentious has been the nature of the mechanism itself. Outlined in Scheme 4.6 are the three limiting cases for the mechanism. The reaction can proceed along a concerted path, passing through a single transition state (**16a**) with no intermediates (path a). This transition state invokes delocalization across all six carbon centers and has been termed an *aromatic* transition state. There are two stepwise possibilities. Following path b, the $\sigma(C_3-C_4)$ bond is cleaved first, creating two noninteracting allyl radical species (**16b**). The ends of these allyl radicals can then combine to give product. The alternative is path c where the bond between C_1 and C_6 forms first, creating cyclohexane-1,4-diyl (**16c**) as a stable intermediate. Cleaving the C_3-C_4 bond then forms the product.

The experimental activation enthalpy for the Cope rearrangement of 1,5-hexadiene is 33.5 kcal mol^{-1}.[62] The cleavage-first pathway (Scheme 4.6 path b) has been discounted for two reasons. First, the estimate for the dissociation energy of 1,5-hexadiene into two allyl radicals is 59.7 kcal mol^{-1}, much higher

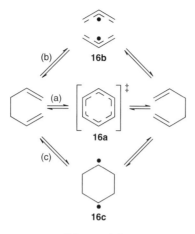

Scheme 4.6

than the activation barrier. Second, experiments show no crossover products, which would be expected if allyl fragments were produced.[63,64] Doering et al.[62] estimated that cyclohexane-1,4-diyl would be 33.7 kcal mol^{-1} above 1,5-hexadiene, essentially identical to the experimental activation barrier, championing path c of Scheme 4.6. However, they utilized a faulty estimate for the bond dissociation energy for forming the *iso*-propyl radical from propane. With current group equivalents[65] and bond energies,[66] the diyl is estimated to be 42 kcal mol^{-1} higher in energy than 1,5-hexadiene, suggesting that it too is unlikely to participate in the Cope rearrangement. Nevertheless, the diyl pathway remained intriguing, as Cope rearrangements with radical-stabilizing substituents might make the diyl competitive.[67–73] This summarizes the environment in which computational chemists came to weigh in on the nature of the Cope rearrangement.

4.2.1 Theoretical Considerations

The implication of the multiple possible reaction pathways shown in Scheme 4.6 is that any computational approach must allow for the possible contribution of at least these three valence bond structures.[74] The simplest approach to the nature of the wavefunction for the Cope rearrangement is to just account for the correlation of the active orbitals of the reactants with those of the products. The σ-bond between C_3 and C_4 of the reactant correlates to $\sigma(C_1-C_6)$ in the product. Assuming that 1,5-hexadiene has C_2 symmetry, both of these orbitals have a symmetry. The in-phase mixing of the two π-bonds of the reactant $(\pi(C_1-C_2)+\pi(C_5-C_6))$ has b symmetry and correlates with $(\pi(C_2-C_3)+\pi(C_4-C_5))$ of the product. The out-of-phase combination of the reactant π-bonds $(\pi(C_1-C_2)-\pi(C_5-C_6))$ has a symmetry and correlates with $(\pi(C_2-C_3)-\pi(C_4-C_5))$ of the product. If the reaction proceeds through a C_{2h} geometry, orbital symmetry demands that these active orbitals of $a^2a^2b^2$ must become $a_g^2a_u^2b_u^2$. So, we may take as the "aromatic"

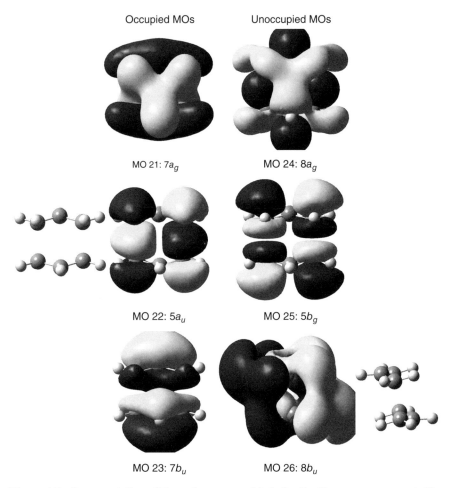

Figure 4.7 Representation of the active space orbitals for the Cope rearrangement. (*See insert for color representation of this figure.*)

configuration the Slater determinant Ψ_{arom}

$$\Psi_{\text{arom}} = |\ldots 7a_g^{\ 2}5a_u^{\ 2}7b_u^{\ 2}|$$

Pictorial representations of these orbitals are displayed in Figure 4.7.

To account for the possible contribution of cyclohexane-1,4-diyl, we must include the configuration arising from the excitation of the electrons from the HOMO to the LUMO. Thus, the diyl Slater determinant Ψ_{diyl} is

$$\Psi_{\text{diyl}} = |\ldots 7a_g^{\ 2}5a_u^{\ 2}7b_u^{\ 2}| - c_1|\ldots 7a_g^{\ 2}5a_u^{\ 2}8a_g^{\ 2}|$$

where the value of the coefficient c_1 is near 1.

The last contribution comes from the bis-allyl supermolecule. This structure arises from removing the electrons from $5a_u$ and placing them into $5b_g$:

$$\Psi_{\text{bis-allyl}} = |\ldots 7a_g{}^2 5a_u{}^2 7b_u{}^2| - c_2 |\ldots 7a_g{}^2 5b_g{}^2 7b_u{}^2|$$

where $c_2 = 1$ when the two allyl groups are infinitely separated.

An HF wavefunction includes only Ψ_{arom}, and therefore minimizes the contribution of any diradical or bis-allyl character. The most reasonable approach would be to use a CASSCF wavefunction where the active space involves the three occupied MOs ($7a_g, 5a_u, 7b_u$) and the three lowest energy unoccupied orbitals ($8a_g, 5b_g, 8b_u$). All possible occupations of the six valence electrons within this set of MOs will capture the three most important valence bond contributors. It will also provide some accounting for the dynamic correlation between these six electrons. What is missing is the dynamic correlation between these six valence electrons and the remaining 28 electrons. One reason why the Cope rearrangement represents such an important test case for computational chemistry is that this dynamic correlation turned out to play a much larger role than ever suspected, a finding that was especially surprising for a reaction as seemingly straightforward as this one.

4.2.2 Computational Results

The computational studies dealing with the Cope rearrangement focus on the nature of the mechanism. Specifically, is there a stable intermediate (the diyl) or just a single transition state? The majority of researchers approached this question by examining a C_{2h} slice through the PES. They scanned the distance, R_{16}, between the formal allylic species, looking for energy minima. These C_{2h} minima were then determined to be either transition states or intermediates on the global energy surface. We will concentrate this discussion solely on the reaction in the chair conformation.

The earliest serious computational studies of the Cope rearrangement were done using the semiempirical methods MINDO/2[75] and MINDO/3.[76] Both of these studies identified a stable C_{2h} intermediate with R_{16} of about 1.6 Å sitting about 3 kcal mol^{-1} below the transition state. A follow-up study using the more sophisticated AM1 method also identified a stable intermediate, but one residing in a well so small (0.1 kcal mol^{-1}) as to not have "any definite significance."[77] We note these results mainly because they, along with Dewar's controversial contention[38] that reactions involving making and breaking many bonds could not be concerted, inspired the application of ever more rigorous ab initio calculations to the Cope rearrangement.

Instead of proceeding with a historical presentation, we discuss the computational results by methodology. Using HF/6-31G*, the only feature located on the C_{2h} slice through the PES is a transition state with $R_{16} = 2.046$ Å.[74] This geometry looks quite reasonable; however, as is found with many other pericyclic reactions, the activation barrier is dramatically overestimated: $\Delta H_a = 55.0$ kcal mol^{-1} versus 33.5 kcal mol^{-1} experiment (see Table 4.6).

TABLE 4.6 Energies and R_{16} for Transition States and Intermediates for the Cope Rearrangement

Method	R_{16} (Å)	$\Delta E^{\ddagger}$ (kcal mol^{-1})	$\Delta H^{\ddagger}_{298}$ (kcal mol^{-1})
Transition state			
RHF/6-31G*[a]	2.046	56.6	55.0
CASSCF(6,6)/6-31G*[a]	2.189	48.7	46.9
CASPT2N/6-31G*[b]	1.745	31.2	30.8
CASPT2N/6-311G(2d,2p)[b]	1.775	33.1	32.2
CCD/6-31G*[c]	1.874	42.1	41.1
CCSD/6-31G**[d]	1.89	41.1	
CCSD(T)/6-311G**[d]	1.82	35.2	
CR-CCSD(T)/6-311G**[d]	1.86	37.7	
B3LYP/6-31G*[a]	1.966	34.4	33.2
B3LYP/6-31+G(d,p)[e]	2.004		34.0
B3LYP/6-311+G*[f]		33.7	32.2
B3PW91/6-31G*[a]	1.877	32.1	31.0
B3PW91/6-311+G(2d,p)[g]		32.5	31.6
CBS-QB3[h]			33.0
MD-CISD(CAS6,6)/6-31G*[i]		40.5	
MR-AQCC(CAS6,6)/6-31G*[i]	1.725	37.3	
MR-AQCC(CAS6,6)/6-311G(2d,1p)[i]	1.902	36.8	
MR-AQCC-ars(CAS6,6)/6-311G(2d,1p)[i]		33.4	
MCQDPT/6-311G**[d]	1.88	28.3	
Intermediate			
UHF/6-31G*[a]	1.558	20.4	19.2
CASSCF(6,6)/6-31G*[a]	1.641	46.8	47.0
MP2/6-31G*[f]	1.784	28.5	28.1
CCSD(T)/6-31G*[d]	1.72	36.2	
CCSD(T)/6-311G**[d]	1.72	35.3	
UB3LYP/6-31G*[a]	1.652	37.4	36.4
B3PW91/6-31G*[a]	1.611	32.3	31.5

[a]Ref. 74.
[b]Ref. 81.
[c]Ref. 82.
[d]Ref. 88.
[e]Ref. 89.
[f]Ref. 86.
[g]Ref. 90.
[h]Ref. 12.
[i]Ref. 84.

A number of different CASSCF computations have been reported, varying in the size of the active space and the basis set.[78–80] Optimization of C_{2h} structures using CASSCF(6,6)/6-31G*[80] revealed both a diyl structure with $R_{16} = 1.641$ Å and a transition state with $R_{16} = 2.189$ Å. The diyl structure is 1.9 kcal mol^{-1} below the transition state. A C_s transition state leading to the diyl was also located;

it lies 1.3 kcal mol^{-1} above the diyl but 0.6 kcal mol^{-1} below the C_{2h} "aromatic" transition state. Further complicating the matters, the energies of both the diyl and transition state give calculated barrier heights that are at least 11 kcal mol^{-1} above the experimental value. These large differences lead to the surprising conclusion that CASSSCF(6,6) calculations are unsuitable for probing the mechanism of the Cope rearrangement.

Resolution came with the development of perturbation methods that use a multireference wavefunction. Of interest here are the techniques of applying second-order Moller–Plesset perturbation theory using a CASSCF wavefunction as the reference. In the mid-1990s, two slightly different formulations of multireference perturbation theory (MRPT) were applied to the Cope rearrangement.[81,82] Both methods find only a single transition state with R_{16} between 1.74 and 1.88 Å, depending on the basis set. There is no diyl intermediate. Furthermore, the activation energies are much improved: $\Delta H^{\ddagger} = 32.2$ kcal mol^{-1} at CASPT2/6-311G(2d,p).

A limitation of these MP-CASSCF calculations was that full geometry optimization could not be performed due to a lack of analytical gradients. In 2003, Lischka[22,83] reported MR-CISD (multireference configuration interaction with single and double excitations) and MRAQCC calculations that include fully optimized structures for the Cope rearrangement of 1,5-hexadiene.[84] Their largest computations (MR-AQCC/6-311G(2d,1p) using the CASSCF(6,6) reference) indicate a single transition state with $R_{16} = 1.902$ Å. The activation enthalpy at this level is 33.4 kcal mol^{-1}, essentially the experimental value.

Why are the CASSCF computations not only quantitatively but also qualitatively incorrect? Inclusion of dynamic correlation decreases the diradical contribution to the wavefunction, which is overestimated by CASSCF. This was a very surprising result at that time. The simple valence bond model described above would imply that the CASSCF approach should be satisfactory. The failure of CASSCF meant that much greater computational resources than anyone had expected would be needed to adequately describe even simple organic reactions, such as the Cope rearrangement.

Since inclusion of dynamic electron correlation increases the importance of the aromatic contribution (Ψ_{arom}) at the expense of the diradical contributions, a single-reference post-HF method might be satisfactory. CCD and QCISD find a transition state with $R_{16} = 1.87$ Å. Both methods predict an activation barrier that is about 7 kcal mol^{-1} too high.[82]

A recent study of the Cope rearrangement using coupled-cluster theory revealed a discouraging result. McGuire and Piecuch first obtained the C_{2h} PES at UB3LYP/6-311G** by optimizing the geometry with a range of fixed values of R_{16}. They then computed the energy of these geometries using the CCSD, CCSD(T), and a renormalized version of CCSD(T), called *CR-CCSD(T)*. The renormalization is a slightly reworked treatment of the perturbative correction of the triple excitations.[85] The CCSD surface shows a single minimum on the C_{2h} slice, corresponding to an aromatic-like transition state. However, CCSD(T) predicts a very different result. CCSD(T)/6-31G* predicts a single minimum with $R_{16} = 1.72$ Å, the diyl structure. With the larger 6-311G** basis set, the CCSD(T) C_{2h} PES has two minima, one with $R_{16} = 1.72$ Å and the second with $R_{16} = 1.82$ Å. These two structures differ by only 0.03 kcal mol^{-1}. Whether these two local minima are real is certainly questionable, but the PES is certainly very flat in the region that differentiates the aromatic transition state from the diyl. Since CCSD(T) is generally considered to be one of the most accurate computational methods, this failure to discern the nature of the Cope rearrangement is disappointing. The completely renormalized variation, CR-CCSD(T), predicts a single minima on the C_{2h} PES with $R_{16} = 1.86$ Å. This result is consistent with the best MP-CAS computations in predicting a single aromatic transition state for the Cope rearrangement.

MP2/6-31G* finds a diyl structure,[86] but the MP4 energy using the CCD geometry differs by less than a kilocalorie per mole from the experimental value.[82] None of these single-configuration methods, however, have reached convergence; higher levels of perturbation theory might actually give poorer agreement with experiment.

Density functional theory has also been applied to the Cope rearrangement. Nonlocal methods, such as BLYP and B3LYP, find a single transition state with R_{16} approximately 2 Å. The barrier height is in excellent agreement with experiment.[87] These first DFT results were extremely encouraging because DFT computations are considerably less resource-intensive than MRPT. Moreover, analytical first and second derivatives are available for DFT, allowing for efficient optimization of structures (particularly transition states) and the computation of vibrational frequencies needed to characterize the nature of the stationary points. Analytical derivatives are not available for MRPT calculations, which means that there is a more difficult optimization procedure and the inability to fully characterize structures.

The competition between the chair and boat transition states for the Cope rearrangement has also been the subject of computations.[81,89] Based on Goldstein's studies of the Cope rearrangement of deuterated 1,5-hexadienes, the chair transition state is estimated to be 11.3 kcal mol^{-1} lower in enthalpy than the boat transition state.[61] Shea and Phillips[91] designed the diastereomeric pair **17c** and **17b**, which can undergo a Cope rearrangement exclusively through a chair transition state (**17cTS**) or a boat transition state (**17bTS**), respectively. Consistent with Goldstein's results, the activation enthalpy for the Cope rearrangement of **17c** is 13.8 kcal mol^{-1} lower than that of **17b**. Dolbier[92] followed these experiments with a study of the difluorinated analogs, **18c** and **18b**. The activation enthalpy for the Cope rearrangement of **18c** is 5.6 kcal mol^{-1} below that of **17c**, but the barrier for reaction of **18b** is 7.9 kcal mol^{-1} above that for **17b**. Perhaps even more intriguing are the experimental activation entropies: -11.3 and -17.5 eu for **17c** and **18c**, respectively, which are in the range of typical values, but $\Delta S^{\ddagger}(\mathbf{17b}) = -0.7$ eu and $\Delta S^{\ddagger}(\mathbf{18b}) = +8.7$ eu! The more positive activation entropies of the boat than chair paths suggest more bond breaking than bond forming in the former. The very positive activation entropy for **18b** suggests that there is essentially no bond making and only bond breaking in this boat transition state. To quote Dolbier directly, "This *(the reaction of 18b)* is a Cope rearrangement which does not want to be pericyclic!"

17c X=H
18c X=F

17b X=H
18b X=F

The B3LYP/6-31+G(d,p) chair and boat transition states[89] for the Cope rearrangement of 1,5-hexadiene (**15**) are shown in Figure 4.8 and their relative energies are listed in Table 4.7. The boat TS (**15bTS**) has a computed activation enthalpy that is higher than the chair TS (**15cTS**) by 7.2 kcal mol^{-1}. (The enthalpy difference is predicted to be 9.2 kcal mol^{-1} at B3PW91/6-311+G(2d,p),[90] in somewhat better agreement with experiment, but still too small.) Since the agreement between experiment and calculation for the chair barrier is excellent, this computational method is underestimating the boat transition state enthalpy. This trend carries over to the computations on **17** and **18**; the chair activation enthalpies are in excellent agreement with experiment while the boat TS enthalpies are underestimated.[89]

Also of concern is that the predicted activation entropies for the boat TSs, while more positive than those computed for the chair TSs, are not nearly positive enough, particularly for **18b**. These results suggest that the transition states for **17b** and especially **18b** are poorly described by the B3LYP/6-31+G(d,p) method. Rather, the transition states are likely to have much larger dissociative character, that is, a strong bisallyl contribution. This same failure was noted in the early ab initio calculations of the boat transition state **15bTS**.[81] It has been suggested that the variational transition state (i.e., the geometry that minimizes the *free energy of activation*) occurs at a looser geometry with higher entropy than the C_{2v} structure that minimizes the activation enthalpy (Table 4.7).

A word of caution with regard to DFT computations is necessary. A spin-restricted wavefunction cannot properly describe either of the dissociation limits for the Cope rearrangement, but using an unrestricted wavefunction can help to mitigate this problem. UB3LYP, UB3PW91, and UB3P86 all identify a C_{2h} cyclohexane-diyl radical for the Cope rearrangement. The aromatic transition state is found using their restricted analogs. One now has to ask which (if either) is correct?

Anytime a wavefunction may have diradical character, care must be taken to insure that the wavefunction is stable with respect to lifting the requirement of spin restriction. This can be simply tested by examining whether an unrestricted wavefunction is lower in energy. For the Cope rearrangement, the

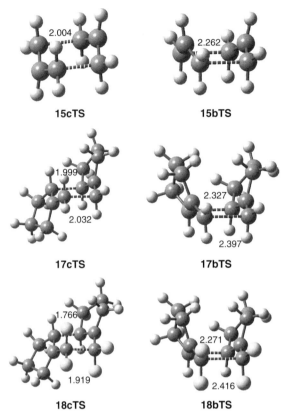

Figure 4.8 Chair and boat transition state geometries for the Cope rearrangements of **15**, **17**, and **18**.[89]

restricted Hartree–Fock (RHF) wavefunction is *always* unstable for the region of R_{16} between 1.4 and 3.4 Å. In contrast, the B3LYP and B3PW91 wavefunctions are both stable in the region of interest for the Cope rearrangement (1.8 Å < R_{16} < 2.7 Å). This means that the DFT prediction of an aromatic transition state is likely to be real.[93] As for the diyl found on the unrestricted density functional theory (UDFT) surfaces, Stavarov and Davidson[94] examined the MRPT energies of points along a C_{2h} cut with varying R_{16} distances, using UB3LYP and UB3PW91 geometries about the predicted diradical intermediate. For both cases, a diyl was not found by MRPT. They concluded that the prediction of a diradical intermediate is an artifact of the unrestricted formalism.

What is the nature of the single transition state for the Cope rearrangement? Examination of the CASSCF wavefunction indicates that the Ψ_{arom} dominates the wavefunction, and becomes even more important when dynamic correlation is included. The C–C distances within the allyl fragments are typical of those for aromatic molecules, and R_{16} is similar to the lengths of forming C–C distances in a variety of pericyclic reactions. Individual gauge for localized orbital (IGLO) calculations for the transition state using either CISD or MP2 predict upfield

TABLE 4.7 Activation Enthalpies (kcal mol^{-1}) and Entropies (eu) for the Cope Rearrangements of 15, 17, and 18[a]

	$\Delta H^{\ddagger}$	$\Delta S^{\ddagger}$	$\Delta H^{\ddagger}$	$\Delta S^{\ddagger}$
15	34.0	−8.1	41.2	−6.1
	33.5	*−13.8*		
17c	27.0	−5.6		
	28.0	*−11.3*		
18b			35.3	−4.3
			41.6	*−0.7*
18c	24.2	−8.5		
	22.4	*−17.5*		
18b			39.8	−3.6
			49.5	*8.1*

[a]Values computed at B3LYP/6-31+G(d,p) are listed in normal text and experimental values are listed in italics. See Ref. 89.

shifts for axial protons and downfield shifts for the equatorial protons.[86] The magnetic susceptibility exaltation of the chair Cope transition state is larger than for benzene and its NICS value is quite negative (−22.7). All these point toward a concerted, aromatic transition state.[31] Bader, on the other hand, argued that topological analysis points toward delocalization of the π-electrons to each of the two allylic-like fragments and that there is no aromatic stabilization energy.[90]

Computational chemistry has converged upon a mechanism for the Cope rearrangement. The reaction is characterized by a single transition state, having C_{2h} symmetry. There are no intermediates; the reaction occurs by a single chemical step. The transition state can be best characterized as having some aromatic character. A representation of the mechanism that captures the nature of the transition state is given in Scheme 4.7; the transition state is composed of resonance structures for the two radical contributors but with a much larger contribution from the delocalized aromatic structure.

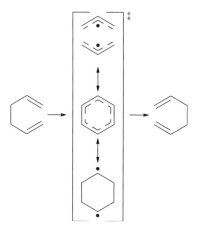

Scheme 4.7

TABLE 4.8 Activation Enthalpies (kcal mol^{-1}) and R_{16} (Å) for the Cope Rearrangements of Cyano- and Phenyl-Substituted 1,5-Hexadienes[a]

Substituents	ΔH_{298} Calc	ΔH_{298} Expt	R_{16}
H	33.2	33.5 ± 0.5[b]	1.965
1-CN	35.5		2.082, 2.131
3-CN	29.3		2.131, 2.082
1,4-diCN	29.9		2.236
1,3,4,6-tetraCN	24.7		2.467
2-CN	28.0		1.825
	27.8		*1.607*
2,5-diCN	24.4	(23.3)[c]	1.752
	20.2		*1.575*
2,4-diCN	26.5		1.915, 1.966
1,2,3-triCN	29.1		2.104
1-phenyl	36.2		2.062, 2.122
3-phenyl	28.4	28.1 ± 0.4[d]	2.122, 2.062
1,4-diphenyl	29.2	29.9 ± 1.6[e]	2.241
1,3,4,6-tetraphenyl	19.1	21.3 ± 0.1[f]	2.649
2-phenyl	30.3	29.3 ± 1.6[d]	1.777, 1.700
	29.4		*1.599*
2,5-diphenyl	24.8	21.3 ± 0.3[d,g]	1.839, 1.667
	21.3		*1.576*
2,4-diphenyl	26.7	24.6 ± 0.8[d]	1.979, 1.900
1,3,5-triphenyl	29.2	27.8 ± 0.2[d]	2.113, 2.106

[a] Values in italics refer to reaction intermediates.
[b] Ref. 62.
[c] Activation enthalpy for the reaction of 2,5-dicyano-3-methyl-1,5-hexadiene.[69]
[d] Ref. 68.
[e] Ref. 102.
[f] Ref. 103.
[g] Ref. 104.

Perhaps the most significant result of these computational studies is in clarifying the decision concerning which is the appropriate methodology for investigating pericyclic reactions. For reactions where multiple configurations may play a role, CASSCF is necessary to provide an unbiased description of the transition structure, but CASSCF is unlikely to be sufficient to adequately describe the reaction surface. Rather, the CASSCF wavefunction is the appropriate *reference wavefunction* but some accounting of dynamic correlation among *all* of the electrons will be necessary.[95] MRPT, usually in its CASPT2[96] or MRMP2[97] formulation, has become the *de facto* standard for high level authoritative computations, particularly for pericyclic reactions. Unfortunately, this methodology is very resource-intensive. Frequently, MRPT is employed simply to benchmark cheaper and more efficient methods. For example, it appears that UB3LYP produces the correct results for the Cope rearrangement. Along with its success with the Diels–Alder reaction described in the previous sections, the wide adoption of this DFT methodology by organic chemists was encouraged by its success in treating the Cope rearrangement. Subsequent studies have uncovered some serious problems that B3LYP has in treating, for example, alkane energies (see Section 3.3.3), and so B3LYP should not be considered as the panacea for all organic systems. Future practitioners of computational chemistry should be hesitant to use the Cope rearrangement as the sole reaction on which to rest their methodological choice; it is always best to try out new computational approaches on a variety of different reactions, preferably ones closely related to those reactions where the methodology is intended to be applied.

4.2.3 Chameleons and Centaurs

The transition state of the Cope rearrangement, as depicted in Scheme 4.7, is a mixture of the three resonance contributors. The C_{2h} slice of the PES is very flat in the range of R_{16} from 1.6 to 2.2 Å. A continuum representation whereby the cyclohexane-1,4-diyl resonance structure becomes more important at small values of R_{16} and the bis-allyl contribution becomes greater at large values of R_{16} has been proposed by a number of people.[68–71] Doering[98] named this model "chameleonic." The chameleonic model suggests that substituents may alter the relative weights of the resonance contributions, although the transition state maintains C_{2h} (or near C_{2h}) symmetry of the six-member ring. Radical-stabilizing groups, such as cyano, phenyl, or vinyl, can then be classified based on their position of attachment. Those attached at C_1, C_3, C_4, or C_6 will stabilize the allylic structure and are termed *active*. Those attached to C_2 or C_5 will stabilize the cyclohexadiyl structure and are termed *nodal*. Multiple substituents of the same type, either all "active" or all "nodal," should be additive and enhance the biallylic or diyl contribution, respectively.

Doering also considered the case where substituents of different types are placed on 1,5-hexadiene. For example, 1,3,5-triphenylhexa-1,5-diene (**19**) has two substituents of the "active" type (the phenyl groups on C_1 and C_3), which would favor the bis-allyl contribution, and one substituent of the "nodal" type (the phenyl on C_5), which would favor the diyl contribution. If the stronger of these two dominates, that would be a "chameleonic" transition state. However, Doering proposed the possibility where each set of substituents acts only on its half of the molecule; for our example, the C_1–C_3 fragment would be allylic-like and the C_4–C_6 fragment would be diyl-like. The transition state would still have equal R_{16} bond lengths; but each half would distort to have bond length appropriate to its substituents, with C_1–C_2 and C_2–C_3 shorter than C_4–C_5 or C_5–C_6. Doering[98] termed such a transition state *centauric* after the mythological creature that is half man, half horse.

Doering examined the Cope rearrangements of 1,3,5-triphenylhexa-1,5-diene[98] (**19**) and 1,3-dicyano-5-phenyl-1,5-hexadiene (**20**).[99] These two molecules are set up to test the "centauric" model: are the stabilizing effects of substituents additive, as the "centauric" model suggests, or does one set dominate, indicative of the "chameleonic" model. Direct comparison of activation barriers between model compounds (such as 1,3-diphenyl-1,5-hexadiene and 1,3-dicyano-1,5-hexdiene and the parent itself) might be inappropriate since the substituents affect the reactants and products and not just the transition state region. Correcting for these stabilizing effects, Doering finds that the activation barriers are lowered more than what is suggested by the "chameleonic" model but shy of what the "centauric" model predicts.

19: Ph, Ph, Ph substituted 1,5-hexadiene
20: NC, CN, Ph substituted 1,5-hexadiene

The Houk and Borden groups[100,101] have collaborated on a couple of important studies to address the effect of multiple substituents on the Cope rearrangement. They examined the effects of cyano, phenyl, and vinyl substituents at various positions on 1,5-hexadiene. All three substituents give similar results, but only the cyano and phenyl cases will be discussed here. The cyano case was also examined by Staroverov and Davidson,[94] coming to the same conclusions as presented here but with a slightly different analysis.

The calculated activation enthalpies for the Cope rearrangements of various cyano- and phenyl-substituted 1,5-hexadienes were calculated at B3LYP/6-31G* and are listed in Table 4.8.

A couple of initial observations are warranted. First, the agreement in the activation enthalpy between the calculated B3LYP and the experimental values is truly outstanding! Second, radical-stabilizing substituents, such as cyano and phenyl, can dramatically decrease the activation barrier of the Cope rearrangement, suggesting a greater participation by the radical contributors to the wavefunction (see Scheme 4.7).

Both cyano and phenyl substituents prefer to conjugate with a double bond, making the Cope rearrangement of 1-cyano and 1-phenyl-1,5-hexadiene endothermic, and making direct comparisons of their activation enthalpies meaningless. However, a simplified Marcus model suggests that in the absence of any special stabilizing effects, the average of the activation enthalpies of 1-cyano and 3-cyano-1,5-hexadienes (or 1-phenyl and 3-phenyl-1,5-hexdienes) should equal that of the parent Cope rearrangement. However, the average for the cyano case is 32.4 kcal mol^{-1}, 0.8 kcal mol^{-1} less than for the parent, suggesting that a cyano group in the "active" position will lower the barrier by 0.8 kcal mol^{-1}. The average for the phenyl case is 32.3 kcal mol^{-1}, indicating that a phenyl group in the "active" position lowers the barrier by 0.9 kcal mol^{-1}.

The 1,4-disubstituted cases are degenerate, so no averaging is needed. The "chameleonic" model suggests that the substituent effects should be additive, or a predicted value of $\Delta H^{\ddagger} = 31.6$ kcal mol^{-1} for 1,4-dicyano-1,5-hexadiene and $\Delta H^{\ddagger} = 31.4$ kcal mol^{-1} for 1,4-diphenyl-1,5-hexadiene. The computed barriers are actually lower than these values. The barrier for the 1,3,4,6-tetrasubstituted cases are lowered by more than four times the effect of a single substituent. Thus, the multiple substituents in the "active" position act cooperatively and not simply additively. Substituents in the "active" positions should act to increase the bis-allyl contribution. This should manifest itself in a longer R_{16} distance as the number of "active" substituents is increased. This trend is true for both the cyano and phenyl cases; in fact, R_{16} is very long in the tetrasubstituted cases: 2.467 Å for the cyano and 2.649 Å for the phenyl.

Placing a substituent in the "nodal" positions should stabilize the diyl contribution. The activation enthalpies for both 2-cyano and 2-phenyl-1,5-hexadiene are lower than for the parent. The stabilizing effects of a cyano or phenyl group placed in a "nodal" position are 5.2 kcal mol^{-1} and 3.0 kcal mol^{-1}, respectively. In fact, the substituents stabilize the diyl contribution so much that an intermediate is found for both cases. The second "nodal" substituent further stabilizes both the transition state and the intermediate. The barrier for 2,5-dicyano-1,5-hexdiene is 24.4 kcal mol^{-1} or 8.8 kcal mol^{-1} less than for the parent. This is less than twice the stabilizing effect of a single cyano, yet this is misleading. First, the intermediate diyl is now much lower in energy, 20.2 kcal mol^{-1} above reactants. Second, both the intermediate and transition state suffer from spin contamination; Houk and Borden estimate that the diyl intermediate is only 18 kcal mol^{-1} above the reactants, for a net stability greater than three times the value of a single "nodal" cyano group. Similar stabilizing effects are seen with the phenyl substituent. The geometric effect of the "nodal" substituents is also clearly seen in the shrinking R_{16} distances with increasing substitution. Both the 2,5-dicyano and 2,5-diphenyl intermediates have C_1–C_6 distances in the range of normal C–C single bonds.

Cooperativity between multiple substituents in "active" or "nodal" sites can be rationalized in the following way. The first substituent substantially alters R_{16}: the first "active" substituent lengthens R_{16}, creating the allyl radicals, while the first "nodal" substituent shortens this distance, creating the diyl. The next substituent(s) benefit from the "work" the first did in changing R_{16} and their effect is mostly just to stabilize the resulting radical. Thus, each subsequent substituent effectively acts to stabilize a radical, while the first substituent largely has to create the radical. Therefore, energetic stabilization afforded by later substituents is greater than that afforded by the first.

The last issue we need to address is what happens when "active" and "nodal" substituents are both present. The "centauric" model predicts an additive substituent effect. Since $\Delta H^{\ddagger} = 29.3$ for 3-cyano-1,5-hexadiene and a "nodal" cyano group offers a stabilizing effect of 5.2 kcal mol^{-1}, the activation enthalpy for 2,4-dicyano-1,5-hexadiene should be 24.1 kcal mol^{-1}. The same argument suggests that $\Delta H^{\ddagger} = 25.5$ kcal mol^{-1} for 2,4-diphenyl-1,5-hexadiene. Both, in fact, have barriers higher than what additivity predicts. Furthermore, the barrier for the Cope rearrangement of 1,3,5-tricyano-1,5-hexadiene is only 0.6 kcal mol^{-1} less than that of 1,3-dicyano-1,5-hexadiene; the "nodal" cyano group here has a stabilizing effect of about 10 percent its expected value. Simple additivity of the effect of three phenyl groups predicts $\Delta H^{\ddagger} = 27.4$ kcal mol^{-1} for 1,3,5-triphenyl-1,5-hexadiene, but B3LYP indicates a barrier of 29.2 kcal mol^{-1}.

These results suggest a competitive interaction between the "active" and "nodal" substituents. The geometries of these transition states support this competition: their R_{16} values are quite similar to the distance found in the parent 1,5-hexadiene. Computational examinations of the substituent effects on the Cope rearrangement conclude that the "centauric" model does not apply. The "chameleonic" model makes a better accounting of the cooperative and competitive ways the substituents affect the Cope rearrangement. Borden[105] has proposed a simple mathematical model that allows for the prediction of the stabilization of the transition state by substituents solely on the change in R_{16}.

In a related study, Borden and coworkers[106] have examined whether other pericyclic reactions might express chameleonic behavior. Using B3LYP/6-31G* calculations, they located transition states for the 1,5-hydrogen shift in phenyl-substituted 1,3-pentadienes (**21–23**). The activation enthalpy for the 1,5-hydrogen shift of the parent 1,3-pentadiene is 32.0 kcal mol^{-1}. The effect of a phenyl group on C_1 or C_5 can be estimated by taking the average activation enthalpies for **22a** and **22b**; this phenyl substituent lowers the barrier by 1.5 kcal mol^{-1}. (This effectively removes the bias due to the fact that **22a** is 8.7 kcal mol^{-1} more stable than **22b**.) Similarly, averaging of the activation enthalpies of **22b** and **22d** gives the effect of a phenyl group on C_2 or C_4, lowering the barrier by 1.0 kcal mol^{-1}. A single phenyl substitution on C_3 increases the barrier by 0.5 kcal mol^{-1} (Figure 4.9).

Diphenyl substitution reduces the barrier, by 3.2 kcal mol^{-1} in **23a** and by 2.1 kcal mol^{-1} in **23b**. The effect of the two substituents is about additive. This is quite different than the effect of phenyl substitution in the Cope rearrangement, where the effect is *cooperative* and *much* more stabilizing of the transition state. The smaller substituent effect on the 1,5-hydrogen migration is also apparent in the very small geometry changes upon substitution.

Borden concludes that 1,5-hydrogen migration is not chameleonic. Rather than having a shift of the dominant resonance contributor with substitution as occurs in the Cope rearrangement, regardless of substitution, the transition state for the 1,5-H

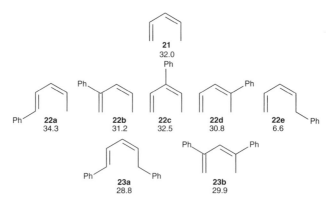

Figure 4.9 B3LYP/6-31G* activation enthalpies for 1,5-hydrogen shifts of substituted 1,3-pentadienes.[106]

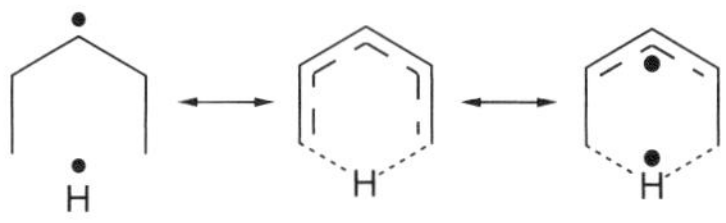

THE COPE REARRANGEMENT 231

Scheme 4.8

shift is dominated by resonance contribution **C**, as shown in Scheme 4.8. Resonance structure **C** can be viewed as an allyl radical and a hydrogen atom bonded to the two terminal carbon atoms. This structure has a node through C_3, which explains why the phenyl substituent on C_3 has very little effect. The geometry of the transition state provides further support for **C**. The two terminal carbon atoms must twist toward the central migrating hydrogen, resulting in disruption of the π-overlap with C_2 and C_3. Doering[107] investigated the [1,5]-H migrations in **24** and **25**, finding that the activation barriers of both are lowered by essentially the same amount relative to **26**. Doering evokes this as evidence of a chameleonic effect, though alternative explanations involving strain relief in the transition states may also account for these energy differences.

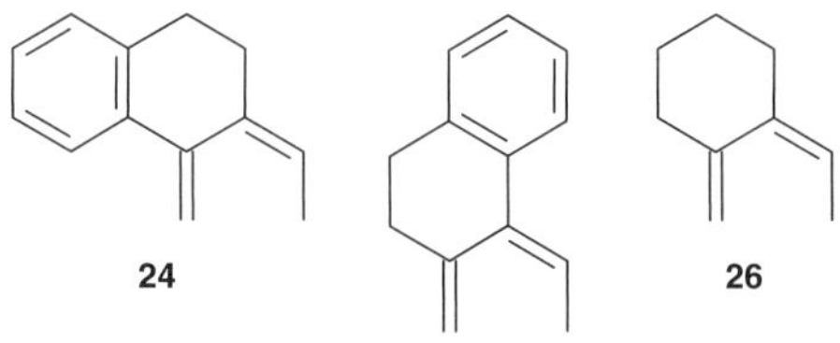

24 25 26

Schreiner[108] has identified three different classes of Cope rearrangements based on the model shown in Scheme 4.7: (1) the reaction proceeds through a single transition state, (2) the reaction proceeds through a diyl intermediate, and (3) the reaction has two competitive routes through either a single TS or through the intermediate diyl. An example of the second class, involving a diyl intermediate, is shown in Scheme 4.9. The energies of the critical points shown in Scheme 4.9 were computed at BD(T)/cc-pVDZ//BLYP/6-31G*.

Though Schreiner examined 64 Cope reactions, none involved in the formation of a bis-allyl intermediate. However, Kertesz[109] found just such a case in the

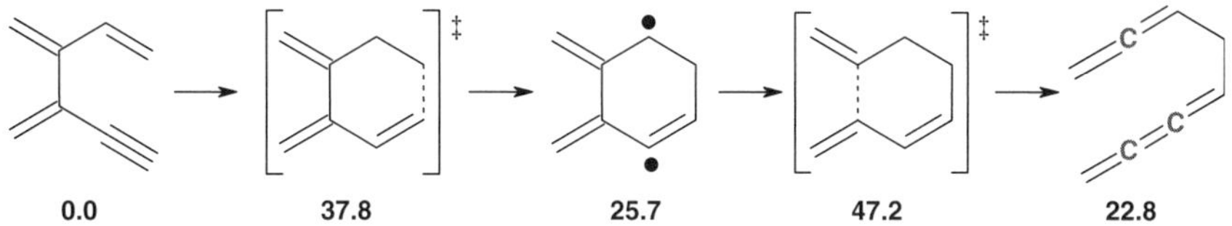

0.0   37.8   25.7   47.2   22.8

Scheme 4.9

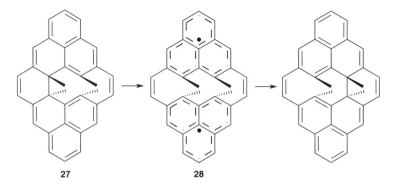

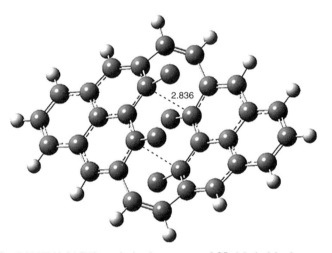

Figure 4.10 B3LYP/6-31G(d) optimized structure of **28**. Methyl hydrogens suppressed for clarity. Distance in angstroms.

rearrangement of **27**. Using B3LYP and BPW91 computations with two different basis sets, he identified the stable diradical intermediate **28**. This structure, shown in Figure 4.10, clearly has very long distances (2.836 Å) separating the ends of the two "allylic" components. A true transition state connects the reactant **27** with the intermediate **28**. The activation energy is 6.3 kcal mol^{-1} and the intermediate **28** lies 3.3 kcal mol^{-1} above **27**.

Why does a stable bis-allyl analog exist on the Cope reaction surface of **27**? In the prototype Diels–Alder reaction of 1,5-hexadiene, the possible bis-allyl intermediate (i.e., two isolated allyl radicals) is about 26 kcal mol^{-1} higher in energy than the Cope transition state. Only with significant radical stabilization might one expect a bis-allyl intermediate to occur. One can consider **28** as composed of two bridged phenalenyl radicals (**29**), which affords a rather dramatic stabilization of each radical.

29

4.3 THE BERGMAN CYCLIZATION

In the 1980s, a series of structurally related naturally occurring antibiotics were discovered and characterized, generating a cottage industry within the chemistry community. Synthetic organic chemists pursued the total synthesis of these complex molecules and then their nonnatural analogs. Biochemists and biologists looked toward understanding the nature of the activity of these molecules. And computational chemists tried to understand the chemical mechanism underlying the biological activity. Of particular interest for us here is that once again a seemingly straightforward mechanism proved to be very difficult to accurately compute. Nonetheless, once computational chemists settled on appropriate methodologies, these computations helped to settle some experimental controversies and provided great insight toward the design of new potential drugs with similar molecular mechanism of biological action.

These novel, extremely potent antibiotics all possessed one very unusual chemical structural feature: an enediyne fragment within a ring.[110–112] The first discovered was neocarzinostatin (NSC), which was found to be composed of a protein (NSC apoprotein) and the neocarzinostatin chromophore **30** in a 1:1 complex.[113] The biological activity is associated with the NSC chromophore. However, it took until the discovery of the calichaemicin[114,115] **31** and esperamicin[116,117] **32** families in 1987 to really garner the broad attention of chemists. Other enediyne antibiotics include the dynemicin[118,119] family, the kedarcidin chromophore,[120,121] namenamicin,[122] the C-1027 chromophore,[123] and uncialamycin.[124]

Neocarzinostatin chromophore
30

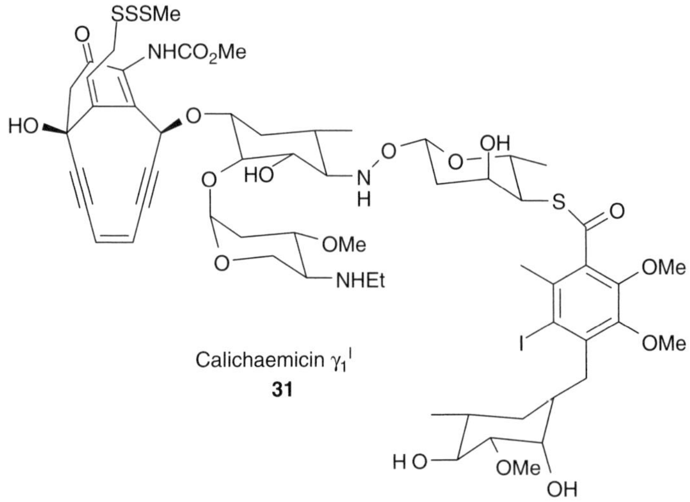

Calichaemicin γ₁¹
31

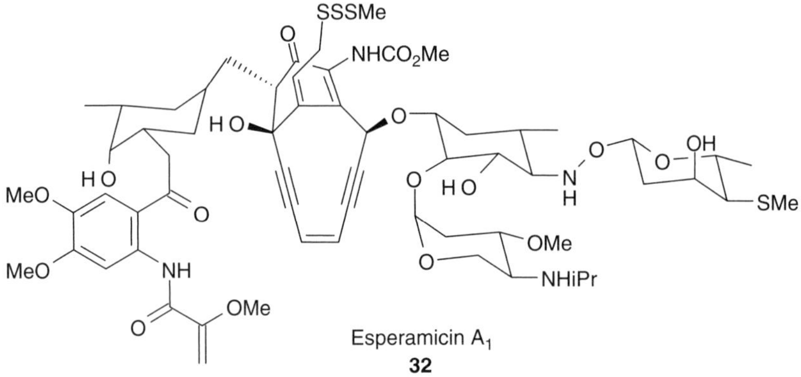

Esperamicin A₁
32

Intensive investigation of these drugs uncovered a very elegant method of action involving the delivery of the molecule to a DNA strand, a triggering step to activate the molecule, followed by abstraction of hydrogen atoms from the DNA molecule leading to its scission. This is diagrammed in Figure 4.11 for calichaemicin as a representative example of the enediyne drugs. First, the sugar moiety binds the molecule to the minor groove of DNA. Next, the molecule is activated by a nucleophilic substitution at the trisulfide fragment, creating a thiolate anion that can undergo a conjugate addition to the bridgehead alkene. The resulting enediyne **33** undergoes a Bergman cyclization to form the aromatic diradical **34**. This diradical then excises hydrogen atoms from the neighboring DNA molecule, leading to the eventual cleavage of the DNA strands and cell death. Using militaristic terms, the sugar is the targeting mechanism, the trisulfide cleavage and conjugate addition act as the triggering mechanism, and the diradical is the warhead, wreaking havoc

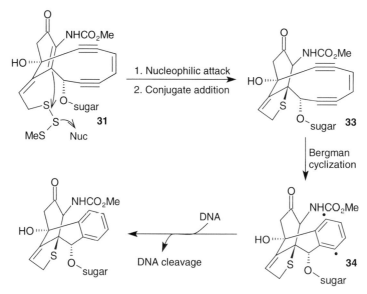

Figure 4.11 Mechanism of action of calichaemicin **31**. Nicolaou, K. C.; Dai, W.-M. "Chemistry and Biology of the Enediyne Anticancer Antibiotics," *Angew. Chem. Int. Ed. Engl.* **1991**, *30*, 1387–1416. Copyright Wiley-VCH Verlag GmbH & Co. KGaA. Reproduced with permission.

upon the target DNA molecule. Computational chemists have explored a number of these critical steps that we will detail in later chapters. This section explores the step that creates the warhead; in other words, the Bergman cyclization converts the enediyne into an aromatic diradical.

The Bergman cyclization derives its name from the studies of the Bergman group that began in 1972. Jones and Bergman[125,126] prepared (3Z)-3-hexene-1,5-diyne deuterated in the acetylenic positions (**35**). Heating **35** up to 200 °C in the gas phase led to rapid scrambling of the deuteriums, but only the two symmetrically substituted analogs **35** and **37** were identified. Heating **38** in solution with a hydrogen donor, such as 1,4-cyclohexadiene, gave benzene, while heating in the presence of CCl_4 gave 1,4-dichlorobenzene. These reactions are typical of free radical behavior, suggesting the intermediacy of *p*-benzyne **40**. They estimated the barrier for the conversion of **38** to **40** to be about 32 kcal mol^{-1}. A careful thermochemical study of the Bergman cyclization of **38** gave $\Delta H^{\ddagger}(470\,K) = 28.2 \pm 0.5$ kcal mol^{-1}.[127] A follow-up study of the thermal reaction of **41** provides additional thermochemical data.[128] The activation barriers are: **41** → **42**, 27.4 kcal mol^{-1}; **42** → **41**, 16 kcal mol^{-1}; and **42** → **43**, 10 kcal mol^{-1}.

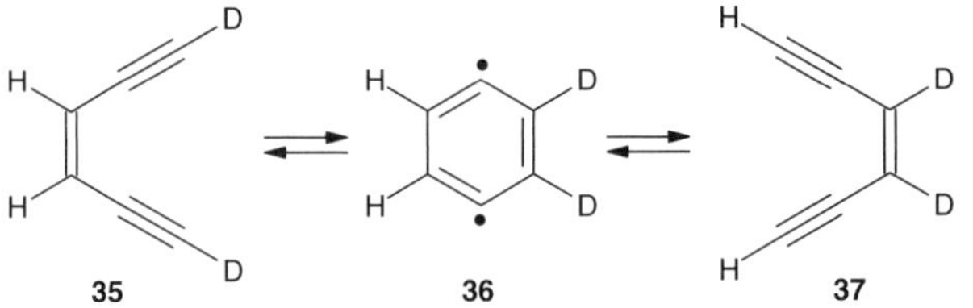

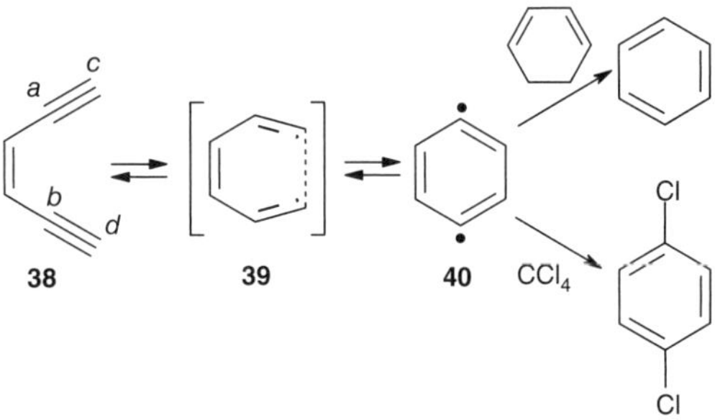

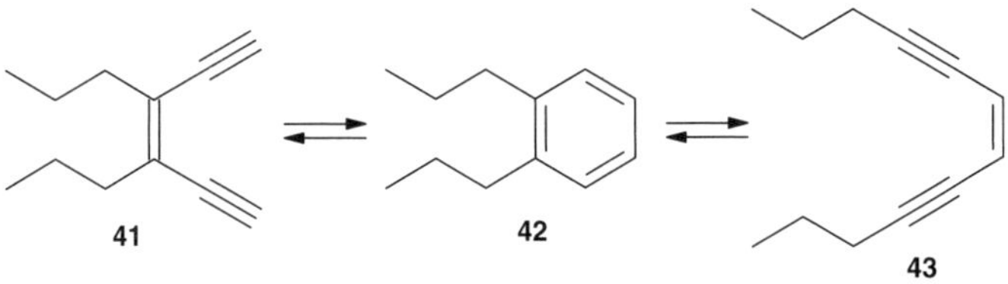

The cyclization of **38** has a large activation barrier and it is in fact stable at 25 °C. Calichaemicin and the other enediyne drugs are also stable at 25 °C, but rapidly undergo cyclization after activation. Nicolaou et al.[129,130] proposed, based on molecular mechanics (MM) calculations, that the critical factor is the distance between the terminal alkynyl carbons, a distance they called *cd*. When the terminal alkynes are far apart ($cd > 3.31$ Å), the enediyne is stable, and when the distance is small ($cd < 3.20$ Å), the cyclization is spontaneous. Based on this model, the 10-membered ring enediyne **44** falls in the intermediate range ($cd = 3.25$ Å). They prepared this molecule and found that its activation energy is 23.8 kcal mol^{-1}, with a half-life at 37 °C of 18 h. This is consistent with the predicted intermediate-type behavior.

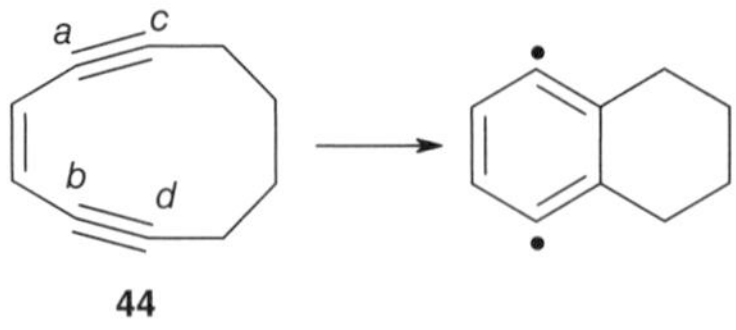

Computational chemists have investigated a number of relevant issues. What is the nature of the mechanism? Can the *cd* distance predict the reactivity of the enediynes or does some other factor dictate reactivity? What factors govern the reactivity of the intermediate *p*-benzyne **40**—will it ring open or abstract hydrogen atoms? What is the ground state of *p*-benzyne? How does the activation take place, especially the nucleophilic substitution at the trisulfide moiety? We address the first three questions in this chapter. The electronic structure and properties of the benzynes are discussed in Chapter 5 and the substitution reaction is presented in Chapter 6.

4.3.1 Theoretical Considerations

Before presenting the application of quantum calculations to the Bergman cyclization, we first discuss the orbital transformations that occur during the course of the reaction. A simple description of the bond changes is that the in-plane π-bonds of the two triple bonds of (3Z)-3-hexene-1,5-diyne **38** are converted into the σ-bond between C_1 and C_6 and two singly occupied sp^2-like orbitals, one on C_2 and the other on C_5, making *p*-benzyne **40**. In terms of molecular orbitals, for the reactant enediyne, MO 18 and MO 19 are the in-phase and out-of-phase combinations of the in-plane π-bonds (see Figure 4.12). These will transform into MO 15, describing the new σ-bond of *p*-benzyne, and MO 20. Note that this latter molecular orbital (MO 20) mostly comprises the out-of-phase combination of the p-orbitals at the *para* positions which can favorably interact with the σ* orbitals at C_1–C_6 and C_3–C_4. The in-phase combination of these p-orbitals appears in MO 21, the LUMO of *p*-benzyne. Here, the p-orbitals interact with the σ orbitals at C_1–C_6 and C_3–C_4.

The appropriate description of the wavefunction of *p*-benzyne comes down to the relative contribution of two configurations that differ only in the occupation of one orbital. The first configuration doubly occupies MO 20, while the second doubly occupies MO 21. The wavefunction can therefore be written as

$$\Psi = c_1 | \ldots 1b_{2g}^2 2b_{1g}^2 5b_{1u}^2 | + c_2 | \ldots 1b_{2g}^2 2b_{1g}^2 6a_g^2 | \qquad (4.3)$$

The pure orbital-symmetry-conserved wavefunction has $c_2 = 0$ and a single-configuration wavefunction would therefore be an acceptable reference. The pure biradical wavefunction has $c_2 = -c_1$, necessitating a multiconfiguration reference.

Here then is the crux of the computational difficulty. The reactant, (3Z)-3-hexene-1,5-diyne, is well described by a single-configuration reference wavefunction. The product, *p*-benzyne, is likely to have appreciable diradical character and necessitates a multiconfiguration wavefunction. The transition state will exist somewhere in between. The choice of computational method suited to describe all three structures equally well is nontrivial, and in the next section we discuss the various approaches employed and results obtained by a number of research groups.

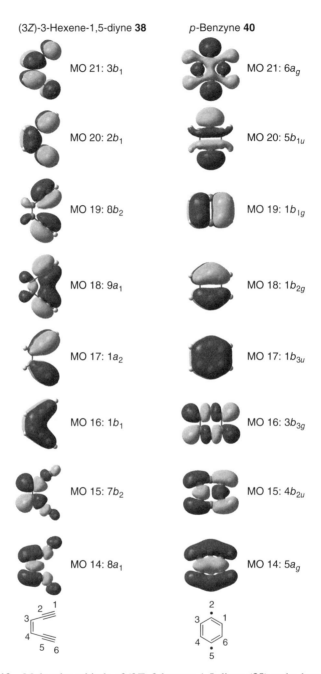

Figure 4.12 Molecular orbitals of (3Z)-3-hexene-1,5-diyne (**38**) and *p*-benzyne (**40**).

4.3.2 Activation and Reaction Energies of the Parent Bergman Cyclization

We first take up the prototype Bergman cyclization, which is the conversion of (3Z)-3-hexene-1,5-diyne (**38**) into *p*-benzyne (**40**). Computational chemists came to the Bergman cyclization problem at about the same time as they came to understand the computational difficulties of the Cope rearrangement discussed in the previous section. These studies of the Cope rearrangement greatly colored the early choice of methodologies to be used for the Bergman cyclization.

Recognizing the inherent multiconfigurational nature of **40** and the likely multiconfigurational nature of the transition state **39**, the CASPT2, large CI, or DFT approaches offered the most promising options. None, however, is without potential difficulties. The CASPT2 method, while offering a true multiconfiguration reference, systematically overestimates the dynamic correlation of unpaired electrons, and this is a real concern for the Bergman cyclization where the number of unpaired electrons is not conserved.[131] The coupled clusters method is a powerful tool for treating dynamic correlation, but it too does not treat the reactant and product equivalently. In describing **40**, the contribution from the second important configuration (see Eq. (4.3)) enters as a double excitation from the reference configuration. This second configuration will be a large contributor, and one should correct it for dynamic correlation. This can be accomplished by taking double excitations from the second configuration, but these are not included within the CCSD(T) ansatz; one would need to include contributions from the T_4 operator. Thus, CCSD(T) will do a better accounting of the correlation energy for **38** than for **40**, thereby overestimating the heat of reaction. With respect to DFT, questions abound concerning whether the single-determinant DFT can describe the multiconfiguration problem, whether a restricted DFT solution will be stable for *p*-benzyne, and, of course, the usual nagging problem of which functional is best suited for this problem.

An additional complication is that the available experimental reaction and activation enthalpies were obtained at different temperatures: $\Delta H(\mathbf{38} \rightarrow \mathbf{40}, 298\,\text{K}) = 8.5 \pm 1.0$ kcal mol^{-1} and: $\Delta H^{\ddagger}(\mathbf{38} \rightarrow \mathbf{40}, 470\,\text{K}) = 28.2 \pm 0.5$ kcal mol^{-1}. These can be corrected for different temperatures if the vibrational frequencies are known. Unfortunately, the experimental values are unavailable, but DFT-computed frequencies can be used in their place.[132] This gives $\Delta H^{\ddagger}(\mathbf{38} \rightarrow \mathbf{40}, 298\,\text{K}) = 28.7 \pm 0.5$ kcal mol^{-1}. One can also obtain relative electronic energies at 0 K with no ZPE corrections for direct comparison to computed values: $\Delta E(\mathbf{38} \rightarrow \mathbf{40}) = 7.8 \pm 1.0$ kcal mol^{-1} and $\Delta E^{\ddagger}(\mathbf{38} \rightarrow \mathbf{40}) = 30.1 \pm 0.5$ kcal mol^{-1}.

The first serious computational study employed a CASSCF wavefunction followed by a CI expansion.[133] However, the active space employed for the geometry optimization only included the in-plane π-orbitals of the triple bond, which neglects the true symmetry of the species involved; the C_1-C_6 σ-bond in **40** is included in the active while the C_3-C_4 σ-bond is not.

Using large CASSCF and CASPT2 calculations, Lindh and coworkers[134,135] examined the parent Bergman cyclization. The CASSCF(12,12) active space included all of the out-of-plane π-orbitals (the occupied orbitals included for **38** are MOs 16, 17, and 20 and for **40** are MOs 17, 18, and 19—see Figure 4.12) and the in-plane π-orbitals and the C_3–C_4 σ-bond (for **38** these are MOs 14, 18, and 19 and for **40** these are MOs 14, 15, and 20). They included dynamic correlation using a couple of different perturbation schemes, CASPT2[0] and CASPT2[g1], along with a number of different basis sets. CASSCF(12,12) predicts a barrier that is too high (>40 kcal mol^{-1}, see Table 4.9) and a reaction enthalpy that is too large; essentially CASSCF is overestimating the stability of the reactant relative to the TS and product. The situation is dramatically improved when dynamic correlation is included. Their best CASPT2 calculation (CASPT2[g1]/ANO//CASSCF(12,12)/ANO) gives the activation enthalpy as 23.87 kcal mol^{-1} and the reaction energy as 3.84 kcal mol^{-1}. Using empirical corrections for the difference in the number of unpaired electrons in reactant, TS and product, and BSSE, they estimate that the reaction enthalpy at 298 K

TABLE 4.9 Computed Energies[a] (kcal mol^{-1}) for the Prototype Bergman Cyclization **38 → 40**

Method	TS	Product	cd^b
CASSCF(4,4)/MIDI1[c]	45.3	20.7	1.95
CAS(10,10)/MIDI4//CASSCF(4,4)/MIDI1[c]	44.7	28.3	
MRSDCI[c]	37.6	22.0	
CCSD(T)/6-31G(d,p)[d]	29.5	5.5	1.993
CCSD(T)/6-31G(d,p) with MBPT(2) frequencies[d]	28.5	8.0	
MBPT(2)/6-31G(d,p)[d]	23.3	−12.3	2.077
CCSD/6-31G(d,p)[d]	37.7	27.2	
CAS(12,12)/DZP[e]	40.40	22.39	1.923
CASPT2// CAS(12,12)/DZP[e]	20.63	−4.22	
CAS(12,12)/TZ2P[e]	43.64	27.64	1.896
CASPT2// CAS(12,12)/TZ2P[e]	0.42	21.55	
CAS(12,12)/ANO[e]	43.64	27.70	1.895
CASPT2[0]//CAS(12,12)/ANO[e]	23.20	2.34	
CASPT2[g1]//CAS(12,12)/ANO[f]	23.87	3.84	
CASPT2(12,12)/ANO[g]	25.0	8.6	1.896
CCSD//CAS(12,12)/ANO[f]	40.03	31.10	
CCSD(T)//CAS(12,12)/ANO[f]	29.65	10.81	
BLYP/6-311+G**//BLYP/6-31G*[h]	28.4	17.6	2.077
BPW91/cc-pVDZ[i]	22.0	0.9	2.116
CASPT2//BPW91/cc-pVDZ[i]	21.7	−2.6	
CCSD(T)//BPW91/cc-pVDZ[i]	25.6	4.8	
BCCD(T)//BPW91/cc-pVDZ[i]	25.6	4.8	
Composite[i]	27.7	11.0	
BPW91/6-311G**[j]	25.50	11.64	2.064
BPW91/6-311G**[j]	24.33	11.62	
BP86/6-311G**//BPW91/6-311G**[j]	24.62	11.01	
BPW91/ANO//BPW91/6-311G**[j]	24.40	10.08	
B3LYP/6-311G**//BPW91/6-311G**[j]	34.35	25.67	
B3LYP/ANO//BPW91/6-311G**[j]	32.81	23.77	

TABLE 4.9 (*Continued*)

Method	TS	Product	cd^b
SVWN/6-31G**[k]	*17.7*	*−4.6*	2.178
SVWN/cc-pVDZ[k]	*19.2*	*0.6*	2.105
BP86/6031G**[k]	*22.4*	*0.6*	2.123
BP86/cc-pvDZ[k]	*24.6*	*6.0*	2.062
BPW91/6-31G**[k]	*23.3*	*0.2*	2.120
BPW91/cc-pvDZ[k]	*25.4*	*5.4*	2.061
BLYP/6-31G**[k]	*25.4*	*6.8*	2.073
BLYP/cc-pVDZ[k]	*28.6*	*13.6*	2.008
B3PW91/6-31G**[k]	*28.9*	*−2.8*	1.929
B3PW91/cc-pVDZ[k]	*31.2*	*2.7*	1.966
B3LYP/6-31G**[k]	*31.2*	*3.3*	1.978
B3LYP/cc-pVDZ[k]	*34.4*	*10.1*	1.925
MPW1PW91/6-31G**[k]	*29.6*	*−5.6*	1.996
MPW1PW91/cc-pVDZ[k]	*31.8*	*−0.1*	1.949
experimental[k]	30.1	7.8	

[a] Energies of the transition state **39** and product **40** relative to reactant **38**. Relative enthalpies at 298 K are given as normal type, and relative electronic energies are given in italics.
[b] cd distance in the transition state.
[c] Ref. 133.
[d] Ref. 137.
[e] Ref. 134.
[f] Ref. 135.
[g] Ref. 136.
[h] Ref. 139.
[i] Ref. 138. The composite enthalpy is defined as BCCD(T)/cc-pvDZP + [CCSD/cc-pVDZ − CCSD/ccpvDZ].
[j] Ref. 140.
[k] Ref. 132.

is 5.4 ± 2.0 kcal mol^{-1}, just below the error limits of the experimental result. They estimate that the activation energy is 25.5 kcal mol^{-1}, well below the corrected experimental value of 30.1 kcal mol^{-1}. Lindh[136] recently examined the reaction with CASPT2 (using a CAS(12,12) reference) optimized geometries and CAS(10,10) vibrational frequencies. The reaction enthalpy is very nicely reproduced (8.6 kcal mol^{-1}), with slightly worse agreement in the reaction enthalpy (25.0 kcal mol^{-1}).

At about the same time, Kraka and Cremer[137] reported a CCSD(T)/6-31G(d,p) study of the parent Bergman cyclization. The CCSD(T) optimized structures are shown in Figure 4.13. Using MP2/6-31G* frequencies with the CCSD(T) electronic energies, they find that $\Delta H(\mathbf{39} \rightarrow \mathbf{41}, 298\,\text{K}) = 8.0$ kcal mol^{-1} and $\Delta H^{\ddagger}(\mathbf{39} \rightarrow \mathbf{41}, 298\,\text{K}) = 28.5$ kcal mol^{-1}, both in excellent agreement with experiment. The role of the triples configurations is key; they allow for correlation between an electron pair and a single electron. Without these contributions (i.e., CCSD), the barrier is 37.7 kcal mol^{-1} and, even worse, the reaction energy is +27.2 kcal mol^{-1}!

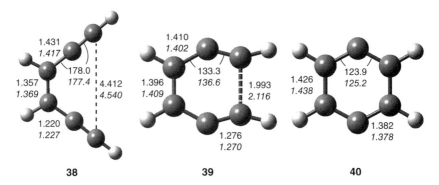

Figure 4.13 Optimized geometries of **38**, **39**, and **40**. CCSD(T)/6-31G(d,p)[137] results in normal text and (U)BPW91/cc-pVDZ[138] results are in italics. All distances are in angstroms and all angles are in degrees.

Cramer[138] has criticized these CCSD(T) results on a number of issues. First, he argues that MP2 frequencies for **40** are likely to be very poor, for reasons we will discuss later in Section 4.4, which deals with the nature of the benzynes. If BPW91/cc-pVDZ frequencies are used with the CCSD(T)/6-31G(d,p) energies, the reaction endothermicity is 5.8 kcal mol^{-1}, still in quite respectable agreement with experiment. Next, Cramer points out that the HF reference does not dominate the CCSD wavefunction, rather $\langle \Psi_{CCSD} | \Psi_{HF} \rangle = 0.520$. This is much lower, for example, than the overlap value for ozone of 0.7–0.8, a molecule possessing significant multireference character. This results in the CCSD(T) energy being inordinately sensitive to geometry since the triples contribution is performed in a perturbative sense. Cramer suggested the usage of Brueckner orbitals, which remove the singles contribution to the CC expansion. The resulting BCCD(T) energies are much less sensitive to geometry. He proposed a composite method to combine the BCCD(T) energy with some accounting of the basis set effect.

$$E_{comp} = BCCD(T)/cc\text{-}pVDZ + [CCSD/cc\text{-}pVTZ - CCSD/cc\text{-}pVDZ] \quad (4.4)$$

This composite approach gives $\Delta H(\mathbf{38} \rightarrow \mathbf{40}, 298\,K) = 11.0$ kcal mol^{-1} and $\Delta H^{\ddagger}(\mathbf{38} \rightarrow \mathbf{40}, 298\,K) = 27.7$ kcal mol^{-1}.

The earliest attempts at using DFT to obtain the energetics of the parent Bergman cyclization found the hybrid methods wanting. Schreiner[139] compared B3LYP with BLYP in their ability to describe some properties of **40**. While both density functionals are far from ideal, the hybrid B3LYP method performs more poorly. Chen and coworkers[140] used B3LYP, BP86, BLYP, and BPW91 to determine the activation and reaction energies for the Bergman cyclization (Table 4.9). B3LYP drastically overestimates both the barrier and overall reaction energies. BLYP overestimates the reaction energy. Both BP86 and BPW91 provide reasonable estimates. Cramer[138] noted that the spin-unrestricted DFT energy is always lower than the spin-restricted energy for **40**, a problem for both of these previous studies that employed restricted solutions only. Use of the UDFT geometry and energy suggests that the reaction is nearly thermoneutral. The UBPW91/cc-pVDZ geometry of **40** is shown in Figure 4.13.

In 2000, Kraka and coworkers[132] presented a large study on DFT methods applied to the Bergman cyclization, which clarified a number of important issues. They compared a local spin density (LSD) functional (SVWN), three generalized gradient approximation (GGA) functionals (BP86, BPW91, and BLYP), and three hybrid functionals (B3LYP, B3PW91, and mPW91PW91). For **40**, the spin-restricted solution is unstable with respect to the unrestricted solution with *all* of the functionals. In contrast, the spin-restricted solution for the reactant **38** and the transition state **39** is stable with all the functionals. Therefore, UDFT must be employed to describe p-benzyne in order to get a reasonable geometry and vibrational frequencies. The LSD and GGA functions produce more stable solutions than the hybrid functions, though again *all* are unstable for **40**. The LSD and GGA functionals underestimate the activation energy, a situation not realized earlier because of the comparison to the inappropriate experimental values. The hybrid methods, previously discounted because of too high energy values, give the best results. In fact, the best results are obtained using RB3LYP for **38** and **39** and UB3LYP (correct for spin contamination) for **40** with the 6-311+G(3df,3pd) basis set: $\Delta H(\mathbf{38}\rightarrow\mathbf{40}, 298\text{ K}) = 8.5 \text{ kcal mol}^{-1}$ (experiment: 8.5 kcal mol^{-1}) and $\Delta H^{\ddagger}(\mathbf{38}\rightarrow\mathbf{40}, 470\text{ K}) = 29.9 \text{ kcal mol}^{-1}$ (experiment: 28.2 kcal mol^{-1}).

To summarize, with sufficient care, the energetics of the Bergman cyclization can be accurately reproduced by theory. The basis set must include polarization functions, preferentially on all atoms, and at least a DZ representation of the valence space. Larger basis sets (TZ or bigger, multiple polarization functions, and higher angular functions) are warranted if very accurate energies are desired. If a CASSCF approach is taken, the active space must include all π-electrons, and followed by accounting for some dynamical correlation, for example, CASPT2. A CC approach must include some accounting of triples configurations. Finally, if a DFT approach is employed, UDFT must be used for **40**. GGA functionals are appropriate though hybrid functionals may give slightly better energies.

The geometry of p-benzyne will be discussed later in Section 5.5. The geometry of the reactant **38** can be adequately modeled by a large variety of methodologies.[141] One interesting feature is that the two alkyne units bend slightly away from each other, in opposite direction of the reaction coordinate for the Bergman cyclization. The geometry of the TS **39** is dependent on the methodology. Listed in Table 4.9 is the critical geometric parameter of the TS, which is the distance between C_1 and C_6 and is also referred to as the *cd* distance. Using CASSCF(12,12) to optimize the structure, the *cd* distance decreases as the basis set improves: 1.923 Å with DZP, 1.896 Å with TZP, and 1.895 Å with ANO. It is slightly longer when optimized at CCSD(T)/6-31G*: 1.993 Å. LSD and GGA DFT optimizations provide even longer *cd* distances, always over 2 Å regardless of method or basis set, up to 2.178 Å using SVWM/6-31G**. Hybrid methods provide *cd* distances between the CASSCF and the CCSD(T) values, for example, 1.925 Å at B3LYP/cc-pVDZ.

It is interesting to note that stability problems occur in computing the product **40** but not for the TS **39**. The instability arises from the need for multiple configurations to adequately describe the electron occupancy. Another viewpoint is that **40** has significant diradical character, which requires a multiconfiguration description. Based on the occupation numbers of the natural orbitals from their CCSD(T) computations, Kraka and Cremer[137] find that while *p*-benzyne has significant diradical character (about 65 percent), both the reactant and TS have essentially no diradical character at all. Lindh and coworkers used two different measures of the diradical character of the CASSCF(12,12) wavefunctions. First, they compare the excess electron occupation of the symmetric representation of the in-plane π-orbitals (which includes the σ radical orbitals of **40**). A pure diradical will have an excess occupation of one electron. The excess occupation in **40** is 0.65, but it is only 0.201 in the TS **39**.[134] Alternatively, they examined the relative percentage of the two configurations that describe the two radical electrons in the two available orbitals. For the reactant **38**, the first configuration's weight is 84 percent while the second does not contribute at all. At the TS, the weights are 73.3 percent and 8.8 percent, while in the product, they are 56.6 percent and 27.6 percent.[135] Similar weights are reported by Cramer.[138] All of these measures come to the same conclusion: that while there is considerable diradical character in *p*-benzyne, there is very little diradical character in the TS. Since the TS has a little diradical character, it has less need for a multiconfigurational description. This result also suggests an early transition state, with the diradical character being generated rather late along the reaction pathway.

A couple of studies have addressed the degree of aromatic character of the Bergman cyclization transition state using the NICS criteria. Both studies find large negative NICS values for **39** (−18.8, −17.9, and −19.5 ppm, depending on computational method), indicating significant aromatic character. However, the first study, using a valence bond method, argued that it is due to σ-aromaticity rather than π-aromaticity.[142] This argument is rejected in the second study that partitions the NICS contributions into σ and π contributions.[143] The π contribution is negative (−15.1 ppm) and dominates the small positive σ contribution.

4.3.3 The *cd* Criteria and Cyclic Enediynes

Nicolau developed the *cd* reaction criteria based on two observations. First, based on MM2 structures, the *cd* distance in calichaemicin is 3.35 Å and the molecule is stable. However, after activation to give **33** (Figure 4.11), the distance contracts to 3.16 Å and cyclization occurs spontaneously.[129] Second, Nicolaou and coworkers[130] prepared a series of cyclic enediynes **45a–h**. Their *cd* distances, as estimated by MM2, and their stabilities are shown in Table 4.10, along with a few related compounds. (It is important to mention that MM2 grossly underestimates the *cd* distance in the parent enediyne **38**: MP2 estimates that this distance is 4.12 Å, it is 4.38 Å at CCSD(T)/cc-pVTZ, and the microwave experimental value is 4.32 Å.[141]) Their conclusion, the *cd* criteria, is that molecules having *cd* less than 3.2 Å will cyclize spontaneously, while those with *cd* greater than 3.3 Å are

THE BERGMAN CYCLIZATION

45a-h, n = 1–8

TABLE 4.10 cd Distances and Stabilities of Cyclic Enediynes[a]

Compound	n	cd (Å)	Stability
45a	1	2.84	Unknown
45b (= 44)	2	3.25	$t_{1/2} = 18$ h at 25 °C
45c	3	3.61	Stable at 25 °C
45d	4	3.90	Stable at 25 °C
45e	5	4.14	Stable at 25 °C
45f	6	4.15	Stable at 25 °C
45g	7	4.33	Stable at 25 °C
45h	8	4.20	Stable at 25 °C
(diol)		3.20	$t_{1/2} = 11.8$ h at 37 °C
(diol)		3.29	$t_{1/2} = 4$ h at 50 °C
(diol)		3.34	$t_{1/2} = 2$ h at 50 °C
(carbonate)		3.42	Stable at 25 °C

[a] From Ref. 144.

stable. This argument implies that the critical feature that initiates (or precludes) Bergman cyclization is the distance between the ends of the two alkynes; if they are close enough together, the reaction occurs, otherwise the enediyne is stable.

An alternative proposal put forward by Magnus[145−147] and Snyder[148,149] is that the amount of strain energy relief afforded by the transition state dictates the relative stabilities of the enediynes. For example,[147] **46** and **47** have similar *cd* distances, with the distances slightly longer in the former (3.391 Å) than in the latter (3.368 Å). Yet, **46** cyclizes 650 times faster than **47**, corresponding to $\Delta\Delta G^{\ddagger}(\mathbf{47}-\mathbf{46}) = 5.1 \pm 0.2$ kcal mol^{-1}. Semiempirical calculations estimate that **46** experiences a drop of about 6 kcal mol^{-1} in going from reactant to transition state, while **47** experiences a net gain in strain energy of 1.5 kcal mol^{-1}.

Kraka and Cremer[137] addressed the *cd* criteria by performing CCSD(T)/6-31G(d,p) constrained optimizations of **38**. When *cd* is 3.3 Å, the reaction becomes thermoneutral and the barrier is lowered by 6 kcal mol^{-1} and further shortening of *cd* to 3.0 Å reduced the barrier by another 3 kcal mol^{-1}. Using CASPT2, Lindh and Persson[134] find similar energetic consequences to shortening the *cd* distance, seemingly in agreement with the *cd* criteria. However, they note that these reductions in activation barriers still leave sizable barriers such that reaction kinetics are much slower than for the natural enediynes. They suggest that the *cd* distance is insufficient to predict reaction rates.

Due to the large size of the cyclic enediynes presented in Table 4.10, computational chemists were forced to use DFT methods in order to study their activation barriers. Schreiner[139] employed BLYP/6-311+G**//BLYP/6-31G* and Chen[140] used BPW91/6-311G**. Both did not appreciate the need for using spin-unrestricted solutions for the *p*-benzyne analogs, and so we will discuss just the activation barriers here. BLYP gives longer *cd* distances and higher barriers than BPW91; however, the trends within each set of results are identical.

Methyl substitution at the termini of **38**, giving (3Z)-3-heptene-1,5-diyne **48** and (4Z)-4-octene-2,6-diyne **49**, only slightly increases the *cd* distance, yet the activation barrier increases by 3–4 kcal mol^{-1} with each methyl addition (Table 4.11).

TABLE 4.11 DFT-Computed *cd* Distances and Activation Enthalpies for Acyclic and Cyclic Enediynes[a]

Compound	cd	$\Delta H^{\ddagger}$
38	4.548	28.4
	4.512	*24.33*
48	4.522	27.07
49	4.571	29.9
	4.541	*30.99*
50	2.512	n.a.
51	2.636	n.a.
45a	2.924	16.3
	2.913	*11.56*
45b (=44)	3.413	25.0
	3.393	*20.21*
45c	3.588	31.9
	3.976	*25.72*
45d	4.353	40.3

[a]BLYP/6-311+G**//BLYP/6-31G*[139] in regular print and BPW91/6-311G**[140] in italics.

Schreiner argues that a methyl group stabilizes an sp carbon over an sp^2 carbon, thereby lowering the energy of the reactant. This is inconsistent with the *cd* criteria.

Schreiner was unable to find either transition states or products for the cyclization of the smallest cyclic enediynes, **50** and **51**. These are simply too strained. The barrier for **45a** is very low and should spontaneously cyclize at room temperature (Table 4.11). The barrier for **45b** is lower than that of **38** and is consistent with it slowly cyclizing at RT. These barriers are consistent with the *cd* criteria. However, the *cd* distance in **45c** (3.588 Å) is only slightly longer than in **45b** (3.413 Å), yet its barrier is 5–7 kcal mol^{-1} greater. Schreiner, therefore, claims that there is no straightforward correlation between *cd* and the activation barrier. Chen and coworkers note that the MM2 strain energies for **45a**, **45b**, and **45c** are 14.8, 11.4, and 9.0 kcal mol^{-1}, respectively. This accounts for about 40 percent of the activation barrier differences between the three compounds, indicating that the differential strain energy alone is not a complete answer. It appears that both models provide limited predictive capacity. An attempt to predict the activation energy as a cubic expansion about the *cd* distance shows only modest correlation.[150] Accurate barriers prediction still requires direct experimental or computational determination.

A theoretical study of the biological activity of dynamicin A (**52**) dealt in part with the question of the *cd* distance. Tuttle et al.[151] examined the Bergman cyclization of **52** to **53** and compared it to cyclization of activated dynamicin A (**54**) to its diradical product (**55**). Given the size of these molecules, they optimized the structures at the relatively small B3LYP/3-21G level, but it is important to note that this level predicts the barrier for the cyclization of the 10-member ring enediyne (**44**) with an error of less than 1 kcal mol^{-1}—clearly a situation of fortuitous cancellation of errors. Energies were then computed at B3LYP/6-31G(d). The activation enthalpy for the cyclization of **52** is 52.3 kcal mol^{-1}, while the barrier for cyclization of **54** is dramatically reduced to 17.9 kcal mol^{-1}. It is even much lower than the barrier for the cyclization of **44** (28 kcal mol^{-1}).

The epoxide ring of **52** forces the two adjoining cyclohexene rings into boat-like conformations. This conformation forces a very long *cd* distance of 3.540 Å, leading to the very large activation energy. However, when dynamicin A is activated, the epoxide ring is broken and the cyclohexane rings can adopt half-chair conformations. This results in a sharp contraction of the *cd* distance to 3.169 Å. An additional factor toward reducing the activation barrier for Bergman cyclization is that the enediyne ring in **54** is strained relative to **55** by about 11 kcal mol^{-1}. The accompanying strain relief in the transition state for the Bergman cyclization lowers the barrier in **54** relative to **55**.

Thus, dynamicin A is carefully tuned to act as a warhead toward DNA. Inactivated dynamicin A will not undergo the Bergman cyclization at body temperature because of its very high activation barrier. However, the activated form **54** has a barrier of only 17 kcal mol^{-1}, sufficiently low enough for efficient reaction within the body. In order to insure that the diradical **55** has sufficient lifetime to abstract protons from DNA, the retrocyclizations back to **54** or to **56** must have a barriers greater than 12 kcal mol^{-1}—the barrier for H-abstraction. These barriers are estimated as 23.5 and 25.6 kcal mol^{-1}, respectively, consistent with potential high biological activity.

4.3.4 Myers–Saito and Schmittel Cyclization

Unlike the other naturally occurring enediyne molecules, neocarzinostatin chromophore (**30**) is activated into an enyne-butatriene. This unusual cumulene **57** undergoes a cyclization analogous to the Bergman cyclization, proposed concurrently by Myers[152] and Saito.[153] This cyclization, now referred to as *Myers–Saito cyclization*, produces the diradical **58** that can abstract hydrogen atoms from DNA. Thus, neocarzinostatin's bioactivity is remarkably similar to calichaemicin (Figure 4.14).

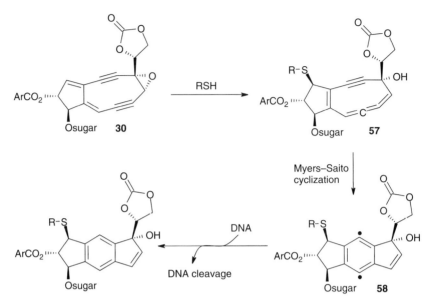

Figure 4.14 Mechanism of action of neocarzinostatin chromophore **30** Nicolaou, K. C.; Dai, W.-M. "Chemistry and biology of the enediyne anticancer antibiotics," *Angew. Chem. Int. Ed. Engl.* **1991**, *30*, 1387–1416. Copyright Wiley-VCH Verlag GmbH & Co. KGaA. Reproduced with permission.

In the mid-1990s, Schmittel[154,155] discovered an alternative reaction to the Myers–Saito cyclization for enyne-allenes. If an aryl group (R = Ph or tolyl) or a sterically bulky substituent (R = tBu or SiMe$_3$) is appended to the alkynyl terminus, then the cyclization is switched from the Myers–Saito pathway (making the C$_1$–C$_6$ bond) to a cyclization where the C$_2$–C$_6$ bond is formed, now termed the *Schmittel cyclization* (Figure 4.15). Interested readers are directed to Schmittel's excellent review, published in 2011.[156] Both the Myers–Saito[157] and Schmittel[158] cyclization reactions lead to molecules that can cleave DNA, suggestive of the formation of intermediate diradicals analogous to the mechanism of the Bergman cyclization.

Limited thermochemical data are available for these two cyclization variants. Myers[152] examined the thermolysis of (4Z)-1,2,4-heptatrien-6-yne (**59**) to give the diradical intermediate **61**, which then abstracts hydrogen atoms from 1,4-cyclohexadiene to give toluene. For this parent Myers–Saito cyclization, $\Delta H^{\ddagger} = 21.8 \pm 0.5$ kcal mol^{-1} and $\Delta S^{\ddagger} = -11.6 \pm 1.5$ eu. The Schmittel cyclization

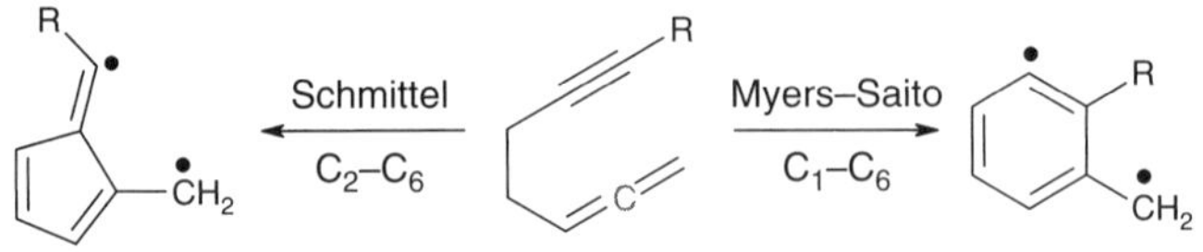

Figure 4.15 Myers–Saito versus Schmittel cyclization.

of **62**, to give the diradical intermediate **63**, which is then trapped by reaction with 1,4-cyclohexadiene, has $\Delta H^{\ddagger} = 24.9$ kcal mol^{-1} and $\Delta S^{\ddagger} = -5.3$ eu.[154]

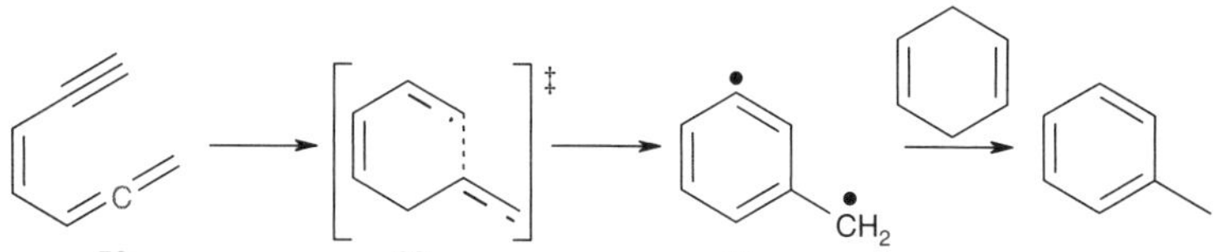

Computational methodology has been utilized to address a number of issues related to the Myers–Saito and Schmittel cyclizations. The first issue is the structure of the transition state. Unlike the product of the Bergman cyclization, where the two radical centers occupy σ-orbitals, the product of both the Myers–Saito and Schmittel cyclizations have one radical in a σ-orbital and one in a π-orbital. The exocyclic π-radical can conjugate with the ring, but this requires the terminal methylene group to rotate 90° in transforming from the reactant to the product. Koga and Morokuma[133] argued that this radical delocalization favors the Myers–Saito product over the Bergman product. Interestingly, the methylene group has not come into significant conjugation with the ring in the transition state (**60**) of the parent Myers–Saito cyclization (Figure 4.16). The methylene group is rotated from about 5° (CASSCF(10,10)) to 45° (DFT) in **60**.[159] On the other hand, the methylene group has almost completely rotated into conjugation in the transition state of the Schmittel cyclization (**64**). Both transition states are early, appearing more reactant-like than product-like (see Figure 4.16). Analysis of the CASSCF configuration weights indicates no contribution of the $|\ldots p_{\sigma}{}^{1}p_{\pi}{}^{1}|$ configuration in either transition state.[159] Furthermore, the restricted DFT solutions for **60** and **64** are stable.[159,160] All of this suggests very little, if any, diradical character in either transition state, just as is the case for the Bergman cyclization. The NICS values for the two transition states are negative, -9.3 in **60** and -5.7 in **64**, indicative of some aromatic character.[143]

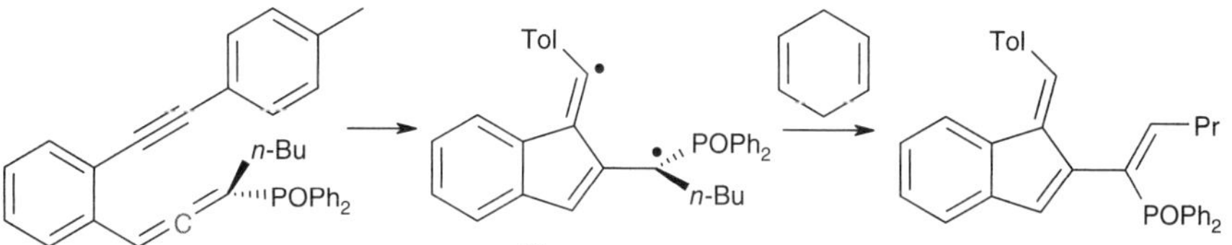

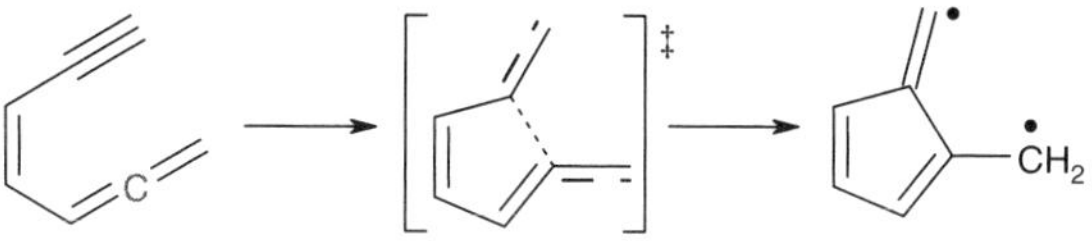

252 PERICYCLIC REACTIONS

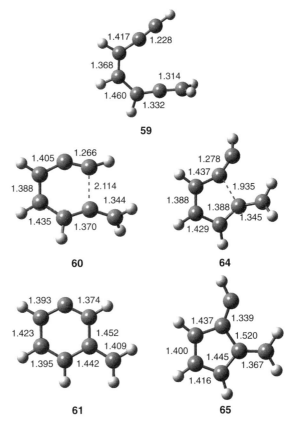

Figure 4.16 BPW91/6-31G* optimized structures of (4Z)-1,2,4-heptatrien-6-yne (**59**) and its Myers–Saito and Schmittel transition states and products. All distances are in angstroms.

The two possible cyclization reactions of (4Z)-1,2,4-heptatrien-6-yne (**59**) have been examined with a number of different computational techniques. Their resulting reaction energies are listed in Table 4.12. The trends in activation energy for the Myers–Saito cyclization mimic those seen with the Bergman cyclization. Namely, CASSCF predicts a barrier that is too large, but a CI computation using configurations based on the CASSCF solution gives a reasonable barrier.[133] CCSD also overestimates the barrier, but inclusion of triples configurations again results in a barrier in close agreement with experiment.[161] All of the DFT variants provide good barrier estimates, though B3LYP tends to be on the high side, with some authors claiming that this hybrid method should be avoided.[160–163] We will defer the discussion of the overall reaction energy to a later chapter (Section 5.5), where we discuss the benzynes and related diradicals, except to say here that restricted DFT is unsatisfactory, but unrestricted DFT produces reaction energies close to the empirical estimate of -15 kcal mol^{-1}.[160,161]

TABLE 4.12 Computed Energies[a] (kcal mol^{-1}) for the Prototype Myers–Saito (59 → 61) and Schmittel (55 → 61) Cyclizations

Method	Myers–Saito TS	Myers–Saito Product	Schmittel TS	Schmittel Product
CASSCF(4,4)/MIDI1[b]	*35.5*	*2.1*		
CAS(10,10)/MIDI4//CASSCF(4,4)/MIDI1[b]	*33.3*	*-1.2*		
MRSDCI[b]	*19.1*	*-0.7*		
CASSCF(10,10)/DZ+P[c]	29	4	37	18
MRCI//CASSCF(10,10)/DZ+P[c]	25	-21	35	12
B3LYP/6-31G*[c]	22	-17	31	9
B3LYP/6-31G*[d]	21.4		29.0	
BLYP/6-31G(d)[e]	23	-2	34	19
BLYP/6-31G(d)[e]	(20.2)	(-7.8)	(31.5)	(14.2)
BLYP/cc-pVDZ// BLYP/6-31G(d)[e]	(22.7)	(-2.1)	(34.4)	(19.3)
CCSD(T)/cc-pVDZ// BLYP/6-31G(d)[e]	23	-24	35	17
CCSD(T)/cc-pVDZ// BLYP/6-31G(d)[e]	(22.2)	(-24.3)	(35.0)	(17.3)
BCCD(T)/cc-pVDZ// BLYP/6-31G(d)[e]	23	-12	35	10
BCCD(T)/cc-pVDZ// BLYP/6-31G(d)[e]	(22.2)	(-11.9)	(34.8)	(10.0)
B3LYP/6-31G*[f]	22.0	-17.9	29.1	11.0
mPWPW91/6-31+G*[g]	19.8	-13.0	26.6	13.9
BPW91/6-311G**[h]	18.3	-13.0		
CCSD/6-31G**//BPW91/6-311G**[h]	27.0	-15.3		
CCSD(T)/6-31G**//BPW91/6-311G**[h]	21.8	-12.1		
BP86/6-311G**[h]	17.7	-12.5		
BLYP/6-311G**[h]	21.2	-5.8		
B3LYP/6-311G**[h]	24.6	-11.9		
CASSCF(10,10)/6-31G(d)[i]	*31.1*	*-3.1*	*39.1*	*22.1*
MRMP2/6-31G(d)//CASSCF(10,10)/6-31G(d)[i]	*17.4*	*-15.0*	*24.0*	*10.9*
MRMP2/6-311+G(d,p)//CASSCF(10,10)/6-31G(d)[i]	*16.9*	*-14.0*	*22.9*	*10.5*

[a] For the Myers–Saito reaction, energies are of the transition state **60** and product **61** relative to reactant **59**. For the Schmittel reaction, energies are of the transition state **64** and product **65** relative to reactant **59**. Relative enthalpies at 298 K are given as normal type, relative electronic energies are given in italics, and relative Gibbs free energies at 298 K are given within parentheses.
[b] Ref. 133.
[c] Ref. 159.
[d] Ref. 166.
[e] Ref. 160.
[f] Ref. 162.
[g] Ref. 163.
[h] Ref. 161.
[i] Ref. 167.

As an aside, Carpenter[164] suggested that the Myers–Saito cyclization of **59** may proceed through a nonadiabatic mechanism. Based on the experiments and computations, Carpenter proposed that a nonadiabatic transition from the ground-state singlet to an excited-state singlet occurs after the transition state, creating a zwitterionic form of the intermediate. While these results certainly complicate the mechanism, it does not significantly color the discussion of the competition between Myers–Saito and Schmittel cyclizations since the surface hopping occurs in the post-transition state. Of more concern is the work of Singleton,[165] which implicates nonstatistical dynamic behavior in the Schmittel cyclization.[156] Further discussion of this nontraditional mechanism is deferred to Chapter 8.

In the experimental thermolysis of **59**, no Schmittel product was detected. Computational estimates for the activation enthalpy for the Schmittel cyclization of **59** range from 31 to 35 kcal mol^{-1}, significantly higher than the barrier for the Myers–Saito cyclization of 20–22 kcal mol^{-1}. Furthermore, the Schmittel cyclization is predicted to be endothermic ($\Delta H = +10 - +19$ kcal mol^{-1}), while the Myers–Saito cyclization is exothermic. Therefore, the Myers–Saito cyclization of **59** is both thermodynamically and kinetically favored over the Schmittel reaction.

The switch from Myers–Saito to Schmittel cyclization occurs with phenyl or bulky groups attached to the alkynyl position of the enyne-allene. Engels[166] compared the Myers–Saito and Schmittel cyclization for the parent reaction with R=H, phenyl, and *t*-butyl (Figure 4.15) at B3LYP/6-31G*. The activation enthalpy for the Myers–Saito reaction increases from 21.4 kcal mol^{-1} when R=H to 26.7 kcal mol^{-1} for R=phenyl, and 27.9 kcal mol^{-1} for R=*t*-butyl. Conversely, the phenyl group decreases the activation enthalpy for the Schmittel reaction from 29.0 to 25.1 kcal mol^{-1}. However, the activation enthalpy for the *t*-butyl substituted case is 31.4 kcal mol^{-1}. They argue that increasing steric demand gives rise to the increasing barrier for both cyclizations, but that the Myers–Saito cyclization is more sensitive to steric demands. The Schmittel cyclization is favored by the phenyl substituents due to mesomeric stabilization of the incipient benzylic radical. Phenyl has the same effect in switching the reaction pathway for the reactions of model eneyne-allenes **66**: the Myers–Saito barrier is increased while the Schmittel barrier is decreased by phenyl substitution (see Table 4.13).[168]

While these model studies have shed light on the nature of the Myers–Saito and Schmittel cyclizations, the molecules examined bear important differences from the neocarzinostatin chromophore: the chromophore is a butadiene not an allene; the enyne-butatriene is embedded within a nine-member ring; this ring is fused with a cyclopentane moiety; and the ring bears an alcohol substituent.

Schreiner and Prall[160] examined a series of cyclic enyne-allenes (**67**) and find that ring strain lowers the barrier for both the Myers–Saito and Schmittel cyclizations. The lowest barrier and most exothermic reaction is found for the Myers–Saito cyclization of the nine-membered ring. Cramer and Squires[169] examined the Myers–Saito cyclization of cyclonona-1,2,3,5-tetraen-7-yne (**68**) using the BD(T)/cc-pVDZ method. The activation enthalpy for **68** is about 10 kcal mol^{-1} less than for **59**, which they attribute to ring strain. They note that the cyclization of **68** leads to a σ,σ-diradical and not the σ,π-diradical formed from cyclization of **59**.

TABLE 4.13 Activation Free Energies (kcal mol^{-1}) for the Cyclizations of Model Enyne-Allenes 66[a]

n	R	Myers–Saito $\Delta G^{\ddagger}$	Schmittel $\Delta G^{\ddagger}$
1	H	25.0	34.1
2	H	22.3	27.8
3	H	20.6	26.6
1	Ph	30.9	30.3
2	Ph	28.1	24.2
3	Ph	26.1	22.2

[a]Ref. 168.

This σ,σ-diradical lacks the benzylic radical stabilization found in **61**, and therefore cyclization of **68** is less exothermic than that of **59**. Musch and Engels[170] note that the Schmittel cyclization ($\Delta G^{\ddagger}$(CCSD(T)/cc-pVDZ) = 18.6 kcal mol^{-1}) of **68** is favored over the Myers–Saito cyclization ($\Delta G^{\ddagger}$(CCSD(T)/cc-pVDZ) = 21.1 kcal mol^{-1}). This is opposite to the case for the open chain analog **69**, where the barrier for Myers–Saito cyclization is 9.7 kcal mol^{-1} below the barrier for the Schmittel cyclization. However, fusing a cyclopentane ring with the eneyne-butatriene (**70**) favors the Myers–Saito cyclization over the Schmittel cyclization, $\Delta G^{\ddagger}$(UB3LYP/6-31G(d)) = 22.2 versus 22.8 kcal mol^{-1}. Since neocarzinostatin chromophore follows the Myers–Saito pathway exclusively, nature has carefully balanced many factors in creating this system.

In an analogy to the relationship between Myers–Saito and Schmittel cyclization, Schreiner[171] proposed another cyclization path of an endiyne, C_1–C_5, as an alternative to the C_1–C_6 Bergman cyclization (Figure 4.17). For the case where the terminal group is a hydrogen atom (**38** → **40**), the barrier is 27.1 kcal mol^{-1} with a reaction enthalpy of 8.3 kcal mol^{-1} at BCCD(T)/cc-pVDZ. However, the reaction to the fulvene diradical (**38** → **72**) is noncompetitve with an activation barrier of 42.1 kcal mol^{-1} and reaction enthalpy of 39.6 kcal mol^{-1}. Interestingly, the transition state **71** has considerable diradical character and must be treated

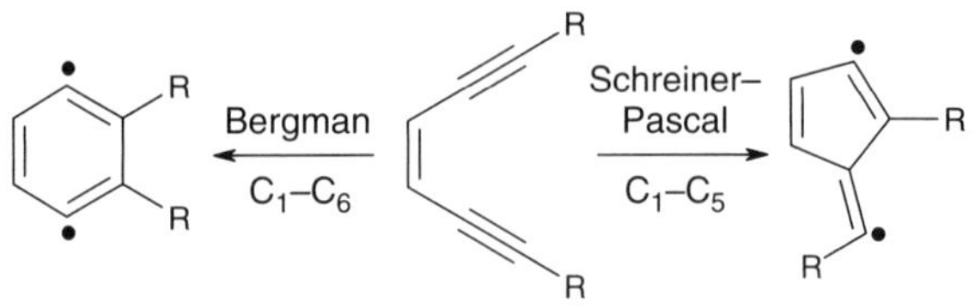

Figure 4.17 Bergman versus Schreiner–Pascal cyclization.

with a broken symmetry spin-unrestricted wavefunction, unlike in the Bergman cyclization.

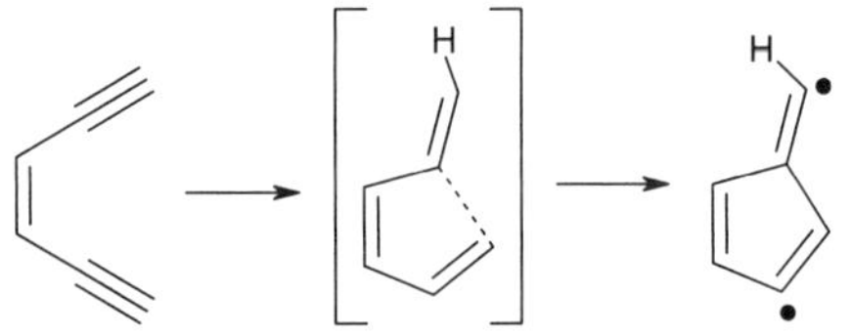

Pascal[172] proposed that terminal aryl substitution on the enediyne would both (1) inhibit the C_1–C_6 cyclization due to steric interactions and (2) enhance the C_1–C_5 cyclization due to stabilization of the radical by the neighboring aryl group. The RBLYP/6-31G(d) activation energies for both the C_1–C_6 and C_1–C_5 cyclizations of a series of analogs are listed in Table 4.14. Phenyl substitution does accomplish both suggestions: the activation barrier for the Bergman cyclization increases by 4 kcal mol^{-1}, while the barrier for the C_1–C_5 cyclization is lowered by nearly 6 kcal mol^{-1}. Further substitution of the phenyl ring by either chloro or methyl groups brings the barriers into near degeneracy, suggesting increased steric inhibition of the Bergman cyclization with the substitution.

For the diphenyl substituted case, BLYP favors the C_1–C_5 cyclization over the Bergman cyclization. This may be an overestimation as the BCCD(T)/cc-pVDZ//-BLYP/6-31G(d) computations predict the opposite energy ordering. However, if one corrects the BLYP energies for this overestimation, the C_1–C_5 cyclization of **73** remains favored over the Bergman cyclization. In fact, thermolysis of **74** does give only indene product, and thermolysis of **73** at 180 °C gives a small amount of the indene product.

4.4 BISPERICYCLIC REACTIONS

Cyclopentadiene **75** can dimerize by a Diels–Alder reaction, whereby each molecule could act as either the diene or the dienophile, that is, in a [4+2] or a [2+4] fashion (Scheme 4.10). One might therefore expect two isoenergetic transition states leading to these two products **76a** and **76b**. Instead, to their surprise, Caramella and coworkers[173] located a single transition state on the B3LYP/6-31G* PES. TS **77**, as shown in Figure 4.18, has C_2 symmetry. The

TABLE 4.14 RBLYP/6-31G(d) Activation Energies (kcal mol^{-1}) for C_1–C_6 and C_1–C_5 Cyclizations[a]

R	R'	$E_a(C_1-C_6)$	$E_a(C_1-C_5)$
H	H	24.6	37.2
Phenyl	H	28.7	31.4
2,6-Dichlorophenyl	H	30.8	31.6
2,6-Dimethylphenyl	H	30.5	30.9
Phenyl	Phenyl (**73**)	38.5	36.3
		(32.9)[b]	(35.1)[b]
2,4,6-Trichlorophenyl	2,4,6-Trichlorophenyl (**74**)	43.2	38.7

[a]Ref. 172.
[b]Computed at BCCD(T)/cc-pVDZ//-BLYP/6-31G(d).

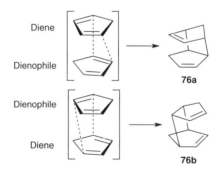

Scheme 4.10

intrinsic reaction coordinate (IRC) pathway forward maintains this symmetry and leads to the Cope TS **78**, also shown in Figure 4.18.

How does one understand this reaction? At first look, the reaction path looks like it connects one TS (**77**) to a second TS (**78**). In fact, the reaction trajectory follows the IRC downward from **77**, maintaining C_2 symmetry until it reaches the valley–ridge inflection (VRI) point. At this point, a vibrational mode perpendicular to the path inflects, meaning that it goes from having positive curvature to having negative curvature. In other words, the path changes from following a valley to suddenly being along the crest of a ridge. The trajectory will bifurcate after this point since moving in the direction perpendicular to the ridge will mean a more rapid decline in energy. The VRI is the origin of two diverging paths, one leading to the well associated with **76a** and one

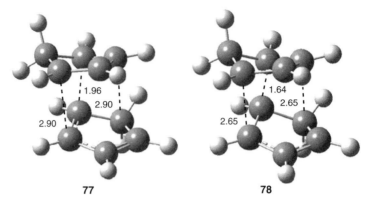

Figure 4.18 Bispericyclic TS **77** and Cope TS **78** for the dimerization of cyclopentadiene **75**.

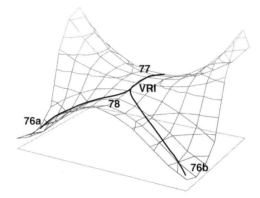

Figure 4.19 Bifurcating PES for the dimerization of cyclopentadiene. Heavy black lines are two trajectories that lead from the Diels–Alder bispericyclic TS **77** to the valley–ridge inflection (VRI) point and then diverge, one into the well of product **76a** and one into the well of product **76b**.

into the well associated with **76b**. This bifurcating PES and two trajectories are shown in Figure 4.19, and we will encounter this type of surface again in Chapter 8.

Caramella called this kind of TS *bispericyclic*, where two independent TSs merge into one. The bispericyclic TS can be thought of as a secondary orbital interactions (SOIs) as defined by Salem[174] and Houk[175] (see **79**) taken to their extreme, that is, the secondary and primary interactions become identical in magnitude.

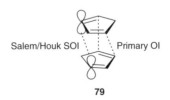

79

The dimerization of cyclopentadiene was studied by Houk[12] with a variety of computational methods. The bispericyclic TS was found using MP2, CASSCF, B3LYP, and MPW1K. Only with the HF method were two different TSs ([4+2] and [2+4]) located. Symmetric C_2 bispericyclic transition states were also located by Caramella for the dimerization of 1,3-butadiene[176] and cyclopentadienone.[177]

An asymmetric yet bispericyclic TS was found in the reaction of cyclopentadiene and cycloheptatriene, but this transition state is not the lowest energy pathway for either of the [4+2] routes.[178] A true asymmetric bispericyclic TS was located in the reaction of cyclopentadiene (**75**) and nitrostyrene (**80**), with an interesting twist concerning its regiochemical control.[179] Using B3LYP/6-31+G(d), a single bispericyclic transition state **81** connects reactants to the two [4+2] products through a VRI. The Cope TS **84** interconverts the two products **82** and **83**. TS **81** is geometrically closer to **82** than **83**, while TS **84** lies closer to **83** than **82**. This topology directs most molecules to traverse a path over **81** and on to **82**. What is most interesting in this study is that the acid-catalyzed reaction, using $SnCl_4$, shifts **81** toward **83** and **84** toward **82**, leading to the opposite product distribution. In fact, this is exactly what is experimentally observed: the uncatalyzed reaction gives **82** while the catalyzed reaction gives **83**.

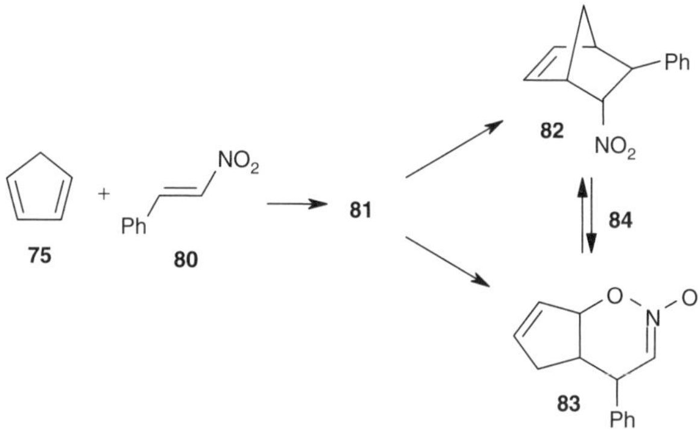

For the dimerization of 1,3-cyclohexadiene **84**, Houk found a concerted endo [4+2] TS (**85**) at B3LYP/6-31G*, along with a second equivalent TS **85'**.[180] These lead to the endo products **86** and **86'**, respectively. A Cope rearrangement interconverts the two products through TS **87**. There is a bispericyclic structure (**88**), but it is a SOSP. The three critical points **85**, **87**, and **88** are shown in Figure 4.20. While the geometries of the [4+2] TS **85** and the SOSP **88** are quite different, **85** is only 0.3 kcal mol^{-1} lower in energy than **88**. The region on the PES from **85** to **88** to **85'** is therefore quite flat, and Houk has termed this situation *quasibispericyclic*. Whether this flat region results in any interesting phenomena will require a molecular dynamics study similar to that to be discussed in Chapter 8.

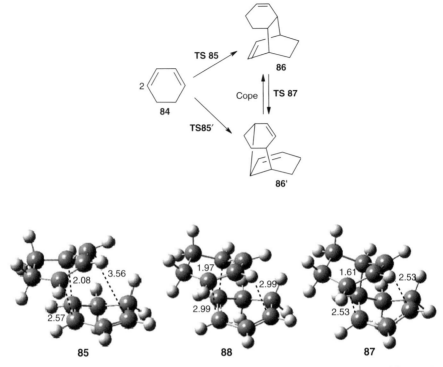

Figure 4.20 B3LYP/6-31G* geometries of the Diels–Alder TS **85**, the SOSP **88**, and the Cope TS **87**.

4.5 PSEUDOPERICYCLIC REACTIONS

Perhaps the most important consequence of the Woodward–Hoffman rules is its predictions of allowed and forbidden reactions. In particular, reactions involving $4n$ ($4n + 2$) electrons are allowed if there are an odd (even) number of antarafacial two-electron components. A consequence of this symmetry property is that changing the number of electrons will alter whether the reaction is allowed or forbidden.

Pericyclic reactions are the ones where the electrons rearrange through a closed loop of interacting orbitals, such as in the electrocyclization of 1,3,5-hexatriene (**88**). Lemal[181] pointed out that a concerted reaction could also take place within a cyclic array, but where the orbitals involved do not form a closed loop. Rather, a disconnection occurs at one or more atoms. At this disconnection, nonbonding and bonding orbitals exchange roles. Such a reaction has been termed *pseudopericyclic*.

The electrocyclization of 5-oxo-2,4-pentadienal (**89**) to pyran-2-one (**90**) is an example of a pseudopericyclic reaction. As shown in Figure 4.21, the disrotatory electrocyclization of **88** occurs with a closed loop of the p-orbitals in the transition state. On the other hand, the electrocyclization of **89** has no such closed loop. Rather, two orbital disconnections interchange the role of bonding and nonbonding orbitals.

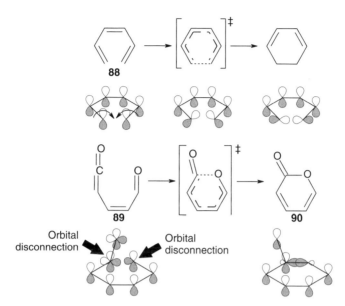

Figure 4.21 Active orbitals in the electrocyclization of **88** and **89**.

The highest six occupied molecular orbitals and the LUMO of **89** and **90** are displayed in Figure 4.22. If a planar transition state connects these two molecules, then the MOs along the reaction path are rigorously separated into those that are symmetric and antisymmetric with respect to the molecular plane. The symmetric orbitals cannot mix with the antisymmetric orbitals and therefore the electron occupancy of each set of MOs will not change. The reaction is allowed because of the orbital symmetry conservation.

Birney, making use of experiments and ab initio and DFT computations, has been the principal champion of pseudopericyclic reactions. He examined the electrocyclization of **89** and could find a transition state only at the HF level; at MP2, all attempts to locate a stable structure of **89** failed, collapsing directly to **90**.[182] This was later confirmed with a B3LYP/6-31G** study that also failed to locate a barrier for this reaction.[183] This lack of a barrier (or more generally, a small barrier) for a pseudopericyclic reaction is not unique.

Cycloaddition reactions can also be pseudopericyclic. Birney examined a number of these and a few examples involving the reactions of formylketene (**91**) are covered here. Formylketene reacts with alcohols to produce β-ketoesters from the enols **92**. Birney[184] examined the model reaction of formylketene with water (Reaction 4.6). The reactants first come together to form a hydrogen-bonded complex (**93**) before passing though the transition state **94** to give the enol product **95**. The activation barrier, defined as the energy for the reaction **93** → **94**, is 6.4 kcal mol^{-1} when computed at MP4(SDQ)/6-31G*//MP2/6-31G* + ZPE. This is a small barrier, much smaller, for example, than for a typical Diels–Alder reactions. The geometry of the transition state is quite unusual: the atoms involved in the formal bond making/breaking process are nearly coplanar. This can be seen in Figure 4.23

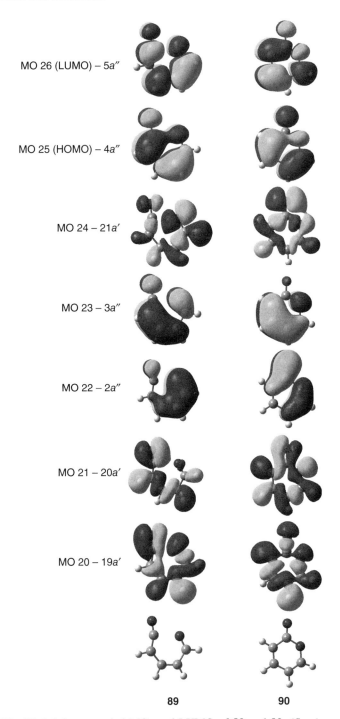

Figure 4.22 High lying occupied MOs and LUMO of **89** and **90**. (*See insert for color representation of this figure.*)

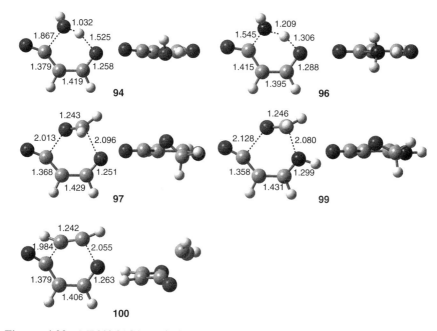

Figure 4.23 MP2/6-31G* optimized geometries of transition states for Reactions 4.6–4.10.

where top and side views of **94** are drawn. Transition states of pericyclic cycloadditions are generally nonplanar in order to maximize the orbital overlaps between the reacting fragments. The planar TS is emblematic of a pseudopericyclic reaction, where orbital disconnections at the sp C and the formyl O permit an in-plane interaction with the incoming water molecule.

(Reaction 4.6)

(Reaction 4.7)

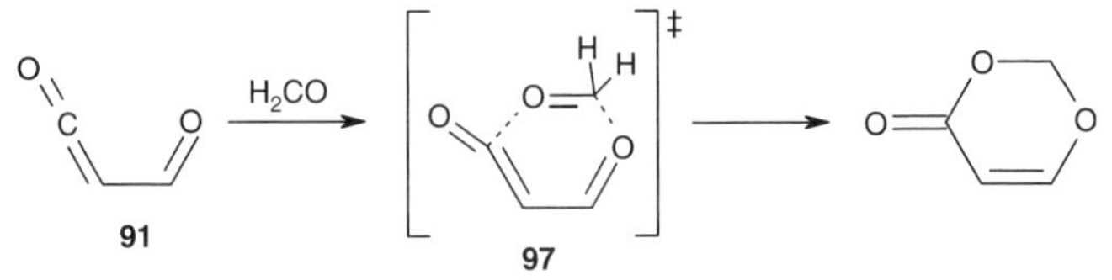

264 PERICYCLIC REACTIONS

(Reaction 4.8)

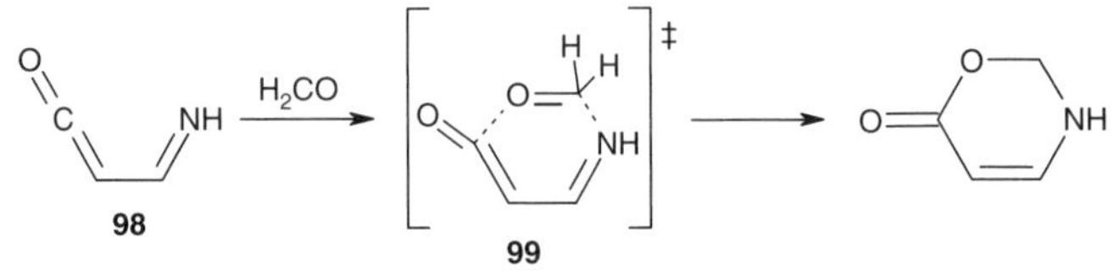

(Reaction 4.9)

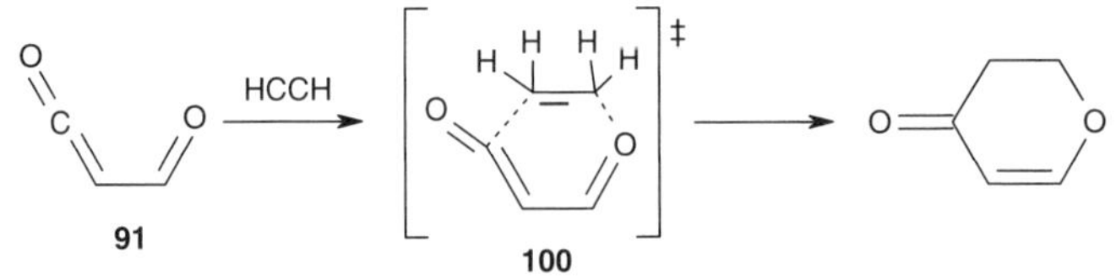

(Reaction 4.10)

The reaction of formylketene with ammonia (Reaction 4.7) follows an analogous pathway as the reaction of formylketene with water. The transition state **96** lies only 1.1 kcal mol^{-1} above the complex at MP2/6-31G* and the nonhydrogen atoms are coplanar.[185] The reaction of **91** with formaldehyde (Reaction 4.8) proceeds without a complex. The reaction barrier is 10.9 kcal mol^{-1} and the transition state (**97**) has nearly coplanar heavy atoms.[184] In the analogous reaction of imidoylketene **98** with formaldehyde (Reaction 4.9), the barrier is slightly lower than in the former case (10.6 kcal mol^{-1}), but both have nearly planar transition states.[186] The transition states for Reactions 4.6–4.9 are drawn in Figure 4.23. Their planar or nearly planar geometry is in striking contract to the transition state for the reaction of **91** with acetylene (Reaction 4.10).[187] This transition state **100** is decidedly nonplanar, with the dienophilic acetylene fragment attacking from above the plane of **91**. This arrangement is like that of normal pericyclic cycloaddition reactions, though the attack is from a shallower angle than what is typically seen. **100** lies 28.3 kcal mol^{-1} above the reactants, a barrier typical of Diels–Alder reactions. Pseudopericyclic reactions, exemplified by Reactions 4.6–4.9, are characterized by their planar transition states and low activation barriers.

An important consequence of the pseudopericyclic mechanism is that the planar (or nearly planar) transition states preclude orbital overlap between the σ- and π-orbitals. This implies that *all* pseudopericyclic reactions are allowed. Therefore, π-electron count, which dictates whether a pericyclic reaction will be allowed or forbidden, disrotatory, or conrotatory, is inconsequential when it comes to pseudopericyclic reactions. Birney[188] demonstrated this "allowedness" for all pseudopericyclic reactions in the study of the electrocyclic reactions 4.11–4.14.

Carbene **101** is expected to be very unstable due to its antiaromatic character and large ring strain. In fact, it could not be located and an energy scan performed by holding the breaking C–O bond fixed at a series of distances reveals a smoothly declining curve with no barrier. Importantly, along this path the $H_1-C_1-C_2-O$ dihedral angle remains close to 180°. A similar situation occurs with carbene **103**; it too is antiaromatic and strained. All attempts to optimize its geometry led to the ring-opened product. A constrained energy scan indicates a barrierless ring opening. Again, the geometry about the breaking C–O bond is planar. The ring opening of **102** occurs with a barrier of 1.2 kcal mol^{-1} (B3LYP/6-31(d,p)) to 6.1 kcal mol^{-1} (MP2/6-311G(d,p)//MP2/6-311G(d)). Furthermore, the transition state is planar. All three systems, involving differing number of electrons, express properties of pseudopericyclic reactions: planar transition states with low energy barriers. This is in sharp contrast to the electrocyclic reaction of **104**. The transition state is nonplanar (the $H_1-C_1-C_2-C_3$ and $C_4-C_5-C_6-H_6$ dihedral angles are 147.9° and 77.3°, respectively) and conrotatory motion is seen at both C_1 and C_6. The electrocyclic opening of Reaction 4.14 is clearly a pericyclic reaction and distinctly different than the other three reactions.

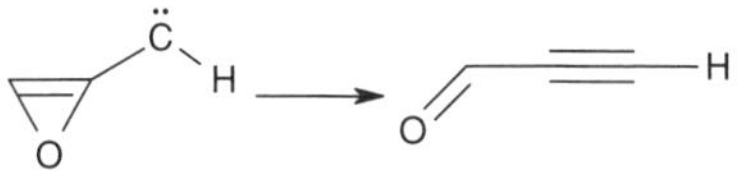

(Reaction 4.11)

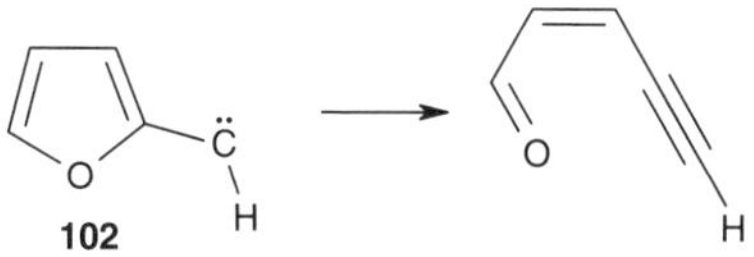

101

(Reaction 4.12)

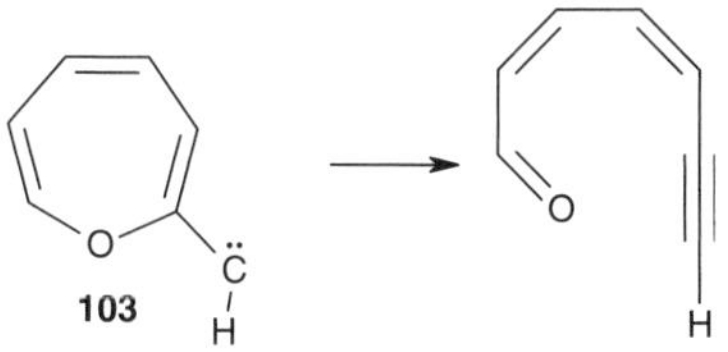

102

(Reaction 4.13)

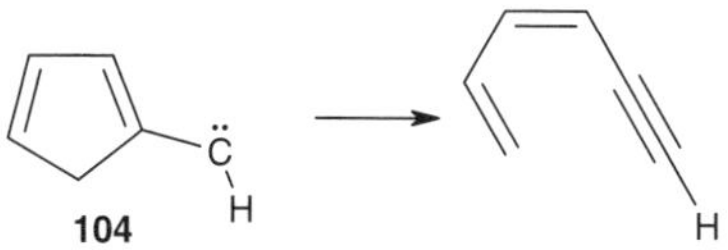

103

(Reaction 4.14)

104

Using CASSCF/6-31G* computations, Duncan[189] examined Reactions 4.15–4.17. The transition state of Reaction 4.15 is much more planar than is typical for a [3,3]-rearrangement. Dihedral angles are about 20° in the TS for Reaction 4.15, while the dihedral angles in the TSs of the other two reactions are about 50°. This is consistent with Birney's contention that pseudopericyclic reactions have nearly planar TSs. Furthermore, the activation barrier for Reaction 4.15 is also quite small, 19.4 kcal mol^{-1}, much lower than for Reactions 4.16 (26.2 kcal mol^{-1}) and 4.17 (33.1 kcal mol^{-1}). Reaction 4.15 is decidedly psueudopericyclic in nature.

266 PERICYCLIC REACTIONS

X = O: (Reaction 4.15)
X = CH$_2$: (Reaction 4.16)
X = NH: (Reaction 4.17)

We end this discussion with a cautionary tale, one concerned with just what distinguishes a pericyclic reaction from a pseudopericyclic reaction. de Lera and Cossío[190] examined the electrocyclic reaction of 2,4,5-hexatrienal and related species (**105–107**) at B3LYP/6-31+G*. They argued that the cyclization of **105** and **106** occurs by a pseudopericyclic path while **107** undergoes a normal pericyclic reaction. The crux of their argument centers on the atomic motion in the imaginary frequency and magnetic properties of the TS. There is no N–H rotation about the C=N bond in the TS for the cyclization of **105**, while there is disrotatory motion in **107TS**. The NICS value for **107TS** is much more negative (−13.6) than for **105TS** (−6.8) or **106TS** (−7.8), indicative of less aromatic character.

105: X = O
106: X = NH
107: X = CH$_2$

Rodríguez-Otero and Caleiro-Lago offer an alternative take on the reactions of **105** and **106**.[183,191] They note that atomic motion in a frequency may be misleading and instead plot the H–X–C$_6$–C$_5$ dihedral angles along the IRC. The plots are quite similar for **106** and **107**, diverging only well past the TS, and, therefore, these proceed via the same pathway. de Lera and Cossío,[192] in contrast, use the IRC to suggest that there is no disrotatory motion in the cyclization of **106**. Rodríguez-Otero and Caleiro-Lago compare the NICS values for the TSs to those of other aromatic TSs and ground states (such as benzene where the NICS value is −9.7) and argue that **105TS** and **106TS**, while having somewhat smaller values are still comparable with other aromatic systems. de Lera and Cossío correctly point out that NICS values can only be compared among closely related systems: TSs with other TSs and systems with the same ring size. A more compelling argument is that the variation in the NICS value along the IRC is similar for the reactions of **105–107**, but is quite different than for the reaction of **89**. Rodríguez-Otero and Caleiro-Lago's last, and perhaps strongest, argument is that **105TS** and **107TS** are not planar. The sum of the dihedral angles about the ring in the TS for the cyclization of **89** is zero, as expected for a pseudopericyclic reaction. In **107TS**, a pericyclic reaction involving a nonplanar TS, this sum is 129.8. The sum is 100.3 in **105TS** and 86.9 in **106TS**, more planar than for **107TS** but still far from planarity. While it is difficult to conclusively come down in favor of one side or another in this matter, perhaps the answer lies somewhere in the middle. For reactions of this type, both pericyclic and pseudopericyclic reactions are possible, and the appropriate electronic configuration exits for each path. Rather than choosing one configuration or the other, the transition state is best described as a mixture of these two configurations, expressing both pericyclic and pseudopericyclic characteristics to some extent (Birney, D. M., personal communication). In fact, one might argue that transition state **100** is dominated by pericyclic character, but has some pseudopericyclic character as well, as seen by the shallow attack angle of the acetylene fragment. In effect, what we have is a continuum from pericyclic to pseudopericyclic character, analogous to the S_N1 to S_N2 continuum for nucleophilic substitution reactions.[193] The cyclization of **105** and **106** thus expresses attributes of both pericyclic and pseudopericyclic mechanisms. B3LYP and CASSCF(10,9) computations performed by Duncan,[194] in fact, led him to conclude that the reaction of **106** is "neither purely pericyclic nor pseudopericyclic."

4.6 TORQUOSELECTIVITY

The conservation of orbital symmetry dictates that electrocyclic reactions involving $4n$ electrons follow a conrotatory pathway while those involving $4n+2$ electrons follow a disrotatory pathway.[1] For each case, two different rotations are possible. For example, 3-substituted cyclobutenes can ring open via two allowed conrotatory but diastereomeric paths, leading to *E*- or *Z*-1,3-butadienes, as shown in Scheme 4.11. Little attention was paid to this fact until Houk and coworkers developed the theory of torquoselectivity in the mid-1980s. They defined torquoselectivity as the preference of one of these rotations over the other.

Actually, the first example of torquoselectivity dates back to the 1960s, though it was not identified using this terminology. The solvolysis of cyclopropyl halides and tosylates proceeds via a two-electron electrocyclic reaction to yield allyl cation.[195,196] While two disrotatory pathways are allowed, only one occurs: when substituents are cis to the leaving group, they rotate inward, while trans substituents rotate outwards (Scheme 4.12). The favored rotation allows for the breaking $\sigma(C_2-C_3)$ bond to overlap with the $\sigma^*(C_1-X)$ orbital, thereby assisting in the simultaneous ejection of the leaving group and opening of the ring (Scheme 4.13).[197]

Surprisingly, the ring opening of cyclopropanes had not been studied using computational techniques until 2004. Using B3LYP/6-311++G** (and the SKBJ psuedopotential for bromine), de Lera[198] examined a series of diastereomeric substituted bromomethylcyclopropanes (**108** and **109**). Diastereomers **108** ring open with the substituents moving inwards; the outward rotation activation energy is 10–17 kcal mol^{-1} greater than the inward activation energy (see Table 4.15). The exception

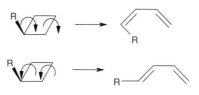

Scheme 4.11

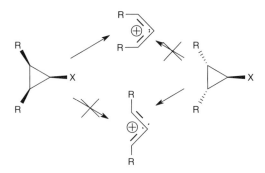

Scheme 4.12

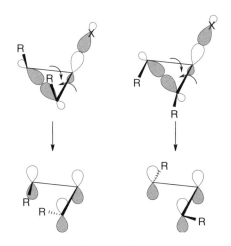

Scheme 4.13

TABLE 4.15 Activation Free Energies for the Electrocyclic Ring Opening of 108 and 109[a]

108	a	b	c	d	e	f	g	h	i
$\Delta G^{\ddagger}_{in}$	36.08	38.82	36.67	30.29	35.33	16.60	28.50	37.04	37.93
$\Delta G^{\ddagger}_{out}$	46.29	51.31	43.56	42.89	49.37	31.66	45.30	53.68	54.82
109	a	b	c	d	e	f	g	h	i
$\Delta G^{\ddagger}_{in}$	60.62	57.57	64.33	55.10	63.53	49.50	55.40		61.58
$\Delta G^{\ddagger}_{out}$	30.26	36.83	26.89	22.98	33.02	17.26	22.91	31.09	35.20

[a]B3LYP/6-311++G**, see Ref. 198.

is **108c**, where the methyl and bulky *t*-butyl groups experience some steric congestion; nevertheless, the inward rotation is favored by almost 7 kcal mol^{-1}. Confirming the stereoselectivity, diastereomers **109** all ring open with the substituents rotating outward. Important in the context of the arguments on torquoselectivity that will follow, varying the substituents does not switch the preference of the favored rotation. We will discuss the origins of this lack of a substituent effect later in this section.

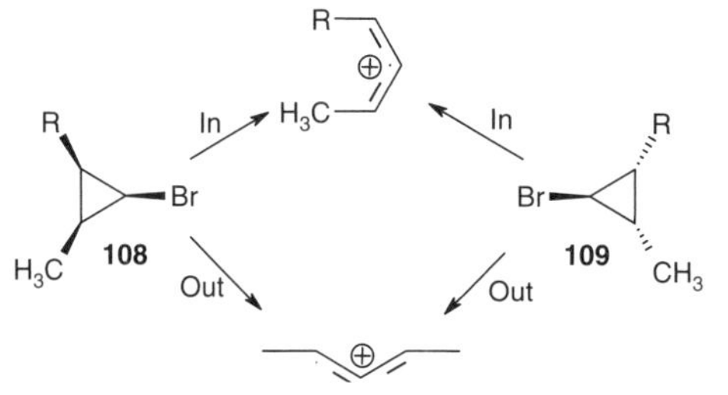

a: R=Me   d: R=CHCH$_2$ g: R=OMe
b: R=H   e: R=F   h: R=CHO
c: R=tBu   f: R=NH$_2$ i: R=NO$_2$

Until the mid-1980s, the electrocyclic ring opening of cyclobutenes had been understood to be under steric control. For example, the ring opening of trans-1,2,3,4-tetramethylcyclobutene (**110**) produces only (2Z,4Z)-3,4-dimethyl-2,4-hexadiene (**111**) without the formation of the E,E-isomer (**112**).[199] As a four-electron electrocyclization, both methyl groups can rotate outwards to form **11** or inwards to form **112**. The transition state for the inward rotation is expected to be sterically more congested and therefore less likely. However, Dolbier[200] discovered that the perflouro analog **113** opens preferentially to the Z,Z-isomer **114**. The activation barrier toward **114** is 18 kcal mol^{-1} lower than the barrier to produce the sterically less congested E,E-isomer **115**.

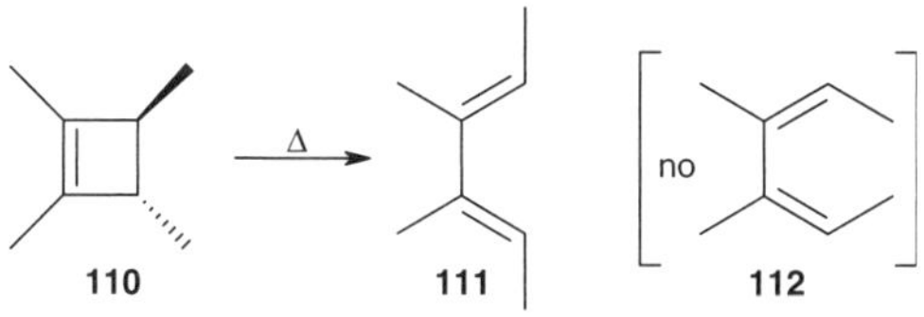

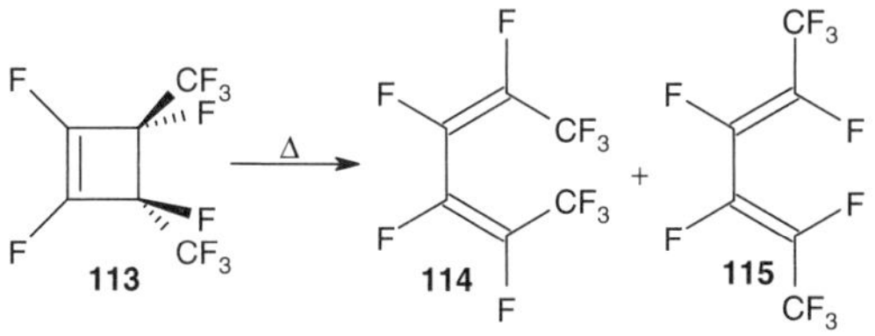

Inspired by these results, Houk and coworkers initiated the computational studies of the electrocyclic ring opening of cyclobutenes. Based on the studies of a broad range of pericyclic reactions,[8] including the Diels–Alder reaction explicitly discussed in the first part of this chapter, their early studies employed HF-optimized structures and occasional single-point energy calculations at the MP2 level. While the HF activation energies are likely to be too large, relative

TS energies are likely to be adequately estimated, especially the trends among related inward versus outward TS for a series of substituted cyclobutenes. They first examined the ring opening of *trans*-3,4-dimethylcyclobutene (**116**) and *trans*-3,4-dihydroxycyclobutene (**117**).[201,202] The outward rotation is favored over the inward for both cases. More importantly, while the inward barriers are similar, the outward barrier for **117** is 17 kcal mol^{-1} lower than for **116**, a difference not readily amenable to rationalization based solely on substituent size.

$$E_a = 40.4 \quad \text{116} \quad E_a = 53.4$$

$$E_a = 23.4 \quad \text{117} \quad E_a = 55.0$$

Houk[203] proposed a molecular orbital model shown in Scheme 4.14 to explain these results. Pictured in Scheme 4.14 are representations of the resulting HOMO and LUMO of the cyclobutene moiety in the transition state for the electrocyclic ring opening. The most energetically favorable situation will occur when this HOMO energy is as low as possible. Both the hydroxyl and methyl groups are electron donors, characterized by having an energetically high lying filled orbital. As a filled–filled orbital interaction, the interaction of this donor substituent orbital with the HOMO of cyclobutene will raise the energy of the latter. Since the donor will act to raise the cyclobutene HOMO energy, the overlap between the donor orbital and the cyclobutene HOMO should be minimized. This minimized overlap is realized when the donor is situated in the outward direction. Inward rotation develops the overlap between the C_4 hybrid and the donor substituent orbital, an interaction that cannot occur when the donor rotates outward. Therefore, a donating substituent will preferentially rotate outward. In addition, the donor orbital will interact with the LUMO of cyclobutene, leading to a stabilization of

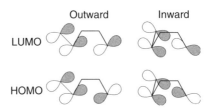

Scheme 4.14 Schematic diagram of the frontier orbitals for the inward and outward rotations of 3-substituted cyclobutene.

the donor orbital energy. This donor stabilization is larger upon outward rotation due to better overlap than inward rotation, where there is a phase mismatch between the donor orbital and the C_4 hybrid.

The strength of the molecular orbital model is the prediction of the effect of an electron-withdrawing group. An electron-withdrawing substituent will stabilize the cyclobutene HOMO. This stabilization can be maximized by *inward* rotation, where the overlap of the substituent orbital with the C_4 hybrid is now favorable. Outward rotation of the substituent will also stabilize the transition state HOMO but not as much as inward rotation. Houk then sought a cyclobutene with an electron-withdrawing substituent, hoping to find inward rotation upon electrocyclization. This is exactly the case for 3-formylcyclobutene. HF/6-31G*//HF/3-21G calculations predict that the barrier for inward rotation is 4.6 kcal mol^{-1} less than the barrier for outward rotation. (Subsequent B3LYP/6-31G* computations also show the inward rotational preference.[204]) These calculations were then confirmed by monitoring the electrocyclization of 3-formylcyclobutene by NMR spectroscopy at 50–70 °C. Only (Z)-pentadienal was observed, indicating that inward rotation is favored by at least 2.7 kcal mol^{-1}.[205] Clearly, steric arguments are insufficient to rationalize these results.

Subsequent theoretical and experimental studies have validated Houk's torquoselectivity model for the electrocyclization of cyclobutenes.[203] Computational results for a broad range of 3-substituted cyclobutenes are listed in Table 4.16. Of particular note is the comparison of the activation energies for cyclobutene-3-carboxylic acid and its conjugate acid and base.[206] The conjugate base (**118b**) is a strong electron-donating group and favors (by 7.3 kcal mol^{-1}) outward rotation. The neutral acid (**118**) is a weaker donor, and it also prefers outward rotation, but the difference in activation barriers is only 2.3 kcal mol^{-1}. The protonated acid (**118a**) is an electron-withdrawing group and it prefers inward rotation by 4.8 kcal mol^{-1}. Clearly, tuning the donating/withdrawing power can alter the stereo-outcome of the electrocyclization.

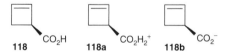

An interesting example of the electronic nature of torquoselectivity is the silyl substituent effect. Murakami[208] reported that the *t*-butyl group rotates exclusively outward in the electrocyclic ring opening of **119**. However, the trimethylsilyl analog **120**, which should have an even greater steric preference for outward rotation than **119**, opens to give the inward rotation product over the outward rotation product in a 69 : 31 ratio. Their B3LYP/6-31G* calculations indicated that the inward TS is 1.67 kcal mol^{-1} lower in energy than the outward TS. Murakami argued that the preference for inward silyl rotation is due to the

TABLE 4.16 Activation Energies (kcal mol⁻¹) for 3-Substituted Cyclobutenes[a]

R	$\Delta E^{\ddagger}_{in}$	$\Delta E^{\ddagger}_{out}$
H	41.6 (46.2) *33.8*	
OLi	45.5 (27.8)	24.4 (27.8)
NH₂	46.5 (52.2) *35.4*	27.0 (34.7) *20.7*
PH₂	*33.8*	*29.7*
OH	48.7 (54.6)	32.3 (37.4)
F	52.8 (58.9) [44.3]	35.5 (42.0) [29.2]
Cl	53.4 (60.2)	38.6 (46.6)
SH	50.0 (55.0)	36.9 (41.3)
CH₃	47.3 (51.3) [38.8] *37.3*	40.4 (44.5) [33.5] *31.4*
SiH₃	*30.6*	*32.2*
SiMe₃	*30.6*	*31.9*
HCCH₂	41.7 (45.3)	37.0 (40.4)
CO₂⁻	42.1 (47.1)	37.2 (39.8)
NH₃⁺	49.8 (54.2) *39.3*	42.1 (46.3) *33.0*
CCH	45.5 (48.2)	38.2 (40.6)
CF₃	47.0 (50.9) *36.8*	45.1 (48.3) *34.5*
SiF₃	*28.3*	*32.4*
CN	43.9 (47.2) [35.2]	39.3 (42.9) [31.0]
CO₂H	40.0 (44.2)	38.5 (41.9)
NO₂	45.2 (50.1)	38.3 (42.8)
COMe	40.5 (42.6)	39.0 (41.4)
CHO	34.7 (38.0) [26.5] *25.3*	39.2 (42.6) [31.2] *29.2*
NO	34.5 (39.1) [25.1]	37.1 (41.7) [29.8]
CO₂H⁺	23.3 (30.4)	29.6 (35.2)
BMe₂	24.6 (29.0)	37.1 (40.5)
BH₂	17.1 (20.8) [10.8] *9.9*	32.2 (39.0) [29.5] *25.8*

[a]HF/3-21G//HF/3-21G in normal text, HF/6-31G*//HF/3-21G in parentheses, MP2/6-31G*//HF/3-21G in italics (see Ref. 207), and B3LYP/6-31G(d)//B3LYP/6-31G(d) in italics (these are $\Delta H^{\ddagger}$ values—see Ref. 204).

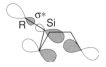

Scheme 4.15

overlap of the σ* orbital of the silyl group with the C₄ hybrid of the HOMO of cyclobutene (see Scheme 4.15). Houk[204] examined the ring opening of a number of methyl- and silyl-substituted cyclobutenes and found a strong correlation between the activation enthalpy for the inward rotation with the energy of the substituent LUMO, consistent with the orbital model proposed by Murakami.

A boryl substituent on cyclobutene, with its vacant p-orbital, should also rotate in the inward direction during the electrocyclic ring opening reaction. Houk first predicted the preference for inward rotation of a boryl group in 1985.[202] Subsequent B3LYP calculations indicate that the inward rotation transition state for 3-dimethylboracyclobutene is 11.5 kcal mol^{-1} lower in energy than the outward transition state.[204] Murakami[209] confirmed this inward preference in 2005: the electrocyclic reaction of **121** results in the inward rotation product exclusively. The experimental activation energy is 24.3 kcal mol^{-1} in fine agreement with computations for simpler systems.

Electronic control of the ring opening can overcome significant steric strains. Both **122a** and **122b** will ring open to give a single diasteromeric product: the isomer where the very bulky *t*-butyl group rotates inward! To examine this stereoselectivity, the Houk group optimized the transition states for the ring opening of 3-hydroxy- (as a model for the ether substituents), 3-methyl- (as a model for the *t*-butyl group), and 3-hydoxy-3-methylcyclobutene at HF/3-21G.[210] What is really examined here is whether torquoselectivity induced by multiple substituents is additive. The hydroxyl group is a stronger donor than the methyl group, reflected in its much greater preference for outward rotation; the difference in the outward versus inward rotation TS is 16.4 kcal for OH and 6.4 kcal mol^{-1} for Me. If the effects are additive, when both substituents are present, the TS with the OH group moving outward and the methyl group moving inward would be favored by 10.0 kcal mol^{-1} over the reverse motion. In fact, this TS is favored by 10.6 kcal mol^{-1}. This result is consistent with the experiments, and demonstrates that torquoselectivity effects are additive. Additivity has been further observed in a variety of other systems.[211,212]

Another example of electronic effect outweighing the effects of sterics is in the electrocyclic ring opening of **123**. Inward conrotatory ring opening leads to the very congested diene **124**, while outward rotation to the much less congested diene **125** would seem to be the likely product. However, the electropositive silicon substituent favors inward rotation, as we saw in the reaction of **120**. At B3LYP/6-31G(d), the inward conrotatory transition state is 0.51 kcal mol^{-1} *below* the transition state for the outward rotation.[213] Murakami also prepared **123** and upon heating it at 110 °C, obtained the two diene products **124** and **125** in a ratio of 78 : 22, which corresponds to a difference in activation barriers of 0.96 kcal mol^{-1}, in fine agreement with the computations. This reaction is an outstanding example of the predictive power of the concept of torquoselectivity and ably demonstrates that electronic effects can dominate steric effects.

Houk and Tang[214] reported a seeming contradiction. The ring opening of **126a** gives *only* the inward product **127a**, but according to the torquoselection model, the phenyl group should rotate outward to give **128a**. B3LYP/6-31G* computations on the ring opening of **126b** indicate that the activation barrier for the outward path (leading to **128b**) is nearly 8 kcal mol^{-1} lower than the barrier for the inward path (leading to **127**). This is consistent with the torquoselectivity rules, but inconsistent with the experiment.

While investigating the isomerization of the outward product to the inward product, they discovered a pyran intermediate **129b**. This led to the proposal of the mechanism shown in Scheme 4.16. The highest barrier in this mechanism is for the electrocyclization that leads to the outward product **128b**. The subsequent barriers—closing to the pyran **129b** and then the torquoselective ring opening to **127b**—are about 13 kcal mol^{-1} lower in energy than for the first step.

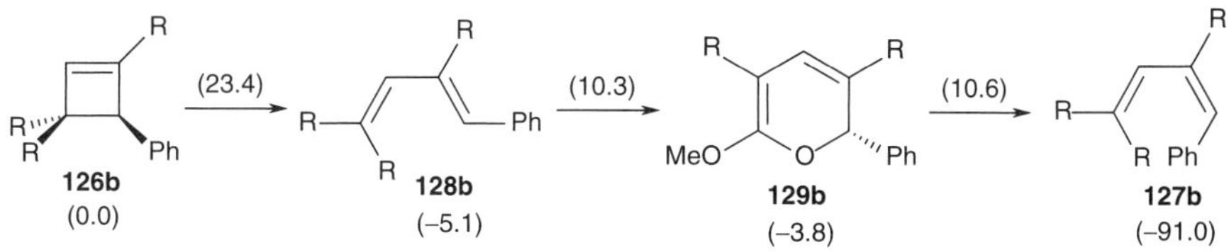

TORQUOSELECTIVITY

Scheme 4.16 (relative energies in kilocalorie per mole, activation energies above arrows).

126b (0.0) → (23.4) → 128b (−5.1) → (10.3) → 129b (−3.8) → (10.6) → 127b (−91.0)

The observed product is the thermodynamic sink. Torquoselection is preserved according to this mechanism, and fully accounts for the experiment.

We now turn our attention to the possibility of torquoselectivity in electrocyclization reactions involving larger rings. The ring opening of cyclopentenyl cation is a four-electron process, which is a conrotatory process involving five carbon atoms. HF/3-21G computations of the parent cation, along with amino and boryl substituents (**130a–c**) indicate distinct torquoselectivity (see Table 4.17).[215] The amino substituent prefers outward rotation by 14.7 kcal mol^{-1}, while the boryl substituent prefers inward rotation by 5.6 kcal mol^{-1}. These are the same rotational preferences (donors rotate outward and acceptors rotate inward) as seen in the ring opening of 3-amino- and 3-borylcyclobutene. However, the barrier energy differences between inward and outward rotations are larger for the cyclobutene system than for the cyclopentenyl cations. Houk argued that the greater geometric flexibility of the five-member ring over the four-member ring allows for the substituents to move to limit the unfavorable filled–filled orbital interactions. Similar conclusions are also found in a study of the silyl substituent effect on the Nazarov cyclization.[216] A very large torquoselectivity was predicted for the retro-Nazarov reaction where inward rotation of the methoxy group is favored by nearly 10 kcal mol^{-1} over outward rotation.[217]

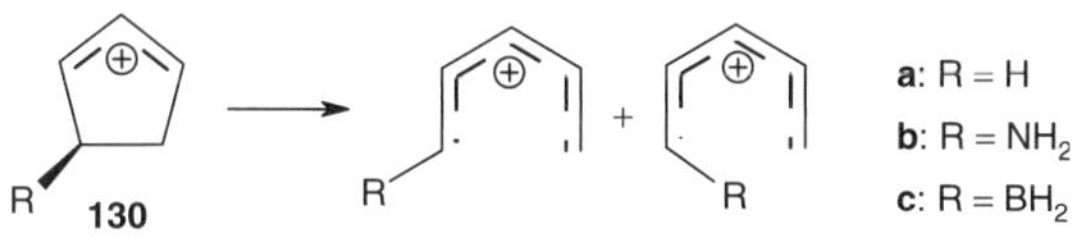

130 a: R = H
b: R = NH$_2$
c: R = BH$_2$

A six-electron cyclization will proceed with disrotatory motion. Torquoselectivity in a six-electron electrocyclic reaction was first examined in the ring opening

TABLE 4.17 Activation Energies (kcal mol^{-1}) for the Ring Opening of 113a–c[a]

R	$\Delta E^{\ddagger}$(Inward)	$\Delta E^{\ddagger}$(Outward)
H	43.5	
NH$_2$	23.6	8.9
BH$_2$	25.1	30.7

[a]Computed at HF/3-21G, Ref. 215.

TABLE 4.18 Activation Energies (kcal mol^{-1}) for the Ring Opening of 131[a]

R	$\Delta E^{\ddagger}$(Inward)	$\Delta E^{\ddagger}$(Outward)
F	51.8	47.1
CH$_3$	52.0	48.1
CN	48.5	45.2
CHO	40.5	40.9
NO	44.3	46.0
BH$_2$	33.1	42.3

[a]Computed at MP2/6-31G*//HF/3-21G + ZPE(HF/3-21G), Ref. 218.

of 5-substituted-1,3-cyclohexadienes (**131**).[218] The MP2/6-31G*//HF/3-21G energies for the inward and outward activation energies due to differing substituents are listed in Table 4.18. Similarly to the ring opening of cyclobutenes, electron-donating groups prefer to rotate outwards, while withdrawing groups show a small preference for inward rotation. (Inward rotation of the formyl group is favored by only 0.4 kcal mol^{-1} at this level, but B3LYP/6-31G* computations suggest that outward rotation is slightly lower than inward rotation. Nonetheless, this withdrawing group certainly does diminish the energy difference relative to a donor, like a methyl group.) As seen with the ring opening of cyclopentyl cations, the substituent effect is smaller here than in the cyclobutenes.

Houk employed the MO model shown in Scheme 4.17 to explain substituent effects in the ring opening of cyclohexadienes. The key orbital interaction is

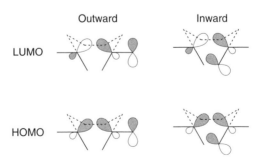

Scheme 4.17 MO model for torquoselectivity in disrotatory six-electron electrocyclizations.

again present within the HOMO for the inward rotation. A donor substituent will have an occupied orbital interacting with the breaking σ-bond of the ring, leading to an unfavorable, destabilizing, filled–filled interaction; this is the cause for electron-donating groups favoring an outward rotation. On the other hand, a withdrawing substituent provides a low lying empty orbital to interact with the breaking σ-bond, a net stabilizing interaction. Withdrawing groups therefore favor inward rotation. However, since the overlap between the empty donor orbital and the C_6 hybrid is poorer than the analogous interaction in the cyclobutene case (see Scheme 4.14), the stabilizing/destabilizing effect for the inward rotation is less in the six-electron case than in the four-electron case. Furthermore, the steric interactions of the inward rotating groups are greater in the six-electron case than the four-electron case, making inward rotation inherently more difficult.

Electrocyclizations involving eight electrons will proceed by a conrotatory transition state. Houk[219] examined the ring opening of three 7-substituted-cycloocta-1,3,5-trienes (**132**) at MP2/6-31G*//HF/3-21G. The ring can open with the substituent rotating inward to give the Z product or outward to give the E product. Since the inward and outward substituents are about equally removed from C_8, torquoselectivity should be minimal. In fact, as seen in Table 4.19, outward rotation is favored for all three substituents, including a strong donor (F) and acceptor (CHO). It appears that steric interactions alone account for the stereochemical outcome; inward rotation is simply too congested to compete with outward rotation.

132

The concept of torquoselectivity is now accepted as an extension of the Woodward–Hoffmann rules. It has been used as a guide for synthetic chemists to prepare the appropriate stereoisomer. Examples include Danishefsky's[220] exploitation of the stereoselective ring opening of *trans*-1,2-disiloxybenzocyclobutenes **133** to prepare idarubicin, Paquette's[221] use of the electrocyclization of **134** that ultimately leads to a very efficient synthesis of pentalene, and Murakami's[222]

TABLE 4.19 Activation Energies (kcal mol⁻¹) for the Ring Opening of 132[a]

R	$\Delta E^{\ddagger}$(Inward)	$\Delta E^{\ddagger}$(Outward)
F	24.9	24.1
CH₃	23.6	20.7
CHO	20.2	17.1

[a]Computed at MP2/6-31G*/HF/3-21G + ZPE(HF/3-21G), Ref. 219.

preparation of a broad spectrum of functionalized 1,3-butadienes prepared by controlled ring opening of cyclobutenes. Torquoselectivity is a premier example of a concept developed by computational chemists that then migrated to the bench chemists for exploitation.

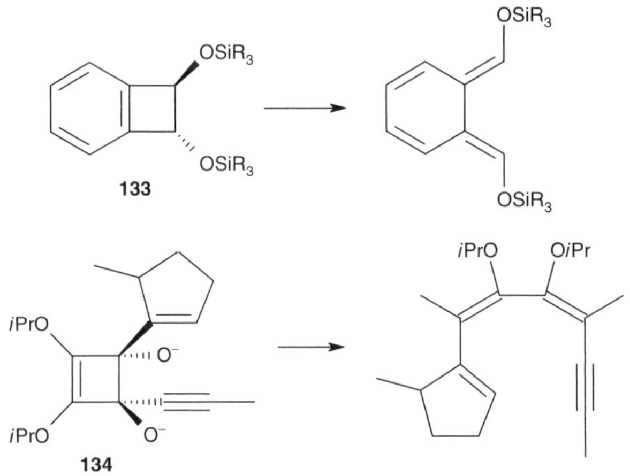

4.7 INTERVIEW: PROFESSOR WESTON THATCHER BORDEN

Interviewed April 1, 2005

Professor Wes Borden holds the Welch Chair at the University of North Texas, which he assumed in 2004. He previously spent the bulk of his academic career at the University of Washington.

Dr. Borden was trained in theoretical chemistry by Professor H. C. Longuet-Higgins, learning the 'proper' way to do theoretical chemistry—only those calculations that could be performed on the back of an envelope. When Borden returned to Harvard as a graduate student, his PhD adviser, E.J. Corey, encouraged him to continue his theoretical work.

After receiving his PhD in 1968, Borden joined the faculty at Harvard. Although he had a largely experimental research program, Borden taught a course in theoretical chemistry. The lecture notes for this course became a textbook, entitled *Modern Molecular Orbital Theory for Organic Chemists*.

Borden was not to conduct his first ab initio calculation until he joined the faculty at the University of Washington in 1973. There he began a very fruitful collaboration with Professor Ernest Davidson. Collaborative research remains a signature feature of Borden's career.

Borden was interested in the PES of trimethylenemethane. Davidson indicated that his MELD program was capable of handling such a molecule. The ensuing collaboration with Davidson led to more than 20 coauthored papers and to one of Borden's major contributions to theoretical chemistry: the Borden–Davidson rules for predicting the ground states of diradicals. Borden considers the research done with Davidson to be the best of his many collaborations because he learned so much from Davidson. In particular, Borden acquired a much more mathematical way of thinking about chemical problems, while in turn he believes that Davidson acquired a more intuitive chemical sense.

The success of the Borden–Davidson rules directly led to enormously fruitful collaboration collaborations with Professors Jerry Berson, Carl Lineberger, and Matt Platz. About the last of these collaborations, Borden says jokingly, "It's all Matt's fault!" Platz had been concerned about size of the singlet–triplet gap in phenylnitrene. Impressed with Borden's success in computing this energy diffcrence in a number of hydrocarbons, Platz asked Borden to tackle the phenylnitrene problem. Borden admits with some embarrassment that his own lack of familiarity with nitrene chemistry resulted in Platz's having to continue to ask for a number of years. More out of a desire to finally placate Platz than anything else, Borden agreed to compute the singlet–triplet gap of phenylnitrene. This eventually led to a realization of the unusual wavefunction of singlet phenylnitrene, as discussed in Section 5.2. Borden says, "Even after this initial discovery, Platz was not satisfied and continued to drive our collaboration forward. Every computational result would trigger further questions from Matt." Eventually, Borden's calculations led to a novel two-step mechanism that helped to rationalize the chemistry of phenylnitrene. Platz has told Borden that this mechanism would not have been found without the help of the calculations.

Although Platz was certainly the driving force behind their collaboration, occasionally Borden did suggest a new direction. For example, the Borden group's calculations of the effects of cyano substituents on the phenylnitrene rearrangement were completed prior to Platz' experimental tests of the computational predictions. "It is more fun to make the prediction before the experiment is done," said Borden. This attitude results from the familiar complaint of experimentalists about computational chemists: "All you ever do is predict things that are already known!"

Borden's interest in the Cope rearrangement dates back to 1973 when, as an Assistant Professor at Harvard, Professor William von Eggers Doering approached Borden with the question about substituents on this rearrangement. Borden utilized an orbital correlation diagram to support an argument for a continuum transition state model. This conclusion was in direct opposition to Professor Michael Dewar contention of two distinct mechanisms. To this day, Borden can still vividly recollect the reviews to his submitted *JACS* paper. The first, signed by Professor Howard Zimmerman, was supportive but warned that Dewar will not approve. The second anonymous review read: *The publication of this manuscript will only serve to provide the gladiatorial spectacle of Borden being torn to shreds in print*. The article never was published. Nonetheless, as discussed in Section 3.2, the calculations of Houk and Borden and Doering's experiments firmly established the variable (chameleonic) nature of the Cope transition state.

Thus began Borden's long involvement with the Cope rearrangement. It included a sabbatical trip to Japan on a Guggenheim Fellowship to collaborate with Professor Morokuma, which, as a side effect, fomented Borden's passionate interest in Japanese culture.

He has worked on the Cope rearrangement with many students and in particular two long-time associates, Dr. David Feller and Dr. Dave Hrovat. Their CASSCF(6,6)/3-21G computations located a single synchronous transition state. Borden recalls the dismay and almost sense of betrayal he felt when Davidson's larger MCSCF results confirmed Dewar's finding of two distinct transition states. However, the poor activation energy predicted by this method inspired Borden to realize the need for both nondynamic and *dynamic* correlation in order to properly describe the Cope transition state. Ultimately, Borden's CASPT2 computations settled the matter in favor of the concerted mechanism. The CASPT2 calculations gave an enthalpy of activation that was in excellent agreement with the experimental value that had been measured by Doering. Borden says, "When we saw the CASPT2 activation enthalpy, we knew immediately that CASPT2 had gotten the Cope rearrangement right."

Borden considers his group's work on the importance of the inclusion of dynamic electron correlation in many different types of problems to be his most significant contribution to computational methodology. In addition, his work on the Cope rearrangement and artifactual symmetry breaking taught him to question the applicability of routine methods to what first appears to be a routine problem but one that later turns out to require high level calculations.

Borden said, "For example, first we thought our CASSCF(2,2) calculations on the Cope rearrangement should have worked. What we failed to realize initially is that CASSCF(2,2) is prejudiced, because it can describe cyclohexane-1,4-diyl, but not the other diradical extreme of two allyl radicals for the Cope transition structure. Then, we thought that CASSCF(6,6) would give the correct answer, but didn't realize the need to include dynamic correlation." The Cope rearrangement opened his eyes to the dangers afoot in computational chemistry. "You could get a wrong answer really easily. It is an interesting problem to challenge methodology."

The last collaboration we discussed was the work on the response of the Cope transition state to substituents. Doering independently contacted Professor Ken Houk and Borden to get them interested in the problem. Both subsequently did pursue the project, only to discover that they were both presenting similar results at a Bartlet session at a Reaction Mechanisms Conference. Following the conference, they began to truly collaborate, working together on calculations, interpretations and writing.

A major influence that Houk had on Borden was insisting on the importance of including lot of figures in the article. Borden claims Houk taught him that "organic chemists look at the pictures; they don't look at the tables."

I asked Borden, "After working on the Cope rearrangement for more than 30 years, what do you think still remains to be done?" He responded, "Though it is always dangerous to reply to such a question about future research, I have told Ken Houk that I have published my last paper on the Cope."

Borden emphasizes the importance of experiment as part of his work. The ratio of his experimental to computational emphasis has shifted over his career, from mostly experiment when he began his career at Harvard to mostly computation during his later years at Washington, to exclusively computation at North Texas. Nevertheless, he still regards himself as an experimentalist, "My group used to do experiments in the laboratory. Now we do them on the computer."

His students have always been encouraged (if not forced) to engage in both experimental and computational research studies. He believes this approach "is great training for a theoretician to know the kinds of questions that experimentalists want answered. And it's wonderful for an experimentalist to be able to say, 'Gee, maybe we should do a calculation to understand an experimental result'. The synergy between experiments and calculations is extraordinary."

Borden has seen the evolution of the discipline of computational chemistry from being the object of friendly ridicule by experimentalists to being a full partner with experimentation. He points to the methylene problem (Section 5.1) and discrepancies in the heats of formation of the isomeric benzynes (Section 5.5) as examples of where computations identified the existence of errors in the interpretation of well-regarded experiments. Nonetheless, he notes that some experimentalists regard computations as second-rate science, partly because calculations are relatively easy to do, partly because (bad) calculations can yield bad results, and partly because many computational chemists still seem only to offer explanations, without also providing predictions.

It is Borden's belief that computational methodology is now good enough to address most problems in organic chemistry. However, he remains somewhat wary about density functional methods, characterizing them as "AM1 for the 21st century." Borden says, "Functionals that work well for most problems, can unexpectedly fail. The same is, of course, also true for *ab initio* methods. However, *ab initio* calculations can systematically be improved, by expanding the basis set and/or including more electron correlation; whereas, there is no systematic way to improve upon a DFT result. Although my group certainly uses DFT methods for many problems, whenever possible, we try to validate the results by comparison with those obtained from *ab initio* calculations."

Borden points to Professor Roald Hoffmann as a great influence on the way he does science. "Roald's early papers, which relied exclusively on Extended Hückel theory, taught me that critical interpretation of calculations is even more important than their numerical accuracy." The key to Borden's success in computational chemistry can be summarized by his statement: "You need to ask the right question and use the right tools to get the right answer. However, the goal of calculations is not just numbers; it's the interpretation that really matters."

> Borden's career is marked by the many research problems that he has tackled with just this approach.

REFERENCES

1. Woodward, R. B.; Hoffmann, R. "The conservation of orbital symmetry," *Angew. Chem. Int. Ed. Engl.* **1969**, *8*, 781–853.
2. Fukui, K. "Recognition of stereochemical paths by orbital interaction," *Acc. Chem. Res.* **1971**, *4*, 57–64.
3. Fukui, K. *Theory of Orientation and Stereoselection*; Springer-Verlag: Berlin, **1975**.
4. Zimmerman, H. "Möbius-Hückel concept in organic chemistry. Application of organic molecules and reactions," *Acc. Chem. Res.* **1971**, *4*, 272–280.
5. Fleming, I. *Pericyclic Reactions*; Oxford University Press: Oxford, **1999**.
6. Miller, B. *Advanced Organic Chemistry: Reactions and Mechanisms*; Prentice Hall: UpperSaddle River, NJ, **2004**.
7. Carroll, F. A. *Perspectives on Structure and Mechanism in Organic Chemistry*; 2nd ed.; Wiley: Hoboken, NJ, **2010**.
8. Houk, K. N.; Li, Y.; Evanseck, J. D. "Transition structures of hydrocarbon pericyclic reactions," *Angew. Chem. Int. Ed. Engl.* **1992**, *31*, 682–708.
9. Houk, K. N.; Gonzalez, J.; Li, Y. "Pericyclic transition states: passion and punctilios, 1935–1995," *Acc. Chem. Res.* **1995**, *28*, 81–90.
10. Wiest, O.; Houk, K. N. "Density functional theory calculations of pericyclic reaction transition structures," *Top. Curr. Chem.* **1996**, *183*, 1–24.
11. Rowley, D.; Steiner, H. "Kinetics of diene reactions at high temperatures," *Discuss. Faraday Soc.* **1951**, 198–213.
12. Guner, V.; Khuong, K. S.; Leach, A. G.; Lee, P. S.; Bartberger, M. D.; Houk, K. N. "A standard set of pericyclic reactions of hydrocarbons for the benchmarking of computational methods: the performance of ab initio, density functional, CASSCF, CASPT2, and CBS-QB3 methods for the prediction of activation barriers, reaction energetics, and transition state geometries," *J. Phys. Chem. A* **2003**, *107*, 11445–11459.
13. Houk, K. N.; Lin, Y. T.; Brown, F. K. "Evidence for the concerted mechanism of the Diels-Alder reaction of butadiene with ethylene," *J. Am. Chem. Soc.* **1986**, *108*, 554–556.
14. Bach, R. D.; McDouall, J. J. W.; Schlegel, H. B. "Electronic factors influencing the activation barrier of the Diels-Alder reaction. An ab initio study," *J. Org. Chem.* **1989**, *54*, 2931–2935.
15. Stanton, R. V.; Merz, K. M., Jr. "Density functional transition states of organic and organometallic reactions," *J. Chem. Phys.* **1994**, *100*, 434–443.
16. Li, Y.; Houk, K. N. "Diels-Alder dimerization of 1,3-butadiene: an ab initio CASSCF study of the concerted and stepwise mechanisms and butadiene-ethylene revisited," *J. Am. Chem. Soc.* **1993**, *115*, 7478–7485.
17. Barone, B.; Arnaud, R. "Diels–Alder reactions: an assessment of quantum chemical procedures," *J. Chem. Phys.* **1997**, *106*, 8727–8732.

18. Isobe, H.; Takano, Y.; Kitagawa, Y.; Kawakami, T.; Yamanaka, S.; Yamaguchi, K.; Houk, K. N. "Extended Hartree–Fock (EHF) theory of chemical reactions VI: hybrid DFT and post-Hartree–Fock approaches for concerted and non-concerted transition structures of the Diels–Alder reaction," *Mol. Phys.* **2002**, *100*, 717–727.

19. Sakai, S. "Theoretical analysis of concerted and stepwise mechanisms of Diels-Alder reaction between butadiene and ethylene," *J. Phys. Chem. A* **2000**, *104*, 922–927.

20. Lischka, H.; Ventura, E.; Dallos, M. "The Diels-Alder reaction of ethene and 1,3-butadiene: an extended multireference ab initio investigation," *ChemPhysChem* **2004**, *5*, 1365–1371.

21. Jorgensen, W. L.; Lim, D.; Blake, J. F. "Ab initio study of Diels-Alder reactions of cyclopentadiene with ethylene, isoprene, cyclopentadiene, acrylonitrile, and methyl vinyl ketone," *J. Am. Chem. Soc.* **1993**, *115*, 2936–2942.

22. Szalay, P. G.; Bartlett, R. J. "Multi-reference averaged quadratic coupled-cluster method: a size-extensive modification of multi-reference CI," *Chem. Phys. Lett.* **1993**, *214*, 481–488.

23. Jursic, B.; Zdravkovski, Z. "DFT study of the Diels–Alder reactions between ethylene with buta-1,3-diene and cyclopentadiene," *J. Chem. Soc., Perkin Trans. 2* **1995**, 1223–1226.

24. Goldstein, E.; Beno, B.; Houk, K. N. "Density functional theory prediction of the relative energies and isotope effects for the concerted and stepwise mechanisms of the Diels-Alder reaction of butadiene and ethylene," *J. Am. Chem. Soc.* **1996**, *118*, 6036–6043.

25. Kraka, E.; Wu, A.; Cremer, D. "Mechanism of the Diels–Alder reaction studied with the united reaction valley approach: mechanistic differences between symmetry-allowed and symmetry-forbidden reactions," *J. Phys. Chem. A* **2003**, *107*, 9008–9021.

26. Guner, V. A.; Khuong, K. S.; Houk, K. N.; Chuma, A.; Pulay, P. "The performance of the Handy/Cohen functionals, OLYP and O3LYP, for the computation of hydrocarbon pericyclic reaction activation barriers," *J. Phys. Chem. A* **2004**, *108*, 2959–2965.

27. Froese, R. D. J.; Humbel, S.; Svensson, M.; Morokuma, K. "IMOMO(G2MS): a new high-level G2-like method for large molecules and its applications to Diels-Alder reactions," *J. Phys. Chem. A* **1997**, *101*, 227–233.

28. Herges, R.; Jiao, H.; Schleyer, P. v. R. "Magnetic properties of aromatic transition states: the Diels-Alder reactions," *Angew. Chem. Int. Ed. Engl.* **1994**, *33*, 1376–1378.

29. Bradley, A. Z.; Kociolek, M. G.; Johnson, R. P. "Conformational selectivity in the Diels-Alder cycloaddition: predictions for reactions of s-trans-1,3-butadiene," *J. Org. Chem.* **2000**, *65*, 7134–7138.

30. Kobko, N.; Dannenberg, J. J. "Effect of basis set superposition error (BSSE) upon ab initio calculations of organic transition states," *J. Phys. Chem. A* **2001**, *105*, 1944–1950.

31. Jiao, H.; Schleyer, P. v. R. "Aromaticity of pericyclic reaction transition structures: magnetic evidence," *J. Phys. Org. Chem.* **1998**, *11*, 655–662.

32. Pieniazek, S. N.; Clemente, F. R.; Houk, K. N. "Sources of error in DFT computations of C-C bond formation thermochemistries: $\pi \to \sigma$ transformations and error cancellation by DFT methods," *Angew. Chem. Int. Ed.* **2008**, *47*, 7746–7749.

33. Pang, L. S. K.; Wilson, M. A. "Reactions of fullerenes C_{60} and C_{70} with cyclopentadiene," *J. Phys. Chem.* **1993**, *97*, 6761–6763.

34. Giovane, L. M.; Barco, J. W.; Yadav, T.; Lafleur, A. L.; Marr, J. A.; Howard, J. B.; Rotello, V. M. "Kinetic stability of the fullerene C_{60}-cyclopentadiene Diels-Alder adduct," *J. Phys. Chem.* **1993**, *97*, 8560–8561.
35. Osuna, S.; Swart, M.; Sola, M. "Dispersion corrections essential for the study of chemical reactivity in fullerenes," *J. Phys. Chem. A* **2011**, *115*, 3491–3496.
36. Doering, W. v. E.; Franck-Neumann, M.; Hasselmann, D.; Kaye, R. L. "Mechanism of a Diels-Alder reaction. Butadiene and its dimers," *J. Am. Chem. Soc.* **1972**, *94*, 3833–3844.
37. Doering, W. v. E.; Roth, W. R.; Breuckmann, R.; Figge, L.; Lennartz, H.-W.; Fessner, W.-D.; Prinzbach, H. "Verbotene reaktionene. [2+2]-cycloreversion starrer cyclobutane," *Chem. Ber.* **1988**, *121*, 1–9.
38. Dewar, M. J. S. "Multibond reactions cannot normally be synchronous," *J. Am. Chem. Soc.* **1984**, *106*, 209–219.
39. Bigeleisen, J.; Mayer, M. G. "Calculation of equilibrium constants for isotopic exchange reactions," *J. Chem. Phys.* **1947**, *15*, 261–267.
40. Storer, J. W.; Raimondi, L.; Houk, K. N. "Theoretical secondary kinetic isotope effects and the interpretation of transition state geometries. 2. The Diels-Alder reaction transition state geometry," *J. Am. Chem. Soc.* **1994**, *116*, 9675–9683.
41. Streitwieser, A., Jr.; Jagow, R. H.; Fahey, R. C.; Suzuki, S. "Kinetic isotope effects in the acetolyses of deuterated cyclopentyl tosylates," *J. Am. Chem. Soc.* **1958**, *80*, 2326–2332.
42. Gajewski, J. J.; Peterson, K. B.; Kagel, J. R.; Huang, Y. C. J. "Transition-state structure variation in the Diels-Alder reaction from secondary deuterium kinetic isotope effects. The reaction of nearly symmetrical dienes and dienophiles is nearly synchronous," *J. Am. Chem. Soc.* **1989**, *111*, 9078–9081.
43. Van Sickle, D. E.; Rodin, J. O. "The secondary deuterium isotope effect on the Diels-Alder reaction," *J. Am. Chem. Soc.* **1964**, *86*, 3091–3094.
44. Singleton, D. A.; Thomas, A. A. "High-precision simultaneous determination of multiple small kinetic isotope effects at natural abundance," *J. Am. Chem. Soc.* **1995**, *117*, 9357–9358.
45. Beno, B. R.; Houk, K. N.; Singleton, D. A. "Synchronous or asynchronous? An "experimental" transition state from a direct comparison of experimental and theoretical kinetic isotope effects for a Diels-Alder reaction," *J. Am. Chem. Soc.* **1996**, *118*, 9984–9985.
46. Singleton, D. A.; Hang, C. "Isotope effects and the experimental transition state for a prototypical thermal ene reaction," *Tetrahedron Lett.* **1999**, *40*, 8939–8943.
47. Singleton, D. A.; Merrigan, S. R.; Beno, B. R.; Houk, K. N. "Isotope effects for Lewis acid catalyzed Diels-Alder reactions. The experimental transition state," *Tetrahedron Lett.* **1999**, *40*, 5817–5821.
48. Singleton, D. A.; Merrigan, S. R. "Resolution of conflicting mechanistic observations in ester aminolysis. A warning on the qualitative prediction of isotope effects for reactive intermediates," *J. Am. Chem. Soc.* **2000**, *122*, 11035–11036.
49. Singleton, D. A.; Hang, C.; Szymanski, M. J.; Meyer, M. P.; Leach, A. G.; Kuwata, K. T.; Chen, J. S.; Greer, A.; Foote, C. S.; Houk, K. N. "Mechanism of ene reactions of singlet oxygen. A two-step no-intermediate mechanism," *J. Am. Chem. Soc.* **2003**, *125*, 1319–1328.

50. Vo, L. K.; Singleton, D. A. "Isotope effects and the nature of stereo- and regioselectivity in hydroaminations of vinylarenes catalyzed by palladium(II)-diphosphine complexes," *Org. Lett.* **2004**, *6*, 2469–2472.

51. Singleton, D. A.; Schulmeier, B. E.; Hang, C.; Thomas, A. A.; Leung, S.-H.; Merrigan, S. R. "Isotope effects and the distinction between synchronous, asynchronous, and stepwise Diels–Alder reactions," *Tetrahedron* **2001**, *57*, 5149–5160.

52. Nagase, S.; Morokuma, K. "An ab initio molecular orbital study of organic reactions. The energy, charge, and spin decomposition analyses at the transition state and along the reaction pathway," *J. Am. Chem. Soc.* **1978**, *100*, 1666–1672.

53. Kitaura, K.; Morokuma, K. "A new energy decomposition scheme for molecular interactions within the Hartree-Fock approximation," *Int. J. Quantum Chem.* **1976**, *10*, 325–340.

54. Bickelhaupt, F. M. "Understanding reactivity with Kohn–Sham molecular orbital theory: E_2-S_N2 mechanistic spectrum and other concepts," *J. Comput. Chem.* **1999**, *20*, 114–128.

55. Ess, D. H.; Houk, K. N. "Distortion/interaction energy control of 1,3-dipolar cycloaddition reactivity," *J. Am. Chem. Soc.* **2007**, *129*, 10646–10647.

56. Ess, D. H.; Houk, K. N. "Theory of 1,3-dipolar cycloadditions: distortion/interaction and frontier molecular orbital models," *J. Am. Chem. Soc.* **2008**, *130*, 10187–10198.

57. Li, X.; Danishefsky, S. J. "Cyclobutenone as a highly reactive dienophile: expanding upon Diels−Alder paradigms," *J. Am. Chem. Soc.* **2010**, *132*, 11004–11005.

58. Paton, R. S.; Kim, S.; Ross, A. G.; Danishefsky, S. J.; Houk, K. N. "Experimental Diels–Alder reactivities of cycloalkenones and cyclic dienes explained through transition-state distortion energies," *Angew. Chem. Int. Ed.* **2011**, *50*, 10366–10368.

59. Gajewski, J. J. *Hydrocarbon Thermal Isomerizations*; Academic Press: New York, **1981**.

60. Doering, W. v. E.; Roth, W. R. "The overlap of two allyl radicals or a four-centered transition state in the Cope rearrangement," *Tetrahedron* **1962**, *18*, 67–74.

61. Goldstein, M. J.; Benzon, M. S. "Boat and chair transition states of 1,5-hexadiene," *J. Am. Chem. Soc.* **1972**, *94*, 7147–7149.

62. Doering, W. v. E.; Toscano, V. G.; Beasley, G. H. "Kinetics of the Cope rearrangement of 1,1-dideuteriohexa-1,5-diene," *Tetrahedron* **1971**, *27*, 5299–5306.

63. Cope, A. C.; Hofmann, C. M.; Hardy, E. M. "The rearrangement of allyl groups in three-carbon systems. II," *J. Am. Chem. Soc.* **1941**, *63*, 1852–1857.

64. Humski, K.; Malojcic, R.; Borcic, S.; Sunko, D. E. "Thermodynamic and kinetic secondary isotope effects in the Cope rearrangement," *J. Am. Chem. Soc.* **1970**, *92*, 6534–6538.

65. Cohen, N.; Benson, S. W. "Estimation of heats of formation of organic compounds by additivity methods," *Chem. Rev.* **1993**, *93*, 2419–2438.

66. Russell, J. J.; Seetula, J. A.; Gutman, D. "Kinetics and thermochemistry of methyl, ethyl, and isopropyl. Study of the equilibrium R + HBr ? R-H + Br," *J. Am. Chem. Soc.* **1988**, *110*, 3092–3099.

67. Dewar, M. J. S.; Wade, L. E. "Possible role of 1,4-cyclohexylene intermediates in Cope rearrangements," *J. Am. Chem. Soc.* **1973**, *95*, 290–291.

68. Dewar, M. J. S.; Wade, L. E., Jr. "A study of the mechanism of the Cope rearrangement," *J. Am. Chem. Soc.* **1977**, *99*, 4417–4424.

69. Wehrli, R.; Schmid, H.; Bellus, D.; Hansen, H. J. "The mechanism of the Cope rearrangement," *Helv. Chim. Acta* **1977**, *60*, 1325–1356.
70. Gajewski, J. J.; Conrad, N. D. "The mechanism of the Cope rearrangement," *J. Am. Chem. Soc.* **1978**, *100*, 6268–6269.
71. Gajewski, J. J.; Conrad, N. D. "Variable transition-state structure in the Cope rearrangement as deduced from secondary deuterium kinetic isotope effects," *J. Am. Chem. Soc.* **1978**, *100*, 6269–6270.
72. Gajewski, J. J.; Conrad, N. D. "Variable transition state structure in 3,3-sigmatropic shifts from α-secondary deuterium isotope effects," *J. Am. Chem. Soc.* **1979**, *101*, 6693–6704.
73. Lutz, R. P.; Berg, H. A. J. "Kinetics of the Cope rearrangement of a 3,4-diphenylhexa-1,5-diene," *J. Org. Chem.* **1980**, *45*, 3915–3916.
74. Staroverov, V. N.; Davidson, E. R. "The Cope rearrangement in theoretical retrospect," *J. Mol. Struct. (THEOCHEM)* **2001**, *573*, 81–89.
75. Komornicki, A.; McIver, J. W. J. "Structure of transition states. 4. MINDO/2 study of rearrangements in the C_6H_{10} system," *J. Am. Chem. Soc.* **1976**, *98*, 4553–4561.
76. Dewar, M. J. S.; Ford, G. P.; McKee, M. L.; Rzepa, H. S.; Wade, L. E. "The Cope rearrangement. MINDO/3 studies of the rearrangements of 1,5-hexadiene and bicyclo[2.2.0]hexane," *J. Am. Chem. Soc.* **1977**, *99*, 5069–5073.
77. Dewar, M. J. S.; Jie, C. "Mechanism of the Cope rearrangement," *J. Am. Chem. Soc.* **1987**, *109*, 5893–5900.
78. Osamura, Y.; Kato, S.; Morokuma, K.; Feller, D.; Davidson, E. R.; Borden, W. T. "Ab initio calculation of the transition state for the Cope rearrangement," *J. Am. Chem. Soc.* **1984**, *106*, 3362–3363.
79. Hrovat, D. A.; Borden, W. T.; Vance, R. L.; Rondan, N. G.; Houk, K. N.; Morokuma, K. "Ab initio calculations of the effects of cyano substituents on the Cope rearrangement," *J. Am. Chem. Soc.* **1990**, *112*, 2018–2019.
80. Dupuis, M.; Murray, C.; Davidson, E. R. "The Cope rearrangement revisited," *J. Am. Chem. Soc.* **1991**, *113*, 9756–9759.
81. Hrovat, D. A.; Morokuma, K.; Borden, W. T. "The Cope rearrangement revisited again. Results of ab initio calculations beyond the CASSCF level," *J. Am. Chem. Soc.* **1994**, *116*, 1072–1076.
82. Kozlowski, P. M.; Dupuis, M.; Davidson, E. R. "The Cope rearrangement revisited with multireference perturbation theory," *J. Am. Chem. Soc.* **1995**, *117*, 774–778.
83. Szalay, P. G.; Bartlett, R. J. "Approximately extensive modifications of the multireference configuration interaction method: a theoretical and practical analysis," *J. Chem. Phys.* **1995**, *103*, 3600–3612.
84. Ventura, E.; Andrade do Monte, S.; Dallos, M.; Lischka, H. "Cope rearrangement of 1,5-hexadiene: full geometry optimizations using analytic MR-CISD and MR-AQCC gradient methods," *J. Phys. Chem. A* **2003**, *107*, 1175–1180.
85. Kowalski, K.; Piecuch, P. "The method of moments of coupled-cluster equations and the renormalized CCSD[T], CCSD(T), CCSD(TQ), and CCSDT(Q) approaches," *J. Chem. Phys.* **2000**, *113*, 18–35.
86. Jiao, H.; Schleyer, P. v. R. "The Cope rearrangement transition structure is not diradicaloid, but is it aromatic?" *Angew. Chem. Int. Ed. Engl.* **1995**, *34*, 334–337.

87. Wiest, O.; Black, K. A.; Houk, K. N. "Density functional theory isotope effects and activation energies for the Cope and Claisen rearrangements," *J. Am. Chem. Soc.* **1994**, *116*, 10336–10337.
88. McGuire, M. J.; Piecuch, P. "Balancing dynamic and nondynamic correlation for diradical and aromatic transition states: a renormalized coupled-cluster study of the Cope rearrangement of 1,5-hexadiene," *J. Am. Chem. Soc.* **2005**, *127*, 2608–2614.
89. Black, K. A.; Wilsey, S.; Houk, K. N. "Dissociative and associative mechanisms of Cope rearrangements of fluorinated 1,5-hexadienes and 2,2'-bis- methylenecyclopentanes," *J. Am. Chem. Soc.* **2003**, *125*, 6715–6724.
90. Brown, E. C.; Bader, R. F. W.; Werstiuk, N. H. "QTAIM study on the degenerate Cope rearrangements of 1,5-hexadiene and semibullvalene," *J. Phys. Chem. A* **2009**, *113*, 3254–3265.
91. Shea, K. J.; Phillips, R. B. "Diastereomeric transition states. Relative energies of the chair and boat reaction pathways in the Cope rearrangement," *J. Am. Chem. Soc.* **1980**, *102*, 3156–3162.
92. Dolbier Jr., W. R.; Palmer, K. W. "Effect of terminal fluorine substitution on the Cope rearrangement: boat versus chair transition state. Evidence for a very significant fluorine steric effect," *J. Am. Chem. Soc.* **1993**, *115*, 9349–9350.
93. Davidson, E. R. "How robust is present-day DFT?" *Int. J. Quantum Chem.* **1998**, *69*, 241–245.
94. Staroverov, V. N.; Davidson, E. R. "Transition regions in the Cope rearrangement of 1,5-hexadiene and its cyano derivatives," *J. Am. Chem. Soc.* **2000**, *122*, 7377–7385.
95. Borden, W. T.; Davidson, E. R. "The importance of including dynamic electron correlation in ab initio calculations," *Acc. Chem. Res.* **1996**, *29*, 67–75.
96. Andersson, K.; Malmqvist, P.-Å.; Roos, B. O. "Second-order perturbation theory with a complete active space self-consistent field reference function," *J. Chem. Phys.* **1992**, *96*, 1218–1226.
97. Hirao, K. "Multireference Moeller-Plesset method," *Chem. Phys. Lett.* **1992**, *190*, 374–380.
98. Doering, W. v. E.; Wang, Y. "Perturbation of Cope's rearrangment: 1,3,5-triphenylhexa-1,5-diene. Chameleonic or centauric transition region?" *J. Am. Chem. Soc.* **1999**, *121*, 10112–10118.
99. Doering, W. v. E.; Wang, Y. "*Crypto*Cope rearrangement of 1,3-dicyano-5-phenyl-4,4-d_2-hexa-2,5-diene. Chameleonic or centauric?" *J. Am. Chem. Soc.* **1999**, *121*, 10967–10975.
100. Hrovat, D. A.; Beno, B. R.; Lange, H.; Yoo, H.-Y.; Houk, K. N.; Borden, W. T. "A Becke3LYP/6-31G* study of the Cope rearrangements of substituted 1,5-hexadienes provides computational evidence for a chameleonic transition state," *J. Am. Chem. Soc.* **1999**, *121*, 10529–10537.
101. Hrovat, D. A.; Chen, J.; Houk, K. N.; Thatcher, B. W. "Cooperative and competitive substituent effects on the Cope rearrangements of phenyl-substituted 1,5-hexadienes elucidated by Becke3LYP/6-31G* calculations," *J. Am. Chem. Soc.* **2000**, *122*, 7456–7460.
102. Doering, W. v. E.; Birladeanu, L.; Sarma, K.; Teles, J. H.; Klaerner, F.-G.; Gehrke, J.-S. "Perturbation of the degenerate, concerted Cope rearrangement by two phenyl groups in active positions of (E)-1,4-diphenylhexa-1,5-diene. Acceleration by high pressure as criterion of cyclic transition states," *J. Am. Chem. Soc.* **1994**, *116*, 4289–4297.

103. Doering, W. v. E.; Birladeanu, L.; Sarma, K.; Blaschke, G.; Scheidemantel, U.; Boese, R.; Benet-Bucholz, J.; Klarner, F.-G.; Gehrke, J.-S.; Zimny, B. U.; Sustmann, R.; Korth, H.-G. "A non-Cope among the Cope rearrangements of 1,3,4,6-tetraphenylhexa-1,5-dienes," *J. Am. Chem. Soc.* **2000**, *122*, 193–203.

104. Roth, W. R.; Lennartz, H. W.; Doering, W. v. E.; Birladeanu, L.; Guyton, C. A.; Kitagawa, T. "A frustrated Cope rearrangement: thermal interconversion of 2,6-diphenylhepta-1,6-diene and 1,5-diphenylbicyclo[3.2.0]heptane," *J. Am. Chem. Soc.* **1990**, *112*, 1722–1732.

105. Hrovat, D. A.; Borden, W. T. "A simple mathematical model for the cooperative and competitive substituent effects found in the Cope rearrangements of phenyl-substituted 1,5-hexadienes," *J. Chem. Theory and Comput.* **2005**, *1*, 87–94.

106. Hayase, S.; Hrovat, D. A.; Borden, W. T. "A B3LYP study of the effects of phenyl substituents on 1,5-hydrogen shifts in 3-(Z)-1,3-pentadiene provides evidence against a chameleonic transition structure," *J. Am. Chem. Soc.* **2004**, *126*, 10028–10034.

107. Doering, W. v. E.; Keliher, E. J.; Zhao, X. "Perturbations by phenyl on the 1,5-hydrogen shift in 1,3(Z)-pentadiene. Another chameleonic transition region?" *J. Am. Chem. Soc.* **2004**, *126*, 14206–14216.

108. Navarro-Vazquez, A.; Prall, M.; Schreiner, P. R. "Cope reaction families: to be or not to be a biradical," *Org. Lett.* **2004**, *6*, 2981–2984.

109. Huang, J.; Kertesz, M. "Stepwise Cope rearrangement of cyclo-biphenalenyl via an unusual multicenter covalent π-bonded intermediate," *J. Am. Chem. Soc.* **2006**, *128*, 7277–7286.

110. Nicolaou, K. C.; Dai, W.-M. "Chemistry and biology of the enediyne anticancer antibiotics," *Angew. Chem. Int. Ed. Engl.* **1991**, *30*, 1387–1416.

111. Borders, D. B.; Doyle, T. W., Eds. *Enediyne Antibiotics as Antitumor Agents*; Marcel Dekker: New York, **1994**.

112. Smith, A. L.; Nicolaou, K. C. "The enediyne antibiotics," *J. Med. Chem.* **1996**, *39*, 2103–2117.

113. Edo, K.; Mizugaki, M.; Koide, Y.; Seto, H.; Furihata, K.; Otake, N.; Ishida, N. "The structure of neocarzionostatin chromophore possessing a novel bicyclo(7,3,0)dodecadiyne system," *Tetrahedron Lett.* **1985**, *26*, 331–334.

114. Lee, M. D.; Dunne, T. S.; Siegel, M. M.; Chang, C. C.; Morton, G. O.; Borders, D. B. "Calichemicin, a novel family of antitumor antibiotics 1. Chemistry and partial structure of calichemicin," *J. Am. Chem. Soc.* **1987**, *109*, 3464–3466.

115. Lee, M. D.; Dunne, T. S.; Chang, C. C.; Ellestad, G. A.; Siegel, M. M.; Morton, G. O.; McGahren, W. J.; Borders, D. B. "Calichemicin, a novel family of antitumor antibiotics 2. Chemistry and structure of calichemicin," *J. Am. Chem. Soc.* **1987**, *109*, 3466–3468.

116. Golik, J.; Clardy, J.; Dubay, G.; Groenewold, G.; Kawaguchi, H.; Konishi, M.; Krishnan, B.; Ohkuma, H.; Saitoh, K.; Doyle, T. W. "Esperamicins, a novel class of potent antitumor antibiotics. 2. Structure of esperamicin X," *J. Am. Chem. Soc.* **1987**, *109*, 3461–3462.

117. Golik, J.; Dubay, G.; Groenewold, G.; Kawaguchi, H.; Konishi, M.; Krishnan, B.; Ohkuma, H.; Saitoh, K.; Doyle, T. W. "Esperamicins, a novel class of potent antitumor antibiotics. 3. Structures of esperamicins A1, A2, and A1b," *J. Am. Chem. Soc.* **1987**, *109*, 3462–3464.

118. Konishi, M.; Ohkuma, H.; Matsumoto, K.; Tsuno, T.; Kamei, H.; Miyaki, T.; Oki, T.; Kawaguchi, H. "Dynemicin a, a novel antibiotic with the anthraquinone and 1,5-diyn-3-ene subunit," *J. Antibiot.* **1989**, *42*, 1449–1452.

119. Konishi, M.; Ohkuma, H.; Tsuno, T.; Oki, T.; VanDuyne, G. D.; Clardy, J. "Crystal and molecular structure of dynemicin A: a novel 1,5-diyn-3-ene antitumor antibiotic," *J. Am. Chem. Soc.* **1990**, *112*, 3715–3716.

120. Leet, J. E.; Schroeder, D. R.; Hofstead, S. J.; Golik, J.; Colson, K. L.; Huang, S.; Klohr, S. E.; Doyle, T. W.; Matson, J. A. "Kedarcidin, a new chromoprotein antitumor antibiotic: structure elucidation of kedarcidin chromophore," *J. Am. Chem. Soc.* **1992**, *114*, 7946–7948.

121. Leet, J. E.; Schroeder, D. R.; Langley, D. R.; Colson, K. L.; Huang, S.; Klohr, S. E.; Lee, M. S.; Golik, J.; Hofstead, S. J.; Doyle, T. W.; Matson, J. A. "Chemistry and structure elucidation of the kedarcidin chromophore," *J. Am. Chem. Soc.* **1993**, *115*, 8432–8443.

122. McDonald, L. A.; Capson, T. L.; Krishnamurthy, G.; Ding, W.-D.; Ellestad, G. A.; Bernan, V. S.; Maiese, W. M.; Lassota, P.; Discafini, C.; Kramer, R. A.; Ireland, C. M. "Namenamicin, a new enediyne antitumor antibiotic from the marine ascidian polysyncraton lithostrotum," *J. Am. Chem. Soc.* **1996**, *118*, 10898–10899.

123. Yoshida, K.; Minami, Y.; Azuma, R.; Saeki, M.; Otani, T. "Structure and cycloaromatization of a novel enediyne, C-1027 chromophore," *Tetrahedron Lett.* **1993**, *34*, 2637–2640.

124. Davies, J.; Wang, H.; Taylor, T.; Warabi, K.; Huang, X.-H.; Andersen, R. J. "Uncialamycin, a new enediyne antibiotic," *Org. Lett.* **2005**, *7*, 5233–5236.

125. Jones, R. R.; Bergman, R. G. "p-benzyne. Generation as an intermediate in a thermal isomerization reaction and trapping evidence for the 1,4-benzenediyl structure," *J. Am. Chem. Soc.* **1972**, *94*, 660–661.

126. Bergman, R. G. "Reactive 1,4-dehydroaromatics," *Acc. Chem. Res.* **1973**, *6*, 25–31.

127. Roth, W. R.; Hopf, H.; Horn, C. "1,3,5-cyclohexatrien-1,4-diyl und 2,4-cyclohexadien-1,4-diyl," *Chem. Ber.* **1994**, *127*, 1765–1779.

128. Lockhart, T. P.; Comita, P. B.; Bergman, R. G. "Kinetic evidence for the formation of discrete 1,4-dehydrobenzene intermediates. Trapping by inter- and intramolecular hydrogen atom transfer and observation of high-temperature CIDNP," *J. Am. Chem. Soc.* **1981**, *103*, 4082–4090.

129. Nicolaou, K. C.; Zuccarello, G.; Ogawa, Y.; Schweiger, E. J.; Kumazawa, T. "Cyclic conjugated enediynes related to calicheamicins and esperamicins: calculations, synthesis, and properties," *J. Am. Chem. Soc.* **1988**, *110*, 4866–4868.

130. Nicolaou, K. C.; Zuccarello, G.; Riemer, C.; Estevez, V. A.; Dai, W. M. "Design, synthesis, and study of simple monocyclic conjugated enediynes. The 10-membered ring enediyne moiety of the enediyne anticancer antibiotics," *J. Am. Chem. Soc.* **1992**, *114*, 7360–7371.

131. Andersson, K.; Roos, B. O. "Multiconfigurational second-order perturbation theory: a test of geometries and binding energies," *Int. J. Quantum Chem.* **1993**, *45*, 591–607.

132. Gräfenstein, J.; Hjerpe, A. M.; Kraka, E.; Cremer, D. "An accurate description of the Bergman reaction using restricted and unrestricted DFT: stability test, spin density, and on-top pair density," *J. Phys. Chem. A* **2000**, *104*, 1748–1761.

133. Koga, N.; Morokuma, K. "Comparison of biradical formation between enediyne and enyn-allene. Ab initio CASSCF and MRSDCI study," *J. Am. Chem. Soc.* **1991**, *113*, 1907–1911.

134. Lindh, R.; Persson, B. J. "Ab initio study of the Bergman reaction: the autoaromatization of hex-3-ene-1,5-diyne," *J. Am. Chem. Soc.* **1994**, *116*, 4963–4969.

135. Lindh, R.; Lee, T. J.; Bernhardsson, A.; Persson, B. J.; Karlstroem, G. "Extended ab initio and theoretical thermodynamics studies of the Bergman reaction and the energy splitting of the singlet *o*-, *m*-, and *p*-benzynes," *J. Am. Chem. Soc.* **1995**, *117*, 7186–7194.

136. Dong, H.; Chen, B.-Z.; Huang, M.-B.; Lindh, R. "The Bergman cyclizations of the enediyne and its N-substituted analogs using multiconfigurational second-order perturbation theory," *J. Comput. Chem.* **2012**, *33*, 537–549.

137. Kraka, E.; Cremer, D. "CCSD(T) investigation of the Bergman cyclization of enediyne. Relative stability of *o*-, *m*-, and *p*-didehydrobenzene," *J. Am. Chem. Soc.* **1994**, *116*, 4929–4936.

138. Cramer, C. J. "Bergman, aza-Bergman, and protonated aza-Bergman cyclizations and intermediate 2,5-arynes: chemistry and challenges to computation," *J. Am. Chem. Soc.* **1998**, *120*, 6261–6269.

139. Schreiner, P. R. "Monocyclic enediynes: relationships between ring sizes, alkyne carbon distances, cyclization barriers, and hydrogen abstraction reactions. Singlet-triplet separations of methyl-substituted p-benzynes," *J. Am. Chem. Soc.* **1998**, *120*, 4184–4190.

140. Chen, W.-C.; Chang, N.-y.; Yu, C.-h. "Density functional study of Bergman cyclization of enediynes," *J. Phys. Chem. A* **1998**, *102*, 2584–2593.

141. McMahon, R. J.; Halter, R. J.; Fimmen, R. L.; Wilson, R. J.; Peebles, S. A.; Kuczkowski, R. L.; Stanton, J. F. "Equilibrium structure of cis-hex-3-ene-1,5-diyne and relevance to the Bergman cyclization," *J. Am. Chem. Soc.* **2000**, *122*, 939–949.

142. Galbraith, J. M.; Schreiner, P. R.; Harris, N.; Wei, W.; Wittkopp, A.; Shaik, S. "A valence bond study of the Bergman cyclization: geometric features, resonance energy, and nucleus-independent chemical shift (NICS) values," *Chem. Eur. J.* **2000**, *6*, 1446–1454.

143. Stahl, F.; Moran, D.; Schleyer, P. v. R.; Prall, M.; Schreiner, P. R. "Aromaticity of the Bergman, Myers-Saito, Schmittel, and directly related cyclizations of enediynes," *J. Org. Chem.* **2002**, *67*, 1453–1461.

144. Nicolaou, K. C.; Smith, A. L.; Yue, E. W. "Chemistry and biology of natural and designed enediynes," *Proc. Natl. Acad. Sci. U. S. A.* **1993**, *90*, 5881–5888.

145. Magnus, P.; Carter, P. A. "A model for the proposed mechanism of action of the potent antitumor antibiotic esperamicin A_1," *J. Am. Chem. Soc.* **1988**, *110*, 1626–1628.

146. Magnus, P.; Lewis, R. T.; Huffman, J. C. "Synthesis of a remarkably stable bicyclo[7.3.1]diynene esperamicin A_1/calicheamicin γ system. Structural requirements for facile formation of a 1,4-diyl," *J. Am. Chem. Soc.* **1988**, *110*, 6921–6923.

147. Magnus, P.; Fortt, S.; Pitterna, T.; Snyder, J. P. "Synthetic and mechanistic studies on esperamicin A_1 and calichemicin γ_1. Molecular strain rather than π-bond proximity determines the cycloaromatization rates of bicyclo[7.3.1]enediynes," *J. Am. Chem. Soc.* **1990**, *112*, 4986–4987.

148. Snyder, J. P. "The cyclization of calichemicin-esperamicin analogs: a predictive biradicaloid transition state," *J. Am. Chem. Soc.* **1989**, *111*, 7630–7632.

149. Snyder, J. P. "Monocyclic enediyne collapse to 1,4-diyl biradicals: a pathway under strain control," *J. Am. Chem. Soc.* **1990**, *112*, 5367–5369.
150. Gaffney, S. M.; Capitani, J. F.; Castaldo, L.; Mitra, A. "Critical distance model for the energy of activation of the Bergman cyclization of enediynes," *Int. J. Quantum Chem.* **2003**, *95*, 706–712.
151. Tuttle, T.; Kraka, E.; Cremer, D. "Docking, triggering, and biological activity of dynemicin a in DNA: a computational study," *J. Am. Chem. Soc.* **2005**, *127*, 9469–9484.
152. Myers, A. G.; Kuo, E. Y.; Finney, N. S. "Thermal generation of α,3-dehydrotoluene from (Z)-1,2,4-heptatrien-6-yne," *J. Am. Chem. Soc.* **1989**, *111*, 8057–8059.
153. Nagata, R.; Yamanaka, H.; Okazaki, E.; Saito, I. "Biradical formation from acyclic conjugated eneyne-allene system related to neocarzinostatin and esperamicin-calichemicin," *Tetrahedron Lett.* **1989**, *30*, 4995–4998.
154. Schmittel, M.; Strittmatter, M.; Kiau, S. "Switching from the Myers reaction to a new thermal cyclization mode in enyne-allenes," *Tetrahedron Lett.* **1995**, *36*, 4975–4978.
155. Schmittel, M.; Keller, M.; Kiau, S.; Strittmatter, M. "A surprising switch from the Myers-Saito cyclization to a novel biradical cyclization in enyne-allenes: formal Diels-Alder and ene reactions with high synthetic potential," *Chem. Eur. J.* **1997**, *3*, 807–816.
156. Schmittel, M.; Vavilala, C.; Cinar, M. E. "The thermal C^2-C^6 (schmittel)/ene cyclization of enyne–allenes – crossing the boundary between classical and nonstatistical kinetics," *J. Phys. Org. Chem.* **2012**, *25*, 182–197.
157. Nicolaou, K. C.; Maligres, P.; Shin, J.; De Leon, E.; Rideout, D. "DNA-cleavage and antitumor activity of designed molecules with conjugated phosphine oxide-allene-ene-yne functionalities," *J. Am. Chem. Soc.* **1990**, *112*, 7825–7826.
158. Schmittel, M.; Kiau, S.; Strittmatter, M. "Steric effects in enyne-allene thermolyses: switch from the Myers-Saito reaction to the C^2-C^6-cyclization and DNA strand cleavage," *Tetrahedron Lett.* **1996**, *37*, 7691–7694.
159. Engels, B.; Hanrath, M. "A theoretical comparison of two competing diradical cyclizations in enyne-allenes: the Myers-Saito and the novel C^2-C^6 cyclization," *J. Am. Chem. Soc.* **1998**, *120*, 6356–6361.
160. Schreiner, P. R.; Prall, M. "Myers-Saito versus C^2-C^6 ("Schmittel") cyclizations of parent and monocyclic enyne-allenes: challenges to chemistry and computation," *J. Am. Chem. Soc.* **1999**, *121*, 8615–8627.
161. Chen, W.-C.; Zou, J.-W.; Yu, C.-H. "Density functional study of the ring effect on the Myers-Saito cyclization and a comparison with the Bergman cyclization," *J. Org. Chem.* **2003**, *68*, 3663–3672.
162. Wenthold, P. G.; Lipton, M. A. "A density functional molecular orbital study of the C^2-C^7 and C^2-C^6 cyclization pathways of 1,2,4-heptatrien-6-ynes. The role of benzannulation," *J. Am. Chem. Soc.* **2000**, *122*, 9265–9270.
163. Cramer, C. J.; Kormos, B. L.; Seierstad, M.; Sherer, E. C.; Winget, P. "Biradical and zwitterionic cyclizations of oxy-substituted enyne-allenes," *Org. Lett.* **2001**, *3*, 1881–1884.
164. Cremeens, M. E.; Hughes, T. S.; Carpenter, B. K. "Mechanistic studies on the cyclization of (Z)-1,2,4-heptatrien-6-yne in methanol: a possible nonadiabatic thermal reaction," *J. Am. Chem. Soc.* **2005**, *127*, 6652–6661.

165. Bekele, T.; Christian, C. F.; Lipton, M. A.; Singleton, D. A. ""Concerted" transition state, stepwise mechanism. Dynamics effects in C^2—C^6 enyne allene cyclizations," *J. Am. Chem. Soc.* **2005**, *127*, 9216–9223.

166. Engels, B.; Lennartz, C.; Hanrath, M.; Schmittel, M.; Strittmatter, M. "Regioselectivity of biradical cyclizations of enyne-allenes: influence of substituents on the switch from the Myers-Saito to the novel C^2—C^6 cyclization," *Angew. Chem. Int. Ed.* **1998**, *37*, 1960–1963.

167. Sakai, S.; Nishitani, M. "Theoretical studies on Myers–Saito and Schmittel cyclization mechanisms of hepta-1,2,4-triene-6-yne," *J. Phys. Chem. A* **2010**, *114*, 11807–11813.

168. Schmittel, M.; Steffen, J.-P.; Maywald, M.; Engels, B.; Helten, H.; Musch, P. "Ring size effects in the C^2—C^6 biradical cyclisation of enyne-allenes and the relevance for neocarzinostatin," *J. Chem. Soc., Perkin Trans. 2* **2001**, 1331–1339.

169. Cramer, C. J.; Squires, R. R. "Quantum Chemical Characterization of the Cyclization of the Neocarzinostatin Chromophore to the 1,5-Didehydroindene Biradical," *Org. Lett.* **1999**, *1*, 215–218.

170. Musch, P. W.; Engels, B. "Which Structural Elements Are Relevant for the Efficacy of Neocarzinostatin?" *Angew. Chem. Int. Ed.* **2001**, *40*, 3833–3836.

171. Prall, M.; Wittkopp, A.; Schreiner, P. R. "Can fulvenes form from enediynes? A systematic high-level computational study on parent and benzannelated enediyne and enyne–allene cyclizations," *J. Phys. Chem. A* **2001**, *105*, 9265–9274.

172. Vavilala, C.; Byrne, N.; Kraml, C. M.; Ho, D. M.; Pascal, R. A. "Thermal C^1—C^5 diradical cyclization of enediynes," *J. Am. Chem. Soc.* **2008**, *130*, 13549–13551.

173. Caramella, P.; Quadrelli, P.; Toma, L. "An unexpected bispericyclic transition structure leading to 4+2 and 2+4 cycloadducts in the endo dimerization of cyclopentadiene," *J. Am. Chem. Soc.* **2002**, *124*, 1130–1131.

174. Salem, L. "Intermolecular orbital theory of the interaction between conjugated systems. II. Thermal and photochemical cycloadditions," *J. Am. Chem. Soc.* **1968**, *90*, 553–566.

175. Birney, D. M.; Houk, K. N. "Transition structures of the Lewis acid-catalyzed Diels-Alder reaction of butadiene with acrolein. The origins of selectivity," *J. Am. Chem. Soc.* **1990**, *112*, 4127–4133.

176. Quadrelli, P.; Romano, S.; Toma, L.; Caramella, P. "Merging and bifurcation of 4+2 and 2+4 cycloaddition modes in the archetypal dimerization of butadiene. A case of competing bispericyclic, pericyclic and diradical paths," *Tetrahedron Lett.* **2002**, *43*, 8785–8789.

177. Quadrelli, P.; Romano, S.; Toma, L.; Caramella, P. "A bispericyclic transition structure allows for efficient relief of antiaromaticity enhancing reactivity and endo stereoselectivity in the dimerization of the fleeting cyclopentadienone," *J. Org. Chem.* **2003**, *68*, 6035–6038.

178. Leach, A. G.; Goldstein, E.; Houk, K. N. "A cornucopia of cycloadducts: theoretical predictions of the mechanisms and products of the reactions of cyclopentadiene with cycloheptatriene," *J. Am. Chem. Soc.* **2003**, *125*, 8330–8339.

179. Celebi-Olcum, N.; Ess, D. H.; Aviyente, V.; Houk, K. N. "Lewis acid catalysis alters the shapes and products of bis-pericyclic Diels-Alder transition states," *J. Am. Chem. Soc.* **2007**, *129*, 4528–4529.

180. Ess, D. H.; Hayden, A. E.; Klärner, F.-G.; Houk, K. N. "Transition states for the dimerization of 1,3-cyclohexadiene: a DFT, CASPT2, and CBS-QB3 quantum mechanical investigation," *J. Org. Chem.* **2008**, *73*, 7586–7592.

181. Ross, J. A.; Seiders, R. P.; Lemal, D. M. "An extraordinarily facile sulfoxide automerization," *J. Am. Chem. Soc.* **1976**, *98*, 4325–4327.

182. Birney, D. M. "Further pseudopericyclic reactions: an ab initio study of the conformations and reactions of 5-oxo-2,4-pentadienal and related molecules," *J. Org. Chem.* **1996**, *61*, 243–251.

183. Rodríguez-Otero, J.; Cabaleiro-Lago, E. M. "Criteria for the elucidation of the pseudopericyclic character of the cyclization of (Z)-1,2,4,6-heptatetraene and its heterosubstituted analogues: magnetic properties and natural bond orbital analysis," *Chem. Eur. J.* **2003**, *9*, 1837–1843.

184. Birney, D. M.; Wagenseller, P. E. "An ab initio study of the reactivity of formylketene. Pseudopericyclic reactions revisited," *J. Am. Chem. Soc.* **1994**, *116*, 6262–6270.

185. Birney, D. M.; Xu, X.; Ham, S.; Huang, X. "Chemoselectivity in the reactions of acetylketene and acetimidoylketene: confirmation of theoretical predictions," *J. Org. Chem.* **1997**, *62*, 7114–7120.

186. Ham, S.; Birney, D. M. "Imidoylketene: an ab initio study of its conformations and reactions," *J. Org. Chem.* **1996**, *61*, 3962–3968.

187. Wagenseller, P. E.; Birney, D. M.; Roy, D. "On the development of aromaticity in cycloadditions: ab initio transition structures for the trimerization of acetylene and for the addition of ethylene and acetylene to formylketene," *J. Org. Chem.* **1995**, *60*, 2853–2859.

188. Birney, D. M. "Electrocyclic ring openings of 2-furylcarbene and related carbenes: a comparison between pseudopericyclic and coarctate reactions," *J. Am. Chem. Soc.* **2000**, *122*, 10917–10925.

189. Forte, L.; Lafortune, M. C.; Bierzynski, I. R.; Duncan, J. A. "CASSCF molecular orbital calculations reveal a purely pseudopericyclic mechanism for a [3,3] sigmatropic rearrangement," *J. Am. Chem. Soc.* **2010**, *132*, 2196–2201.

190. de Lera, A. R.; Alvarez, R.; Lecea, B.; Torrado, A.; Cossío, F. P. "On the aromatic character of electrocyclic and pseudopericyclic reactions: thermal cyclization of (2Z)-hexa-2,4-5-trienals and their schiff bases," *Angew. Chem. Int. Ed.* **2001**, *40*, 557–561.

191. Rodríguez-Otero, J.; Cabaleiro-Lago, E. M. "Electrocyclization of (Z)-1,2,4,6-Heptatetraene and its Heterosubstituted Analogues: Pericyclic or Pseudopericyclic?" *Angew. Chem. Int. Ed.* **2002**, *41*, 1147–1150.

192. de Lera, A. R.; Cossio, F. P. "Reply," *Angew. Chem. Int. Ed.* **2002**, *41*, 1150–1152.

193. Lowry, T. H.; Richardson, K. S. *Mechanism and Theory in Organic Chemistry*; 3rd ed.; Harper and Row: New York, **1987**.

194. Duncan, J. A.; Calkins, D. E. G.; Chavarha, M. "Secondary orbital effect in the electrocyclic ring closure of 7-azahepta-1,2,4,6-tetraene – a CASSCF molecular orbital study," *J. Am. Chem. Soc.* **2008**, *130*, 6740–6748.

195. DePuy, C. H.; Schnack, L. G.; Hausser, J. W. "Chemistry of cyclopropanols. IV. The solvolysis of cycopropyl tosylates," *J. Am. Chem. Soc.* **1966**, *88*, 3343–3346.

196. Schleyer, P. v. R.; Su, T. M.; Saunders, M.; Rosenfeld, J. C. "Stereochemistry of allyl cations from the isomeric 2,3-dimethylcyclopropyl chlorides. Stereomutations of allyl cations," *J. Am. Chem. Soc.* **1969**, *91*, 5174–5176.

197. DePuy, C. H. "The chemistry of cyclopropanols," *Acc. Chem. Res.* **1968**, *1*, 33–41.
198. Faza, O. N.; Lopez, C. S.; Alvarez, R.; de Lera, A. R. "The Woodward-Hoffmann-De Puy rule revisited," *Org. Lett.* **2004**, *6*, 905–908.
199. Criegee, R.; Seebach, D.; Winter, R. E.; Boerretzen, B.; Brune, H. A. "Cyclobutenes. XXI. Valency isomerization of cyclobutenes," *Chem. Ber.* **1965**, *98*, 2339–2352.
200. Dolbier, W. R., Jr.; Koroniak, H.; Burton, D. J.; Bailey, A. R.; Shaw, G. S.; Hansen, S. W. "Remarkable, contrasteric, electrocyclic ring opening of a cyclobutene," *J. Am. Chem. Soc.* **1984**, *106*, 1871–1872.
201. Kirmse, W.; Rondan, N. G.; Houk, K. N. "Stereoselective substituent effects on conrotatory electrocyclic reactions of cyclobutenes," *J. Am. Chem. Soc.* **1984**, *106*, 7989–7991.
202. Rondan, N. G.; Houk, K. N. "Theory of stereoselection in conrotatory electrocyclic reactions of substituted cyclobutenes," *J. Am. Chem. Soc.* **1985**, *107*, 2099–2111.
203. Dolbier, W. R., Jr.; Koroniak, H.; Houk, K. N.; Sheu, C. "Electronic control of stereoselectivities of electrocyclic reactions of cyclobutenes: a triumph of theory in the prediction of organic reactions," *Acc. Chem. Res.* **1996**, *20*, 471–477.
204. Lee, P. S.; Zhang, X.; Houk, K. N. "Origins of inward torquoselectivity by silyl groups and other σ-acceptors in electrocyclic reactions of cyclobutenes," *J. Am. Chem. Soc.* **2003**, *125*, 5072–5079.
205. Rudolf, K.; Spellmeyer, D. C.; Houk, K. N. "Prediction and experimental verification of the stereoselective electrocyclization of 3-formylcyclobutene," *J. Org. Chem.* **1987**, *52*, 3708–3710.
206. Buda, A. B.; Wang, Y.; Houk, K. N. "Acid-base-controlled torquoselectivity: theoretical predictions of the stereochemical course of the electrocyclic reactions of cyclobutene-3-carboxylic acid and the conjugate base and acid," *J. Org. Chem.* **1989**, *54*, 2264–2266.
207. Niwayama, S.; Kallel, E. A.; Spellmeyer, D. C.; Sheu, C.; Houk, K. N. "Substituent effects on rates and stereoselectivities of conrotatory electrocyclic reactions of cyclobutenes. A theoretical study," *J. Org. Chem.* **1996**, *61*, 2813–2825.
208. Murakami, M.; Miyamoto, Y.; Ito, Y. "Stereoselective synthesis of isomeric functionalized 1,3-dienes from cyclobutenones," *J. Am. Chem. Soc.* **2001**, *123*, 6441–6442.
209. Murakami, M.; Usui, I.; Hasegawa, M.; Matsuda, T. "Contrasteric stereochemical dictation of the cyclobutene ring-opening reaction by a vacant boron p orbital," *J. Am. Chem. Soc.* **2005**, *127*, 1366–1367.
210. Houk, K. N.; Spellmeyer, D. C.; Jefford, C. W.; Rimbault, C. G.; Wang, Y.; Miller, R. D. "Electronic control of the stereoselectivities of electrocyclic reaction of cyclobutenes against incredible steric odds," *J. Org. Chem.* **1988**, *53*, 2125–2127.
211. Niwayama, S.; Houk, K. N. "Competition between ester and formyl groups for control of torquoselectivity in cyclobutene electrocyclic reactions," *Tetrahedron Lett.* **1992**, *33*, 883–886.
212. Niwayama, S.; Wang, Y.; Houk, K. N. "The torquoselectivity of electrocyclic reactions of 3-donor-3-acceptor-substituted cyclobutenes," *Tetrahedron Lett.* **1995**, *36*, 6201–6204.
213. Murakami, M.; Hasegawa, M. "Synthesis and thermal ring opening of trans-3,4-disilylcyclobutene," *Angew. Chem. Int. Ed.* **2004**, *43*, 4874–4876.

214. Um, J. M.; Xu, H.; Houk, K. N.; Tang, W. "Thermodynamic control of the electrocyclic ring opening of cyclobutenes: C-X substituents at C-3 mask the kinetic torquoselectivity," *J. Am. Chem. Soc.* **2009**, *131*, 6664–6665.
215. Kallel, E. A.; Houk, K. N. "Theoretical predictions of torquoselectivity in pentadienyl cation electrocyclizations," *J. Org. Chem.* **1989**, *54*, 6006–6008.
216. Smith, D. A.; Ulmer, C. W., II "Theoretical studies of the Nazarov cyclization 3. Torquoselectivity and hyperconjugation in the Nazarov cyclization. The effects of inner versus outer β-methyl and β-silyl groups," *J. Org. Chem.* **1993**, *58*, 4118–4121.
217. Harmata, M.; Schreiner, P. R.; Lee, D. R.; Kirchhoefer, P. L. "Combined computational and experimental studies of the mechanism and scope of the retro-Nazarov reaction," *J. Am. Chem. Soc.* **2004**, *126*, 10954–10957.
218. Evanseck, J. D.; Thomas IV, B. E.; Spellmeyer, D. C.; Houk, K. N. "Transition structures of thermally allowed disrotatory electrocyclizations. The prediction of stereoselective substituent effects in six-electron pericyclic reactions," *J. Org. Chem.* **1995**, *60*, 7134–7141.
219. Thomas IV, B. E.; Evanseck, J. D.; Houk, K. N. "Electrocyclic reactions of 1-substituted 1,3,5,7-octatetraenes. An ab initio molecular orbital study of torquoselectivity in eight-electron electrocyclizations," *J. Am. Chem. Soc.* **1993**, *115*, 4165–4169.
220. Allen, J. G.; Hentemann, M. F.; Danishefsky, S. J. "A powerful *o*-quinone dimethide strategy for intermolecular Diels-Alder cycloadditions," *J. Am. Chem. Soc.* **2000**, *122*, 571–575.
221. Paquette, L. A.; Feng Geng, F. "A highly abbreviated synthesis of pentalenene by means of the squarate ester cascade," *Org. Lett.* **2002**, *4*, 4547–4549.
222. Murakami, M.; Miyamoto, Y.; Ito, Y. "Stereoselective Synthesis of Isomeric Functionalized 1,3-Dienes from Cyclobutenones," *J. Am. Chem. Soc.* **2001**, *123*, 6441–6442.

CHAPTER 5

Diradicals and Carbenes

Reactive intermediates play a critical role in the chemist's understanding of many reaction mechanisms.[1] Given their often fleeting appearance, experimental observation, let alone acquiring any property data on them, can be a great challenge. It is not, therefore, surprising that computational chemistry has been a strong partner with experiment in gaining knowledge of a broad spectrum of reactive intermediates.

In this chapter, we focus on the class of reactive intermediates that bear at least two unpaired electrons: diradicals and carbenes. The exact definition of a diradical is somewhat in the eye of the beholder. Salem and Rowland[2] provided perhaps the most general, yet effective, definition—a diradical is a molecule that has two degenerate or nearly degenerate orbitals occupied by two electrons. With this definition, carbenes can be considered as a subcategory of diradicals. In a carbene, the two degenerate molecular orbitals are localized about a single carbon atom.

The chemistry of radicals has been described many times.[3-6] Likewise, the chemistry of diradicals and carbenes has been the subject of many reviews.[7-16] Two issues will guide our presentation in this chapter. First and foremost, we discuss examples of diradicals and carbenes, where computational chemistry has greatly aided in understanding the properties, structure, and chemistry of these intermediates. A theme that will emerge here is the strong collaborative relationship between experimental and computational chemists that greatly aided in resolving the controversies and discrepancies that arose in trying to understand these unusual species. The second aspect of this chapter is that computational studies of diradicals and (especially) carbenes helped to establish the discipline of computational chemistry. In fact, the first example we discuss in this chapter is the nature of methylene, which is the topic that many claim to have single handedly won over skeptics to the advantages and power of computational chemistry.

Following the section on methylene, we present the chemistry of phenylcarbene and phenylnitrene and describe how computational chemistry helped detail why these two closely related molecules behave so differently. A discussion of tetramethyleneethane (TME) and oxyallyl diradicals explores how theories of apparently simple molecules may be quite complicated. Next, we discuss the chemistry of

---

Computational Organic Chemistry, Second Edition. Steven M. Bachrach
© 2014 John Wiley & Sons, Inc. Published 2014 by John Wiley & Sons, Inc.

the benzynes. These reactive diradicals became of heightened interest with the emergence of enediyne chemistry in the 1990s. This section furthers our discussion of the application of the Bergman cyclization discussed in Section 4.3. This chapter concludes with a discussion of tunneling in the rearrangement of hydroxycarbenes, which led to the development of a new control mechanism to complement kinetic and thermodynamic control.

5.1 METHYLENE

Computations of the properties of methylene played the central role in legitimizing the discipline of computational chemistry. Through the 1960s, algorithmic insufficiencies and the lack of computer power limited the scope of computations in extreme ways. Computations could be performed only with very small basis sets and, at best, limited the accounting of electron correlation. Typically, geometries were assumed or only restricted optimizations could be run. These constraints inevitably led to calculations that were erroneous and misguided, and consequently the entire field was painted with a broad brush of general neglect by most chemists.

This was all to change in 1970 when the first chink in the armor of the inherent superiority of experiment over computation appeared. Within a few years, computational results pointed out serious experimental errors and thereby secured a place for computational chemistry as an equal partner with experiment in exploring the chemical sciences.[17] All of these centered on the simple molecule methylene, CH_2, establishing what Schaefer[18] has called the "paradigm of computational quantum chemistry".

5.1.1 Theoretical Considerations of Methylene

Methylene is the simplest example of a carbene, a molecule containing a carbon formally bearing only six valence electrons. Of these, four electrons are involved in the C–H bonds. The orbital occupation of the last two electrons defines the specific electronic state of methylene. If we assume a bent structure, we can use the simple model of an sp^2-hybridized carbon. The four bonding electrons occupy two of these sp^2 hybrids. This leaves the third sp^2 hybrid ($3a_1$) and the p-orbital ($1b_1$) available for the last two electrons (see Figure 5.1). Placing one electron in each of these orbitals with their spins aligned creates a triplet state. The electronic configuration of this triplet state ($^3B_1$) is

$$\Psi(^3B_1) = |1a_1^2 2a_1^2 1b_2^2 1b_1^1 3a_1^1|$$

The singlet state can be formed by doubly occupying the $3a_1$ orbital. This, however, neglects the possibility of double occupation of the $1b_1$ orbital, an orbital not too much higher in energy. Thus, a two-configuration wavefunction is necessary to adequately describe the lowest singlet state ($^1A_1$)

$$\Psi(^1A_1) = c_1|1a_1^2 2a_1^2 1b_2^2 3a_1^2| + c_2|1a_1^2 2a_1^2 1b_2^2 1b_1^2|$$

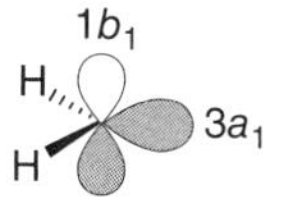

Figure 5.1 Highest occupied orbitals of methylene.

All configuration interaction (CI) based treatments must utilize at least the one-configuration reference for the triplet and the two-configuration reference for the singlet.

5.1.2 The H–C–H Angle in Triplet Methylene

The year 1970 proved to be the watershed for methylene. It started out with a general consensus that triplet methylene was linear. Though recognizing some small internal discrepancies, Herzberg[19–21] concluded that triplet methylene is linear based on a series of spectroscopic examinations. This conclusion was in contrast to the results from one of the first serious computation of a polyatomic molecule, a single-*zeta* near self-consistent field (SCF) followed by limited CI computation of the triplet methylene molecule.[22] The optimized geometry had an angle of 129°. Unfortunately, the strength of Herzberg's reputation and a devastating critique by the theorist Longuet-Higgins[23] effectively negated the computational work.

In 1970, Bender and Schaefer[24] reported *ab initio* computations of triplet methylene. Employing the CISD/DZ method, they computed the energy of triplet methylene at 48 different geometries, varying the C–H distance and H–C–H angle. Fitting this surface to a quadratic function, they predicted that the H–C–H angle is 135.1°, and emphatically concluded that the molecule is not linear.

This report was quickly followed by two solid-state electron spin resonance (ESR) studies, which also indicated a bent structure with an angle of 136°.[25,26] By the end of the year, Herzberg[27] reexamined his UV spectra and conceded that in fact triplet methylene is bent. His estimate of 136° is remarkably similar to the computed value. Theory had established a beachhead!

Experimentalists and theorists continue to refine the potential energy surface (PES) of triplet methylene. Using a Morse oscillator-rigid bender internal dynamics Hamiltonian, a broad array of experimental data were fit to a PES, giving the best experimental value of the H–C–H angle as 133.9308°.[28] A sampling of some of the computed values of the H–C–H angle of triplet methylene is listed in Table 5.1. Hartree–Fock (HF) optimizations consistently underestimate the angle (about 129.5°), while CISD with at least a TZ basis set brings it to about 133°. Density functional theory (DFT) methods overestimate the angle by 1–1.5°. Both MRCISD and RCCSD(T) optimization with a TZ or better basis sets give structures in excellent agreement with experiment; their estimates of the H–C–H angle range from 133.42° to 133.82°.

TABLE 5.1 H–C–H Angle of $^3B_1$ Methylene

Method	H–C–H Angle
HF/DZP[a]	129.36
HF/TZ2P[a]	129.47
HF/TZ3P[a]	129.44
HF/TZ3P((2f,2d)+2diff[a]	129.50
CASSCF/TZP(f)[b]	133.06
limited CI/SZ[c]	129
CISD/DZ[d]	135.1
CISD/DZP[a]	131.79
CISD/TZ2P[a]	132.84
CISD/TZ3P[a]	132.68
CISD/TZP((2f,2d)+2diff[a]	132.94
RCCSD(T)/cc-pVTZ[e]	133.46
RCCSD(T)/cc-pVQZ[e]	133.67
RCCSD(T)/cc-pV5Z[e]	133.79
RCCSD(T)/d-aug-cc-pV6Z[f]	133.82
MRCISD/cc-pVTZ[e]	133.42
MRCISD/cc-pVQZ[e]	133.59
MRCISD/cc-pV5Z[e]	133.70
MRCISD/d-aug-cc-pV6Z[f]	133.66
BVWN5/cc-pVTZ[g]	134.8
BLYP/cc-pVTZ[g]	135.4
B3LYP/aug-cc-pVDZ[h]	135.5
B3LYP/aug-cc-pVTZ[h]	135.2
B3PW91/aug-cc-pVDZ[h]	135.1
Expt.[i]	133.9308

[a] Ref. 29.
[b] Ref. 30.
[c] Ref. 22.
[d] Ref. 24.
[e] Ref. 31.
[f] Ref. 32.
[g] Ref. 33.
[h] Ref. 34.
[i] Ref. 28.

5.1.3 The Methylene and Dichloromethylene Singlet–Triplet Energy Gap

Following directly on the success of determining the bond angle in triplet methylene, computational chemists turned to the size of the methylene singlet–triplet energy gap.[17,18,35] Methylene is a ground-state triplet ($X\ ^3B_1$) and its first excited state is the $\tilde{a}\ ^1A_1$ state. Prior to 1972, experiments placed the value of their energy difference (ΔE_{ST}) at either 1–2 kcal mol^{-1} or 8–9 kcal mol^{-1}.[35] Appearing nearly coincident in 1972, Goddard and Schaefer independently computed ΔE_{ST} for methylene. Goddard[36] used a GVB-CI wavefunction with a DZ+P basis set and estimated the energy gap as 11.5 kcal mol^{-1}. Schaefer[37] used a first-order CI with a DZ+P basis set to predict that the triplet is 11.0 kcal mol^{-1} below the singlet. Their work also pointed out some of the computational difficulties in this problem. The gap is 29.3 kcal mol^{-1} at HF/DZ, which reduces to 24.9 kcal mol^{-1} upon inclusion of d-functions on carbon. Therefore, ΔE_{ST} is highly sensitive to both the basis set and electron correlation. Theory seemed to have helped settle the experimental discrepancies in favor of the higher (~9 kcal mol^{-1}) barrier.

This calm was shattered by the first direct measurement of the singlet–triplet gap of methylene by Lineberger[38] in 1976. Laser photoelectron spectrometry of CH_2^- provided a spectrum interpreted as transitions to singlet and triplet CH_2. Despite some difficulty in arriving at the 0–0 transition, Lineberger concluded that ΔE_{ST} was equal to 19.6 kcal mol^{-1}, far above the theoretical (or previous experimental) predictions. However, since this was the first direct observation of the gap, it was considered to be the most accurate determination yet.

A number of computational papers appeared in the following year responding to Lineberger's ΔE_{ST} value. Goddard[39] performed a larger GVB-POL-CI computation and obtained a value of 11.1 kcal mol^{-1}. Schaefer used a larger basis set than before with a CI expansion and still found a gap of 13.5 kcal mol^{-1}.[40] Addition of diffuse functions or expanding the carbon sp basis had little effect, though estimation of the effect of higher order excitations reduced the ΔE_{ST} value to 11.3 kcal mol^{-1}. Roos[41] employed a multireference-CI wavefunction and a slightly larger than DZ+2P basis set, finding ΔE_{ST} = 10.9 kcal mol^{-1}. He then estimated a zero-point vibrational energy correction that lowers the gap by another 0.4 kcal mol^{-1}.

One year later, Harding and Goddard[42] computed the PES of CH_2^- along with the singlet and triplet CH_2 surfaces to model the photoelectron spectrum. They suggested that three, and only three, "hot bands"—originating from vibrationally excited CH_2^-—appear in the experimental spectrum, resulting in the too large singlet–triplet measurement. Theoreticians had dug in their heals, held to their convictions and offered an explanation for the faulty experiment.

Resolution came in the early 1980s. Using far-IR laser magnetic resonance, McKellar and coworkers[43] were able to directly measure the singlet–triplet energy difference. Their result was 9.05 ± 0.06 kcal mol^{-1}. By estimating the zero-point vibrational energy, they were also able to offer a value of 9.08 ± 0.18 kcal mol^{-1} for T_e, the energy difference of the minimum on each surface. This value can be directly compared to computation. A year later, Lineberger[44] reported a new PES experiment, this one introducing CH_2^- from a flowing afterglow apparatus. This technique provides much cooler reactant than the previous PES experiments. The resulting PES is notable for its absence of the previously suggested "hot bands," leading Lineberger to conclude that the gap is only 9 kcal mol^{-1}. The experimental and theoretical values for the methylene singlet–triplet energy gap were now in accord. The theorists were vindicated. Computational chemistry had earned its place as a significant tool in interpreting results and understanding chemistry.

A compilation of computed methylene singlet–triplet energy gaps is presented in Table 5.2. The best experimental value is 9.215 kcal mol^{-1} for T_e and 8.998 kcal mol^{-1} for T_0.[28] The HF method grossly overestimates the gap, due to the necessity of including the second configuration in the reference description of the $^1A_1$ state. This overestimation persists at MP2, where again the insufficient single-configuration reference is used.[45] A simple two-configuration SCF treatment brings the gap to about 11 kcal mol^{-1}. A large basis-set effect is

TABLE 5.2 Singlet–Triplet Energy Gap (kcal mol^{-1}) of Methylene

Method	ΔE_{ST}[a]
HF/DZP[b]	26.39 (25.99)
HF/TZ2P[b]	25.13 (24.74)
HF/TZ3P[b]	25.03 (24.63)
TC-SCF/DZP[b]	13.58 (13.12)
TC-SCF/TZ2P[b]	11.33 (10.90)
TC-SCF/TZ3P[b]	10.94 (10.52)
HF/TZ3P(2f,2d)+2diff[b]	10.49 (10.06)
CISD(TC-SCF)/DZP[b]	12.61 (12.13)
CISD(TC-SCF)/TZ2P[b]	11.02 (10.54)
CISD(TC-SCF)/TZ3P[b]	10.50 (10.05)
CISD(TC-SCF)/TZ3P(2f,2d)+2diff[b]	9.25 (8.84)
CASSCF/DZP[b]	13.50 (13.02)
CASSCF/TZ2P[b]	11.13 (10.65)
CASSCF/TZ3P[b]	10.75 (10.30)
CASSCF/TZ3P(2f,2d)+2diff[b]	10.25 (9.84)
CASSCF/TZP(f)[c]	10.8
CASSCF-SOCI/DZP[b]	12.65 (12.17)
CASSCF-SOCI/TZ2P[b]	11.12 (10.64)
CASSCF-SOCI/TZ3P[b]	10.62 (10.17)
CASSCF-SOCI/TZ3P(2f,2d)+2diff[b]	9.44 (9.02)
UMP2/6-31G*[d]	(20.1)
UMP3/6-31G*//UMP2/6-31G*[d]	(18.0)
UMP4/SDTQ/6-31G*//UMP/6-31G*[d]	(16.7)
RCCSD/cc-pVTZ[e]	11.67
RCCSD/cc-pVQZ[e]	11.01
RCCSD(T)/cc-pVTZ[f]	10.33
RCCSD(T)/cc-pVQZ[f]	9.80
RCCSD(T)/cc-pV5Z[f]	9.58
RCCSD(T)/CBS[g]	9.2
RCCSDT/cc-pVTZ[e]	10.54
MRCISD/cc-pVTZ[f]	10.05
MRCISD/cc-pVQZ[f]	9.50
MRCISD/cc-pV5Z[f]	9.32
MRCISD/d-aug-cc-pV6Z[h]	9.178 (8.822)
MR-BWCCSD/cc-pVTZ[e]	10.18
MR-BWCCSD/cc-pVQZ[e]	9.33

TABLE 5.2 (*Continued*)

Method	$\Delta E_{ST}{}^a$
MR-BWCCSDT/cc-pVQZ[i]	9.51
MR-ccCA-AQCC/S_DT[j]	8.7
BVWN5/cc-pVTZ[k]	(9.22)
BLYP/cc-pVTZ[k]	(9.68)
B3LYP/aug-cc-pVDZ[l]	11.15 (10.77)
B3LYP/aug-cc-pVTZ[l]	11.38 (10.92)
B3LYP/6-311+G(3df,2p)[l]	11.27 (10.92)
B3PW91/aug-cc-pVDZ[l]	15.95 (15.54)
B3PW91/6-311+G(3df,2p)[l]	16.28 (15.88)
G3MP2[m]	9.4
Expt.[n]	9.215 (8.998)

[a] T_e values in regular text, T_0 values within parentheses.
[b] Ref. 29.
[c] Ref. 30.
[d] Ref. 45.
[e] Ref. 46.
[f] Ref. 31.
[g] Ref. 50.
[h] Ref. 32.
[i] Ref. 47.
[j] Ref. 48.
[k] Ref. 33.
[l] Ref. 34.
[m] Ref. 49.
[n] Ref. 28.

seen at this level; the gap decreases from 13.6 kcal mol^{-1} with a DZP basis to 11.3 kcal mol^{-1} with a TZ2P basis to 10.9 with a TZ3P basis. Further expansion with a set of two diffuse functions along with two f-functions on C and two d-functions on H reduces the gap by another 0.5 kcal mol^{-1}.[29] This basis set effect on ΔE_{ST} is seen with *all* ab initio methods. With at least at TZP basis set, all of the methods predict a gap of 9–11 kcal mol^{-1}. DFT gives the values in this range as well, except for B3PW91. The best calculation to date (MRCISD/d-aug-cc-pV6Z) predicts a gap that differs with experiment by only 0.1 kcal mol^{-1}.[32]

A similar controversy, only recently resolved, arose with Lineberger's negative ion photoelectron spectrum of CCl_2^- used to obtain the singlet–triplet gap of dichloromethylene.[51] The ground state of CCl_2 is $^1A_1$ and it first excited state is $^3B_1$. The PES results suggest that the singlet is only 3 kcal mol^{-1} below the triplet state. This is in contradiction with a wide variety of theoretical studies that predict a difference of 20–24 kcal mol^{-1}.[52–55] The largest computations are those of Schaefer.[56] At CCSD(T)/cc-pVQZ, ΔE_{ST} is -20.1 kcal mol^{-1}. (The negative sign indicates a ground-state singlet.) Adding a set of diffuse s-, p-, d-, f- , and g-functions increased the gap to -20.3 kcal mol^{-1}. Using the core-valence basis set (CCSD(T)/cc-pCVQZ//CCSD(T)/cc-pVQZ) reduced the gap to -19.3 kcal mol^{-1}, leading to a best estimate of -19.5 kcal mol^{-1}.

McKee and Michl[57] performed a series of G3 computations to rationalize the disagreement between the computations and experiment. Their computed energy for the electron detachment $^2CCl_2^- \rightarrow {}^1CCl_2$ of 35 kcal mol^{-1} is in fine agreement with Lineberger's experimental value of 37 kcal mol^{-1}. On the other hand, their computed value for $^2CCl_2^- \rightarrow {}^3CCl_2$ (89 kcal mol^{-1}) disagrees with the value assigned by Lineberger (62 kcal mol^{-1}). Rather, they claim that this signal arises from either the electron detachment from the excited quartet state ($^4CCl_2^- \rightarrow {}^3CCl_2$: 65 kcal mol^{-1}) or from the quartet state of its isomeric species ($^4C=Cl-Cl^- \rightarrow {}^3C=Cl-Cl^-$: 59 kcal mol^{-1}).

In fact, the original spectroscopic experiment was contaminated with a small amount of $CHCl_2^-$. Lineberger[58] repeated the experiment using CD_2Cl_2 as the source material to better aid in separating out CCl_2^- from $CDCl_2^-$ and also subtracting out the spectrum of authentic $CHCl_2^-$. This led to clear identification of the $^1A_1$ state of CCl_2. The onset energy for the formation of the $^3B_1$ state could only be estimated based on the comparison with simulations. Nonetheless, the revised best experimental estimate of the singlet–triplet of CCl_2 is 20.8 kcal mol^{-1}, which is in fine agreement with all of the computational estimates.

5.2 PHENYLNITRENE AND PHENYLCARBENE

The development of the chemistry of aryl nitrenes has occurred primarily over the last 20 years, even though initial forays were made nearly a century ago.[59] This slow development can be attributed to the fact that photolysis or pyrolysis of typical aryl azides, such as phenylnitrene **1**, produces polymeric tar (Reaction 5.1). This is in sharp contrast to the rich chemistry of phenylcarbene (**2**), which exhibits insertion into C–H bonds, addition to π-bonds, reaction with oxygen, and so on.[14]

(Reaction 5.1)

Instead of intermolecular chemistry, phenylnitrene (**1**) rapidly undergoes intramolecular rearrangement. Substituted azepine (**3**) is the product of both thermolysis[60,61] and photolysis[62] of phenyl azide in the presence of amine. On the other hand, substituted aniline (**4**) is obtained when phenyl azide is photolyzed with ethanthiol.[63]

Intermolecular chemistry of phenylnitrene does occur if the ortho positions are occupied. The first such demonstration was the thermolysis of perfluorophenyl azide, producing perfluorophenylnitrene, which inserts into C–H bonds;[64,65] a representative example of which is shown in Reaction 5.2. These ortho-protected phenylnitrenes have been exploited as crosslinking reagents in photoresists[66] and precursors for conducting polymers.[67] Perhaps their greatest use has been as photoaffinity labels,[68,69] where they have been employed in the exploration of the estrogen[70] and progesterone[71] receptors, to label LSD[72] and map protein–protein interaction in bacteriophage T4 DNA polymerase holoenzyme.[73]

(Reaction 5.2)

Significant progress has been made in the past 10 years toward understanding the properties and chemistry of phenylnitrene, particularly in its relationship to phenylcarbene. What are the low lying states of each and what gives rise to their differences? Why does phenylcarbene participate in bimolecular chemistry over rearrangement while the opposite is true for phenylnitrene? How does the rearrangement of **1** to **3** occur? What is the nature of the ortho-substituent effect upon phenylnitrene? These questions have been answered by a symbiotic approach of computational and experimental sciences. This section highlights the computational component of this collaboration, though always pointing out the important and necessary contributions made by the experimentalists. After reading this section, the reader will appreciate the power of computation and experiment working in concert to fully explicate complicated reaction chemistries.

5.2.1 The Low Lying States of Phenylnitrene and Phenylcarbene

The electronic structure of phenylcarbene is analogous to that of methylene. To obtain the lowest energy states, the two nonbonding electrons can be placed into two molecular orbitals that are similar to the two MOs of methylene shown in Figure 5.1. Again, the in-plane MO (*21a'*) has significant s-character and is lower in energy than the p-orbital (*4a"*). Unlike the case in methylene, the neighboring phenyl group can donate electron density into this p-orbital, especially in the singlet state where this orbital is empty. This orbital interaction is clearly seen in MO 25 (*4a"*) of Figure 5.2. This donation will stabilize the singlet state more than the triplet state, thus reducing the singlet–triplet energy gap in phenylcarbene from that in methylene.

The results of a number of ab initio computations of **2** are listed in Table 5.3. In all the cases, the triplet is predicted to be the ground state, in accord with experiment.[14] CASSCF predicts an energy difference of about 10 kcal mol^{-1}; and correction for dynamic correlation increases the gap.[45] These values

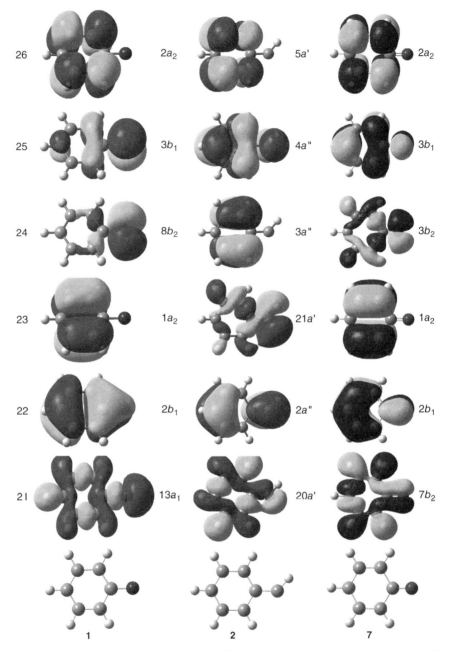

Figure 5.2 Molecular orbitals of **1**, **2**, and **7**. (*See insert for color representation of this figure.*)

TABLE 5.3 Singlet–Triplet Energy Gap (kcal mol^{-1}) of Phenylcarbene (2)

Method	ΔE_{ST}
R(O)HF/6-31G*[a]	17.3
MP2/cc-pVDZ//BLYP/6-31G*[b]	7.02
MP2/cc-pVTZ//BLYP/6-31G*[b]	5.52
CASSCF(8,8)/6-31G*[a]	13.3
CASSCF(8,8)/6-311G(2d,p)//CASSCF(8,8)/6-31G*[a]	8.5
CASPT2N(8,8)/6-31G*[a]	14.1
CASPT2N(8,8)/6-311G(2d,p)// CASSCF(8,8)/6-31G*[a]	12.5
BLYP/6-31G*[b]	5.54
BLYP/6-311+G(3df,2p)//BLYP/6-31G*[c]	3.6
(U)B3LYP/6-31G*[a]	7.4
(U)B3LYP/6-311+G*//(U)B3LYP/6-31G*[a]	5.0
BVWN5/cc-pVTZ//BVWN5/cc-pVDZ[d]	4.3
CISD/DZP[a]	10.2
CCSD/cc-pVDZ//BLYP/6-31G*[b]	5.80
CCSD/DZP//CISD/DZP[a]	7.7
CCSD(T)/cc-pVDZ//BLYP/6-31G*[b]	4.63
CCSD(T)/DZP//CISD/DZP[a]	6.2
CCSD(T)/CBS//B3LYP/6-31+G(d,p)[e]	3.5
G2(RMP2,SVP)//QCI[c]	1.4
G3MP2[f]	2.32

[a]Ref. 45.
[b]Ref. 74.
[c]Ref. 75.
[d]Ref. 77.
[e]Ref. 76.
[f]Ref. 49.

are not in accord with any of the other high level computations. DFT[45,74,75] and CCSD(T)[45,74,76] calculations place the gap at about 5 kcal mol^{-1}. A G2(RMP2/SVP)[75] calculation suggests an even smaller gap of 1.4 kcal mol^{-1}. Unfortunately, singlet phenylcarbene has not been spectroscopically observed, and the singlet–triplet gap has not been measured.[59] Nevertheless, the best calculations clearly indicate that the phenyl substituent reduces the ΔE_{ST} from 9 kcal mol^{-1} in methylene to 2–5 kcal mol^{-1} in **2**.

To understand the electronic structure of phenylnitrene, we first consider the simpler compound NH, an analog of methylene. In NH, one pair of nonbonding electrons on nitrogen occupies a hybrid orbital that is largely sp-like. The other two nonbonding electrons are distributed into the two perpendicular and degenerate p-orbitals on nitrogen. In phenylnitrene, this degeneracy is lifted, and we can consider the p-orbital in the plane of the phenyl ring and the p-orbital perpendicular to the plane (see Figure 5.3). Unlike the case in phenylcarbene, where the in-plane orbital has significant s character that stabilizes it relative to the perpendicular p-orbital, the two p-orbitals in **1** are only slightly different in energy. The triplet

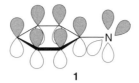

Figure 5.3 Schematic of the AOs of **1**.

state $^3A_2$ is therefore the ground state and has occupancy

$$\Psi(^3A_2) = |13a_1{}^2 2b_2{}^2 1a_2{}^2 8b_2{}^1 3b_1{}^1|$$

and its molecular orbitals are shown in Figure 5.2.

There are a number of possible singlet configurations of **1**. The first possibility, analogous to methylene and phenylcarbene, is to doubly occupy the in-plane nitrogen p-orbital ($8b_2$, see Figure 5.2), producing the $^1A_1$ state. A first clue that this state might not be the lowest singlet state is that unlike the case in **2**, the in-plane p-orbital of **1** ($8b_2$) is not much lower in energy than the perpendicular p-orbital ($3b_1$). In fact, distributing one electron into each of these p-orbitals proves to be energetically favorable in **1**, making the lowest singlet the $^1A_2$ state, which has the same electron occupancy as the $^3A_2$ state but with singlet spin coupling. A third possibility is to doubly occupy the perpendicular p-orbital, which gives another $^1A_1$ configuration. In fact, one might expect that a two-configuration wavefunction is needed to properly describe both the lower and higher energy $^1A_1$ states:

$$\Psi(^1A_1) = c_1|\ldots 13a_1{}^2 2b_2{}^2 1a_2{}^2 8b_2{}^2| + c_2|\ldots 13a_1{}^2 2b_2{}^2 1a_2{}^2 3b_1{}^2|$$

where c_1 and c_2 have opposite signs in the lower $^1A_1$ state and the same sign in the upper state.

An often-overlooked aspect of molecular orbital theory is that orbital shapes and energies will differ between the singlet and triplet states of a molecule. The highest lying occupied MOs of triplet phenylnitrenes are shown in Figure 5.2. MOs 21, 22, and 23 are doubly occupied, and MOs 24 and 25 are singly occupied with their spins aligned. The highest occupied orbitals for $^1A_1$ phenylnitrene are shown in Figure 5.4. These are all singly occupied orbitals, three with α spin and three with β spin, generated from the unrestricted Hartree–Fock (UHF) wavefunction. The α MO 22 and β MO 23 correspond with MO 22 of triplet **1**, however their shapes differ. Note that the node in the two orbitals of singlet **1** isolates the C–N fragment from the rest of the ring, while the node bisects the ring in MO 22 of the triplet. The in-plane singly occupied orbital of the triplet (MO 24) drops in relative energy ordering in the singlet due to more delocalization of density about the ring. The largest difference in MOs between the two electronic states is with the MO that describes the perpendicular p-orbital. In the triplet state, this is MO 25, which has its greatest density at the nitrogen. In the singlet state, this is α MO 24, which has its greatest density at the ring carbon para to the nitrogen. The consequences of these different MO shapes will be discussed later.

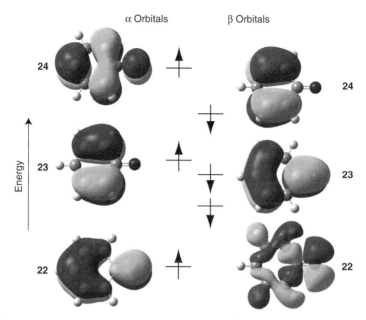

Figure 5.4 Highest lying occupied UHF orbitals of the $^1A_1$ state of phenylnitrene (**1**). (*See insert for color representation of this figure.*)

CISD,[78] CASSCF,[79,80] CASPT2N,[80] and MR-AQCC[81] (multireference averaged quadratic coupled-cluster[82]) computations all agree on the nature and energies of the low lying states of **1**. The ground state is the triplet, in accord with experiments.[59] The lowest lying singlet state is $^1A_2$. The best calculations (Table 5.4) predict that the singlet–triplet gap is 18.3 kcal mol^{-1} (CCSD+Q[78]), 18.5 kcal mol^{-1} (CASPT2N[80]), or 15.9 kcal mol^{-1} (MR-AQCC[81]). The initial experimental determination *via* negative ion photoelectron spectroscopy[83] suggested a singlet–triplet gap of 18 ± 2 kcal mol^{-1}. Wenthold[84] reexamined the spectra of phenylnitrene anion and suggests that the gap is somewhat smaller, 14.8 ± 0.5 kcal mol^{-1}. The more recent CCSD and MR-AQCC computations are in good accord with this revised estimate. The next lowest singlet state is the $^1A_1$ state dominated by the configuration that doubly occupies the in-plane orbital ($8b_2$); this state is about 30 kcal mol^{-1} above the triplet.[85]

The two lowest states of **1** differ in their electronic configuration only by their spin coupling, yet their geometries show significant differences (Figure 5.5). In particular, at CASSCF(8,8)/6-31G*, the C–N distance is 1.338 Å in the triplet but only 1.276 Å in the singlet $^1A_2$ state.[80] (A similar geometric difference is found at CISD/DZ+d[78] and MR-AQCC.[81]) The singlet state has considerable double-bond character between N and C, and in fact the singlet is better represented by the resonance structure **5**, which delocalizes the electron in the perpendicular p-orbital off of nitrogen and into the phenyl π-system. This delocalization acts to physically separate the two nonbonding electrons, localizing one onto nitrogen in the in-plane orbital and the second into the phenyl π-orbitals. This singlet is thereby stabilized,

TABLE 5.4 Singlet–Triplet Energy Gap (kcal mol⁻¹) of Phenylnitrene (1)

Method	ΔE_{ST}
HF/DZ+d[a]	24.8
CSISD/DZ+d[a]	20.1
CSISD+Q/DZ+d[a]	18.3
σ-S,π-SDCI/6-31G*//CASSCF(8,8)/3-21G[b]	18.3
CASSCF(8,8)/6-31G*[c]	17.5
CASSCF(8,8)/6-311G(2d,p)//CASSCF(8,8)/6-31G*[c]	17.1
CASPT2N/6-31G*//CASSCF(8,8)/6-31G*[c]	19.1
CASPT2N/6-311G(2d,p)//CASSCF(8,8)/6-31G*[c]	18.5
CASPT2/cc-pVDZ//CASSCF(8,8)/cc-pVDZ[d]	19.3
BPW91/cc-pVDZ[d]	14.3
SF-CCSD(T)/cc-pVTZ//SF-CCSD/cc-pVDZ[e]	15.4
MR-AQCC/cc-pVTZ//CASSCF(8,8)/cc-pVTZ[f]	17.0
MR-AQCC/CBS//CASSCF(8,8)/cc-pVTZ[f]	15.9
Expt.	18±2[g]
	14.8±0.5[h]

[a]Ref. 78.
[b]Ref. 79.
[c]Ref. 80.
[d]Ref. 86.
[e]Ref. 87.
[f]Ref. 81.
[g]Ref. 83.
[h]Ref. 84.

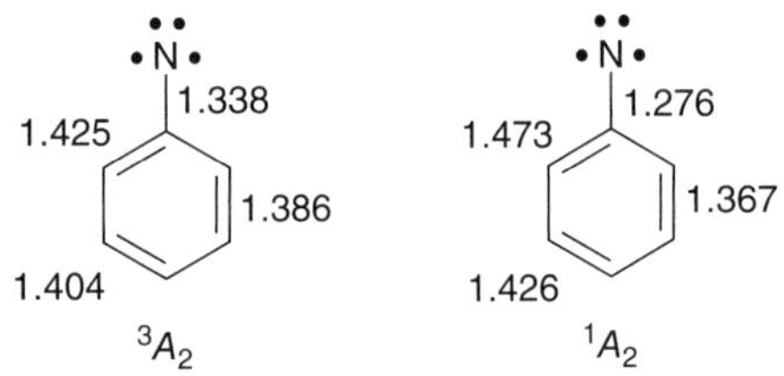

Figure 5.5 CASSCF(8,8)/6-31G* bond lengths in the lowest singlet and triplet states of **1**.

relative to the lowest singlet state of NH, by the reduced repulsion energy between these two electrons.[88] The stabilization afforded by the phenyl group is substantial. The singlet–triplet gap of NH is 36.4 kcal mol⁻¹.[89] The phenyl substituent in **1** reduces this splitting by 20 kcal mol⁻¹. This is in contrast to the circa 5 kcal mol⁻¹ stabilization, relative to methylene, of the singlet state of **2** afforded by its phenyl group.

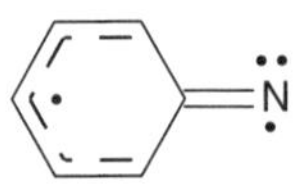

Quantum computations assisted in the spectroscopic identification of the triplet and especially the singlet state of phenylnitrene.[90] The electronic absorption spectrum of triplet **1** is dominated by a strong band at 308 nm and a broad band at 370 nm. CASPT2 was used to obtain the vertical excitation energies from the ground triplet state of **1**. The transitions to the second and third $^3B_1$ states are predicted to be at 301 and 299 nm, respectively, both with strong oscillator strength, in excellent agreement with experiment. The transition to the second $^3A_2$ state is calculated to be at 393 nm, in reasonable agreement with experiment. The strongest computed (CASPT2) excitation for singlet **1** is to the second $^1B_1$ state, with a frequency of 368 nm. This absorption is predicted to be at 217 nm with CASSCF(12,12), demonstrating the critical need for inclusion of dynamical correlation in computing excited state energies. The experimental spectra of singlet **1**, obtained by laser flash photolysis of phenyl azide, shows a very strong absorption at 350 nm, with the computed spectra verifying this assignment.

With the large differences in ΔE_{ST} between **1** and **2** and the dramatic differences in their reactivities, the suggestion had been made that triplet phenylnitrene is thermodynamically more stable than triplet phenylcarbene.[91,92] Borden[93] used 3-pyridylcarbene (**6**), an isomer of **1**, in order to make a direct comparison in the stability of **1** versus an appropriate carbene. First, Borden established that 3-methylpyridine and its isomer N-methylaniline have nearly identical enthalpies (within 3 kcal mol^{-1}), as computed with CASPT2N/6-31G*//CASSCF(8,8)/6-31G*, and BVWN5/aug-cc-pVDZ. Further, the radical derived from these two molecules by the loss of a hydrogen (3-PyCH$_2$ and PhNH) is also nearly identical in enthalpy (within 3 kcal mol^{-1}). This implies that (1) the N–H bond dissociation energy of primary amines is nearly identical to the C–H bond dissociation energy of primary alkanes and (2) phenyl and 3-pyridyl are comparable in stabilizing a π-electron.

6

Next, he demonstrated that **1** is about 25–26 kcal mol^{-1} lower in enthalpy than **6**. Combining these results in Reaction 5.3 allows for an alternate way of comparing the stabilities of **1** with **6**. The enthalpy of Reaction 5.3 ranges from −21.9 (BVWN5/aug-cc-pVDZ) to −24.8 kcal mol^{-1} (CASPTSN/6-31G*//CASSCF(8,8)/6-31G*). Therefore, it is clear that triplet nitrenes are about 20 kcal mol^{-1} more stable than triplet carbenes. Given that the singlet–triplet separation in **1** is about 13 kcal mol^{-1} greater than in **2**, singlet phenylnitrene is about 7 kcal mol^{-1} more stable than singlet phenylcarbene. The stability of the nitrenes over the carbenes derives mostly from the stabilization of the σ-lone pair on nitrogen. In moving from amine, to amine radical, to nitrene (just as in moving from alkane to alkyl radical to carbene), the hybridization of nitrogen (or carbon) changes by increasing its s-character. While the C–H hybrid orbital is stabilized in progressing along this series, the nitrogen lone pair is affected to a much greater extent, since this pair of electrons resides solely on the nitrogen, rather than being shared between the carbon and hydrogen atoms. The net result is that nitrenes are thermodynamically more stable than carbenes, which will help explain some of the differences in their reactivities.

(Reaction 5.3)

An interesting comparison is also made with phenyloxenium cation **7**. It is isoelectronic with both **1** and **2**. As expected, the MOs of **7** are similar to those of **1** and **2**, as shown in Figure 5.2. There are, however, some important differences. The more electronegative oxygen atom in **7** instead of the nitrogen atom in **1** (along with the positive charge on the former) lowers the energy of the p-orbital on oxygen relative to nitrogen. This results in a higher energy for the out-of-plane p molecular orbital $3b_1$ in **7** compared to this orbital in **1**, leading to a greater break in the degeneracy of MOs 24 and 25 in **7**. One might therefore expect that the lowest energy singlet state of **7** might be described by just a single configuration $\Psi(^1A_1) = c_1 |\ldots 7b_2^2 2b_2^2 1a_2^2 3b_2^2|$, in contrast to the two-configuration wavefunction needed for the singlet state of **1**.

In fact, CASPT2/pVTZ//CASSCF(8,8)/pVTZ computations indicate that the ground state of **7** is the $^1A_1$ state, lying 22.1 kcal mol^{-1} below the lowest triplet state ($^3A_2$).[94] This is in excellent agreement with the ultraviolet photoelectron spectrum[95] if a reassignment of the bands is made. With the large split in the degeneracy of the two p-orbitals on oxygen, the triplet state requires a single-electron occupation of the more energetic MO 25 instead of doubly occupying the much lower in energy MO 24.

The geometry of $^1A_1$ **7** shows a strong quinoidal structure, indicating a large contribution of the **7q** resonance structure. **7** displays greater quinoidal character than **1**. The greater electronegativity of oxygen prefers the positive charge in **7** be delocalized into the ring rather than residing on oxygen.

5.2.2 Ring Expansion of Phenylnitrene and Phenylcarbene

Both phenylcarbene and phenylnitrene can undergo intramolecular rearrangement; however, they do so under very different conditions. Phenylcarbene will participate in normal intermolecular carbene reactions (cycloadditions and insertions[14]) except

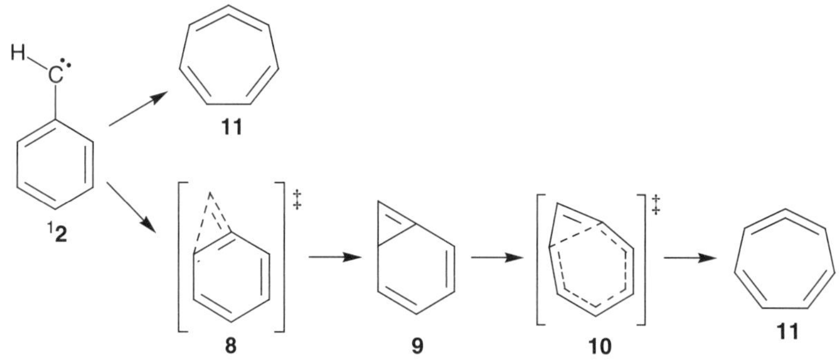

Scheme 5.1

in the gas phase[96] or when sequestered from other reagents.[97–99] Phenylnitene, on the other hand, preferentially undergoes intramolecular rearrangements[59] and substituents must be placed upon the ring to inhibit this rearrangement in order to observe intermolecular chemistry.

Ring expansion of phenylcarbene had been implicated in the gas-phase rearrangement of p-tolylcarbene,[100,101] but it was not until matrix isolation studies were carried out that conclusive evidence was obtained that indicated the product of the expansion was 1,2,4,6-cycloheptatetraene (**11**).[97,98] No intermediates have ever been detected in the rearrangement of phenylcarbene, leading to speculation over which of two mechanisms (Scheme 5.1) might be operating: a single-step ring-expansion that is unprecedented in carbene chemistry or a two-step path with an unobserved intermediate, bicycle[4.1.0]hepta-2,4,6-triene (**9**). Experimental support for the latter (two-step) mechanism is provided by the spectroscopic observation of benzannulated analogs of **9** obtained from 1- and 2-naphthylcarbene.[102,103] Ring opening of these benzannulated derivatives (**12** and **13**) does not occur since this would involve further loss of aromaticity.

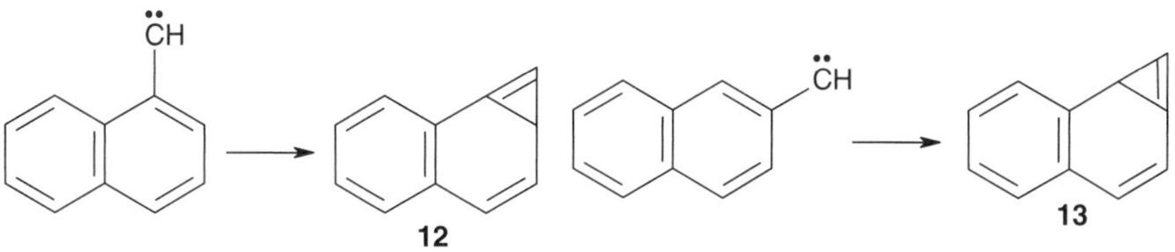

Computational chemists weighed in on this problem in 1996, when three independent reports concurred on the nature of the mechanism.[45,74,75] Using a variety of computational methods, including CASSCF, CASPT2N, CCSD, DFT, and G2(MP2,SPV), the rearrangement of ^{1}**2** → **11** was found to proceed via the two-step mechanism of Scheme 5.1. (A subsequent CCSD(T)/CBS study[76] provided identical conclusions as these three earlier computational reports.) The relative energies of all of the critical points along this reaction pathway are listed in Table 5.5 and the B3LYP/6-31G*-optimized structures are drawn in Figure 5.6.

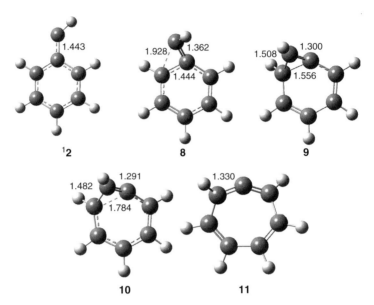

Figure 5.6 B3LYP/6-31G*-optimized geometries of the critical points for the rearrangement of ^{1}**2**.

A few important points can be discerned from Table 5.5. The first step is the rate-determining step, with a barrier of 13–15 kcal mol^{-1}, leading to the bicyclic intermediate **9** that is energetically slightly below **2**. The second step is the Woodward–Hoffmann-allowed, disrotatory, electrocyclic opening of a cyclohexadiene ring. This step has a much lower barrier (1–2 kcal mol^{-1} relative

TABLE 5.5 Relative Energies (kcal mol^{-1}) for the Critical Points along the Reaction of ^{1}2 → 11

	^{1}2	8	9	10	11
CASSCF(8,8)/cc-pVDZ//CASSCF(8,8)/6-31G*[a]	0.0	20.3	5.3	11.5	−15.8
CASSCF(8,8)/6-311G(2d,p)//CASSCF(8,8)/6-31G*[a]	0.0	20.1	5.5	11.5	−15.7
CASPT2N/cc-pVDZ// CASSCF(8,8)/6-31G*[a]	0.0	13.2	−2.9	−1.4	−17.8
CCSD/cc-pVDZ//BLYP/6-31G*[b]	0.0	17.5	0.5	4.8	−14.0
CCSD(T)/cc-pVDZ//BLYP/6-31G*[b]	0.0	15.7	−0.7	2.5	−16.3
CCSD(T)/CBS//B3LYP/6-31+G(d,p)[c]	0.0	12.8	−3.8	−1.1	−17.3
MP2/cc-pVDZ//BLYP/6-31G*[b]	0.0	13.6	−5.8	−4.7	−18.2
MP2/cc-pVTZ//BLYP/6-31G*[b]	0.0	15.1	−7.9	−7.2	−19.5
BLYP/6-31G*[b]	0.0	12.4	−0.8	−0.2	−18.2
B3LYP/6-311+G*//B3LYP/6-31G*[a]	0.0	14.9	2.5	3.6	−13.5
G2(MP2,SVP)[d]	0.0	13.1	−3.3	−1.4	−17.4

[a]Ref. 45.
[b]Ref. 74.
[c]Ref. 76.
[d]Ref. 75.

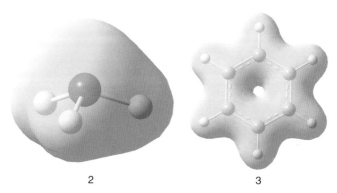

Figure 1.10 Isoelectronic surface of the total electron density of ammonia **2** and benzene **3**. Note the lack of lone pairs or a π-cloud.

Computational Organic Chemistry, Second Edition. Steven M. Bachrach.
© 2014 John Wiley & Sons, Inc. Published 2014 by John Wiley & Sons, Inc.

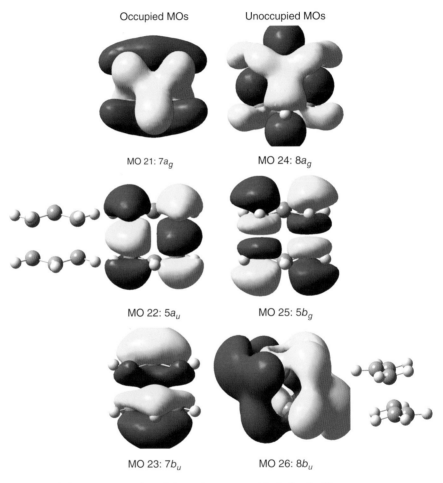

Figure 4.7 Representation of the active space orbitals for the Cope rearrangement.

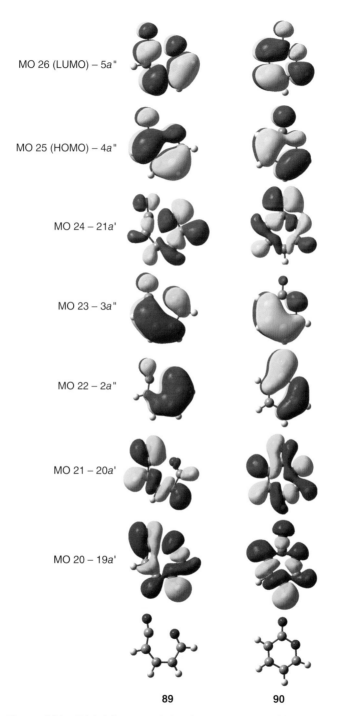

Figure 4.22 High lying occupied MOs and LUMO of **89** and **90**.

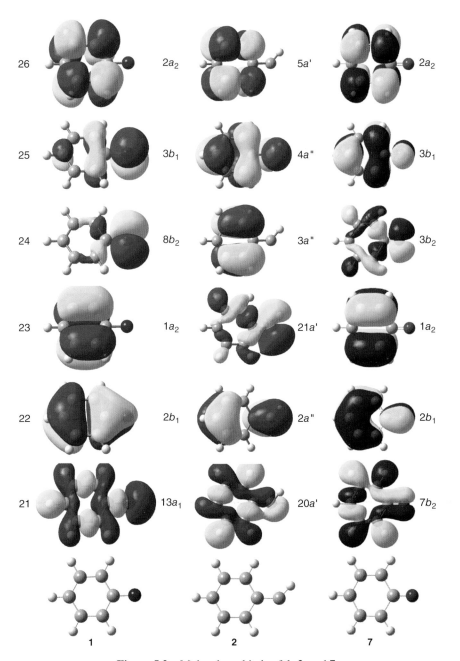

Figure 5.2 Molecular orbitals of **1**, **2**, and **7**.

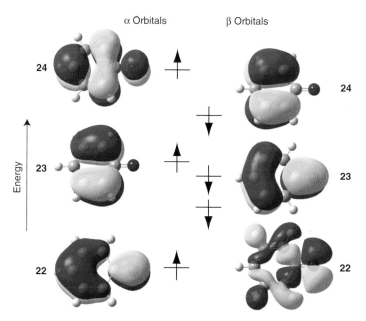

Figure 5.4 Highest lying occupied UHF orbitals of the $^1A_1$ state of phenylnitrene (**1**).

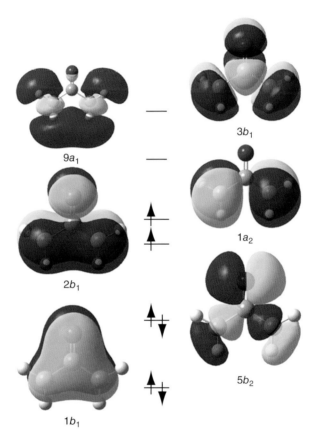

Figure 5.11 MOs of oxyallyl diradical **40**.

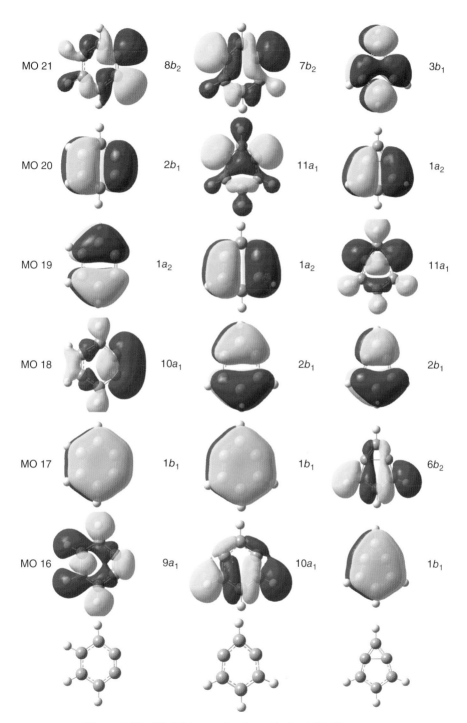

Figure 5.12 High-lying molecular orbitals of **41**, **42**, and **44**.

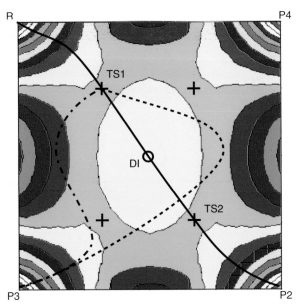

Figure 8.4 Contour plot of the PES in Figure 8.3, looking down from above. The heavy line is the direct trajectory from **TS1** across **DI** and over **TS2** to **P2**. The long-short dashed line is the semidirect trajectory from **TS1** to **P3**. The dashed line is the rebound trajectory from **TS1** to **P3**.

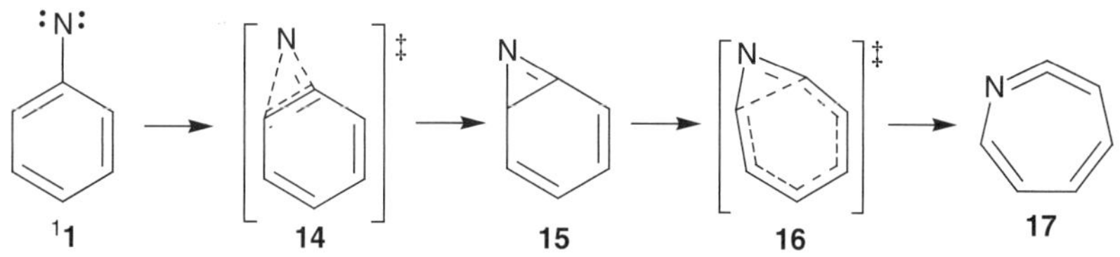

Scheme 5.2

to **9**) and is very exothermic, with **11** lying about 14–17 kcal mol^{-1} below **9**. These computations suggest that once the first barrier (**8**) is crossed, sufficient energy is available to readily cross the second barrier (**10**) and therefore little concentration of **9** will ever accumulate. Further, the large exothermicity of the overall reaction indicates that the reverse reaction is highly unlikely, consistent with the lack of observation of this process, except at high temperatures.[100,101]

Borden[80] examined the rearrangement of phenylnitrene using CASSCF and CASPT2 methods. Just as in the rearrangement of phenylcarbene, the calculations found a two-step mechanism involving a bicyclic intermediate (in this case the azirine **15**), as shown in Scheme 5.2. The energies of the critical points along this reaction path obtained by Borden, along with some later DFT[86,104,105] results, are listed in Table 5.6. The CASSSCF(8,8)/6-31G* structures are drawn in Figure 5.7.

The computed barrier for the first step is about 8–9 kcal mol^{-1}. DFT computation of the singlet state of **1** and the transition state **14** is difficult due to spin contamination and both restricted and unrestricted solutions for open shell singlets must be treated with great caution. Note here that Cramer was unable to locate an unrestricted solution of **14** and the restricted solution gives a barrier that is much

TABLE 5.6 Relative Energies (kcal mol^{-1}) for the Critical Points along the Reaction of ^{1}**1** → **17**

	^{1}**1**	**14**	**15**	**16**	**17**
CASSCF(8,8)/6-31G*^{a}	0.0	8.9	4.7	24.2	3.7
CASSCF(8,8)/cc-pVDZ//CASSCF(8,8)/6-31G*^{a}	0.0	9.0	5.3	23.7	3.2
CASSCF(8,8)/6-311G(2d,p)//CASSCF(8,8)/6-31G*^{a}	0.0	8.5	4.6	23.4	2.4
CASPT2/6-31G*// CASSCF(8,8)/6-31G*^{a}	0.0	8.6	1.6	6.8	−1.3
CASPT2/cc-pVDZ// CASSCF(8,8)/6-31G*^{a}	0.0	9.3	3.5	6.7	−1.0
CASPT2/6-311G(2d,p)// CASSCF(8,8)/6-31G*^{a}	0.0	9.2	4.1	7.2	−1.6
CASPT2/cc-pVDZ//CASSCF(8,8)/cc-pVDZ^b	0.0	8.5	2.7	5.8	−1.9
B3LYP/6-31G*^{c}			0.0	4.7	−5.1
BPW91/cc-pVDZ^b	0.0	19.4^d	5.0	7.3	1.4
BLYP/cc-pVDZ^b	0.0	11.2^d	2.5	4.1	−4.4
(U)B3LYP/6-311+G(2d,p)//(U)B3LYP/6-31G*^{e}	0.0	8.0	0.5	3.5	−6.7

^{a}Ref. 80.
^{b}Ref. 86.
^{c}Ref. 104.
^{d}Unrestricted calculations did not converge. The restricted result is listed here.
^{e}Ref. 105.

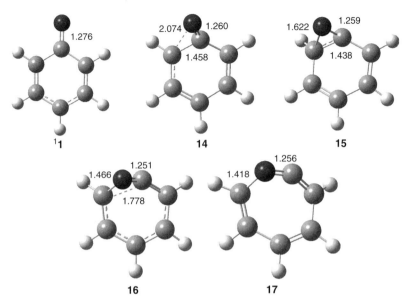

Figure 5.7 CASSCF(8,8)/6-31G*-optimized geometries of the critical points for the rearrangement of **1**.[80]

too large.[86] The intermediate **15** is about 2 kcal mol^{-1} higher in energy than **1**. As in the rearrangement of **2**, the first step for the rearrangement of **1** is predicted to be the rate-limiting step. The barrier for the second step is lower than the first, though CASSCF erroneously predicts a very large activation barrier for the electrocyclic reaction in the second step. As noted in Chapter 3, without the inclusion of dynamic electron correlation, barriers for Woodward–Hoffmann-allowed reactions are predicted to be much too high.

Laser flash photolysis of phenyl azide provided the experimental barrier for the reaction of singlet **1**: 5.6 ± 0.3 kcal mol^{-1}.[90] The best computational result (CASPT2) overestimates this barrier. However, CASPT2 tends to systematically overestimate the stability of open-shell species relative to closed-shell analogs,[106,107] and so the computed barrier, corrected for this error (estimated[85] as about 3 kcal mol^{-1}), is in fine agreement with experiment.

The barrier for the reverse reaction, **17** → **1**, has been obtained by photolysis of **1** encapsulated in a hemicarcerand and then monitoring the disappearance of **17**.[108] This experimental value for the reverse reaction is 12.3 ± 0.6 kcal mol^{-1}. Combining the experimental values for the forward and reverse activation barriers indicates that the forward reaction is exothermic by 6.7 kcal mol^{-1}. The CASPT2 calculations predict that the reaction is overall slightly exothermic. Since this method overstabilizes **1** relative to **17** by about 3 kcal mol^{-1}, the reaction is actually more exothermic than predicted by CASPT2; and the DFT values of −5 to −6 kcal mol^{-1} are quite reasonable and in excellent agreement with experiment. Thus, **17** is predicted to dominate at normal temperatures; but some singlet phenylnitrene will be present, and it can convert to the triplet ground state by intersystem crossing (ISC).

This should lead to the irreversible production of triplet **1** from **17**, which has been observed.[109,110]

How do the calculations account for the differing chemistries of **1** and **2**? The rate-limiting step for both rearrangments is the first step: formation of the bicyclic intermediate. The barrier for this step is computed to be much higher for the reaction of **2** (13–15 kcal mol^{-1}) than for **1** (8–9 kcal mol^{-1}). This larger barrier for **2** impedes the rearrangement, providing more time and a higher concentration of singlet phenylcarbene to partake in intermolecular chemistry than for **1**, which will much more readily undergo this rearrangement. The electronic structure of singlet phenylnitrene facilitates its closure to **15**. The diradical structure of singlet **1**, represented as **5**, requires only movement of nitrogen out of the plane for the radical orbitals to overlap, resulting in the closure of the three-member ring. On the other hand, singlet **2** is a closed-shell configuration. Tipping the exocyclic carbon out of the ring plane and then closing the angle to form the three-member ring requires increasing contribution of the higher energy ionic resonance structure **18**.[88]

Another difference between **1** and **2** is the rate of the intersystem crossing from the singlet to the triplet. The rate constant for ISC in **1** is $3.2 \pm 0.3 \times 10^6$ s^{-1}, as measured by laser flash photolysis.[90] Unfortunately, no such measurement is available for **2**, but ISC in **1** is about three orders of magnitude slower than in diphenylcarbene[111] or fluorenylidene.[112] Gritsan and Platz[90] proposed three reasons to account for the slow ISC of phenylnitrene. First, the singlet–triplet gap is much larger for **1** than **2**. The smaller is this gap, the faster is the rate of intersystem crossing. Second, phenylcarbene possesses a bending mode about the divalent carbon that can couple the singlet and triplet surfaces; this mode is absent in phenylnitrene. Finally, the differing electronic configurations of singlets **1** and **2** affect their rates of ISC. The transition from singlet to triplet phenylcarbene involves moving an electron from the σ orbital to the π orbital, creating orbital angular momentum. This enhances the spin-orbital coupling and encourages the change in the spin state. In contrast, the singlet and triplet states of phenylnitrene have identical orbital occupancies, and so no orbital angular momentum change occurs to assist in the intersystem crossing process.

5.2.3 Substituent Effects on the Rearrangement of Phenylnitrene

Abramovitch[64] and Banks[65] discovered that polyfluorinated phenylnitrene participates in intermolecular chemistry. Interestingly, perfluoro- and 2,6-difluorophenylnitrene react with pyridine but 4-fluorophenylnitrene undergoes rearrangement prior to reacting with pyridine.[69] The key to intermolecular phenylnitrene chemistry appears to be di-ortho substitution. Platz[92,113] has shown that fluorine substitution into both the ortho positions increases the barrier for rearrangement to about 8 kcal mol^{-1}, about 3 kcal mol^{-1} greater than the barrier for the parent phenylnitrene. Due to the higher barrier to intramolecular rearrangement, the substituted singlet phenylnitrene has sufficient lifetime to react with another molecule.

Platz[92] originally suggested that the rearrangement proceeds in one step to ketenimine from the configuration where both carbene electrons occupy the p-orbital perpendicular to the ring plane, or in other words, from the 2 $^1A_1$ state. He suggested that the 1 $^1A_1$ state is stabilized by ortho fluorine substitution at the expense of destabilizing the 2 $^1A_1$ state, thereby making the rearrangement less likely. BLYP/cc-pVDZ calculations confirm that the 2 $^1A_1$ state is raised with ortho-F substitution.[114]

However, this explanation necessitates a one-step mechanism, and, as discussed above, the rearrangement of phenylnitrene proceeds by a two-step mechanism (Scheme 5.2). In fact, a year after the review article of Platz appeared, Sander[115] detected the difluoroazirine **19** produced by the irradiation of 2,6-difluorophenyl azide in an argon matrix.

Sundberg[116] has observed products of the rearrangement of 2-alkylsubstituted phenylnitrenes that indicated that the initial cyclization is toward the unsubstituted *ortho* carbon (the presumed intermediate **20** in Reaction 5.4). Matrix isolation studies of 2,4-dimethylphenylnitrene find very little rearrangement product.[117]

(Reaction 5.4)

These experimental results set the stage for Karney and Borden's[118] seminal computational study of the role of ortho substituents on the rearrangement of phenylnitrenes. They computed the CASSCF(8,8)/6-31G* structures of a variety of fluoro-, chloro-, and methylphenylnitrenes (**21**), the transition states (**22**), and products of the first step (**23**) of their rearrangement (Scheme 5.3). Single-point energies were obtained with CASPT2. At this computational level, the barrier for

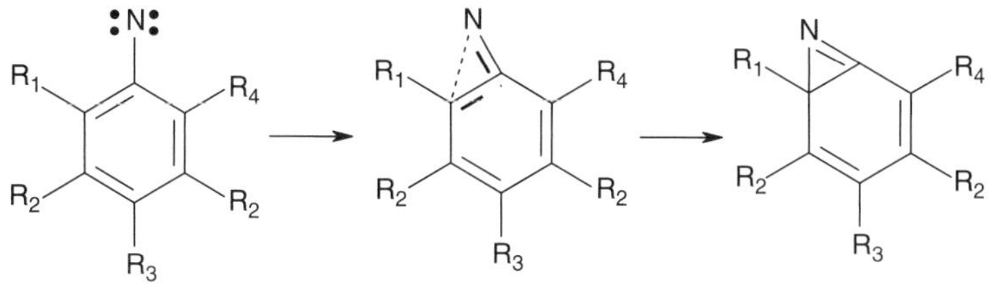

a: $R_1=R_4=F$, $R_2=R_3=H$
b: $R_2=F$, $R_1=R_3=R_4=H$
c: $R_3=F$, $R_1=R_2=R_4=H$
d: $R_4=F$, $R_1=R_2=R_3=H$
e: $R_1=F$, $R_2=R_3=R_4=H$
f: $R_4=Cl$, $R_1=R_2=R_3=H$
g: $R_1=Cl$, $R_2=R_3=R_4=H$
h: $R_4=Me$, $R_1=R_2=R_3=H$
i: $R_1=Me$, $R_2=R_3=R_4=H$

Scheme 5.3

the rearrangement of the parent phenylnitrene is 8.6 kcal mol^{-1} (Table 5.6), which overestimates the barrier by about 3 kcal mol^{-1}. Fluoro substitution into both meta positions (**21b**) or into the para position (**21c**) is calculated to have very little effect on the activation barrier relative to **1** (see Table 5.7), which is a prediction that is consistent with the experimental observation of this rearrangement for **21c**.[69] Laser flash photolysis studies by Gritsan and Platz[104] confirmed this prediction; the experimental activation barriers are 5.5 ± 0.3 and 5.3 ± 0.3 kcal mol^{-1} for **21b** and **21c**, respectively, little different from the barrier for **1**. If we assume that the

TABLE 5.7 Relative Energies (kcal mol^{-1}) for the Rearrangement 21 → 25[a]

x	21x[b]	22x[b]	23x[b]	24x[c]	25x[c]
a	0.0 (0.0)	13.0 (13.4)	−0.5 (1.0)	(6.6)	(−2.8)
b	0.0 (0.0)	7.9 (8.6)	−0.7 (1.1)	(6.3)	(−1.6)
c	0.0 (0.0)	8.5 (9.1)	1.6 (3.3)	(7.8)	(−1.0)
d	0.0 (0.0)	9.5 (9.9)	3.6 (4.8)	(11.8)	(3.2)
e	0.0 (0.0)	12.3 (13.0)	−2.4 (−0.3)	(2.1)	(−6.3)
f	0.0	8.0	0.8		
g	0.0	11.7	−2.9		
h	0.0	6.4	−1.3		
i	0.0	8.4	−1.3		

[a]Energies computed at CASPT2N/6-31G*//CASSCF(8,8)/6-31G*, CASPT2N/cc-pVDZ//, and CASSCF(8,8)/6-31G* values are within parentheses.
[b]Ref. 118.
[c]Ref. 104.

CASPT2 error in computing the barrier for **1** is the same as for **21b** and **21c**, then the agreement with experiment is quite good.

On the other hand, fluorine substitution at both ortho positions (**21a**) raises the CASPT2/6-31G* barrier from 8.6 to 13.0 kcal mol^{-1}. The increase in the barrier height inhibits the rearrangement reaction and allows **21a** to participate in intermolecular chemistry. Subsequent experiments determined that its activation barrier is 7.3–8.0 kcal mol^{-1} (depending on the solvent).[104] Calculations predict a 4.4 kcal mol^{-1} increase in E_a for the rearrangement of **21a** compared to **1**, which is slightly larger than the experimental value, but the predicted trend is correct: ortho fluoro-substitution increases E_a while substitution at the meta or para position has no effect.

Since **21a** is computed to be 3.8 kcal mol^{-1} higher in energy than **21b**, the larger activation barrier of the former cannot be attributed to fluorine stabilization of phenylnitrene. Instead, the ortho fluorines in **21a** must destabilize the transition state for rearrangement.

The rearrangement of 2-fluorophenylnitrene (**21d=21e**) can proceed with the nitrogen migration away from the fluorine substituent (**22d** and **23d**) or toward the fluorine substituent (**22e** and **23e**). The barrier for migration away from the fluorine is only 1 kcal mol^{-1} higher than for the parent phenylnitrene, which is a difference subsequently confirmed by the experiment.[104] However, the barrier for migration toward the fluorine substituent is 3 kcal mol^{-1} greater than that for migration away from the fluorine. Thus, 2-fluorophenylnitrene can undergo rapid rearrangement with the nitrogen migrating away from fluorine, consistent with experimental observation.[119] Since **21d** is higher in energy than **21c**, ortho-fluorine substitution destabilizes the former phenylnitrene, yet raises the barrier for its rearrangement.

Next, Karney and Borden examined the rearrangement of 2-chlorophenylnitrene (**21f=21g**) and 2-methylphenylnitrene (**21h=21i**). For both cases, the barrier for migration away from the substituent is smaller than for migration toward the substituent. Again, this result is consistent with previous experiments.[116] Shortly after the computations were published, Gritsan and Platz[120] reported on their laser flash photolysis of methyl derivatives of phenylazide. They found that the activation barriers for the rearrangement of phenylnitrene (**1**), 2-methylphenylnitrene (**21h**), and 4-methylphenylnitrene are identical (about 5.6 kcal mol^{-1}), within their error limits. However, the rearrangement barriers for 2,6-dimethyl- and 2,4,6-trimethylphenylnitrene are higher (about 7.3 kcal mol^{-1}). The experimental barriers are systematically smaller than the CASPT2 barriers, but when corrected for the tendency of CASPT2 to overestimate the barrier heights, the agreement between experiment and computation is excellent.

Based on their computational results, Karney and Borden argued that an ortho substituent sterically hinders migration. When phenylnitrene is substituted at one ortho position, rearrangement occurs away from the blocked position with a barrier little different from that of phenylnitrene itself. However, when both ortho positions are occupied, the barrier for rearrangement is raised by 2–3 kcal mol^{-1}, allowing for alternate reactions, especially intermolecular chemistry, to take place.

Platz and coworkers performed computations that complete the rearrangement pathway for the fluoro-substituted nitrenes **21a–e** by locating the transition states (**24a–e**) and product ketenimines (**25a–e**).[104] The relative energies for these species are listed in Table 5.7. In all cases, the reaction proceeds by two chemical steps, passing through the intermediate azirine **23**. For all of the cases examined except **21d**, the first step is rate limiting. These computations are consistent with the experiment where the rate of formation of ketenimine equals the rate of disappearance for **21a–c**, but for **21d**, its rate of decay is faster than the rate of appearance of ketenimine **25e**.

a: $R_1=R_4=F$, $R_2=R_3=H$
b: $R_2=F$, $R_1=R_3=R_4=H$
c: $R_3=F$, $R_1=R_2=R_4=H$
d: $R_4=F$, $R_1=R_2=R_3=H$
e: $R_1=F$, $R_2=R_3=R_4=H$

While generally agreeing with the contention that the ortho effect is steric in origin, Platz and coworkers noted that the barrier for **21d** → **23d** is 1 kcal mol^{-1} larger than the barrier for phenylnitrene and that the lifetime of **21d** is greater than either **1** or **21c**. The *ortho* fluorine has a small effect that cannot be steric. Using natural population analysis (NPA) charges, they argue that fluorine polarizes the phenyl ring such that the carbon bearing the fluorine becomes very positively charged. In progressing from **21** to transition state **22**, the carbon bearing the nitrogen becomes more positively charged. In the case of ortho-fluoro substitution (**22d** and **22e**), this leads to neighboring positively charged carbon atoms, which is significantly destabilized relative to the *para*-fluoro case (**22c**), where the positively charged carbon atoms are farther apart. Simple electrostatic arguments thus account for the small *ortho*-fluoro effect in increasing the activation barrier.

The steric effect of ortho substituents was further examined in experimental and theoretical studies of 2,6-dialkylphenylnitrenes (**26**).[105] The rearrangement was computed at B3LYP/6-311+G(2d,p)//B3LYP/6-31G*. Singlet phenylnitrenes are poorly described at this level due to spin contamination. Singlet phenylnitrene energies were obtained by computing the energy of the triplet at this DFT level and then adding in the singlet–triplet separation computed with the more reliable CASPT2 method. The barrier for the first step (creating the azirine) is little affected by methyl or ethyl substituents. This barrier is 1.5 kcal mol^{-1} lower for **26c** than **1**, and is dramatically lower still for **26d**—only 4.1 kcal mol^{-1} compared to 8.0 kcal mol^{-1} for **1**. The barrier for the second step is affected in the opposite way, increasing with greater steric bulk: 3.0 kcal mol^{-1} for **1**, 3.8 kcal mol^{-1} for **26a–c**, and 4.9 kcal mol^{-1} for **26d**. Platz argued that strain relief explains the decreasing barrier of the first step with increasing steric bulk. In this step, the nitrogen swings toward one *ortho* carbon, thereby increasing its distance from the other ortho substituent and diminishing the steric interactions with it. Perhaps of more importance is the fact that cyclization requires that the nitrogen atom moves out of the plane that contains both ortho substituents, which is an action that diminishes its steric interactions with both substituents.

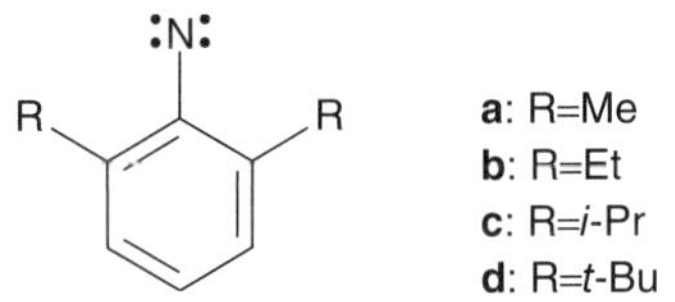

a: R=Me
b: R=Et
c: R=*i*-Pr
d: R=*t*-Bu

26

Perhaps their most significant result is that the opposite trends in changes in the barrier heights for the two steps with increasing steric bulk infers that the intermediate azirine might be observable if a sufficiently bulky substituted phenylnitrene could be produced. Just such a molecule is 2,4,6-tri-*t*-butylphenylnitrene. Computations indicate a barrier of 4.1 kcal mol^{-1} for the first step and 6.3 kcal mol^{-1} for the second step. Laser flash photolysis detected the azirine intermediate **27** having a lifetime of 62 ns and a barrier to ring open to the azepine **28** of 7.4 ± 0.2 kcal mol^{-1}.

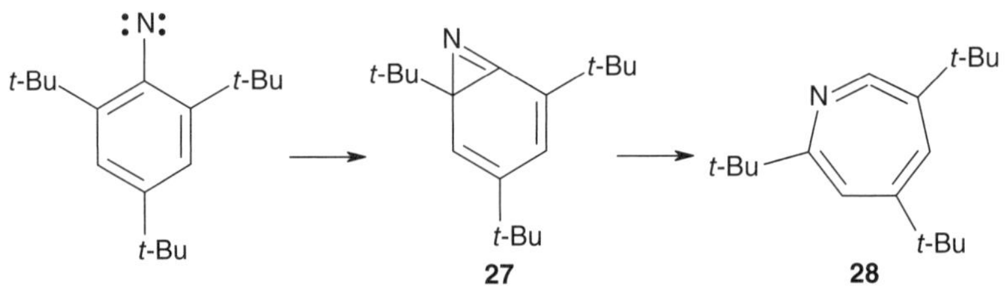

27   **28**

If singlet phenylcarbene is best represented by resonance structure **5**, radical-stabilizing substituents may alter its propensity to rearrange. A cyano group in the ortho position of phenylnitrene (**29**) should localize the ring π-radical onto that position, making rearrangement *toward* the substituted carbon more favorable than rearrangement away from the substituted carbon. The cyano substituent might outweigh the steric effect observed with fluoro, chloro, and methyl substituents. Since there is little π-radical density at the meta positions, the barrier for rearrangement of **30** should be comparable with phenylnitrene. Finally, placing a cyano group in the para position (**31**) should localize the π-radical onto the *para* carbon, thereby diminishing its presence at the ortho positions. This should result in a higher activation barrier for rearrangement.

[Structures 29, 30, 31, 32: cyano-substituted phenylnitrenes — 29 (ortho-CN), 30 (meta-CN), 31 (para-CN), 32 (2,6-diCN)]

In a combined experimental and computational study, Gritsan and coworkers[121] examined the rearrangement of the cyano-substituted phenylnitrenes **29–32**. They optimized the critical points for the reaction of these nitrenes analogous to the those shown in Scheme 5.3. Geometries were obtained with CASSCF(8,8)/6-31G*, and single-point energies were computed at CASPT2/6-31G* to account for dynamical correlation. As discussed above, these computed activation barriers, listed in Table 5.8, are likely to be too large by about 2–3 kcal mol^{-1}. In addition, laser flash photolysis was used to obtain the experimental activation energies, which is also listed in Table 5.8.

For **29**, the CASPT2 barrier for rearrangement toward the ortho-substituted carbon is slightly less than that for the rearrangement away. Both barriers are very close to the barrier for the rearrangement of unsubstituted phenylnitrene, and the experiments also show that the barriers for **1** and **29** are nearly identical. This is quite different than the *o*-fluoro, *o*-chloro, or *o*-methyl cases where rearrangement away from the substituted center is decidedly favored. As already noted, this reversal for the cyano substituent is readily understood in terms of the radical-stabilizing effect of the cyano group essentially balancing the steric effect.

The barrier for the rearrangement of **32** is slightly greater than for **29**. In this case, both cyano groups in **32** act to localize radical density onto the ortho ring carbon atoms, so each *ortho* carbon ends up with less radical density than the cyano-bearing carbon in the monosubstituted ring **29**. A greater change in the distribution of nonbonding p-electron density must occur in order to reach the TS in the reaction of **32**, hence the higher barrier for **32** than for **29**.

TABLE 5.8 Activation Energy (kcal mol^{-1}) for the Formation of Azirine Derivatives from **29–32**[a]

	Mode	CASSCF[b]	CASPT2[c]	Expt.
29	Away	8.3	8.6	5.5 ± 0.3
	Toward	8.4	7.5	
30	Away	8.6	8.2	
	Toward	8.1	7.6	
31		9.4	9.8	7.2 ± 0.8
32		8.2	8.0	6.5 ± 0.4

[a]Ref. 121.
[b]CASSCF(8,8)/6-31G*.
[c]CASPT2/6-31G*//CASSCF(8,8)/6-31G*.

The meta-substituted isomer **30** has a barrier height quite similar to **1**. In this case, there is neither any steric inhibition nor any radical concentration and so the net effect of the *meta*-cyano group is nil. Finally, the barrier for the para-substituted isomer **31** is much higher than for any of the other cyano isomers, and also higher than the barrier for rearrangement of **1**. The *para*-cyano group in **31** localizes the charge onto the *para* ring carbon, thereby depleting the radical density on the *ortho* carbon atoms that are involved in making the new bond to nitrogen.

In conclusion, the barriers for the rearrangements of the cyano derivatives of **1** are completely rationalized by the model of singlet phenylnitrene being described as a diradical: one electron localized to nitrogen and the second electron delocalized about the phenyl ring. The predictions of this qualitative model are neatly confirmed by this combined computational and experimental study.

5.3 TETRAMETHYLENEETHANE

Hund's rules provide a framework for predicting the relative energies of electronic states of atoms and molecules where multiple electrons occupy degenerate orbitals.[122] Hund's first rule is that energy increases with decreasing spin S. The second rule is that for states with the same spin, the energy increases with decreasing orbital angular momentum L. As with any rule, interesting chemistry can be found when the rule is violated or nearly violated. From the 1970s onward, computational chemists have been actively involved in finding molecules that violate Hund's rules.

In 1977, Borden and Davidson[123] developed the rules for predicting the spin for the ground-state diradicals. Diradicals are the molecules that have two electrons populating two (near) degenerate orbitals, oftentimes two nonbonding molecular orbitals (NBMOs). Hund's rule suggests a triplet ground state for such a system. The triplet state forces the two electrons to occupy different regions of space in order not to violate the Pauli's exclusion principle. The key insight provided by Borden and Davidson is that the two NBMOs might be spatially separated and then singlet spin coupling could be in the ground state.

Orbital separation comes about if the two NMBOs are disjoint, meaning that one NMBO has density on one set of atoms and the second NMBO has density only on atoms not in the first set. An example they gave of disjoint NMBOs is for tetramethyleneethane (TME) **33** and a nondisjoint example is trimethylenemethane (TMM) **34**. A schematic of their respective NMBOs is given in Figure 5.8. Notice that the left NMBO of **33** is localized to the atoms of the left allyl fragment, while the right NMBO is localized to the atoms of the right allyl fragment; they share no common atoms. On the other hand, the two NMBOs of **34** have orbital coefficients on two atoms in common.

On the basis of the simple Hückel theory, the two NMBOs of both **33** and **34** will be degenerate. Hund's rule therefore dictates a triplet state. However, Borden and Davidson argued that since the electrons in the two NBMOs of **33** are specially isolated and have reduced the electrostatic repulsion, the singlet state will not violate the Pauli's exclusion principle and will be energy competitive with the triplet

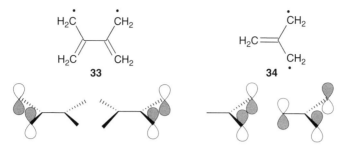

Figure 5.8 NBMOs of **33** and **34**.

state. In fact, they suggested that electron correlation will lower the singlet state below the triplet.

Borden and Davidson provided a quick way to decide if the NMBOs are disjoint or not for alternate hydrocarbons. Using the standard technique of starring alternate carbons (see Figure 5.9), if the difference between the number of starred (n^*) and unstarred (n) carbons is zero, then the molecule will have disjoint NMBOs. Otherwise, when $n^* - n = 2$, the molecule will be nondisjoint. The so-called Davidson–Borden rule predicts that disjoint alternate hydrocarbons will have small singlet–triplet energy gaps with the singlet likely to be the ground state. Nondisjoint alternate hydrocarbons will have large singlet–triplet energy gaps with a triplet ground state.[124] A similar conclusion can be derived from valence bond (VB) theory.[125] Accurate prediction of spin states is critical toward the development of organic magnetic materials, particularly organic ferromagnets.[8,126,127]

Over the following 25 years, significant computational and experimental effort has been expended to determine if TME is in fact a triplet, and therefore conforms

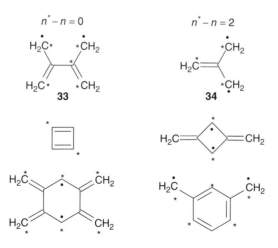

Figure 5.9 Representative alternate hydrocarbons with starred and unstarred carbon atoms.

to Hund's rule, or a singlet, and thereby violates Hund's rule. We will first present a more detailed description of the MOs of TME before telling this story in chronological order.

5.3.1 Theoretical Considerations of Tetramethyleneethane

Tetramethyleneethane can rotate about the central C–C bond, giving rise to three distinct conformations. These can be distinguished by the angle (τ) between the planes of the two allyl fragments (Scheme 5.4). The planar conformation ($\tau = 0°$) has D_{2h} symmetry, the perpendicular conformation ($\tau = 90°$) has D_{2d} symmetry, and any intermediate value of τ will have D_2 symmetry.

The NMBOs of **33** drawn in Figure 5.8, while being proper orbitals, do not transform according to the symmetry of the molecule. Rather, a linear combination of the two gives the correct symmetry-adapted orbitals. The shapes of the four lowest π-orbitals, which comprise the two highest lying molecular orbitals and the two NMBOs, are drawn in Figure 5.10. The orbitals for both the planar (D_{2h}) and perpendicular (D_{2d}) structures are shown.

The triplet state of **33** can be described by a single configuration. The $^3B_1$ state, using the D_{2d} designation, has the wavefunction

$$\Psi(^3B_1) = |\ldots 5e^4 1b_1{}^1 1a_2{}^1|$$

The singlet $^1A_1$ state will require at least two configurations since the NBMOs (MO 22 and 23) are close in energy. Thus, a reasonable reference wavefunction for the singlet is

$$\Psi(^3B_1) = c_1|\ldots \infty 5e^4 1b_1{}^2| + c_2|\ldots 5e^4 1a_2{}^2|$$

Any computation that does not account for the multiconfigurational nature of singlet TME will run into difficulties, analogous to the difficulties discussed previously with methylene and phenylnitrene and with the benzynes discussed later in this chapter.

5.3.2 Is TME a Ground-State Singlet or Triplet?

Controversy arose almost from the start in this story of the ground state of TME. Dowd,[128] in 1986, prepared TME by photolysis of 4,5-bis(methylene)-3,4,5,6-tetrahydropyridazine **35** in a methyltetrahydrofuran glass at 10 K. The ESR spectrum

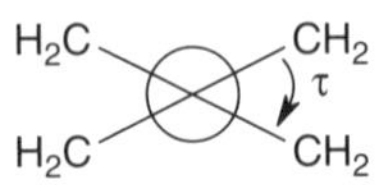

Scheme 5.4

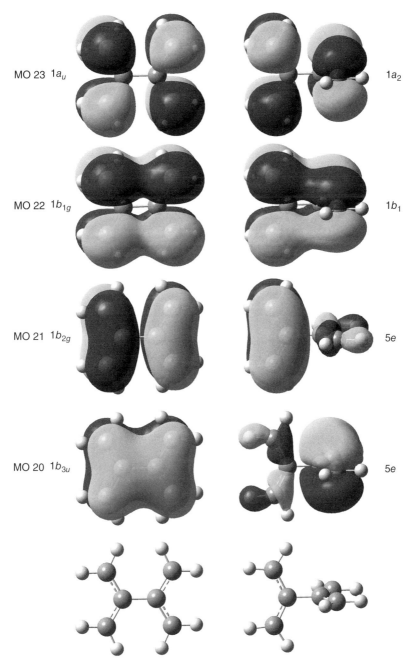

Figure 5.10 Lowest four p-MOs of **33** in its D_{2h} and D_{2d} conformations.

for the $\Delta m = 2$ line showed hyperfine splitting into nine lines, consistent with the eight hydrogens of **33** if all are approximately magnetically equivalent. The Curie–Weiss law plot for the $\Delta m = 2$ line is linear. Dowd concluded that **33** is a ground-state triplet, conflicting with the prediction of simple molecular orbital (MO) and VB rules.

The following year, Borden[129] reported computations of TME at three geometries: D_{2h}, D_{2d}, and D_2. The singlet state was optimized with a TC-SCF/3-21G wavefunction and the triplet was optimized at UHF/3-21G. Dynamic electron correlation was treated by obtaining the energies at CISD/DZP using the above geometries. The lowest energy structure was the D_{2d} singlet. The lowest energy triplet has D_2 symmetry. For all the three structures, the singlet is lower in energy; the singlet–triplet gap is predicted to be 2.8 kcal mol^{-1} for the planar structure, 0.4 for the D_2 structure, and 2.2 kcal mol^{-1} for the perpendicular conformer. These computations are incompatible with the previous experimental results.

Borden suggested that though planar TME should be a ground-state singlet, the triplet state, in a twisted conformation, might be lower in energy than the singlet. To test this hypothesis, Dowd[130] prepared 2,3-dimethylene–1,4-cyclohexanediyl **36** and Roth[131] prepared 2,2-dimethyl-4,5-dimethylene–1,3-cyclopentanediyl **37**. Both of these TME derivatives were anticipated to be planar or nearly so, and theory predicts that they should be singlets. However, the ESR spectra of both indicate that they are triplet ground states having linear Curie–Weiss plots.

These experiments, however, were not definitive. **36** is not planar; the computed value of τ is 25°.[132] A configuration interaction wave function including all singles, doubles, triples, and quadruples excitations (CISDTQ) computation where all excitations were confined to a minimal basis set comprising the π-orbitals predicted that the singlet state lies 1.4 kcal mol^{-1} below the triplet of **38**.[133] CASSCF(6,6)/DZP computation of **36** finds a singlet–triplet gap of less than 0.2 kcal mol^{-1} and a triplet ground state for **37**.[134] Hyperconjugation involving the orbitals to the methyl groups reduces the energy difference between the NMBOs, thereby making the triplet favorable. The same near-NMBO degeneracy occurs in **36**, induced by the nonzero τ angle. A linear Curie–Weiss plot can occur if the energy difference between the singlet and triplet states is very small (<50 cal mol^{-1}) and so the experiments may not be definitive tests of the ground state of TME. In fact, a later ESR and superconducting quantum interference device (SQUID) study of **36** can be best interpreted as resulting from a near-zero singlet–triplet energy gap.

In 1992, Nachtigall and Jordan[135] reinvestigated TME by first optimizing its structure using a CASSCF(6,6)/3-21G wavefunction. Both the lowest energy singlet and triplet have D_2 symmetry; the angle τ is 58.0° for the singlet and 59.0° for the triplet. Single-point energies computed using the larger DZP basis set suggest that the singlet is lower in energy for all values of τ: $\Delta E_{ST} = 3.84$ for $\tau = 0°$, 1.92 for $\tau = 180°$, and 1.36 kcal mol^{-1} for $\tau = 59°$. They noted that the most serious omission in these calculations is the lack of correlation among the σ-electrons. To address this omission, Nachtigall and Jordan[136] recomputed the energies using a CISD method, with a two-configuration reference for the singlet and the Davidson correction to accounted for higher order excitations. Their results with the TZ2P basis set, the largest they used, predict that the lowest energy singlet has D_{2d} symmetry. For both the D_{2d} and D_{2h} conformers, the singlet lies below the triplet. However, the lowest energy structure is the D_2 triplet. This triplet is only 0.1 kcal mol^{-1} lower in energy than the D_{2d} singlet; however, it is 0.5 kcal mol^{-1} lower than the singlet at the geometry of the triplet. Given the low temperature of the experimental observation of triplet TME, their computations support the observation of a triplet. These results, however, conflict with the ESR experiment,[128] which indicates a D_{2d} triplet conformation.

A dramatic reinterpretation of the nature of TME was introduced by the work of Lineberger coworkers.[137] They reported the photoelectron spectrum of the anion **39** (TME$^-$). The spectrum revealed electron loss to give two different states of TME. Based largely on the population distribution, they characterized the lower energy state as the singlet. The measured singlet–triplet energy gap is 3 kcal mol^{-1}, but this requires correction for Franck–Condon factors. Basically, TME$^-$ has D_{2d} symmetry, but triplet TME is D_2. Subtracting the difference in energy between triplet TME in its D_{2d} and ground state D_2 conformations from the direct experimental energy gap yields $\Delta E_{ST} = 2.0$ kcal mol^{-1}. These authors proposed that triplet TME may energetically lie below the singlet at the geometry of the triplet. Further, in the glass at 10 K, the triplet may be locked into its twisted D_2 conformation, creating a metastable species that will only slowly revert to the singlet state.

39

Over the next 3 years, three computational attempts at reconciling these contradictory results appeared. The first, by Filatov and Shaik,[138] utilized a REKS/6-311G* calculation of singlet and triplet TMEs. The singlet minimum energy structure has D_{2d} symmetry, while the minimum triplet structure has d_2 symmetry with $\tau = 50.1°$. The singlet is lower in energy for all values of τ except between 38° and 51°. The spin-orbit-coupling matrix elements were computed to be very small. Therefore, Filatov and Shaik conclude that triplet TME is metastable, consistent with the arguments of Lineberger et al.[137]

On the other hand, Rodriguez[139] reported the CISD/TZ+P//CASSCF(6,6)/6-31+G* PES for the singlet and the difference-dedicated configuration interaction (DDCI) PES for the triplet. For all values of t, the singlet is lower in energy than the triplet. However, at the triplet energy minimum ($t = 47°$), the singlet is only 0.29 kcal mol^{-1} below the triplet. Therefore, the two states are very close in energy.

Finally, Pittner and coworkers[140] examined the PESs of singlet and triplet TMEs using two multireference coupled cluster methods (two-determinant CCSD and multireference Brillouin–Wigner CCSD) with the cc-pVDZ basis set. With both methods, the singlet is always below the triplet for all values of τ. The smallest singlet–triplet gap, 1.3 kcal mol^{-1}, occurs at an angle $\tau = 45°$ and this is reduced when a near-cc-pVTZ basis set is used.

So what is the answer to our initial question? Just what is the ground state of TME? The gas-phase PES experiment agrees with virtually all of the computations that TME is a ground-state singlet. The singlet–triplet gap is undoubtedly quite small, especially in the vicinity of the triplet energy minimum. This allows for the possibility of a metastable triplet being trapped within a glass, explaining the ESR results of Dowd. Final resolution will likely require extremely accurate treatment of electron correlation along with a large basis set and perhaps additional experiments.

5.4 OXYALLYL DIRADICAL

Related to TME is TMM **34**. The tale of TMM also entails a synergistic interplay of theory and experiment similar to that of TME. This story is nicely retold in the review article of Lineberger and Borden.[16] Instead of reviewing this TMM story, we present the case of the analogous oxyallyl diradical (OXA) **40**. This diradical has been proposed as an intermediate in the ring opening of cyclopropanone[141] and other reactions.[142,143]

40

The frontier MOs of OXA **40** are drawn in Figure 5.11. The electron occupation of the $^3B_2$ state, which according to Hund's rule should be the ground state, is shown in this figure, with the wavefunction $\Psi(^3B_2) = |\ldots 1b_1{}^1 5b_2{}^2 2b_1{}^1 1a_2{}^1|$. The singlet ground state may again require at least a two-configuration description because the $2b_1$ and $1a_2$ orbitals are close in energy, that is, $\Psi(^1A_1) = c_1$

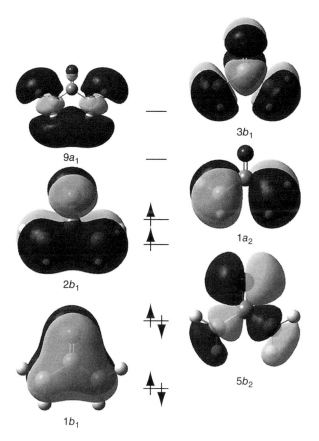

Figure 5.11 MOs of oxyallyl diradical **40**. *(See insert for color representation of this figure.)*

$|\ldots 1b_1^1 5b_2^2 2b_1^2| + c_2 |\ldots 1b_1^1 5b_2^2 1a_2^2|$. The oxygen atom removes the degeneracy of these two orbitals (relative to TMM) and so the coefficients c_1 and c_2 will not be identical; in fact, the greater electronegativity of oxygen versus carbon suggests that the first configuration will dominate.

The first study of OXA, by Morokuma and Borden,[144] performed at a very low level: the geometry was optimized at CASSCF(5,6)/3-21G, imposing C_{2v} symmetry. The $^3B_2$ state was predicted to be 12.3 kcal mol^{-1} below the $^1A_1$ state, though this gap was reduced by more than 50 percent when single-point energies were computed at 3-21G* (see Table 5.9). This suggested that reoptimization with a large basis set might further change the gap, and their next study[145] did show a smaller singlet–triplet gap at CASSCF(4,4)/DZP and MR-SDQ/DZP of 4.4 and 1.6 kcal mol^{-1}, respectively. Clearly, this is a system that is very sensitive to computational methodology and further refinement was delayed until computers and computation algorithms had significantly improved. However, these computations were good enough to indicate that the $^1A_1$ state is dominated

TABLE 5.9 Singlet–Triplet Gap (kcal mol⁻¹) of 40

Computational Method	$\Delta E_{ST}{}^a$
CASSCF(6,5)/3-21G[b]	12.3
CASSCF(6,5)/3-21G*//CASSCF(6,5)/3-21G[b]	5.6
CASSCF(4,4)/DZP[c]	4.4
MR-CISDQ/DZP//CASSCF(4,4)/DZP[c]	1.6
EOM-SF-CCSD(dT)/aug-cc-pVTZ[d]	−1.4
CASPT2(4,4)/aug-cc-pVTZ//CASSCF(4,4)/cc-pVTZ[e]	−0.92
MkCCSD(T)$_u$/vv-pVQZ//CASSCF(2,2)/cc-pVTZ[f]	−1.3
Expt.[g]	−1.3

[a] A positive value indicates that the triplet state is lower in energy than the singlet state.
[b] Ref. 144.
[c] Ref. 145.
[d] Ref. 146.
[e] Ref. 147.
[f] Ref. 148.
[g] Ref. 149.

by the diradical resonance structure, and that the zwitterion resonance structure is only a very minor component.

In 2009, Lineberger and Borden[149] reported the photoelectron spectrum of the oxyallyl radical anion, with more complete analysis provided in 2011.[147] They observed two sets of peaks in the spectrum, one originating at 1.942 eV and the second at 1.997 eV. This second set has rather sharp peaks with a progression of 405 cm⁻¹. B3LYP/6-311++G(d,p) computations of the OXA radical anion and triplet diradical show a large difference in the C–C–C angle of these two structures (121.9° and 114.4°, respectively). The vibrational frequency of the C–C–C bend is computed to be 408 cm⁻¹, in excellent agreement with the PES.

The other set of peaks in the PES is broad and shows a progression of about 1660 cm⁻¹. CASSCF(4,4)/cc-pVTZ computations of the singlet oxyallyl diradical indicate that while this $^1A_1$ state is a local energy minimum, when zero-point vibrational energy is included, it is actually higher in energy than the geometrically nearby transition state for disrotatory ring opening of cyclopropane. The C–O bond is predicted to shorten by 0.057 Å in going from the OXA radical anion to the single OXA diradical, and the C–O vibrational frequency should be similar to that observed in this progression.

The interpretation of the PES leads to two important conclusions. First the ground state of **40** is in fact the singlet state, about 0.13 kcal mol⁻¹ below the triplet. This is corroborated by the subsequent high level computations (Table 5.9). Secondly, the broad peaks associated with the $^1A_1$ ground state arise from the transition from the OXA radical anion to the $^1A_1$ state, which is actually a transition state. The overtones observed in the spectrum are observations of the structure of this *transition state*. Wavepacket dynamics computations[146] indicate a lifetime of about 100 fs for the $^1A_1$ transition state, which corroborates the observed peak broadening.

5.5 BENZYNES

Loss of two hydrogen atoms from benzene results in the reactive biradical species called *benzynes* or *arynes*. There are three possible isomers: *o*-benzyne (**41**), *m*-benzyne (**42**), and *p*-benzyne (**43**). Of these three, *o*-benzyne is the best known and studied.[150] It was first demonstrated to be a reactive intermediate by Roberts[151] in 1953 and later characterized by UV,[152] MS[153], IR,[154] and microwave[155] spectroscopies. *o*-Benzyne is a widely utilized reactive intermediate, particularly as the dienophile component in a Diels–Alder reaction. The other two isomers have only recently become of interest. Due to the rise of enediyne chemistry, especially the Bergman cyclization discussed in Section 4.3, *p*-benzyne has been the subject of many theoretical and experimental studies. Interest in *m*-benzyne largely grew out of the desire to compare and contrast the whole set of benzyne isomers, though as it turned out, *m*-benzyne brings its own interesting challenges.

Computational chemistry has been thoroughly involved in the study of the benzyne family over the past 30 years. This section will focus on three aspects of benzyne chemistry. First, computational studies disclosed a serious error in the experimental determination of the heats of formation of the benzynes and directly led to the corrected values. Second, a combination of matrix isolation spectroscopy and computations led to the characterization of the structure of *m*-benzyne, which is a process complicated by the extreme sensitivity of the geometry to computational method. Finally, the propensity for the benzynes to abstract hydrogen atoms, which is a key step for the biological activity of enediyne drugs, can be understood by a simple model based on the singlet–triplet gap of benzyne. Computations played an important role in determining the singlet–triplet separation, modeling the hydrogen abstraction step, and proposing potential new drug candidates.

5.5.1 Theoretical Considerations of Benzyne

The difficulty in computing the wavefunction, structure, energy, and properties of the benzynes is how to properly treat the interaction between the two separated electrons. If the interaction between them is strong, a single configuration describing them as a bonding pair is satisfactory. However, as the interaction becomes weaker, and the biradical character increases, the contribution of a second (antibonding) configuration becomes more and more important.

For **41**, the bonding interaction between the ortho radicals is described by MO 18 ($10a_1$) and the corresponding antibonding orbital is MO 21 ($8b_2$)—see Figure 5.12. The HF wavefunction is therefore

$$\Psi_{\text{HF}}(\mathbf{41}) = |\ldots\ 1b_1{}^2 10a_1{}^2 1a_2{}^2 2b_1{}^2| \tag{5.1}$$

but if there is some biradical character, the two-configuration self-consistent field (TC-SCF) wavefunction might be more apropos

$$\Psi_{\text{TC-SCF}}(\mathbf{41}) = c_1|\ldots\ 1b_1{}^2 10a_1{}^2 1a_2{}^2 2b_1{}^2| + c_2|\ldots\ 1b_1{}^2 1a_2{}^2 2b_1{}^2 8b_2{}^2| \tag{5.2}$$

Schaefer examined the geometry, vibrational frequencies, and singlet–triplet gap of **41**, comparing the results of a HF wavefunction with that of the TC-SCF wave function.[156] As expected with the inclusion of the antibonding configuration, the distance between the two radical carbon atoms is longer in the optimized TC-SCF/DZ+P geometry (1.263 Å) than in the HF/DZ+P structure (1.225 Å). The experiments reporting IR spectra of **41** at that time suggested a frequency of around 2080 cm^{-1} or 1860 cm^{-1}.[154,157,158] The frequency corresponding to the stretch between these two radical carbon centers is 2184 cm^{-1} at HF and reduces to 1931 cm^{-1} with the TC-SCF wavefunction, indicating that the two configurations improve the description of **41**.

Most damning for the single-configuration approach was the evaluation of the singlet–triplet gap. The experimental measurement of this separation is 37.7 kcal mol^{-1}[158] with a singlet ground state. The HF treatment actually predicts a ground-state triplet. Using a CISD expansion from the HF reference improves matters: the singlet is 17.0 kcal mol^{-1} below the triplet, but inclusion of the Davidson correction increases the gap to 48.2 kcal mol^{-1}. This large correction indicates a poor reference wavefunction. Describing the singlet with a TC-SCF wavefunction dramatically improves this situation. The gap is 27.7 kcal mol^{-1} at TC-SCF/DZ+P, 32.2 kcal mol^{-1} when CISD is used with the TC-SCF reference, and inclusion of the Davidson correction increases the gap to 33.3 kcal mol^{-1}. This is still 4 kcal mol^{-1} less than the experiment, but clearly these results indicate the necessity of using a multiconfiguration wavefunction for **41**.

With the two radical centers farther apart in **42** than in **41**, it is reasonable to expect the meta isomer to express greater biradical character than the ortho isomer. Therefore, a multiconfiguration wavefunction will be necessary to adequately describe **41**. The two configurations that doubly occupy either the radical bonding orbital ($11a_1$) or antibonding orbital ($7b_2$)

$$\Psi_{\text{TC-SCF}}(\mathbf{42}) = c_1|\ldots\ 1b_1{}^2 2b_1{}^2 1a_2{}^2 11a_1{}^2| + c_2|\ldots\ 1b_1{}^2 2b_1{}^2 1a_2{}^2 7b_2{}^2| \tag{5.3}$$

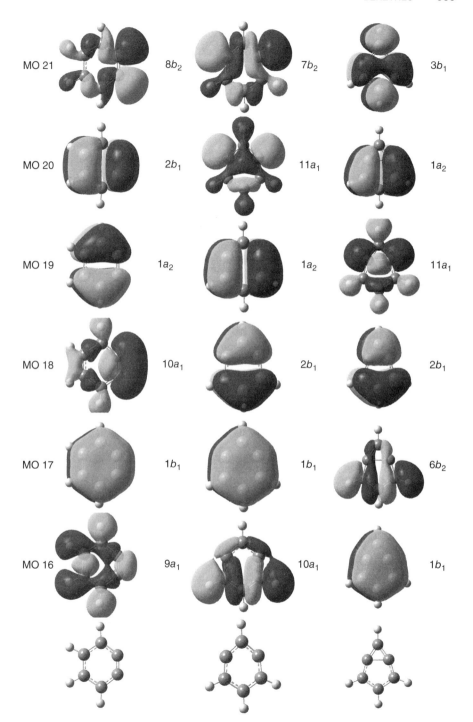

Figure 5.12 High-lying molecular orbitals of **41**, **42**, and **44**. *(See insert for color representation of this figure.)*

should dominate the configuration expansion. The MOs of **42** are drawn in Figure 5.12.

Again, a multiconfiguration wavefunction will be necessary to describe the expected large biradical character of *p*-benzyne **43**. This wavefunction will be dominated by the two configurations that define the bonding and antibonding interactions between the radical centers.

$$\Psi_{TC-SCF}(\mathbf{43}) = c_1 | \ldots 1b_{3u}1b_{2g}^{2}2b_{1g}^{2}5b_{1u}^{2}| + c_2 | \ldots 1b_{3u}1b_{2g}^{2}2b_{1g}^{2}6a_{g}^{2}| \quad (5.4)$$

Kraka and Cremer[159] have estimated the biradical character of the benzynes in two ways. The first involves the occupation number of the natural orbitals (n_i) obtained from the CCSD(T) wavefunction. The biradical character is then defined as $\Sigma n_i - \Sigma n_i(\text{reference})$ where the reference here is benzene. With this method, the biradical character is 11 percent, 20 percent, and 65 percent for **41**, **42**, and **43**, respectively. Alternatively, one can use the largest amplitude for a doubles excitation within the CCSD(T) wavefunction to indicate how strongly the next most important configuration (after the HF configuration) contributes to the total wavefunction. This amplitude is 0.24 for **41**, 0.34 for **42**, and 0.71 for **43**. Yet another measure of biradical character is the ratio C_S^2/C_A^2, where these are the coefficients of the configurations involving symmetric and antisymmetric combination of the radical orbitals. A pure biradical will have a value of 1 for this ratio. From the CASSCF(8,8)/3-21G wavefunction, Squires[160] finds that this ratio is 11.1 for **41**, 4.3 for **42**, and 0.6 for **43** (the value less than 1 simply means that the antisymmetric configuration dominates the wavefunction). Biradical character clearly increases for the benzynes as the radical centers become farther apart, and one might expect a concomitant increasing need for a multiconfiguration wavefunction. The fact that the biradical character differs among the three benzynes presents the underlying difficulty in computing the geometries and energies of these compounds—selecting the appropriate computational method that will treat them equivalently is a challenge for all chemists wishing to take on the benzynes.

Also presented in Figure 5.12 are the molecular orbitals for the bicyclic isomer of *m*-benzyne, namely, bicycle[3.1.0]hexa-1,3,5-triene (**44**). This structure has a bond between the *meta*-dehydro carbon atoms. A single configuration is likely to be sufficient to describe **44** as long as the C_1-C_3 distance is not too long.

44

5.5.2 Relative Energies of the Benzynes

The first contribution to benzyne chemistry produced by computational chemists concerns the relative energies of the isomeric benzynes. (Wenthold[161] has published an excellent summary of the thermochemical properties of the benzynes.)

Squires[162] reported collision-induced dissociation (CID) of benzyne precursors (Reactions 5.5–5.7) to obtain their heats of formation. The heat of formation of **41** was determined to be 106 ± 3 kcal mol^{-1}, in fine accord with other measurements.[163,164] The values of the other two isomers were the first to be recorded: $\Delta H_{f,298}$ (**42**) = 116 ± 3 kcal mol^{-1} and $\Delta H_{f,298}$ (**43**) = 128 ± 3 kcal mol^{-1}. Bergman[165] had previously made a rough estimate of the heat of formation of **43** as 140 kcal mol^{-1}, but this discrepancy with the CID results went unmentioned by Squires.

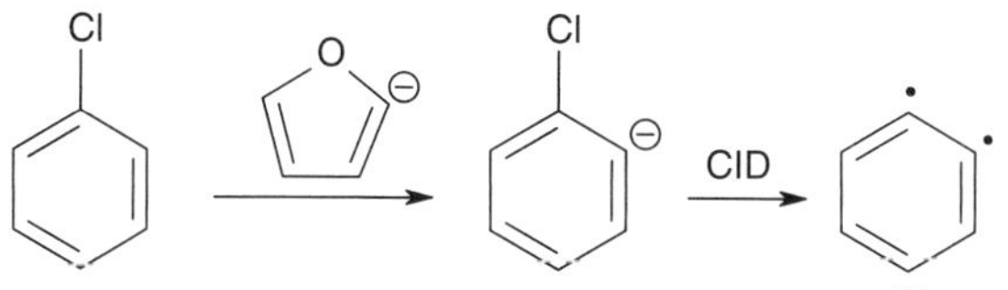

(Reaction 5.5)

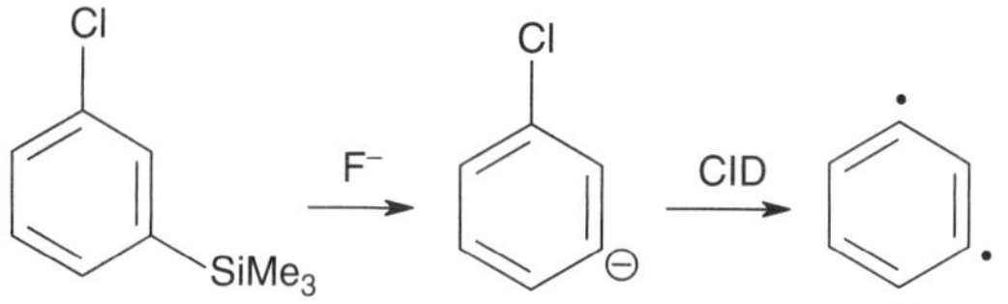

(Reaction 5.6)

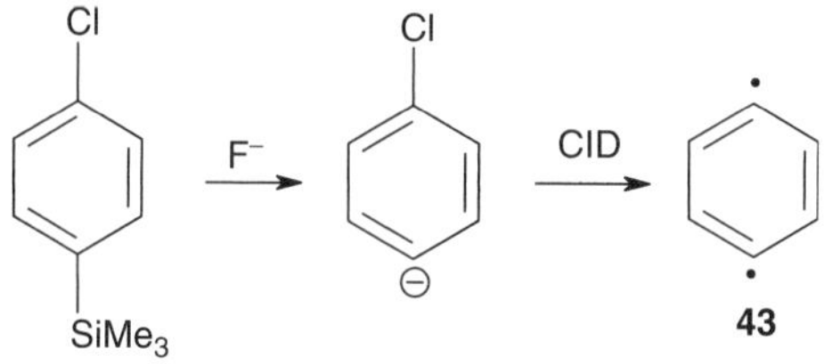

(Reaction 5.7)

Within 2 years, computational reports appeared, which questioned these values of the experimental heats of formation of **42** and **43**. In back-to-back articles, Borden[166] and Squires[160] computed the relative energies of **41**–**43** using different methods. The experiment indicates that their relative energies are **41**, 0.0; **42**, 10.0; and **43**, 22.0 kcal mol^{-1}. Borden first optimized the geometries using the TC-SCF/6-31G* method. At this level, **42** is 12.6 kcal mol^{-1} higher than **41**, and **43** is 22.7 kcal mol^{-1} higher than **41**. In order to account for some of the correlation energy, they performed single-point energy computations including all singles and doubles configurations from the π-orbitals and the bonding and antibonding radical orbitals (π-SDCI/6-311G**). This increases the separation between the benzynes: the energies of **42** and **43** relative to **41** are 15.8 and 28.4 kcal mol^{-1}, respectively.

Squires[160] estimated the heats of formation of the benzynes with Eq. (5.5), making use of the biradical separation energy (BSE) defined in Reaction 5.8. The BSE is computed with different computational methods, and experimental heats of formation are utilized for benzene and phenyl radical. Their best estimate for the heats of formation was obtained using the correlation-consistent CI method:

$\Delta H_{f,298}$ (**41**) = 107 kcal mol^{-1}, $\Delta H_{f,298}$ (**42**) = 125 kcal mol^{-1}, and $\Delta H_{f,298}$ (**43**) = 138 kcal mol^{-1}.

$$\Delta H_{f,298}(\textbf{41, 42, or 43}) = 2\Delta H_{f,298}(C_6H_6) - \Delta H_{f,298}(C_6H_5) - \text{BSE}(\textbf{41, 42, or 43})$$
(5.5)

[Reaction scheme: benzene + **41** (or **42** or **43**) → 2 phenyl radical] (Reaction 5.8)

Shortly thereafter, a CCSD(T) examination of the benzynes appeared.[159] Heats of formation were obtained from the hydrogenation of each benzyne, using the experimental ΔH_f of benzene. Using the CCSD(T)/6-31G* energies, the computed heats of formation are 110.8, 123.9, and 135.7 kcal mol^{-1} for **41**, **42**, and **43**, respectively. The optimized geometries of the benzynes at this level appear in Figure 5.13.

These three studies employing different computational methodologies all provided the same conclusion: the experimental heats of formation for **42** and **43** are too low by about 5–7 kcal mol^{-1}! This discrepancy is well outside the error bars of the experiments. The unanimity among the computational work compelled Squires[167] to reassess his experimental values. Close examination of the CID results indicated that less than a 13 percent impurity of the *ortho*-benzyne isomer would lead to the apparently too low appearance energy for *meta*- or *para*-benzyne. Trace amounts of water can cause rearrangement of **42** and **43** into **41**. Use of deuterated precursors and careful mass selection allowed Squires to obtain revised values for authentic samples of **42** and **43**. These new values (including some minor corrections made a few years later[168]) are $\Delta H_{f,298}$ (**41**) = 105.9 ± 3.3, $\Delta H_{f,298}$ (**42**) = 121.9 ± 3.1, and $\Delta H_{f,298}$ (**43**) = 137.8 ± 3.1 kcal mol^{-1}, all of which are in agreement with the computational estimates.

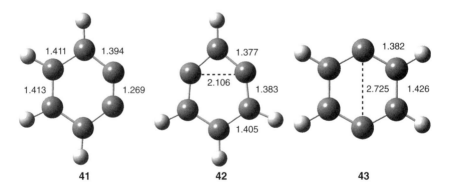

Figure 5.13 CCSD(T)/6-31G(d,p)-optimized geometries of **41–43**.[159]

Over the following years, a few additional computational assessments of the relative energies of the benzynes have appeared. A collection of the computational results is found in Table 5.10. The TC-SCF[166] and CASSCF(2,2)[160] methods underestimate the energetic separation between **42** or **43** from **41**. Dynamic correlation must be included in the computation. MP2 is unsuitable since it uses the HF reference and cannot recover the nondynamic correlation.[159] The HF solution is also at the heart of the failure of G3 to adequately describe **43**; G3 makes use of the RHF/6-31G(d) zero-point energy and thermal corrections, and this method gives three imaginary frequencies for **43**.[169] CASSCF(8,8) provides good agreement for **42** but underestimates the energy of **43**.[160,170] Unfortunately, including dynamic correlation using CASPT2 stabilizes both **42** and **43** relative to **41** due to inherent overstabilization[106] of unpaired electrons.

The CI methods build upon the HF wavefunction and so there is concern that limited expansions might not be able to recover enough nondynamic correlation to provide reasonable treatment of the multiconfigurational nature of the benzynes, especially for **43**. Nevertheless, CISD provides reasonable results, and the CCCI results are in excellent agreement with experiment.[160] Both **42** and **43** are predicted to be too energetic relative to **41** with CCSD, but inclusion of triples configurations using CCSD(T) lowers this separation.[159,169−171]

DFT is also a single-configuration method and again raises doubt about its applicability to the benzynes. Nonetheless, both BPW91 and B3LYP provide fair agreement with experiment. BPW91 underestimates the energy gaps between **41** and **42** or **43**,[171] while B3LYP overestimates the energy difference between **41** and **43**.[169]

The energies listed in Table 5.10 do not tell the whole story. A more critical examination of *m*-benzyne will be presented in the next section as computational results played a significant role in the characterization of this reactive species. We close this section with further discussion of problems associated with computing the energy and geometry of *p*-benzyne.

Cremer and coworkers published two in-depth analyses of computational problems associated with *p*-benzene. The first deals with HF and post-HF methods,[172] while the second focused on DFT methods.[173]

The restricted Hartree–Fock (RHF) wavefunction allows for no biradical character and the resulting optimized geometry has three imaginary frequencies. UHF overplays the biradical character, partly through spin contamination from multiple triplet states. RMP2 does provide an optimized structure with only real frequencies, but many of these frequencies are erroneous. RMP4 gives a structure with one imaginary frequency and only slightly improved frequencies relative to RMP2. Spin-restricted CCSD again gives a stable structure with erroneous frequencies, and adding in the effect of triple excitations (RHF-CCSD(T)) produces a structure with two imaginary frequencies. Increasing the basis set from 6-31G(d,p) to cc-pVTZ made little improvement for many of these methods. However, the CCD(T) method using Brueckner orbitals as the reference resolves most of the problems: the optimized structure is a local minimum with good agreement between its computed and experimental vibrational frequencies. Even though the UHF wavefunction suffers from triplet contamination, inclusion of dynamic correlation does reduce the

TABLE 5.10 Relative Energies (kcal mol^{-1}) of 41–43[a]

Methods	41	42	43
TC-SCF/6-31G*[b]	0.0	12.0	23.2
CASSCF(2,2)/6-311G**[c]	0.0	13	24
CASSCF(8,8)/cc-pVDZ//CASSCF(8,8)/3-21G[c]	0.0	16	27
CASSCF(8,8)/aANO[d]	0.0	16.6	27.5
CASPT2[0]//CASSCF(8,8)/aANO[d]	0.0	10.9	23.4
CASPT2/aANO//CASSCF(8,8)/cc-pVDZ[e]	0.0	9.8	21.8
		10.0	*21.9*
MP2/6-31G(d,p)[f]	(0.0)	(7.5)	(10.3)
π-SDCI/6-311G**//TC-SCF/6-31G*[b]	0.0	15.8	28.4
CISD/6-31G*//CASSCF(2,2)/6-31G*[c]	0.0	13	25
CISD/cc-pVDZ//CASSCF(8,8)/3-21G[c]	0.0	13	25
QCISD/cc-pVTZ[d]	0.0	13.0	40.1
		13.1	*40.4*
CCCI/cc-pVTZ//CASSCF(8,8)/3-21G[c]	0.0	18	31
CCSD/6-31G(d,p)//CCSD(T)/6-31G(d,p)[f]	(0.0)	(18.0)	(42.5)
CCSD/aANO//CASSCF(8,8)/aANO[d]	0.0	19.8	44.3
CCSD(T)/6-31G(d,p)[f]	0.0	13.7	25.3
		13.1	*24.9*
CCSD(T)/aANO//CASSCF(8,8)/aANO[d]	0.0	14.3	27.2
CCSD(T)/cc-pVDZ//BPW91/cc-pVDZ[e]	0.0	13.5	25.5
		13.7	*25.6*
CCSD(T)/cc-pVTZ[g]	0.0	14.8	26.2
		14.9	*26.4*
BPW91/cc-pVDZ[e]	0.0	7.6	28.1
		7.8	*28.2*
BPW91/cc-pVTZ//BPW91/cc-pVDZ[e]	0.0	8.7	29.3
		8.9	*29.4*
B3LYP/cc-pVTZ[g]	0.0	12.1	37.0
		12.2	*37.2*
G3[g]	0.0	13.0	22.2
		12.8	*22.1*
Expt.[h,i]	0.0	16.0	31.9
	105.9±3.3[j]	121.9±3.1[j]	137.8±3.1[j]

[a] Entries in normal text are energies including ZPE, while entries within parentheses omit the ZPE and entries in italics are corrected for 298 K.
[b] Ref. 166.
[c] Ref. 160.
[d] Ref. 170.
[e] Ref. 171.
[f] Ref. 159.
[g] Ref. 169.
[h] Ref. 167.
[i] Ref. 168.
[j] $\Delta H_{f,298}$.

contamination. In fact, UHF-CCSD(T) appears to be the most reliable of all of the traditional ab initio methods.

In their examination of a broad range of functionals, Cremer[173] found that the spin-restricted solution for *all* of the examined functionals is unstable for **43**. (The same is true also for the M06-2x and ωB97X-D functionals.) The spin-restricted solution for **43** using a local spin density functional, such as SVWN, is more stable than the generalized gradient approach, such as BLYP and BPW91, which in turn is more stable than the hybrid functionals, such as B3LYP, B3PW91, or mPW1PW91. The unrestricted DFT solution, regardless of the functional, is always lower in energy. While RB3LYP suffers from the most problems, in part due to its incorporation of a HF term, UB3LYP performs the best of all the functionals in terms of geometry and prediction of the singlet–triplet energy gap. The generalized gradient approximation (GGA) functionals also perform quite well. Unrestricted density functional theory (UDFT) includes some nondynamic correlation effects, and the exchange-correlation terms account for dynamic correlation effects. The net result is that UDFT, especially the hybrid functional UB3LYP and also the GGA functionals UBP91 and UBLYP, provides perfectly reasonable descriptions of *p*-benzyne.

5.5.3 Structure of *m*-Benzyne

The next controversy concerning the benzynes is the structure of *m*-benzyne. Does it exist as the monocyclic biradical **42** or as the bicyclic closed-shell species **44**? Answering this question with a computational approach will take some care. While the biradical character of **42** is small, a multiconfiguration wavefunction (Eq. (5.3)) is likely to be necessary for adequate description of its electronic structure. On the other hand, **44** is a closed-shell species and its electronic configuration can be expressed by a single Slater determinant made from the molecular orbitals shown in Figure 5.12:

$$\Psi(\mathbf{44}) = | \ldots 6b_2{}^2 2b_1{}^2 11a_1{}^2 1a_2{}^2 | \quad (5.6)$$

Comparing the geometries and energetics of these two isomers requires appropriate treatment of their respective electronic natures.

A critical distinction between **42** and **44** is the distance between the two radical centers, $r(C_1-C_3)$. Selected results for computed values of this distance in **42** and **44** are presented in Table 5.11. The results can be generally divided into three categories. First, the HF and restricted B3LYP and B3PW91 methods predict a single local minimum: **44**. (The RB3LYP/cc-pVDZ-optimized structure of **44** is shown in Figure 5.14.) CASSCF, MP2, CCSD(T), and BLYP predict that **42** is the sole local minimum. Finally, unrestricted B3LYP and B3PW91 locate both **42** and **44** with the latter slightly lower in energy.

Sander and coworkers[174] reported a detailed examination of how the energy varies with $r(C_1-C_3)$ for a number of computational methods. They optimized the structure of **42/44** holding the $r(C_1-C_3)$ distance fixed and then plotted the energy against this varying distance. Figure 5.15a displays this PES computed at

RB3LYP/cc-pVDZ, RBLYP/cc-pVDZ, and CASSCF(8,8)/cc-pVTZ. The RB3LYP curve shows a minimum at 1.615 Å corresponding to **44**, but the surface remains very shallow until a separation of 2 Å. The UB3LYP curve deviates from the RB3LYP potential above 2 Å, with a very shallow second well at 2.15 Å. The CASSCF(8,8) surface shows a sharper minimum at 2.174 Å, corresponding to **42**. RBLYP shows a broad flat bottom surface with a minimum near 2 Å. Figure 5.15b shows the energy surface computed at RCCSD(T) using either the UB3LYP or CASSCF(8,8) structures. The point of this plot is that even though the underlying geometries are different, the shape of the two curves is quite comparable. Therefore, the choice of geometry is not critical. What is important

TABLE 5.11 Computed $r(C_1-C_3)$ in angstroms for 42 and 44

Method	$r(C_1-C_3)$
44	
HF/cc-pVTZ[a]	1.479
QCISD/cc-pVTZ[b]	1.565
RB3LYP/6-31G(d,p)[c]	1.598
RB3LYP/6-311++G(3df,3pd)[c]	1.606
RB3LYP/cc-pVDZ[a]	1.615
RB3LYP/cc-pVTZ[a]	1.603
42	
RMP2/cc-pVTZ[a]	2.083
TC-SCF/6-31G*[d]	2.198
CASSCF(2,2)/6-311G**[e]	2.190
CASSCF(8,8)/3-21G[e]	2.251
CASSCF(8,8)/6-31G(d)[c]	2.198
CASSCF(8,8)/cc-pVTZ[a]	2.174
CASSCF(8,8)/aANO[f]	2.178
UB3LYP/6-31G(d,p)[g]	2.136
UBPW91/cc-pVDZ[f]	1.874
UB3LYP/cc-pVDZ[a]	2.147
BLYP/cc-pVDZ[a]	2.021
BLYP/cc-pVTZ[a]	1.997
CCSD(T)/6-31G(d,p)[h]	2.106
CCSD(T)/6-311G(2dp,2p)[i]	2.101

[a]Ref. 174.
[b]Ref. 169.
[c]Ref. 175.
[d]Ref. 166.
[e]Ref. 160.
[f]Ref. 171.
[g]Ref. 176.
[h]Ref. 159.
[i]Ref. 177.

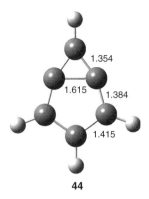

Figure 5.14 B3LYP/cc-pVDZ-optimized geometry of **44**.[174]

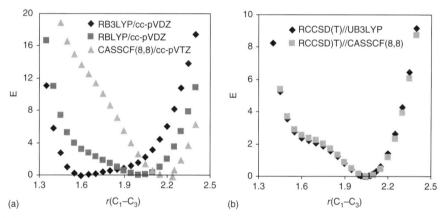

Figure 5.15 PES for the conversion of **44** into **42**. (a) Comparison of surfaces computed at RB3LYP/cc-pVDZ, RBLYP/cc-pVDZ and CASSCF(8,8)/cc-pVTZ. (b) Comparison of surfaces computed at RCCSD(T)/cc-pVTZ//UB3LYP/cc-pVTZ and RCCSD(T)/cc-pVTZ//CASSCF(8,8)/cc-pVTZ. Reproduced with permission from Winkler, M.; Sander, W. *J. Phys. Chem. A.* **2001**, *105*, 10422–10432. Copyright 2001 American Chemical Society.

to note is that in all cases, the PES near the minimum is quite flat, with variation of $r(C_1-C_3)$ by 0.5 Å resulting in less than a 2 kcal mol^{-1} change in energy.

While most of the computational methods indicate that just a single isomer exits on the PES, that being **42**, confirmation of this prediction requires comparison to experiment. The IR spectra of **42** has been obtained in a number of ways: UV photolysis of matrix-isolated **45** or flash vacuum pyrolysis of **46**, 1,3-diiodobenzene, or 1,3-dinitrobenzene followed by trapping in argon.[177,178] In the first of these experiments, an intense band was observed at 547 cm^{-1}, with weaker bands at 751, 824, and 936 cm^{-1}. All of these matched well with the computed CCSD(T)/6-31G(d,p) spectrum (545, 743, 818, and 975 cm^{-1}), which is a level that predicts the open structure of **42**.[178] A subsequent computed spectrum using BLYP/cc-pVTZ,[174] which predicts structure **42**, is also in excellent agreement with the experimental spectrum.

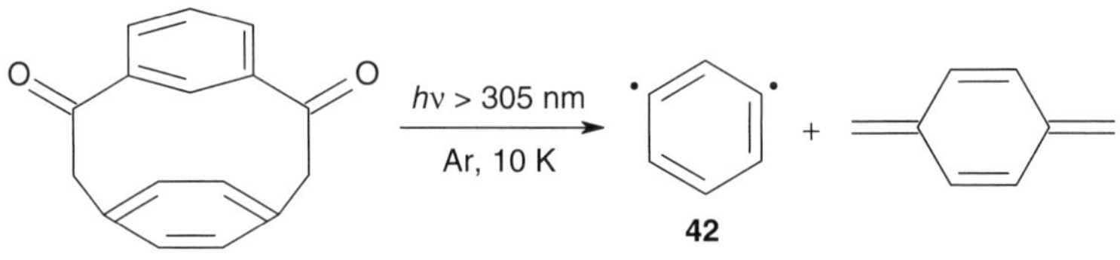

Hess, however, proposed that structure **44** is also consistent with the experimental IR spectra.[179] B3LYP indicates that **44** is lower in energy than **42**. The B3LYP/cc-pVTZ-computed spectrum of **44** predicts the most intense band at slightly higher frequency (564 cm^{-1}) than experiment. In the region of 600–1000 cm^{-1}, the agreement between experiment and the CCSD(T) spectrum is better than with the B3LYP spectrum, yet between 1200 and 1600 cm^{-1} B3LYP predicts four bands, seemingly in agreement with experiment and in conflict with CCSD(T), which predicts only two bands. Furthermore, Hess noted a strong band at 363 cm^{-1}, which is absent in the CCSD(T) spectrum, and this frequency region was not recorded in the previous experiments.

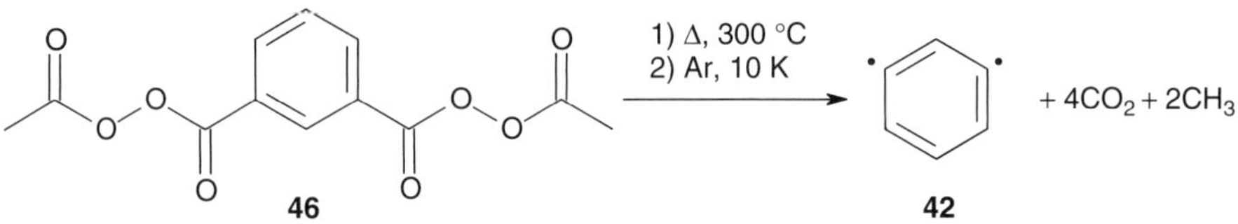

Kraka and Cremer[175] answered Hess' criticisms by first arguing that the UB3LYP description of **42** includes appreciable triplet contamination. When this is corrected for using the sum method, the open ring isomer **42** is now more stable than **44**. Furthermore, they note that Hess misassigned the experimental IR spectrum and there are actually only two bands in the 1200–1600 cm^{-1} region.

Finally, Sander revisited *m*-benzyne in 2002, repeating and extending his experiments and performing additional computations.[177] The experimental and computed IR spectroscopic data are presented in Table 5.12. CCSD(T)/6-311G(2d,2p) computations find only one minimum on the PES, corresponding to **42** with $r(C_1-C_3) = 2.101$ Å. The IR spectrum predicted at this level of theory is in excellent accord with experiment, both in terms of the frequencies and their relative intensities. The computed B3LYP/cc-pVDZ spectrum of **44** differs markedly from the experimental IR spectrum. It predicts that the second most intense band appears at 363 cm^{-1}, while the experiment shows only a very weak absorption in this vicinity. B3LYP predicts three absorptions in the region between the most intense peak and 1100 cm^{-1}, while four are found in the experiment. As Hess

TABLE 5.12 Comparison of Spectroscopic Data of 42 (CCSD(T)) and 44 (B3LYP) with Experiment

Expt.[a]		CCSD(T)/6-311G(2d,2p)[a]		B3LYP/cc-pVDZ[b]	
ν (cm^{-1})	I^c	ν (cm^{-1})	I^c	ν (cm^{-1})	I^c
				316	12
367		405	6	363	60
547	100	535	100	564	100
561	10	541	7		
751	45	748	49	762	34
824	20	827	10	815	42
936	25	944	27	1064	12
				1286	7
1402	15	1425	12	1396	18
				1416	14
1486	15	1504	16	1573	19
		3152	18	3159	30
		3192	7	3162	14
		3197	6	3211	19
				3216	14

[a]Ref. 177.
[b]Ref. 174.
[c]Relative intensity based on the most intense absorption.

claimed, B3LYP indicates that there are four absorptions in the 1200–1600 cm^{-1} region, but only two appear in the experiment.

Based on the experimental IR spectrum and appropriate computations, it is clear that the structure of *m*-benzyne is that of **42**. The bicyclic isomer **44** is an artifact of computational methods that fail to adequately account for the multiconfigurational nature of the wavefunction for **42**.

5.5.4 The Singlet–Triplet Gap and Reactivity of the Benzynes

In the early 1990s, Chen proposed a model to describe the thermochemistry of biradicals and carbenes.[180,181] The heat of formation of an organic biradical or carbene can be produced by sequential cleavage of two C–H bonds. A first approximation is that these bond dissociation energies are equivalent and additive, leading to a hypothetical, noninteracting biradical/carbene. Chen's model suggests that this corresponds with the triplet state, and that the singlet state is stabilized below this additivity approach by the energy of the singlet–triplet gap.

The key to the antitumor/antibiotic activity of the enediyne drugs is the cleavage of hydrogen atoms from DNA by substituted *p*-benzynes. We presented this activity in Chapter 4 (see in particular, Figure 4.10). The enediyne molecules typically suffer from being too reactive and nonselective. Chen[182] applied his model toward understanding the behavior of *p*-benzyne **43** and proposed a scheme for producing potentially more selective, and thereby more suitable, target drug molecules.

Triplet **43** can be envisaged as two isolated, noninteracting radical centers. If this is true, then triplet **43** should abstract hydrogen atoms at the same rate as does phenyl radical. Singlet **43** is more stable than the triplet and at least some of this stabilization energy will need to be released in the transition state for hydrogen abstraction. The singlet–triplet gap is the upper bound for the difference in hydrogen abstraction activation barriers between the "noninteracting" triplet and singlet. Singlet **43** should therefore abstract hydrogen atoms more slowly than does phenyl radical. Even though the singlet–triplet gap is small for **43** (about 2 kcal mol^{-1}, see below), it is enough to introduce a rate decrease by a factor of 30. Simple model computations in fact suggest that singlet **43** abstracts hydrogen with a rate 14 times slower than phenyl radical.[183] Chen[184] later demonstrated that the related *p*-biradical 9,10-dehydroanthracene **47** abstracts hydrogen atoms about 100 times slower than either phenyl radical or 9-anthracenyl radical. Chen[182] proposed that enhanced selectivity of *p*-benzyne analogs could be produced by generating molecules with larger singlet–triplet energy gaps than **43**.

47

Before we discuss potential new enediyne drugs based on this proposal, we first present computational results on the singlet–triplet energy gap of **41–43**, which are presented in Table 5.13. As already discussed, a single-configuration treatment of **41** predicts that the ground state is a triplet,[156] when, in fact, the ground state of all three benzynes is the singlet.[168] While a two-configuration wavefunction does place the singlet energetically below the triplet, the energy gap is underestimated by 10 kcal mol^{-1}.[156,166] A larger multiconfiguration wavefunction, such as CASSCF(8,8) dramatically improves ΔE_{ST} for **41**,[160,171] but inclusion of dynamic electron correlation is needed for **42**.[171,185] CI methods provide very good estimates of the singlet–triplet gap for all three benzynes.[160] CCSD(T) proves to be extraordinarily good in evaluating ΔE_{ST} for the three benzynes: it underestimates the gap by less than 2 kcal mol^{-1}.[171] The results are even better with the MkCCSD method.[186] BPW91 underestimates ΔE_{ST} for **41** but gets the gap for **42** within 2 kcal mol^{-1}. Restricted DFT predicts a ground-state triplet for *p*-benzyne,[171,185] but spin-unrestricted DFT gives very reasonable estimates of the singlet–triplet gap[187,188]. So, while the most accurate results are obtained by the very computationally intensive CCSD(T) method, DFT, properly done, can provide perfectly reasonable values at a fraction of the computational expense.

Chen first proposed 2,5-didehydropyridine **48** as a potential selective biradical that might be useful as drug target.[189] **48** can act as a weak base, and since tumor cells are generally more acidic than normal healthy cells, it would exist in its conjugate acid form **49** in cancer cells. Chen conjectured that if the singlet–triplet gaps of **48** and **49** are different, then a drug selective for tumor cells might be possible. While Chen's calculations were supportive of this idea, he used a CASSCF(6,6) wavefunction that did not allow for correlation of the π-electrons and an untested, unusual method for adding in dynamic electron correlation.

Cramer[187] reexamined the singlet–triplet energy gaps of **48** and **49** with a variety of computational methods. His results, along with results by Cremer and

TABLE 5.13 Singlet–Triplet Energy Gap (ΔE_{ST}) for 41–43[a]

Method	41	42	43
HF/DZ+P[b]	−2.9		
TC-SCF/DZ+P[b]	27.7		
TC-SCF/6-31G*[c]	27.7	13.0	0.8
CASSCF(2,2)/6-311G**[d]	28.6	13.1	0.6
CASSCF(8,8)/cc-pVDZ//CASSCF(8,8)/3-21G[d]	33.6	15.2	2.5
CASSCF(8,8)/aANO[e]	*35.1*	*16.1*	*3.8*
CASPT2/aANO//CASSCF(8,8)/aANO[e]	*32.6*	*19.0*	*5.8*
CASPT2/cc-pVDZ//BPW91/cc-pVDZ[f]	30.4	18.0	5.8
π-SDCI/6-311G**//TC-SCF/6-31G*[c]	34.4		
TC-CISD/DZ+P[b]	32.2	16.0	1.6
CISD/cc-pVDZ//CASSCF(8,8)/3-21G[d]	31.2	15.8	1.8
CCCI/cc-pVTZ//CASSCF(8,8)/3-21G[d]	36.9	17.1	2.2
CCSD(T)/cc-pVTZ[e, g]	*35.3*	*20.7*	*2.3*
CCSD(T)/cc-pVTZ[h]	36.0	19.9	4.9
BCCD(T)/cc-pVDZ//UBPW91/cc-pVDZ[i]			4.5
MkCCSD/cc-pVDZ//(U)CCSD(T)/cc-pVDZ[j]	38.5	19.5	2.4
RBPW91/cc-pVDZ[f]	31.3	19.4	−2.0
RBPW91/cc-pVTZ[e]	*33.2*	*20.1*	*−1.6*
RB3LYP/cc-pVTZ[h]	31.4	14.8	−11.3
UBPW91/cc-pVDZ[i]			4.0
UB3LYP/6-31G(d,p)[k]			2.5
G3[h]	38.5	23.3	11.1
Expt.	37.7±0.6[l]		
	7.5±0.3[m]	21.0±0.3[m]	3.8±0.5[m]

[a]All energies are in kilocalorie per mole. Positive values indicate a ground-state singlet. Values in italics are for 298 K, otherwise 0 K is assumed.
[b]Ref. 156.
[c]Ref. 166.
[d]Ref. 160.
[e]Ref. 171
[f]Ref. 185.
[g]Estimated as E(CCSD/cc-pVT) + E(CCSD(T)/cc-pVDZ)-E(CCSD/cc-pVDZ).
[h]Ref. 169.
[i]Ref. 187.
[j]Ref. 186.
[k]Ref. 188.
[l]Ref. 158.
[m]Ref. 168.

TABLE 5.14 Singlet–Triplet Energy Gap (ΔE_{ST}) for 48–49[a]

Method	48	49
CASPT2/cc-pVDZ//BPW91/cc-pVDZ[b]	14.1	5.6
CCSD(T)/cc-pVDZ//BPW91/cc-pVDZ[b]	12.9	29.8
BCCD(T)/cc-pVDZ//BPW91/cc-pVDZ[b]	11.6	5.6
RBPW91/cc-pVDZ[b]		−2.3
UBPW91/cc-pVDZ[b]	14.0	4.3
UB3LYP/6-31(d,p)[c]	8.2	2.8

[a] All energies are in kilocalorie per mole. Positive values indicate a ground-state singlet.
[b] Ref. 187
[c] Ref. 188.

Kraka,[188] are listed in Table 5.14. Once again we see problematic results with many methods. Consistent with the previous results, restricted DFT erroneously predicts a ground-state triplet for **49**. More discouraging is the gross overestimate of ΔE_{ST} for **49** obtained with the CCSD(T) method. The origins of this problem are with its single-configuration reference and low symmetry, which allows for many single and double configurations to strongly mix in. The high symmetry of **42** fortuitously eliminates this problem. The general solution is to use the BSSD(T) method, which uses Brueckner orbitals as the reference and eliminates contributions from single-electron excitations.

Nevertheless, computations clearly distinguish **48** and **49** in terms of their singlet–triplet energy gaps. The gap in **49** is about 4–6 kcal mol^{-1}, somewhat larger than the gap in **43**. However, the gap in **48** is much larger, 8–12 kcal mol^{-1}. Based on Chen's reactivity model, **48** should be a very poor hydrogen abstracter, while **49** should behave very similar to **43**. Therefore, if the two species could partition between normal and cancer cells based on their acidity, **49** should be a potent antitumor agent while not appreciably damaging normal cells. Kraka and Cremer,[188] however, point out that the proton affinity of **48** is only 214 kcal mol^{-1}, indicating that it is too weakly basic to be protonated in tumor cells. They propose amidine **50** as a better target molecule. It has a singlet–triplet gap of 2.8 kcal mol^{-1}, which reduces to 1.8 kcal mol^{-1} upon protonation (**51**). While the difference between **50** and **51** is not as great as that between **48** and **49**, the proton affinity of **50** is 249 kcal mol^{-1}, making it much more basic and likely to be protonated in tumor cells.

A number of computational studies have been performed examining different didehydoropyridines[190,191] and analogs containing more than one-ring nitrogen.[169] Kraka and Cramer[192] have examined amidine-substituted dynamicin analogs. Also, numerous studies of substituted benzynes have been reported.[193–197] A range of singlet–triplet energy gaps have been found, including some cases where the ground state is the triplet.[196] Nevertheless, no drug based on this variation upon enediyne chemistry has been developed. The only marketed enediyne-based drug is mylotarg,[198] a derivatized version of calichaemicin. It was, however, removed from the marketplace in 2010 due in part to a high death rate among trial patients using the drug, though a recent study suggests that a low dose regimen may prove useful.[199]

5.6 TUNNELING OF CARBENES

Carbenes, as species bearing a carbon atom with an unfilled octet, are especially reactive. Their rich chemistry is dominated by insertion reactions and rearrangements.[10–12,200] Over the past five years, Schreiner has uncovered just how reactive carbenes can be, opening up a new method of chemical control.[201]

Though speculated upon its existence for some time, hydroxymethylene **53** was not truly identified until 2008.[202] Schreiner and coworkers developed a technique of extruding CO_2 from appropriate precursors using high vacuum flash pyrolysis (HVFP) to obtain hydroxylated carbenes. Their first example was the preparation of **53** by HVFP of glyoxylic acid **52** with capture of the product stream in an argon matrix at 11 K (Reaction 5.9).

(Reaction 5.9)

Previous computations at CCSD(T)/CBS[50] and Schreiner's[202] own computations at CCSD(T)/cc-pCVQZ indicate that **53** has a large singlet–triplet gap of 25.3 and 28.0 kcal mol^{-1}, respectively, with a singlet ground state. (The hydroxycarbenes discussed here are ground-state singlets with large singlet–triplet gaps, and so only the singlet PESs need be considered.) The production of **53** is confirmed by comparison of the experimental and computed IR vibrational frequencies; using anharmonic corrected frequencies leads to agreement between the experimental and computed frequencies with a standard deviation of only 6 cm^{-1}.

The PES for the rearrangement of hydroxymethylene **53**, with computed focal point energies, is shown in Scheme 5.5. **53** appears to be trapped in a deep well: the [1,2]-hydrogen shift to give formaldehyde **54** requires surmounting a barrier of 29.7 kcal mol^{-1} and fragmentation into CO_2 and H_2 has a barrier of 47.0 kcal mol^{-1}. It seems that **53** should exist indefinitely in the argon matrix.

Astonishingly, **53** disappears with a half-life of about 2 h at 11–20 K, with little rate change with the increase in temperature. As the IR signature of **53** disappears, signals corresponding to formaldehyde **54** appear. The deuterated analog HCOD (**d-53**) is completely stable at these temperatures. Apparently, **53** rearranges to **54** via quantum mechanical tunneling through the high barrier.

This is supported by computed rate constants with tunneling contributions using the semiclassical Wentzel–Kramers–Brillouin (WKB) theory.[203,204] The WKB procedure requires a carefully described intrinsic reaction path, and

350 DIRADICALS AND CARBENES

Scheme 5.5 PES of rearrangement of **53**. Relative energies in kilocalorie per mole.

Schreiner et al. computed this path at CCSD(T)/cc-pCVQZ. The computed WKB half-life of **53** is 122 min, in excellent accord with the experimental value of about 2 h. (A WKB estimate was performed using a slightly less rigorous method; the path was obtained at MP2/aug-cc-pvTZ and energies then recomputed at CCSD(T)/aug-cc-pVTZ gave a very similar half-life of 126 min.[205]) Furthermore, the computed half-life of **d-53** is over 1200 years, which also agrees with the experimental observation. There can be no doubt that **53** rearranges rapidly at very low temperatures by a tunneling mechanism.

How prevalent is tunneling among the hydroxycarbenes? Is there a continuum of tunneling rates? Schreiner answered these questions in a series of outstanding papers that present a combination of experimental and computational results.

Phenylhydroxycarbene **56** was prepared by HVFP of phenylglyoxylic acid **55** (Reaction 5.10).[206] The presence of **56** is confirmed by comparison of its experimental IR spectrum and computed (CCSD(T)/cc-pVDZ and corrected for anharmonicity) vibrational frequencies.

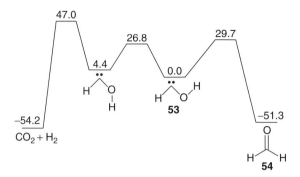

(Reaction 5.10)

Again, it appears as if **56** lies in a deep potential energy well (see Scheme 5.6). The computed barrier (CCSD(T)/cc-pVQZ//M06-2x/cc-pVDZ) for the [1,2]-H shift to form benzaldehyde **57** is 28.8 kcal mol^{-1}, and the alternative exit channel leading to benzene and CO has an even higher barrier of 55.0 kcal mol^{-1}.

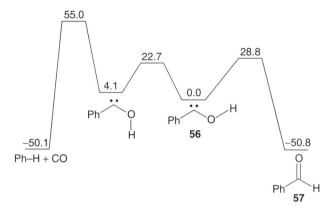

Scheme 5.6 PES of rearrangement of **56**. Relative energies in kilocalorie per mole.

However, **56** disappears at 11 K with a half-life of 2.46 h, while the deuterated analog is stable. The WKB-computed half-life (using the path computed at MP2/aug-cc-pVDZ and energies at CCSD(T)/cc-pVTZ) of **56** is 3.3 h and 8700 years for ***d*-56**. This fast tunneling rate also accounts for the lack of rearrangement of **56** into the tropone **58**, a reaction analogous to the chemistry of phenyl carbene and phenyl nitrene discussed earlier in this chapter (Section 5.2).

Cyclopropylhydroxycarbene **60** has an even longer tunneling half-life.[207] **60** was produced by HVFP of cyclopropylglyoxylic acid **59** (Reaction 5.11). The structure of **60** was confirmed by the agreement between its experimental IR spectrum and the CCSD(T)/cc-pVTZ-computed vibrational frequencies. Despite a high computed barrier of 30.4 kcal mol^{-1} (CCSD(T)/cc-pVTZ//M06-2x/6-311++G(d,p)) for [1,2]-hydrogen rearrangement to cyclopropanecarbaldehyde **61**, the half-life for this rearrangement is 17.8 h at 11 K. The deuterated analog ***d*-60** does not rearrange. The WKB-computed half-lives are again in excellent agreement with those of the experiment: 16.6 h for **60** and 10^5 years for ***d*-60**.

(Reaction 5.11)

In contrast to the above examples, dihydroxycarbene **63** does not undergo a tunneling rearrangement.[208] **63** was made by the HVFP of oxalic acid **62** (Reaction 5.12), and its structure was confirmed by the comparison of the experimental and computed (CCSD(T)//cc-pVTZ) vibrational frequencies. **63** persists for days at 11 K, as does its dideuterated analog d_2-**63**. The carbene is stabilized by π-donation from each oxygen lone pair into the empty p-orbital on carbon. This donation manifests in the very short C–O distance of 1.325 Å in **63**, compared with the C–O distance of 1.427 Å in methanol. Similarly, the carbenes **64** and **65** do not rearrange at low temperature.

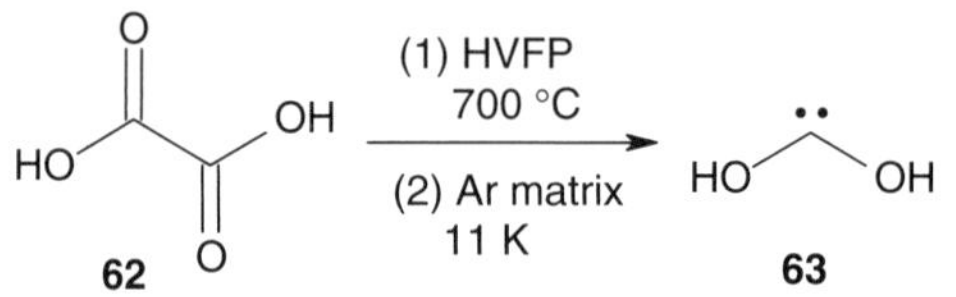

$$\underset{\mathbf{62}}{\text{HO-C(=O)-C(=O)-OH}} \xrightarrow[\text{(2) Ar matrix 11 K}]{\text{(1) HVFP 700 °C}} \underset{\mathbf{63}}{\text{HO-}\ddot{\text{C}}\text{-OH}} \quad \text{(Reaction 5.12)}$$

$$\underset{\mathbf{64}}{\text{MeO-}\ddot{\text{C}}\text{-OH}} \quad \underset{\mathbf{65}}{\text{H}_2\text{N-}\ddot{\text{C}}\text{-OH}}$$

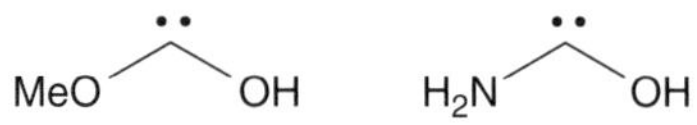

The fastest tunneling rate for rearrangement of a hydroxycarbene that Schreiner[209] has observed is for methylhydroxycarbene **67**. This carbene is produced by HVFP of pyruvic acid **66** (Reaction 5.13). As with the other carbenes, its structure is confirmed by comparison of the experimental and computed vibrational frequencies. The observed rate of rearrangement of **67** into acetaldehyde **68** is 66 min at 11 K. The WKB-computed rate (at CCSD(T)/cc-pCVQZ) is 71 min, which is in excellent accord with experiment.

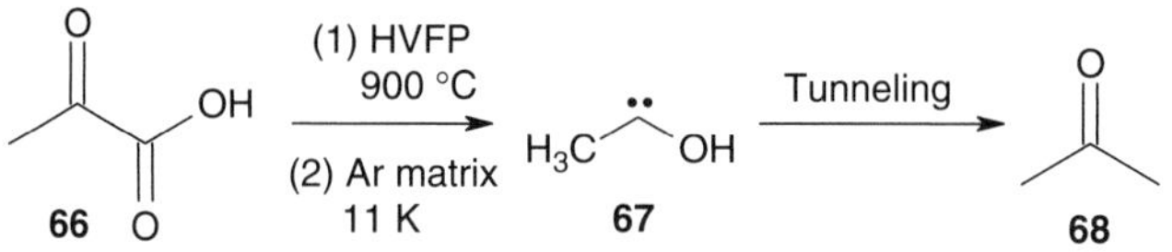

$$\underset{\mathbf{66}}{\text{CH}_3\text{-C(=O)-C(=O)-OH}} \xrightarrow[\text{(2) Ar matrix 11 K}]{\text{(1) HVFP 900 °C}} \underset{\mathbf{67}}{\text{H}_3\text{C-}\ddot{\text{C}}\text{-OH}} \xrightarrow{\text{Tunneling}} \underset{\mathbf{68}}{\text{CH}_3\text{-CHO}} \quad \text{(Reaction 5.13)}$$

Schreiner[207] offers a correlation between the stabilities of the carbenes and their tunneling rates. Both the phenyl and cyclopropyl groups of **56** and **60** act as weak π-donors into the adjacent empty p-orbital of the carbene carbon. This acts to stabilize the carbene, but a much greater stabilization is afforded when the atom adjacent to the carbene carbon has a lone pair. In both **63** and **64**, the second adjacent oxygen atom donates significant π-density to the carbene carbon, leading to a much more stable carbene with short C–O bonds. The adjacent nitrogen in **65** is an even better donor.

The stability of the hydroxycarbenes can be judged using an isodesmic reaction that compares a particular carbene to methylhydroxycarbene **67** (Reaction 5.14). The computed energy of Reaction 5.14 associated with each carbene and its tunneling half-life is listed in Table 5.15. As the carbene becomes more stable, the half-life increases, with a threshold around 20 kcal mol^{-1} leading to carbenes that do not tunnel at all.

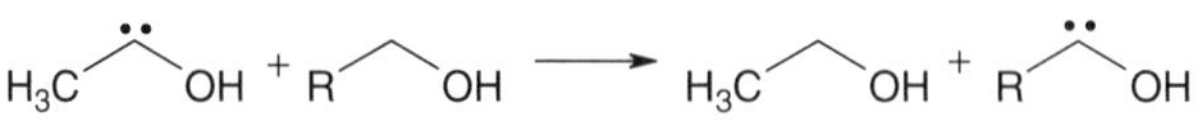

$$\text{H}_3\text{C-}\ddot{\text{C}}\text{-OH} + \text{R-CH-OH} \longrightarrow \text{H}_3\text{C-CH-OH} + \text{R-}\ddot{\text{C}}\text{-OH} \quad \text{(Reaction 5.14)}$$

R = Ph, *cyc*-C$_3$H$_5$, OH, OCH$_3$, NH$_2$

TABLE 5.15 Tunneling Half-Life and Stabilization Energy (kcal mol^{-1}) for Hydroxycarbenes

Carbene	R	Half-life	$E_{stabilization}$[a]
67	CH_3	1.1 h	0.0
56	Ph	2.5 h	−2.3
60	cyc-C_3H_5	17.8 h	−4.1
63	OH	No tunneling	−23.8
64	OCH_3	No tunneling	−25.2
65	NH_2	No tunneling	−28.7

[a]Energy of Reaction 5.14 computed at CCSD(T)/cc-pVDZ//M06-2x/6-311++G(d,p).[207]

5.6.1 Tunneling control

The chemistry of methylhydroxycarbene **67** merits further consideration. Early studies of the pyrolysis of pyruvic acid indicated that **67** is an intermediate (though not specifically detected) that rearranged directly to acetaldehyde **68**, and that vinyl alcohol **69** was not an intermediate nor a product.[210–212] This interpretation is in conflict with G1 computational studies,[213] which suggest that the barrier for conversion of **67** to **69** is 4.8 kcal mol^{-1} lower in energy than the barrier to **68**, and that computations at higher levels would not be likely to change the relative order of these two TSs. Schreiner and Allen[209] recomputed the PES for the rearrangement of **67** using the focal point method (Scheme 5.7). It also shows a lower barrier leading to **69** (22.6 kcal mol^{-1}) than the barrier leading to **68** (28.0 kcal mol^{-1}).

At very low temperature, it is impossible for **67** to cross over either barrier. Following their previous studies, Schreiner and Allen proposed that **67** tunnels through the barrier to form **68**. Their WKB computation (CCSD(T)/cc-PCVQZ) of the half-life for the rearrangement of **67** → **68** is 71 min, which is in excellent

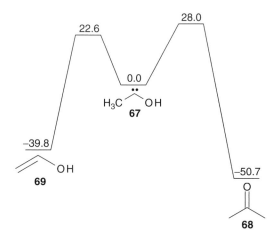

Scheme 5.7 PES of rearrangement of **67**. Relative energies in kilocalorie per mole.

accord with the experimental value of 66 min. The WKB barrier for the rearrangement of the deuterated analog **d-67** is 4000 years, which is again consistent with the experimental observation that the deuterated species does not rearrange. The computed half-life for the rearrangement **67** → **69** is 190 days, and this slow rate explains why no vinyl alcohol is observed: the rate to form **68** is simply much faster than the rate to form **69**.

But why does **66** prefer to tunnel through the *higher* barrier? The tunneling rate is dependent on three factors: (1) the square root of the height of the barrier, (2) the width of the barrier, and (3) the square root of the mass of the atom(s) moving through the barrier. So while the lower barrier height favors the rearrangement to **69**, the barrier width for the rearrangement to **68** is much smaller; its path is 19 percent shorter than for the path to **69**. The path taking **67** to **68** involves mostly motion of the migrating hydrogen, which is a distance of about 2.1 Å. However, in going from **67** to **69**, the migrating hydrogen has a longer distance to travel (2.4 Å) and a second methyl hydrogen must move down into the plane of the forming carbon–carbon double bond. As Schreiner and Allen[209] put it, "barrier width trumps barrier height".

A similar situation prevails in the rearrangement of cyclopropylhydroxycarbene **60**. This hydroxycarbene can rearrange via a hydrogen shift to **61** or via a carbon shift to **70**. The PES, computed at CCSD(T)/cc-pVTZ//M06-2x/6-311+g(d,p), for these paths is shown in Scheme 5.8.[207] The (much) lower barrier is for the rearrangement to **70**; however, this reaction is not observed. Rather, the rearrangement that involves the shorter path and the lower effective mass, **60** → **61**, is the sole reaction that occurs.

The implication of these studies is of critical importance. Chemists generally think of the product distribution of a chemical reaction being controlled by kinetics or thermodynamics. Under kinetic control, the distribution favors the product that results from crossing the lowest activation barrier. Under thermodynamic control, the distribution favors the lowest energy product. Schreiner and Allen now add

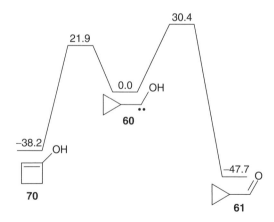

Scheme 5.8 PES of rearrangement of **60**. Relative energies in kilocalorie per mole.

a third option, *tunneling control*, where the product formation is dictated by the factors critical to tunneling (barrier height, barrier width, and effective mass).

5.7 INTERVIEW: PROFESSOR HENRY "FRITZ" SCHAEFER

Interviewed March 28, 2006

Professor Henry Schaefer is the Graham Perdue Professor of Chemistry and director of the Center for Computational Chemistry at the University of Georgia. Schaefer moved to Georgia in 1987 from the University of California at Berkeley, where he had been since 1969.

Schaefer's interest in computational chemistry began as an undergraduate at MIT. While trying to muster the nerve to knock on the door of Professor Richard Lord's office to pursue a senior thesis in spectroscopy, Schaefer asked his advisor Professor Walter Thorson for advice. Instead, Thorsen sufficiently flattered young Schaefer to join his own research group. Schaefer worked on cubic H_8, successfully determining its molecular orbital solution but failing to generate the VB solution. Nonetheless, Schaefer's honors thesis was accepted, and he then obtained his doctorate at Stanford under the direction of Professor Frank Harris.

Just prior to joining the faculty at Berkeley, Schaefer met with Dr. Charlie Bender at the computational chemistry Gordon Conference in June 1968. Bender was about to join University Computing in Palo Alto. They decided to work together, making use of what Schaefer calls "a magnificent device", a state-of-the-art CI code written with Professor Isiah Shavitt, and the Univac 1108, a state-of-the-art computer, at University Computing. They had to settle on a molecule to study. Their guidelines were amazingly simple—"It was a matter of deciding what would fit. Triatomics were large systems at that time, and non-linear triatomics were considered to be difficult. Water seemed to be uninteresting, so it was methylene. Herzberg seemed like he was going to get a Nobel Prize, and it looks like methylene is going to be the feather in his cap. Let's go for it. Who knew what we were going to find? Basically, we had the hammer; we're just looking for something to pound."

What they found was surprising—triplet methylene was *not* linear as Herzberg had suggested, and that everyone accepted. But Schaefer and Bender trusted their computations, clearly and plainly arguing that triplet CH_2 is bent. This despite previous computations by Pople and Allen *had also indicated a bent geometry*, but these authors deferred to Herzberg and his experiments. What gave Schafer his confidence to challenge the iconic Herzberg? "Well, we were younger. We had nothing to lose. Our calculations were a whole lot better and we were rambunctious. It was the only molecule we had, and we had to run with it," Schafer recalls with a hint of modesty and a twinkle in his eyes.

Schaefer readily remembers his reaction when Lineberger's PES study of methylene appeared, indicating a very large singlet–triplet gap, much larger than his computational prediction. His first response was that Lineberger was wrong. Schaefer then "went to talk to wiser heads, like Yuan Lee and Brad Moore. Their response was 'You made a splash on triplet methylene, so what if you got the singlet wrong'." For 6 months, Schaefer was convinced that his calculation must be wrong, but he didn't know why. So he returned to the problem with a vengeance, along with many others. "Methylene became a cause," Schaefer recalls.

Schaefer's improved calculations continued to indicate a gap of only about 10 kcal mol^{-1}, and Goddard's calculation indicated the possible presence of hot bands in the PES spectrum. Schaefer recalled a conversation with Lineberger at that time (about 1977–1978): "These preprints all show the computational results indicating a much smaller singlet-triplet gap than your experiments. Aren't you going to be embarrassed about this? Lineberger said, I'm not worried about that at all. In fact, I'm feeling pretty good about it. Before, yours was the only career I destroyed. Now, I'm taking down all of quantum chemistry!" Lineberger did eventually retract, but only after achieving a much better experimental design. Schaefer continues to admire Lineberger's resolve and faith in his own research, and finished with the comment "All of us are still friends."

Schaefer holds to his assertion made in his *Science* article that methylene remains the paradigm for computational chemistry. He notes that one could point to Kolos' paper on H_2 or Davidson's doctoral thesis that demonstrated a double minimum in an excited state of H_2. But he maintains the prominence of methylene: "Ours was the first non-linear triatomic. Organic chemists were interested in methylene; lots of experiments were available." The computational studies on methylene corrected two serious interpretations of experimental data, and clearly demonstrated that computational chemistry can hold its own with experiment. Because of the impact of this work, Schaefer considers his methylene studies to be his most important contribution to science. "We were at the right place, at the right time. There's a brashness about it that set it apart."

Schaefer has a long interest in aromaticity, including studies of paracylophanes and N_6. "[10]-annulene just seemed obvious. I read Masamune's paper and said 'We could probably help these guys out'." He was surprised, however, in the difficulties that occurred in dealing with [10]-annulene. Never being a great fan of DFT or perturbation methods, he does find a sense of satisfaction that these fail to properly treat [10]-annulene and that the coupled cluster method arrives at the proper answer. One technique utilized in tackling the [10]-annulene story has piqued Schaefer's interest. He was very much taken by Monte Carlo and other broad search techniques, whereby "you put in the atoms and it goes and finds everything." He calls this 'mindless chemistry' and has used this term as the title of a collaborative paper with Professor Schleyer describing stochastic searches used to locate all local energy structures of a series of pentatomic molecules. This paper has resonated within the chemistry community. Though rejected by the *Journal of the American Chemical Society*, it was published in 2006 by the *Journal of Physical Chemistry A* and was that journal's most frequently accessed paper for the first half of 2006!

Collaborations have played a major role in his scientific pursuits, but he points to the very asynchronous nature of such collaborations with experimentalists. Early on, experimentalists would have nothing to do with him or his computations—"They were all skeptics." Nowadays, Schaefer is frequently contacted by experimentalists looking for computational assistance or confirmations. However, experimentalists are much less receptive to his calls for suggested experiments based upon his computations. He believes that this reluctance is not due to mistrust of the computations, but rather to the fundamental differences in the nature of doing computations and doing experiments. "Experimentalists have several tricks in their hats and that's what they know how to do." If the suggested chemistry is not within that scope, they hesitate to take on the suggestion. "Computational chemistry has much greater coverage. We don't have universal coverage but it's heading that way." He qualifies this statement, "A spectroscopist can measure something to 10^{-4} wavenumbers. They may not be exactly sure of what they are measuring, but it is what it is. And we're not going to compute *anything* to 10^{-4} wavenumbers."

Schaefer has few regrets when it comes to missed scientific opportunities. The one in particular that he notes is C_{60}. He met with Professor Rick Smalley days after the discovery of C_{60}, but didn't get excited about it.

When asked to speculate as to what the computational chemistry landscape might look like a decade from now, Schaefer became a bit pessimistic. He notes that computer performance has stagnated recently, calling it "depressing that PCs are about as fast now as they were two years ago". He feels that parallel processing will be the key to future progress, a prospect he finds discouraging when contemplating the coding efforts that will be required to maximize parallel performance.

Nonetheless, he continues to actively pursue new projects, welcoming new students into the group and pushing forward. "Why do we study chemistry?" he asks. "We want to know answers to qualitative questions. That is what distinguishes us from physics."

5.8 INTERVIEW: PROFESSOR PETER R. SCHREINER

Interviewed September 6, 2012

Peter Schreiner is the Professor of Organic Chemistry at the Justus-Liebig-University Giessen (Germany). He was introduced to computational chemistry in graduate school at Erlangen working with Paul Schleyer. Wanting to understand more about what the quantum chemical computer programs were actually doing, he took a fellowship at the University of Georgia, where in addition to pursuing an experimental project with Dick Hill, he worked with Fritz Schaefer. Computational chemistry "just sucked me in," he said. "The more you know, the better you can answer organic chemistry questions. It turns out to be a great tool: you can ask questions about why this bond is a bit long or this angle is odd and it doesn't tell you the answer at first glance but as you pour over it, there is more to see. Computational chemistry is such a creative tool; it can trigger your thinking."

Combining experiments and computations, Schreiner's research has always exhibited a symbiotic character. His research in carbenes exemplifies this interplay. Despite a great deal of previous work on carbenes, Schreiner discovered that the chalcogene carbenes were not well known. Knowing that the simplest example, hydroxymethylene, had not been made or characterized, he was sure that its preparation would be difficult. "How does nature make carbenes?" he asked. "The pyruvate cycle—CO_2 extrusion from alpha-ketocarboxylic acids. She figured it all out!" So, the group tried oxalic acid and "Bam! Extrusion gave us dihydroxycarbene smoothly! The surprise was it was so simple." But it was not all to be so simple.

The next case they examined was the pyrolysis of glyoxylic acid, and hydroxymethylene was observed, but it disappeared overnight. This offered a clue as to why it might not have been previously prepared. Schreiner thought that perhaps someone might have done this previously but did not spectroscopically identify it right away. "Let it sit too long and it's gone!" he noted. After many repeated tries to isolate hydroxymethylene, Schreiner had to face this problem. He discounted a matrix or a stereoelectronic effect because the molecule is too small. He was at a loss for explaining the rate, but "then as a good organic chemist and the rate doesn't fit with what you'd expect, you say 'something's fishy and it's a tunneling effect.'" His students measured the rate of disappearance as 2 h, which was way too fast for any normal reaction at 11 K. When they increased the temperature to 20 K and the rate remained 2 h, "that sort of confirmed the hunch," he said. Worried that no one would accept tunneling through a barrier as large as 30 kcal mol^{-1}, he contacted his colleague Wesley Allen, whom he had met at the University of Georgia. Schreiner asked Allen to compute the proper IRC, and from it the barrier height and width, and ultimately a tunneling rate. Allen responded, "apart from the fact that you're crazy, I'll do it." Of particular importance is that Schreiner did not tell Allen the value of the experimental rate. In about 3 weeks, Allen called Schreiner and told him that the computed rate was 2 h! (Actually, Allen computed a rate of 122 min, "a bit more accuracy than I needed," said a sheepish Schreiner.) Both Schreiner and Allen were in shock: this was a clear case of tunneling.

Next was the study of methylhydroxycarbene, which could potentially tunnel from either the C—H or O—H direction. The kinetic product is vinyl alcohol while the thermodynamic product is acetaldehyde. Which direction would tunneling choose? The previous examples all had the tunneling selection through the lower barrier. "We got lucky, and it tunnels through the higher barrier," said Schreiner. "I always thought of tunneling as a correction to the rate but it's not. It *is* the rate." Thus, was born the expression *tunneling control*. "Now we have several cases. It's a general phenomenon, not unusual at all! Practicing organic chemists should care about this."

The outcome of this endeavor was really quite a surprise to Schreiner. "Tunneling was not something that I thought about. In fact it was so far back in my mind I was hoping that I would never have to unbury it," said Schreiner. He noted that the theory of tunneling is rigorous but "so non-classically difficult for a paper-and-pencil organic chemist to think about it. Qualitative models are not at hand. And this is our goal now—that's why we did the Hammett plots, to give a handle on the qualitative understanding of the electronic effects."

There has been a recent shift in how new projects are approached within the Schreiner group—computations are now leading experiments. "We simply can't put in the effort for weeks to months to characterize these species without knowing beforehand if the molecule will be interesting," he noted. The group is expanding into other areas in the pursuit of tunneling effects. "Who would have thought about a *JACS* paper on benzoic acid in 2010?" he exclaimed. "We have looked at many acids now and haven't found one that does not undergo tunneling-assisted conformational change. How is it possible this was not broadly known?" He noted that all discussions on the conformations of amino acids in space are "pointless since tunneling will be fast even at low temperature."

Schreiner continues to collaborate with Wesley Allen, who brings the ability to carry out very high level computations, which are especially valuable in structure identification *via* vibrational frequency matching—computations that require anharmonic corrections that Allen is expert in performing. They hope to develop in the near future an add-on to the *Gaussian* and *CFOUR* computer packages that will allow nonspecialists to do simple tunneling computations. A "friendly competitor" group, led by Magyarfalvi in Hungary, has spent time in Giessen and has now scooped Schreiner on tunneling in glycine. Schreiner takes this as a good development: "You have to have good competitors. It keeps you on your toes." His other collaborations are with surface and solid-state physicists, because "the questions they ask are so different."

Schreiner's interest in dispersion effects led to collaboration with Stefan Grimme. Schreiner always felt uncomfortable with the usual textbook explanations for some basic organic facts, like the fact that branched alkanes are more stable than linear ones. The culprit is dispersion. "Dispersion is hard to grasp because it is so dispersed," he complained. "Typically we just wipe it under the carpet. The most blatant examples are the hydrocarbon dimers with very long C—C bonds that you can heat to 300 °C without decomposing. What keeps them together? It's a balance of things: repulsion and attraction." Schreiner discussed this problem with Grimme, and they joined forces on the hexaphenylethane riddle. Ultimately, they found that "dispersion is a huge component of the total binding energy. It's so large you can no longer overlook it." He believes that we have just begun to recognize the importance of dispersion. "With modern density functionals, we can cleanly dissect out the role of dispersion and now we can look at larger molecules—and dispersion plays a larger role in larger molecules."

Schreiner points to the C–H activation problem as one that has so far eluded his group's solution. "Methane to methanol conversion, if we manage, that that would be an enormous breakthrough—and by manage I mean large scale with simple affordable methods—but it's such a difficult nut to crack," he notes with disappointment. "I didn't think we would solve it, but I thought we would learn something about it. You keep running into a wall and your head is bloody, you heal and run again—but the wall isn't getting any thinner."

In thinking about the future of computational chemistry, Schreiner, like Stefan Grimme, noted the conformation and multiple isomer problems. He explains that "we scribble a mechanism and come up with a feeling for the reactive conformation, but science is not just a feeling. So we have to probe large conformational spaces for large molecules, and what is missing are filters that would rule out the conformations that are nonsense. How do we input our chemical knowledge into the optimization process?" What Schreiner is hoping for is a way of incorporating chemical intuition into the computational chemistry toolset. Synthetic chemists seem to have this intuition—"K.C. Nicolaou, he knows how to cut a large molecule into pieces in a few minutes; it's amazing. How does he know this?" Schreiner wistfully mused. *"That logic* coupled with quantum mechanics would be amazing."

REFERENCES

1. Moss, R. A.; Platz, M. S.; Jones, M., Jr., Eds. *Reactive Intermediate Chemistry*; Wiley-Interscience: Hoboken, NJ, **2004**.
2. Salem, L.; Rowland, C. "The electronic properties of diradicals," *Angew. Chem. Int. Ed. Engl.* **1972**, *11*, 92–111.
3. Giese, B. *Radicals in Organic Synthesis : Formation of Carbon-Carbon Bonds*; Pergamon Press: Oxford, **1986**.
4. Henry, D. J.; Parkinson, C. J.; Mayer, P. M.; Radom, L. "Bond dissociation energies and radical stabilization energies associated with substituted methyl radicals," *J. Phys. Chem. A* **2001**, *105*, 6750–6756.
5. Renaud, P.; Sibi, M. P., Eds; *Radicals in Organic Synthesis*; Wiley-VCH: Weinheim, **2001**.
6. Newcomb, M. "Radicals," in *Reactive Intermediate Chemistry*; Moss, R. A.; Platz, M. S., Jones, M., Jr., Eds.; Wiley-Interscience: Hoboken, NJ, **2004**, p 122–163.
7. Borden, W. T., Ed. *Diradicals*; John Wiley & Sons: New York, **1982**.
8. Rajca, A. "Organic Diradicals and Polyradicals: From Spin Coupling to Magnetism?," *Chem. Rev.* **1994**, *94*, 871–893.
9. Berson, J. A. "Diradicals: conceptual, inferential, and direct methods for the study of chemical reactions," *Science* **1994**, *266*, 1338–1339.
10. Brinker, U. H., Ed. *Advances in Carbene Chemistry*; JAI Press: Greenwich, CT, **1994**; Vol. *1*.

11. Brinker, U. H., Ed. *Advances in Carbene Chemistry*; JAI Press: Greenwich, CT, **1998**; Vol. *2*.
12. Brinker, U. H., Ed. *Advances in Carbene Chemistry*; Elsevier Science: Amsterdam, **2001**; Vol. *3*.
13. Berson, J. A. "Non-Kekule Molecules as Reactive Intermediates," in *Reactive Intermediate Chemistry*; Moss, R. A., Platz, M. S., Jones, M. Jr., Eds.; Wiley-Interscience: Hoboken, NJ, **2004**, p 165–203.
14. Jones, M., Jr.; Moss, R. A. "Singlet Carbenes," in *Reactive Intermediate Chemistry*; Moss, R. A., Platz, W. S., Jones, M. Jr., Eds.; Wiley-Interscience: Hoboken, NJ, **2004**, p 273–328.
15. Tomioka, H. "Triplet Carbenes," in *Reactive Intermediate Chemistry*; Moss, R. A., Platz, W. S., Jones, M., Jr., Eds.; Wiley-Interscience: Hoboken, NJ, **2004**, p 375–461.
16. Lineberger, W. C.; Borden, W. T. "The synergy between qualitative theory, quantitative calculations, and direct experiments in understanding, calculating, and measuring the energy differences between the lowest singlet and triplet states of organic diradicals," *Phys. Chem. Chem. Phys.* **2011**, *13*, 11792–11813.
17. Goddard, W. A., III "Theoretical chemistry comes alive: full partner with experiment," *Science* **1985**, *227*, 917–923.
18. Schaefer, H. F., III "Methylene: a paradigm for computational quantum chemistry," *Science* **1986**, *231*, 1100–1107.
19. Herzberg, G.; Shoosmith, J. "Spectrum and structure of the free methylene radical," *Nature* **1959**, *183*, 1801–1802.
20. Herzberg, G. "The spectra and structure of free methyl and free methylene," *Proc. R. Soc. A* **1961**, *262*, 291–317.
21. Herzberg, G.; Johns, J. W. C. "The spectrum and structure of singlet CH_2," *Proc. R. Soc. A* **1966**, *295*, 107–128.
22. Foster, J. M.; Boys, S. F. "Quantum variational calculations for a range of CH_2 configurations," *Rev. Mod. Phys.* **1960**, *32*, 305–307.
23. Jordan, P. C. H.; Longuet-Higgins, H. C. "The lower electronic levels of the radicals CH, CH_2, CH_3, NH, NH_2, BH, BH_2, and BH_3," *Mol. Phys.* **1962**, *5*, 121–138.
24. Bender, C. F.; Schaefer, H. F., III "New theoretical evidence for the nonlinearlity of the triplet ground state of methylene," *J. Am. Chem. Soc.* **1970**, *92*, 4984–4985.
25. Bernheim, R. A.; Bernard, H. W.; Wang, P. S.; Wood, L. S.; Skell, P. S. "Electron paramagnetic resonance of triplet CH_2," *J. Chem. Phys.* **1970**, *53*, 1280–1281.
26. Wasserman, E.; Yager, W. A.; Kuck, V. J. "EPR of CH_2: a substantially bent and partially rotating ground state triplet," *Chem. Phys. Lett.* **1970**, *7*, 409–413.
27. Herzberg, G.; Johns, J. W. C. "On the structure of CH_2 in its triplet ground state," *J. Chem. Phys.* **1971**, *54*, 2276–2278.
28. Jensen, P.; Bunker, P. R. "The potential surface and stretching frequencies of X^3B_1 methylene (CH_2) determined from experiment using the morse oscillator-rigid bender internal dynamics hamioltonian," *J. Chem. Phys.* **1988**, *89*, 1327–1332.
29. Yamaguchi, Y.; Sherrill, C. D.; Schaefer, H. F., III "The X^3B_1, a^1A_1, b^1B_1, and c^1A_1 electronic states of CH_2," *J. Phys. Chem.* **1996**, *100*, 7911–7918.
30. Apeloig, Y.; Pauncz, R.; Karni, M.; West, R.; Steiner, W.; Chapman, D. "Why is methylene a ground state triplet while silylene is a ground state singlet?," *Organometallics* **2003**, *22*, 3250–3256.

31. Woon, D. E.; Dunning, T. H., Jr. "Gaussian basis sets for Use in correlated molecular calculations. V. Core-valence basis sets for boron through neon," *J. Chem. Phys.* **1995**, *103*, 4572–4585.

32. Kalemos, A.; Dunning, T. H., Jr.; Mavridis, A.; Harrison, J. F. "CH_2 revisited," *Can. J. Chem.* **2004**, *82*, 684–693.

33. Worthington, S. E.; Cramer, C. J. "Density functional calculations of the influence of substitution on singlet-triplet gaps in carbenes and vinylidenes," *J. Phys. Org. Chem.* **1997**, *10*, 755–767.

34. Das, D.; Whittenburg, S. L. "Performance of the hybrid density functionals in the determination of the geometric structure, vibrational frequency and singlet-triplet energy separation of CH_2, CHF, CF_2, CCl_2 and CBr_2," *J. Mol. Struct. (THEOCHEM)* **1999**, *492*, 175–186.

35. Shavitt, I. "Geometry and singlet-triplet energy gap in methylene. A critical review of experimental and theoretical determinations," *Tetrahedron* **1985**, *41*, 1531–1542.

36. Hay, P. J.; Hunt, W. J.; Goddard III, W. A. "Generalized valence bond wavefunctions for the low lying states of methylene," *Chem. Phys. Lett.* **1972**, *13*, 30–35.

37. Bender, C. F.; Schaefer, H. F., III; Franceschetti, D. R.; Allen, L. C. "Singlet-triplet energy separation, Walsh-Mulliken diagrams, and singlet d-polarization effects in methylene," *J. Am. Chem. Soc.* **1972**, *94*, 6888–6893.

38. Zittel, P. F.; Ellison, G. B.; O'Neil, S. V.; Herbst, E.; Lineberger, W. C.; Reinhardt, W. P. "Laser photoelectron spectrometry of CH_2^-. singlet-triplet splitting and electron affinity of CH_2," *J. Am. Chem. Soc.* **1976**, *98*, 3731–3732.

39. Harding, L. B.; Goddard, W. A., III "Ab initio studies on the singlet-triplet splitting of methylene (CH_2)," *J. Chem. Phys.* **1977**, *67*, 1777–1779.

40. Lucchese, R. R.; Schaefer, H. F., III "Extensive configuration interaction studies of the methylene singlet-triplet separation," *J. Am. Chem. Soc.* **1977**, *99*, 6765–6766.

41. Roos, B. O.; Siegbahn, P. M. "Methylene singlet-triplet separation. An ab initio configuration interaction study," *J. Am. Chem. Soc.* **1977**, *99*, 7716–7718.

42. Harding, L. B.; Goddard, W. A., III "Methylene: ab initio vibronic analysis and reinterpretation of the spectroscopic and negative Ion photoelectron experiments," *Chem. Phys. Lett.* **1978**, *55*, 217–220.

43. McKellar, R. W.; Bunker, P. R.; Sears, T. J.; Evenson, K. M.; Saykally, R. J.; Langhoff, S. R. "Far infrared laser magnetic resonance of singlet methylene: singlet-triplet perturbations, singlet-triplet transitions, and the singlet-triplet splitting," *J. Chem. Phys.* **1983**, *79*, 5251–5264.

44. Leopold, D. G.; Murray, K. K.; Lineberger, W. C. "Laser photoelectron sprectoscopy of vibrationally relaxed CH_2^-: a reinvestigation of the singlet-triplet splitting in merthylene," *J. Chem. Phys.* **1984**, *81*, 1048–1050.

45. Schreiner, P. R.; Karney, W. L.; von Rague Schleyer, P.; Borden, W. T.; Hamilton, T. P.; Schaefer, H. F., III "Carbene rearrangements unsurpassed: details of the C_7H_6 potential energy surface revealed," *J. Org. Chem.* **1996**, *61*, 7030–7039.

46. Demel, O.; Pittner, J.; Carsky, P.; Hubac, I. "Multireference Brillouin-Wigner coupled cluster singles and doubles study of the singlet-triplet separation in alkylcarbenes," *J. Phys. Chem. A* **2004**, *108*, 3125–3128.

47. Demel, O.; Pittner, J. "Multireference Brillouin–Wigner coupled cluster method with singles, doubles, and triples: efficient implementation and comparison with approximate approaches," *J. Chem. Phys.* **2008**, *128*, 104108–104111.

48. Oyedepo, G. A.; Wilson, A. K. "Multireference correlation consistent composite approach [MR-ccCA]: toward accurate prediction of the energetics of excited and transition state chemistry," *J. Phys. Chem. A* **2010**, *114*, 8806–8816.

49. Gronert, S.; Keeffe, J. R.; More O'Ferrall, R. A. "Stabilities of carbenes: independent measures for singlets and triplets," *J. Am. Chem. Soc.* **2011**, *133*, 3381–3389.

50. Matus, M. H.; Nguyen, M. T.; Dixon, D. A. "Heats of formation and singlet–triplet separations of hydroxymethylene and 1-hydroxyethylidene," *J. Phys. Chem. A* **2006**, *110*, 8864–8871.

51. Schwartz, R. L.; Davico, G. E.; Ramond, T. M.; Lineberger, W. C. "Singlet-triplet splittings in CX_2 (X = F, Cl, Br, I) dihalocarbenes via negative Ion phoetoelectron spectroscopy," *J. Phys. Chem. A* **1999**, *103*, 8213–8221.

52. Gutsev, G. L.; Ziegler, T. "Theoretical study on neutral and anionic halocarbynes and halocarbenes," *J. Phys. Chem.* **1991**, *95*, 7220–7228.

53. Russo, N.; Sicilia, E.; Toscano, M. "Geometries, singlet-triplet separations, dipole moments, ionization potentials, and vibrational frequencies in methylene (CH_2) and halocarbenes (CHF, CF_2, CCl_2, CBr_2, and CCl_2)," *J. Chem. Phys.* **1992**, *97*, 5031–5036.

54. Gobbi, A.; Frenking, G. "The singlet-triplet gap of the Halonitrenium ions NHX^+, NX_2^+ and the halocarbenes CHX, CX_2 (X=F, Cl, Br, I)," *J. Chem. Soc., Chem. Commun.* **1993**, 1162–1164.

55. Garcia, V. M.; Castell, O.; Reguero, M.; Caballol, R. "Singlet-triplet energy gap in halogen-substituted carbenes and silylenes: a difference-dedicated configuration interaction calculation," *Mol. Phys.* **1996**, *87*, 1395–1404.

56. Barden, C. J.; Schaefer, H. F., III "The singlet-triplet separation in dichlorocarbene: a surprising difference between theory and experiment," *J. Chem. Phys.* **2000**, *112*, 6515–6516.

57. McKee, M. L.; Michl, J. "A possible reinterpretation of the photoelectron spectra of $[CCl_2]^{-\cdot}$, $[CBr_2]^{-\cdot}$ and $[CI_2]^{-\cdot}$: a role for quartet isodihalocarbene radical anions?," *J. Phys. Chem. A* **2002**, *106*, 8495–8497.

58. Wren, S. W.; Vogelhuber, K. M.; Ervin, K. M.; Lineberger, W. C. "The photoelectron spectrum of CCl_2^-: the convergence of theory and experiment after a decade of debate," *Phys. Chem. Chem. Phys.* **2009**, *11*, 4745–4753.

59. Platz, M. S. In *Reactive Intermediate Chemistry*; Moss, R. A., Platz, W. S., Jones, M., Jr., Eds.; Wiley-Interscience: Hoboken, NJ, **2004**, p 501–560.

60. Huisgen, R.; Vossius, D.; Appl, M. "Thermolysis of phenyl azide in primary amines; the constitution of dibenzamil," *Chem. Ber.* **1958**, *91*, 1–12.

61. Huisgen, R.; Appl, M. "The mechanism of the ring enlargement in the decomposition of phenyl azide in aniline," *Chem. Ber.* **1958**, *91*, 12–21.

62. Doering, W. v. E.; Odum, R. A. "Ring enlargement in the photolysis of phenyl azide," *Tetrahedron* **1966**, *22*, 81–93.

63. Carroll, S. E.; Nay, B.; Scriven, E. F. V.; Suschitzky, H.; Thomas, D. R. "Decomposition of aromatic azides in ethanethiol," *Tetrahedron Lett.* **1977**, *18*, 3175–3178.

64. Abramovitch, R. A.; Challand, S. R.; Scriven, E. F. V. "Mechanism of intermolecular aromatic substitution by arylnitrenes," *J. Am. Chem. Soc.* **1972**, *94*, 1374–1376.

65. Banks, R. E.; Sparkes, G. R. "Azide chemistry. V. Synthesis of 4-azido-2,3,5,6-tetrafluoro-, 4-azido-3-chloro-2,5,6-trifluoro-, and 4-azido-3,5-dichloro-2,6-difluoropyridine,

and thermal reactions of the tetrafluoro compound," *J. Chem. Soc., Perkin Trans. 1* **1972**, 2964–2970.

66. Cai, S. X.; Glenn, D. J.; Kanskar, M.; Wybourne, M. N.; Keana, J. F. W. "Development of highly efficient deep-UV and electron beam mediated cross-linkers: synthesis and photolysis of bis(perfluorophenyl) azides," *Chem. Mater.* **1994**, *6*, 1822–1829.

67. Meijer, E. W.; Nijhuis, S.; van Vroonhoven, V. C. B. M. "Poly-1,2-azepines by the photopolymerization of phenyl azides. Precursors for conducting polymer films," *J. Am. Chem. Soc.* **1988**, *110*, 7209–7210.

68. Keana, J. F. W.; Cai, S. X. "New reagents for photoaffinity labeling: synthesis and photolysis of functionalized perfluorophenyl azides," *J. Org. Chem.* **1990**, *55*, 3640–3647.

69. Schnapp, K. A.; Poe, R.; Leyva, E.; Soundararajan, N.; Platz, M. S. "Exploratory photochemistry of fluorinated aryl azides. Implications for the design of photoaffinity labeling reagents," *Bioconjugate Chem.* **1993**, *4*, 172–177.

70. Kym, P. R.; Anstead, G. M.; Pinney, K. G.; Wilson, S. R.; Katzenellenbogen, J. A. "Molecular structures, conformational analysis, and preferential modes of binding of 3-aroyl-2-arylbenzo[b]thiophene estrogen receptor ligands: LY117018 and aryl azide photoaffinity labeling analogs," *J. Med. Chem.* **1993**, *36*, 3910–3922.

71. Kym, P. R.; Carlson, K. E.; Katzenellenbogen, J. A. "Evaluation of a highly efficient aryl azide photoaffinity labeling reagent for the progesterone receptor," *Bioconjugate Chem.* **1995**, *6*, 115–122.

72. Kerrigan, S.; Brooks, D. E. "Optimization and immunological characterization of a photochemically coupled lysergic acid diethylamide (LSD) immunogen," *Bioconjugate Chem.* **1998**, *9*, 596–603.

73. Alley, S. C.; Ishmael, F. T.; Jones, A. D.; Benkovic, S. J. "Mapping protein-protein interactions in the bacteriophage T4 DNA polymerase holoenzyme using a novel trifunctional photo-cross-linking and affinity reagent," *J. Am. Chem. Soc.* **2000**, *122*, 6126–6127.

74. Matzinger, S.; Bally, T.; Patterson, E. V.; McMahon, R. J. "The C_7H_6 potential energy surface revisited: relative energies and IR assignment," *J. Am. Chem. Soc.* **1996**, *118*, 1535–1542.

75. Wong, M. W.; Wentrup, C. "Interconversions of phenylcarbene, cycloheptatetraene, fulvenallene, benzocyclopropene. A theoretical study of the C_7H_6 surface," *J. Org. Chem.* **1996**, *61*, 7022–7029.

76. Polino, D.; Famulari, A.; Cavallotti, C. "Analysis of the reactivity on the C_7H_6 potential energy surface," *J. Phys. Chem. A* **2011**, *115*, 7928–7936.

77. Cramer, C. J.; Dulles, F. J.; Falvey, D. E. "Ab initio characterization of phenylnitrenium and phenylcarbene: remarkably different properties for isoelectronic species," *J. Am. Chem. Soc.* **1994**, *116*, 9787–9788.

78. Kim, S.-J.; Hamilton, T. P.; Schaefer, H. F., III "Phenylnitrene: energetics, vibrational frequencies, and molecular structure," *J. Am. Chem. Soc.* **1992**, *114*, 5349–5355.

79. Hrovat, D. A.; Waali, E. E.; Borden, W. T. "Ab initio calculations of the singlet-triplet energy difference in phenylnitrene," *J. Am. Chem. Soc.* **1992**, *114*, 8698–8699.

80. Karney, W. L.; Borden, W. T. "Ab initio study of the ring expansion of phenylnitrene and comparison with the ring expansion of phenylcarbene," *J. Am. Chem. Soc.* **1997**, *119*, 1378–1387.

81. Winkler, M. "Singlet–triplet energy splitting and excited states of phenylnitrene," *J. Phys. Chem. A* **2008**, *112*, 8649–8653.

82. Szalay, P. G. "Multireference averaged quadratic coupled-cluster (MR-AQCC) method based on the functional of the total energy," *Chem. Phys.* **2008**, *349*, 121–125.
83. Travers, M. J.; Cowles, D. C.; Clifford, E. P.; Ellison, G. B. "Photoelectron spectroscopy of the phenylnitrene anion," *J. Am. Chem. Soc.* **1992**, *114*, 8699–8701.
84. Wijeratne, N. R.; Fonte, M. D.; Ronemus, A.; Wyss, P. J.; Tahmassebi, D.; Wenthold, P. G. "Photoelectron spectroscopy of chloro-substituted phenylnitrene anions," *J. Phys. Chem. A* **2009**, *113*, 9467–9473.
85. Karney, W. L.; Borden, W. T. "Differences between Phenylcarbene and Phenylnitrene and the Ring Expansion Reactions They Undergo," in *Advances in Carbene Chemistry*; Brinker, U. H., Ed.; *Elsevier Science: Amsterdam*, **2001**; Vol. 3, p 205-251.
86. Johnson, W. T. G.; Sullivan, M. B.; Cramer, C. J. "*Meta* and *para* substitution effects on the electronic state energies and ring-expansion reactivities of phenylnitrenes," *Int. J. Quantum Chem.* **2001**, *85*, 492–508.
87. Wenthold, P. G. "Spin-state dependent radical stabilization in nitrenes: the unusually small singlet–triplet splitting in 2-furanylnitrene," *J. Org. Chem.* **2011**, *77*, 208–214.
88. Borden, W. T.; Gritsan, N. P.; Hadad, C. M.; Karney, W. L.; Kemnitz, C. R.; Platz, M. S. "The interplay of theory and experiment in the study of phenylnitrene," *Acc. Chem. Res.* **2000**, *33*, 765–771.
89. Engelking, P. C.; Lineberger, W. C. "Laser photoelectron spectrometry of NH−: electron affinity and intercombination energy difference in NH," *J. Chem. Phys.* **1976**, *65*, 4323–4324.
90. Gritsan, N. P.; Zhu, Z.; Hadad, C. M.; Platz, M. S. "Laser flash photolysis and computational study of singlet phenylnitrene," *J. Am. Chem. Soc.* **1999**, *121*, 1202–1207.
91. Wentrup, C. "Rearrangements and interconversions of carbenes and nitrenes," *Top. Curr. Chem.* **1976**, *62*, 173–251.
92. Platz, M. S. "Comparison of phenylcarbene and phenylnitrene," *Acc. Chem. Res.* **1995**, *28*, 487–492.
93. Kemnitz, C. R.; Karney, W. L.; Borden, W. T. "Why are nitrenes more stable than carbenes? An ab initio study," *J. Am. Chem. Soc.* **1998**, *120*, 3499–3503.
94. Hanway, P. J.; Winter, A. H. "Phenyloxenium ions: more like phenylnitrenium ions than isoelectronic phenylnitrenes?," *J. Am. Chem. Soc.* **2011**, *133*, 5086–5093.
95. Dewar, M. J. S.; David, D. E. "Ultraviolet photoelectron spectrum of the phenoxy radical," *J. Am. Chem. Soc.* **1980**, *102*, 7387–7389.
96. Joines, R. C.; Turner, A. B.; Jones, W. M. "The rearrangement of phenylcarbenes to cycloheptatrienylidenes," *J. Am. Chem. Soc.* **1969**, *91*, 7754–7755.
97. West, P. R.; Chapman, O. L.; LeRoux, J. P. "1,2,4,6-cycloheptatetraene," *J. Am. Chem. Soc.* **1982**, *104*, 1779–1782.
98. McMahon, R. J.; Abelt, C. J.; Chapman, O. L.; Johnson, J. W.; Kreil, C. L.; LeRoux, J. P.; Mooring, A. M.; West, R. P. "1,2,4,6-Cycloheptatetraene: the key intermediate in arylcarbene interconversions and related C_7H_6 rearrangements," *J. Am. Chem. Soc.* **1987**, *109*, 2456–2469.
99. Warmuth, R. "Inner-phase stabilization of reactive intermediates," *Eur. J. Org. Chem.* **2001**, 423–437.
100. Vander Stouw, G.; Kraska, A. R.; Shechter, H. "Rearrangement and insertion reactions of 2-methylbenzylidenes," *J. Am. Chem. Soc.* **1972**, *94*, 1655–1661.

101. Baron, W. J.; Jones Jr., M.; Gaspar, P. P. "Interconversion of o-, m- and p-tolylcarbene," *J. Am. Chem. Soc.* **1970**, *92*, 4739–4740.
102. Albrecht, S. W.; McMahon, R. J. "Photoequilibration of 2-naphthylcarbene and 2,3-benzobicyclo[4.1.0]hepta-2,4,6-triene," *J. Am. Chem. Soc.* **1993**, *115*, 855–859.
103. Bonvallet, P. A.; McMahon, R. J. "Photoequilibration of 1-naphthylcarbene and 4,5-benzobicyclo[4.1.0]hepta-2,4,6-triene," *J. Am. Chem. Soc.* **1999**, *121*, 10496–10503.
104. Gritsan, N. P.; Gudmundsdottir, A. D.; Tigelaar, D.; Zhu, Z.; Karney, W. L.; Hadad, C. M.; Platz, M. S. "A laser flash photolysis and quantum chemical study of the fluorinated derivatives of singlet phenylnitrene," *J. Am. Chem. Soc.* **2001**, *123*, 1951–1962.
105. Tsao, M.-L.; Platz, M. S. "Photochemistry of ortho, ortho' dialkyl phenyl azides," *J. Am. Chem. Soc.* **2003**, *125*, 12014–12025.
106. Andersson, K.; Roos, B. O. "Multiconfigurational second-order perturbation theory: a test of geometries and binding energies," *Int. J. Quantum Chem.* **1993**, *45*, 591–607.
107. Andersson, K. "Different forms of the zeroth-order Hamiltonian in second-order perturbation theory with a complete active space self-consistent field reference function," *Theor. Chim. Acta* **1995**, *91*, 31–46.
108. Warmuth, R.; Makowiec, S. "The phenylnitrene rearrangement in the inner phase of a hemicarcerand," *J. Am. Chem. Soc.* **2005**, *127*, 1084–1085.
109. Schrock, A. K.; Schuster, G. B. "Photochemistry of phenyl azide: chemical properties of the transient intermediates," *J. Am. Chem. Soc.* **1984**, *106*, 5228–5234.
110. Li, Y. Z.; Kirby, J. P.; George, M. W.; Poliakoff, M.; Schuster, G. B. "1,2-Didehydroazepines from the photolysis of substituted aryl azides: analysis of their chemical and physical properties by time-resolved spectroscopic methods," *J. Am. Chem. Soc.* **1988**, *110*, 8092–8098.
111. Sitzmann, E. V.; Langan, J.; Eisenthal, K. B. "Intermolecular effects on intersystem crossing studied on the picosecond timescale: the solvent polarity effect on the rate of singlet-to-triplet intersystem crossing of diphenylcarbene," *J. Am. Chem. Soc.* **1984**, *106*, 1868–1869.
112. Grasse, P. B.; Brauer, B. E.; Zupancic, J. J.; Kaufmann, K. J.; Schuster, G. B. "Chemical and physical properties of fluorenylidene: equilibration of the singlet and triplet carbenes," *J. Am. Chem. Soc.* **1983**, *105*, 6833–6845.
113. Marcinek, A.; Platz, M. S. "Deduction of the activation parameters for ring expansion and intersystem crossing in fluorinated singlet phenylnitrenes," *J. Phys. Chem.* **1993**, *97*, 12674–12677.
114. Smith, B. A.; Cramer, C. J. "How do different fluorine substitution patterns affect the electronic state energies of phenylnitrene?," *J. Am. Chem. Soc.* **1996**, *118*, 5490–5491.
115. Morawietz, J.; Sander, W. "Photochemistry of fluorinated phenyl nitrenes: matrix isolation of fluorinated azirines," *J. Org. Chem.* **1996**, *61*, 4351–4354.
116. Sundberg, R. J.; Suter, S. R.; Brenner, M. "Photolysis of *o*-substituted aryl azides in diethylamine. Formation and autoxidation of 2-diethylamino-1H-azepine intermediates," *J. Am. Chem. Soc.* **1972**, *94*, 513–520.
117. Dunkin, I. R.; Donnelly, T.; Lockhart, T. S. "2,6-Dimethylphenylnitrene in low-temperature matrices," *Tetrahedron Lett.* **1985**, *26*, 59–362.
118. Karney, W. L.; Borden, W. T. "Why does *o*-fluorine substitution raise the barrier to ring expansion of phenylnitrene?," *J. Am. Chem. Soc.* **1997**, *119*, 3347–3350.

119. Levya, E.; Sagredo, R. "Photochemistry of fluorophenyl azides in diethylamine. Nitrene reaction versus ring expansion," *Tetrahedron* **1998**, *54*, 7367–7374.

120. Gritsan, N. P.; Gudmundsdottir, A. D.; Tigelaar, D.; Platz, M. S. "Laser flash photolysis study of methyl derivatives of phenyl azide," *J. Phys. Chem. A* **1999**, *103*, 3458–3461.

121. Gritsan, N. P.; Likhotvorik, I.; Tsao, M.-L.; Celebi, N.; Platz, M. S.; Karney, W. L.; Kemnitz, C. R.; Borden, W. T. "Ring-expansion reaction of cyano-substituted singlet phenyl nitrenes: theoretical predictions and kinetic results from laser flash photolysis and chemical trapping experiments," *J. Am. Chem. Soc.* **2001**, *123*, 1425–1433.

122. Hund, F. "The interpretation of complicated spectra," *Z. Physik* **1925**, *22*, 345–371.

123. Borden, W. T.; Davidson, E. R. "Effects of electron repulsion in conjugated hydrocarbon diradicals," *J. Am. Chem. Soc.* **1977**, *99*, 4587–4594.

124. Borden, W. T. In *Diradicals*; Borden, W. T., Ed.; John Wiley & Sons: New York, **1982**, p 1–72.

125. Ovchinnikov, A. A. "Multiplicity of the ground state of large alternant organic molecules with conjugated bonds (do organic ferromagnets exist?)," *Theor. Chim. Acta* **1978**, *47*, 297–304.

126. Miller, J. S.; Epstein, A. J. "Organic and organometallic molecular magnetic materials—designer magnets," *Angew. Chem. Int. Ed. Engl.* **1994**, *33*, 385–415.

127. Rajca, A. "From high-spin organic molecules to organic polymers with magnetic ordering," *Chem. Eur. J.* **2002**, *8*, 4834–4841.

128. Dowd, P.; Chang, W.; Paik, Y. H. "Tetramethyleneethane, a ground-state triplet," *J. Am. Chem. Soc.* **1986**, *108*, 7416–7417.

129. Du, P.; Borden, W. T. "Ab initio calculations predict a singlet ground state for tetramethyleneethane," *J. Am. Chem. Soc.* **1987**, *109*, 930–931 (errata *J. Am. Chem. Soc.*, **1992**, *114*, 4949).

130. Dowd, P.; Chang, W.; Paik, Y. H. "2,3-Dimethylenecyclohexa-1,3-diene diradical is a ground-state triplet," *J. Am. Chem. Soc.* **1987**, *109*, 5284–5285.

131. Roth, W. R.; Kowalczik, U.; Maier, G.; Reisenauer, H. P.; Sustmann, R.; Müller, W. "2,2-Dimethyl-4,5-dimethylene-l,3-cyclopentanediyl," *Angew. Chem. Int. Ed. Engl.* **1987**, *26*, 1285–1287.

132. Choi, Y.; Jordan, K. D.; Paik, Y. H.; Chang, W.; Dowd, P. "Ab initio calculations of the geometries and IR spectra of two derivatives of tetramethyleneethane," *J. Am. Chem. Soc.* **1988**, *110*, 7575–7576.

133. Du, P.; Hrovat, D. A.; Borden, W. T. "Ab initio calculations of the singlet-triplet energy separation in 3,4-dimethylenefuran and related diradicals," *J. Am. Chem. Soc.* **1986**, *108*, 8086–8087.

134. Nash, J. J.; Dowd, P.; Jordan, K. D. "Theoretical study of the low-lying triplet and singlet states of diradicals: prediction of ground-state multiplicities in cyclic analogs of tetramethyleneethane," *J. Am. Chem. Soc.* **1992**, *114*, 10071–10072.

135. Nachtigall, P.; Jordan, K. D. "Theoretical study of the low-lying triplet and singlet states of diradicals. 1. Tetramethyleneethane," *J. Am. Chem. Soc.* **1992**, *114*, 4743–4747.

136. Nachtigall, P.; Jordan, K. D. "Theoretical study of the low-lying triplet and singlet states of tetramethyleneethane: prediction of a triplet below singlet state at the triplet equilibrium geometry," *J. Am. Chem. Soc.* **1993**, *115*, 270–271.

137. Clifford, E. P.; Wenthold, P. G.; Lineberger, W. C.; Ellison, G. B.; Wang, C. X.; Grabowski, J. J.; Vila, F.; Jordan, K. D. "Properties of tetramethyleneethane (TME) as

revealed by ion chemistry and ion photoelectron spectroscopy," *J. Chem. Soc., Perkin Trans 2* **1998**, 1015–1022.

138. Filatov, M.; Shaik, S. "Tetramethyleneethane (TME) diradical: experiment and density functional theory reach an agreement," *J. Phys. Chem. A* **1999**, *103*, 8885–8889.

139. Rodriguez, E.; Reguero, M.; Caballol, R. "The controversial ground state of tetramethyleneethane. An ab initio CI study," *J. Phys. Chem. A* **2000**, *104*, 6253–6258.

140. Pittner, J.; Nachtigall, P.; Carsky, P.; Hubac, I. "State-specific Brillouin-Wigner multireference coupled cluster study of the singlet-triplet separation in the tetramethyleneethane diradical," *J. Phys. Chem. A* **2001**, *105*, 1354–1356.

141. Cordes, M. H. J.; Berson, J. A. "Thermal interconversion of a pair of diastereomeric cyclopropanones. An upper limit for a cyclopropanone-oxyallyl energy separation," *J. Am. Chem. Soc.* **1992**, *114*, 11010–11011.

142. Chan, T. H.; Ong, B. S. "Chemistry of allene oxides," *J. Org. Chem.* **1978**, *43*, 2994–3001.

143. Schuster, D. I. "Mechanisms of photochemical transformations of cross-conjugated cyclohexadienones," *Acc. Chem. Res.* **1978**, *11*, 65–73.

144. Osamura, Y.; Borden, W. T.; Morokuma, K. "Structure and stability of oxyallyl. An MCSCF study," *J. Am. Chem. Soc.* **1984**, *106*, 5112–5115.

145. Coolidge, M. B.; Yamashita, K.; Morokuma, K.; Borden, W. T. "Ab initio MCSCF and CI calculations of the singlet-triplet energy differences in oxyallyl and in dimethyloxyallyl," *J. Am. Chem. Soc.* **1990**, *112*, 1751–1754.

146. Mozhayskiy, V.; Goebbert, D. J.; Velarde, L.; Sanov, A.; Krylov, A. I. "Electronic structure and spectroscopy of oxyallyl: a theoretical study," *J. Phys. Chem. A* **2010**, *114*, 6935–6943.

147. Ichino, T.; Villano, S. M.; Gianola, A. J.; Goebbert, D. J.; Velarde, L.; Sanov, A.; Blanksby, S. J.; Zhou, X.; Hrovat, D. A.; Borden, W. T.; Lineberger, W. C. "Photoelectron spectroscopic study of the oxyallyl diradical," *J. Phys. Chem. A* **2011**, *115*, 1634–1649.

148. Šimsa, D.; Demel, O.; Bhaskaran-Nair, K.; Hubač, I.; Mach, P.; Pittner, J. "Multireference coupled cluster study of the oxyallyl diradical," *Chem. Phys.* **2012**, *401*, 203–207.

149. Ichino, T.; Villano, S. M.; Gianola, A. J.; Goebbert, D. J.; Velarde, L.; Sanov, A.; Blanksby, S. J.; Zhou, X.; Hrovat, D. A.; Borden, W. T.; Lineberger, W. C. "The lowest singlet and triplet states of the oxyallyl diradical," *Angew. Chem. Int. Ed.* **2009**, *48*, 8509–8511.

150. Wenk, H. H.; Winkler, M.; Sander, W. "One century of aryne chemistry," *Angew. Chem. Int. Ed.* **2003**, *42*, 502–528.

151. Roberts, J. D.; Simmons Jr., H. E.; Carlsmith, L. A.; Vaughan, C. W. "Rearrangement in the reaction of chlorobenzene-1-C^{14} with potassium amide," *J. Am. Chem. Soc.* **1953**, *75*, 3290–3291.

152. Berry, R. S.; Spokes, G. N.; Stiles, M. "The absorption spectrum of gaseous benzyne," *J. Am. Chem. Soc.* **1962**, *84*, 3570–3577.

153. Berry, R. S.; Clardy, J.; Schafer, M. E. "Benzyne," *J. Am. Chem. Soc.* **1964**, *86*, 2738–2739.

154. Chapman, O. L.; Mattes, K.; McIntosh, C. L.; Pacansky, J.; Calder, G. V.; Orr, G. "Photochemical transformations. LII. Benzyne," *J. Am. Chem. Soc.* **1973**, *95*, 6134–6135.

155. Kukolich, S. G.; Tanjaroon, C.; McCarthy, M. C.; Thaddeus, P. "Microwave spectrum of o-benzyne produced in a discharge nozzle," *J. Chem. Phys.* **2003**, *119*, 4353–4359.
156. Scheiner, A. C.; Schaefer III, H. F.; Bowen Liu, B. "The X^1A_1 and a^3B_2 states of o-benzyne: a theoretical characterization of equilibrium geometries, harmonic vibrational frequencies, and the singlet-triplet energy gap," *J. Am. Chem. Soc.* **1989**, *111*, 3118–3124.
157. Wentrup, C.; Blanch, R.; Briehl, H.; Gross, G. "Benzyne, cyclohexyne, and 3-azacyclohexyne and the problem of cycloalkyne versus cycloalkylideneketene genesis," *J. Am. Chem. Soc.* **1988**, *110*, 1874–1880.
158. Leopold, D. G.; Miller, A. E. S.; Lineberger, W. C. "Determination of the singlet-triplet splitting and electron affinity of o-benzyne by negative Ion photoelectron spectroscopy," *J. Am. Chem. Soc.* **1986**, *108*, 1379–1384.
159. Kraka, E.; Cremer, D. "*Ortho-*, *meta-*, and *para*-benzyne. A comparative CCSD (T) investigation," *Chem. Phys. Lett.* **1993**, *216*, 333–340.
160. Wierschke, S. G.; Nash, J. J.; Squires, R. R. "A multiconfigurational SCF and correlation-consistent CI study of the structures, stabilities, and singlet-triplet splittings of *o*-, *m*-, and *p*-benzyne," *J. Am. Chem. Soc.* **1993**, *115*, 11958–11967.
161. Wenthold, P. G. "Thermochemical properties of the benzynes," *Aus. J. Chem.* **2010**, *63*, 1091–1098.
162. Wenthold, P. G.; Paulino, J. A.; Squires, R. R. "The absolute heats of formation of *o*-, *m*-, and *p*-benzyne," *J. Am. Chem. Soc.* **1991**, *113*, 7414–7415.
163. Riveros, J. M.; Ingemann, S.; Nibbering, N. M. M. "Formation of gas phase solvated bromine and iodine anions in ion/molecule reactions of halobenzenes. Revised heat of formation of benzyne," *J. Am. Chem. Soc.* **1991**, *113*, 1053–1053.
164. Guo, Y.; Grabowski, J. J. "Reactions of the benzyne radical anion in the gas phase, the acidity of the phenyl radical, and the heat of formation of o-benzyne," *J. Am. Chem. Soc.* **1991**, *113*, 5923–5931.
165. Jones, R. R.; Bergman, R. G. "p-Benzyne. Generation as an intermediate in a thermal isomerization reaction and trapping evidence for the 1,4-benzenediyl structure," *J. Am. Chem. Soc.* **1972**, *94*, 660–661.
166. Nicolaides, A.; Borden, W. T. "CI calculations on didehydrobenzenes predict heats of formation for the *meta* and *para* Isomers that are substantially higher than previous experimental values," *J. Am. Chem. Soc.* **1993**, *115*, 11951–11957.
167. Wenthold, P. G.; Squires, R. R. "Biradical thermochemistry from collision-induced dissociation threshold energy measurements. Absolute heats of formation of *ortho-*, *meta-*, and *para*-benzyne," *J. Am. Chem. Soc.* **1994**, *116*, 6401–6412.
168. Wenthold, P. G.; Squires, R. R.; Lineberger, W. C. "Ultraviolet photoelectron spectroscopy of the o-, m-, and p-benzyne negative ions. Electron affinities and singlet-triplet splittings for *o*-, *m*-, and *p*-benzyne," *J. Am. Chem. Soc.* **1998**, *120*, 5279–5290.
169. Cioslowski, J.; Szarecka, A.; Moncrieff, D. "Energetics, electronic structures, and geometries of didehydroazines," *Mol. Phys.* **2003**, *101*, 839–858.
170. Lindh, R.; Lee, T. J.; Bernhardsson, A.; Persson, B. J.; Karlstroem, G. "Extended ab initio and theoretical thermodynamics studies of the bergman reaction and the energy splitting of the singlet *o*-, *m*-, and *p*-benzynes," *J. Am. Chem. Soc.* **1995**, *117*, 7186–7194.

370 DIRADICALS AND CARBENES

171. Cramer, C. J.; Nash, J. J.; Squires, R. R. "A reinvestigation of singlet benzyne thermochemistry predicted by CASPT2, coupled-cluster and density functional calculations," *Chem. Phys. Lett.* **1997**, *277*, 311–320.

172. Crawford, T. D.; Kraka, E.; Stanton, J. F.; Cremer, D. "Problematic *p*-benzyne: orbital instabilities, biradical character, and broken symmetry," *J. Chem. Phys.* **2001**, *114*, 10638–10650.

173. Gräfenstein, J.; Hjerpe, A. M.; Kraka, E.; Cremer, D. "An accurate description of the Bergman reaction using restricted and unrestricted DFT: stability test, spin density, and on-top pair density," *J. Phys. Chem. A* **2000**, *104*, 1748–1761.

174. Winkler, M.; Sander, W. "The structure of *meta*-benzyne revisited-a close look into σ-bond formation," *J. Phys. Chem. A* **2001**, *105*, 10422–10432.

175. Kraka, E.; Angladab, J.; Hjerpea, A.; Filatova, M.; Cremer, D. "*m*-benzyne and bicyclo[3.1.0]hexatriene—which isomer is more stable?—a quantum chemical investigation," *Chem. Phys. Lett.* **2001**, *348*, 115–125.

176. Kraka, E.; Cremer, D.; Bucher, G.; Wandel, H.; Sander, W. "A CCSD(T) and DFT investigation of *m*-benzyne and 4-hydroxy-*m*-benzyne," *Chem. Phys. Lett.* **1997**, *268*, 313–320.

177. Sander, W.; Exner, M.; Winkler, M.; Balster, A.; Hjerpe, A.; Kraka, E.; Cremer, D. "Vibrational spectrum of *m*-benzyne: a matrix isolation and computational study," *J. Am. Chem. Soc.* **2002**, *124*, 13072–13079.

178. Marquardt, R.; Sander, W.; Kraka, E. "1,3-Didehydrobenzene (*m*-benzyne)," *Angew. Chem. Int. Ed. Engl.* **1996**, *35*, 746–748.

179. Hess Jr., B. A. "Do bicyclic forms of m- and p-benzyne exist?," *Eur. J. Org. Chem.* **2001**, 2185–2189.

180. Clauberg, H.; Minsek, D. W.; Chen, P. "Mass and photoelectron spectroscopy of C_3H_2. ΔH_f of singlet carbenes deviate from additivity by their singlet-triplet gaps," *J. Am. Chem. Soc.* **1992**, *114*, 99–107.

181. Blush, J. A.; Clauberg, H.; Kohn, D. W.; Minsek, D. W.; Zhang, X.; Chen, P. "Photoionization mass and photoelectron spectroscopy of radicals, carbenes, and biradicals," *Acc. Chem. Res.* **1992**, *25*, 385–392.

182. Chen, P. "Design of diradical-based hydrogen abstraction agents," *Angew. Chem. Int. Ed. Engl.* **1996**, *35*, 1478–1480.

183. Logan, C. F.; Chen, P. "Ab initio calculation of hydrogen abstraction reactions of phenyl radical and *p*-benzyne," *J. Am. Chem. Soc.* **1996**, *118*, 2113–2114.

184. Schottelius, M. J.; Chen, P. "9,10-Dehydroanthracene: *p*-benzyne-type biradicals abstract hydrogen unusually slowly," *J. Am. Chem. Soc.* **1996**, *118*, 4896–4903.

185. Cramer, C. J.; Squires, R. R. "Prediction of singlet-triplet splittings for aryne biradicals from ^{1}H hyperfine interactions in aryl radicals," *J. Phys. Chem. A* **1997**, *101*, 9191–9194.

186. Evangelista, F. A.; Allen, W. D.; Schaefer III, H. F. "Coupling term derivation and general implementation of state-specific multireference coupled cluster theories," *J. Chem. Phys.* **2007**, *127*, 024102–024117.

187. Cramer, C. J. "Bergman, aza-bergman, and protonated aza-Bergman cyclizations and intermediate 2,5-arynes: chemistry and challenges to computation," *J. Am. Chem. Soc.* **1998**, *120*, 6261–6269.

188. Kraka, E.; Cremer, D. "The *para*-didehydropyridine, *para*-didehydropyridinium, and related biradicals—a contribution to the chemistry of enediyne antitumor drugs," *J. Comput. Chem.* **2001**, *22*, 216–229.

189. Hoffner, J.; Schottelius, M. J.; Feichtinger, D.; Chen, P. "Chemistry of the 2,5-didehydropyridine biradical: computational, kinetic, and trapping studies toward drug design," *J. Am. Chem. Soc.* **1998**, *120*, 376–385.

190. Cramer, C. J.; Debbert, S. "Heteroatomic substitution in aromatic small σ biradicals: the six pyridynes," *Chem. Phys. Lett.* **1998**, *287*, 320–326.

191. Winkler, M.; Cakir, B.; Sander, W. "3,5-Pyridyne-a heterocyclic *meta*-benzyne derivative," *J. Am. Chem. Soc.* **2004**, *126*, 6135–6149.

192. Kraka, E.; Tuttle, T.; Cremer, D. "Design of a new warhead for the natural enediyne dynemicin a. An increase of biological activity," *J. Phys. Chem. B* **2008**, *112*, 2661–2670.

193. Johnson, W. T. G.; Cramer, C. J. "Influence of hydroxyl substitution on benzyne properties. Quantum chemical characterization of the didehydrophenols," *J. Am. Chem. Soc.* **2001**, *123*, 923–928.

194. Johnson, W. G.; Cramer, C. J. "Substituent effects on benzyne electronic structures," *J. Phys. Org. Chem.* **2001**, *14*, 597–603.

195. Amegayibor, F. S.; Nash, J. J.; Lee, A. S.; Thoen, J.; Petzold, C. J.; Kenttamaa, H. I. "Chemical properties of a *para*-benzyne," *J. Am. Chem. Soc.* **2002**, *124*, 12066–12067.

196. Clark, A. E.; Davidson, E. R. "*p*-benzyne derivatives that have exceptionally small singlet-triplet gaps and even a triplet ground state," *J. Org. Chem.* **2003**, *68*, 3387–3396.

197. Price, J. M.; Kenttamaa, H. I. "Characterization of two chloro-substituted m-benzyne isomers: effect of substitution on reaction efficiencies and products," *J. Phys. Chem. A* **2003**, *107*, 8985–8995.

198. Fenton, C.; Perry, C. M. "Spotlight on gemtuzumab ozogamicin in acute myeloid leukaemia," *BioDrugs* **2006**, *20*, 137–139.

199. Castaigne, S.; Pautas, C.; Terré, C.; Raffoux, E.; Bordessoule, D.; Bastie, J.-N.; Legrand, O.; Thomas, X.; Turlure, P.; Reman, O.; de Revel, T.; Gastaud, L.; de Gunzburg, N.; Contentin, N.; Henry, E.; Marolleau, J.-P.; Aljijakli, A.; Rousselot, P.; Fenaux, P.; Preudhomme, C.; Chevret, S.; Dombret, H. "Effect of gemtuzumab ozogamicin on survival of adult patients with de-novo acute myeloid leukaemia (ALFA-0701): a randomised, open-label, phase 3 study," *Lancet*, *379*, 1508–1516.

200. Bertrand, G., Ed. *Carbene Chemistry: From Fleeting Intermediate To Powerful Reagents*; Marcel Dekker: New York, **2002**.

201. Ley, D.; Gerbig, D.; Schreiner, P. R. "Tunnelling control of chemical reactions - the organic chemist's perspective," *Org. Biomol. Chem.* **2012**, *10*, 3781–3790.

202. Schreiner, P. R.; Reisenauer, H. P.; Pickard Iv, F. C.; Simmonett, A. C.; Allen, W. D.; Matyus, E.; Csaszar, A. G. "Capture of hydroxymethylene and its fast disappearance through tunnelling," *Nature* **2008**, *453*, 906–909.

203. Miller, W. H.; Handy, N. C.; Adams, J. E. "Reaction path Hamiltonian for polyatomic molecules," *J. Chem. Phys.* **1980**, *72*, 99–112.

204. Razavy, M. *Quantum Theory of Tunneling*; World Scientific: Singapore, **2003**.

205. Kiselev, V. G.; Swinnen, S.; Nguyen, V. S.; Gritsan, N. P.; Nguyen, M. T. "Fast reactions of hydroxycarbenes: tunneling effect versus bimolecular processes," *J. Phys. Chem. A* **2010**, *114*, 5573–5579.
206. Gerbig, D.; Reisenauer, H. P.; Wu, C.-H.; Ley, D.; Allen, W. D.; Schreiner, P. R. "Phenylhydroxycarbene," *J. Am. Chem. Soc.* **2010**, *132*, 7273–7275.
207. Ley, D.; Gerbig, D.; Wagner, J. P.; Reisenauer, H. P.; Schreiner, P. R. "Cyclopropylhydroxycarbene," *J. Am. Chem. Soc.* **2011**, *133*, 13614–13621.
208. Schreiner, P. R.; Reisenauer, H. P. "Spectroscopic identification of dihydroxycarbene," *Angew. Chem. Int. Ed.* **2008**, *47*, 7071–7074.
209. Schreiner, P. R.; Reisenauer, H. P.; Ley, D.; Gerbig, D.; Wu, C.-H.; Allen, W. D. "Methylhydroxycarbene: tunneling control of a chemical reaction," *Science* **2011**, *332*, 1300–1303.
210. Rosenfeld, R. N.; Weiner, B. "Energy disposal in the photofragmentation of pyruvic acid in the gas phase," *J. Am. Chem. Soc.* **1983**, *105*, 3485–3488.
211. Weiner, B. R.; Rosenfeld, R. N. "Pyrolysis of pyruvic acid in the gas phase. A study of the isomerization mechanism of a hydroxycarbene intermediate," *J. Org. Chem.* **1983**, *48*, 5362–5364.
212. Yamamoto, S.; Back, R. A. "The photolysis and thermal decomposition of pyruvic acid in the gas phase," *Can. J. Chem.* **1985**, *63*, 549–554.
213. Smith, B. J.; Nguyen Minh, T.; Bouma, W. J.; Radom, L. "Unimolecular rearrangements connecting hydroxyethylidene (CH_3–C-OH), acetaldehyde (CH_3–CH=O), and vinyl alcohol (CH_2=CH–OH)," *J. Am. Chem. Soc.* **1991**, *113*, 6452–6458.

CHAPTER 6

Organic Reactions of Anions

This chapter presents computational studies of organic reactions that involve anions. These reactions are usually not grouped together in textbooks. However, these reactions are fundamentally variations on a theme. Anions, acting as nucleophiles, can attack sp^3 carbon atoms; we call these as nucleophilic substitution reactions that follow either the S_N1 or S_N2 mechanism. Reactions where the nucleophile attacks sp^2 or sp carbon atoms are addition reactions. The 1,2- and 1,4-addition reactions follow the classic addition mechanism, where the nucleophile adds first followed by the addition of an electrophile. Other nucleophilic reactions at carbonyl compounds, especially carboxylic acid derivatives, follow the addition–elimination pathway.

In a sense, these reaction mechanisms differ simply in the timing of the critical steps: the formation of the new C-nucleophile bond and the cleavage of a carbon-leaving group (C-LG) bond. In the S_N1 mechanism, cleavage occurs prior to bond formation. The opposite order characterizes the addition–elimination mechanism—bond formation precedes bond breaking. Lastly, the two bond changes occur together in one step in the S_N2 mechanism.

In this chapter, we present the contributions of computational chemistry toward understanding the mechanism and chemistry for three reactions involving nucleophilic attack. The S_N2 reaction, with emphasis on the gas versus solution phase, is presented first. Next, we describe the critical contribution that computational chemists made in developing the theory of asymmetric induction at carbonyl and vinyl compounds. The chapter concludes with a discussion on the collaborative efforts of synthetic and computational chemists in developing organic catalysts, especially proline and proline-related molecules, for the aldol, Mannich and Michael reaction, and other related reactions.

6.1 SUBSTITUTION REACTIONS

Nucleophilic substitution reactions are among the first synthetic transformations introduced to beginners learning organic chemistry. The mechanisms for these

---

Computational Organic Chemistry, Second Edition. Steven M. Bachrach.
© 2014 John Wiley & Sons, Inc. Published 2014 by John Wiley & Sons, Inc.

reactions—S_N1 and S_N2—provide the paradigm for mechanistic understanding of organic reactions. The level of detailed understanding of how nucleophilic substitution occurs in solution is unprecedented, providing the ability to predict regioselectivity and stereoselectivity and relative rates and solvent effects with great accuracy.

Most quantum chemical calculations model the gas phase. The S_N1 reaction is essentially unknown in the gas phase. The first step of this mechanism is the heterolytic cleavage of the C–X bond, forming a carbocation and an anionic leaving group. In the gas phase, this separation is strongly opposed by the electrostatic attraction of these two ions, creating a very large barrier. Therefore, this chapter focuses on the S_N2 reaction: the nature of its potential energy surface (PES), the effect of α- and β-branching, and a brief survey of approaches taken toward incorporating solvent into the calculation.

6.1.1 The Gas Phase S_N2 Reaction

The mechanism for the gas phase S_N2 reaction is well understood. The extensive experimental studies by Brauman provide the general PES shown in Figure 6.1.[1,2] The nucleophile and substrate first come together purely by electrostatic attraction to form the entrance ion–dipole complex (**ID1**). Next, the nucleophile approaches the carbon atom from the backside, pushing out the leaving group, through the classic S_N2 transition state (**TS**). As the leaving group exits, the exit ion–dipole complex (**ID2**) is formed prior to dissociating to separated products. This double-well potential is ubiquitous in gas-phase substitution chemistry. The features that characterize the surface are the depths of the wells associated with each ion–dipole complex and the height of its central barrier.

Computational chemistry has played a large role in helping to determine these energies. Early computational studies confirmed the double-well nature of the PES.[3–6] Coming to an agreement on the well depths for the ion–dipole complexes and the activation barrier remained a challenge into this century.

We begin by examining how the computational predictions for the complexation energy and the activation barrier for the simple identity substitutions, Reactions

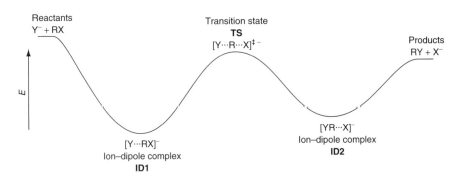

Figure 6.1 Potential energy surface for the gas phase S_N2 reaction.

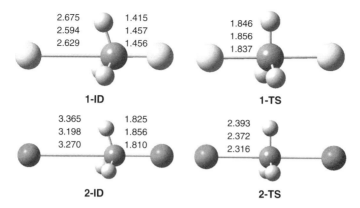

Figure 6.2 Optimized geometries of the ion dipole complexes and transition states for Reactions 6.1 and 6.2. Distances (Å) are for the HF/6-31+G*, B3LYP/6-31+G*, and MP2/6-31+G* geometries reading top to bottom.

6.1 and 6.2, have evolved with increasing computational rigor. The complexation energy is defined as the energy change for the reaction $X^- + RY \rightarrow$ **ID**, and the activation barrier is defined as the energy change for the reaction $X^- + RY \rightarrow$ **TS**.

$$F^- + CH_3F \rightarrow FCH_3 + F^- \qquad \text{(Reaction 6.1)}$$

$$Cl^- + CH_3Cl \rightarrow ClCH_3 + Cl^- \qquad \text{(Reaction 6.2)}$$

The ion–dipole complexes (**1-ID**, **2-ID**) and the TSs (**1-TS**, **2-TS**) for Reactions 6.1 and 6.2 are shown in Figure 6.2. The geometries are only slightly sensitive to computational level. Inclusion of electron correlation induces greater interaction between the nucleophile and the substrate in the ion–dipole complex, namely, shorter nucleophile-carbon distances and longer C-LG distances. For the TSs, the C–X distance is shortest in the HF structures and longest in the MP2 structures. However, the variations in any of the distances are less than 0.05 Å.

The PES for Reaction 6.2 was first mapped out by Jorgensen[6] using the HF/6-31G* method. This PES is recreated in Figure 6.3 with four computational methods: HF/6-31+G*, B3LYP/6-31+G*, MP2/6-31+G*, and ωB97X-D/6-31+G*. The reaction coordinate is defined as the difference between the two C–Cl distances. The gas-phase PES clearly shows the two wells associated with the ion–dipole complexes, separated by a single TS. While the location of these critical points is generally independent of the computational method, the relative energies, especially the barrier height, are strongly dependent on methodology. The HF and MP2 methods predict a barrier that is 6–8 kcal mol^{-1} above the energy of the reactants, while B3LYP predicts a barrier that is *below* the energy of the reactants. ωB97X-D predicts a barrier that is above the reactants (3.4 kcal mol^{-1}) and an ion–dipole that is more strongly bound than the other methods.

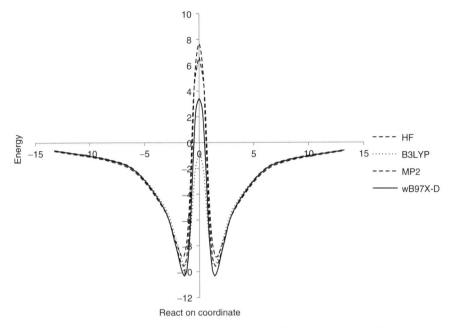

Figure 6.3 Potential energy surface for Reaction 6.2. The energies are in kilocalories per mole and the reaction coordinate (defined as the difference between the two C–Cl distances) is in angstrom.

Reactions 6.1 and 6.2 have been subjects of many computational studies using a variety of methods. The relative energies of the ion–dipole complexes and TSs for Reactions 6.1 and 6.2 are listed in Tables 6.1 and 6.2, respectively. A few systematic trends can be discerned.

The composite methods designed to produce very accurate energies predict very similar energetics for these two reactions. The focal point analysis of Schaefer, probably the most accurate method applied to date, indicates that the complexation energy for the formation of **ID-1** is 13.73 kcal mol^{-1} and the TS lies 0.80 kcal mol^{-1} below separated reactants.[14] All of the energies produced by the other composite methods (the G, W, and CBS methods) are within 1 kcal mol^{-1}. For Reaction 6.2, the experimental values for the relative energies of **ID-2**[18] and **TS-2**[19] are −10.4 and 2.5 kcal mol^{-1}, respectively. The energies predicted by the composite methods differ from experiment by less than 0.7 kcal mol^{-1}.

Of the other more traditional computational methods, only CCSD(T), a very computationally demanding method, is in close agreement with the composite methods. HF overestimates the barrier height and underestimates the complexation energy. MP2 does well with the ion–dipole energetics but grossly overestimates the activation barrier. Perhaps most disheartening is the performance of the DFT methods, which consistently underestimate the activation barrier. The DFT method that provides the most consistent results is the mPW1PW91 functional, but if high accuracy is required, none of the DFT methods deliver.[10,17]

TABLE 6.1 Energies (in kcal mol^{-1}) of the Ion–Dipole Complex and Transition State Relative to Separated Reactants for Reaction 6.1

Method	ID-1	TS-1
HF/6-31++G**[a]	−12.84	5.69
MP2/6-31++G**[a]	−13.92	−0.98
MP2/6-31+G*+ZPE[b]	−13.63	−1.11
MP4/6-31+G*//HF/6-31+G*[b]	−14.19	−4.00
CCSD/QZ(+)(2d,2p)[c]	−13.03	1.95
CCSD(T)/QZ(+)(2d,2p)//CCSD/QZ(+)(2d,2p)[c]	−13.31	−0.48
CCSD(T)/13s8p6d4f/8s6p4d// CCSD/QZ(+)(2d,2p)+ZPE[c]	−13.58	−0.77
CCSD(T)/ TZ2Pf+dif+ZPE(MP2)[d]	−13.28	−0.53
B3LYP/6-31+G*[e]	−13.70	−4.27
B3LYP/cc-pVTZ[f]	−12.72	−2.58
B3LYP/TZ2Pf+dif+ZPE [d]	−12.74	−3.05
BLYP/ TZ2Pf+dif+ZPE[d]	−12.91	−6.97
BP86/ TZ2Pf+dif+ZPE [d]	−12.81	−6.39
mPW1PW91/cc-pVTZ[f]	−12.49	−0.95
ZORA-OLYP/TZ2p[i]	−15.7	−7.6
G1[f]	−13.01	−1.37
G2[f]	−13.34	−1.15
G2(+)[g]	−13.50	−1.91
G3[f]	−14.23	−1.97
W1'[f]	−13.66	−0.37
W2h[f]	−13.72	−0.34
CBS-QB2[f]	−13.46	−0.85
Focal point analysis[h]	−13.726	−0.805

[a]Ref. 7.
[b]Ref. 8.
[c]Ref. 9.
[d]Ref. 10.
[e]Ref. 11.
[f]Ref. 12.
[g]Ref. 13.
[h]Ref. 14.
[i]Ref. 15.

Similar behavior of the computation methods is found for nonidentity S_N2 reactions. The optimized geometries of the entrance and exit ion–dipole complexes (**ID1** and **ID2**) and TSs for Reaction 6.3–6.5 are shown in Figure 6.4. The entrance ion–dipole complex for Reaction 6.4 is unusual among the reactions discussed here. The ion–dipole complexes are characterized by the electrostatic interaction of the nucleophile with the most acidic hydrogen(s). In all of the other examples, the nucleophile associates with the methyl protons, but in **4-ID1**, the most acidic proton is the alcohol hydrogen, so fluoride forms a hydrogen bond to it. For all three reactions, the TS displays the normal S_N2 features, namely, backside attack of the nucleophile accompanied by Walden inversion. They differ in how symmetric the

TABLE 6.2 Energies (in kcal mol^{-1}) of the Ion–Dipole Complex and Transition State Relative to Separated Reactants for Reaction 6.2

Method	ID-2	TS-2
HF/6-31G*[a]	−10.3	3.6
HF/6-31++G**[b]	−8.88	6.59
MP2/6-31G*[c]	−10.96	4.55
MP2/6-31+G(d)+ZPE[d]	−9.20	7.12
MP2/6-31++G**[b]	−9.66	7.68
MP2/6-311+G(3df,2p)//MP2/6-31+G(d)+ZPE[d]	−10.47	4.95
MP4/6-311+G(d,p)//MP2/6-31+G(d)+ZPE[d]	−9.75	6.14
B3LYP/6-31+G*[e]	−9.52	−0.85
B3LYP/cc-pVTZ[f]	−9.50	−0.48
mPW1PW91/cc-pVTZ[f]	−9.59	1.23
ZORA-OLYP/TZ2p[j]	−9.0	−0.2
G1[f]	−10.52	1.79
G2[f]	−10.77	3.06
G2(+)[g]	−10.52	2.75
G3[f]	−11.15	1.79
W1'[f]	−10.54	3.07
W2h[f]	−10.94	2.67
CBS-QB3[f]	−10.65	2.47
Experimental	−10.4[h]	2.5[i]

[a]Ref. 6.
[b]Ref. 7.
[c]Ref. 16.
[d]Ref. 17.
[e]Ref. 11.
[f]Ref. 12.
[g]Ref. 13.
[h]Ref. 18.
[i]Ref. 19.
[j]Ref. 15.

TS is as evidenced by the extent of their inversion: **3-TS** is early, **4-TS** is late, and **5-TS** is nearly symmetric.

$$F^- + CH_3Cl \rightarrow FCH_3 + Cl^- \quad \text{(Reaction 6.3)}$$

$$F^- + CH_3OH \rightarrow FCH_3 + HO^- \quad \text{(Reaction 6.4)}$$

$$Cl^- + CH_3Br \rightarrow ClCH_3 + Br^- \quad \text{(Reaction 6.5)}$$

These nonidentity S_N2 reactions will have asymmetric PESs, reflecting the overall reaction energies. As an example, the PES for Reaction 6.3, computed at B3LYP/6-31+G*, is given in Figure 6.5. The PES is computed in an analogous manner to that displayed in Figure 6.3: the geometry is optimized for a series of points where either the C–Cl or C–F distance is held fixed. The reaction

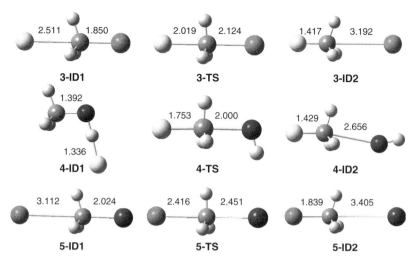

Figure 6.4 Optimized critical points on the reaction surface for Reactions 6.3–6.5. Geometries optimized at CCSD/TZ2pf+dif[10] for Reactions 6.3 and 6.4 and at B3LYP/aug-ccpVTZ[12] for Reaction 6.5.

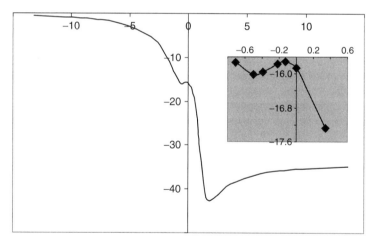

Figure 6.5 Potential energy surface at B3LYP/6-31+G* for Reaction 6.3. The energies are in kilocalories per mole and the reaction coordinate (defined as the difference between the C–F and C–Cl distances) is in angstrom. The inset is a close-up of the region about **3-TS**.

coordinate is defined as the difference between the C–F and C–Cl distances. The reaction is overall exothermic, reflecting the greater basicity of fluoride over chloride. The Hammond Postulate states that an exothermic reaction should have an early TS. This is seen in both the geometry of the TS and the shorter distance between **3-ID1** and **3-TS** than between **3-TS** and **3-ID2**.

TABLE 6.3 Energies (in kcal mol^{-1}) of the Ion–Dipole Complexes and Transition State Relative to Separated Reactants for Reactions 6.3–6.5

Method	Reaction 6.3 3-ID1	3-TS	3-ID2	Products
B3LYP/cc-pVTZ[a]	−15.37	−14.69	−40.86	−32.77
B3LYP/TZ2Pf+dif+ZPE[b]	−15.85	−15.43	−40.39	−32.19
mPW1PW91/cc-pVTZ[a]	−15.06	−13.43	−41.21	−33.08
CCSD(T)/ TZ2Pf+dif+ZPE(MP2)[b]	−14.85	−11.16	−38.12	−28.51
W1' core[a]	−15.43	−12.54	−42.16	−32.65
Focal point analysis[c]	−15.595	−12.118	−40.203	−30.693
Method	Reaction 6.4 4-ID1	4-TS	4-ID2	Products
B3LYP/TZ2Pf+dif+ZPE[b]	−30.05	14.15	6.30	18.26
CCSD(T)/ TZ2Pf+dif+ZPE(MP2)[b]	−30.63	16.47	5.68	18.69
Focal point analysis[c]	−30.609	16.058	4.848	17.866
Method	Reaction 6.5 5-ID1	5-TS	5-ID2	Products
MP2/6-31+G(d)+ZPE[d]	−10.1	0.2		−10.6
MP2/6-311+G(3df,2p)//MP2/6-31+G(d)+ZPE[d]	−10.9	0.6		−7.5
MP4/6-311+G(d,p)//MP2/6-31+G(d)+ZPE[d]	−10.3	0.2		−10.0
B3LYP/6-31+G(d)[d]	−10.0	−5.3	−15.3	−7.0
B3LYP/6-311+G(3df,2p)//B3LYP/6-31+G(d)[d]	−10.5	−5.5	−16.3	−8.0
B3LYP/cc-pVTZ[a]	−10.24	−5.25	−16.43	−8.01
mPW1PW91/cc-pVTZ[a]	−10.32	−3.99	−17.11	−8.57
G1[a]	−11.17	−3.45	−19.35	−10.25
G2[a]	−11.35	−1.82	−17.79	−8.43
G2(+)[d]	−11.1	−1.6	−17.0	−7.7
W1'-core[a]	−11.91	−1.82	−18.88	−8.56
Experimental		−2.5[e]		
		−2.2[f]		

[a]Ref. 12.
[b]Ref. 10.
[c]Ref. 14.
[d]Ref. 17.
[e]Ref. 20.
[f]Ref. 21.

The inset of Figure 6.5 shows a close-up of the region about **3-TS**, especially the very small barrier in the forward direction. As with the identity S_N2 reactions, the barrier heights and well depths associated with the ion–dipole complexes varies with computational method (Table 6.3). DFT continues to underestimate activation barriers, although the mPW1PW91 method performs adequately. MP2 overestimates the activation barrier for Reaction 6.4. The composite methods again are quite consistent with each other. For Reaction 6.5, in which the experimental barrier has been measured (-2.2^{21} and -2.5^{20} kcal mol^{-1}), G2 and W1′ predict the activation energy to within 1 kcal mol^{-1} of this value. Clearly, the composite methods provide a very accurate description of the reaction surface for gas-phase S_N2 reactions.

All of these gas-phase S_N2 reactions (Reaction 6.1–6.5) have topologically identical PESs: the double-well potential. However, there is a great variation in one critical feature: the relative energy of the TS. The energy of the TS can be above, nearly equal to, or below the energy of the reactants. **4-TS** lies about 16 kcal mol^{-1} above the reactants, imposing a significant enthalpic reaction barrier. The TSs for Reactions 6.1, 6.2, and 6.5 are close in energy to their respective reactants. Reaction 6.3 provides an example of a case in which the TS lies energetically well below the reactants. At first glance, it might appear that the activation barrier for Reaction 6.3 is -12.1 kcal mol^{-1}! This is not a logical inconsistency. The TS connects in the reverse direction to **3-ID1**, not to the reactants. The activation barrier is really the energy difference between **3-TS** and **3-ID1**, in other words, $+3.5$ kcal mol^{-1}. Furthermore, it should be mentioned that gas-phase S_N2 reactions with low energy TSs, even those as low as -8 kcal mol^{-1} relative to separated reactants can have rates well below that of the collision-controlled limit because of the entropic barriers.[22,23]

Branching at the α- or β-carbon is well known to retard the S_N2 reaction in solution.[24,25] In fact, the diminution of reaction rate with increasing α-substitution is one of the hallmarks of the S_N2 reaction, distinguishing it from the S_N1 mechanism. The argument is basically one of the steric hindrance: α-substitution blocks access to the backside of the C-LG bond, while the β-substitution precludes a linear attack of the nucleophile along the C-LG bond.

Recent experimental studies have indicated that α- or β-substitution may alter the gas-phase substitution reactions in different ways than in solution. Kebarle[20] has determined that the activation barrier for the gas-phase reaction of chloride with alkyl bromides increases as Me (-1.9 kcal mol^{-1}) < Et (-1.3 kcal mol^{-1}) < n-Bu (0.0 kcal mol^{-1}) < i-Pr (5.6 kcal mol^{-1}). While α-branching does increase the barrier for the substitution reaction, β-branching appears to lower the barrier. This is even more apparent in the essentially equivalent rates for the reaction of fluoride with ethyl and neopentyl chlorides in the gas phase.[26] Another example is the unexpectedly small difference (1.9 kcal mol^{-1}) in the activation energies for the reactions of chloride with methyl- and *tert*-butyl chlororoacetonitriles,[27] which was subsequently expanded to include the ethyl and *i*-propyl analogs (Table 6.4).[28]

Jensen,[29] Gronert,[30] and Streitwieser[11] have examined the effect of α- or β-branching using computational techniques. Streitwieser examined the reactions of chloride with methyl, ethyl, *n*-propyl, neopentyl, and iso-propyl chlorides at B3LYP and MP2. The MP2/6-31+G* optimized TSs (**6–10**) are shown in Figure 6.6 and their respective activation energies are listed in Table 6.4. To complete the series, we have computed the MP2/6-31+G* energies for the reaction of chloride with *tert*-butyl chloride, and this TS (**11**) is shown in Figure 6.6.

TABLE 6.4 Activation Energies (in kcal mol⁻¹) for the Reactions of Chloride with Various Alkylchlorides

Reaction	MP2[a]	CBS-QB3[b,c]	M06-2x[d]	Experimental[c]
Cl⁻ + CH₃Cl	7.68	2.0		
Cl⁻ + CH₃CH₂Cl	11.10	4.5		
Cl⁻ + CH₃CH₂CH₂Cl	9.71	3.2		
Cl⁻ + (CH₃)₂CHCH₂Cl		4.8		
Cl⁻ + (CH₃)₃CCH₂Cl	17.18	10.3		
Cl⁻ + i-PrCl	13.99			
Cl⁻ + t-BuCl	23.01			
Cl⁻ + CH₃CH(CN)Cl		−0.7	−0.9	−1.6
Cl⁻ + CH₃CH₂CH(CN)Cl		−1.8		−1.8
Cl⁻ + (CH₃)₂CHCH(CN)Cl		0.3		−0.6
Cl⁻ + (CH₃)₃CCH(CN)Cl		5.7	5.6	0.3

[a]Ref. 11.
[b]Ref. 31
[c]Ref. 28.
[d]Ref. 32.

Two important trends resulting from α-branching can be readily discerned. First, the C–Cl distance in the TS increases with increasing α-branching: 2.32 Å in **6**, approximately 2.36 Å in **7**, 2.44 Å in **10**, and approximately 2.76 Å in **11**. This systematic increase with increasing α-branching reflects the steric crowding of the backside. This is manifested also in the increasing activation energy with α-branching. The barrier for the unsubstituted case, Cl⁻ + MeCl, is 7.7 kcal mol⁻¹. Adding one methyl group increases the barrier by 3.4 kcal mol⁻¹ and a second methyl group increases the barrier by another 2.9 kcal mol⁻¹. Addition of the third methyl group, that is, the reaction of *tert*-butyl chloride, raises the barrier by another 9 kcal mol⁻¹, giving rise to an enormous barrier of 23 kcal mol⁻¹. These results are completely consistent with the kinetics of solution-phase S_N2 reactions, where S_N2 reactions with tertiary alkyl halides are exceedingly rare.

In the case of β-branching, the situation is a bit more complicated. While the C–Cl distances are very similar in **7** and **8**, indicating little increase in congestion with one methyl group in the β-position, the C–Cl distances are quite long in the neopentyl case (**9**), longer even than in the *iso*-propyl case. If steric strain is the sole factor then one might expect the activation barrier for the reactions of ethyl- and *n*-propyl chloride to be quite similar. However, the barrier for the latter reaction is actually lower by 1.4 kcal mol⁻¹. Gronert[30] argued that this reflects the greater polarizability in the larger TS **8** than **7**, allowing for stabilization of the excess negative charge.

The structure of **9** and the large activation energy through this TS (17.2 kcal mol⁻¹) are consistent with solution-phase studies of S_N2 reactions, which show that β-branching substantially retards the rate of reaction. However, this is in

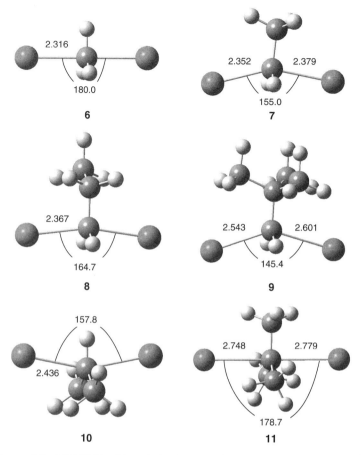

Figure 6.6 MP2/6-31+G* optimized geometries of transition states **6–11**.[11]

conflict with Brauman's gas-phase experiments, which suggest that the *tert*-butyl group in the β-position raises the barrier by only 1.9 kcal mol^{-1} relative to hydrogen.[27,28] Houk et al.[28] re-examined the PES for the reaction of chloride ion with the increasingly β-branched acetonitrilechlorides using the large composite method CBS-QB3. The activation energies, listed in Table 6.4, show the proper trend for the four compounds. However, the barrier for the neo-pentyl case is computed to be substantially larger than that observed in the experiment. A similar result was also obtained by Cramer and Truhlar[32] using M06-2x. Houk et al. attribute this discrepancy to an unknown contaminant in the experiment. The computational methods consistently indicate that β-branching increases the activation barrier, and this is largely a steric effect.

The classic S_N2 mechanism invokes backside attack of the nucleophile, leading to inversion at the carbon center. The alternative attack from the frontside, the S_NF mechanism, has been examined by Glukhovtsev et al.[33] and Bickelhaupt.[15]

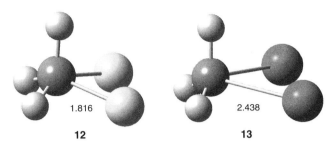

Figure 6.7 MP2/6-31+G* optimized S_NF transition states for Reactions 6.1 and 6.2.[33]

Computational chemistry is a uniquely qualified tool for examining this mechanism, which is both unlikely to occur in reality and if it did occur, distinguishing it from other possible mechanisms would be extraordinarily difficult. Since one can compute and examine the TS for many different pathways using theoretical methods, comparisons of these mechanisms can be directly made.

The MP2/6-31+G*-optimized TSs for the S_NF TSs for Reaction 6.1 (**12**) and Reaction 6.2 (**13**) are drawn in Figure 6.7. The C–X distances are similar to those in the S_N2 TSs (**1-TS** and **2-TS**) even though both partially negatively charged halide atoms are closer in these S_NF TSs. This close approach of the halides, however, manifests itself in drastically higher activation barriers for the S_NF path. The G2(+) barriers for the S_NF pathway for Reactions 6.1 and 6.2 are 44.1 and 46.3 kcal mol^{-1}, respectively. These barriers are about 40 kcal mol^{-1} higher than that for the classic S_N2 reaction. Similar large increases in the activation barriers in going from the backside to frontside attack are computed using the relativistic functional ZORA-OLYP.[15] While the S_NF reaction appears to be noncompetitive with the S_N2 reactions, an interesting counterexample is Reaction 6.6. The S_N2 (**14**) and S_NF (**15**) TSs, computed at B3LYP/6-31G*, are displayed in Figure 6.8.[34] The two TSs are very close in energy, with **15** being 1.1 kcal mol^{-1} *lower* in energy. There is experimental evidence in support of these computations and the S_NF mechanism.[35] Frontside attack may also be competitive with backside attack in the reaction of

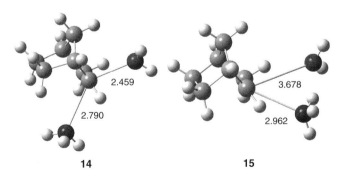

Figure 6.8 B3LYP/6-31G* optimized S_N2 and S_NF transition states for Reaction 6.6.[34]

water with protonated *t*-butanol, where the barriers for the two processes differ by about 10 kcal mol^{-1} at G3.^{36}

$$NH_4^+ + \text{[norbornane-OH]} \longrightarrow \text{[norbornane-NH}_3^+\text{]} + H_2O \quad \text{(Reaction 6.6)}$$

6.1.2 Effects of Solvent on S$_N$2 Reactions

Solvent plays a critical role in substitution reactions, often dictating which mechanism will play out. For the S$_N$2 reaction, the rate in the gas and solution phases may differ by up to 20 orders of magnitude.^{22,37,38} Gas-phase studies of nucleophilic substitution in the presence of a small number ($n = 1-5$) of water molecules have shown that the rate slows with each water molecule associated in the reaction.$^{2,39-41}$ The rate can in fact be attenuated by three orders of magnitude when three water molecules are present.^{39}

Virtually all methods for treating solvent within quantum mechanical computations have been applied to nucleophilic substitution reactions. We present here just a few studies that demonstrate the utility and limitations of solvation treatments.

A strong argument that the combined quantum mechanics (QM)–Monte Carlo (MC) simulation of Reaction 6.2 by Chandrasekhar et al.^{6} is one of the most important papers in the field of computational chemistry can be made. They first generated the PES for the reaction at HF/6-31G*, similar to that presented in Figure 6.3. The solvent (water) was treated as 250 water molecules treated as described by the TIP4P^{42} model. Water in the TIP4P approximation is treated as a rigid molecule, with a potential function having an electrostatic term (based on a partial positive charge assigned to each hydrogen atom and a partial negative charge assigned to a point on the bisector of the H–O–H angle) and Lennard–Jones 6–12 terms. Potential terms to describe the water–solute interactions were developed, with parameters that depend on the geometry of the solute; in other words, the parameters adjust as the reaction proceeds along the reaction coordinate. A MC simulation of the solute and 250 water molecules was run to determine the energy surface along the reaction coordinate, as shown in Figure 6.9. The calculated free energy barrier of the solution is 26.3 kcal mol^{-1}, in excellent agreement with the experimental^{43} value of 26.6 kcal mol^{-1}.

The effect of the solvent is readily apparent. The activation barrier is much higher in solution than that in the gas phase. This can be ascribed to strong solvation of the small chloride anion relative to the much weaker association between the solvent and the delocalized charge in the TS. The solution PES in the neighborhood of the gas-phase ion–dipole is quite flat. While there may be some minima corresponding to ion–dipole-like complexes, these minima are very shallow and unlikely to play a role in the chemistry. This flat surface leading up to the very sharp rise in the barrier suggests that the desolvation of the nucleophile is balanced by the ion–dipole attraction until the charge delocalization within the substrate about the TS dramatically reduces the solute–solvent attraction. This paper linked the long-standing

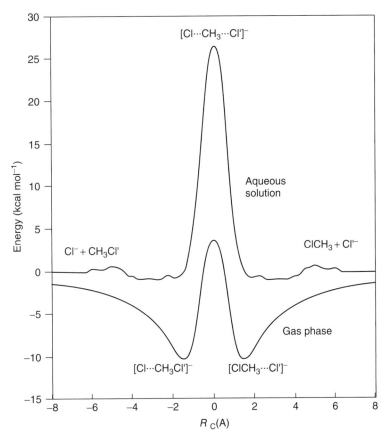

Figure 6.9 Calculated HF/6-31G* gas-phase energies (dashed curve) and potential of mean force in aqueous solution (solid curve) for Reaction 6.2. Adapted from Ref. 6. Copyright 1984 American Chemical Society.

single-hill PES ascribed to solution-phase S_N2 reactions with the double-well PES of the gas-phase reaction.

A second approach to modeling solvation effects is the continuum methods, which treat the solvent as a bulk field characterized as a dielectric or conductor medium. One of the earliest applications of the continuum method to the S_N2 reaction that incorporated geometry optimization of the solute was the work by Cossi et al.[44] They examined Reaction 6.2 at mPW1PW/6-311+G(2df,2p) for the gas phase and then reoptimized all critical points with this method and the conductor-like polarized continuum method (C-PCM) treatment for water. Their results are listed in Table 6.5. As in Jorgensen's QM/MC study, the barrier is much larger in the solution phase (26.5 kcal mol^{-1}) than that in the gas phase (0.96 kcal mol^{-1}). The geometry of the TS is slightly more compact in solution than in the gas phase, probably because of the greater charge concentration resulting in

TABLE 6.5 mPW1PW/6-311+G(2df,2p) Geometries (Distances in Angstrom) and Relative Energies (in kcal mol^{-1}) for the Ion–Dipole Complex and Transition State of Reaction 6.2[a]

	Vacuum	C-PCM
Ion–Dipole Complex		
$r_{C-Cl'}$	3.1614	3.5386
r_{C-Cl}	1.8198	1.7895
r_H	1.0807	1.0837
ΔE_C	−9.81	+5.01
Transition State		
r_{C-Cl}	2.3108	2.2955
r_H	1.0696	1.0708
ΔE_a	+0.96	+26.46

[a]Ref. 44.

TABLE 6.6 Relative Activation Energies (in kcal mol^{-1}) for X$^-$ + RY

	Computed		Experiment		
	$\Delta\Delta E^{\ddagger}$ X = Y = Cl		$\Delta\Delta E^{\ddagger}_{gas}$[b]	$\Delta\Delta E^{\ddagger}_{acetone}$[c]	$\Delta\Delta E^{\ddagger}_{acetone}$[d]
R	Gas[a]	PCM[a]	X = Cl, Y = Br	X = Cl, Y = Br	X = Y = Br
Me	0.0	0.0	0.0	0.0	0.0
Et	3.6	4.3	3.3	1.9	1.7
i-Pr	6.1	6.6	7.6	3.1	3.9
t-Bu	12.6	12.6		7.6	6.0

[a]Theoretical values calculated at B3LYP/6-31+G*, Ref. 45.
[b]Ref. 20.
[c]Ref. 46.
[d]Ref. 47

stronger interaction with the bulk continuum. An ion–dipole structure was located with the C-PCM method, but its energy is more than that of separated reactants. Since all of frequencies for this ion–dipole complex are real, a TS must exits, connecting separated reactants to this complex. The authors did not search for this TS. Nonetheless, these results are in general good agreement with the QM/MC study in terms of the shape of the solution-phase S$_N$2 PES.

Solvent tends to attenuate the α-branching effect in S$_N$2 reactions (see Table 6.6). The gas-phase barrier increases twice as rapidly with the addition of each methyl group compared to that in acetone. Jensen[45] performed B3LYP/6-31+G* optimizations of the reactants and TSs in the gas phase and then obtained free energies in water using the polarized continuum method (PCM). These results are also listed in Table 6.6. The gas-phase computational values match up well with the experimental values. However, the relative activation barriers computed with PCM are quite similar to the gas-phase values and, therefore, overestimate the solution-phase barriers.

A concern with the continuum methods is that local effects, particularly interactions between the solute and the first-shell solvent, may be poorly represented by the bulk treatment inherent in the method. In the case of the S_N2 reaction where the nucleophile may be a small densely charged anion, strong hydrogen bonding may not be appropriately treated, especially across the whole reaction coordinate where the strength of this hydrogen bonding will change. As early as 1982,[48] computations were made by microsolvating the reaction coordinate for the S_N2 reaction, placing a small number of water molecules about the reagents, and including them explicitly within the full quantum mechanical treatment. This approach is fairly computationally demanding; each additional water molecule increases the execution time and disk space required for the energy computation and also increases the configuration space that must be sampled. These demands restrict the number of explicit water molecules that can be reasonably accommodated. A further problem is that it is unknown how many solvent molecules are needed to reproduce bulk solvent effects.

Mohamed and Jensen[45] took an interesting tack by combining both explicit water molecules and PCM, the former providing QM treatment of the closest interacting solute–solvent molecules and the later incorporating the effect of bulk solvent. Examining the identity reactions Cl^- + RCl, with R = Me, Et, *i*-Pr, and *t*-Bu, they first optimized the structures of the reactants, ion–dipole complexes, and TSs with one to four water molecules at B3LYP/6-31+G*. They did not rigorously sample all possible water configurations, especially when three or four solvent molecules were present. Rather, they located representative examples of the ensemble of microsolvated structures. These microsolvated TSs for the reaction of chloride with methyl chloride are shown in Figure 6.10. These microsolvated structures are then used for a PCM calculation without further optimization, providing a means for capturing both the local and the bulk interactions of the solute with the solvent.

These microsolvation structures fall into three general categories: (a) where the water molecules are distributed (more or less) equally between the two chlorine atoms, (b) where the water cluster bridges the two chlorine atoms, and (c) where the water molecules are asymmetrically distributed to the two chlorine atoms. The type **b** structures are lower in energy than the others because of their additional hydrogen bonding between the water molecules. Jensen suggests that the type **a** structures are the best model since these more open water structures can better hydrogen bond with the bulk solution. However, the type **b** structures have the lowest energy at both the B3LYP and PCM levels (Table 6.7), a situation also seen for microsolvated thiolate–disulfide TSs.[49]

It is difficult to draw many conclusions from this work, given the incomplete sampling of the water configurations. However, it does appear that microsolvation does decrease the range of activation barriers with increased α-branching. When four water molecules are included in the cluster, the relative activation barriers for reaction with methyl, ethyl, *iso*-propyl, and *tert*-butyl chlorides are 0.0, 2.5, 6.7, and 7.3, respectively. This computed compression of the activation barrier range (especially relative to that observed in the gas phase) is not as great as what is observed in experiment (see Table 6.6), but the trend is correct. Furthermore, PCM alone does

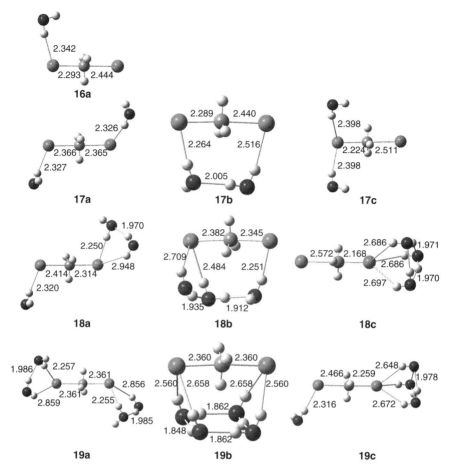

Figure 6.10 B3LYP/6-31+G* optimized hydrated transition states for the reaction $Cl^- + RCl$, R = Me, Et, i-Pr, and t-Bu. All structures are from Ref. 45 except **19a** optimized for this work.

not reproduce this trend, and incorporating PCM with the microsolvated structures works to increase the range. Jensen cautions that PCM should therefore be used with care. Further development of solvation techniques are needed to adequately address this problem.

Within the studies of the effect of β-substitution on the gas-phase S_N2 reaction described in the previous section was an examination of the solution-phase effect. Listed in Table 6.8 are the experimental values of the relative activation barriers of two simple substitution reactions with increasing β-substitution. The solution-phase barriers for the chloride exchange reactions were computed by adding the C-PCM/B3LYP/6-31G(d) solvation energies to the CBS-QB3 gas-phase energies or using the SM8/M06-2x/6-31+G(d,p) method. These computed relative activation barriers are also listed in Table 6.8. The trend of

TABLE 6.7 Activation Energy (in kcal mol⁻¹) for the Substitution Reactions Through Transition States 31–34[a]

	Me	Et	Relative E_a i-Pr	t-Bu
16a	4.5	3.2	5.6	11.2
	20.6	*3.9*	*5.9*	*11.4*
17a	10.1	3.1	5.0	9.9
	20.8	*3.9*	*6.0*	*10.2*
17b	7.8	2.5	4.6	
	19.2	*2.9*	*5.9*	
17c	11.8			
	19.9			
18a	15.4	2.8	4.7	8.7
	19.2	*4.1*	*6.3*	*9.0*
18b	11.1			
	18.3			
18c	12.9			
	16.3			
19a	16.6[b]		6.7	7.3
19b	6.2	2.5		
	19.9	*2.6*		
19c	14.6			
	23.7			

[a]The activation energy is given for the reaction Cl⁻ + MeCl, and the activation barriers relative to it are given for the other three reactions. Values in normal text are at B3LYP/6-31+G* and those in italics are at PCM B3LYP/6-31+G*. See Ref. 45.
[b]Computed for this work.

TABLE 6.8 Relative Experimental and Computed Activation Barriers (in kcal mol⁻¹) for β-Substitution Effects in Simple S_N2 Reactions in Acetone

R	Br⁻ + RCH₂Br[a]	Cl⁻ + RCH₂Br[a]	Cl⁻ + RCH₂Cl[b]	Cl⁻ + RCH₂Br[c]
Methyl	0.0	0.0	0.0	0.0
Ethyl	−0.1	0.3	0.5	
i-Propyl	1.6	1.4	2.6	
t-Butyl	4.0	3.8	8.1	8.0

[a]Ref. 50.
[b]C-PCM/B3LYP/6-31G(d) solvation energies added to CBS-QB3 gas-phase energies, Ref. 28.
[c]SM8/M06-2x/6-31+G(d,p), Ref. 32.

increasing barrier with increasing β-substitution is replicated in the computations, although the computations tend to overestimate the effect.

What is somewhat surprising is that the computed relative activation barriers for the β-substitution effect in the gas and solution phases are so similar—compare the values in Table 6.4 with those in Table 6.8. Given the strong solvent effect in other substitution reactions, and the fact that steric bulk is increasing so much in this series, one might have expected a larger solution effect. For example, the computed CPCM solvation energies of the alkoxides RO$^-$ (R = methyl, ethyl, i-propyl, and t-butyl) span a range of 8.2 kcal mol^{-1}, while solvation energies of the TSs of the reactions listed in Table 6.8 vary by only 1.1 kcal mol^{-1}. One might have expected the increasing alkyl substitution to screen the negative charge in the TSs just as it does in the alkoxides, but this seems to be significantly attenuated.

The explanation is found by examining the geometries of the TSs. As the β-substitution increases, the C–Cl distances increase as well. This manifests also in the greater charge on the chlorines as the β-substitution increases. The screening effect, which diminishes the solvation with increasing β-substitution, is offset by an increasing ionic character with β-substitution.[28,32]

6.2 ASYMMETRIC INDUCTION VIA 1,2-ADDITION TO CARBONYL COMPOUNDS

Addition of a nucleophile to a carbonyl group can create a new chiral center. This type of asymmetric induction has been of active interest for more than 50 years. While countless examples of 1,2-addition to carbonyl compounds resulting in enantiomeric or diasteriomeric excess are known, the origin of this asymmetric induction remains in dispute.

Cram[51] proposed the earliest model, shown in Scheme 6.1, for predicting and understanding this asymmetric induction. The three substituents on the α-carbon are ranked in terms of size and are labeled large (L), medium (M), and small (S). Cram proposed that the largest substituent is placed antiperiplanar to the carbonyl oxygen in order to minimize the steric repulsion between these two groups. The incoming nucleophile will then preferentially attack at the side of the smallest substituent. Karabatsos[52] noted the failure of this model to explain an increase in selectivity with increasing size of the R group. He proposed the model whereby either the medium or large substituent is placed synperiplanar to the carbonyl. The nucleophile then attacks from the side of the smallest substituent.

Felkin[53] objected to the Karabatsos model for having no physical basis for these two conformations. Instead, he proposed the model whereby the largest substituent is placed antiperiplanar to the incoming nucleophile. Felkin argued that the repulsion between the incipient C–Nuc bond and the largest substituent should be minimized. The preferred pathway leading to the major product is through the conformation that places the medium group near the carbonyl, the so-called inside position, and the small group occupies the outside position. For many reactions, all three models predict the same major product, but the Felkin model is the most broadly successful.

Anh and Eisenstein[54] published their first quantum chemical contribution to this issue in 1977. They first constructed a model of the TS for 1,2-addition of hydride to 2-chloropropanal. They placed the hydride ion 1.5 Å away from the carbonyl carbon and perpendicular to the CO bond (**20**). By rotating about the C–C bond in

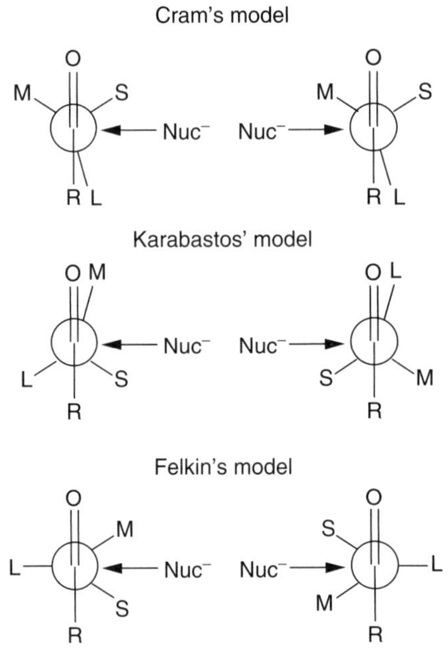
Cram's model
O
M
S
Nuc⁻
Nuc⁻
R L
Karabastos' model
O M
O L
L
S
R
M
Felkin's model
O
M
L
S
R

Scheme 6.1

30° increments for the two diastereomers, they generated 24 geometries, for which HF/STO-3G energies were computed.

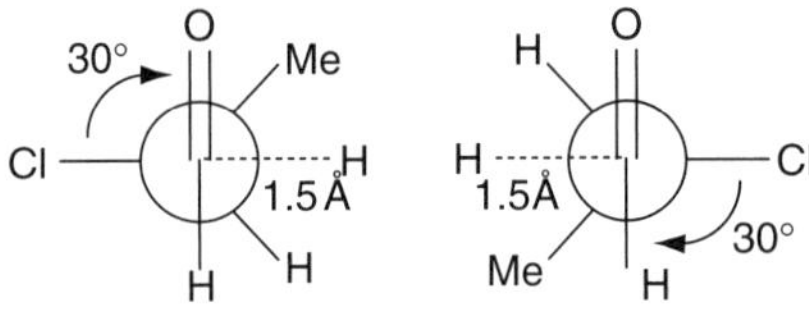
30°
O
Me
Cl
H
1.5Å
H
H

20

Figure 6.11 displays the relative energies of these 24 geometries against the rotation angle about the C_1-C_2 bond. These energies used in this plot were obtained in a method slightly modified from the Anh–Eisenstein procedure. First, the geometry of the lowest energy TS leading to the major and minor products were fully optimized at B3LYP/6-31++G(d). Next, 12 conformers of each structure were obtained by rotating by 30° about the C_1-C_2 bond, holding all other geometric parameters fixed except the $O-C-H_{incoming}$ angle that was allowed to relax. This process produces more rigorously defined rotation energy TS surface than that performed by Anh and Eisenstein but the results are qualitatively very similar.

The lowest energy TS, which leads to the major product, has just the conformation predicted by Felkin. The lowest energy TS leading to the minor product is also that predicted by Felkin. In other words, Felkin's assumption of placing the largest

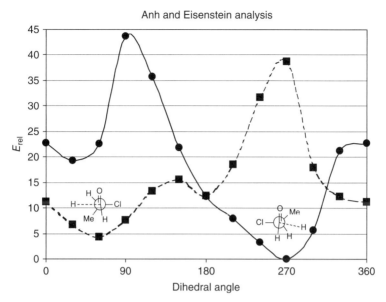

Figure 6.11 Model transition state rotational energy surface for the reaction of H⁻ with 2-chloropropanal. The full line represents transition states leading to the major product and the dashed line represents those leading to the minor product. Modified from the original model proposed by Anh and Eisenstein.[54]

group perpendicular to the carbonyl and antiperiplanar to the incoming nucleophile is correct.

Anh and Eisenstein also supplied a reason for the preference of the M group being in the "inside" position. The incoming hydride does not enter along a path perpendicular to the carbonyl, but rather on a path making an angle of about 103° (**21**). Therefore, the path past the smallest group (S) in the "outside" position will be sterically less congested than that if the M group were in the "outside" position.

Felkin's model relied on reducing steric interactions and strain energy about the carbonyl group. Anh and Eisenstein, however, supplied an alternative reason for the preferred transition-state conformation. The dominant orbital interaction in this reaction is between the highest occupied molecular orbital (HOMO) of the nucleophile and the lowest unoccupied molecular orbital (LUMO) of the carbonyl

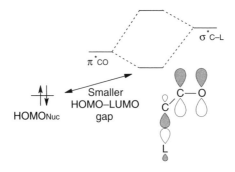

Scheme 6.2

(π^*_{CO}). When the L group is located antiperiplanar to the incoming nucleophile, the antibonding orbital between carbon and the L group (σ^*_{C-L}) can effectively lower the π^*_{CO} energy, allowing for greater interaction with the incoming nucleophile (Scheme 6.2). The definition of the L group, therefore, becomes the group that has the lowest σ^*_{C-L} orbital. So while the Felkin model relies on reducing steric interactions, the Anh–Eisenstein variation relies on maximizing hyperconjugative stabilization in the TS. Heathcock[55] has modified this viewpoint, arguing that in fact, both electronic (Anh's hyperconjugation argument) and steric (Felkin's argument) effects determine which group is the "largest."

Early ab initio computations fully supported the arguments of Anh and Eisenstein. Schleyer and Houk[56] optimized the transition structures for the reaction of formaldehyde with LiH (**22**), methyllithium (**23**), and methyllithium dimer (**24**) at HF/3-21G. These TSs, reoptimized at HF/6-31G*, are shown in Figure 6.12. The nucleophile adds from the face perpendicular to the carbonyl group. Wong and Paddon-Row[57] examined the reaction of cyanide with 2-fluoropropanal. They located six TSs at HF/3-21G, differing by which group is anti to the incoming cyanide nucleophile. The two lowest energy TSs (**25** and **26**, Figure 6.13) correspond to the two Felkin–Anh conformations, where the fluorine atom is anti to the incoming cyanide. TS **25**, the one predicted by the Felkin–Anh model to lead to

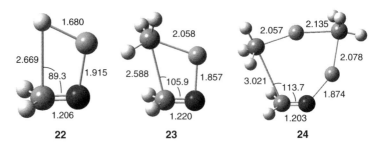

Figure 6.12 HF/6-31G* optimizations of the transition states for the reaction of formaldehyde with HLi, CH$_3$Li, and the CH$_3$Li dimer.

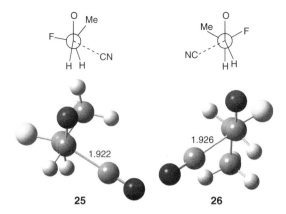

Figure 6.13 HF/6-31G* optimized transition states for the reaction of cyanide with 2-chloropropanal.

the major product, is 6.9 kcal mol^{-1} lower in energy at MP2/6-31+G*//HF/3-21G than the alternative TS **26**. While the $C_{Nuc}-C_1-O$ angle is nearly identical in the two TSs (about 113°), the $C_N-C_1-C_2-$(outside group) dihedral angle is 12° larger in **26** than in **25**, reflecting the poorer steric interactions when the nucleophile must pass by the medium group rather than by the smallest group.

Unquestionably, the work of Anh and Eisenstein is seminal. Even though their computations were performed at what is today considered an extremely rudimentary level, and as we will see, it is not the last word on the subject, the predictive power of the Felkin–Anh model has become a standard tool in organic chemistry.[58] As argued by Houk,[59] the Ahn–Eisenstein "was important not only because it solved an important general problem in stereoselectivity but also because it demonstrated the power of quantum mechanical calculations to solve important problems on real organic systems."

Cieplak[60] countered the Anh explanation with an alternative orbital model. He noted that reductions of cyclohexanones and other additions at carbonyls occasionally resulted in the major product coming from the Felkin–Anh *minor* TS. Arguing that since the incipient bond was electron deficient—a partial bond lacks the full two-electron occupation—it is donation of density from the σ_{C2-L} into the σ^*_{C-Nuc} orbital that will stabilize the TS (Scheme 6.3). Support for the Cieplak model was provided by experimental results for nucleophilic addition to 3-substituted cyclohexanones,[61] reductions of 2,2-diarylcyclopentanones,[62] and especially le Noble's extensive studies of reductions of adamantanones.[63]

Negative reaction to the Cieplak model by many computational chemists was quickly forthcoming, resulting in modifications to the simple Felkin–Anh model. Houk[64] reported the three TSs for the reaction of NaH with propanal (the HF/6-31G* optimizes structures are shown in Figure 6.14). The lowest energy TS is **27a**, where the methyl group is in the "inside" position. Assuming that the methyl group is larger than a hydrogen atom, the Felkin–Anh TS, **27b**, is 1.0 kcal

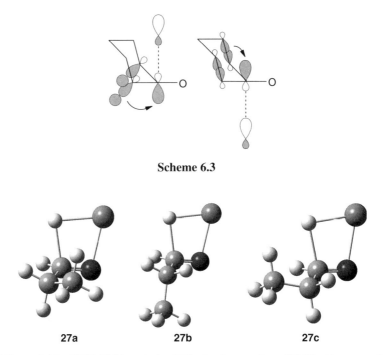

Scheme 6.3

Figure 6.14 HF/6-31G* optimized TSs for the reaction of NaH with propanal.

mol^{-1} above **27a**, and the TS where the methyl group is in the outside position (**27c**) is 0.9 kcal mol^{-1} above **27a**. For the Cieplak model to work, the CH bond would have to be a better donor than the CC bond, an argument Cieplak[60,61] and others[65,66] have made, but this contention has been disputed.[67] Houk maintained that the CH bond is a poorer donor than the CH bond and rejects the Cieplak model. Rather, he argued that the methyl group, being a better donor, destabilizes **27b** relative to **27a**. Keep in mind that the Anh model is based on the anti group being the best acceptor, that is, having the lowest σ^* orbital.

Frenking has also criticized the Cieplak model. His initial argument[68] is with the conceptual footing of the model itself. Stabilization of the Cieplak TS comes by donation of density into the vacant antibonding orbital for the forming bond. In other words, this is a HOMO(σ_{C-L})–LUMO(σ^*_{C-Nuc}) interaction. This is *not* the interaction favored by frontier molecular orbital (FMO) theory: HOMO(σ_{C-Nuc})–LUMO(π^*_{C-O}). Instead of using the Cieplak model, Frenking argued that the preferred orientation of 1,2-addition is understood in terms of the Felkin model, the effect of the conformation of the aldehyde, and the shape of the carbonyl LUMO.

Frenking examined the TSs for the reaction of LiH with propanal, 2-chloroethanal, and 2-chloropropanal.[69] He found three TSs (**28a–c**) very similar to **27a–c**. **28a** is the lowest energy of the three, while **28b** and **28c** are 1.3 and 1.6 kcal mol^{-1} higher in energy, respectively. If the LiH fragment is removed,

ASYMMETRIC INDUCTION VIA 1,2-ADDITION TO CARBONYL COMPOUNDS   397

but the structure of the remaining aldehyde is kept unchanged, **28a** remains the lowest in energy, **28b** is 1.2 kcal mol^{-1} higher and **28c** is 0.7 kcal mol^{-1} above **28a**. Frenking argued that this indicates that the energy of the conformers of the aldehyde alone determine much of the energy difference between the TSs.

For the reaction of lithium hydride with 2-chloroethanal, the lowest energy TS is **29a** (Figure 6.15), with **29b** and **29c** lying 1.7 and 0.3 kcal mol^{-1} above **29a**, respectively. When the lithium cation is removed from each structure and its energy recomputed, **29c** has the lowest energy, with **29b** and **29a** are 3.4 and 6.4 kcal mol^{-1}, respectively, above it. Frenking suggested that this indicates strong complexation energy in **29a**, reminiscent of the Cram chelation model where the cation associates with both the carbonyl oxygen and the electronegative substituent on C_2. This chelation stabilizes **29a** more than the Anh orbital interaction available in **29c**. Paddon-Row[70] noticed the same favorable chelation in the most favorable TS for the analogous reaction of LiH with 2-fluoroethanal.

Lastly, in the reaction of LiH with 2-chloropropanal, five TSs were located, **30a–e** (Figure 6.16). **30a** and **30b** lead to the major product, while the other three lead to the minor product. The two lowest energy structures correspond with the Felkin–Anh major (**30a**) and minor (**30c**) TSs. Their energy difference corresponds with the energy difference for their aldehyde fragments when LiH is removed. **30b** and **30d** display the chelation effect, but unlike with **29**, chelation is not enough to make up for the favorable orbital interactions in the Felkin–Anh approach.

Frenking next examined the reaction of LiH with cyclohexanone.[68] The Hartree–Fock (HF)/3-21G transition structures for axial (**31ax**) and equatorial (**31eq**) attack are shown in Figure 6.17. Axial attack is lower than equatorial

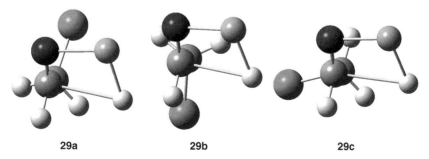

Figure 6.15 HF/6-31G* optimized transition structures for the reaction of LiH with 2-chloroethanal.

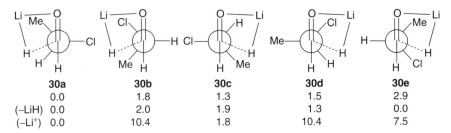

	30a	30b	30c	30d	30e
	0.0	1.8	1.3	1.5	2.9
(−LiH)	0.0	2.0	1.9	1.3	0.0
(−Li⁺)	0.0	10.4	1.8	10.4	7.5

Figure 6.16 Transition states for the reaction of LiH with 2-chloropropanal. (a) HF/6-31G* relative energies; (b) relative energies without LiH; (c) relative energies without Li⁺.

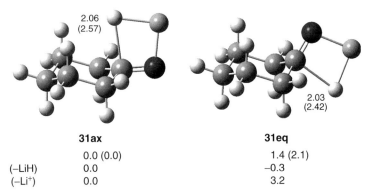

	31ax	31eq
	0.0 (0.0)	1.4 (2.1)
(−LiH)	0.0	−0.3
(−Li⁺)	0.0	3.2

Figure 6.17 HF/3-21G optimized structures for the reaction of LiH with cyclohexanone.[68] (a) MP2/6-31G*//HF/3-21G relative energies; (b) relative energies without LiH; (c) relative energies without Li⁺. Values in parenthesis are for the B3LYP/6-31G** optimized structures.[72]

attack by 1.4 kcal mol^{-1}. The energy difference between the two is only 0.3 kcal mol^{-1} when the LiH fragment is removed, indicating that two ketone fragments have similar energies. This is counter to Houk's[71] explanation for the preference for axial attack. Houk argues that the equatorial TS is more strained, based on geometrical arguments; however, the equatorial TS is actually slightly *less* strained than the axial TS. The large difference in energies when Li⁺ is removed from the structures suggests that the preference for axial attack must come from the interaction of the nucleophile (H⁻) with cyclohexanone. Frenking argued that the carbonyl LUMO is distorted by hyperconjugation making the carbon lobe larger on the side of axial attack. Very similar conclusions have been drawn by Luibrand in a more recent B3LYP/6-31G** study of the reaction of cyclohexanone with LiH or LiAlH$_4$.[72] The B3LYP TSs are much earlier than the HF ones, but their energy differences are comparable—both the methods predict the axial attack to be lower in energy than the equatorial attack.

Frenking also examined the reaction of LiH with 3-fluorocyclohexanone, for which four TSs were located. Fluorine can occupy either the equatorial (**32**) or the axial (**33**) position and then attack can come from the axial or equatorial faces. The Cieplak model does not distinguish between **32** and **33**, predicting axial attack for both. While axial attack is preferred in **32**, equatorial attack is favored by 2.3 kcal mol^{-1} over axial attack in **33**. Frenking suggests that this difference is reflected in the orbital coefficients of the LUMO of **32** and **33**. A simpler explanation, one that will be further explored next, is that electrostatic interactions between the hydride and the axial fluorine destabilize **33ax** relative to **33eq**.

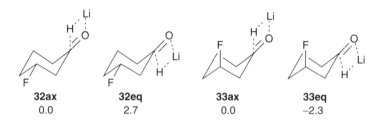

32ax	**32eq**	**33ax**	**33eq**
0.0	2.7	0.0	−2.3

Paddon-Row and Houk[73] were the first to strongly advocate for the role of electrostatics in determining the stereo-outcome of 1,2-addition reactions. They examined the addition of LiH to a number of substituted-7-norbornanones, of which **34** is representative. The addition can come from the same side of substituents (**34syn**) or from the opposite side (**34anti**). These two transitions states are shown in Figure 6.18. At MP2/6-31G*//HF/3-21G, attack is favored from the syn face by 4.0 kcal mol^{-1} over anti attack. When the LiH fragment is removed, the two TSs differ by only 0.8 kcal mol^{-1}, with **34anti** lower in energy. Natural population

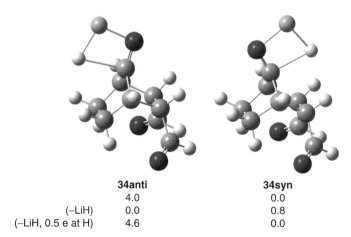

	34anti	**34syn**
	4.0	0.0
(−LiH)	0.0	0.8
(−LiH, 0.5 e at H)	4.6	0.0

Figure 6.18 HF/6-31G*-optimized transition states for the reaction of LiH with **34**. (a) Relative energies at MP2/6-31G*//HF/6-31G*; (b) relative energies without LiH; (c) relative energies with LiH and a −0.5 point charge at the position of the hydride.[73]

analysis suggests that the charge on the hydride is about −0.5. When the LiH fragment is removed and a -0.5 charge is placed at the position of the hydride, the *syn* TS structure is 4.6 kcal mol^{-1} lower in energy than the anti one, almost perfectly matching the energy difference between **34syn** and **34anti**. Since the formyl group will polarize C_2 and C_3 (the substituted ring carbons) and will make them partially positively charged, stronger electrostatic stabilization will occur in **34syn**, where the hydride atom is closer to these positive centers than in **34anti**.

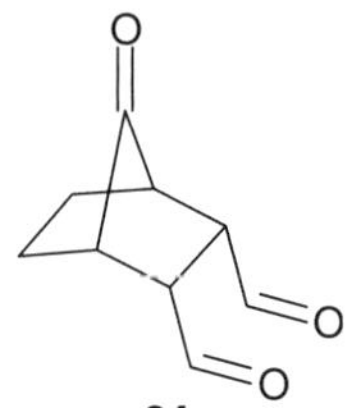

34

Paddon-Row extended this work with the investigation of the reaction of disubstituted norbornen-7-ones (**35**) with LiH or methyllithium.[74] Again, there are two pathways, one from the same side as the substituents (syn) and one from the side of the alkene (anti). The energy difference between the TSs for these two paths with either LiH or MeLi are listed in Table 6.9.

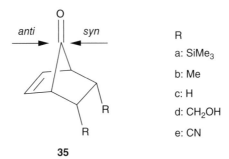

R

a: SiMe$_3$
b: Me
c: H
d: CH$_2$OH
e: CN

35

Syn attack of LiH is favored for all five cases (**35a–e**). This is contrary to the predictions of the Cieplak model, which would favor anti attack for compounds with electron-donating substituents, such as **35a**. The Anh–Eisenstein model fairs no better, as it predicts anti attack for compounds with electron-withdrawing substituents, such as **35e**. Paddon-Row, however, argued that electrostatic effect can account for this uniform attack direction. Minimization of steric interactions would argue for the anti approach. However, the anti approach takes the nucleophile over the electron-rich double bond. The resulting electrostatic repulsion disfavors the anti path and the syn approach dominates. The very large energy difference between the *syn* and *anti* TSs for LiH addition to **35e** is understood also in terms of electrostatics. Here, the strongly electron-withdrawing cyano groups build up positive

TABLE 6.9 MP2/6-31G*//HF/6-31G* Energy Difference between the *anti* and *syn* Transition States ($E_{anti} - E_{syn}$) for the Reaction of 35 with LiH or MeLi[a]

	LiH	MeLi
35a	10.14	−14.56
35b	8.32	−14.06
35c	12.38	−7.88
35d	7.83	−11.28
35e	34.78	7.19

[a]Ref. 74.

charge on their neighboring carbon atoms, generating favorable electrostatic interactions with the incoming nucleophile for the syn face.

For the reactions of **35a–e** with methyllithium, steric interactions overcome the electrostatic effects and the *anti* TSs are lower in energy than the *syn* TSs for all cases but **35e**. Here, the strong electron-withdrawing nature of the cyano substituents generates enough favorable electrostatic interaction to outweigh the steric demands, and the reaction through the *syn* TS is the preferred path.

Marsella[75] has proposed a scheme for predicting the preferred direction of attack at carbonyls. He suggests visualizing the computed electrostatic potential on an isosurface of the carbonyl π^*_{CO} orbital (the LUMO). The lobe that is more positive accurately predicts the correct face that is attacked for 13 different reactions.

The interplay between favorable electrostatic interactions and steric demands was further demonstrated in the computational study of the reaction of LiH with 2-silylethanal (Reaction 6.7) and 2-trimethylsilylethanal (Reaction 6.8). For both the reactions, Fleming et al.[76] located three TSs corresponding to different rotamers about the C–C bond. The Felkin model, in which the largest (most sterically demanding) group is placed anti to the nucleophile, suggests that TSs **36c** and **37c** should be the best. The Anh–Eisenstein model favors **36b** and **37c**; here, the strongest donor is placed anti to the nucleophile and the larger silyl group is placed in the "inside" position.

$$RCH_2CHO + LiH \rightarrow RCH_2CHOH \quad R = SiH3, \quad \text{(Reaction 6.7)}$$

$$R = SiMe3, \quad \text{(Reaction 6.8)}$$

For Reaction 6.7, the lowest energy (MP2/6-31G*//HF/6-31G*) TS is **36a**, with the Anh TS lying 0.56 kcal mol^{-1} higher, and the Felkin TS 0.05 kcal mol^{-1} higher still. In contrast, the relative ordering of the TSs is reversed for Reaction 6.8; the Felkin TS is the lowest in energy, followed by the Anh TS lying 0.98 kcal mol^{-1} higher, and **37a** is 1.15 kcal mol^{-1} above **37c**. **36a** is stabilized by electrostatic attraction between the nearby electropositive silyl group and the hydride; these electrostatics dominate any steric repulsions, and **36a** is the preferred TS for Reaction 6.7. In **37a**, the much larger trimethylsilyl group does not allow for the two oppositely charged groups to closely approach each other. The sheer bulk of the trimethylsilyl group makes **37c** the most favorable TS for Reaction 6.8.

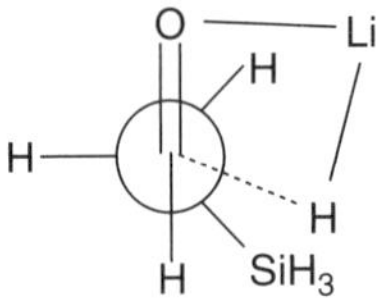

36a 0.0	**36b** 0.56	**36c** 0.61
37a 1.15	**37b** 0.98	**37c** 0.0

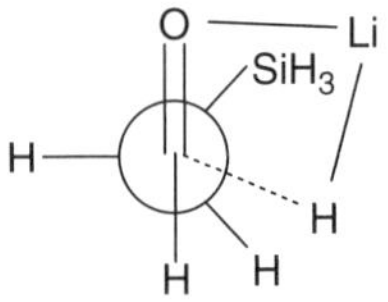

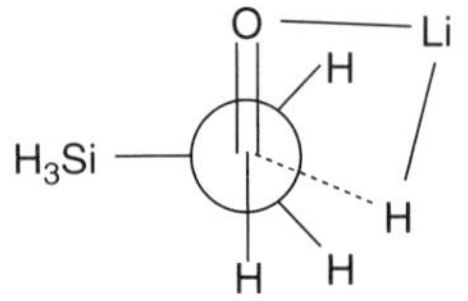

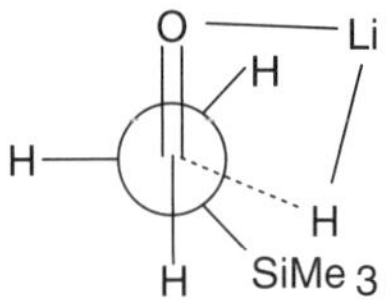

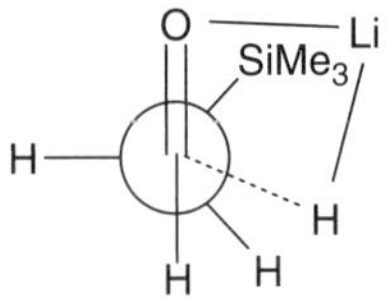

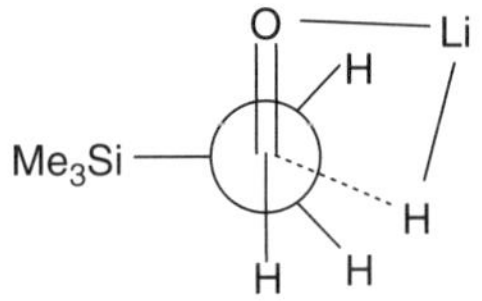

Another interesting variation on the Felkin–Anh model is the reduction of ketones with very bulky α-substituents: phenyl (**38**), cyclohexyl (**39**), and *t*-butyl (**40**). The TSs for the reactions of these ketones with LiH were examined at MP2/6-311+G**//HF/6-31G*.[77] The optimized geometries for the lowest energy TS leading to the Cram and anti-Cram products are shown in Figure 6.19. For the reactions of **38** and **39**, the computations predict the Cram product to be the major product, consistent with the experimental reduction of these ketones with LiAlH$_4$. At first sight, this result appears consistent with the Felkin–Anh model: TSs **38cram** and **39cram** have the largest group anti to the incoming hydride and the methyl group occupies the "inside" position. However, the TSs leading to the minor, anti-Cram, product (**38anticram** and **39anticram**) have the methyl group in the anti position and the large group in the "inside" position.

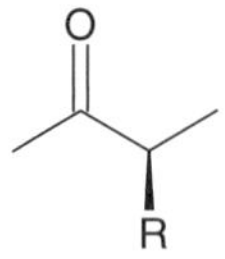

38: R = Phenyl
39: R = Cyclohexyl
40: R = *t*-Butyl

More unusual are the results for the reaction of **40**. The two TSs, **40cram** and **40anticram**, are geometrically consistent with the Felkin–Anh model. Both have the largest group anti to the hydride. The Felkin–Anh model predicts that the TS with the methyl group in the "inside" position (**40cram**) should be lower than that when the methyl group is in the "outside" position (**40anticram**). The MP2 energies are in fact opposite: **40anticram** is 3.42 kcal mol^{-1} lower in energy than **40cram**, and the major product found in the experimental reduction of **40** with LiAlH$_4$ is the anti-Cram product!

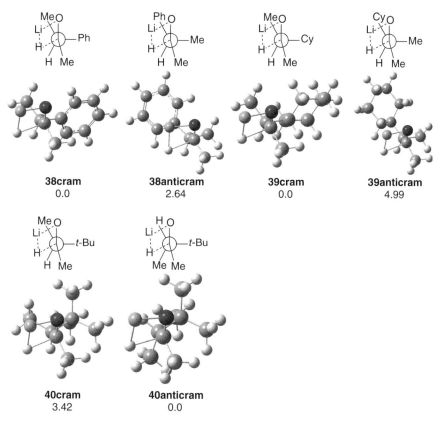

Figure 6.19 HF/6-31G* optimized transition states for the reactions of **38–40** with LiH. Relative energies at MP2/6-311+G**//HF/6-31G* are listed below each structure.

Scheme 6.4

Smith et al.[77] argue that a modification of the Felkin–Anh model is needed to accommodate these results (Scheme 6.4). They suggest that neither the large nor the medium groups should occupy the "outside" positions because of the steric interactions with the incoming nucleophile. Thus, for the reactions of **38** and **39**, the usual lowest-energy Felkin–Anh TS is fine (**41**), but the Felkin–Anh model for the TS leading to the minor product places the methyl group into the "outside" position. This can be avoided by adopting the conformation in which the medium group is in the "anti" position and the large group is in the "outside" position (**42**).

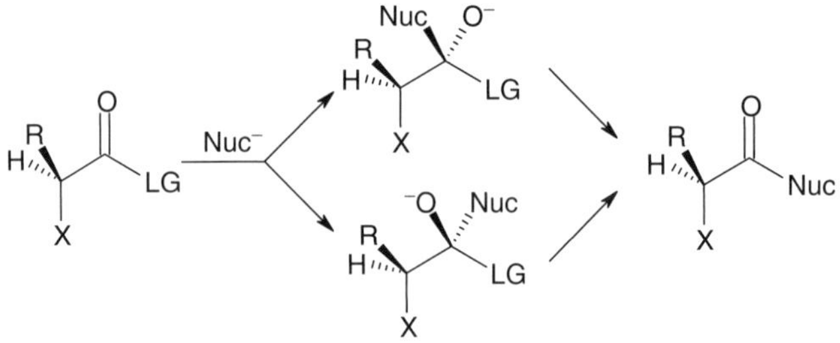

Scheme 6.5

When the large group is very bulky, such as the *t*-butyl group in **40**, then the usual Felkin–Anh conformations (**41** and **43**) are appropriate, thereby minimizing the repulsion between this extremely large group and the gegenion associated with the oxygen atom. Of the two competing conformations, the preferred TS is the one with the least interaction with the "inside" group. So for the reactions of **38** and **39**, one needs to compare TS models **41** and **42**, with the former being the preferred TS since its "inside" group is smaller. For the reaction of **40**, the two competing TS models are **41** and **43**, with the latter preferred since its "inside" group is smaller.

Stereocontrol in nucleophilic acyl substitution might at first appear to be not an issue, since the carbonyl carbon is achiral. However, if the incoming nucleophile or the leaving group are chiral, then the TS and intermediates could be diasterotopic, and selectivity can occur (see Scheme 6.5). Houk and Birman[78] address this in a computational study of kinetic resolution of carboxylic acids. They found that the Felkin–Anh model applies, extending its applicability to the nucleophilic acyl substitution.

What conclusions can be drawn from the computations on 1,2-addition? The Felkin–Anh rules seem to apply fairly well across a broad spectrum of reactions. Computations clearly indicate that the stereoselectivity is based on a number of competing factors—sterics, orbital interactions, and electrostatic interactions—that can be subtly balanced. Perhaps most critical is that relatively simple computations can offer real predictive power, providing guidance to the synthetic chemist in their pursuit of high enantioselectivity and diastereoselectivity.

6.3 ASYMMETRIC ORGANOCATALYSIS OF ALDOL REACTIONS

The Aldol reaction is one of the most powerful methods for creating the C–C bond. Typical conditions involve the formation of an enolate, usually with a stoichiometric equivalent of base. Stereoinduction is usually accomplished with chiral enolates, aldehydes, or auxiliaries.[79,80] Nature, however, is much more efficient, having created enzymes that both catalyze the aldol reaction and produce stereospecific product. These enzymes, called *aldolases*, are of two types.[81] The type II aldolases make use of a zinc enolate. Of interest for this section are the type I aldolases, which make use of enamine intermediates. Sketched in Scheme 6.6 is

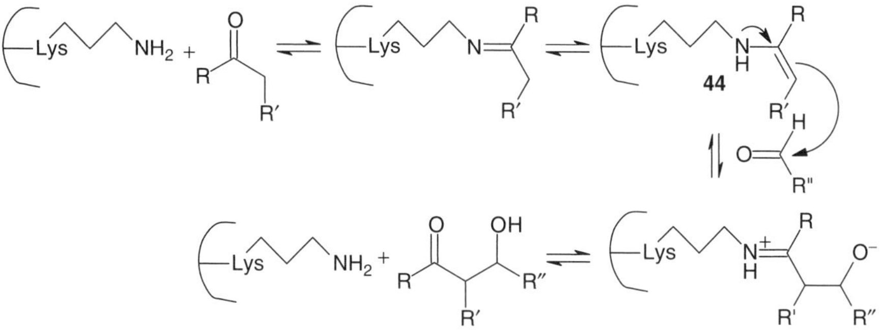

Scheme 6.6

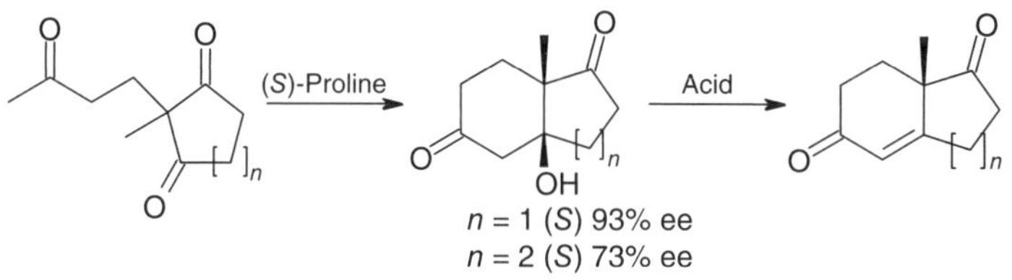

Scheme 6.7

the mechanism of action of type I aldolases. First, a ketone reacts with a lysine residue of the enzyme to form an enamine intermediate **44**. The key step involves this enamine intermediate acting as a nucleophile (a masked enolate) in attacking the carbonyl carbon. The resulting iminium is hydrolyzed, releasing the final aldol product and regenerating the enzyme. Catalytic antibodies based on this model have been developed to catalyze aldol reactions with high enantioselectivity.[82]

Prior to the determination of the aldolase mechanism and the development of catalytic antibodies for the aldol reaction, Hajos and Parrish[83] and independently Wiechert et al.[84] discovered that (S)-proline catalyzes the intramolecular aldol reaction of cyclic triketones (Scheme 6.7). This is not only a catalytic effect: the reaction proceeds with high yields and large enantiomeric excess.

Surprisingly, little follow-up work on this idea of small molecule asymmetric catalysis appeared for the next 25 years. In the late 1980s, Agami[85] reported the asymmetric intramolecular aldol reaction of acyclic diketones with (S)-proline as the catalyst. It was not until the twenty-first century, however, when this notion of organocatalysts became fully exploited.[86,87] List[88,89] and Barbas[90] pioneered enamines as catalysts for aldol and Mannich and related reactions. MacMillan has developed a variety of iminium-based catalysts producing large asymmetric induction for Diels–Alder chemistry,[91] Friedel–Crafts alkylations,[92] Mukaiyama–Michael[93] and cyclopropanation[94] reactions.

Examples of amine-catalyzed aldol reactions are shown in Table 6.10. For all entries, the mole percent of amine used was generally 10–20 percent. While, in some cases, yields may be poor, generally the yields are quite acceptable.

TABLE 6.10 Examples of Amine-Catalyzed Aldo Reactions

Reaction	Yield, %	d.r., %	ee, %	Reactions
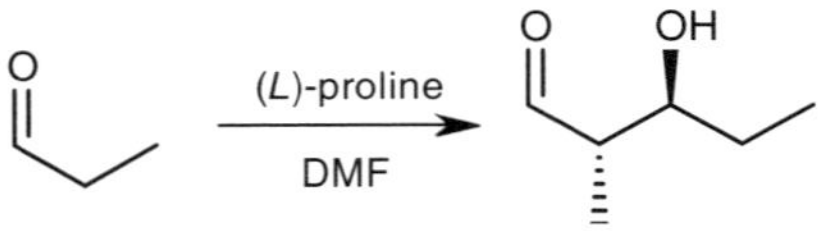	91	75	99	Reaction 6.9[97]
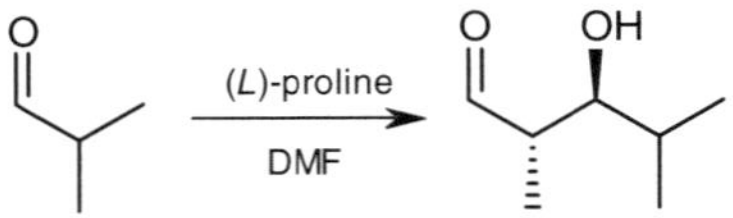	82	96	>99	Reaction 6.10[97]
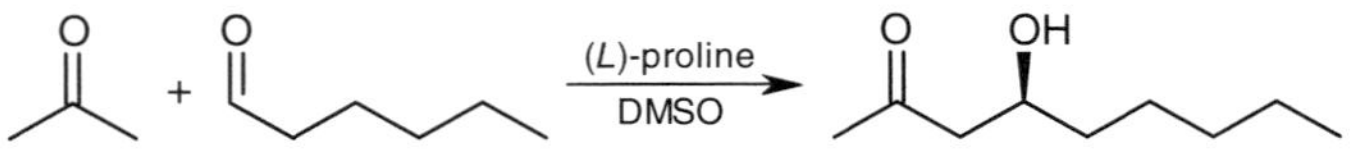	35		73	Reaction 6.11[98]
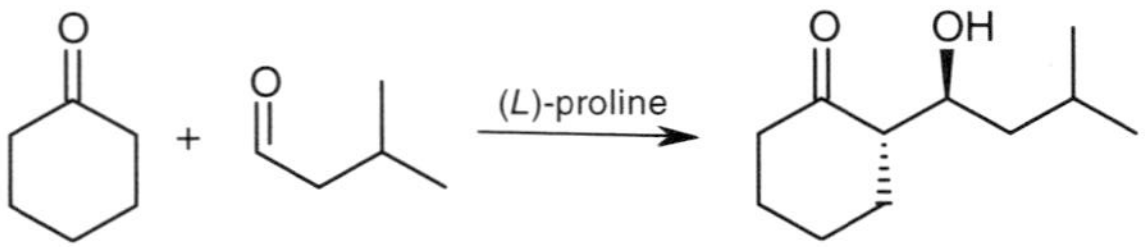		88	86	Reaction 6.12[98]
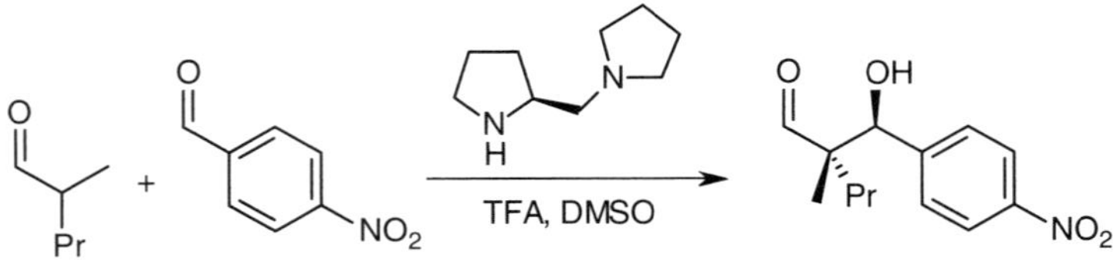	92	32	89 (anti) 66 (syn)	Reaction 6.13[99]

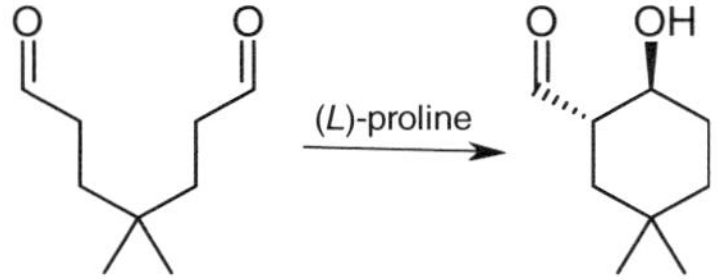 75	>95	97	Reaction 6.14[100]
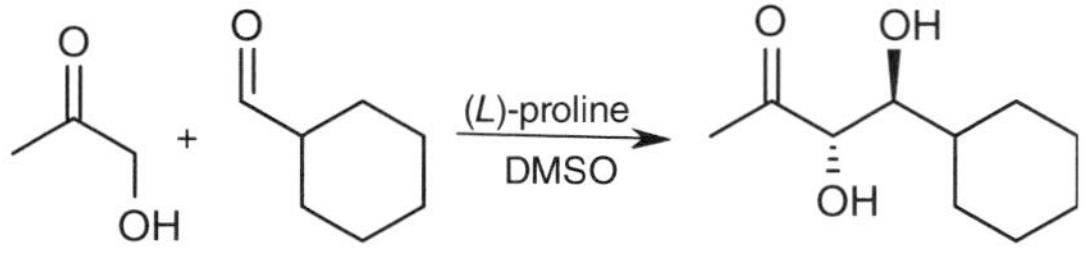 60	>95	>99	Reaction 6.15[101]
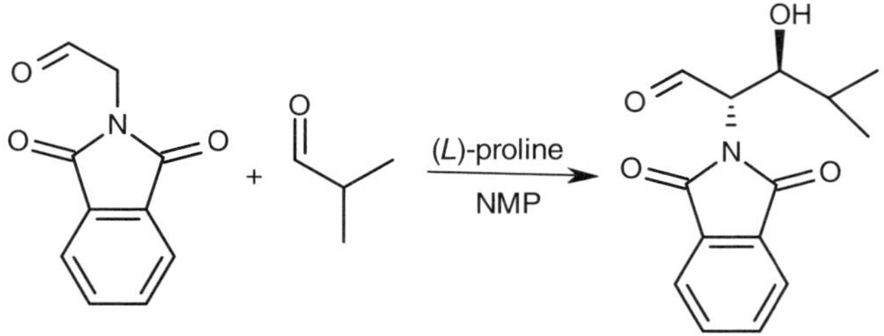 91	>99	>99	Reaction 6.16[95]

In all cases, the enantiomeric excess is outstanding, typically greater than 90 percent. Reactions 6.9 and 6.10 demonstrate asymmetric aldol reactions involving only aldehydes, and Reactions 6.11 and 6.12 involve ketones adding to aldehydes. Reaction 6.20 demonstrates an aldol that establishes a quaternary α-carbon center. The intramolecular aldol can also be catalyzed (Reaction 6.14). The last two entries, Reactions 6.15 and 6.16, demonstrate the ability to create chiral α,β-dihydroxyketones and α,β-aminohydroxyketones, respectively. These latter compounds can be precursors to chiral α-amino acids.[95] Proline catalysis is fast becoming a standard synthetic stratagem, as witnessed by its critical use in the recent synthesis of the natural products brasoside and littoralizone.[96]

Theoretical chemistry has played a major role in elucidating the mechanism whereby proline and other amines catalyze the aldol reaction. Houk has reviewed the contribution of computational chemistry towards understanding the actions of organcatalysts in the aldol and other reaction.[102] Before detailing the calculations and their implications, we describe some of the key experiments that provided data relevant to the mechanism.

List[103] and Barbas[104] investigated the effect of different potential catalysts on the reaction of acetone with *p*-nitrobenzaldehyde (Table 6.11). The important conclusions drawn from this work are that primary and acyclic secondary amino acids do not catalyze the reaction. Cyclic secondary amino acids do catalyze the aldol reaction, the best being proline. Converting proline into either a tertiary amine (*N*-methylproline) or an amide destroys its catalytic behavior. It is clear that the catalyst must provide both basic and acidic sites.

Experiments conducted in the mid-1980s by Agami indicated a small nonlinear effect in the asymmetric catalysis in the Hajos–Parrish–Wiechert–Eder–Sauer reaction (Scheme 6.7). Agami proposed that two proline molecules were involved in the catalysis: the first proline forms an enamine with the side chain ketone and the second proline molecule facilitates a proton transfer. Hajos and Parrish[83] reported that the proline-catalyzed cyclization shown in Scheme 6.7 did not incorporate ^{18}O when run in the presence of labeled water. While both of these results have since been discredited—the catalysis is first order in catalyst and ^{18}O is incorporated into the diketone product[105,106]—these erroneous results did stand at the time of the first computational approaches toward understanding amine-catalyzed aldol reactions.

Prior to 2001, when the first serious computational approaches to the problem appeared in print, four mechanistic proposals had been offered for understanding the Hajos–Parrish–Wiechert–Eder–Sauer reaction (Scheme 6.8). Hajos and Parrish[83] proposed the first two mechanisms: Mechanisms **A** and **B**. Mechanism **A** is a nucleophilic substitution reaction where the terminal enol attacks the carbinolamine center, displacing proline. The other three mechanisms start from an enamine intermediate. Mechanism **B** invokes an enaminium intermediate, which undergoes C–C formation with proton transfer from the aminium group. Mechanism **C**, proposed by Agami[107,108] to account for the nonlinear proline result, has the proton transfer assisted by the second proline molecule. Lastly, Mechanism **D**, proffered by Jung,[109] proposed that the proton transfer that accompanies C–C bond formation is facilitated by the carboxylic acid group of proline.

TABLE 6.11 Effect of Catalyst on the Direct Asymmetric Aldol Reaction[a]

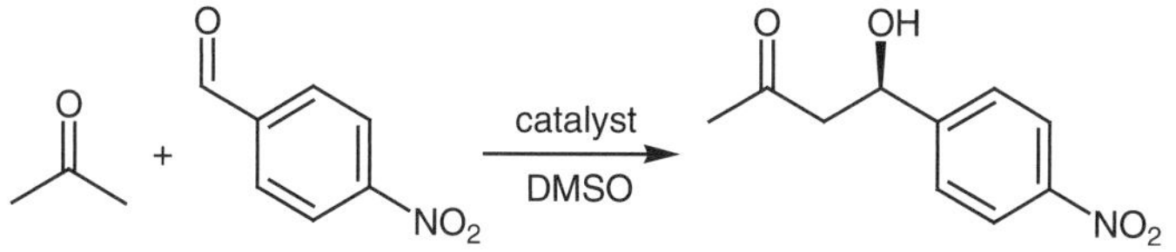

Catalyst	Yield, %	ee, %	Catalyst	Yield, %	ee, %
(L)-His, (L)-Val, (L)-Tyr, (L)-Phe	<10		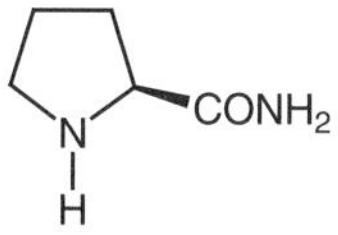 pyrrolidine-CONH$_2$	<10	
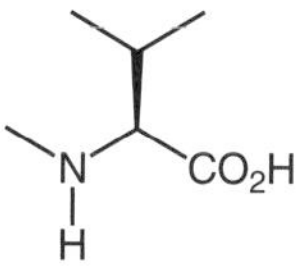 N-iPr Val-CO$_2$H	<10		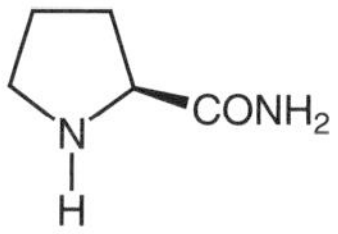 pyrrolidine-CONH$_2$	26	61
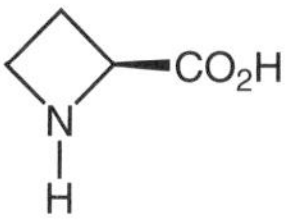 azetidine-CO$_2$H	55	40	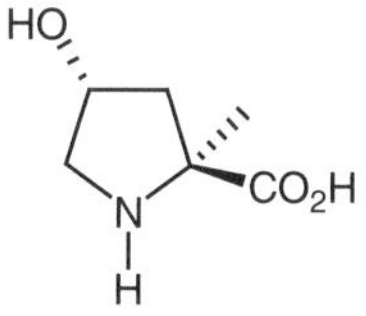 4-HO-pyrrolidine-CO$_2$H	85	78
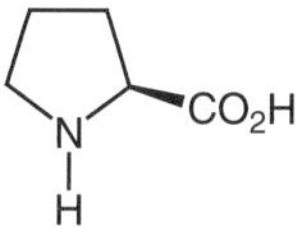 pyrrolidine-CO$_2$H	68	76	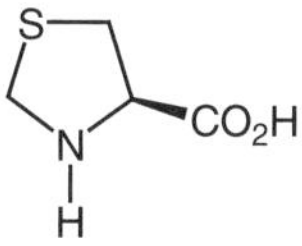 thiazolidine-CO$_2$H	67	73
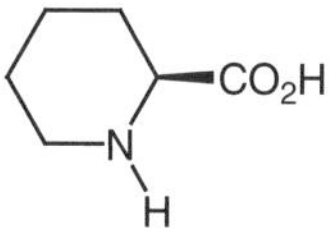 piperidine-CO$_2$H	<10		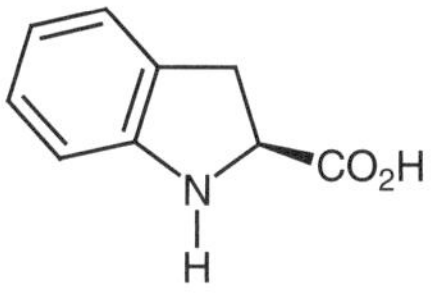 indoline-CO$_2$H	<10	
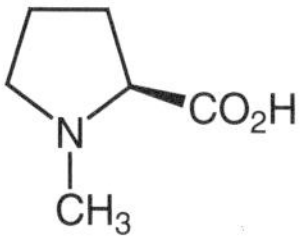 N-CH$_3$-pyrrolidine-CO$_2$H	<10				

[a]Refs. 103 and 104.

6.3.1 Mechanism of Amine-Catalyzed Intermolecular Aldol Reactions

We begin by examining the computational results for the intermolecular amine-catalyzed aldol reaction. The most likely mechanisms posit an enamine intermediate that then attacks an aldehyde to produce the aldol product. Boyd and coworkers[110] examined this entire reaction pathway for the reaction of acetone and

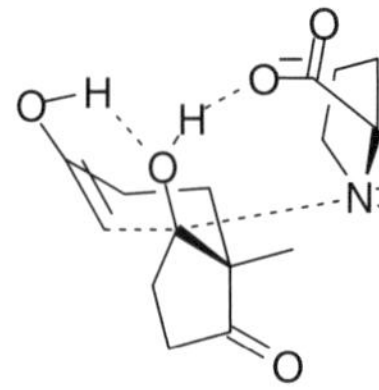

A
Nucelophilic substitution TS

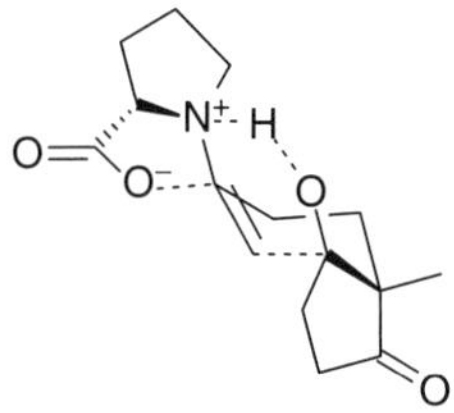

B
Enaminium-catalyzed TS

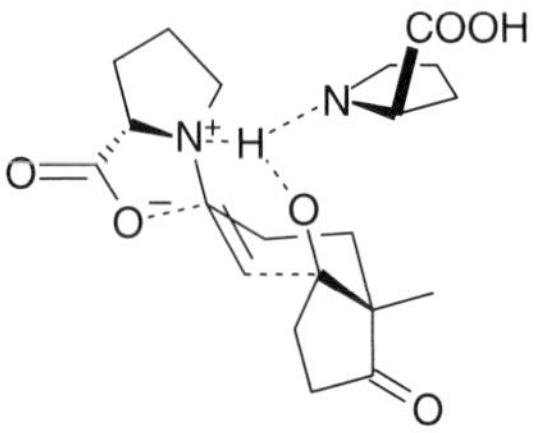

C
Dual proline-catalyzed TS

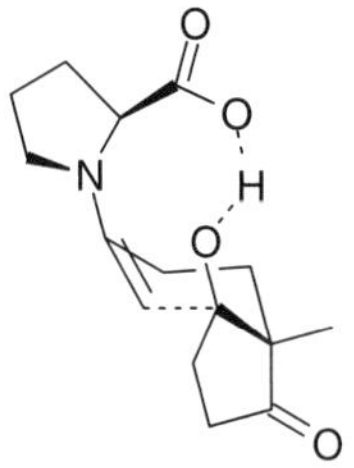

D
Carboxylic-acid-catalyzed TS

Scheme 6.8

acetaldehyde with (*S*)-proline as the catalyst. Their reaction profile, computed at B3LYP/6-311+G(2df,p)//B3LYP/6-31G(d,p) with zero-point vibrational energy, is shown in Figure 6.20. They also modeled the effect of solvent (DMSO) with a simple Onsager model; these energies are shown in italics below the gas-phase energies.

For the gas-phase reaction, surprisingly, the step where the new C–C is formed is *not* rate limiting. The barrier for this step (**49** → **50**) is 13.7 kcal mol^{-1}. (It should be noted that proton transfer from the carboxylic acid group accompanies the C–C bond formation in this step.) Rather, it is the initial addition of proline to acetone (**45** → **46**) that has the highest barrier: 41.0 kcal mol^{-1}. Boyd noted that many charged species exist along this pathway and that a polar solvent, such as DMSO, might significantly perturb the energy surface. Keeping in mind that the Onsager model rather poorly accounts for solvation, the relative barriers of the key steps are radically altered by solvent. What was the most difficult step, **45** → **46**, has a barrier in solution of 9.8 kcal mol^{-1}. Formation of the enamine, **47** → **48**, and the C–C addition step, **49** → **50**, have larger barriers, 12.0 and 11.1 kcal mol^{-1}, respectively. There are two important conclusions to be carried forward: first, that solvation is likely to be of critical importance and should be included in the computations, and second, that focusing on the C–C forming step is appropriate. This C–C forming step is critical in accounting for the stereochemical outcome of the aldol reaction.

Houk and Bahmanyar[111] began a series of computational studies of organocatalyzed aldol reactions with an examination of simple intermolecular aldol reactions catalyzed by small amines. They first looked at the methylamine-catalyzed aldol

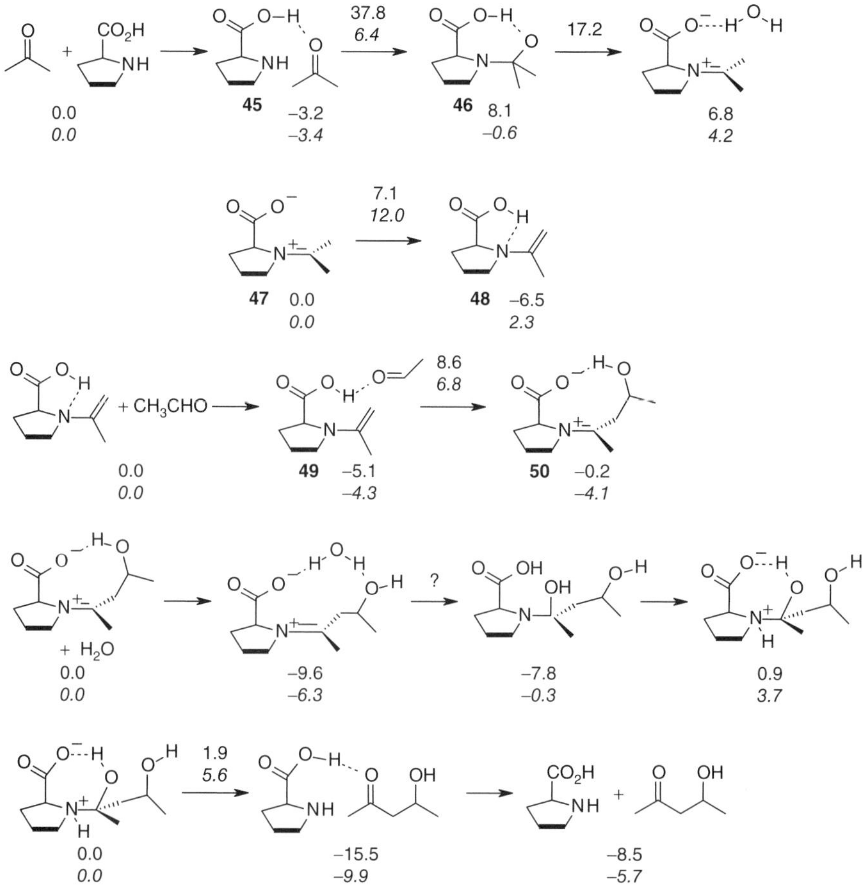

Figure 6.20 B3LYP/6-311+G(2df,p)//B3LYP/6-31G(d,p) energies (in kcal mol^{-1}) for the reaction of acetone and acetaldehyde with (S)-proline as catalysts. Energies in DMSO (Onsager model) are in italics below the gas-phase energies.

reaction of acetaldehyde. This can occur with (Reaction 6.17) or without (Reaction 6.18) concomitant proton transfer. At B3LYP/6-31G*, Reaction 6.17 is exothermic: $\Delta E = -14$ kcal mol^{-1} in the gas phase and -16 kcal mol^{-1} in solution phase (computed for water using the conductor-like screening model (COSMO) continuum model). Reaction 6.18 is much less exothermic: -3.6 and -9.0 kcal mol^{-1} in the gas and solution phases, respectively. Even though the zwitterion is formed in Reaction 6.18, which should be stabilized in a polar medium, Reaction 6.17 is thermodynamically favorable.

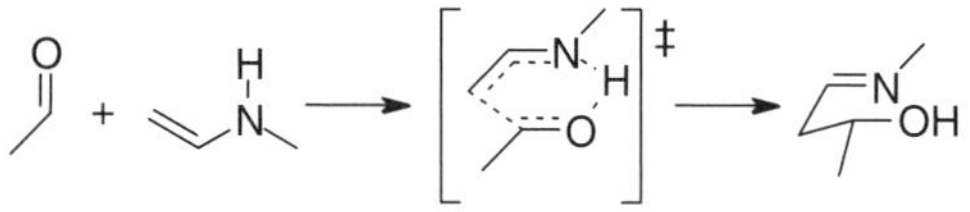

(Reaction 6.17)

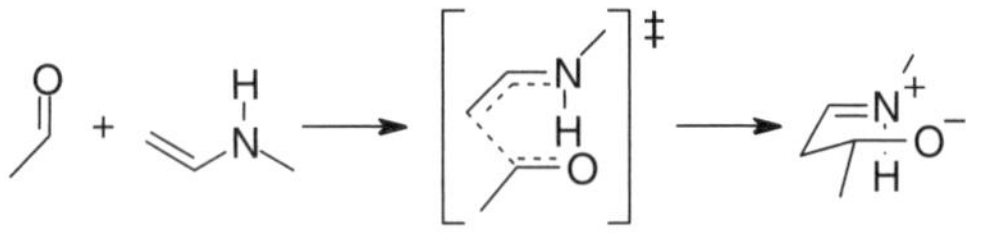

(Reaction 6.18)

More importantly, the barriers for Reaction 6.17 are much lower than that for Reaction 6.18. The two diastereomeric TSs for Reaction 6.17 (**51a** and **51b**) are 14 (gas phase) and 25 (solution) kcal mol^{-1} above reactants. The barriers for Reaction 6.18 are 33.2 and 35.7 kcal mol^{-1} in the gas phase. These are greatly reduced, as expected for increasingly polar structures, in solution to 16.0 and 17.5 kcal mol^{-1}, respectively, but Reaction 6.17 nevertheless has the lower barriers. This aldol reaction prefers to form the new C–C bond with a simultaneous proton transfer, avoiding zwitterionic intermediates.

The structures of **51a** and **51b** are drawn in Figure 6.21. The geometries are half-chairs with s-trans arrangement of the enamine. The c-cis alternative does not allow the nitrogen lone pair to conjugate with the enamine π-bond, leading to a barrier of 30 kcal mol^{-1}.

To examine the syn/anti selectivity, Houk examined the aldol reaction of acetaldehyde and propanal with methanamine as catalyst, Reaction 6.19. The E or Z enamine can react to give syn or anti product. They located four TSs **52a–d**, shown in Figure 6.22. Again, these TSs are in a half-chair conformation with internal proton transfer. The TSs involving the E isomer are lower than those with the Z isomer. The E isomer prefers to give the *anti* isomer (**52a** is 0.7 kcal mol^{-1} below **52b**), while the Z isomer favors the syn product (**52c** is 1.4 kcal mol^{-1} below **52d**). These results are consistent with experiments that show preference for the anti product.

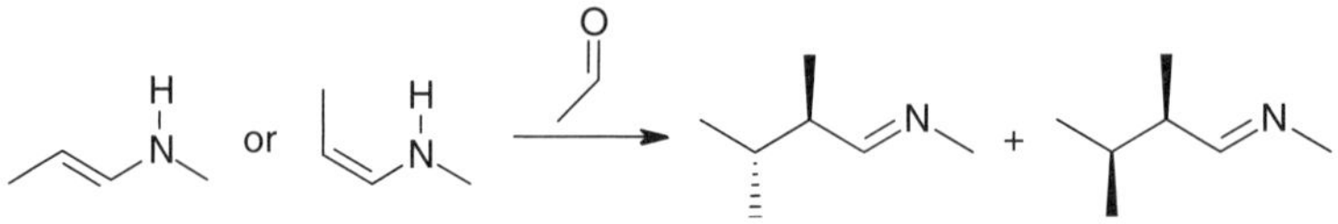

(Reaction 6.19)

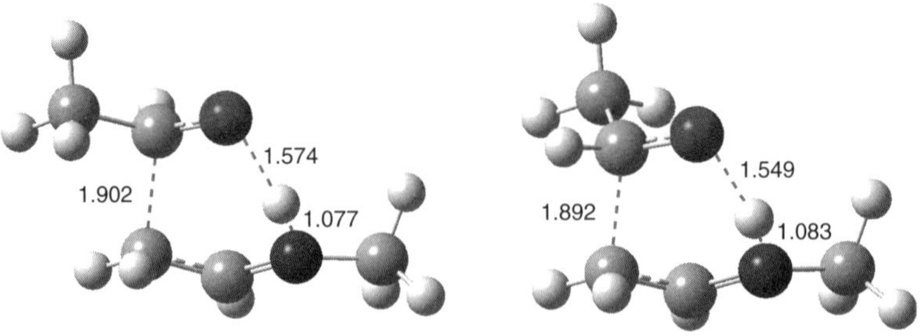

Figure 6.21 B3LYP/6-31G* optimized transition states for Reaction 6.17.[111]

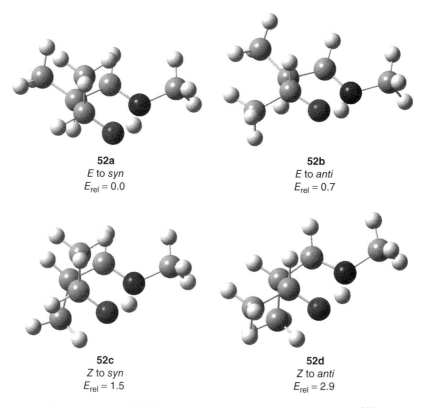

Figure 6.22 B3LYP/6-31G* transition states for Reaction 6.29.[111]

Houk[112] next examined the aldol reaction of cyclohexanone with benzaldehyde (Reaction 6.20) and isobutyraldehyde (Reaction 6.21) with (S)-proline as the catalyst. Four diastereomeric TSs starting from the enamine formed from cyclohexanone and proline were optimized at B3LYP/6-31G* for each reaction. These transitions states, **53** and **55**, are shown in Figure 6.23. In all of these TSs, proton transfer from the carboxylic acid group to the carbonyl oxygen accompanies the formation of the new C–C bond, creating a carboxylate and alcohol product.

(Reaction 6.20): R=Ph
(Reaction 6.21): R=i-Pr

The TSs can be readily understood by viewing them via the Newman projection down the forming C–C bond. The TSs involving the *anti*-enamine (**53a**, **53b**, **55a**, **55b**) are lower in energy than those with the *syn*-enamine. The aldehyde substituent preferentially occupies the site anti to the enamine carbon. Therefore, the lowest

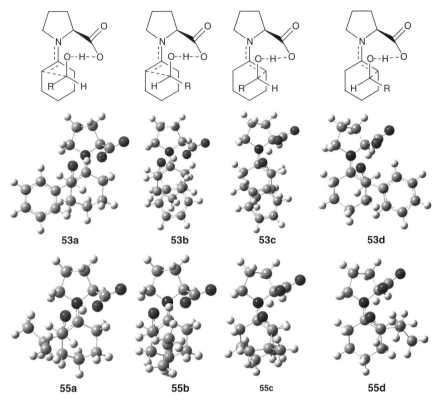

Figure 6.23 B3LYP/6-31G* transition states for Reactions 6.20 and 6.21. Adapted from Bahmanyar, S.; Houk, K. N.; Martin, H. J.; List, B. *J. Am. Chem. Soc.* **2003**, *125*, 2475–2479. Copyright 2003 American Chemical Society.

TS is **53a** and **55a** for the two reactions. These TSs lead to the anti products **54a** and **56a**, respectively, which should then be the major products produced. Table 6.12 presents the computed and experimental product ratios for Reactions 6.20 and 6.21. The agreement is outstanding—computation correctly identifies the major product and that Reaction 6.21 is more selective than Reaction 6.20.

The preference for the amine-catalyzed aldol reaction to go through a TS having the features of **53a** or **55a** is now called the *Houk–List model*. This type of TS has three major characteristics: (1) proton transfer from the carboxylic acid to the incipient alcohol concomitant with the formation of the C–C bond (TS **D** of Scheme 6.8); (2) the enamine is in the anti orientation; and (3) the aldehyde substituent is anti to the enamine carbon.

Houk[113] has further explored conformational factors that may play a role in determining selectivity. The proline ring can be puckered in two orientations, up **57** or down **58**. Consideration of the four TSs with the up orientation and the four with the down orientation for the aldol reaction of acetone with *p*-nitrobenzaldehyde is necessary. The up TSs predict an ee that is too low, while the down TSs predict an

TABLE 6.12 Computed and Experimental Product Ratios for Reactions 6.20 and 6.21[a]

	54: R = Ph		56: R = i-Pr	
	Computed	Experimental	Computed	Experimental
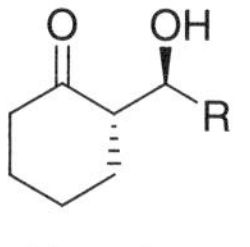 54a, 56a	50–80% ($\Delta H_{298} = 0.0$)	45–47% ($\Delta G_{exp} = 0.0$)	>99% ($\Delta H_{298} = 0.0$)	97–100% ($\Delta G_{exp} = 0.0$)
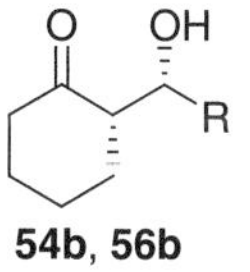 54b, 56b	20–50% ($\Delta H_{298} = 0.4$)	43–45% ($\Delta G_{exp} = 0.03 \pm 0.05$)	<1% ($\Delta H_{298} = 6.7$)	<1% ($\Delta G_{exp} = 4.1 \pm 0.03$)
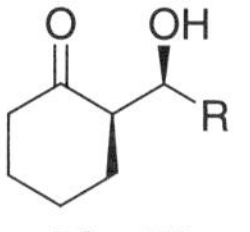 54c, 56c	<1% ($\Delta H_{298} = 3.6$)	5–7% ($\Delta G_{exp} = 1.2 \pm 0.05$)	<1% ($\Delta H_{298} = 7.8$)	<1% ($\Delta G_{exp} = 4.1 \pm 0.03$)
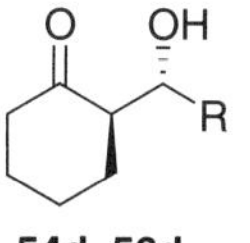 54d, 56d	1–4% ($\Delta H_{298} = 2.3$)	3–5% ($\Delta G_{exp} = 1.4 \pm 0.05$)	<1% ($\Delta H_{298} = 4.6$)	0–3% ($\Delta G_{exp} = 2.5 \pm 0.03$)

[a]Ref. 112.

ee that is too high; including all eight TSs results in a predicted ee quite close to the experimental value. Similar results are obtained for structurally related catalysts.

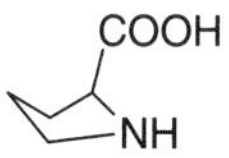

Primary amino acids also catalyze the aldol reaction. For example, the aldol between cyclohexanone and 4-nitrobenzaldehyde is catalyzed by alanine, valine, and isoleucine with enantiomeric excess greater than 99 percent.[114] Himo and Córdova[115] computationally examined a model of this reaction, namely, Reaction 6.20 with alanine as the catalyst. At B3LYP/6-311+G(2d,p), they located 32 different TSs. The lowest energy TS leads to the (S,R) product, and the next lowest energy TS, lying 3.2 kcal mol^{-1} higher in energy, leads to the (R,S) product. These computations are in accord with the experiment, including the ee. The TS geometry also agrees with the Houk–List model.

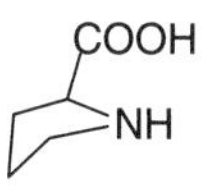

Sunoj[116] also predicted the enantioselectivity offered by a series of bicyclic proline analogs in the aldol reaction of acetone with *p*-nitrobenzaldehyde. The two best performing catalysts are **59** and **60**; both are predicted to give an ee greater than 90 percent, which exceeds the selectivity afforded by proline itself. The TS geometries all display the Houk–List characteristics.

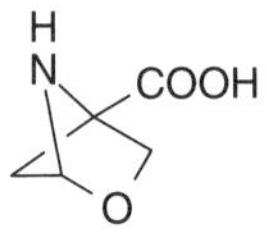

59 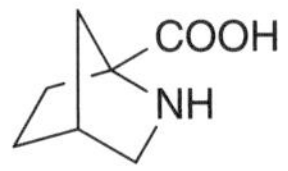   **60**

An interesting twist in the amino-acid-catalyzed aldol reaction is the use of histidine as catalyst. The key element of the Houk–List model is the proton transfer that accompanies the C–C bond formation. Histidine in aqueous solution can supply two sites for the proton transfer: the carboxylic acid group (analogous to proline) or the imidazole. Experimental studies of the histidine-catalyzed aldol reaction demonstrates appreciable selectivity as shown in Reactions 6.22 and 6.23.[117]

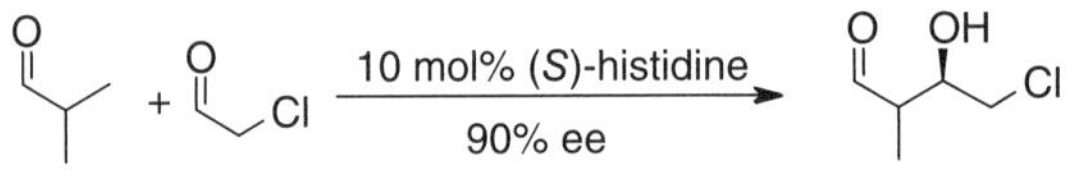

(Reaction 6.22)

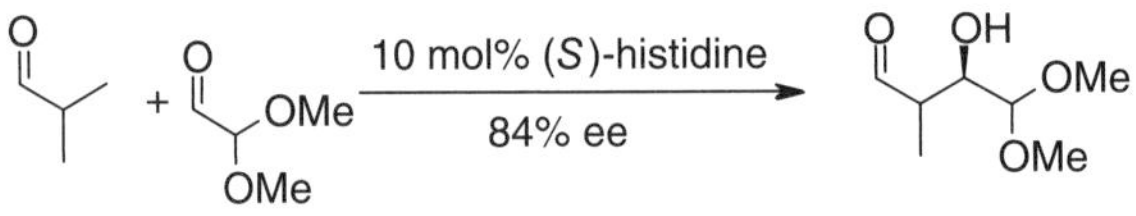

(Reaction 6.23)

Houk[117] employed SMD(water)/M06-2x/6-31+G(d,p) computations to model these histidine-catalyzed reactions. For the reaction of isobutyraldehyde with formaldehye, eight TSs were located, four with the carboxylic acid group as the proton donor and four with the imidazole as the proton donor. The lowest energy TS for each case has the Houk–List geometry with the proton transfer occurring with the C–C formation. In this case, the two lowest TSs are of equal energy.

In the computation of Reaction 6.22, the lowest energy TS involves the imidazole as the proton donor and it leads to the major product. The lowest energy TS that leads to the minor product involves the carboxylic acid as the donor. The predicted ee is 75 percent, in fine agreement with the experiment. This study should spur further exploration of tuning the stereoselectivity through catalysts with multiple binding opportunities.

Seebach[118] has proposed the participation of an oxazolidinone intermediate along the reaction pathway of the proline-catalyzed aldol reaction, driven by the NMR observation of this species in the reaction mixture. Sharma and Sunoj[119] employed DFT and MP2 computations to assess whether the oxazolidinone is actually on the reaction pathway. They examined the proline-catalyzed self-aldol reaction of propanal, for which MacMillan has experimentally shown an ee of

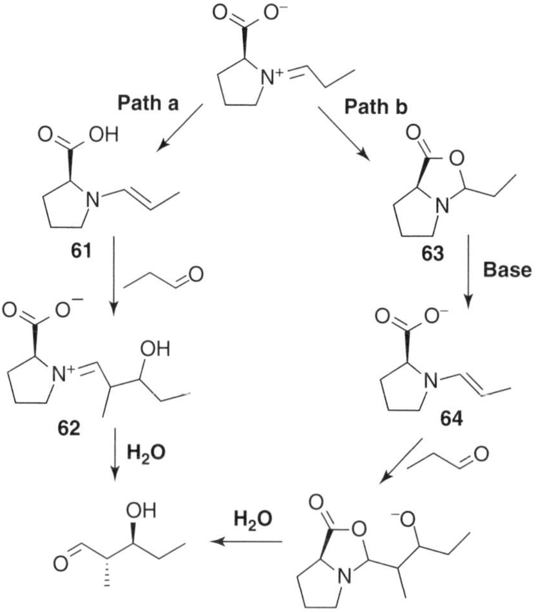

Scheme 6.9

more than 99 percent and a 4:1 preference of the anti product over the syn product.[97] Two distinct reaction paths were investigated (Scheme 6.9). The first (path a) follows the Houk–List model, where the key step is the addition of propanal to **61** to give the adduct **62**. The alternative is path b, where the oxazolidinone **63** is formed and the key step is the addition of propanal to **64**.

For path a, gas-phase B3LYP/6-31+G** predicts an *anti* : *syn* ratio of 5.4 : 1, but PCM computations modeling acetonitrile as solvent reduces the ratio to 2 : 1. MP2 computations predict an even smaller ratio of 1.2 : 1. Nonetheless, all methods do predict the correct stereoproduct. The oxazolidinone **63** is found to be lower in energy than the enamine **61**, and the barrier for reversion is fairly high (13 kcal mol^{-1}). Thus, it is not unreasonable that oxazolidinones are observed in the reaction mixture. However, path b must traverse the barrier **63** → **64**, and the free energy of this barrier is predicted to be 12 kcal mol^{-1} higher than the key barrier on path a (**61** → **62**). Furthermore, the computations for path b predict that syn stereochemistry would dominate, in direct conflict with experiment. Thus, the oxazolidinone pathway (path b) can be discounted.

6.3.2 Mechanism of Proline-Catalyzed Intramolecular Aldol Reactions

We next turn our attention to the intramolecular aldol reaction, typified by the Hajos–Parrish–Wiechert–Eder–Sauer reaction (Scheme 6.7). Houk[120] first examined the intramolecular proline-catalyzed aldol reaction of 4-methyl-2,6-hexadione

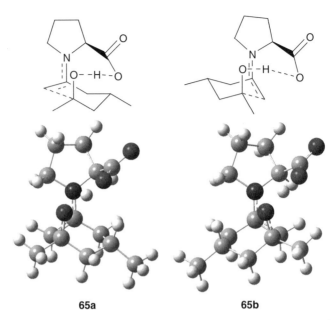

Figure 6.24 B3LYP/6-31G* optimized structures of the transition states for Reaction 6.24.[120]

(Reaction 6.24) by locating the two diastereomeric TSs **65a** and **65b** at B3LYP/6-31G*. These two geometries are drawn in Figure 6.24. **65a** lies 1.0 kcal mol^{-1} below **65b**, attributed to two factors. First, the forming iminium bond is more planar in the former structure. Second, the distance between the partial positively charged nitrogen and the forming alkoxide is shorter in **65a** than in **65b**, providing the former with greater electrostatic stabilization. Both TSs have the two characteristics noted from the TSs of the intermolecular aldol reactions, namely, the half-chair conformation and the proton transfer. **65a** also has the other characteristic, that of an anti arrangement of the double bond relative to the carboxylic group.

The lower barrier through **65a** leads to the isomeric product that is found in excess in the experiment, with reasonable agreement with the experimental ee of 42 percent.[85] The two diastereomeric TSs for Reaction 6.25 differ in energy by only 0.1 kcal mol^{-1}. No enantiomeric excess was observed in this experiment.[85]

(Reaction 6.24): R=Me
(Reaction 6.25): R=i-Bu

Clemente and Houk[121] examined the mechanistic possibilities for the Hajos–Parrish–Wiechert–Eder–Sauer reaction (Scheme 6.7). They determined the energies of the different TSs and intermediates at B3LYP/6-31+G(d,p)//B3LYP/ 6-31G* and added in the free energy of solvation in DMSO computed with PCM at HF/6-31+G(d,p). The lowest energy TS is for the carboxylic-acid-catalyzed enamine route (Scheme 6.8 D). This barrier is 10.7 kcal mol^{-1} smaller than that for the barrier without proline acting as a catalyst. The transition state for the enaminium-catalyzed route (Scheme 6.8 B) is 29.0 kcal mol^{-1} higher than the TS for Scheme 6.8 D. All attempts to locate a TS for the nucleophilic substitution route (Scheme 6.8 **A**) failed. However, the carbinolamine intermediate that proceeds the TS on path **A** lies 12.7 kcal mol^{-1} above the TS for Scheme 6.8 **D**. Since the pathway involving two proline molecules is inconsistent with new experiments that indicate a reaction that is first order in proline, they did not pursue **C**. Thus, computation indicates that the Hajos–Parrish–Wiechert–Eder–Sauer reaction proceeds by the carboxylic-acid-catalyzed enamine mechanism **D**, which is consistent with all of the computations for intermolecular proline-catalyzed aldol examples.

The two TSs of Reaction 6.26, computed at B3LYP/6-31G*, are shown in Figure 6.25.[120] Both TSs conform to mechanism **D**—proton transfer accompanying the C–C formation within a half-chair conformation. The barrier for Reaction 6.26 passing through TS **66a** to product **67a** is only 9.1 kcal mol^{-1}. It is 3.4 kcal mol^{-1} (3.1 kcal mol^{-1} in DMSO using CPCM) lower in energy than the barrier through **66b**. The enamine double bond is anti to the carboxylic acid in **66a**, which

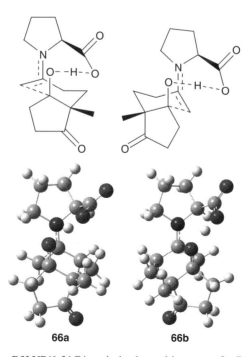

Figure 6.25 B3LYP/6-31G* optimized transition states for Reaction 6.26.

we have seen is preferable to the syn arrangement. This preference arises from a much more planar geometry about the forming iminium bond in **66a** than in **66b**. Again, these computations are consistent with experimental product distributions, where the ee is 93 percent after hydrolysis to give the diketones.

(Reaction 6.26)

Recent experiments and computations, however, suggest that the rate-limiting step of the Hajos–Parrish–Eder–Sauer–Wiechert reaction is not the step forming the new C–C bond. Meyer and Houk[122] determined the $^{12}C/^{13}C$ kinetic isotope effects (KIEs) for the parent reaction. The only KIE that differs from unity is at the acyclic carbonyl. B3LYP/6-31+G(d,p) computations were then performed to obtain the theoretical KIEs for three potential rate-limiting steps. The predicted KIEs for the step where the carbinolamine is formed **67** and the step for the iminium formation **68** are both in agreement with the experiments. However, the computed KIEs for the C–C forming step are not in accord with the experiment. Although a definitive statement about what is the rate-limiting step cannot be decided, it appears that the rate-limiting step occurs prior to the C–C forming step.

Houk[123] has also computationally examined a number of proline derivatives as catalysts for the Hajos–Parrish–Eder–Sauer–Wiechert reaction. The enantioselectivity resides largely in the ability to adopt a planar enamine. Proton transfer from the sulfur of **69** or **70** requires a longer distance because sulfur is larger than oxygen. This results in large nonplanar distortion of the enamine in the *syn* TS, increasing the enantioselectivity. The catalyst **71** actually favors the syn approach, leading to the opposite enantioselectivity than when proline is used. This results from the more flexible ammonium side chain being able to accommodate proton transfer in the syn conformer with little nonplanar distortion about the enamine.

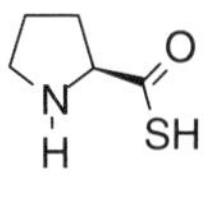

69 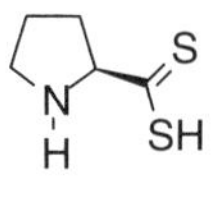 **70** 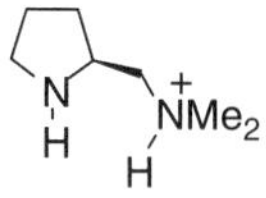 **71**

6.3.3 Comparison with the Mannich Reaction

The Mannich reaction is a close relative of the aldol, whereby an imine replaces the aldehyde acceptor. Proline has been demonstrated to be an excellent catalyst of the Mannich reaction, inducing high enantiomeric excess, as shown in Table 6.13.[104,124] Pertinent to this discussion is that the stereochemical outcome of the proline-catalyzed Mannich reaction is *opposite* to that of the proline-catalyzed aldol reaction.

List[124] suggested that the proline-catalyzed Mannich reaction proceeds in close analogy to the proline-catalyzed aldol reaction. As detailed in Scheme 6.10, the ketone and proline combine to form an enamine. The aldehyde reacts with a primary amine (usually an aniline derivative) giving an imine. The enamine and imine then combine to produce, after hydrolysis, the Mannich product.

In order to explain the high enantiomeric excess, and one that is opposite to that of the aldol reaction, List[124] postulated that the two reactions proceed through similar TSs (Scheme 6.11). As detailed earlier, Houk's computations indicate that the TS is in a half-chair conformation with proton transfer synchronous with C–C bond formation. List suggested that the (*E*)-configuration would be the most stable form of the imine. The imine would then position itself in order to minimize the interactions between proline and its phenyl ring. This places the alkyl group of the aldehyde in the interior pseudoaxial position, while in the aldol TS, this group preferentially occupies the exterior pseudoequatorial position. This then accounts for the differing stereochemical outcomes for the two reactions.

Houk and Bahmnyar[125] located the TSs for the Mannich reaction of the enamine of acetone and proline with *N*-ethylidine-*N*-phenylamine (Reaction 6.27) at B3LYP/6-31G*. They looked only for the TSs that allowed for proton transfer from

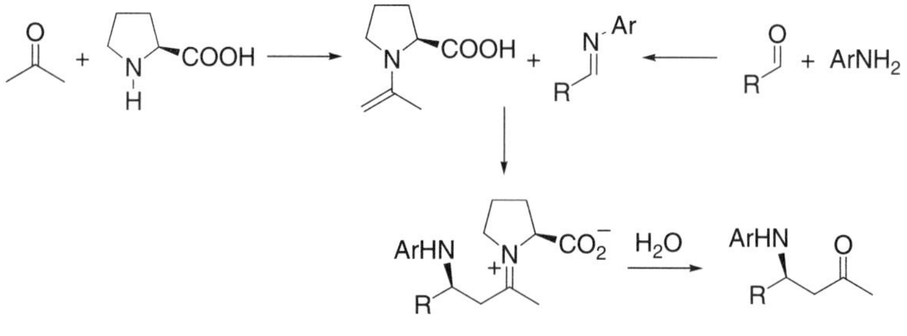

Scheme 6.10

TABLE 6.13 Comparison of the Proline-Catalyzed Mannich and Aldol Reactions

Reaction	Yield, %	ee, %	References
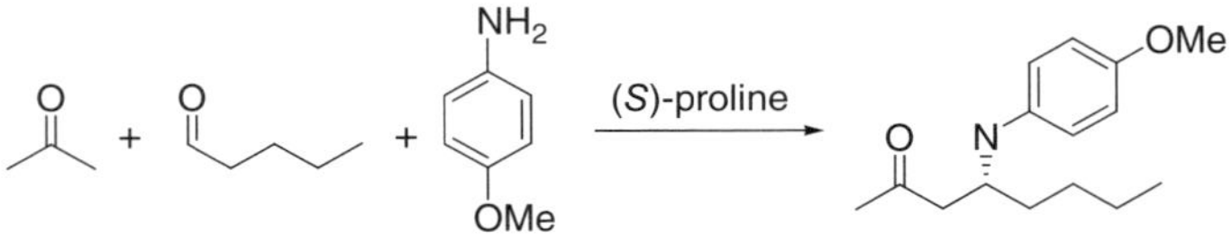 Mannich (pentanal + acetone + p-anisidine → β-amino ketone)	74	73	124
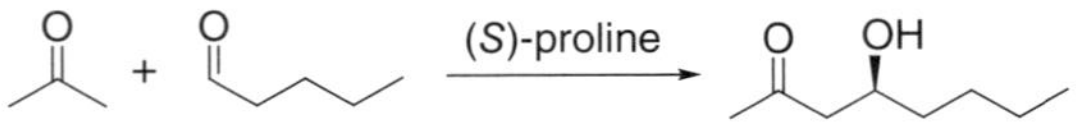 Aldol (pentanal + acetone → β-hydroxy ketone)	31	67	98
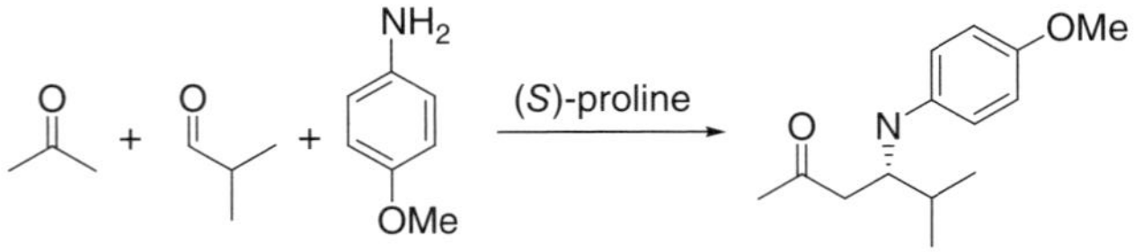 Mannich (isobutyraldehyde + acetone + p-anisidine → β-amino ketone)	56	70	124
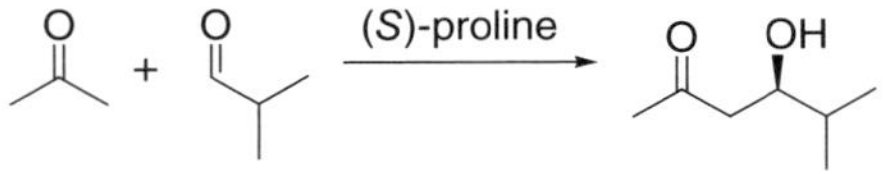 Aldol (isobutyraldehyde + acetone → β-hydroxy ketone)	97	96	103

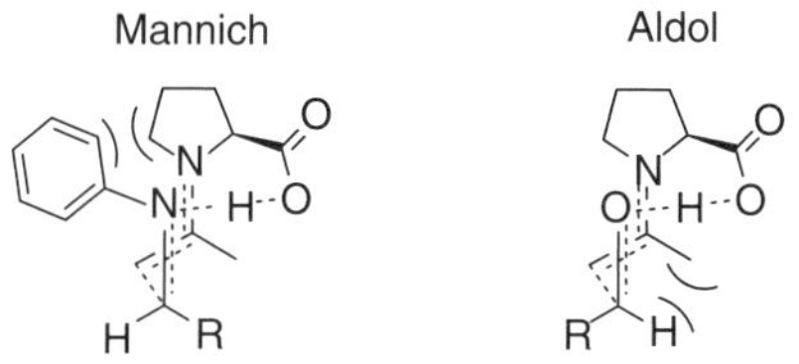

Scheme 6.11

the carboxylic acid group to the imine nitrogen with either the (E)- or (Z)-enamine; seven such TSs were found, shown in Figure 6.26.

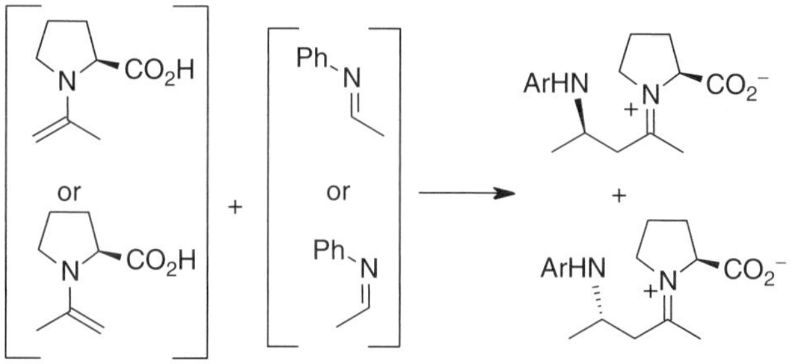

(Reaction 6.27)

Unlike the aldol TSs, in all of these Mannich TSs, the proton has completely transferred to the imine nitrogen. The lowest energy TS is **72a**. In close analogy with the aldol TS model, **72a** has the enamine double bond anti to the carboxylic acid group and the atoms involved in the bond changes occupy the half-chair conformation. The phenyl group is positioned away from the proline ring, all in accord with List's suggested TS model (Scheme 6.11). TS **72b** adopts the same general conformation as in **72a** except with the (Z)-imine; it lies 1.6 kcal mol^{-1} above **72a**. The two TSs where the enamine double bond is cis to the carboxylic acid group, **72c** and **72d**, lie 1.7 and 4.9 kcal mol^{-1} above **72a**, respectively. The last three TSs, **72e–g**, suffer from poor steric interactions, resulting in large distortion from planarity about the forming iminium bond.

The lowest energy TS **72a** does correspond with the experimentally observed major product, and the energy differences with the other TSs are consistent with the typically large ee found in experiment. These computations are consistent with List's mechanistic proposal for the proline-catalyzed Mannich reaction. Further confirmation was provided by a computational study of Reaction 6.28.

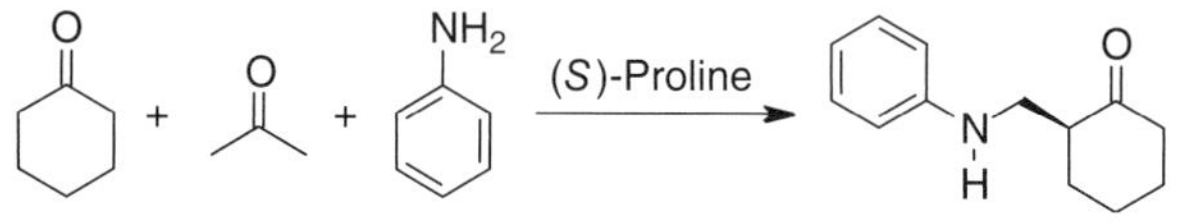

(Reaction 6.28)

Two important aspects of the mechanism of this reaction were probed in detail using MP2/6-31++G(d,p)//B3LYP/6-31++G(d,p) computations.[126] The first

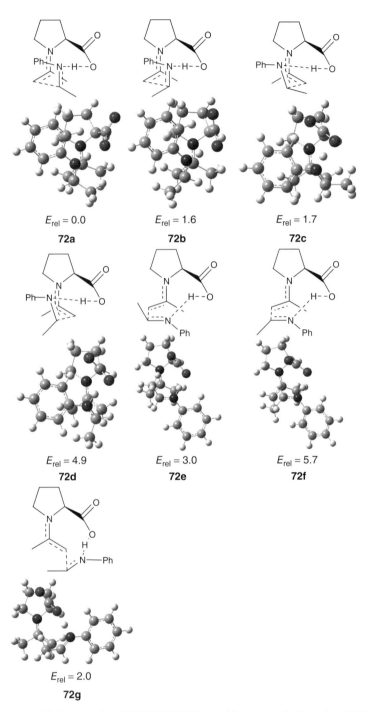

Figure 6.26 Optimized B3LYP/6-31G* transition states for Reaction 6.27.

question deals with the selectivity of the formation of the *syn*-enamine versus *anti*-enamine **73**. The *syn* isomer involves proton transfer from a nearby position; this path has a relatively low barrier of 10.2 kcal mol^{-1}, and the resulting enamine is lower in energy than the iminium precursor. Formation of the *anti*-enamine requires proton transfer from a remote site, one that is too far away to occur intramolecularly. Instead, Parasuk and Parasuk[126] incorporate a water molecule to act to shuttle the proton. Nonetheless, the barrier to form the *anti* isomer is 7.7 kcal mol^{-1} higher than the barrier to form the *syn* isomer, and the *anti*-enamine is higher in energy than the *syn*-enamine. This all suggests that the *syn* isomer would be preferred, but the computed syn to anti rotational barrier is only 4.2 kcal mol^{-1}. So while the *syn*-enamine is predominant, the *anti* isomer is accessible.

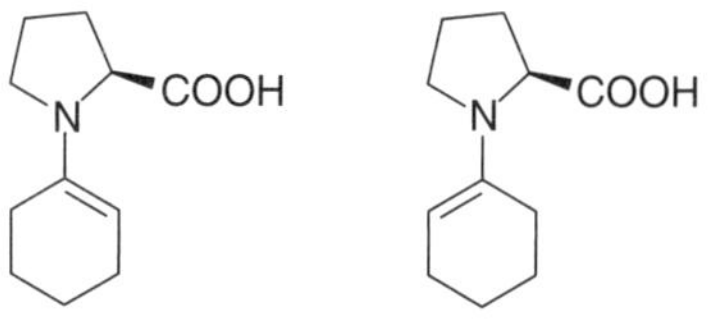

syn-**73**   anti-**73**

The second part of this study examined the barriers for the addition of the enamine to the imine, the step that determines the stereochemistry of the product. The pathway leading to the *S*-product involves the *anti*-enamine, while the pathway to the *R*-product uses the more stable *syn*-enamine. Nonetheless, the pathway leading to the *S*-product is both thermodynamically and kinetically preferred at MP2. The TS for the *S*-pathway is consistent with the List model (Scheme 6.11) and also corresponds with the experimental result.

Application of these principles for stereoinduction via the catalyzed Mannich reaction led Barbas and Houk to develop **74** as a catalyst to affect the *anti*-Mannich reaction, to complement the use of proline as a catalyst for the *cis*-Mannich reaction (Scheme 6.12).[127] In the larger context, Houk's extensive

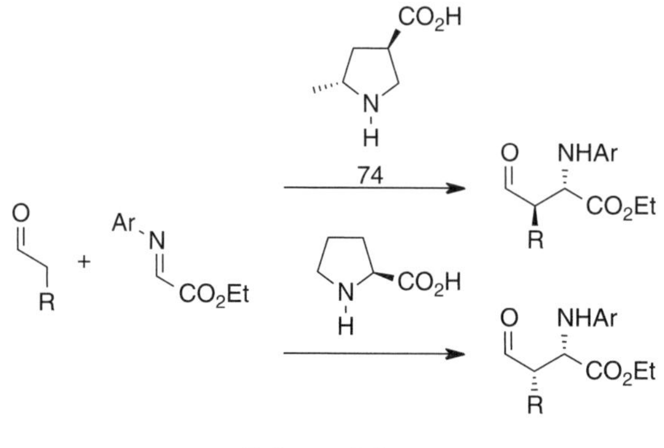

Scheme 6.12

studies of proline organocatalysis demonstrate the power of DFT and computational studies in general, in providing insight and understanding of complex real-world synthetic organic chemistry.

It is also worth noting that B3LYP and related functionals provide very poor reaction energetics for the aldol and Mannich reactions.[128,129] The more modern functionals, such as M06-2x and the long-range-corrected functionals CAM-B3LYP and LC-ωPBE, perform much better, with mean errors of less than 2 kcal mol^{-1}.

6.3.4 Catalysis of the Aldol Reaction in Water

The catalyzed aldol reaction in pure water offers a cautionary tale on overinterpretation of computationally derived mechanisms. The term *catalysis* may seem to be inappropriately applied to the role of water in the aldol reaction. The aldol reaction is in fact usually much slower in water than in organic solvents. Rather, as will be demonstrated, the catalytic role of water is to create an alternate pathway with a lower barrier than that for the noncatalytic, but aqueous, reaction.

In an effort to understand the role of nornicotine in catalyzing aqueous aldol reactions,[130] Noodleman and Janda examined the aldol reaction of acetaldehyde and acetone in an organic solvent (THF, tetrahydrofuran) and water.[131] The calculations were performed at B3LYP/6-311+G(2d,2p)//B3LYP/6-311(d,p) and were corrected for ZPVE at HF/3-21G and the effects of solvent using the COSMO model at B3LYP/6-311(d,p).

For the reaction in an organic solvent (THF), they obtained a single TS **75** (Figure 6.27). In this TS, the new C–C bond is formed in conjunction with the proton transfer from the enol to the aldehyde oxygen (Reaction 6.29). The activation energy for this reaction is 18.6 kcal mol^{-1}.

(Reaction 6.29)

For the aqueous reaction, they proposed an alternative reaction scheme, one that is stepwise and involves two water molecules (Reaction 6.30). In the first step, a water molecule transfers a hydrogen atom to acetaldehyde as its oxygen atom bonds to the enol carbon all the while forming the new C–C bond. The TS for this step (**76**) is 32.9 kcal mol^{-1} above the reactants. Proceeding forward from **76** gives the *gem*-diol **77**, which is 8.6 kcal mol^{-1} more stable than the reactants. The second chemical step involves the second water molecule assisting in the proton transfer from one hydroxyl group of the *gem*-diol to the other, generating the ketone product and two water molecules. The cyclic TS for this second step (**78**) is 18.6 kcal mol^{-1} above **77** and the overall reaction is exothermic, $\Delta E = -12.7$ kcal mol^{-1}. The structures of the two TSs (**76** and **78**) are drawn in Figure 6.27, along with a reaction energy diagram. Although the authors noted that the activation energy for Reaction

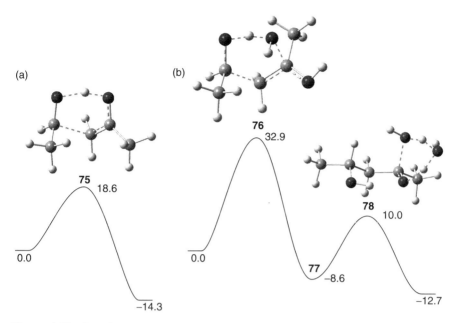

Figure 6.27 Reaction energy diagram and transition structures for Reactions 6.29 and 6.30. Energies (in kcal mol^{-1}) computed at B3LYP/6-311+G(2d,2p)//B3LYP/6-311(d,p) with solvation energy using COSMO.[131]

6.30 is quite large, they downplayed it by suggesting that the computational method might have an error of 2–3 kcal mol^{-1} for *each* critical point and that perhaps an alternative mechanism exists. They did note that the barrier for Reaction 6.30 in water is greater than that for Reaction 6.29 in THF, consistent with experiment.

(Reaction 6.30)

In the following year, Houk[132] provided three alternate mechanisms for the aldol reaction in water, computing the critical points at B3LYP/6-311++G(3d,3p)//B3LYP/6-31G(d) with the CPCM solvation model. Houk noted that the proposed mechanism, Reaction 6.30, begins not with acetone but with its enol, produced by some unmentioned mechanism. Since acetone enol lies 9 kcal mol^{-1} above acetone, the reaction barrier is actually 42 kcal mol^{-1}! Furthermore, the slow step involves three molecules, so in addition to its high activation enthalpy, there would be a large entropic barrier.

Houk's first alternative mechanism, Reaction 6.31, posits the direct proton transfer from acetone to acetaldehyde to form an ion pair. This process requires a great deal of energy, 59 kcal mol^{-1}, and even though the subsequent formation of the C–C bond and the ultimate aldol product occur without further barrier, this mechanism requires too much energy to be competitive.

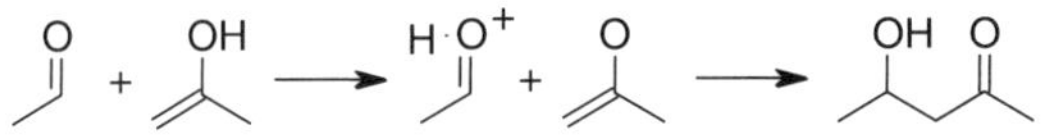

(Reaction 6.31)

The second mechanism, Reaction 6.32, proposes that water acts as a base to create acetone enolate anion. This enolate next adds to acetaldehyde through a TS that is 49.1 kcal mol^{-1} above reactants. The final aldol product is obtained by proton transfer. While this mechanism require less energy than the first of Houk's mechanisms, it too is unlikely to be competitive, given its large barrier for formation of the acetone enolate.

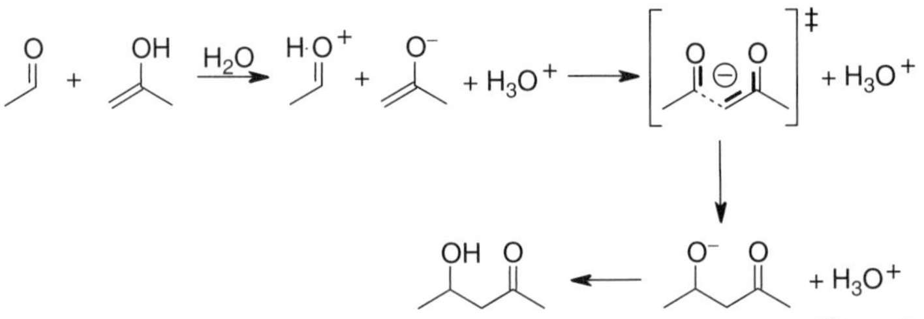

(Reaction 6.32)

The problem with Reaction 6.32 is that water is simply not a strong enough base to bring about the first proton transfer. Instead of using water as the base, in Houk's third mechanism, Reaction 6.33, autoionization of water produces hydroxide, a much stronger base. The ionization of water is endergonic ($\Delta G = 24.7$ kcal mol^{-1}). Deprotonation of acetone by hydroxide and reprotonation from hydronium gives acetone enol. This enol can then directly react with acetaldehyde to give the aldol product via the path proposed by Janda and Noodleman for the reaction in THF (Reaction 6.29). This three-step process has an activation barrier of only 28.1 kcal mol^{-1} (see Figure 6.28), a much more reasonable overall reaction barrier than any of the alternative mechanisms.

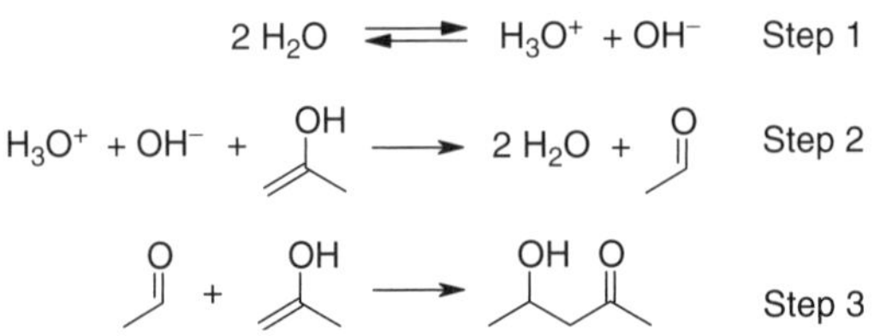

(Reaction 6.33)

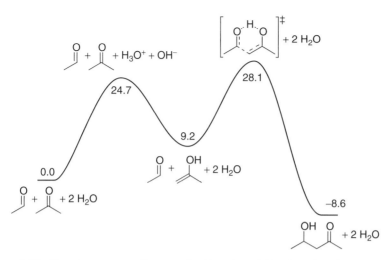

Figure 6.28 Reaction energy diagram for Reaction 6.33. Energies (in kcal mol^{-1}) computed at B3LYP/6-311++G(3d,3p)//B3LYP/6-311(d) with solvation energy using CPCM.[132]

6.3.5 Another Organocatalysis Example—The Claisen Rearrangement

A particularly active area of study of synthetic chemistry in this century has been in organocatalysis, and as we have discussed earlier, computational chemistry has been a strong partner in elucidating the mechanism and understanding the stereoinduction. To give a sense of the breadth of activity in this area, the chapter ends with a discussion of a very different reaction, the Claisen rearrangement.

Jacobsen[133,134] has pioneered the use of urea and thiourea derivatives as organocatalysts. The interesting difference in this approach compared to the use of proline and related compounds discussed previously is that the ureas and thioureas interact with substrate through hydrogen bonds, as opposed to a covalent attachment in the proline cases. In the first example that we discuss here, the guanidinium salt **79** provides significant catalysis of some simple Claisen rearrangements, such as Reactions 6.34 and 6.35.[135] Use of the chiral gaunidinium catalyst **80** provides outstanding enantioselectivity. (Jacobsen[136] recently employed a similar approach using a combination of computational and experimental to develop a thiourea catalyst for the hydroxyamination reaction.)

Uncatalyzed: 0% yield
20 mol% **79**: 72% yield   (Reaction 6.34)

Uncatalyzed: 12% yield
20 mol% **79**: 76% yield
20 mol% **80**: 86% yield
92% ee (S)
(Reaction 6.35)

The Claisen rearrangement of Reaction 6.35 was modeled at B3LYP/6-31G(d) using dimethylguanidinium as the catalyst. The TS and the hydrogen bonding between the catalyst and the two carbonyl carbons are shown in Figure 6.29. This association reduces the activation enthalpy from 25.7 kcal mol^{-1} in the uncatalyzed version to 21.1 kcal mol^{-1} with the guanidinium.

In a follow-up paper, Jacobsen[137] showed that the chiral guanidinium **80** provides catalytic power and enantioselectivity in the Claisen rearrangement systems such as Reaction 6.36. Here the reaction proceeded in 81 percent yield with an ee of 84 percent. The reaction is also diasteroselective: for Reaction 6.37, the diastereoselectivity increases from 11 : 1 without catalyst to 16 : 1 with **80**.

20 mol% **80**   (Reaction 6.36)

5 mol% **80**   (Reaction 6.37)

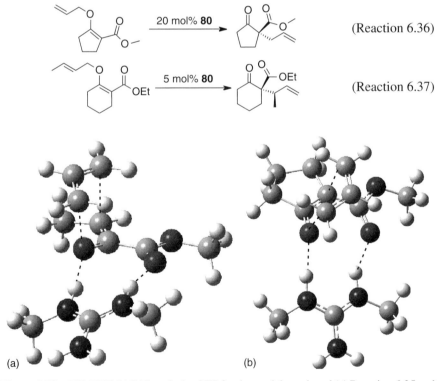

Figure 6.29 B3LYP/6-31G(d) optimized TS for the model-catalyzed (a) Reaction 6.35 and (b) Reaction 6.36.

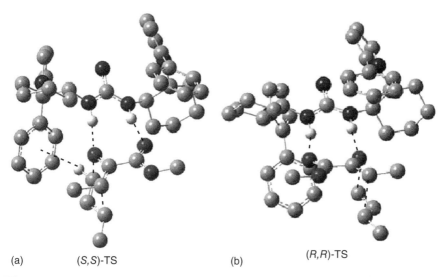

(a) (S,S)-TS (b) (R,R)-TS

Figure 6.30 B3LYP/6-31G(d) optimized geometries of the enantiomeric TSs for Reaction 6.38. All hydrogen atoms have been suppressed except those involved in hydrogen bonding or special electrostatic interactions.

B3LYP/6-31G(d,p) computations provide some insight. The uncatalyzed Reaction 6.36 has a predicted activation energy of 25.9 kcal mol^{-1}. Using N,N'-dimethylguanidnium as a model for the catalyst (and with no counteranion and no treatment of solvent), the complexation energy between the reactant and the catalyst is 27 kcal mol^{-1}. The complexed TS, shown in Figure 6.29, has two hydrogen bonds, with an activation energy of 20.6 kcal mol^{-1} above the complexed state. Thus, the catalyst lowers the barrier by about 5 kcal mol^{-1}.

To address the enantioselectivity, Jacobsen[138] computed the critical points for Reaction 6.38 with catalyst **80** (neglecting the counterion and solvent) at B3LYP/6-31G(d). Experimentally, the reaction goes with 85 percent yield and with an ee of 73 percent. The TSs leading to the two different enantiomeric products are shown in Figure 6.30. The TS leading to the (S,S) product is computed to be 3.0 kcal mol^{-1} lower in energy than the TS leading to the (R,R) product; this agrees with the experiment, although the computations overestimate the selectivity. Jacobsen argues that the (S,S) TS is preferred because of the favorable electrostatic interaction between the C–H bond of the allyl group and a phenyl ring (Shown in Figure 6.30a), an interaction that is absent in the other TS. Experiments and computations of phenyl-substituted analogs support the notion of this electrostatic interaction in the (S,S) TS.

(Reaction 6.38)

6.4 INTERVIEW: PROFESSOR KENDALL N. HOUK

Interviewed September 11, 2006

Professor Ken Houk is a Distinguished Professor in the Department of Chemistry at the University of California at Los Angeles. Houk began his academic career at the Louisiana State University, rising to become professor before moving to the University of Pittsburgh. After spending 6 years there, he relocated to the UCLA in 1986.

Professor Roy Olofson (now at the Penn State University) introduced Houk to research as an undergraduate and inspired him to pursue a career in chemistry. Olofson was a graduate from the Woodward group, and he encouraged Houk to continue at Harvard for his doctorate. Houk met with Professor R. B. Woodward three times to decide on a thesis project. The first time, Woodward proposed synthesizing dodecahedrane, a proposal Houk is, in retrospect, quite thankful he did not take on. With his third visit, they settled on seeking out examples of [6 + 4] cycloadditions, a reaction type that Woodward and Hoffmann predicted should occur. Woodward suggested looking at the reaction of cycloheptatriene with cyclopentadiene. The experimental work proved to be quite difficult, complicated by the multiple products coming from both [6 + 4] and undesired [4 + 2] cycloadditions. Houk returned to this reaction almost 40 years later, publishing a computational study of the reaction. Houk now claims "to have finally completed my thesis project!"

Woodward was clearly one of Houk's great mentors. "He was such a towering presence," Houk recollects. "It was an amazing experience to see him think through a problem. I vividly remember a series of three consecutive group meetings where he and Hoffman were trying to understand ketene cycloadditions." However, Houk maintains that "I could count on two hands the number of times I actually talked to him. I recall once when I thought I had some results, a post-doc, Bali Singh, suggested 'Don't try to see Woodward; just leave a note on the secretary's desk that you found something interesting.' So I did. The next morning when I came in, Woodward was at my desk pouring through my notebooks. On the whole it was a good situation for me. I could work with the post-docs and fellow students and I had a lot of independence." Houk's experiences in graduate school distinctly influence how he runs his own research group. He schedules biweekly individual meetings with each student. "It gives good students a chance to go running with their project, and I can be a cheerleader," he says. "And with other students, I try to make sure that they don't get lost."

The focus of Houk's early academic career was on pericyclic reactions. He began as an experimental chemist, but the relatively strong computational staff and facilities at LSU allowed Houk to first begin testing the ability of semiempirical methods in predicting regiochemistry. Studies of pericyclic reactions, especially from a computational perspective, span his entire academic career, continuing even to this day. It is this long-standing interest in pericyclic reactions that explains Houk's interest in Dolbier's report of the ring opening of perfluoro-*trans*-dimethylcyclobutene (compound **100** of Chapter 3). As Houk explains, "Dolbier generously interpreted the unusual stereoselectivity in terms of distortions of the π-system based on some of our earlier work. This especially piqued my interest, since I didn't think it was really correct. So we decided to do some calculations. What was unusual here is that we entered without really knowing what was going on. Usually, you really have it all figured out first and then do the calculations to confirm your notions. Nelson Rondan took on this project, at first doing HF/STO-3G and 3-21G computations. Our major realization was that the usual orbital correlation diagram, where one draws a straight line for the orbitals as you go from reactant to product, wasn't working. Once we had the basic explanation, the idea of specific orbital interactions that cause torquoselectivity, we realized that if we had a good enough acceptor, we would see inward rotation."

"About the time when I moved to UCLA, we decided to do the experimental study. So first we did a computational study on a series of acceptors and we identified the formyl group as particularly good. Then we synthesized the formyl-substituted cyclobutene and the reaction didn't work! It gave the *trans* product. But upon closer analysis, we realized that it might have rearranged in acid. So we repeated the experiments with the exclusion of acid and then it did give the *cis* product, which slowly isomerizes to the more stable *trans*. Computations guided us to see that our initial experiments were wrong. Furthermore, computation preceded experiment—that's always nice!"

Houk's involvement in proline-catalyzed asymmetric induction began in a similar way. "Occasionally something very interesting appears in the literature, especially so if it is amendable to elucidation by computational methods," Houk recalls. "We are especially interested in synthetic methods where the experimentalist is unclear of just how things work. This was the case with the proline work. We saw the work of Barbas and List on the aldol reaction, and we were aware of the Hajos–Parrish reaction. So we started computations on simple models, and then looked at the Hajos–Parrish reaction, and we came upon an explanation for its stereoselectivity. This was all done without communicating with any of the experimentalists."

After these studies were published, Professors Barbas and List independently contacted Houk, and the List collaboration developed. Houk remembers that "List suggested 'Why don't you try the reaction of cyclohexanone plus benzaldehyde. We'll do the experiments but I won't tell you the results—so you can make a real prediction.' And that's the way it developed. That took quite some time—3 to 4 months to do all the computations—since we had to *predict* something. That's a more arduous task than *explaining* something. We were concerned about missing a TS, so we went about a systematic search. And then we told List our results and there was really spectacular agreement."

The story, however, is not yet complete. A new student re-examined the problem and located additional TSs. "It turns out that there are two conformations of proline," Houk explains. "We had tested this out in model systems, and one conformation is preferred over the other. But it turns out that in some of the transition states, the second conformer is better. So now when we take into account all of the results, our predictions don't fit experiments so well. List then redid the experiments—and he got different results, too! These reactions turn out to be extremely sensitive to the amount of water present. List is trying to get an experiment as close to anhydrous conditions as possible. The final picture is not yet clear."

When asked about the difficulties in modeling reactions such as the proline-catalyzed aldol, Houk points to the difficulties in insuring that all TSs have been located. He had favored a systematic search over the random Monte-Carlo-like procedure, but he has been on the losing side of a wager concerning just such a matter. Houk was "skeptical about the Monte Carlo strategy that Professor Martin Saunders had developed. We made a bet that a systematic search for locating all minima within 3 kcal mol^{-1} of the minimum energy structure of $C_{17}H_{34}$ would be more efficient than Saunders' Monte Carlo approach. So Marty set his MicroVax doing MC searches, and we did a systematic search, and Professor Clark Still tried a number of different approaches. We didn't have the ground rules well set out, but Frank Anet was given the job of deciding who won. Well, Marty was the winner. That was a bummer, but Marty and I have had several excellent dinners at Spago as a result. And Marty insisted on losing a second bet so he could pay, too!" Houk did mention that his group is toying with doing a MC search using a semiempirical method to locate TSs and then follow this with ab initio searches starting with the best structures for stereoselectivity predictions.

I asked Houk about the role that computational chemistry may play in assisting the development of new synthetic methods. He notes "what makes the whole reaction prediction thing daunting is that if you try to imagine all the things that can happen when you mix two compounds together, you have a huge combinatorial problem. If you're just studying stereochemistry, then you're OK, because you're not concerned if that reaction will dominate over all others. Rather, the question is 'if this reaction occurs, which stereoproduct will dominate?' This is an area where computational chemistry can really help. But it's not good for questions like 'will this reaction occur in preference to *all other possibilities?*'."

Collaboration has played a consistent role in his research. His first collaborative project was with Professor Bob Moss, who approached Houk about help in interpreting carbene selectivity. Houk recalls a collaboration with Professor Al Meyers concerning a failed anionic amino-Cope rearrangement. "He had this reaction that didn't work and we did some calculations to explain why," he remembers, "and this appeared as a *JACS* communication. Then for some time afterwards, Al and others would call to say 'Hey, I've got this reaction that doesn't work! Can we get a *JACS* communication?'."

Collaborations also played a role in Houk's career decisions. The move to the University of Pittsburgh was made in part to facilitate working with Professors John Pople and Ken Jordan. "Having that connection," Houk explains, "was critical. Instead of just following the literature as a guide and doing routine calculations, I was able to do more sophisticated things, especially by getting access to state-of-the-art tools, namely the developmental versions of *Gaussian*." Houk enjoys active collaborations with many of his UCLA colleagues, including sharing a number of graduate students. These collaborations are crucial since, over his career, the focus of his research has shifted from exclusively experimental work to almost exclusively computational work.

Houk considers his greatest research challenge to be a collaborative project currently underway. "We are involved in a highly collaborative DARPA project to design new enzymes. We try to design active sites that we think would be good for catalysis. Our collaborator, David Baker, then takes this active site design and tries to appropriately fold up a protein to accommodate this site. We have done the first part a number of times; the difficult part is assessing the many different sequences that might be reasonable for the second part. The first catalyst we tried was to do the Hajos–Parish reaction. We made the design all the way through the actual protein sequence. The enzyme was synthesized, but it doesn't work. In fact, nothing has worked yet. It is a worthy challenge!"

When asked to name his most important scientific discovery, Houk first names, as a single discovery, his development of FMO explanations of selectivity in pericyclic reactions, most notably expressed in the idea of torquoselectivity. Houk believes that more important has been the general impact his research has had on science. "Our work has helped establish computational methods as a real strong partner with experiment," he says. His claim to fame, be believes, is that he's "the synthetic chemist's 'go-to-guy' when it comes to computations."

Houk's credo can be summed up in this characterization of his own work: "There is an intimate inter-relationship of experiment and theory in my work. If there's an interesting phenomenon, we want to explain it!"

REFERENCES

1. Moylan, C. R.; Brauman, J. I. In *Advances in Classical Trajectory Methods*; Hase, W., Ed.; JAI Press: Greenwich, CT, **1994**; Vol. 2, p 95–114.
2. Chabinyc, M. L.; Craig, S. L.; Regan, C. K.; Brauman, J. I. "Gas-phase ionic reactions: Dynamics and mechanism of nucleophilic displacements," *Science* **1998**, *279*, 1882–1886.
3. Dedieu, A.; Veillard, A. "Comparative study of some S_N2 reactions through ab initio calculations," *J. Am. Chem. Soc.* **1972**, *94*, 6730–6738.
4. Keil, F.; Ahlrichs, R. "Theoretical study of S_N2 reactions. Ab initio computations on HF and CI level," *J. Am. Chem. Soc.* **1976**, *98*, 4787–4793.

5. Wolfe, S.; Mitchell, D. J.; Schlegel, H. B. "Theoretical studies of S_N2 transition states. 1. Geometries," *J. Am. Chem. Soc.* **1981**, *103*, 7692–7694.
6. Chandrasekhar, J.; Smith, S. F.; Jorgensen, W. L. "S_N2 reaction profiles in the gas phase and aqueous solution," *J. Am. Chem. Soc.* **1984**, *106*, 3049–3050.
7. Shi, Z.; Boyd, R. J. "An ab initio Study of model S_N2 reactions with inclusion of electron correlation effects through second-order Moeller-Plesset perturbation calculations," *J. Am. Chem. Soc.* **1990**, *112*, 6789–6796.
8. Wolfe, S.; Kim, C.-K. "Secondary H/D isotope effects in methyl-transfer reactions decrease with increasing looseness of the transition structure," *J. Am. Chem. Soc.* **1991**, *113*, 8056–8061.
9. Wladkowski, B. D.; Allen, W. D.; Brauman, J. I. "The S_N2 identity exchange reaction $F^- + CH_3F$ .fwdarw. $FCH_3 + F^-$: Definitive ab initio predictions," *J. Phys. Chem.* **1994**, *98*, 13532–13540.
10. Gonzales, J. M.; Cox, R. S., III; Brown, S. T.; Allen, W. D.; Schaefer, H. F., III "Assessment of density functional theory for model S_N2 reactions: $CH_3X + F^-$ (X = F, Cl, CN, OH, SH, NH_2, PH_2)," *J. Phys. Chem. A.* **2001**, *105*, 11327–11346.
11. Streitwieser, A.; Choy, G. S.-C.; Abu-Hasanayn, F. "Theoretical study of ion pair S_N2 reactions: Ethyl vs methyl reactivities and extension to higher alkyls," *J. Am. Chem. Soc.* **1997**, *119*, 5013–5019.
12. Parthiban, S.; de Oliveira, G.; Martin, J. M. L. "Benchmark ab initio energy profiles for the gas-phase S_N2 Reactions $Y^- + CH_3X$ $CH_3Y + X^-$ (X,Y = F,Cl,Br). Validation of hybrid DFT methods," *J. Phys. Chem. A.* **2001**, *105*, 895–904.
13. Glukhovtsev, M. N.; Pross, A.; Radom, L. "Gas-phase identity S_N2 reactions of halide anions with methyl halides: A high-level computational study," *J. Am. Chem. Soc.* **1995**, *117*, 2024–2032.
14. Gonzales, J. M.; Pak, C.; Cox, R. S.; Allen, W. D.; Schaefer, H. F. I.; Császár, A. G.; Tarczay, G. "Definitive ab initio studies of model S_N2 reactions $CH_3X + F^-$ (X = F, Cl, CN, OH, SH, NH_2, PH_2)," *Chem. Eur. J.* **2003**, *9*, 2173–2192.
15. Bento, A. P.; Bickelhaupt, F. M. "Nucleophilicity and leaving-group ability in frontside and backside S_N2 reactions," *J. Org. Chem.* **2008**, *73*, 7290–7299.
16. Tucker, S. C.; Truhlar, D. G. "Ab initio calculations of the transition-state geometry and vibrational frequencies of the S_N2 reaction of Cl^- with CH_3Cl," *J. Phys. Chem.* **1989**, *93*, 8138–8142.
17. Glukhovtsev, M. N.; Bach, R. D.; Pross, A.; Radom, L. "The performance of B3-LYP density functional theory in describing S_N2 reactions at saturated carbon," *Chem. Phys. Lett.* **1996**, *260*, 558–564.
18. Li, C.; Ross, P.; Szulejko, J. E.; McMahon, T. B. "High-pressure mass spectrometric investigations of the potential energy surfaces of gas-phase S_N2 reactions," *J. Am. Chem. Soc.* **1996**, *118*, 9360–9367.
19. Wladkowski, B. D.; Brauman, J. I. "Application of Marcus theory to gas-phase S_N2 reactions: Experimental support of the Marcus theory additivity postulate," *J. Phys. Chem.* **1993**, *97*, 13158–13164.
20. Caldwell, G.; Magnera, T. F.; Kebarle, P. "S_N2 reactions in the gas phase. Temperature dependence of the rate constants and energies of the transition states. Comparison with solution," *J. Am. Chem. Soc.* **1984**, *106*, 959–966.
21. Knighton, W. B.; Bognar, J. A.; O'Connor, P. M.; Grimsrud, E. P. "Gas-phase S_N2 reactions of chloride ion with alkyl bromides at atmospheric pressure. Temperature

dependence of the rate constants and energies of the transition states," *J. Am. Chem. Soc.* **1993**, *115*, 12079–12084.
22. Olmstead, W. N.; Brauman, J. I. "Gas-phase nucleophilic displacement reactions," *J. Am. Chem. Soc.* **1977**, *99*, 4219–4228.
23. Pellerite, M. J.; Brauman, J. I. "Intrinsic barriers in nucleophilic displacements. A general model for intrinsic nucleophilicity toward methyl centers," *J. Am. Chem. Soc.* **1983**, *105*, 2672–2680.
24. Lowry, T. H.; Richardson, K. S. *Mechanism and Theory in Organic Chemistry*; 3rd ed.; Harper and Row: New York, **1987**.
25. Carroll, F. A. *Perspectives on Structure and Mechanism in Organic Chemistry*; 2nd ed.; John Wiley & Sons, Inc.: Hoboken, NJ, **2010**.
26. DePuy, C. H.; Gronert, S.; Mullin, A.; Bierbaum, V. M. "Gas-phase S_N2 and E_2 reactions of alkyl halides," *J. Am. Chem. Soc.* **1990**, *112*, 8650–8655.
27. Regan, C. K.; Craig, S. L.; Brauman, J. I. "Steric effects and solvent effects in ionic reactions," *Science* **2002**, *295*, 2245–2247.
28. Chen, X.; Regan, C. K.; Craig, S. L.; Krenske, E. H.; Houk, K. N.; Jorgensen, W. L.; Brauman, J. I. "Steric and solvation effects in ionic S_N2 reactions," *J. Am. Chem. Soc.* **2009**, *131*, 16162–16170.
29. Jensen, F. "A theoretical study of steric effects in S_N2 reactions," *Chem. Phys. Lett.* **1992**, *196*, 368–376.
30. Gronert, S. "Theoretical studies of elimination reactions. 3. Gas-phase reactions of fluoride ion with $(CH_3)_2CHCl$ and $CH_3CH_2CH_2Cl$. The effect of methyl substituents," *J. Am. Chem. Soc.* **1993**, *115*, 652–659.
31. Vayner, G.; Houk, K. N.; Jorgensen, W. L.; Brauman, J. I. "Steric retardation of S_N2 reactions in the gas phase and solution," *J. Am. Chem. Soc.* **2004**, *126*, 9054–9058.
32. Kim, Y.; Cramer, C. J.; Truhlar, D. G. "Steric effects and solvent effects on S_N2 reactions," *J. Phys. Chem. A* **2009**, *113*, 9109–9114.
33. Glukhovtsev, M. N.; Pross, A.; Schlegel, H. B.; Bach, R. D.; Radom, L. "Gas-phase identity S_N2 reactions of halide anions and methyl halides with retention of configuration," *J. Am. Chem. Soc.* **1996**, *118*, 1258–1264.
34. Sauers, R. R. "Inversion vs retention of configuration in gas-phase ammonium ion/alcohol reactions," *J. Org. Chem.* **2002**, *67*, 1221–1226.
35. Despeyroux, D.; Cole, R. B.; Tabet, J. C. "Ion-molecule reactions in the gas phase. XVIII. Nucleophilic substitution of diastereomeric norborneols, norbornyl acetates and benzoates under ammonia chemical ionization," *Org. Mass Spectrom.* **1992**, *27*, 300–308.
36. Laerdahl, J. K.; Uggerud, E. "Nucleophilic identity substitution reactions. The reaction between water and protonated alcohols," *Org. Biomol. Chem.* **2003**, *1*, 2935–2942.
37. Perham, R. N. "Domains, motifs, and linkers in 2-oxo acid dehydrogenase multienzyme complexes: A paradigm in the design of a multifunctional protein," *Biochemistry* **1991**, *30*, 8501–8512.
38. Tanaka, K.; Mackay, G. I.; Payzant, J. D.; Bohme, D. K. "Gas-phase reactions of anions with halogenated methanes at $297 \pm 2°K$," *Can. J. Chem.* **1976**, *54*, 1643–1659.
39. Bohme, D. K.; Raksit, A. B. "Gas-phase measurements of the influence of stepwise solvation on the kinetics of nucleophilic displacement reactions with chloromethane and bromomethane at room temperature," *J. Am. Chem. Soc.* **1984**, *106*, 3447–3452.

40. Viggiano, A. A.; Arnold, S. T.; Morris, R. A.; Ahrens, A. F.; Hierl, P. M. "Temperature dependences of the rate constants and branching ratios for the reactions of OH^- $(H_2O)_{0-4}$ + CH_3Br," *J. Phys. Chem.* **1996**, *100*, 14397–14402.

41. Seeley, J. V.; Morris, R. A.; Viggiano, A. A. "Temperature dependences of the rate constants and branching ratios for the reactions of $F^-(H_2O)_{0-5}$ with CH_3Br," *J. Phys. Chem. A* **1997**, *101*, 4598–4601.

42. Jorgensen, W. L.; Chandrasekhar, J.; Madura, J. D.; Impey, R. W.; Klein, M. L. "Comparison of simple potential functions for simulating liquid water," *J. Chem. Phys.* **1983**, *79*, 926–935.

43. McLennan, D. J. "Semiempirical calculation of rates of S_N2 Finkelstein reactions in solution by a quasi-thermodynamic cycle," *Aust. J. Chem.* **1978**, *31*, 1897–1909.

44. Cossi, M.; Adamo, C.; Barone, V. "Solvent effects on an S_N2 reaction profile," *Chem. Phys. Lett.* **1998**, *297*, 1–7.

45. Mohamed, A. A.; Jensen, F. "Steric effects in S_N2 reactions. The influence of microsolvation," *J. Phys. Chem. A.* **2001**, *105*, 3259–3268.

46. Hughes, E. D.; Ingold, C. K.; Mackie, J. D. H. "Mechanism of substitution at a saturated carbon atom. XLIII. Kinetics of the interaction of chloride ions with simple alkyl bromides in acetone," *J. Chem. Soc.* **1955**, 3173–3177.

47. De la Mare, P. B. D. "Mechanism of substitution at a saturated carbon atom. XLV. Kinetics of the interaction of bromide ions with simple alkyl bromides in acetone," *J. Chem. Soc.* **1955**, 3180–3187.

48. Morokuma, K. "Potential energy surface of the S_N2 reaction in hydrated clusters," *J. Am. Chem. Soc.* **1982**, *104*, 3732–3733.

49. Hayes, J. M.; Bachrach, S. M. "Effect of micro and bulk solvation on the mechanism of nucleophilic substitution at sulfur in disulfides," *J. Phys. Chem. A* **2003**, *107*, 7952–7961.

50. DeTar, D. F.; McMullen, D. F.; Luthra, N. P. "Steric effects in S_N2 reactions," *J. Am. Chem. Soc.* **1978**, *100*, 2484–2493.

51. Cram, D. J.; Kopecky, K. R. "Studies in stereochemistry. XXX. Models for steric control of asymmetric induction," *J. Am. Chem. Soc.* **1959**, *81*, 2748–2755.

52. Karabatsos, G. J. "Asymmetric induction. A model for additions to carbonyls directly bonded to asymmetric carbons," *J. Am. Chem. Soc.* **1967**, *89*, 1367–1371.

53. Chérest, M.; Felkin, H.; Prudent, N. "Torsional strain involving partial bonds. The stereochemistry of the lithium aluminium hydride reduction of some simple open-chain ketones," *Tetrahedron Lett.* **1968**, *9*, 2199–2204.

54. Anh, N. T.; Eistenstein, O. "Theoretical interpretation of 1–2 asymmetric induction – Importance of antiperiplanarity," *Nouv. J. Chim.* **1977**, *1*, 61–70.

55. Lodge, E. P.; Heathcock, C. H. "Acyclic stereoselection. 40. Steric effects, as well as σ^*-orbital energies, are important in diastereoface differentiation in additions to chiral aldehydes," *J. Am. Chem. Soc.* **1987**, *109*, 3353–3361.

56. Kaufmann, E.; Schleyer, P. v. R.; Houk, K. N.; Wu, Y.-D. "Ab initio mechanisms for the addition of CH_3Li, HLi, and their dimers to formaldehyde," *J. Am. Chem. Soc.* **1985**, *107*, 5560–5562.

57. Wong, S. S.; Paddon-Row, M. N. "Theoretical evidence in support of the Anh–Eisenstein electronic model in controlling π-facial stereoselectivity in

nucleophilic additions to carbonyl compounds," *J. Chem. Soc. Chem. Commun.* **1990**, 456–458.

58. Gung, B. W. "Diastereofacial selection in nucleophilic additions to unsymmetrically substituted trigonal carbons," *Tetrahedron* **1996**, *52*, 5263–5301.

59. Houk, K. N. "Perspective on "Theoretical interpretation of 1–2 asymmetric induction. The importance of antiperiplanarity": Anh NT, Eisenstein O (1977) Nouv J Chim 1: 61–70," *Theor. Chem. Acc. 2000*, 103, 330–331.

60. Cieplak, A. S. "Stereochemistry of nucleophilic addition to cyclohexanone. The importance of two-electron stabilizing interactions," *J. Am. Chem. Soc.* **1981**, *103*, 4540–4552.

61. Cieplak, A. S.; Tait, B. D.; Johnson, C. R. "Reversal of π-facial diastereoselection upon electronegative substitution of the substrate and the reagent," *J. Am. Chem. Soc.* **1989**, *111*, 8447–8462.

62. Halterman, R. L.; McEvoy, M. A. "Diastereoselectivity in the reduction of sterically unbiased 2,2-diarylcyclopentanones," *J. Am. Chem. Soc.* **1990**, *112*, 6690–6695.

63. Kaselj, M.; Chung, W.-S.; le Noble, W. J. "Face selection in addition and elimination in sterically unbiased systems," *Chem. Rev.* **1999**, *99*, 1387–1414.

64. Wu, Y. D.; Houk, K. N. "Electronic and conformational effects on π-facial stereoselectivity in nucleophilic additions to carbonyl compounds," *J. Am. Chem. Soc.* **1987**, 908–910.

65. Adcock, W.; Abeywickrema, A. N. "Substituent effects in the bicyclo[2.2.2]octane ring system. A carbon-13 and fluorine-19 nuclear magnetic resonance study of 4-substituted bicyclo[2.2.2]oct-1-yl fluorides," *J. Org. Chem.* **1982**, *47*, 2957–2966.

66. Laube, T.; Ha, T. K. "Detection of hyperconjugative effects in experimentally determined structures of neutral molecules," *J. Am. Chem. Soc.* **1988**, *110*, 5511–5517.

67. Rozeboom, M. D.; Houk, K. N. "Stereospecific alkyl group effects on amine lone-pair ionization potentials: Photoelectron spectra of alkylpiperidines," *J. Am. Chem. Soc.* **1982**, *104*, 1189–1191.

68. Frenking, G.; Koehler, K. F.; Reetz, M. T. "The origin of π-facial diastereofacial selectivity in addition reactions to cyclohexane-based systems," *Angew. Chem. Int. Ed. Engl.* **1991**, *30*, 1146–1149.

69. Frenking, G.; Kohler, K. F.; Reetz, M. T. "On the origin of π-facial diastereoselectivity in nucleophilic additions to chiral carbonyl compounds. 2. Calculated transition state structures for the addition of nucleophiles to propionaldehyde 1, chloroacetyldehyde 2, and 2-chloropropionaldehyde 3," *Tetrahedron* **1991**, *47*, 9005–9018.

70. Wong, S. S.; Paddon-Row, M. N. "The importance of electrostatic effects in controlling π-facial stereoselectivity in nucleophilic additions to carbonyl compounds: An ab initio MO study of a prototype chelation model," *J. Chem. Soc. Chem. Commun.* **1991**, 327–330.

71. Wu, Y.-D.; Tucker, J. A.; Houk, K. N. "Stereoselectivities of nucleophilic additions to cyclohexanones substituted by polar groups. Experimental investigation of reductions of trans-decalones and theoretical studies of cyclohexanone reductions. The influence of remote electrostatic effects," *J. Am. Chem. Soc.* **1991**, *113*, 5018–5027.

72. Luibrand, R. T.; Taigounov, I. R.; Taigounov, A. A. "A theoretical study of the reaction of lithium aluminum hydride with formaldehyde and cyclohexanone," *J. Org. Chem.* **2001**, *66*, 7254–7262.

73. Paddon-Row, M. N.; Wu, Y.-D.; Houk, K. N. "Electrostatic control of the stereochemistry of nucleophilic additions to substituted 7-norbornanones," *J. Am. Chem. Soc.* **1992**, *114*, 10638–10639.

74. Williams, L.; Paddon-Row, M. N. "Electrostatic and steric control of π-facial stereoselectivity in nucleophilic additions of LiH and MeLi to endo-5,6-disubstituted bornen-7-ones: an ab initio MO study," *J. Chem. Soc. Chem. Commun.* **1994**, 353–355.

75. Wilmot, N.; Marsella, M. J. "Visualization method to predict the nucleophilic asymmetric induction of prochiral electrophiles," *Org. Lett.* **2006**, *8*, 3109–3112.

76. Fleming, I.; Hrovat, D. A.; Borden, W. T. "The origin of Felkin–Anh control from an electropositive substituent adjacent to the carbonyl group," *J. Chem. Soc. Perkin Trans.* **2001**, *2*, 331–338.

77. Smith, R. J.; Trzoss, M.; Bühl, M.; Bienz, S. "The Cram rule revisited once more – Revision of the Felkin–Anh model," *Eur. J. Org. Chem.* **2002**, *2002*, 2770–2775.

78. Yang, X.; Liu, P.; Houk, K. N.; Birman, V. B. "Manifestation of Felkin–Anh control in enantioselective acyl transfer catalysis: Kinetic resolution of carboxylic acids," *Angew. Chem. Int. Ed.* **2012**, *51*, 9638–9642.

79. Heathcock, C. H. In *Asymmetric Synthesis*; Morrison, J. D., Ed.; Academic Press: Orlando, FL, **1984**; Vol. 3, p 111–212.

80. Atkinson, R. S. *Stereoselective Synthesis*; John Wiley & Sons, Ltd: Chichester, UK, **1995**.

81. Machajewski, T. D.; Wong, C.-H. "The catalytic asymmetric aldol reaction," *Angew. Chem. Int. Ed.* **2000**, *39*, 1352–1375.

82. Tanaka, F.; Barbas, C. F., III. In *Modern Aldol Reactions*; Mahrwald, R., Ed.; Wiley-VCH Verlag: Weinheim, Germany, **2004**; Vol. 1, p 273–310.

83. Hajos, Z. G.; Parrish, D. R. "Asymmetric synthesis of bicyclic intermediates of natural product chemistry," *J. Org. Chem.* **1974**, *39*, 1615–1621.

84. Eder, U.; Sauer, G.; Wiechert, R. "New type of asymmetric cyclization to optically active steroid CD partial structures," *Angew. Chem. Int. Ed. Engl.* **1971**, *10*, 496–497.

85. Agami, C.; Platzer, N.; Sevestre, H. "Enantioselective cyclizations of acyclic 1,5-diketones," *Bul. Soc. Chim. Fr.* **1987**, *2*, 358–360.

86. List, B. "Introduction: Organocatalysis," *Chem. Rev.* **2007**, *107*, 5413–5415.

87. List, B. "Emil Knoevenagel and the roots of aminocatalysis," *Angew. Chem. Int. Ed.* **2010**, *49*, 1730–1734.

88. List, B. "Proline-catalyzed asymmetric reactions," *Tetrahedron* **2002**, *58*, 5573–5590.

89. List, B. "Enamine catalysis is a powerful strategy for the catalytic generation and use of carbanion equivalents," *Acc. Chem. Res.* **2004**, *37*, 548–557.

90. Notz, W.; Tanaka, F.; Barbas, C. F., III "Enamine-based organocatalysis with proline and diamines: The development of direct catalytic asymmetric aldol, Mannich, Michael, and Diels–Alder reactions," *Acc. Chem. Res.* **2004**, *37*, 580–591.

91. Northrup, A. B.; MacMillan, D. W. C. "The first general enantioselective catalytic Diels–Alder reaction with simple α,β-unsaturated ketones," *J. Am. Chem. Soc.* **2002**, *124*, 2458–2460.

92. Paras, N. A.; MacMillan, D. W. C. "The enantioselective organocatalytic 1,4-addition of electron-rich benzenes to α,β-unsaturated aldehydes," *J. Am. Chem. Soc.* **2002**, *124*, 7894–7895.

93. Brown, S. P.; Goodwin, N. C.; MacMillan, D. W. C. "The first enantioselective organocatalytic Mukaiyama–Michael reaction: A direct method for the synthesis of enantioenriched butenolide architecture," *J. Am. Chem. Soc.* **2003**, *125*, 1192–1194.

94. Kunz, R. K.; MacMillan, D. W. C. "Enantioselective organocatalytic cyclopropanations. The identification of a new class of iminium catalyst based upon directed electrostatic activation," *J. Am. Chem. Soc.* **2005**, *127*, 3240–3241.

95. Thayumanavan, R.; Tanaka, F.; Barbas, C. F., III; "Direct organocatalytic asymmetric aldol reactions of amino aldehydes: Expedient syntheses of highly enantiomerically enriched *anti*-hydroxy-amino acids," *Org. Lett.* **2004**, *6*, 3541–3544.

96. Mangion, I. K.; MacMillan, D. W. C. "Total synthesis of brasoside and littoralisone," *J. Am. Chem. Soc.* **2005**, *127*, 3696–3697.

97. Northrup, A. B.; MacMillan, D. W. C. "The first direct and enantioselective cross-aldol reaction of aldehydes," *J. Am. Chem. Soc.* **2002**, *124*, 6798–6799.

98. List, B.; Pojarliev, P.; Castello, C.; "Proline-catalyzed asymmetric aldol reactions between ketones and α-unsubstituted aldehydes," *Org. Lett.* **2001**, *3*, 573–575.

99. Mase, N.; Tanaka, F.; Barbas, C. F., III "Synthesis of hydroxyaldehydes with stereogenic quaternary carbon centers by direct organocatalytic asymmetric aldol reactions," *Angew. Chem. Int. Ed.* **2004**, *43*, 2420–2423.

100. Pidathala, C.; Hoang, L.; Vignola, N.; List, B. "Direct catalytic asymmetric enolexo aldolizations," *Angew. Chem. Int. Ed.* **2003**, *42*, 2785–2788.

101. Notz, W.; List, B. "Catalytic asymmetric synthesis of *anti*-1,2-diols," *J. Am. Chem. Soc.* **2000**, *122*, 7386–7387.

102. Cheong, P. H.-Y.; Legault, C. Y.; Um, J. M.; Çelebi-Ölçüm, N.; Houk, K. N. "Quantum mechanical investigations of organocatalysis: Mechanisms, reactivities, and selectivities," *Chem. Rev.* **2011**, *111*, 5042–5137.

103. List, B.; Lerner, R. A.; Barbas, C. F., III; "Proline-catalyzed direct asymmetric aldol reactions," *J. Am. Chem. Soc.* **2000**, *122*, 2395–2396.

104. Sakthivel, K.; Notz, W.; Bui, T.; Barbas, C. F., III; "Amino acid catalyzed direct asymmetric aldol reactions: A bioorganic approach to catalytic asymmetric carbon-carbon bond-forming reactions," *J. Am. Chem. Soc.* **2001**, *123*, 5260–5267.

105. Hoang, L.; Bahmanyar, S.; Houk, K. N.; List, B. "Kinetic and stereochemical evidence for the involvement of only one proline molecule in the transition states of proline-catalyzed intra- and intermolecular aldol reactions," *J. Am. Chem. Soc.* **2003**, *125*, 16–17.

106. List, B.; Hoang, L. J.; Martin, H. J. "New mechanistic studies on the proline-catalyzed aldol reaction," *Proc. Nat. Acad. Sci. USA* **2004**, *101*, 5839–5842.

107. Agami, C.; Puchot, C.; Sevestre, H. "Is the mechanism of the proline-catalyzed enantioselective aldol reaction related to biochemical processes?," *Tetrahedron Lett.* **1986**, *27*, 1501–4150.

108. Puchot, C.; Samuel, O.; Dunach, E.; Zhao, S.; Agami, C.; Kagan, H. B. "Nonlinear effects in asymmetric synthesis. Examples in asymmetric oxidations and aldolization reactions," *J. Am. Chem. Soc.* **1986**, *108*, 2353–2357.

109. Jung, M. E. "A review of annulation," *Tetrahedron* **1976**, *32*, 3–31.

110. Rankin, K. N.; Gauld, J. W.; Boyd, R. J. "Density functional study of the proline-catalyzed direct aldol reaction," *J. Phys. Chem. A.* **2002**, *106*, 5155–5159.
111. Bahmanyar, S.; Houk, K. N. "Transition states of amine-catalyzed aldol reactions involving enamine intermediates: Theoretical studies of mechanism, reactivity, and stereoselectivity," *J. Am. Chem. Soc.* **2001**, *123*, 11273–11283.
112. Bahmanyar, S.; Houk, K. N.; Martin, H. J.; List, B. "Quantum mechanical predictions of the stereoselectivities of proline-catalyzed asymmetric intermolecular aldol reactions," *J. Am. Chem. Soc.* **2003**, *125*, 2475–2479.
113. Allemann, C.; Um, J. M.; Houk, K. N. "Computational investigations of the stereoselectivities of proline-related catalysts for aldol reactions," *J. Mol. Cat. A: Chem.* **2010**, *324*, 31–38.
114. Cordova, A.; Zou, W.; Ibrahem, I.; Reyes, E.; Engqvist, M.; Liao, W.-W. "Acyclic amino acid-catalyzed direct asymmetric aldol reactions: Alanine, the simplest stereoselective organocatalyst," *Chem. Commun.* **2005**, 3586–3588.
115. Bassan, A.; Zou, W.; Reyes, E.; Himo, F.; Córdova, A. "The origin of stereoselectivity in primary amino acid catalyzed intermolecular aldol reactions," *Angew. Chem. Int. Ed.* **2005**, *44*, 7028–7032.
116. Shinisha, C. B.; Sunoj, R. B. "Bicyclic proline analogues as organocatalysts for stereoselective aldol reactions: An *in silico* DFT study," *Org. Biomol. Chem.* **2007**, *5*, 1287–1294.
117. Lam, Y.-h.; Houk, K. N.; Scheffler, U.; Mahrwald, R. "Stereoselectivities of histidine-catalyzed asymmetric aldol additions and contrasts with proline catalysis: A quantum mechanical analysis," *J. Am. Chem. Soc.* **2012**, *134*, 6286–6295.
118. Seebach, D.; Beck, A. K.; Badine, D. M.; Limbach, M.; Eschenmoser, A.; Treasurywala, A. M.; Hobi, R.; Prikoszovich, W.; Linder, B. "Are oxazolidinones really unproductive, parasitic species in proline catalysis? – Thoughts and experiments pointing to an alternative view," *Helv. Chim. Acta* **2007**, *90*, 425–471.
119. Sharma, A.; Sunoj, R. "Enamine versus oxazolidinone: What controls stereoselectivity in proline-catalyzed asymmetric aldol reactions?," *Angew. Chem. Int. Ed.* **2010**, *49*, 6373–6377.
120. Bahmanyar, S.; Houk, K. N. "The origin of stereoselectivity in proline-catalyzed intramolecular aldol reactions," *J. Am. Chem. Soc.* **2001**, *123*, 12911–12912.
121. Clemente, F. R.; Houk, K. N. "Computational evidence for the enamine mechanism of intramolecular aldol reactions catalyzed by proline," *Angew. Chem. Int. Ed.* **2004**, *43*, 5766–5768.
122. Zhu, H.; Clemente, F. R.; Houk, K. N.; Meyer, M. P. "Rate limiting step precedes C–C bond formation in the archetypical proline-catalyzed intramolecular aldol reaction," *J. Am. Chem. Soc.* **2009**, *131*, 1632–1633.
123. Cheong, P. H.-Y.; Houk, K. N. "Origins and predictions of stereoselectivity in intramolecular aldol reactions catalyzed by proline derivatives," *Synthesis* **2005**, 1533–1537.
124. List, B.; Pojarliev, P.; Biller, W. T.; Martin, H. J. "The proline-catalyzed direct asymmetric three-component Mannich reaction: Scope, optimization, and application to the highly enantioselective synthesis of 1,2-amino alcohols," *J. Am. Chem. Soc.* **2002**, *124*, 827–833.
125. Bahmanyar, S.; Houk, K. N. "Origins of opposite absolute stereoselectivities in proline-catalyzed direct mannich and aldol reactions," *Org. Lett.* **2003**, *5*, 1249–1251.

126. Parasuk, W.; Parasuk, V. "Theoretical investigations on the stereoselectivity of the proline catalyzed Mannich reaction in DMSO," *J. Org. Chem.* **2008**, *73*, 9388–9392.
127. Mitsumori, S.; Zhang, H.; Ha-Yeon Cheong, P.; Houk, K. N.; Tanaka, F.; Barbas, C. F. "Direct asymmetric anti-Mannich-type reactions catalyzed by a designed amino acid," *J. Am. Chem. Soc.* **2006**, *128*, 1040–1041.
128. Wheeler, S. E.; Moran, A.; Pieniazek, S. N.; Houk, K. N. "Accurate reaction enthalpies and sources of error in DFT thermochemistry for aldol, Mannich, and α-aminoxylation reactions," *J. Phys. Chem. A* **2009**, *113*, 10376–10384.
129. Singh, R.; Tsuneda, T.; Hirao, K. "An examination of density functionals on aldol, Mannich and α-aminoxylation reaction enthalpy calculations," *Theor. Chem. Acc.* **2011**, *130*, 153–160.
130. Dickerson, T. J.; Janda, K. D. "Aqueous aldol catalysis by a nicotine metabolite," *J. Am. Chem. Soc.* **2002**, *124*, 3220–3221.
131. Dickerson, T. J.; Lovell, T.; Meijler, M. M.; Noodlcman, L.; Janda, K. D. "Nornicotine aqueous aldol reactions: Synthetic and theoretical investigations into the origins of catalysis," *J. Org. Chem.* **2004**, *69*, 6603–6609.
132. Zhang, X.; Houk, K. N. "Acid/base catalysis by pure water: The aldol reaction," *J. Org. Chem.* **2005**, *70*, 9712–9716.
133. Taylor, M. S.; Jacobsen, E. N. "Asymmetric catalysis by chiral hydrogen-bond donors," *Angew. Che, Int. Ed.* **2006**, *45*, 1520–1543.
134. Doyle, A. G.; Jacobsen, E. N. "Small-molecule H-bond donors in asymmetric catalysis," *Chem. Rev.* **2007**, *107*, 5713–5743.
135. Uyeda, C.; Jacobsen, E. N. "Enantioselective Claisen rearrangements with a hydrogen-bond donor catalyst," *J. Am. Chem. Soc.* **2008**, *130*, 9228–9229.
136. Brown, A. R.; Uyeda, C.; Brotherton, C. A.; Jacobsen, E. N. "Enantioselective thiourea-catalyzed intramolecular cope-type hydroamination," *J. Am. Chem. Soc.* **2013**, *135*, 6747–6749.
137. Uyeda, C.; Rötheli, A. R.; Jacobsen, E. N. "Catalytic enantioselective Claisen rearrangements of *O*-allyl β-ketoesters," *Angew. Chem. Int. Ed. Engl.* **2010**, *49*, 9753–9756.
138. Uyeda, C.; Jacobsen, E. N. "Transition-state charge stabilization through multiple non-covalent interactions in the guanidinium-catalyzed enantioselective claisen rearrangement," *J. Am. Chem. Soc.* **2011**, *133*, 5062–5075.

CHAPTER 7

Solution-Phase Organic Chemistry

Standard quantum chemical computations are performed on a single molecule or complex. This isolated species represents a molecule in the gas phase. While gas-phase chemistry comprises an important chemical subdiscipline, the vast majority of chemical reactions occur in solution. Perhaps most critical is that all of biochemistry takes place in an aqueous environment, and so if computational chemistry is to be relevant for biochemical applications, treatment of the solvent is imperative.

Neglecting solvent effects is extremely hazardous. Equilibria and kinetics can be dramatically altered by the nature of the solvent. For example, the rate of nucleophilic substitution reactions spans 20 orders of magnitude in going from the gas phase to polar and nonpolar solvents.[1–3] A classical example of a dramatic solvent effect on equilibrium is the tautomerism between **1** and **2**. In the gas phase, the equilibrium lies far to the left, while in the solution phase, **2** dominates because of its much larger dipole moment.[4] Another classical example is that the trend in gas-phase acidity of aliphatic alcohols is reverse of the well-known trend in the solution phase; in other words, in the solution phase, the relative acidity trend is $R_3COH < R_2CHOH < RCH_2OH$, but the opposite is true in the gas phase.[5,6]

$$N\!\!\!\diagup\!\!\!\diagdown\!\!\!-OH \quad \rightleftharpoons \quad H-N\!\!\!\diagup\!\!\!\diagdown\!\!\!=O$$

1   2

Through the 1980s, computational chemists did neglect solvent effects, because they either hoped it would not matter or had no real way to effectively treat solvation. In terms of the former case, modeling reactions involving nonpolar molecules with nonpolar transition states (TSs) in nonpolar solvents as gas-phase chemistry might be appropriate. This, however, comprises a very small subset of chemistry, and even small changes in charge distribution can manifest themselves in large solvent effects.

Computational Organic Chemistry, Second Edition. Steven M. Bachrach.
© 2014 John Wiley & Sons, Inc. Published 2014 by John Wiley & Sons, Inc.

Over the past 25 years, a number of significant theoretical and algorithmic advances have been proposed towards incorporating solvent effects into quantum chemical computations. These methodologies were presented in Section 1.4. Interested readers looking for further computational details are referred to the monographs by Cramer[7] and Jensen[8] and comprehensive reviews by Tomasi,[9,10] Cramer and Truhlar,[11–13] and Mennucci.[14] This chapter presents representative case studies of aqueous-phase chemistry analyzed using quantum mechanical computations.

7.1 AQUEOUS DIELS–ALDER REACTIONS

With its concerted mechanism implying little charge distribution change along the pathway, the Diels–Alder reaction has been understood to have little rate dependence on solvent choice.[15] For example, the relative rate of cyclopentadiene dimerization increases only by a factor of 3 when carried out in ethanol.[15] The relative rate for the Diels–Alder reaction of isoprene with maleic anhydride (Table 7.1) varies by only a factor of 13 with solvents whose dielectric constants vary by almost a factor of 10, but the rate acceleration is not a simple function of the solvent polarity.[16] Furthermore, the dimerizations of cyclopentadiene and 1,3-butadiene proceed at essentially identical rates in the gas and solution phases.[17]

In this context, the surprise brought on by Breslow's publication of a study of the Diels–Alder in water is understandable.[18] Breslow noted that the reaction of cyclopentadiene with acrylonitrile is twice as fast in methanol than that in isooctane, but 30 times faster in water (Table 7.2). An even larger acceleration was found for the reaction of cyclopentadiene with butenone (Reaction 7.1): the reaction is 741 times faster in water than that in isooctane. Larger still is the rate acceleration for Reaction 7.2; it is over 12,000 times faster in water than in hexane![19] The effect of water is not limited to rate acceleration. Water also produces an enhanced

TABLE 7.1 Relative Rates for the Diels–Alder Reaction of Maleic Anhydride with Isoprene at 30.3°C[a]

Solvent	Dielectric Constant	k_{rel}
o-Dichlorobenzene	7.5	13.3
Nitrobenzene	36	10.5
Benzonitrile	26	6.84
Nitromethane	39	6.63
m-Dichlorobenzene	5.0	6.59
Chlorobenzene	5.6	5.03
Anisole	4.3	5.03
Benzene	2.3	3.49
Isopropyl ether	4.3	1

[a]Ref. 16.

TABLE 7.2 Relative Rates for the Diels–Alder Reaction of Cyclopentadiene with Acrylonitrile or Butenone[a]

Solvent	k_{rel}
Cyclobutadiene + Acrylonitrile	
Isooctane	1.0
Methanol	2.1
Water	31
Cyclobutadiene + Butenone	
Isooctane	1.0
Methanol	12.7
Water	741

[a]Ref. 18.

TABLE 7.3 Endo : Exo Product Ratios for Reaction 7.1 in Different Solvents[a]

Solvent	Endo : Exo
Cyclopentadiene	3.85
Ethanol	8.5
Water	21.4

[a]Ref. 20.

selectivity for the endo over the exo product: a greater than 20 : 1 ratio for Reaction 7.1 (Table 7.3).[20]

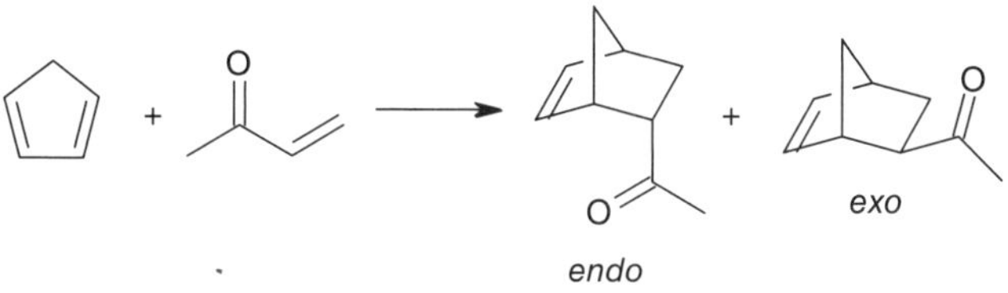

(Reaction 7.1)

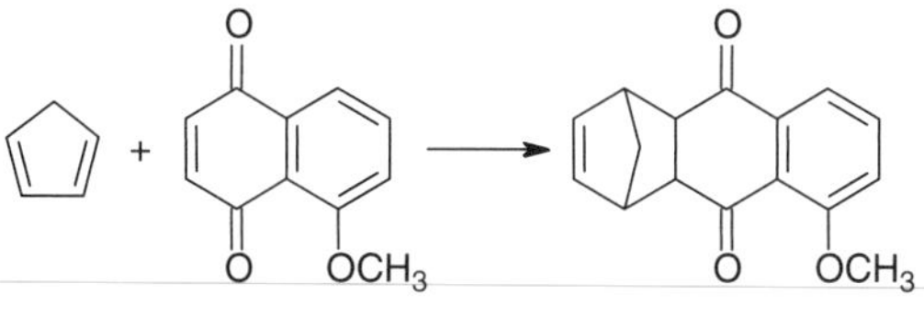

(Reaction 7.2)

Breslow attributed the enhanced rate for the Diels–Alder reaction in water to the hydrophobic effect.[18,21] In an aqueous environment, nonpolar molecules will aggregate to minimize the unfavorable interaction between the hydrocarbon and water. Engberts argued that there is very little evidence for aggregation of diene/dieneophiles in solution of typical concentration levels. Rather, he invoked a minor modification of the hydrophobic effect. The exposed surface area of the reactants is reduced in the TS.[19] With less surface area, the unfavorable hydrocarbon–water interactions in the TS are smaller than that in the reactants, leading to the rate enhancement. This has been called the *enforced hydrophobic interaction*.

Grieco[21] first argued that micellar catalysis explains the rate enhancement in water. Later, he suggested that it is the high internal pressure of water, analogous to the effect of high pressure with ordinary organic solvents on the Diels–Alder reaction, that is responsible for the rate enhancement.[23] Both Engberts[19] and Schneider and Sangwan[24,25] find no evidence of micelles. The latter pair argued that solvophobicity, a parameter that combines the effects of hydrophobicity and lipophilicity, correlates very well with Diels–Alder reaction rates in a number of solvents, including water.[24,25]

The computational work of Jorgensen's group was instrumental in bringing critical insight into the nature of the aqueous Diels–Alder reaction. Their first foray into this question utilized Monte Carlo (MC) simulations of the reaction of butenone with cyclopentadiene (Reaction 7.1).[26] They first optimized the geometry of the four possible TSs (Scheme 7.1) at Hartree–Fock (HF)/3-21G followed by single-point energy calculations at HF/6-31G(d). The lowest energy TS has the endo-cis conformation, consistent with the reaction of butadiene and acrolein.[27] Using this *endo-cis* TS, the minimum energy reaction path connecting reactant to product was produced, creating a total of 65 different geometries. An MC simulation was run to position the solvent molecules at about 43 of these individual geometries in order to obtain their solvation free energies. The solvents used were water (500 molecules), methanol (260 molecules), or propane (250 molecules). The TIP4P model was used for water and optimized potential for liquid simulations (OPLS) was used for the other two solvents to describe their intermolecular potential. For the solute, a Lenard–Jones and electrostatic potential was used, employing the Mulliken charges at HF/6-31G(d)//HF/3-21G for each of the 43 geometries.

As expected, the solvent effect of propane on Reaction 7.1 is negligible. On the other hand, both methanol and water strongly stabilize the TS. In methanol, the TS is stabilized by 2.4 kcal mol^{-1}, while in water, its stabilization is 4.2 kcal mol^{-1}. This is similar to the experimental value (3.8 kcal mol^{-1}) for the lowering of the activation barrier in water relative to isooctane.[18]

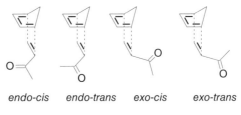

Scheme 7.1

Arbitrarily fixing the atomic charges to zero, eliminating any hydrogen bonding effect, results in an approximation of the free energy owing to hydrophobicity alone: 4.4 kcal mol^{-1}. This, however, overestimates the hydrophobic effect. A better estimate of the hydrophobic effect is about -1.5 kcal mol^{-1}, which is approximately the value of the change in free energy of hydration for Reaction 7.1, butadiene + ethylene, and isoprene + ethylene.

Their most important result concerns what effect could be responsible for the remaining stabilization energy (4.2 kcal mol^{-1} total less 1.5 kcal mol^{-1} because of the hydrophobic effect). Jorgensen noted that the number of hydrogen bonds to the carbonyl oxygen was fairly constant throughout the reaction (about 2–2.5 on average). However, each hydrogen bond was *strongest* in the neighborhood of the TS. This is consistent with slightly more polar C–O bonds, as determined by the Mulliken charges, in the TS than in the reactant or product.

To follow up on this assertion, Jorgensen computed the interaction energy of one water molecule with butenone and with the *endo-cis* TS of Reaction 7.1.[28] Only the six intermolecular degrees of freedom were optimized at HF/6-31G(d). He located two configurations for the butenone–water complex, with binding energies of -6.50 and -6.29 kcal mol^{-1}. The two configurations for the TS–water complex had binding energies of -8.52 and -7.76 kcal mol^{-1}. Figure 7.1 presents the *fully* optimized HF/6-31G(d) structures for the complex of one water molecules with (a) butenone, (b) the TS of Reaction 7.1, and (c) the product of Reaction 7.1. The binding energy of water to the reactant or product is about -6.5 kcal mol^{-1}, but the strength of the water–TS interaction is much larger, -8.8 kcal mol^{-1}. The Mulliken charge on the carbonyl oxygen changes from -0.58 in the water–butenone complex to -0.64 in the water–TS complex. This stronger hydrogen bond in the TS could account for the dramatic acceleration of the Diels–Alder reaction in water.

A couple of other simulations support these conclusions. The MC simulation by Gao used the AM1 model to describe the reaction of cyclopentadiene with either butenone or isoprene.[29] The solvent, 500 water molecules, was described using the TIP3P model, and the solute–solvent interaction term included both the electrostatic and the empirical van der Waals potentials. For the isoprene reaction, the TS is stabilized by 4.6 kcal mol^{-1}. This stabilization is attributable to the (enforced) hydrophobic effect. For the butenone reaction, the TS is stabilized by 3.5 kcal mol^{-1} in water. Separating out the hydrophobic effect from a hydrogen bonding effect in this case is difficult. Gao estimates the electrostatic contribution is about 1.5 kcal mol^{-1} (which should be mostly associated with the hydrogen bonding effect), suggesting that the two effects are about equally important. The O–O$_w$ radial distribution function for the reaction of butenone suggests that roughly 2.5–3 water molecules are associated with the carbonyl oxygen in the reactant and that the coordination decreases to about 2.5 in the TS. However, the pair energy distribution function clearly indicates that the hydrogen bonds are about 1.5–2.0 kcal mol^{-1} *stronger* to the TS than to either reactant or product.

Very similar results were found in Jorgensen's AM1-OPLS MC simulation of the reaction of cyclobutadiene with butenone (Reaction 7.1), acrylonitrile, and 1,4-naphthoquinone.[30] The computed decreases in the activation free energy

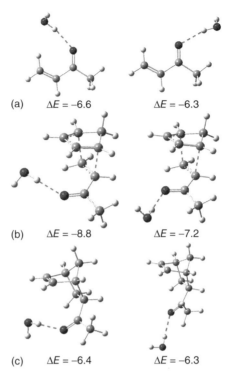

Figure 7.1 HF/6-31G(d) optimized structures[28] of two configurations of the each complex formed between water and (a) butenone, (b) the transition state of Reaction 7.1), and (c) the product of Reaction 7.1).

because of water (Table 7.4) are in reasonable agreement with experiment, but more importantly correctly predict their trend. The changes in solvent accessible surface areas are very similar for the three reactions, with no change seen after the TS. This suggests that the hydrophobic effect should be responsible for about 1 kcal mol^{-1} reduction in activation energy. The enhanced hydrogen bonding in the TS, observed in the solute–solvent pair energy distribution, accounts for the remaining stabilization energy. These results are fully corroborated in a later study by Jorgensen using the PDDG/PM3 method in a free energy perturbation MC simulation of water and other polar solvents.[31]

Evanseck's computational study of the reaction of butadiene with acrolein helped to clarify many issues.[32] Of the four stereoisomeric TSs (analogous to those in Scheme 7.1), the *endo-cis* is again found to be the lowest in energy at B3LYP/6-31G(d) in the gas phase. The activation energy (19.6 kcal mol^{-1}) is in excellent agreement with the experimental value of 19.7 kcal mol^{-1}.[33] The TS geometry is shown in Figure 7.2. Based on analogy with the reaction of cyclopentadiene with methylacrylate, Evanseck estimates that the activation enthalpy in water/methanol for butadiene with acrolein is about 10.9 kcal mol^{-1}, or about 8.9 kcal mol^{-1} lower than the gas-phase barrier.

TABLE 7.4 Difference in the Gas-Phase and Aqueous Activation Free Energy (in kcal mol^{-1}) for the Diels–Alder Reactions of Cyclopentadiene and Three Different Dienophiles[a]

Dienophile	Solvent Effect[b]
Acrylonitrile	−1.5 (−2.1)
Butenone	−2.8 (−3.8)
1,4-Naphthoquinone	−4.4 (−5.0)

[a] See Ref. 30.
[b] Experimental values in parentheses, see Ref. 18.

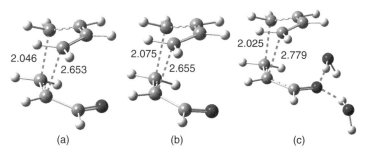

Figure 7.2 B3LYP/6-31G(d) optimized structure of the transition state for the Diels–Alder reaction of butadiene with acrolein: (a) gas phase, (b) PCM, and (c) gas phase incorporating two explicit water molecules.[32]

The reaction was then examined with polarized continuum method (PCM). The optimized PCM TS is shown in Figure 7.2. The barrier is reduced by 3.8 kcal mol^{-1} from the gas-phase value (Table 7.5). Reoptimization with one or two explicit water molecules also leads to a reduced barrier: 14.9 kcal mol^{-1} with two water molecules (Figure 7.2). The addition of a third explicit water has a very small additional effect. Thus, neither the continuum nor discrete microsolvation models recovers all of the barrier reduction produced by an aqueous solution. Combining the two methods—a PCM computation using the geometries of the solute with two explicit water molecules—leads to a reduction of the barrier by 7.8 kcal mol^{-1}, or about 87 percent of the experimental difference. It is interesting to note that the TS with the two explicit water molecules is much more asynchronous than the gas-phase TS. This increased asynchronicity induces a larger polarity in the TS. Clearly, both local and bulk effects are important in understanding this system. The local solvent effect is associated with enhanced hydrogen bonding and accounts for about

TABLE 7.5 Activation Enthalpy (in kcal mol⁻¹) for the Reaction of Butadiene with Acrolein Computed for the Gas Phase, PCM, and Explicit Water Models[a]

TS	$\Delta H^{\ddagger}_{298}$
Vacuum	
endo-s-cis	19.1
exo-s-cis	19.2
PCM	
endo-s-cis	15.3
exo-s-cis	16.4
One Water	
endo-s-cis	16.3
exo-s-cis	16.6
Two Water	
endo-s-cis	14.9
exo-s-cis	16.0

[a]Ref. 32.

half of the barrier reduction, while the hydrophobic (bulk) effect accounts for the other half.

The endo/exo selectivity is also well predicted by the PCM and explicit water computations. The difference in the *s-cis endo* and *exo* gas-phase activation enthalpies is only 0.1 kcal mol⁻¹ (Table 7.5). This difference increases to 1.1 kcal mol⁻¹ with the PCM calculation or explicit water computations.

7.2 GLUCOSE

The structures of carbohydrates have fascinated chemists since the dawn of modern organic chemistry with the work of Fischer.[34,35] While Fischer determined the stereochemical relationship of the simple sugars, controversy still remains as to some of the more subtle features of their three-dimensional structures. In particular, the origins of the anomeric and exo-anomeric effects remain topics of debate.[36,37] Complicating these discussions is the role that solvent may play in perhaps preferentially stabilizing some conformer(s) over others. We will concentrate our attention in this section to structure of D-glucose and the role that aqueous solvent has in altering its conformational preference.

The structure of D-glucose is fascinating because of its complexity. The molecule may appear as a straight chain aldehyde or in one of two cyclic hemiacetal forms, the furanose or pyranose ring. The pyranose ring is likely to exist as a chair conformation, but here are two different chair arrangements. On top of this, many hydroxyl groups can participate in intramolecular hydrogen bonding that can stabilize some conformations better than others. A significant complication is that in aqueous solution, one needs to consider whether the hydroxyl groups will prefer to form intramolecular hydrogen bonds or form hydrogen bonds with the solvent molecules. These intermolecular hydrogen bonds might significantly alter the conformational distribution of glucose.

In order to consider the effect of solvation on the conformational preference of D-glucose, we will first examine two simple analogs: 1,2-ethanediol and 1,2,3-propanetriol. These molecules possess fewer hydroxyl groups than glucose, and so one hopes that some simplifying trends can be observed in terms of the competition between intra- and intermolecular hydrogen bonding. We then discuss the many computational studies of glucose that have addressed the nature of its conformational distribution in the gas and, especially, the aqueous solution phase.

7.2.1 Models Compounds: Ethylene Glycol and Glycerol

7.2.1.1 Ethylene Glycol 1,2-Ethanediol (ethylene glycol) is the simplest analog of a sugar that captures at least some of its critical features, namely, the possibility of intramolecular hydrogen bonding. Even a molecule as small as 1,2-ethanediol possesses a rather complicated conformational profile. Since there are three energy minima associated with torsional rotation about the C–C and each C–O bond, there are, in principle, 27 conformational isomers. However, many of these isomers are identical because of the symmetry, leaving a total of 10 unique conformers. These conformers are labeled by g^+, t, or g^- to indicate gauche clockwise, trans or gauche counterclockwise about the C–O bond and G^+, T, or G^- for the analogous relationship about the C–C bond. Newman projections of the 10 conformational isomers of 1,2-ethanediol are drawn in Figure 7.3.

A number of studies on the conformational isomers of 1,2-ethandiol have been reported, and the relative energies of the isomers from some of these studies are listed in Table 7.6. The structures of these isomers (shown in Figure 7.3) were optimized at MP2/6-31+G** by Hadad.[38] Inspection of Table 7.6 shows relatively uniform agreement among the different computational methods. The range in values is compressed in going from relative energies to relative free energies (compare columns B with C and D with G). The only serious discrepancy is that the g^-G^+g conformer is a first-order saddle point at MP2/6-31+G**.[38] All other computations, performed with basis sets that omit diffuse functions, indicate that the g^-G^+g conformer is a local minimum, although in a shallow well.[39]

All the computational methods concur as to which are the lowest two isomers in the gas phase: tG^+g^- and $g^+G^+g^-$. These two isomers appear to possess an internal hydrogen bond: the alcohol proton of one hydroxyl group is donated to the oxygen of the other hydroxyl group. The stabilization afforded by this hydrogen bond does rationalize why these two isomers are the most stable. A molecular dynamics study casts some doubt as to the presence of this interaction.[42] The O–H distribution function of the terminal hydroxide and the hydrogen-bonded hydroxide are very similar, suggesting that there is no intramolecular hydrogen bond. On the other hand, the O–O distances are consistent with the presence of such an interaction. No bond critical point exists corresponding to the O–H···H bond, suggesting a lack of

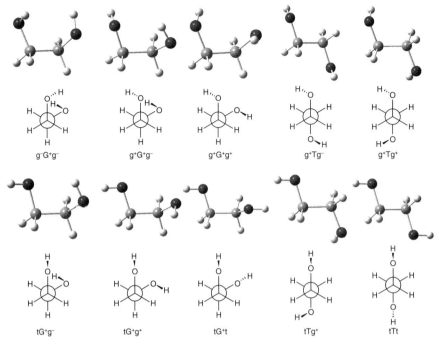

Figure 7.3 MP2/6-31+G** optimized conformers of 1,3-ethanediol.[38]

hydrogen bonding.[43,44] However, Kjaergarrd[41] reported the OH-stretching overtone spectrum of 1,2-ethanediol, noting the redshift of the OH stretches involved in the hydrogen bond. This shift is weaker than that seen in the water dimer, but nonetheless is demonstrative of hydrogen bonding in both the tG$^+$g$^-$ and g$^+$G$^+$g$^-$ isomers.

The equilibrium population of each conformational isomer, computed according to the standard Boltzmann distribution, accounts for the differing degeneracies of the various conformers and allows for the comparison with experiment. These populations are listed in Table 7.7, using the free energies obtained with different computational methods. Experimental microwave spectroscopy indicates that the dominant conformer is tG$^+$g$^-$.[45] Infrared (IR) spectroscopic results indicate that the g$^+$G$^+$g$^-$ isomer is present as a small fraction.[46] The computational results are roughly in accord with these experiments.

Since it appears that internal hydrogen bonds play a role in dictating the gas-phase conformational energy landscape of 1,2-ethanediol, moving to aqueous phase poses an interesting question: how will water compete in making hydrogen bonds to the hydroxyl groups versus this intramolecular hydrogen bond? In other words, will any of the higher energy conformers be better solvated by water, presumably through more and/or better hydrogen bonds, than will the tG$^+$g$^-$ or g$^+$G$^+$g$^-$ isomers, which can make fewer hydrogen bonds with the solvent water molecules? A number of different approaches toward accounting for the solvent effects on the conformational energy profile of 1,2-ethanediol have been taken, in

TABLE 7.6 Computed Gas-Phase Relative Energies^a (in kcal mol^{-1}) of the Conformers of 1,3-Ethanediol

Conformer	Method						
	A	B	C	D	E	F	G
tG$^+$g$^-$	0.00	0.00	0.00	0.0	0.0	0.0	0.00
g$^+$G$^+$g$^-$	0.56	−0.12	0.42		0.8	0.6	0.48
g$^-$G$^+$g$^-$	1.53	1.39	0.84	1.24	1.5	1.3	1.01
tTg$^+$	2.37	3.14	2.22		2.6	2.1	1.93
g$^+$Tg$^-$	2.35	2.60	2.64		2.9	2.3	2.44
g$^+$G$^+$g$^+$	3.22	3.11	2.87		3.1	2.4	2.54
tG$^+$g$^+$	3.94	4.20	3.03	4.15	3.7	3.0	2.55
g$^+$Tg$^+$	2.69	2.85	3.21		3.6	3.0	3.10
tTt	2.08	3.34	3.36		2.3	1.9	2.86
tG$^+$t	3.61	4.27	3.43	4.23	3.6	3.1	2.56

^{a}A, relative energies at HF/cc-pVDZ[40]; B, relative energies at CCSD(T)//cc-pVDZ//MP2/cc-pVDZ[40]; C, relative free energies defined as E(MP2/cc-pVTZ//MP2/cc-pvDZ)-E(MP2/ccpVDZ) +E(CCSD(T)//cc-pVDZ//MP2/cc-pVDZ)+$\Delta G_{vib\text{-}rot}$-RTln$\omega^{40}$; D, relative energies at B3LYP/6-311G(d,p)[39]; E, relative free energies at MP2/6-31+G**//MP2/6-31+G**[38]; F, relative free energies at B3LYP/6-311+G**//MP2/6-31+G**[38]; G, relative free energies at B3LYP/aug-cc-pVTZ.[41]

particular the SM and microsolvation models. An NMR study of 1,2-ethanediol in deuterated water indicated that 88 percent of the alcohol is in the gauche conformation about the C–C bond.[47]

Cramer and Truhlar[40] examined the aqueous distribution of 1,2-ethanediol conformations by using the PM3-SM3 solvation model. They noted the poor performance of semiempirical methods to predict the conformer populations in the gas phase. So, the PM3-SM3 solvation energy for each conformer with the geometry obtained at MP2/cc-pVDZ was added to the free energy computed at CCSD(T)//cc-pVDZ//MP2/cc-pVDZ. A population was then determined from a standard Boltzmann distribution, and these results are listed in Table 7.7. Their prediction that the population of all conformers with a gauche arrangement about the C–C is 84 percent is in fine agreement with the experimental estimate of 88 percent. They also suggest that the conformers with an internal hydrogen bond, namely, tG$^+$g$^-$ and g$^+$G$^+$g$^-$, remain the dominant population even in solution. Its population decreases from about 90 percent in the gas phase to 70 percent in water. Thus, the inherent stability of intramolecular hydrogen bonding outweighs the benefit of (perhaps) more intermolecular hydrogen bonding between the other conformations of 1,2-ethanediol and the solvent.

Microsolvation studies of 1,2-ethanediol with water have also been performed. The first study examined the complex with one to three water molecules using B3LYP/6-311++G*.[48] This study suggested that cyclic structures where the water molecule(s) bridge the two hydroxyl groups were more stable than acyclic structures. This result can be understood simply from the viewpoint that these cyclic

TABLE 7.7 Computed Gas-Phase Populations of the Conformational Isomers of 1,2-Ethanediol[a]

Conformer	Gas-Phase Method			Aqueous-Phase Method	
	A	B	C	D	E
tG$^+$g$^-$	48.7	55.9	77.3	68.5	44.8
g$^+$G$^+$g$^-$	43.2	27.4	20.4	26.5	25.6
g$^-$G$^+$g$^-$	6.2	13.4	0	0	11.9
tTg$^+$	0.6	1.3	1.0	2.1	8.0
g$^+$Tg$^-$	0.6	0.6	0.3	0.7	4.8
g$^+$G$^+$g$^+$	0.4	0.4	0.2	0.6	0.5
tG$^+$g$^+$	0.1	0.3	0.2	0.5	0.9
g$^+$Tg$^+$	0.2	0.2	0.1	0.2	1.9
tTt	0.1	0.2	0.4	0.7	1.2
tG$^+$t	0.0	0.2	0.1	0.2	0.5
Total C–C gauche	98.6	97.6	98.2	96.3	84.3
Total internal H-bond	92.0	83.4	97.7	95.0	70.4

[a] A, derived from E(CCSD(T)//cc-pVDZ//MP2/cc-pVDZ) $+\Delta G_{\text{vib-rot}}$-RTln$\omega$[40]; B, derived from E(MP2/cc-pVTZ//MP2/cc-pvDZ)-E(MP2/ccpVDZ)+E(CCSD(T)//cc-pVDZ//MP2/cc-pVDZ)+ $\Delta G_{\text{vib-rot}}$-RTln$\omega$[40]; C, derived from free energies at B3LYP/6-311+G**//MP2/6-31+G**[38]; D, derived from free energies at B3LYP/aug-cc-pVTZ[41]; E, derived from method A + PM3-SM3.[40]

structures contain *more* hydrogen bonds than their acyclic partners. The study is impaired, however, by the very limited number of configurations examined.

Haddad[38] examined the microsolvation of 1,2-ethanediol with one or two water molecules with a much broader range of configurations. First, 1000 complexes were generated by a pseudo-MC search using the AMBER force field. All of the unique complexes were then optimized at HF/6-31G*, and then all of the resulting unique structures were reoptimized at MP2/6-31+G*. The converged structures were then confirmed to be local energy minima by frequency analysis. The population of any given complex was determined by the Boltzmann distribution using the free energy computed at B3LYP/6-311+G**//MP2/6-31+G*. They were able to identify 21 complexes of 1,2-ethanediol with one water and 57 unique complexes with two water molecules. The four lowest energy complexes with one (**3**) or two water (**4**) molecules are shown in Figure 7.4. Subramanian and coworkers[49] performed a more limited configuration sampling procedure at M06-2X/6-311++G**//M06-2X/6-31+G* but found the same lowest energy configuration for **3** and **4** as shown in Figure 7.4. They also located the lowest energy complexes of 1,2-ethanediol with three or four water molecules.

The four lowest energy complexes of 1,2-ethanediol with one water molecule, **3a–d** comprise 73.4 percent of the total complex population. The two dominant complexes, **3a** and **3b**, have the same underlying tG$^+$g$^-$ conformation of 1,2-ethanediol with water bridging across the two hydroxyl groups, differing slightly in the orientation of the free O–H bond of water. Although this tG$^+$g$^-$ conformation could have an intramolecular hydrogen bond, the O $\cdots$ H distance

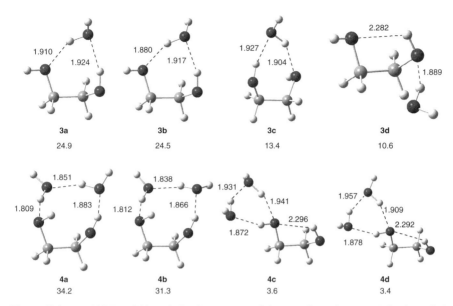

Figure 7.4 MP2/6-31+G** optimized structures of the complexes between 1,2-ethanediol and one (**3**) or two (**4**) water molecules and their relative populations.[38]

is too long (over 2.8 Å) to comprise a hydrogen bond. The same is true for **3c**, which has the g$^+$G$^+$g$^-$ conformation. But the last complex, **3d**, with the tG$^+$g$^-$ conformation, does appear to maintain a weak intramolecular hydrogen bond. In this complex, the water molecule acts as the hydrogen donor to the hydroxyl group that is itself the donor in the intramolecular hydrogen bond.

More than 65 percent of the population of the complexes formed between 1,2-ethanediol and two water molecules is due to **4a** and **4b**. In both the complexes, the 1,2-ethanediol adopts the g$^-$G$^+$g$^-$ conformation and the two water molecules bridge the hydroxyl groups in a nine-member ring. They differ slightly in the orientations of the O–H bonds not involved in the hydrogen bonding. In the other two low energy complexes, **4c** and **4d**, the two water molecules combine with one hydroxyl group to make up a six-member ring with three hydrogen bonds, while the other hydroxyl appears to donate its hydrogen in making a weak intramolecular hydrogen bond. The intermolecular hydrogen bonds of **4a** and **4b** are shorter and stronger than those in **4c** and **4d**, and these outweigh the added weak intramolecular hydrogen bond of the latter two complexes.

A cursory analysis of the SM results of Cramer and Truhlar and the microsolvation results of Hadad might suggest that the two are at odds with each other. The SM results suggest that the intramolecular hydrogen bond persists in solution, while the microsolvation results suggest that complexes that maximize the number of hydrogen bonds to the solvent are preferred. It is important, however, to keep in mind the limitations of each model. The SM approach underplays the importance

458   SOLUTION-PHASE ORGANIC CHEMISTRY

of individual hydrogen bonds between the solvent and the solute. The microsolvation model neglects hydrogen bonds between the water molecules at the surface of the solute and the bulk water.

7.2.1.2 Glycerol   The next larger homolog is 1,2,3-propanetriol (glycerol) **5** and can also be used as a model for the importance of intramolecular hydrogen bonding in sugars. As with 1,2-ethanediol, the conformational space of 1,2,3-propanetriol is extensive and rife with degeneracies. There are 486 possible conformations assuming only staggered arrangements about each C–C and C–O bond, but only 126 of these are unique. These isomers can be designated along the same lines as described for 1,2-ethanediol, but a simplification can be made by focusing on the arrangements about the two C–C bonds. As shown in Scheme 7.2, there are three conformations about each C–C that can be designated as α, β, or γ. There are then six backbone conformations, designated as αα, αβ, αγ, ββ, βγ, and γγ.

Hadad[50] has performed an extensive computational examination of glycerol, for both the gas and solution phases. All 126 conformational isomers were generated and then optimized at HF/6-31G* and B3LYP/6-31G*, resulting in 75 and 76 unique gas-phase energy minima, respectively. The structures of eight of the lowest energy conformers are drawn in Figure 7.5, and their energies are listed in Table 7.8. While density functional theory (DFT) predicts the relative energies for

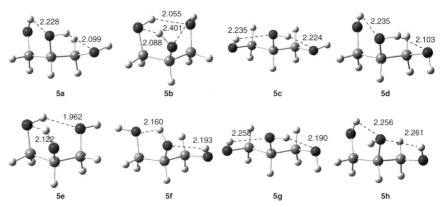

Figure 7.5   HF/6-31G* optimized low energy conformers of 1,2,3-propanediol **5**.[50]

TABLE 7.8 Computed Relative Energies (in kcal mol^{-1}) of Low Lying Conformers of 1,2,3-Propanetriol[a,b]

	Backbone	A	B	C	D	E	F
5a	αγ	0.00	0.00	0.00	0.11	0.00	0.10
5b	γγ	0.27	0.51	0.15	0.00	0.60	0.48
5c	αα	0.80	0.11	0.27	0.16	0.43	0.03
5d	αγ	0.74	0.46	0.30	0.75	0.17	0.02
5e	γγ	1.04	0.71	0.72	0.76	0.83	0.62
5f	αγ	1.33	0.79	0.81		0.62	0.39
5g	αα	1.40	0.55	1.09		0.64	0.00
5h	γα	1.58	1.12	1.50		0.36	0.18

[a]See Ref. 50.
[b]Computational methods are A, HF/6-31G*; B, B3LYP/6-311+G(3df,2p)//B3LYP/6-31G*; C, G2MP2; D, M06-2X/cc-pVTZ[51]; E, SM5.42/HF/6-31G*; F, B3LYP/6-31+G**//SM5.42/HF/6-31G*.

some conformers in a somewhat different order than does the well-tested G2MP2 composite method, these energy differences are relatively small and become smaller with increasing basis set size. Jeong, et al.[51] reproduced this study using M06-2X/cc-pVTZ. They located 78 unique gas-phase structures, and the lowest energy structures coincide with those found by Hadad.

The lowest energy gas-phase conformer is **5a**, which adopts the same conformation about each C–C bond as in the lowest energy conformation of 1,2-ethanediol. Conformer **5b**, the second most stable according to G2MP2 but the lowest energy conformer according to M06-2x, is the only conformer that possesses three intramolecular hydrogen bonds. The other low lying conformations possess two intramolecular hydrogen bonds.

The populations of the conformers of glycerol were computed using standard Boltzmann distributions and are listed in Table 7.9. A gas-phase electron diffraction study indicated that the most likely configurations of glycerol have the αα and αγ backbones. The G2MP2 and DFT computations agree that these two configurations account for more than 50 percent of the total population.

How does the conformational distribution change in going from the gas-phase to aqueous phase? Hadad optimized the structures of the 75 conformations of glycerol using the SM5.42/HF/6-31G* method. DFT single-point energies were also obtained as shown in Eq. (7.1). The aqueous energies are shown in Table 7.8. The largest change in the relative ordering of the low energy conformers is with **5b**. In the gas phase, this conformer is the second lowest in energy, stabilized by its three intramolecular hydrogen bonds. In solution phase, it is the fifth lowest energy conformer because all three hydroxyl groups are involved in these intramolecular hydrogen bonds and are thus precluded from hydrogen bonding with the solvent. Nonetheless, it remains low in energy, as do many conformers with two internal hydrogen bonds. Conformers that do not possess any internal hydrogen bonds and therefore can maximize their participation in making hydrogen bonds with the solvent are among the highest energy conformers in solution, with relative energies

TABLE 7.9 Computed Populations of the Lying Conformers of 1,2,3-Propanetriol[a,b]

	A	B	C	D	E	F	G
5a	12	10					
5b	12	6					
5c	7	10					
5d	8	6					
5e	7	4					
5f	5	5					
5g	5	6					
5h	3	3					
αα	19	27	11	18	30	24	20
αγ	31	29	51	27	28	32	29
αβ	12	11	4	23	20	20	21
ββ	2	2	0	3	2	1	5
βγ	18	21	5	24	16	13	15
γγ	18	9	28	4	4	9	10

[a] See Ref. 50.
[b] Computational methods are A, G2MP2; B, B3LYP/6-311+G(3df,2p)//B3LYP/6-31G*; C, M06-2x/cc-pVTZ[51]; D, SM5.42/HF/6-31G*; E, B3LYP/6-31+g**//SM5.42/HF/6-31G*; F, SMD/M06-2x/cc-pVTZ[51]; G, Average values from two experimental studies.[52,53]

of 4–6 kcal mol^{-1} above **5a**. Solvation does compress the energy scale of the conformers relative to the gas phase, but low energy gas-phase conformers remain low energy structures in the solution phase.

$$E_{\text{B3LYP/6-31+G**//SM5.42/HF/6-31G*}}(\text{aq}) = E_{\text{B3LYP/6-31+G**//SM5.42/HF/6-31G*}}(\text{gas}) \\ + E_{\text{SM5.42/HF/6-31G*}} - E_{\text{HF/6-31G*}}(\text{gas}) \quad (7.1)$$

The computed solution-phase backbone populations are listed in Table 7.9. The principal difference between the gas and solution phase is that there is greater percentage of αβ in solution at the expense of the βγ and γγ configurations. The computed distributions are in fine agreement with the experimental distributions. Again, the αα and αγ configurations dominate the distribution in solution, with little ββ or γγ present.

The main conclusion to be drawn from the studies of the two model compounds (ethylene glycol and glycerol) is that intramolecular hydrogen bonding persists in the solution phase. Obtaining a reasonable description of a sugar in aqueous solution will demand that both the intramolecular and intermolecular hydrogen bonding be treated equally well.

7.2.2 Solvation Studies of Glucose

D-Glucose is the most ubiquitous carbohydrate in nature, found in table sugar, starch, and cellulose.[34,35] Its broad array of configurations and conformations makes it a natural target of experimental and computational studies. D-Glucose can appear as an acyclic aldehyde, in a six-member ring pyranose arrangement with the anomeric hydroxyl group in the α (**6**) or β (**7**) position, or in a five-member furanose arrangement with either α (**8**) or β (**9**) positioning of the anomeric hydroxyl group. In aqueous solution, D-glucose is almost entirely found in its pyranose form, with a 36 : 64 ratio of α : β isomers. In this section, we concentrate on computational approaches toward predicting the structure of D-glucose in the gas and aqueous phases. The structures of the configurations and conformations of D-glucose discussed in this section are shown in Figure 7.6. The geometries were optimized at B3LYP/6-31G(d,p), following on the structures reported by da Silva.[54]

As with any saturated six-member ring, the ring-flip process interconverts two-chair conformations. For D-glucose, these chair conformations are designated $^4C_1$ for **6** (also for **7**) and $^1C_4$ for its ring-flipped conformer **10**. Introductory organic chemistry texts instruct us that equatorial substitution is favored over axial, and therefore, **6** is more stable than **10**.[55,56] In fact, little computational work has been reported on the relative energies of the $^4C_1$ conformers of D-glucose. Momany[57] reported that **10** (the lowest energy conformation among the six examined) is 4.8 kcal mol^{-1} higher in energy (at B3LYP/6-311++G**) than the lowest energy conformer of **6** (Table 7.10). Cramer and Truhlar,[58] using the SM4 procedure, examined two $^4C_1$ and two $^1C_4$ conformations and found the latter two to be at least 5 kcal mol^{-1} higher in free energy in water than the $^4C_1$ conformations.[58] We will, therefore, not concern ourselves any further with the population of the $^1C_4$ structures.

Since almost no furanose form of D-glucose is observed in solution, it is expected that the energies of the furanose conformational isomers will be much greater than those of the pyranose form. In fact, the four furanose isomers examined are predicted to be about 4–5 kcal mol^{-1} higher than the lowest pyranose conformers **6a** and **6c** at HF/6-31G**.[59] Quite remarkable, however, was the B3LYP/6-31G**

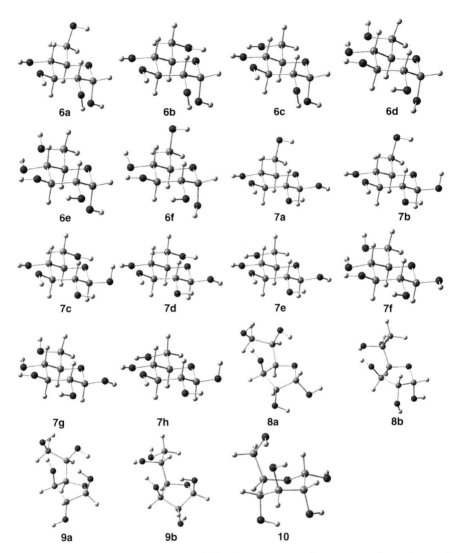

Figure 7.6 B3LYP/6-31G(d,p) optimized structures of some conformations of D-glucose.[54]

results,[59] which predict furanose **8b** to be the lowest energy isomer, 0.2 kcal mol^{-1} below the pyranose **6c** (Table 7.10). This failure was attributed to a basis set superposition error (BSSE), but it is probably more likely that diffuse functions are necessary to adequately treat molecules with many intramolecular hydrogen bonds.[60,61] B3LYP/311++G**//B3LYP/6-31G* computations place the four furanose isomers of D-glucose just over 4 kcal mol^{-1} above the most stable pyranose form,[61] although CCSD(T)/complete basis set (CBS) computations suggest that the energy difference is much smaller, that is, only 1.3 kcal mol^{-1}.[62]

TABLE 7.10 Computed Relative Energies (in kcal mol^{-1}) of D-Glucose Configurations^a

	A	B	C	D	E	F	G	H	I
6a	0.0	0.0	0.0	0.1	0.6	0.2	0.1	0.1	0.0
6b	0.3	0.1	0.0	0.2	0.8	0.0	0.0	0.0	
6c	0.3	0.2	0.5	0.0	0.0	0.4	0.5	0.4	
6d	1.4	1.3	1.2			1.4			
6e	3.3	3.1	2.8			3.0			
6f						1.1			
7a	1.4	0.9	0.4	0.9	2.2	0.5	0.3	0.4	1.1
7b	2.4	1.9	1.3			1.4	1.2		
7c				1.9	2.8	0.3	0.9		
7d	1.7	1.1	0.4				0.4	0.3	
7e	1.8	1.3	1.0	1.0	1.8	0.8	0.8	0.7	
7f	4.3	3.8	3.1			3.2			
7g	4.7	4.1	3.3			3.3			
7h						1.6			
8a				4.5	2.6	4.4			1.3
8b				4.7	5.5	4.2			
9a				3.7	−0.2	4.6			
9b				5.0	2.7	4.7			
10								4.88	6.8

^{a}Computational methods: A, relative energy at MP2/cc-pVDZ[63]; B, relative energy as E(MP2/cc-pVTZ//MP2/cc-pvDZ)+[E(CCSD/6-31G*//MP2/6-31G*)-E(MP2/6-31G*)]+[E(HF/cc-p^{T}TQZ//MP2/cc-pvDZ)-E(HF/cc-pVTZ//MP2/cc-pvDZ)][63]; C, relative free energy as B+ΔG(rot-vib)[63]; D, relative energy at HF/6-31G**//HF/6-31G**[59]; E, relative energy at B3LYP/6-31G**//B3LYP/6-31G**[59]; F, relative free energies at B3LYP/311++G**//B3LYP/6-31G*[61]; G, relative free energy at B3LYP/6-311G++(2d,2p)//B3LYP/6-31G(d)[54]; H, relative free energy at B3LYP/6-311++G**[57]; I, Relative energies at CCSD(T)/CBS//SCS-MP2/def2-TZVPP.[62]

Since the pyranose forms of D-glucose have been computed many times over, representative energies of various conformers of **6** and **7** are listed in Table 7.10. While there is some variation in the ordering of the particular conformers with basis set and computational method, all of the computations confirm that in the gas phase, the α conformers are lower in energy than the β conformers. The lowest energy conformer is either **6a** or **6b**, depending on method, and generally they differ in energy by less than 0.2 kcal mol^{-1}. The only exception is with the B3LYP/6-31G** method,[59] which predicts **6c** to be the most stable pyranose isomer, but keep in mind that this method also erroneously predicts the furanose form to be the most stable.

Much of the computational literature of glucose pertains to the ratio of the two anomers. This is formulated as the anomeric effect: the stability of the α anomer, with its axial group, over the expectedly more stable β anomer, where the hydroxyl group occupies the equatorial position. This has been rationalized in two ways: (1) minimizing the dipole–dipole repulsion because of the C–O bonds or (2) hyperconjugative stabilization by donation of the ring oxygen lone pair into the axial C–O antibond of the α anomer. The latter argument is consistent with certain geometric trends, notably the lengthened C_1-O_1 bond, the shortened C_1-O_6 bond, and the widened O–C–O angle. We will not address the origins of the anomeric effect here, instead focusing our attention on the relative populations of the two pyranose anomers in the gas and solution phases, especially in water where intermolecular hydrogen bonds between glucose and water can compete with the intramolecular hydrogen bonds in glucose.

Inspection of the results of the energies of the pyranose conformers shown in Table 7.10 shows that there is little dependence on the computational method. We therefore focus on the study by Cramer and Truhlar,[63] the first that attempted a systematic review of the conformational space of D-glucose. They selected six conformers from among the 18 lowest energy β-D-glucose structures and five conformers from among the 11 lowest energy α-D-glucose structures obtained by an exhaustive MM3 search. These conformers were selected as representative of the rotational isomers about the C_5-C_6 and C_6-O_5 bonds. Later studies added structures that include other conformations about the C_1-O_1 bond.

All of the conformations exhibit intramolecular hydrogen bonding. The most stable α anomers (**6a–c**) have a counterclockwise arrangement of the OH hydrogen-bond donor groups on C_2, C_3, and C_4, as do the lowest energy β anomers (**7a–e**). The other isomers have a clockwise arrangement of the OH groups. Momany[57,64] has argued that the shorter (and presumably stronger) hydrogen bond between the proton of the 2-hydroxyl group and the oxygen of the 1-hydroxyl group in the α anomer (~2.25 Å) than in the β anomer accounts for the stability of the former structure.

Using the Boltzmann distribution, the populations of the glucose isomers can be computed. These populations, computed with a few different methods, are listed in Table 7.11. These population values must be taken with some caution. Only a small handful of conformations are included of the over 3000 potentially available. It is possible that some low energy structures that might alter the distribution have been omitted. With this caveat, the computed populations using four different methods are in reasonable agreement with each other, with perhaps the MP2 result indicating a slightly high population of the α anomer. The important result is that in the gas phase, the α-D-glucose configurations will dominate the distribution; Cramer and Truhlar's best estimate is that the α-to-β ratio is 57 : 43.[63] Evidently, the anomeric effect is an inherent component of glucose.

This gas-phase distribution is in direct conflict with the solution population where the β anomer is favored by about 2 : 1. As we presented with the studies of ethylene glycol and glycerol, the key issue for computing the solvation energy of glucose will be the competition between forming intramolecular hydrogen bonds and intermolecular hydrogen bonds. Continuum methods may not be able to capture the full importance of hydrogen bonding between the sugar and solvent molecules, while microsolvation approaches might overestimate their importance.

Unfortunately, no one has taken on a study that attempts to combine these two methods—optimize the structures of glucose with a small number of explicit water molecules and then perform a continuum computation on this hydrated sugar

TABLE 7.11 Computed Populations of D-Glucose Conformations

	Gas				Sol			
	A	B	C	D	E	F	G	H
6a	25	21	19	30	11	15	7	14
6b	25	25	26	27	29	20	25	32
6c	12	11	13	9	2	4	2	2
6d	3		2	3	0		0	0
6e	0		0	0	0		0	0
6f			4	8			1	3
7a	14	15	11	10	7	15	6	7
7b	3	3	2	2	9	6	7	7
7c		5	14	9		9	45	32
7d	13	13			37	26		
7e	5	7	6	2	3	5	3	1
7f	0		0	0	0		0	0
7g	0		0	0	0		0	0
7h			2	0			2	1
α:β	65:35	57:43	64:35	77:23	42:56	39:61	35:63	51:48

A, computed using the free energies as described in Method C of Table 7.10[63]; B, computed using the free energy at B3LYP/6-311G++(2d,2p)//B3LYP/6-31G(d)[54]; C, computed using the free energies at B3LYP/311++G**//B3LYP/6-31G*[61]; D, computed using the free energies at MP2/6-311++G**//B3LYP/6-31G*[61]; E, computed using the free energies at SM5.4/AM1//MP2/cc-pVDZ[63]; F, computed using the free energies at IEF-PCM/B3LYP/6-311G++(2d,2p)//B3LYP/6-31G(d)[54]; G, computed by adding the free energy of solvation at SM5.4/AM1 to method C[61]; H, computed by adding the free energy of solvation at SM5.4/AM1 to Method D.[61]

complex. This will be a rather daunting task, given the large number of glucose conformations, the large number of water configurations possible, especially as the number of explicit water molecules increases, and the necessity of using a post-HF method or a density functional that can account for dispersion. Rather, we discuss now a few continuum studies and a few microsolvated studies.

The self-consistent isodensity polarized continuum model (SCIPCM) study by Wladkowski and Brown[65] demonstrate one of the pitfalls of computing the α-to-β ratio of glucopyranose. They compared the energies of **6c** and **7e** to assess the difference in anomeric energies in the gas and solution phases. At MP2/6-31+G(d), **6c** is 2.6 kcal mol^{-1} lower in energy than **7e**, and this difference is reduced to 1.8 kcal mol^{-1} with the inclusion of solvation through the SCIPCM treatment. Solvation is, therefore, favoring the β anomer, however, not enough to make it the lower energy form. They concluded that this failure to reproduce the experimental preference of the β anomer over the α anomer means that the continuum method is failing and that explicit solvent molecules need to be incorporated into the calculation. However, drawing a conclusion about the population of anomers that are composed of many conformers on the basis of a single conformer neglects the possibility that the relative ordering of the conformations in the gas phase may not be the same in solution. In fact, as we shall see, neither of these two conformers comprises the major fraction of the solution population.

In Table 7.11, we list the results of four computational studies that address the glucose anomer population in solution. All four studies use a number of conformers of each anomer and treat the solvent with either SM or PCM. As with their prediction for gas-phase populations, these four methods are generally in good agreement; the MP2[61] results again appear to overestimate the α population.

More importantly, the computed anomer populations are in fine agreement with the experimental value of the ratio of α to β anomer is 36 : 64. Especially good agreement is produced using B3LYP, a basis set that includes both diffuse and polarization functions, and either PCM[54] or SM[61] to treat the effect of water. These calculations show that the solvation energy is more negative for β conformers than for the α conformers. In particular, conformers **7c** and **7d** are dramatically stabilized in water; their population growth in water versus that in the gas is at the expense, primarily, of **6a**. The SM procedure allows for the solvation energy to be apportioned to atoms or groups; Cramer and Truhlar[63] have shown that the solvation contribution of the anomeric hydroxyl group is much greater in the equatorial position than that in the axial position, accounting for the shift in population from α to β when solvated.

Da Silva et al.[54] computed the optical rotation of three α and five β conformations of D-glucose in aqueous solution using a TDDFT/GIAO approach with either the aug-cc-pVDZ or 6-31++G(d,p) basis sets. The α conformations all have large positive values of $[\alpha]_D$, while the β conformations have values that range from −10° to +80°. The Boltzmann-weighted average value is 62.6° or 58.8° with the aug-cc-pVDZ or 6-31++G(d,p) basis sets, respectively. These values are reasonably close to the experimental value of 52.7°.[66] For those interested, Mennucci et al.[67] have reviewed the application of computational models of solvent effects on chiroptical properties.

The agreement in the anomeric populations between experimental and computational studies is certainly encouraging. The computations correctly predict the inversion of anomeric preference between the gas and solution phases. However, there are some underlying concerns. Solution-phase NMR studies have determined the populations of the glucose conformations. The experimental populations of the gg (**6a, 6f, 7a, 7b**), gt (**6b, 7c, 7d**), and tg (**6c, 7e, 7h**) conformations is 51 : 47 : 2 for the α anomer and 53 : 45 : 2 for the β anomer.[68,69] The computational studies of Table 7.11 predict that the gt conformations will dominate the populations for both the α and β anomers.

To address the conformational sampling problem, Momany[70] performed a density functional theory molecular dynamics (DFTMD) whereby trajectories were generated from 36 energetically low lying glucose conformations. The dynamics were run at B3LYP/6-31+G* with the conductor-like screening model COSMO for treating the solvent (water). The resulting computed α-to-β anomeric ratio is 37 : 63, in fine agreement with experiment. The continuum solvent model is inverting the anomeric ratio relative to the gas-phase distribution.

Perhaps more disconcerting is the contention, based on the NMR studies, that sugars are not conformationally stationary in solution.[71–75] Rather, there appears to be free rotation about the anomeric C—O bond in the fructose ring of sucrose and all of the hydroxyl groups exchange protons at equivalent rates, suggesting that there is no *persistent* intramolecular hydrogen bonding in solution. This does not preclude *transient* intramolecular hydrogen bonding. These studies call into question the exclusive use of low energy *gas-phase* conformations in predicting the populations of sugars *in aqueous solution*. The ethylene glycol and glycerol studies indicate that these intramolecular hydrogen-bonded low lying gas-phase structures do dominate in solution but not to the same extent that they do in the gas phase. In other words, structures that possess fewer intramolecular hydrogen bonds become more populated in the solution because of more favorable hydrogen bonding with the solvent. The solution-phase computational studies of glucose to date (Table 7.11) neglect to incorporate any conformations that have some dangling hydroxyl groups, conformations that are not energetically competitive in the gas phase but may be important in solution, where these hydroxyl groups may participate in hydrogen bonds to water.

A few computational studies have been carried out on microsolvated glucose. The conformation/configuration space is much more extensive when explicit water molecules are included, and so restrictions must be placed to limit the space to be explored. Momany[64] has reported 26 configurations of the complex D-glucopyranose with one water molecule. The three lowest-energy α and β anomers complexed with one water (**11a–f**) are shown in Figure 7.7. Once again, the α anomers are lower in energy than the β anomers. Any attempt to

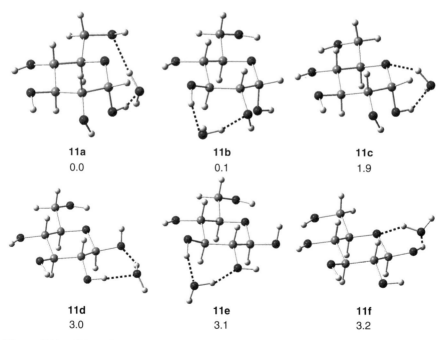

Figure 7.7 B3LYP/6-31G(d) optimized geometries of six low energy configurations of D-glucose with one water molecule (**11**).[64]

Figure 7.8 Lowest energy structure of the pentahydrate of D-glucose.[76]

produce a population ratio must be understood to have limited value because of the very small number of configurations actually utilized in the Boltzman distribution. Nonetheless, using these 26 configurations, the α-to-β ratio is 58 : 42 using electronic energies and 52 : 48 using free energies. The difference in these populations is primarily due to entropy terms. Especially important here are some configurations where the water molecule is involved in only one hydrogen bond; these configurations are entropically favored over those where the water is involved in two hydrogen bonds but are enthalpically disfavored by participating in fewer stabilizing hydrogen bonds. What is unclear is how these complexes would be affected by bulk water. What would be the arrangement of hydrogen bonds involving these complexing water molecules? What would be the competition between hydrogen bonding with neighboring water molecules versus with the sugar?

To begin to address this question, Momany[76] computed 37 different complexes of D-glucopyranose with five water molecules. These 37 configurations in no way span the complexity of the configurational space of this system. However, they do include examples where the water molecules are distributed to have hydrogen-bonding interactions at different locations about the glucose molecule, along with the opportunity for water molecules to hydrogen bond among themselves. In fact, the low energy configurations possess chains or rings formed by the water molecules, and the lowest free energy configuration (**12**), shown in Figure 7.8, exhibits this water chain structure. These results may indicate that water prefers to maintain its hydrogen bond network in preference to extensive incorporation into the intramolecular hydrogen bonding of glucose. This is consistent with Cramer and Trulhar's[58,63] result that the most populated conformations of aqueous-phase glucose retain their internal hydrogen bonds. Modeling of glucose in water clearly remains an ongoing challenge to computational chemists.

7.3 NUCLEIC ACIDS

The determination of the structure[77] of the DNA molecule is arguably the greatest achievement in structural chemistry of the twentieth century.[78] A key feature of this structure is the base pairing that joins the two strands. The base complementarity, guanine with cytosine and adenine with thymine, occurs by hydrogen bonding involving the appropriate tautomer of each base.

The structures and energies of the tautomers of the bases of the nucleic acids have been subject to numerous computational studies. Since DNA is found in the aqueous environment of the cell interior, it is essential that the base tautomers, and the base pairs, be computed in the aqueous phase. In this section, we discuss the structures of the tautomers of each of the four DNA bases and uracil, found in RNA, focusing on the differences in the gas and solution phases.[79] We then take on the structure of the base pairs in solution.

7.3.1 Nucleic Acid Bases

In this section, we present the results of computational studies of the five nucleic acid bases: cytosine **13**, guanine **14**, adenine **15**, thymine **16**, and uracil **17**. The canonical structures, those that are involved in the Watson–Crick base pairing within DNA, are drawn below. Other tautomers for each base can be energetically competitive with the canonical structure, and these other tautomers are invoked in some models of DNA mutations[80–83] and anomalous DNA structures.[84,85] The ensuing discussion focuses on the relative energies of the tautomers, in both the gas and solution phases. Structural changes that accompany this phase change are also noted.

7.3.1.1 Cytosine As with all of the nucleic acid bases, the cytosine tautomers have been the subject of many computational studies. We focus here on two recent studies that examined the relative tautomer energies in both the gas phase and the aqueous solution, noting some additional gas-phase studies as well. Hobza[86] examined the tautomers **13a–g** and Queralt[87] looked at **13a–b, 13e–h** (see Figure 7.9). The relative gas-phase energies were computed with a variety of computational methods; we list in Table 7.12 the results using RI-MP2/TZVPP,[86] CCSD(T)/cc-pVTZ,[86] CCSD(T)/cc-pVQZ,[88] B3LYP/6-31++G**,[87] and MPWB1K/aug-cc-pVDZ,[89] representing post-HF and DFT treatments. Tautomers **13a–e** are energetically clustered together, differing by less than 5 kcal mol^{-1}. (As will be seen with the other nucleic acid bases, Hobza has shown here that the RI-MP2 and CCSD(T) methods provide very comparable energetics for the tautomers.) Both MP2 and CCSD(T) predict that **13b** is lower in energy than the canonical tautomer **13a**, while B3LYP reverses the energy of these two

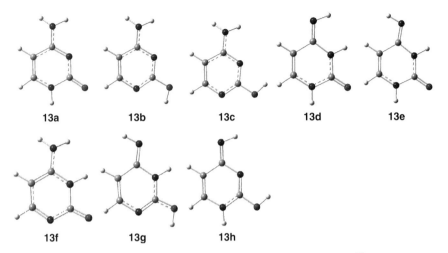

Figure 7.9 Structures of the cytosine tautomers **13a–h**.[86]

isomers. The MPWB1K functional, however, agrees with the wavefunction-based methods in putting **13b** lower in energy than **13a**. Extrapolation to infinite basis sets at HF, MP2, and CCSD(T) all predict that **13b** is the lowest energy tautomer. In fact, tautomer **13c** is also lower in energy than **13a** in the gas phase. However, since five cytosine tautomers are so close in energy, hydration can very well alter their relative energies.

The amino groups of **13a–c** are not planar. The inversion barrier at RI-MP2/TZVPP is 0.2, 0.2, and 0.4 kcal mol^{-1} for **13a–c**, respectively. For each tautomer, the sum of the angles about the amine nitrogen is about 350°.

All of the nucleic acid bases have multiple sites for hydrogen bond formation. Explicit microsolvation may be appropriate for describing the important interactions between the base and its first solvation shell. The complicating factor, as is always true when attempting to carry out microsolvation studies, is how many solvent molecules to place about the substrate and what configurations should be sampled. For the nucleic acid bases, the many tautomeric forms available further add to the number of potential geometry optimizations that need to be performed.

Hobza[86] identified the three lowest energy structures of the monohydrated cytosine tautomers **13a–e** using the RI-MP2/TZVPP method. Optimization of the many configurations that need to be sampled here would be extremely time consuming with the CCSD(T) method; however, the close agreement between these two methods for the gas-phase tautomer energies indicates that the more computationally efficient RI-MP2 method should suffice. (This same logic will apply to the other nucleic acid bases discussed later on.) Shown in Figure 7.10 are these monohydrated complexes for **13a** and **13b** (**13a-1W** and **13b-1W**). These structures differ in the location of the hydrogen bonds between cytosine and water. The lowest energy configuration involving tautomer **13a** has the water donating a hydrogen

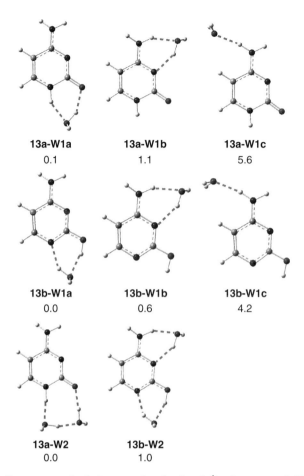

Figure 7.10 Structures and relative energies (kcal mol^{-1}) of mono- and dihydrated cytosine tautomers **13a** and **13b**.[86]

to the carbonyl oxygen and accepting the aromatic N–H hydrogen. Slightly higher in energy is the configuration where water donates a hydrogen to the aromatic N and accepts a hydrogen from the amine group. Much higher in energy is configuration **13a-W1c** where only one hydrogen bond is formed between water and **13a**. Similar structures are found for **13b**. The dihydrated tautomers follow similar sort of trends, along with the possibility of hydrogen bonding between the water molecules themselves. The relative ordering of the configurations for any given tautomer is a function of the strength of the maximum number of hydrogen bonds that can be formed, subject to the geometric constraints of the ability of the water molecule(s) to bridge across hydrogen-bonding sites. (With the other bases, we simply present the lowest energy microsolvated tautomers and refer the readers to the original papers if there is interest in the other arrangements of water about a particular tautomer.) The relative energies of the lowest energy configuration of

TABLE 7.12 Relative Energies (in kcal mol^{-1}) of Cytosine Tautomers (13a–h)

Tautomer	Gas Phase					Solution Phase			
	$\Delta E^{a,b}$	$\Delta E^{a,c}$	$\Delta E^{d,e}$	$\Delta G^{f,g}$	$\Delta G^{h,i}$	$\Delta E(W1)^{a,j}$	$\Delta E(W2)^{a,k}$	Bulk^{a,l}	Bulk^{d,m}
13a	1.9	1.6	0.0	0.3	0.2	0.1	0.0	0.0	0.0
13b	0.0	0.0	0.4	0.0	0.0	0.0	1.0	7.1	5.5
13c	0.72	0.7			0.7	2.0	2.7	6.5	
13d	4.91	3.4				5.2	5.4	4.1	
13e	3.21	1.9	1.8	4.0	0.7	3.5	3.3	4.6	5.8
13f			7.0						6.3
13g			13.6						20.1
13h			18.7						21.9

[a] Ref. 86.
[b] Calculated at RI-MP2/TZVPP//RI-MP2/TZVPP.
[c] Calculated at CCSD(T)/cc-pVTZ// RI-MP2/TZVPP.
[d] Ref. 87.
[e] Calculated at B3LYP/6-31++G**.
[f] Ref. 89.
[g] Calculated at MPWB1K/aug-cc-pVDZ//MPWB1K/aug-cc-pVDZ.
[h] Ref. 88.
[i] Calculated at CCSD(T)/cc-pVQZ//CCSD//cc-pVTZ.
[j] Tautomer and one water molecule calculated at R1-MP2/TZVPP//RI-MP2/TZVPP.
[k] Tautomer and two water molecules calculated at R1-MP2/TZVPP//RI-MP2/TZVPP.
[l] Relative free energies in water using the MST variation of PCM.
[m] Calculated at PCM/B3LYP/6-31++G**.

each monohydrated tautomers are listed in Table 7.12. Other studies[90,91] of monohydrated cytosine have also been reported, but limited to just the hydration of the canonical tautomer **13a**, with similar relative energies as in Figure 7.10. The most interesting result is that although **13b** remains the lowest energy tautomer, **13a** is now only 0.1 kcal mol^{-1} higher in energy. Adding a second explicit water molecule inverts the energy order. The most stable dihydrated complex of **13a** (**13a-2W**, see Figure 7.10) is 1.0 kcal mol^{-1} lower in energy than the dihydrated complex of **13b** (**1b-2W**).

Shishkin and Leszczynski[92] have examined the canonical cytosine **13a** surrounded by up to 14 water molecules. They noted some interesting changes (Table 7.13) in the structure of **13a** as it becomes more and more hydrated. First, the amine nitrogen becomes planar. As hydration increases, the C_2-O_7 distance increases, the C_2-N_3 distance decreases, the N_3-C_4 distance lengthens, and the C_4-N_8 distances shortens. The NBO charge goes from -0.62 e in **13a** to -0.80 in **13a-14W**. They argued that these structural changes are consistent with strong contribution of both resonance structures **A** and **B** shown in Scheme 7.3. The oxyanion nature allows it to act as the acceptor of three hydrogen bonds from three different water molecules. So while the structure of hydrated cytosine appears to be canonical, Shishkin and Leszczynski argue that its electron distribution is not; rather, it possesses a good bit of zwitterion character from resonance structure **B**.

TABLE 7.13 Geometric Parameters (B3LYP/6-31G(d)) of Gas-Phase and Hydrated 13a

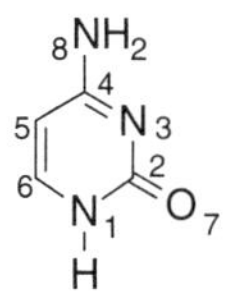

	13a[a]	13a-W1a	13a-W2	13a-W14[a]
$r(N_1-C_2)$	1.430	1.416	1.410	1.377
$r(C_2-N_3)$	1.373	1.365	1.365	1.339
$r(N_3-C_4)$	1.319	1.323	1.324	1.358
$r(C_4-N_8)$	1.366	1.364	1.364	1.329
$r(C_2-O_7)$	1.220	1.236	1.238	1.281
$\Sigma(NH_2)$[b]	353.2	353.9	353.8	359.7

[a]Ref. 92.
[b]Sum of the angles about the amine nitrogen.

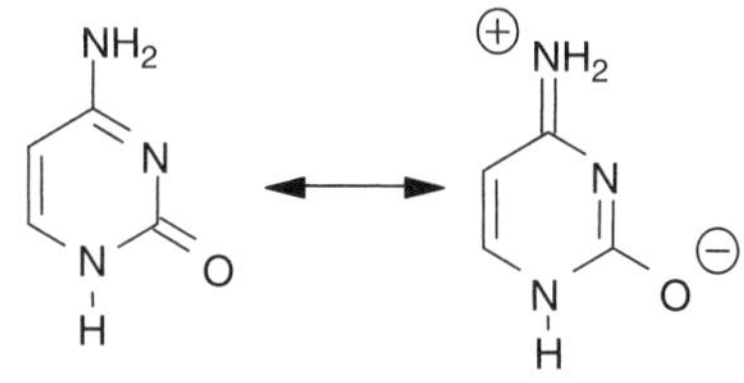

Scheme 7.3

The tautomers of cytosine have also been examined with continuum solvation methods. Both Miertus–Scrocco–Tomasi (MST) and PCM computations predict that **13a** is the lowest energy conformer in aqueous solution.[86,87] This is consistent with the explicit water computations. Aqueous solvation, therefore, substantially alters the relative energies of the tautomers of cytosine. The canonical representation of cytosine **13a**, the tautomer invoked in Watson–Crick base pairing, is in fact the most favorable tautomer in solution, but not in the gas phase.

7.3.1.2 Guanine
While there are many theoretical studies of the guanine tautomers, we focus on three recent studies that compared their gas-phase and aqueous-phase relative energies: that of Hobza,[93] who examined the tautomers **14a–h**; Goddard,[94] who examined **14a–j**; and Orozco,[95] who examined **14a–d** and **14f** (See Figure 7.11). For the gas phase (see Table 7.14), all of the computational studies indicate five low lying tautomers (**14a–f**). In agreements with these computational results, Mons[96] identified four guanine tautomers using gas-phase IR and UV spectroscopies: **14a**, **14b**, **14f**, and **14d** (but they actually could not

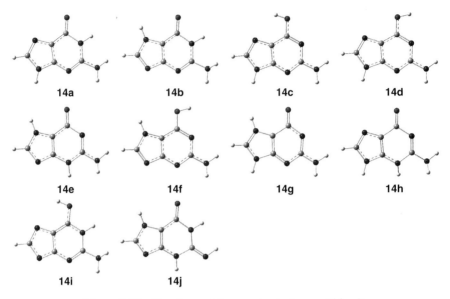

Figure 7.11 Structures of the guanine tautomers (**14a–j**).

distinguish **14c** from **14d**). The remaining structures are much higher in energy in the gas phase.

An interesting feature of the gas-phase guanine tautomers is that their amino group is nonplanar. The sum of the angles about the amine nitrogen is less than 348° for the tautomers **14a–i**, and specifically 340.4° and 337.4° in **14a** and **14b**, respectively.[93] This is more pyramidal than for any of the other nucleic acid bases.

The canonical guanine tautomer is **14a**, the structure invoked in Watson–Crick base pairing. Most of the computation methods predict that it is not the most stable tautomer in the gas phase. Rather, tautomer **14b** is slightly lower in energy. However, **14b** possesses the smallest dipole moment (1.87 D) of all of the tautomers. The dipole moment of **14a** is more than three times greater (6.29 D), and the very high lying tautomers **14g** and **14h** have dipole moments of 9.12 and 10.54 D, respectively. Since the dominant interaction in solvation is dependent on the size of the dipole moment, energy reordering of the tautomers in aqueous solution is very likely.

Hobza[93] optimized the microsolvation cluster of the guanine tautomers with one or two water molecules; the lowest energy configurations involving **14a–c** are shown in Figure 7.12. With one explicit water molecule, both **14b** and **14c** are below the energy of **14a** (Table 7.14). However, with two water molecules, **14a** is the most stable tautomer. The continuum solvation computations, performed with three differing methods, all indicate that **14a** is favored by about 1 kcal mol^{-1} over **14b**. The energy ordering of the other tautomers is dramatically altered by aqueous solvation. The two very high energy tautomers, **14g** and **14h**, which are disfavored by about 20 kcal mol^{-1} in the gas phase, are now only 5–9 kcal mol^{-1} higher in

TABLE 7.14 Relative Energies (in kcal mol^{-1}) of Guanine Tautomers (14a–j)

Tautomer	Gas Phase					Solution Phase				
	ΔE^{ab}	ΔE^{ac}	ΔG^{de}	$\Delta G^{f,g}$	ΔG^{hi}	$\Delta E(W1)^{aj}$	$\Delta E(W2)^{ak}$	ΔG^{al}	ΔG^{dm}	$\Delta G^{f,n}$
14a	0.0	0.0	0.3	0.0	0.4	0.0	0.0	0.0	0.0	0.0
14b	−0.5	−0.7	0.0	0.2	0.0	−1.8	0.6	0.45	1.0	1.0
14c	0.3	0.6	1.5	1.8	0.3	−0.7	1.22		9.7	8.0
14d	0.1	0.2	0.9	1.1	0.1	1.4	2.4	5.1	8.7	7.2
14e	5.8		5.9			2.9	3.8	3.7	3.0	
14f	3.1	3.0	3.6	4.4	3.0	3.8	5.5	6.3	9.9	8.8
14g	20.1		18.9			15.8	13.6	8.5	5.1	
14h	19.0	18.1	18.9			19.3	18.2	9.4	5.2	
14i			21.1						17.9	
14j			5.7		6.2				9.8	

[a] Ref. 93.
[b] Calculated at RI-MP2/TZVPP//RI-MP2/TZVPP.
[c] Calculated at CCSD(T)/cc-pVTZ// RI-MP2/TZVPP.
[d] Ref. 94.
[e] Calculated at B3LYP/6-31++G** with ZPE corrections at B3LYP/6-31G*.
[f] Ref. 95.
[g] Calculated at MP4/6-31++G(d,p)//MP2/6-31G(d).
[h] Ref. 89.
[i] Calculated at MPWB1K/aug-cc-pVDZ//MPWB1K/aug-cc-pVDZ.
[j] Tautomer and one water molecule calculated at R1-MP2/TZVPP//RI-MP2/TZVPP.
[k] Tautomer and two water molecules calculated at R1-MP2/TZVPP//RI-MP2/TZVPP.
[l] Calculated at COSMO/6-31G*.
[m] Calculated using a Poisson–Boltzman equation with nonelectrostatic effects modeled by a linear solvent accessible surface area dependence with B3LYP/6-31++G**.
[n] Calculated by adding the hydration free energy (MST/HF/6-31G(d)) to the values in column 5.

energy than **14a**. Nonetheless, as with cytosine, while the canonical structure of guanine is disfavored in the gas phase, it is the lowest energy tautomer in solution.

7.3.1.3 Adenine

Continuing his study of the gas-phase and aqueous-phase computational study of the nucleic acid bases, Hobza examined 14 different tautomers of adenine (Figure 7.13). The lowest energy (Table 7.15) gas-phase tautomer is **15a**, which is the canonical tautomer in DNA. The next two low energy tautomers, **15b** and **15c**, are about 7.4 kcal mol^{-1} higher in free energy at RI-MP2. Similar energy differences are found at B3PW91/6-311+G(d,p).[97] The dipole moment of **15a** is rather small (2.8 D), smaller than that for the canonical forms of the other nucleic acid bases, and much smaller than that for many of the adenine tautomers; for example, the dipole moment of **15b** and **15c** is 4.7 and 6.8 D, respectively. This might open the possibility for other tautomers to be lower in energy than the canonical form in aqueous solution.

The one- and two-water molecule microsolvated structures of **15a** and **15b** are shown in Figure 7.14. Microsolvation does reduce the energy gap between these two tautomers, such that with two water molecules, **15a-2W** is 5.8 kcal mol^{-1} more

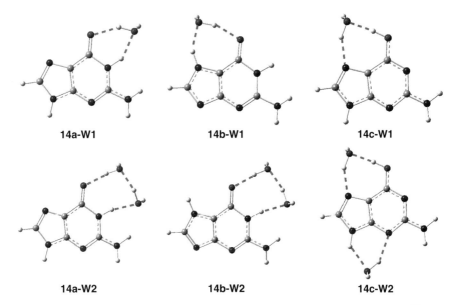

Figure 7.12 Structures of mono- and dihydrated guanine tautomers **14a–c**.[93]

stable than **15b-2W**. Similar results were found with B3PW91 as well.[97] However, the relative energies of the other tautomers remain much higher in energy. Shishkin and Leszczynski[99] optimized the complex of **15a** with 12–16 water molecules, and the structure with 12 water molecules is shown in top and edge view in Figure 7.14. Polyhydration of cytosine resulted in the amine group becoming planar and shortening of the C–N distance. While the C–N bond again contracts on polyhydration of adenine, the amine group remains pyramidal. Since the water molecules are arranged over one face of the adenine, this forces the nitrogen hydrogen atoms to remain out of plane to insure strong hydrogen bonding to water.

Since the explicit solvation results bring **15a** and **15b** closer in energy, it is possible that bulk water might further this trend. Hobza performed two continuum computations, the first with conductor-like polarized continuum method (C-PCM) and the other a hybrid computation involving C-PCM of the microsolvated tautomer. Both the methods further close the gap between these two tautomers, with **15a** now only about 2 kcal mol^{-1} more stable than **15b**. In addition, the explicit solvent models reduce the energy of **15c**, such that it is now the second lowest energy form. Molecular dynamics simulation using 412 water molecules also indicates that the three lowest energy tautomers, **15a–c**, are closely clustered, differing by less than 3 kcal mol^{-1}. In all cases, however, the canonical tautomer **15a** is the lowest energy form. An NMR[100] study in DMSO revealed the presence of these three isomers. A C-PCM study modeling DMSO as solvent altered the relative energies of the **15a–c** tautomers by less than 1 percent, providing very good agreement between computations and experiment.

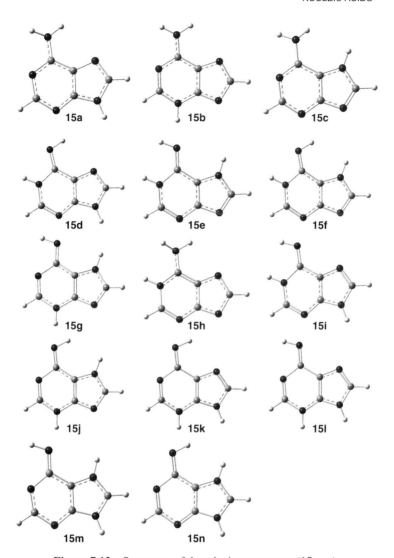

Figure 7.13 Structures of the adenine tautomers (**15a–n**).

7.3.1.4 Uracil and Thymine

Once again, Hobza[101] has examined all of the tautomers of these two related nucleic acid bases. Each can potentially be found in 13 different tautomeric forms, **16a–m** (uracil, see Figure 7.15) and **17a–m** (thymine, see Figure 7.16). Examination of the relative gas-phase energies of the uracil (Table 7.16) and thymine (Table 7.17) tautomers clearly indicates that the canonical structures, **16a** and **17a**, are much more stable than any other tautomer. This is true using both the RI-MP2[101] and B3LYP[102] methods. Since their dipole moments are fairly large (5.0 D for **16a** and 4.0 D for **17a**), both should be substantially stabilized in an aqueous environment. Although some tautomers have even

TABLE 7.15 Relative Energies (in kcal mol⁻¹) of Adenine Tautomers (15a–n)

Tautomer	Gas Phase			Solution Phase					
	$\Delta E^{a,b}$	$\Delta G_0^{298\,a,c}$	$\Delta E^{d,e}$	$\Delta E^{a,f}$	$\Delta E^{a,g}$	$\Delta G^{a,h}$	$\Delta G^{a,i}$	$\Delta G^{a,j}$	$\Delta G^{d,k}$
15a	0.0	0.0	0.0	0.0	0.0	0.0	0.0	0.0	0.0
15b	8.0	7.4	8.2	5.8	4.9	2.9	2.0	2.5	3.9
15c	7.6	7.5	8.4	9.0	11.4	0.69	1.9	2.8	1.9
15d	12.1	12.0		11.0	11.3	8.4	7.9	8.0	
15e	16.1	15.8							
15f	16.6	16.0		16.3	17.0	8.7	8.6	8.2	
15g	17.5	16.9		12.3	13.4	12.9	13.3	11.8	
15h	17.7	17.4		15.0	17.5	3.0	3.5	5.9	
15i	18.5	18.1							
15j	24.3	23.1							
15k	31.6	29.9		33.4	35.8	15.1	19.8	17.7	
15l	32.0	30.2							
15m	35.5	35.1		25.2	23.6	20.9	17.2	13.8	
15n	45.0	43.8							

[a]Ref. 98.
[b]Calculated at RI-MP2/TZVPP//RI-MP2/TZVPP.
[c]Calculated using the relative energies of Method A and thermodynamic quantities computed at MP2/6-31G**.
[d]Ref. 97.
[e]Calculated at B3PW91/6-311+G(d,p).
[f]Tautomer and one water molecule calculated at RI-MP2/TZVPP//RI-MP2/TZVPP.
[g]Tautomer and two water molecules calculated at RI-MP2/TZVPP//RI-MP2/TZVPP.
[h]Calculated at C-PCM/B3LYP/6-31G*.
[i]Tautomer and two water molecules calculated at C-PCM/B3LYP/6-31G*
[j]Molecular dynamics simulation with 412 TIP3P water molecules.
[k]Calculated at IEFPCM/B3PW91/6-311+G(d,p).

larger dipole moments (10.0 D for **16m**, 9.6 D for **17m**), this difference is unlikely to be large enough such that differential hydration energy can make up for a severe gas-phase instability.

The mono- and dihydrated complexes of the two lowest energy tautomers of uracil and thymine are displayed in Figure 7.17. Even though some of the very high lying gas-phase tautomers are particularly well stabilized by complexation with one or two water molecules (Tables 7.15 and 7.16), the hydrates of the canonical tautomers of each base, **16a** and **17a**, remain the lowest energy form. Continuum treatment of the aqueous environment using the C-PCM procedure also predicts the canonical tautomers to be the most stable forms. Tautomers **16d** and **17d** are now the next lowest energy form, but they are at least 9 kcal mol⁻¹ higher in energy than **16a** and **17a**. Thus, the canonical tautomers of uracil and thymine are predicted to dominate the population in both the gas and solution phases. Experiments[101,103] on the aqueous-phase distribution of uracil tautomers are fraught with problems, but the computations strongly indicate that only **16a** will be present in any appreciable concentration.

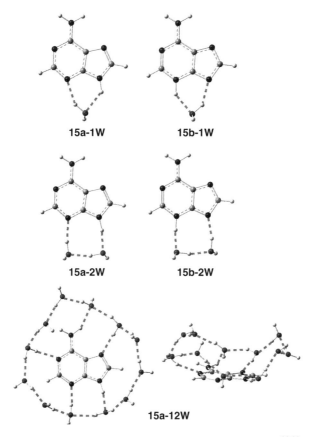

Figure 7.14 Microsolvated structures of **15a** and **15b**.[98,99]

7.3.2 Base Pairs

The key element of the structure of DNA is the encoding of genetic information by the complementary base pairs. Because of the importance of the recognition of guanine for cytosine (the G–C pair) and adenine for thymine (the A–T pair in DNA) or uracil (the A–U pair in RNA), extensive computational study of the hydrogen bonding between bases has been performed. Also critical to the structure of DNA is the base-stacking interactions. These stacking interactions have also been subject of numerous computational studies. Readers interested in an overview of these computational studies of base pairs are referred to the review articles by Hobza and Sponer.[104–106] We address here a few recent articles that discuss hydrogen-bonded and stacked base pairs with emphasis on the role that solvent plays in differentiating these two fundamentals interactions.

The hydrogen-bonded base pairs, of which the Watson–Crick forms are most commonly found in native DNA and RNA, were examined by Sponer and Hobza using high-level computations.[107] The geometries were optimized at

480 SOLUTION-PHASE ORGANIC CHEMISTRY

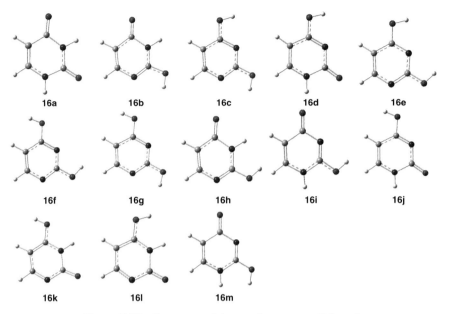

Figure 7.15 Structures of the uracil tautomers (**16a–m**).

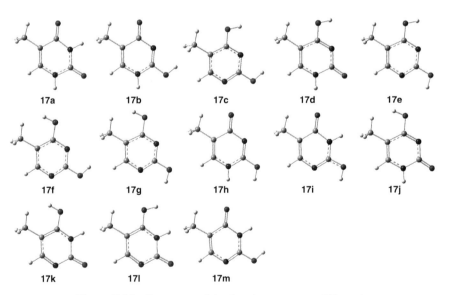

Figure 7.16 Structures of the thymine tautomers (**17a–m**).

TABLE 7.16 Relative Energies (in kcal mol^{-1}) of Uracil Tautomers (16a–m)

	$\Delta E^{a b}$	$\Delta G_0^{298 a c}$	$\Delta E^{d e}$	W1^{af}	W2^{ag}	C-PCM^{ah}
16a	0.0	0.0	0.0	0.0	0.0	0.0
16b	9.9	11.1	11.0	8.3	8.5	11.5
16c	9.4	11.7	12.8	9.1	9.8	13.8
16d	10.8	12.5	11.8	9.3	9.0	9.4
16e	10.6	12.8	13.8	13.3	16.3	14.1
16f	14.5	17.2	18.0			
16g	14.4	17.3	18.0			
16h	17.7	19.0	18.6			
16i	17.7	19.2	19.0	14.6	14.9	13.5
16j	17.5	19.8	18.5			
16k	20.6	21.6	20.6	19.7	19.9	14.3
16l	23.6	24.5	23.3			
16m	28.0	29.7	29.1			

[a]Ref. 101.
[b]Calculated at RI-MP2/TZVPP//RI-MP2/TZVPP.
[c]Calculated using the relative energies in (a) and thermodynamic quantities computed at MP2/6-31G**.
[d]Ref. 102.
[e]Calculated at B3LYP/6-31+G(d,p)
[f]Tautomer and one water molecule calculated at RI-MP2/TZVPP//RI-MP2/TZVPP.
[g]Tautomer and two water molecules calculated at RI-MP2/TZVPP//RI-MP2/TZVPP.
[h]Calculated at C-PCM/B3LYP/6-31G*.

TABLE 7.17 Relative Energies (in kcal mol^{-1}) of Thymine Tautomers (5a–m)^a

	ΔE^b	ΔG_0^{298c}	W1^d	W2^e	C-PCM^f
17a	0.0	0.0	0.0	0.0	0.0
17b	9.4	10.6	7.9	7.2	10.7
17c	10.6	13.1	13.4	15.0	13.2
17d	11.8	13.3	10.0	8.0	9.3
17e	9.4	13.5	9.3	8.7	14.8
17f	15.5	18.2			
17g	15.5	18.2			
17h	17.2	18.5			
17i	17.2	18.7	16.7	11.1	13.0
17j	19.6	21.3			
17k	21.7	24.6	21.2	15.9	18.5
17l	24.2	26.5			
17m	27.4	30.8			

[a]Ref. 101.
[b]Calculated at RI-MP2/TZVPP//RI-MP2/TZVPP.
[c]Calculated using the relative energies in footnote b and thermodynamic quantities computed at MP2/6-31G**.
[d]Tautomer and one water molecule calculated at RI-MP2/TZVPP//RI-MP2/TZVPP.
[e]Tautomer and two water molecules calculated at RI-MP2/TZVPP//RI-MP2/TZVPP.
[f]Calculated at C-PCM/B3LYP/6-31G*.

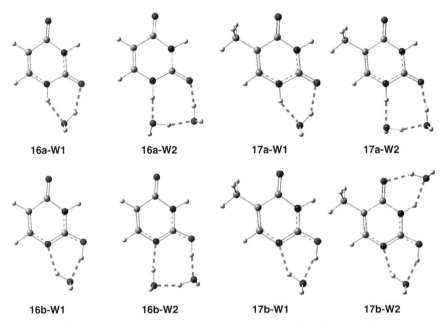

Figure 7.17 Microsolvated structures of **16a**, **16b**, **17a**, and **17b**.

RI-MP2/cc-pVTZ and then single point energies computed using the aug-cc-pVQZ basis set. The difference in energy between the optimized structure of the base pair and each individual base provided the interaction energy. These values were then corrected for BSSE. At this level, the interaction energy for the G–C pair **18** is −27.7 kcal mol^{-1} and for the A-T pair **19** is −15.1 kcal mol^{-1} (see Figure 7.18). These hydrogen-bonded Watson–Crick pairs were also optimized at B3LYP/6-31G** and PW91/6-31G**, and the interaction energies are listed in Table 7.18. The latter DFT method provides values in excellent accord with the RI-MP2 energies, while B3LYP underestimates the interaction energy by a couple of kilocalories per mole. The meta-GGA functionals, those that include kinetic energy density such as MPW1B95[108] and PWB6K,[109] also perform very well. The DFT methods can be very useful (and inexpensive) for obtaining interaction energies. The interaction energy for the A–U pair **20** is −12.3 kcal mol^{-1}, computed at MP2/6-31G*//HF/6-31G*.[110]

A systematic survey of 10 stacked nucleic acid base pairs was reported by Sponer et al.[113] by using the MP2-optimized base geometries and an empirical force field to locate the dimer structures. Drawings of the C–G stacked pair **21**, the A–T stacked pair **22**, and the A–U stacked pair **23** are shown in Figure 7.18. The energy of these optimized pairs were then obtained at MP2/6-31G* and in a follow-up study at RI-MP2 with an extrapolated CBS and corrections for higher order correlation using CCSD(T).[114] The stacking energies are listed in Table 7.19. The best estimates for the G–C **21** and the A–T **22** stacked pairs are −16.9 and −11.6 kcal mol^{-1}, respectively.[115] These values were obtained by optimizing the structures at

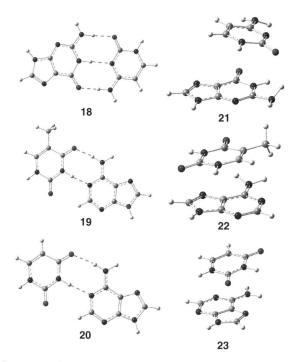

Figure 7.18 Structures of the C–G **18**, A–T **19**, and A–U **20** Watson–Crick pairs and the C–G **21**, A–T **22**, and A–U **23** stacked pairs.

TABLE 7.18 Interaction Energies of the Watson–Crick Base Pairs

	G–C, **18**	A–T, **19**
RI-MP2/cc-pVQZ//RI-MP2/cc-pVTZ[a]	−27.7	−15.1
B3LYP/6-31G**[a]	−25.5	−12.3
PW91/6-31G**[a]	−27.7	−14.5
MPW1B95/6-31+G(d,p)[b]	−26.8	−13.1
PWB6K/6-31+G(d,p)[b]	−28.4	−14.2
X3LYP/cc-pVTZ[c]	−30.5	−28.9

[a]Ref. 107.
[b]Ref. 111
[c]Ref. 112.

RI-MP2/TZVPP, extrapolating to the complete basis set and including corrections from CCSD(T).

B3LYP and B3P86 fail to locate a face-to-face dimer of benzene,[116] leading to the standard thought that DFT fails to appropriately treat dispersion. This would make DFT unsuitable for computing the stacked nucleic acid bases. In fact, both B3LYP/6-31+G(d,p)[111] and X3LYP/cc-pVTZ[112] fail to locate any stacked A–T, C–G, C–C, or U–U pairs.

TABLE 7.19 Stacking Energy (in kcal mol^{-1}) for Nucleic Acid Base Pairs

Stacked Pair	MP2/6-31G*^{a}	RI-MP2^{hc}	CBS(T)^{hd}	CBS(T)^e	MPW1B95^f	PWB6K^f
A–A	−8.8	−11.1	−8.5			
G–G	−11.3	−14.5	−12.7			
A–C	−9.5	−11.9	−10.2			
G–A	−11.2	−13.8	−11.4			
C–C	−8.3	−10.7	−10.0		−8.9	−10.9
A–U 23	−9.1	−11.3	−9.8			
A–T 22				−11.6	−7.5	−9.5
G–C 21	−9.3	−11.7	−10.6	−16.9	−12.8^g	−14.9
C–U	−8.5	−10.8	−10.4			
U–U	−6.5	−8.3	−7.5		−6.0	−7.9
G–U	−10.6	−13.1	−12.1			

^{a}Empirical force field geometries, Ref. 113.
^{b}Ref. 114.
^{c}RI-MP2/aug-ccPVTZ using geometries of Ref. 113.
^{d}Complete basis set extrapolation at RI-MP2 with CCSD(T) correction.
^{e}Geometry optimized at RI-MP2/TZVPP, energy from a complete basis set extrapolation with CCSD(T) correction, Ref. 115.
^{f}Optimized using the 6-31+G(d,p) basis set, Ref. 111.
^{g}Optimized structure not located, PWB6K structure used.

Truhlar[111] has developed hybrid meta functionals that are parameterized in order to properly treat a variety of difficult molecular environments, including $\pi-\pi$ stacking. Table 7.19 shows the results using two of these meta functionals, MPW1B95 and PWB6K. The latter functional locates optimized stacked pairs for all cases, while the former fails to find a G–C stacked pair. The interaction energies, which were not corrected for BSSE, predicted by these DFT methods are quite reasonable. This bodes well that in the future, proper DFT functionals that will allow for accurate computation of base stacking can be constructed.

In fact, Sponer and coworkers have examined more than 100 geometries of the stacked uracil[117] and stacked adenine[118] dimers in order to compare their potential energy surface computed with a variety of methods. For both systems, SCS-MP2 mimics quite nicely the reference PES computed at CCSD(T). DFT-D (in this case, the TPSS functional) is a bit worse in the short stacking region, although the overall agreement is acceptable. Given the much more modest computational cost of DFT-D over the wavefunction methods, these authors recommend its use. Even the molecular mechanics approach of AMBER works well, except in regions of repulsion between the bases because of short separations.

In order for computations of nucleic acid bases to have relevancy to biological questions, the role of solvent (water) must be considered. Computational solvation studies of base pairs are fairly limited.

We begin by considering the stacked base pair. The individual bases are strongly solvated, and when they stack, some solvent molecules must be removed. Thus, differential solvation free energy will be positive for the stacked base pairs. Sponer et al.[114] computed the solvation energy of the individual bases and the stacked pairs using MST/B3LYP/6-31G(d). The solvation free energy ranges from +3.0 to +7.0 kcal mol^{-1} for the 10 pairs listed in Table 7.19. The G–C pair **21** has a large positive solvation free energy (6.8 kcal mol^{-1}) resulting from their large polarity, which favors solvation of them individually.

Hobza examined the microsolvation of both G–C[119] and A–T[120] pairs. First, a molecular dynamics/quenching study using an empirical force field was performed to locate the local energy minima corresponding to planar and nonplanar hydrogen-bonded pairs, T-shaped pairs, and stacked pairs with one or two explicit water molecules. The lowest energy structures were then reoptimized with the RI-MP2/cc-pVDZ method with single-point energy evaluated at RI-MP2/TZVPP. Interaction energies were obtained with these latter energies, corrected for BSSE.

Low energy hydrogen-bonded and stacked G–C pairs with one or two water molecules are shown in Figure 7.19. The interaction energies for these complexes are listed in Table 7.20. The two most stable mono- (**24**, **25**) and dihydrated (**27**, **28**) complexes involve the G–C pair in the Watson–Crick hydrogen-bonded form. The third most stable monohydrated complex (**26**) and the fourth most stable dihydrated complex (**29**) involve a non-Watson–Crick hydrogen-bonding motif. The stacked pairs (**30**, **31**) become relatively more stable with increasing water participation. This is more evident by examining the populations of the various G–C pairs obtained from the molecular dynamics simulation with the AMBER[121] force field (Table 7.21). The hydrogen-bonded planar and nonplanar structures account for 79 percent of the population of the G–C pairs with one water. The stacked pairs account for less than 2 percent. With two water molecules, the hydrogen-bonded pair population is decreased to 54 percent. The stacked pair population increases to 9 percent.

Representative low energy complexes of the A–T pair with one or two water molecules are shown in Figure 7.20. Unlike the G–C pairs, the lowest energy hydrogen-bonded pairs do not have the Watson–Crick motif. For example, **32** and **34** have two hydrogen bonds between the two bases, but not in the positions of the Watson–Crick motif, while in **33**, a water molecule mediates a bridge between the two bases. The Watson–Crick hydrogen-bonding complexes, **35** and **39**, are the ninth lowest energy structures among the mono- and dihydrated A–T pairs, respectively (Table 7.22).

The relative energies of hydrogen bonded versus stacked arrangements of the A–T pair are sensitive to solvent. The stacked monohydrated complex **40** is the fifth lowest energy complex, but for the dihydrated complexes, the stacked form **41** is the most stable. The population of the stacked complexes increases in going from the monohydrated to dihydrated forms. Compared to the G–C pairs, the A–T population is much more favored toward the stacked and T-shaped forms. The preference for the stacked arrangement over the hydrogen-bonded ones in aqueous solution is consistent with experiments.

Bickelhaupt[122] examined 51 hydrogen-bonded base pairs, including the Watson–Crick pairing, mismatched pairs, and incorporating nonnatural

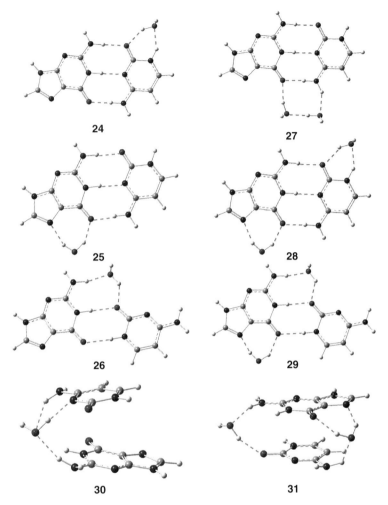

Figure 7.19 Structures of the mono- (**24–26**) and dihydrated (**27–29**) G–C hydrogen bonded pairs and the mono- (**30**) and dihydrated (**31**) G–C stacked pairs.

nucleotides. The binding energy of each pair was computed in the gas phase at BP86-D/TZ2P and in aqueous phase using the COSMO approach for the solvent continuum. Although the most stable gas-phase pair is the Watson–Crick GC pair **18**, the Watson–Crick AT pair is not the lowest energy pair. All of the hydrogen-bonded pairs are destabilized in water, with **18** remaining the most strongly bound pair.

Bickelhaupt et al.[123] also examined some of the factors that lead to the incredible selectivity in DNA replication. They examined the binding of any of the four nucleotide bases with a template–primer complex. The template–primer complex

TABLE 7.20 Interaction Energy (in kcal mol^{-1}) of the G–C Pairs[a]

	RI-MP2/TZVPP[b]	CBS(T)[c]
24	−34.9	−37.7
25	−33.6	
26	−32.6	
30	−29.0	−30.1
27	−45.1	−48.1
28	−43.4	
29	−41.0	
31	−42.6	−44.9

[a]Ref. 119.
[b]RI-MP2/TZVPP//RI-MP2/cc-pVDZ.
[c]Extrapolated to the complete basis set limit and corrected for CCSD(T).

TABLE 7.21 Population of the G–C[119] and A–T[120] Mono- and Dihydrated Pairs

	N^a	Population %	N^a	Population, %
	GC-1W		AT-1W	
Planar	81	68.3	67	30.9
Nonplanar	35	10.7	38	10.8
T-shape	126	17.8	115	30.2
Nonstacked	16	1.5	20	2.1
Stacked	30	1.7	67	25.6
	GC-2W		AT-2W	
Planar	257	36.5	281	33.1
Nonplanar	259	17.7	259	10.1
T-shape	625	30.7	671	23.8
Nonstacked	114	6.0	106	2.2
Stacked	204	9.1	402	30.6

[a]Total number of local energy minima.

consists of a Watson–Crick pair, and then 3.4 Å above this is placed one of the four nucleobases. So for the [X]-G/C-G system, the C–G pair has a G placed above it and then the four bases are bound to this G. The C–G/C–G structure is shown in Figure 7.21. The binding of the nucleotide with the template–primer complexes were computed at BP86-D/TZ2P and the solvent effect included using COSMO. In the gas phase, C–G binding is strongly preferred, but T–A is not the lowest energy pairing. However, once the solvent effect is included, both the G–C and A–T pairings are favored. Interestingly, stacking does enhance the binding energies.

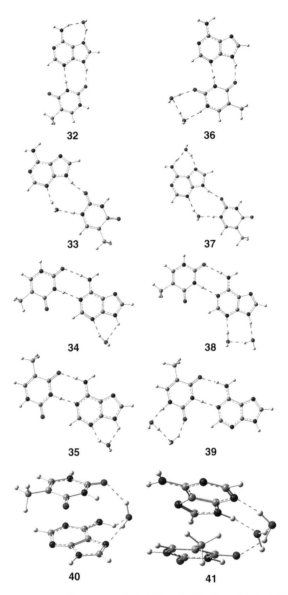

Figure 7.20 Structures of the mono- (**32–35**) and dihydrated (**36–39**) A–T hydrogen bonded pairs and the mono- (**40**) and dihydrated (**41**) G–C stacked pairs.

This section demonstrates that computational methods are suitable for computing the nucleobase components of DNA and RNA. Study of oligomeric or larger DNA or RNA strands remains a major computational challenge. The ongoing development of new functionals, work on linear scaling algorithms, and continued improvement of computer hardware offer hope for continued significant contributions in this area in the future.

TABLE 7.22 Interaction Energy (in kcal mol^{-1}) of the A–T Pairs[a]

	RI-MP2/TZVPP[b]
32	−28.0
33	−26.2
34	−25.3
35	−22.5
40	−23.7
36	−36.6
37	−36.0
38	−35.4
39	−35.0
41	−37.2

[a]Ref. 120.
[b]RI-MP2/TZVPP//RI-MP2/cc-pVDZ.

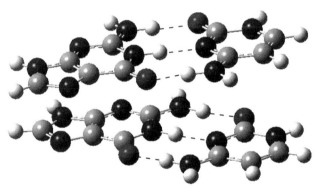

Figure 7.21 Structure of the C–G/C–G model from the [X]–G/C–G template–primer complex interacting with C.

7.4 AMINO ACIDS

Amino acids, the building blocks of proteins, can exist in two principal tautomeric forms. In aqueous solution, the naturally occurring α-amino acids are found in their zwitterionic form **42Z**, but in the gas phase, the neutral form **42N** is the only structure observed. This substantial solvent effect is understood in terms of the dipole moments of the tautomers. The dipole moment of the zwitterion is very large and is favorably stabilized by electrostatic interactions with the polar water molecules. In the absence of the polar environment (the gas phase), the charge separation of the zwitterion is unsustainable and only the neutral form exists.[124,125] In fact, the gas-phase optimization of the zwitterion of glycine **43Z** with many different computational methods all result in the neutral form **43N**.[126,127]

$$\text{R-CHCO}_2^- \; (^+\text{NH}_3) \quad \longleftrightarrow \quad \text{R-CHCO}_2\text{H} \; (\text{NH}_2)$$

$$\mathbf{4_2Z} \qquad\qquad\qquad \mathbf{4_2N}$$

$$\text{H}_3\overset{+}{\text{N}}\frown\text{CO}_2^- \quad \longleftrightarrow \quad \text{H}_2\text{N}\frown\text{CO}_2\text{H}$$

$$\mathbf{4_3Z} \qquad\qquad\qquad \mathbf{4_3N}$$

There are two obvious and interesting questions that arise from this situation. First, how many water molecules are needed to make the zwitterion form of the amino acid stable? Second, how many water molecules are needed to make the zwitterion equal in energy to the neutral form? Both of these questions have been addressed with a variety of computational methods.

Alonso et al.[128] examined the glycine–one water complex using laser ablation molecular beam Fourier-transform microwave spectroscopy. By comparison of the observed rotational parameters and those predicted from an MP2/6-311+G(d,p) optimization, they concluded that the only observed isomer is that formed of the neutral isomer and water, **44a**. Balabin[129] examined the glycine–one water complex using IR spectroscopy. While he too found **44a**, he was able to detect a small amount of **44b** and **44c**. An interesting side note is that anharmonic corrections were necessary in order to match up the computed (MP2) frequencies with the experimental values. These experiments indicate that more than one water molecule is needed to stabilize the zwitterion tautomer.

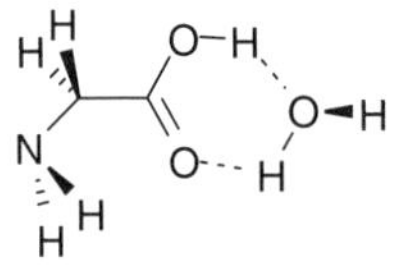

44a 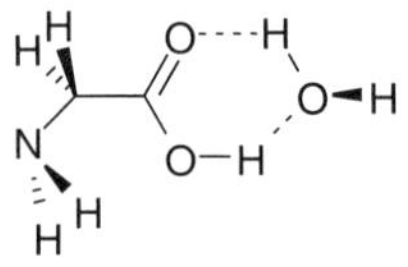   **44b** 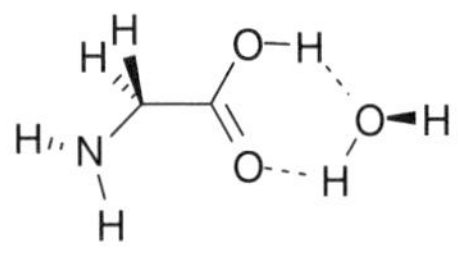   **44c**

Jensen and Gordon[130] approached the question of how many water molecules are required to stabilize the glycine zwitterion by optimizing a variety of configurations of the neutral or zwitterion tautomer with one or two water molecules at HF/DZP and single-point energies were obtained at MP2/DZP++. No stable energy minimum was found for the zwitterion with one water molecule. However, with two water molecules, a stable minimum was located, although it is more than 10 kcal mol^{-1} above the neutral tautomer with two water molecules. A later study by Gordon[131] indicated that a glycine zwitterion:1 water cluster could be found at HF/6-31++G(d,p), but it remains more than 15 kcal mol^{-1} higher in energy than the lowest energy neutral glycine:1 water structure.

Wang[132] determined that the glycine zwitterion:1 water cluster is an artifact of method and basis set; HF yields a local energy minimum, as does B3LYP and MP2 so long as the basis set does not include polarization functions on

TABLE 7.23 Relative Energy (in kcal mol^{-1}) of the Water Clusters of Glycine Zwitterion Minus the Neutral Tautomer

Number of Water Molecules	MP2[a]	PBE0[b]
1	15.3	
2	10.8	7.4
3	5.2	8.4
4	2.6	6.0
5	0.4	2.7
6	0.8	1.7
7	−0.3	0.0
8	−1.8	

[a] MP2/6-311+G(d,p)//HF/6-31++G(d,p), Ref. 131.
[b] PBE0/6-311+G(d,p) + ZPE, Ref. 133.

hydrogen. Bachrach[133] found no stable glycine zwitterion:1 water cluster at PBE0/6-311+G(d,p), but that two water molecules were sufficient to stabilize the zwitterion tautomer. The consensus is that two waters are needed to stabilize the glycine zwitterion.

The second question asks how many water molecules are needed to make the zwitterion structure isoenergetic with the neutral form. Gordon[131] optimized the structures of the two tautomers with up to eight water molecules at HF/6-311++G(d,p) and single-point energies computed at MP23/6-31++G(d,p); zero-point energies (ZPEs) were not computed. Bachrach[133] optimized these clusters with one to seven water molecules at PBE0/6-311+G(d,p) and added in the zero-point vibrational energy. This functional was benchmarked against large MP2 and CCSD(T) computations for the glycine neutral and zwitterion tautomers with two water molecules. The energy differences between the lowest energy neutral cluster and zwitterion cluster with increasing degree of microsolvation are listed in Table 7.23. Both computations come to the same conclusion: seven water molecules are needed to bring the zwitterion tautomer into energy equivalence with the neutral form. Similar results have been found for microsolvation studies of other amino acids.[134–136]

A cautionary note in conclusion: these microsolvation studies depend on two critical factors: (1) the number of solvent molecules to include and (2) selection of the configurations. The two issues are intertwined in that determining the number of solvent molecules that need to be included requires sampling a large number of configurations with many different number of solvent molecules. This problem has both software and hardware components. As computers become faster, it becomes easier to sample more configurations, especially since this type of problem is inherently parallelizable. However, with current technologies, there are real serious limitations as to the size of the clusters that can be examined. On the software side remains the question of how to sample configuration space. An MC or simulated annealing approach means that many highly improbable configurations will be sampled, while a more logical or intuitive approach opens up the possibility of missed configurations. This is an area where opportunities abound for significant breakthroughs and we trust that computational approaches to microsolvation will be dramatically improved in the future.

7.5 INTERVIEW: PROFESSOR CHRISTOPHER J. CRAMER

Interviewed September 10, 2006

Professor Christopher Cramer is the Distinguished McKnight and University Teaching Professor of Chemistry, Chemical Physics, and Scientific Computation at the University of Minnesota. Cramer has been at Minnesota for his entire professional career. Before coming to Minnesota, he served in the US Army, including a tour in Iraq during Operation Desert Storm.

Cramer began his graduate education in synthetic organic chemistry with Professor Scott Denmark at the University of Illinois. In his third year at graduate school, Cramer was very frustrated with his research. "An issue of *Science* crossed my desk with two articles," Cramer recalls, "one by Fritz Schafer on the history of methylene and the second by Ken Houk on modeling transition states. I thought these were fascinating. I was really impressed by this theory thing - that you could just calculate the TS. I went to Professor Denmark and said 'I'm bored with what I'm doing and I would like to incorporate something computational.' Scott agreed and told me to talk with Professor Cliff Dykstra. We started by looking at the 1,2-H shift from phosphine oxide to phosphinous acid. I really liked that experience. My Ph.D. thesis included synthesis, NMR analysis and computations; I was very proud of this. I basically didn't look back and wanted to do more molecular modeling."

Following graduate school, Cramer served in the Army, first in Korea, and then at the Aberdeen Proving Ground. Instead of joining a research group, the post's Chief-of-Staff wanted Cramer to serve as a General's Aide. According to Cramer, "The prize job for a young officer interested in a military career is to be a general's aide." Given a night to think it over, Cramer realized he was at one of life's proverbial crossroads. "You reach a point in your life when you have to make a decision. You burn one set of bridges or another set, and I decided to burn the military career bridges." At his meeting with the General, he told him "I'd be a great staff officer, but I'd be a better scientist."

Looking to establish, as he puts it, "theoretical bona fides in order to pursue an academic career," Cramer wrote a letter to Professor Don Truhlar indicating an interest in collaborating with him. During his first visit to Minnesota, they agreed to pursue a project following on work by Professor Clark Still. Still had combined the Generalized Born (GB) approximation with force fields producing a mathematical form that Cramer describes as "worthy of a chemical physicist". Cramer returned to Aberdeen and began coding GB into AMPAC, a semiempirical quantum chemical program. The work involved "a lot of coding and a lot of parameterization, and checking against a database of solvation free energies.

This was way back before spreadsheets and so I worked on $8'' \times 20''$ paper that I self-ruled, and hand-entered data using four colors and in what must have been 4 point font. I was about ready to write up the work when I got orders to report to Fort Benning in a week and then go to Iraq. I packed up all the result in a large envelope and sent it to Don, hoping that Don would be able to follow it all. In fact, Don did write it up, and it was published while I was in the Gulf."

Their early SM model was relatively crude. "It had some good aspects in terms of accounting for polarization in the wavefunction," Cramer says. "A key feature that nobody else had done was to include terms to account for the other components of solvation. Our model was semiempirical since we fit to experimental data, but that's what people wanted to know—they wanted the free energy of solvation. The PCM models of Tomasi and Rivail were underutilized because it was hard to make contact with particular experimental observables, but we offered that opportunity." Thus began what has been a very fruitful collaboration between Cramer and Truhlar.

Cramer recalls the story of how they got involved with glucose, "When we parameterized SM we were always worried about outliers and one significant outlier was ethylene glycol. While visiting the University of Georgia, I mentioned to Professor Butch Carrera that ethylene glycol was a pain. He indicated that Professor Chuck Eckert at Georgia State had just measured some diols and triols and thought that a lot of prior work was wrong. The value that Eckert got was much closer to what we were predicting. So we thought 'Let's do ethylene glycol seriously.' We nailed down the ab initio gas phase equilibrium population and then solvated each of these with SM and predicted an equilibrium average solvation free energy that was within 0.1 kcal of Eckert's measurement. That was done without readjusting any parameters. I then played with glycerol, but that was kind of boring. But what was not boring was glucose. Is the $\alpha:\beta$ ratio dictated by solvation or the anomeric effect?" This led to a series of papers, including a strong collaboration with Dr. Al French of the USDA.

Cramer believes that there is really only one open question pertaining to glucose. "I think," he says, "everyone agrees that the gas phase equilibrium is what the anomeric effect says it should be and that everyone agrees that solvation inverts it. The last remaining issue is 'are the hydroxyl groups internally bound or unbound?' Our calculations show that when we un-hydrogen bond one of the intramolecular H-bonds the increase in solvation energy is roughly equal to the loss of the internal hydrogen bond. My gut feeling is that intramolecular H-bonds in glucose don't exist in water. The daisy chain of intramolecular H-bonds doesn't actually have a huge effect on the global equilibrium population."

Cramer was drawn into the benzyne problem through collaborations with Professor Bob Squires. Cramer had long been interested in carbenes and related species, having worked on nitreniums with Professor Dan Falvery and trimethylenemethane with Professor Paul Dowd. In 1994, Cramer had applied DFT to XH_2 species where X was B, C, and N. But the collaboration with

Squires first began because of Cramer's paper discussing apicophilicity of hydroxyl groups in phosphoranes. Squires contacted Cramer regarding his own experiments concerning siliconates. Cramer recalls, "I had a little time on my hands, so I ran some calculations, spent about two weeks on this, and instead of sending him a letter in reply, I sent him a manuscript with both of our names on it. He was very impressed with that, and it put me on his map! Bob had been doing some tricky calculations on benzyne. I volunteered to do some DFT and CASPT2 calculations. Our resulting *CPL* paper cleaned up the literature and set a nice standard for comparing CC, DFT, and CASPT2."

Cramer considers the appreciation of how DFT performs with diradicals as the major theoretical contribution of the work on benzynes. "DFT accommodated much more multireference character in singlet wavefunctions of diradicals than did HF. But restricted DFT breaks down with *p*-benzyne. People were just beginning to play with UDFT and recognize the interpretation of S^2. Professor Dieter Cremer deserves more credit for this than anyone else. My contribution was recognizing that the high symmetry of *p*-benzyne allows for CC to work properly. But with lower symmetry systems, like pyridynium, CCSD gives wacky results but Bruckner orbitals remove some instabilities and so it works well."

Cramer believes his work with Squires on hyperfine coupling has been underappreciated. "We recognized that the singlet–triplet splitting measures the interaction of one electron with another electron at some remote site. We also recognized that ESR hyperfine coupling also measures spin–spin interaction. Bob and I thought that if we capped a diradical with a hydrogen, and looked at the hyperfine coupling of the remaining electron to that proton, maybe the magnitude would correlate with the singlet–triple splitting. What we found is that the correlation is fantastic and dirt cheap to compute."

Scientific collaboration is an important part of how Cramer pursues science. "I am very proud that 90 percent of my papers have another senior author on them," Cramer remarks. Of his collaboration with Squires, Cramer says "I got a lot out of Bob's insights. He was a great chemist. Our collaboration worked in both directions, thinking about new molecules that we could look at computationally and that he could get into the flowing afterglow. Bob was really good at making other people do better chemistry."

Cramer's most fruitful collaboration is with his Minnesota colleague Professor Don Truhlar. "The greatest strength of my collaboration with Don is that we have such orthogonal backgrounds. He's a classic chemical physicist and I'm an organic chemist. We would simply argue for some time; he would go in one direction and I'd go in another. Eventually we would wear each other down to the point where we'd settle on a vector roughly the sum of our two ideas. It prevented the solvation model, for example, from being too semiempirical, which might have been the way I was tempted to go, but also from being too disconnected from reality, which is sometimes the way it seemed to me where Don wanted to go. The result is a model that a nice compromise of robustness and applicability that's not available in any other model." One exception to this give-and-take was the first SM paper on glucose. Cramer performed all the computations and wrote up the paper without informing Truhlar. He then dropped it on Truhlar's desk over winter break and told Truhlar it was his Christmas present.

Cramer considers the development of the SMX models as his most important scientific contribution. "This work drove the field forward. Tomasi and Barone moved to put in non-electrostatic components in a more accurate way," Cramer claims. He is somewhat disappointed in the limited use of the SM models by other computational chemists. "What's driving the train is that people use what's in the code that they bought. To the extent that PCM is a very successful solvation model, it's in part because it is in *Gaussian*. The SMX models sadly have never been available in *Gaussian*."

Cramer and Truhlar are currently developing SM7. "We now have a handle on how to treat solvation free energies for ions in organic solvents. We believe we have a more physical way to construct the model. This is probably the last thing that we will do in terms of SM development. Where solvation models will move to is to non-homgeneous solutions: liquid crystals, interfaces, etc.," Cramer predicts.

Cramer looks to inorganic chemistry for his challenges in the future. "I've largely left organic chemistry," he states. "I consider most of the problem in organic chemistry these days to be too easy. They don't need someone with specialized expertise. There are a lot of other elements in the periodic table. For example, I have been working with Professor Bill Tolman on the activation of molecular oxygen by copper catalysts. The modeling here is fantastically difficult; 40 different theoretical methods differ in predictions by 100 kcal mol^{-1}!"

Cramer has never regretted his switch to computational chemistry. "Computational tools allow you to address an enormous breadth of chemistry. Experiments impose more of a focus." His enthusiasm for his research is readily apparent in any of his lectures, where he tries to relate his results back to experiment, "because I was one [an experimentalist]. I really try to emphasize chemistry and not get caught up in the arcana of the calculations." Cramer also brings a sense of whimsy to his science. Take for example, the title "When Anomeric Effects Collide" (a paper by the way he feels has been overlooked) from one of his papers or his computational study of the molecule Y_2K that appeared in 1999 in the journal *Science* as celebration of the upcoming millennium!

We finished our interview by discussing why he does computational chemistry. "Computational chemistry excites me," Cramer says, "because it lets you look at 'reality' all along a pathway, yet can't be probed experimentally. I can look at a TS structure and I know that my experimental colleagues are going to have a hell of a time getting any information on it, except indirectly. Having a microscope the whole way—I like that vision."

REFERENCES

1. Tanaka, K.; Mackay, G. I.; Payzant, J. D.; Bohme, D. K. "Gas-phase reactions of anions with halogenated methanes at 297 ± 2°K," *Can. J. Chem.* **1976**, *54*, 1643–1659.
2. Olmstead, W. N.; Brauman, J. I. "Gas-phase nucleophilic displacement reactions," *J. Am. Chem. Soc.* **1977**, *99*, 4219–4228.
3. Pellerite, M. J.; Brauman, J. I. "Intrinsic barriers in nucleophilic displacements. A general model for intrinsic nucleophilicity toward methyl centers," *J. Am. Chem. Soc.* **1983**, *105*, 2672–2680.
4. Beak, P. "Energies and alkylations of tautomeric heterocyclic compounds: Old problems – New answers," *Acc. Chem. Res.* **1977**, *10*, 186–192.
5. Brauman, J. I.; Blair, L. K. "Gas-phase acidities of alcohols. Effects of alkyl groups," *J. Am. Chem. Soc.* **1968**, *90*, 6561–6562.
6. Brauman, J. I.; Blair, L. K. "Gas-phase acidities of alcohols," *J. Am. Chem. Soc.* **1970**, *92*, 5986–5992.
7. Cramer, C.J. (**2002**) *Essentials of Computational Chemistry: Theories and Models*, John Wiley & Sons, Inc., New York.
8. Jensen, F. (**1999**) *Introduction to Computational Chemistry*, John Wiley & Sons, Ltd, Chichester, England.
9. Tomasi, J.; Persico, M. "Molecular interactions in solution: An overview of methods based on continuous distributions of the solvent," *Chem. Rev.* **1994**, *94*, 2027–2094.
10. Tomasi, J.; Mennucci, B.; Cammi, R. "Quantum mechanical continuum solvation models," *Chem. Rev.* **2005**, *105*, 2999–3094.
11. Cramer, C. J.; Truhlar, D. G. "Continuum solvation models: Classical and quantum mechanical implementations," *Rev. Comput. Chem.* **1995**, *6*, 1–72.
12. Cramer, C. J.; Truhlar, D. G. "Implicit solvation models: Equilbiria, structure, spectra, and dynamics," *Chem. Rev.* **1999**, *99*, 2161–2200.
13. Cramer, C. J.; Truhlar, D. G. "A universal approach to solvation modeling," *Acc. Chem. Res.* **2008**, *41*, 760–768.
14. Mennucci, B. "Polarizable continuum model," *WIREs Comput. Mol. Sci.* **2012**, *2*, 386–404.
15. Sauer, J.; Sustmann, R. "Mechanistic aspects of Diels-Alder reactions: A critical survey," *Angew. Chem. Int. Ed. Engl.* **1980**, *19*, 779–807.
16. Dewar, M. J. S.; Pyron, R. S. "Nature of the transition state in some Diels-Alder reactions," *J. Am. Chem. Soc.* **1970**, *92*, 3098–3103.
17. Beltrame, P. "Addition of unsaturated compounds to each other," *Compr. Chem. Kinet.* **1973**, *9*, 87–162.
18. Rideout, D. C.; Breslow, R. "Hydrophobic acceleration of Diels-Alder reactions," *J. Am. Chem. Soc.* **1980**, *102*, 7816–7817.
19. Engberts, J. B. F. N. "Diels-Alder reactions in water: Enforced hydrophobic interaction and hydrogen bonding," *Pure Appl. Chem.* **1995**, *67*, 823–828.
20. Breslow, R.; Maitra, U.; Rideout, D. "Selective Diels–Alder reactions in aqueous solutions and suspensions," *Tetrahedron Lett.* **1983**, *24*, 1901–1904.
21. Breslow, R. "Hydrophobic effects on simple organic reactions in water," *Acc. Chem. Res.* **1991**, *24*, 159–164.

22. Grieco, P. A.; Garner, P.; He, Z.-M. "'"Micellar" catalysis in the aqueous intermolecular Diels–Alder reaction: Rate acceleration and enhanced selectivity," *Tetrahedron Lett.* **1983**, *24*, 1897–1900.

23. Grieco, P. A.; Nunes, J. J.; Gaul, M. D. "Dramatic rate accelerations of Diels–Alder reactions in 5 M lithium perchlorate-diethyl ether: The cantharidin problem reexamined," *J. Am. Chem. Soc.* **1990**, *112*, 4595–4596.

24. Schneider, H.-J.; Sangwan, N. K. "Diels–Alder reactions in hydrophobic cavities: A quantitative correlation with solvophobicity and rate enhancements by macrocycles," *J. Chem. Soc. Chem. Commun.* **1986**, 1787–1789.

25. Schneider, H.-J.; Sangwan, N. K. "Changes of stereoselectivity in Diels–Alder reactions by hydrophobic solvent effects and by β-cyclodextrin," *Angew. Chem. Int. Ed. Engl.* **1987**, *26*, 896–897.

26. Blake, J. F.; Jorgensen, W. L. "Solvent effects on a Diels–Alder reaction from computer simulations," *J. Am. Chem. Soc.* **1991**, *113*, 7430–7432.

27. Birney, D. M.; Houk, K. N. "Transition structures of the Lewis acid-catalyzed Diels–alder reaction of butadiene with acrolein. The origins of selectivity," *J. Am. Chem. Soc.* **1990**, *112*, 4127–4133.

28. Blake, J. F.; Lim, D.; Jorgensen, W. L. "Enhanced hydrogen bonding of water to Diels–Alder transition states. Ab initio evidence," *J. Org. Chem.* **1994**, *59*, 803–805.

29. Furlani, T. R.; Gao, J. "Hydrophobic and hydrogen-bonding effects on the rate of Diels–Alder reactions in aqueous solution," *J. Org. Chem.* **1996**, *61*, 5492–5497.

30. Chandrasekhar, J.; Shariffskul, S.; Jorgensen, W. L. "QM/MM simulations for Diels-Alder reactions in water: Contribution of enhanced hydrogen bonding at the transition state to the solvent effect," *J. Phys. Chem. B* **2002**, *106*, 8078–8085.

31. Acevedo, O.; Jorgensen, W. L. "Understanding rate accelerations for Diels–Alder reactions in solution using enhanced QM/MM methodology," *J. Chem. Theor. Comput.* **2007**, *3*, 1412–1419.

32. Kong, S.; Evanseck, J. D. "Density functional theory study of aqueous-phase rate acceleration and endo/exo selectivity of the butadiene and acrolein Diels–Alder reaction," *J. Am. Chem. Soc.* **2000**, *122*, 10418–10427.

33. Kistiakowsky, G. B.; Lacher, J. R. "The kinetics of some gaseous Diels–Alder reactions," *J. Am. Chem. Soc.* **1936**, *58*, 123–133.

34. El Khadem, H.S. *Carbohydrate Chemistry: Monosaccharides and their Oligomers*, Academic Press, San Diego, CA, **1988**.

35. Collins, P. and Ferrier, R. *Monosaccharides: Their Chemistry and their Roles in Natural Products*, John Wiley & Sons, Ltd, Chichester, UK, **1995**.

36. Pierson, G. O.; Runquist, O. A. "Conformational analysis of some 2-alkoxytetrahydropyrans," *J. Org. Chem.* **1968**, *33*, 2572–2574.

37. Kirby, A.J. *The Anomeric Effect and Related Stereoelectronic Effects at Oxygen*, Springer-Verlag, Berlin, **1982**.

38. Hommel, E. L.; Merle, J. K.; Ma, G.; Hadad, C. M.; Allen, H. C. "Spectroscopic and computational studies of aqueous ethylene glycol solution surfaces," *J. Phys. Chem. B* **2005**, *109*, 811–818.

39. Csonka, G. I.; Csizmadia, I. G. "Density functional conformational analysis of 1,2-ethanediol," *Chem. Phys. Lett.* **1995**, *243*, 419–428.

40. Cramer, C. J.; Truhlar, D. G. "Quantum chemical conformational analysis of 1,2-ethanediol: Correlation and solvation effects on the tendency to form internal hydrogen bonds in the gas phase and in aqueous solution," *J. Am. Chem. Soc.* **1994**, *116*, 3892–3900.

41. Howard, D. L.; Jorgensen, P.; Kjaergaard, H. G. "Weak intramolecular interactions in ethylene glycol identified by vapor phase OH-stretching overtone spectroscopy," *J. Am. Chem. Soc.* **2005**, *127*, 17096–17103.

42. Crittenden, D. L.; Thompson, K. C.; Jordan, M. J. T. "On the extent of intramolecular hydrogen bonding in gas-phase and hydrated 1,2-ethanediol," *J. Phys. Chem. A* **2005**, *109*, 2971–2977.

43. Klein, R. A. "Ab initio conformational studies on diols and binary diol-water systems using DFT methods. Intramolecular hydrogen bonding and 1:1 complex formation with water," *J. Comput. Chem.* **2002**, *23*, 585–599.

44. Mandado, M.; Graña, A. M.; Mosquera, R. A. "Do 1,2-ethanediol and 1,2-dihydroxybenzene present intramolecular hydrogen bond?," *Phys. Chem. Chem. Phys.* **2004**, 4391–4396.

45. Caminati, W.; Corbelli, G. "Conformation of ethylene glycol from the rotational spectra of the nontunneling O-monodeuterated species," *J. Mol. Spectrosc.* **1981**, *90*, 572–578.

46. Buckley, P. D.; Giguere, P. A. "Infrared studies on rotational isomerism. I. Ethylene glycol," *Can. J. Chem.* **1967**, *45*, 397–407.

47. Pachler, K. G. R.; Wessels, P. L. "Rotational isomerism. X. A nuclear magnetic resonance study of 2-fluoro-ethanol and ethylene glycol," *J. Mol. Struct.* **1970**, *6*, 471–478.

48. Chaudhari, A.; Lee, S.-L. "A computational study of microsolvation effect on ethylene glycol by density functional method," *J. Chem. Phys.* **2004**, *120*, 7464–7469.

49. Kumar, R. M.; Baskar, P.; Balamurugan, K.; Das, S.; Subramanian, V. "On the perturbation of the H-bonding interaction in ethylene glycol clusters upon hydration," *J. Phys. Chem. A* **2012**, *116*, 4239–4247.

50. Callam, C. S.; Singer, S. J.; Lowary, T. L.; Hadad, C. M. "Computational analysis of the potential energy surfaces of glycerol in the gas and aqueous phases: Effects of level of theory, basis set, and solvation on strongly intramolecularly hydrogen-bonded systems," *J. Am. Chem. Soc.* **2001**, *123*, 11743–11754.

51. Jeong, K.-H.; Byun, B.-J.; Kang, Y.-K. "Conformational preferences of glycerol in the gas phase and in water," *Bull. Lorean Chem. Soc.* **2012**, *33*, 917–924.

52. Sheppard, N.; Turner, J. J. "High-resolution nuclear magnetic resonance (nmr) spectra of hydrocarbon groupings. II. Internal rotation in substituted ethanes and cyclic ethers," *Proc. Roy. Soc. (London)* **1959**, *A252*, 506–519.

53. Gutowsky, H. S.; Belford, G. G.; McMahon, P. E. "NMR studies of conformational equilibria in substituted ethanes," *J. Chem. Phys.* **1962**, *36*, 3353–3368.

54. da Silva, C. O.; Mennucci, B.; Vreven, T. "Density functional study of the optical rotation of glucose in aqueous solution," *J. Org. Chem.* **2004**, *69*, 8161–8164.

55. Carey, F. A. *Organic Chemistry*; 5th ed.; McGraw-Hill: Boston, **2003**.

56. Solomons, T. W. G.; Fryhle, C. B. *Organic Chemistry*; 10th ed.; John Wiley & Sons, Inc.: Hoboken, NJ, **2011**.

57. Appell, M.; Strati, G.; Willett, J. L.; Momany, F. A. "B3LYP/6-311++G** study of α- and β-D-glucopyranose and 1,5-anhydro-D-glucitol: $^4C_1$ and $^1C_4$ chairs, $^{3,O}B$ and $B_{3,O}$ boats, and skew-boat conformations," *Carbohydr. Res.* **2004**, *339*, 537–551.

58. Barrows, S. E.; Dulles, F. J.; Cramer, C. J.; French, A. D.; Truhlar, D. G. "Relative stability of alternative chair forms and hydroxymethyl conformations of α-D-glucopyranose," *Carbohydr. Res.* **1995**, *276*, 219–251.
59. Ma, B.; Schaefer, H. F.; Allinger, N. L. "Theoretical studies of the potential energy surfaces and compositions of the D-aldo- and D-ketohexoses," *J. Am. Chem. Soc.* **1998**, *120*, 3411–3422.
60. Lii, J.-H.; Ma, M.; Allinger, N. L. "Importance of selecting proper basis set in quantum mechanical studies of potential energy surfaces of carbohydrates," *J. Comput. Chem.* **1999**, *20*, 1593–1603.
61. Hoffmann, M.; Rychlewski, J. "Effects of substituting a OH group by a F atom in D-glucose. Ab initio and DFT analysis," *J. Am. Chem. Soc.* **2001**, *123*, 2308–2316.
62. Sameera, W. M. C.; Pantazis, D. A. "A hierarchy of methods for the energetically accurate modeling of isomerism in monosaccharides," *J. Chem. Theory Comput.* **2012**, *8*, 2630–2645.
63. Barrows, S. E.; Storer, J. W.; Cramer, C. J.; French, A. D.; Truhlar, D. G. "Factors controlling relative stability of anomers and hydroxymethyl conformers of glucopyranose," *J. Comput. Chem.* **1998**, *19*, 1111–1129.
64. Momany, F. A.; Appell, M.; Strati, G.; Willett, J. L. "B3LYP/6-311++G** study of monohydrates of α- and β-D-glucopyranose: Hydrogen bonding, stress energies, and effect of hydration on internal coordinates," *Carbohydr. Res.* **2004**, *339*, 553–567.
65. Wladkowski, B. D.; Chenoweth, S. A.; Jones, K. E.; Brown, J. W. "Exocyclic hydroxymethyl rotational conformers of β- and α-D-glucopyranose in the gas phase and aqueous solution," *J. Phys. Chem. A* **1998**, *102*, 5086–5092.
66. *The Merck Index*; 11th ed.; Budavari, S., Ed.; Merck & Co.: Rahway, NJ, **1989**.
67. Mennucci, B.; Cappelli, C.; Cammi, R.; Tomasi, J. "Modeling solvent effects on chiroptical properties," *Chirality* **2011**, *23*, 717–729.
68. Nishida, Y.; Ohrui, H.; Meguro, H. "^{1}H-NMR studies of (6R)- and (6S)-deuterated D-hexoses: Assignment of the preferred rotamers about C5—C6 bond of D-glucose and D-galactose derivatives in solutions," *Tetrahedron Lett.* **1984**, *25*, 1575–1578.
69. Rockwell, G. D.; Grindley, T. B. "Effect of solvation on the rotation of hydroxymethyl groups in carbohydrates," *J. Am. Chem. Soc.* **1998**, *120*, 10953–10963.
70. Schnupf, U.; Willett, J. L.; Momany, F. "DFTMD studies of glucose and epimers: Anomeric ratios, rotamer populations, and hydration energies," *Carbohydr. Res.* **2010**, *345*, 503–511.
71. Poppe, L.; van Halbeek, H. "The rigidity of sucrose: Just an illusion?," *J. Am. Chem. Soc.* **1992**, *114*, 1092–1094.
72. Adams, B.; Lerner, L. "Observation of hydroxyl protons of sucrose in aqueous solution: No evidence for persistent intramolecular hydrogen bonds," *J. Am. Chem. Soc.* **1992**, *114*, 4827–4829.
73. Engelsen, S. B.; du Penhoat, C. H.; Perez, S. "Molecular relaxation of sucrose in aqueous solutions: How a nanosecond molecular dynamics simulation helps to reconcile NMR data," *J. Phys. Chem.* **1995**, *99*, 13334–13351.
74. Batta, G.; Kövér, K. E. "Heteronuclear coupling constants of hydroxyl protons in a water solution of oligosaccharides: Trehalose and sucrose," *Carbohydr. Res.* **1999**, *320*, 267–272.

75. Venable, R. M.; Delaglio, F.; Norris, S. E.; Freedberg, D. I. "The utility of residual dipolar couplings in detecting motion in carbohydrates: Application to sucrose," *Carbohydr. Res.* **2005**, *340*, 863–874.

76. Momany, F. A.; Appell, M.; Willett, J. L.; Bosma, W. B. "B3LYP/6-311++G** geometry-optimization study of pentahydrates of α- and β-D-glucopyranose," *Carbohydr. Res.* **2005**, *340*, 1638–1655.

77. Watson, J. D.; Crick, F. H. C. "A structure for deoxyribose nucleic acid," *Nature* **1953**, *171*, 737–738.

78. Judson, H.F. (**1996**) *The Eighth Day of Creation: Makers of the Revolution in Biology*, Cold Spring Harbor Press, Plainview, NY.

79. Kabelac, M.; Hobza, P. "Hydration and stability of nucleic acid bases and base pairs," *Phys. Chem. Chem. Phys.* **2007**, *9*, 903–917.

80. Topal, M. D.; Fresco, J. R. "Complementary base pairing and the origin of substitution mutations," *Nature* **1976**, *263*, 285 289.

81. Morgan, A. R. "Base mismatches and mutagenesis: How important is tautomerism?," *Trends Biochem. Sci.* **1993**, *18*, 160–163.

82. Vonborstel, R. C. "Origins of spontaneous base substitutions," *Mutation Res.* **1994**, *307*, 131–140.

83. Harris, V. H.; Smith, C. L.; Cummins, W. J.; Hamilton, A. L.; Adams, H.; Dickman, M.; Hornby, D. P.; Williams, D. M. "The effect of tautomeric constant on the specificity of nucleotide incorporation during DNA replication: Support for the rare tautomer hypothesis of substitution mutagenesis," *J. Mol. Biol.* **2003**, *326*, 1389–1401.

84. Zhanpeisov, N. U.; Sponer, J.; Leszczynski, J. "Reverse Watson-Crick isocytosine-cytosine and guanine-cytosine base pairs stabilized by the formation of the minor tautomers of bases. An ab initio study in the gas phase and in a water cluster," *J. Phys. Chem. A* **1998**, *102*, 10374–10379.

85. Barsky, D.; Colvin, M. E. "Guanine-cytosine base pairs in parallel-stranded DNA: An ab initio study of the keto-amino wobble pair versus the enol-imino minor tautomer pair," *J. Phys. Chem. A* **2000**, *104*, 8570–8576.

86. Trygubenko, S. A.; Bogdan, T. V.; Rueda, M.; Orozco, M.; Luque, F. J.; Sponer, J.; Slavíek, P.; Hobza, P. "Correlated *ab initio* study of nucleic acid bases and their tautomers in the gas phase, in a microhydrated environment and in aqueous solution. Part 1. Cytosine," *Phys. Chem. Chem. Phys.* **2002**, *4*, 4192–4203.

87. Sambrano, J. R.; de Souza, A. R.; Queralt, J. J.; Andrés, J. "A theoretical study on cytosine tautomers in aqueous media by using continuum models," *Chem. Phys. Lett.* **2000**, *317*, 437–443.

88. Bazso, G.; Tarczay, G.; Fogarasi, G.; Szalay, P. G. "Tautomers of cytosine and their excited electronic states: a matrix isolation spectroscopic and quantum chemical study," *Phys. Chem. Chem. Phys.* **2011**, *13*, 6799–6807.

89. Kosenkov, D.; Kholod, Y.; Gorb, L.; Shishkin, O.; Hovorun, D. M.; Mons, M.; Leszczynski, J. "Ab initio kinetic simulation of gas-phase experiments: Tautomerization of cytosine and guanine," *J. Phys. Chem. B* **2009**, *113*, 6140–6150.

90. Alemán, C. "Solvation of cytosine and thymine using a combined discrete/SCRF model," *Chem. Phys. Lett.* **1999**, *302*, 461–470.

91. Hunter, K. C.; Rutledge, L. R.; Wetmore, S. D. "The hydrogen bonding properties of cytosine: A computational study of cytosine complexed with hydrogen fluoride, water, and ammonia," *J. Phys. Chem. A* **2005**, *109*, 9554–9562.

92. Shishkin, O. V.; Gorb, L.; Leszczynski, J. "Does the hydrated cytosine molecule retain the canonical structure? A DFT study," *J. Phys. Chem. B* **2000**, *104*, 5357–5361.
93. Hanus, M.; Ryjacek, F.; Kabelac, M.; Kubar, T.; Bogdan, T. V.; Trygubenko, S. A.; Hobza, P. "Correlated ab initio study of nucleic acid bases and their tautomers in the gas phase, in a microhydrated environment and in aqueous solution. Guanine: Surprising stabilization of rare tautomers in aqueous solution," *J. Am. Chem. Soc.* **2003**, *125*, 7678–7688.
94. Jang, Y. H.; Goddard, W. A.; Noyes, K. T.; Sowers, L. C.; Hwang, S.; Chung, D. S. "pK_a values of guanine in water: Density functional theory calculations combined with Poisson-Boltzmann continuum-solvation model," *J. Phys. Chem. B* **2003**, *107*, 344–357.
95. Colominas, C.; Luque, F. J.; Orozco, M. "Tautomerism and protonation of guanine and cytosine. Implications in the formation of hydrogen-bonded complexes," *J. Am. Chem. Soc.* **1996**, *118*, 6811–6821.
96. Mons, M.; Dimicoli, I.; Piuzzi, F.; Tardivel, B.; Elhanine, M. "Tautomerism of the DNA base guanine and its methylated derivatives as studied by gas-phase infrared and ultraviolet spectroscopy," *J. Phys. Chem. A* **2002**, *106*, 5088–5094.
97. Kim, H.-S.; Ahn, D.-S.; Chung, S.-Y.; Kim, S. K.; Lee, S. "Tautomerization of adenine facilitated by water: Computational study of microsolvation," *J. Phys. Chem. A* **2007**, *111*, 8007–8012.
98. Hanus, M.; Kabelac, M.; Rejnek, J.; Ryjacek, F.; Hobza, P. "Correlated ab initio study of nucleic acid bases and their tautomers in the gas phase, in a microhydrated environment, and in aqueous solution. Part 3. Adenine," *J. Phys. Chem. B* **2004**, *108*, 2087–2097.
99. Sukhanov, O. S.; Shishkin, O. V.; Gorb, L.; Podolyan, Y.; Leszczynski, J. "Molecular structure and hydrogen bonding in polyhydrated complexes of adenine: A DFT study," *J. Phys. Chem. B* **2003**, *107*, 2846–2852.
100. Laxer, A.; Major, D. T.; Gottlieb, H. E.; Fischer, B. "(^{15}N$_5$)-Labeled adenine derivatives: Synthesis and studies of tautomerism by ^{15}N NMR spectroscopy and theoretical calculations," *J. Org. Chem.* **2001**, *66*, 5463–5481.
101. Rejnek, J.; Hanus, M.; Kabelá, M.; Ryjáek, F.; Hobza, P. "Correlated ab initio study of nucleic acid bases and their tautomers in the gas phase, in a microhydrated environment and in aqueous solution. Part 4. Uracil and thymine," *Phys. Chem. Chem. Phys.* **2005**, 2006–2017.
102. Kryachko, E. S.; Nguyen, M. T.; Zeegers-Huyskens, T. "Theoretical study of uracil tautomers. 2. Interaction with water," *J. Phys. Chem. A* **2001**, *105*, 1934–1943.
103. Morsy, M. A.; Al-Somali, A. M.; Suwaiyan, A. "Fluorescence of thymine tautomers at room temperature in aqueous solutions," *J. Phys. Chem. B* **1999**, *103*, 11205–11210.
104. Hobza, P.; Sponer, J. "Structure, energetics, and dynamics of the nucleic acid base pairs: Nonempirical *ab initio* calculations," *Chem. Rev.* **1999**, *99*, 3247–3276.
105. Sponer, J.; Hobza, P. "Molecular interactions of nucleic acid bases. A review of quantum-chemical studies," *Coll. Czech. Chem. Commun.* **2003**, *68*, 2231–2282.
106. Sponer, J.; Riley, K. E.; Hobza, P. "Nature and magnitude of aromatic stacking of nucleic acid bases," *Phys. Chem. Chem. Phys.* **2008**, *10*, 2595–2610.
107. Sponer, J.; Jurecka, P.; Hobza, P. "Accurate interaction energies of hydrogen-bonded nucleic acid base pairs," *J. Am. Chem. Soc.* **2004**, *126*, 10142–10151.

108. Zhao, Y.; Truhlar, D. G. "Hybrid meta density functional theory methods for thermochemistry, thermochemical kinetics, and noncovalent interactions: The MPW1B95 and MPWB1K models and comparative assessments for hydrogen bonding and van der Waals interactions," *J. Phys. Chem. A* **2004**, *108*, 6908–6918.

109. Zhao, Y.; Truhlar, D. G. "Design of density functionals that are broadly accurate for thermochemistry, thermochemical kinetics, and nonbonded interactions," *J. Phys. Chem. A* **2005**, *109*, 5656–5667.

110. Zhanpeisov, N. U.; Leszczynski, J. "The specific solvation effects on the structures and properties of adenine-uracil complexes: A theoretical *ab initio* study," *J. Phys. Chem. A* **1998**, *102*, 6167–6172.

111. Zhao, Y.; Truhlar, D. G. "How well can new-generation density functional methods describe stacking interactions in biological systems?," *Phys. Chem. Chem. Phys.* **2005**, *7*, 2701–2705.

112. Cerny, J.; Hobza, P. "The X3LYP extended density functional accurately describes H-bonding but fails completely for stacking," *Phys. Chem. Chem. Phys.* **2005**, *7*, 1624–1626.

113. Sponer, J.; Leszczynski, J.; Hobza, P. "Nature of nucleic acid–base stacking: Nonempirical ab initio and empirical potential characterization of 10 stacked base dimers. Comparison of stacked and h-bonded base pairs," *J. Phys. Chem.* **1996**, *100*, 5590–5596.

114. Sponer, J.; Jurecka, P.; Marchan, I.; Luque, F. J.; Orozco, M.; Hobza, P. "Nature of base stacking: Reference quantum-chemical stacking energies in ten unique B-DNA base-pair steps," *Chem. Eur. J.* **2006**, *12*, 2854–2865.

115. Jurecka, P.; Hobza, P. "True stabilization energies for the optimal planar hydrogen-bonded and stacked structures of guanine⋯cytosine, adenine⋯thymine, and their 9- and 1-methyl derivatives: complete basis set calculations at the MP2 and CCSD(T) levels and comparison with experiment," *J. Am. Chem. Soc.* **2003**, *125*, 15608–15613.

116. Hobza, P.; Sponer, J.; Reschel, T. "Density functional theory and molecular clusters," *J. Comput. Chem.* **1995**, *16*, 1315–1325.

117. Morgado, C. A.; Jurečka, P.; Svozil, D.; Hobza, P.; Šponer, J. i. "Balance of attraction and repulsion in nucleic-acid base stacking: CCSD(T)/complete-basis-set-limit calculations on uracil dimer and a comparison with the force-field description," *J. Chem. Theor. Comput.* **2009**, *5*, 1524–1544.

118. Morgado, C. A.; Jurecka, P.; Svozil, D.; Hobza, P.; Sponer, J. "Reference MP2/CBS and CCSD(T) quantum-chemical calculations on stacked adenine dimers. Comparison with DFT-D, MP2.5, SCS(MI)-MP2, M06-2X, CBS(SCS-D) and force field descriptions," *Phys. Chem. Chem. Phys.* **2010**, *12*, 3522–3534.

119. Zendlová, L.; Hobza, P.; Kabelác, M. "Potential energy surfaces of the microhydrated guanine⋯cytosine base pair and its methylated analogue," *ChemPhysChem* **2006**, *7*, 439–447.

120. Kabelac, M.; Zendlova, L.; Reha, D.; Hobza, P. "Potential energy surfaces of an adenine-thymine base pair and its methylated analogue in the presence of one and two water molecules: Molecular mechanics and correlated ab initio study," *J. Phys. Chem. B* **2005**, *109*, 12206–12213.

121. Cornell, W. D.; Cieplak, P.; Bayly, C. I.; Gould, I. R.; Merz, K. M.; Ferguson, D. M.; Spellmeyer, D. C.; Fox, T.; Caldwell, J. W.; Kollman, P. A. "A second generation

force field for the simulation of proteins, nucleic acids, and organic molecules," *J. Am. Chem. Soc.* **1995**, *117*, 5179–5197.

122. Poater, J.; Swart, M.; Guerra, C. F.; Matthias Bickelhaupt, F. "Solvent effects on hydrogen bonds in Watson–Crick, mismatched, and modified DNA base pairs," *Comput. Theor. Chem.* **2012**, *998*, 57–63.

123. Poater, J.; Swart, M.; Guerra, C. F.; Bickelhaupt, F. M. "Selectivity in DNA replication. Interplay of steric shape, hydrogen bonds, π-stacking and solvent effects," *Chem. Commun.* **2011**, *47*, 7326–7328.

124. Godfrey, P. D.; Brown, R. D. "Shape of glycine," *J. Am. Chem. Soc.* **1995**, *117*, 2019–2023.

125. McGlone, S. J.; Elmes, P. S.; Brown, R. D.; Godfrey, P. D. "Molecular structure of a conformer of glycine by microwave spectroscopy," *J. Mol. Struct.* **1999**, *485–486*, 225–238.

126. Ding, Y.; Krogh-Jespersen, K. "The glycine zwitterion does not exist in the gas phase: Results from a detailed ab initio electronic structure study," *Chem. Phys. Lett.* **1992**, *199*, 261–266.

127. Kasalová, V.; Allen, W. D.; Schaefer III, H. F.; Czinki, E.; Császár, A. G. "Molecular structures of the two most stable conformers of free glycine," *J. Comput. Chem.* **2007**, *28*, 1373–1383.

128. Alonso, J. L.; Cocinero, E. J.; Lesarri, A.; Sanz, M. E.; López, J. C. "The glycine–water complex," *Angew. Chem. Int. Ed.* **2006**, *45*, 3471–3474.

129. Balabin, R. M. "The first step in glycine solvation: The glycine–water complex," *J. Phys. Chem. B* **2010**, *114*, 15075–15078.

130. Jensen, J. H.; Gordon, M. S. "On the number of water molecules necessary to stabilize the glycine zwitterion," *J. Am. Chem. Soc.* **1995**, *117*, 8159–8170.

131. Aikens, C. M.; Gordon, M. S. "Incremental solvation of nonionized and zwitterionic glycine," *J. Am. Chem. Soc.* **2006**, *128*, 12835–12850.

132. Wang, W.; Pu, X.; Zheng, W.; Wong, N.-B.; Tian, A. "Some theoretical observations on the 1:1 glycine zwitterion–water complex," *J. Mol. Struct. (THEOCHEM)* **2003**, *626*, 127–132.

133. Bachrach, S. M. "Microsolvation of glycine: A DFT study," *J. Phys. Chem. A* **2008**, *112*, 3722–3730.

134. Bachrach, S. M.; Nguyen, T. T.; Demoin, D. W. "Microsolvation of cysteine: A density functional theory study," *J. Phys. Chem. A* **2009**, *113*, 6172–6181.

135. Blom, M. N.; Compagnon, I.; Polfer, N. C.; vonHelden, G.; Meijer, G.; Suhai, S.; Paizs, B.; Oomens, J. "Stepwise solvation of an amino acid: The appearance of zwitterionic structures," *J. Phys. Chem. A* **2007**, *111*, 7309–7316.

136. Mullin, J. M.; Gordon, M. S. "Alanine: Then there was water," *J. Phys. Chem. B* **2009**, *113*, 8657–8669.

CHAPTER 8

Organic Reaction Dynamics

In the previous chapters, we have focused our discussion of reaction mechanisms on the critical points along the reaction path. These critical points—local energy minima (reactants, intermediates, and products) and transition states (TSs)—were discussed in terms of their geometries and energies. We discussed the nature of the mechanism in terms of the topology of the potential energy surface (PES), the structure of the TSs, and the presence or absence of intermediates. We think of reactants progressing over TSs on to intermediates, perhaps through multiple intermediates and ending at products. This path is called the *reaction coordinate* and corresponds to the intrinsic reaction coordinate (IRC) or the minimum energy path (MEP) (see Section 1.6 for discussion on these two paths).

This standard mechanistic analysis has a long successful history. Organic chemistry textbooks are filled with PESs and discussions of the implication of single-step versus multiple-step mechanisms, concerted TSs, and so on.[1,2] Transition state theory (TST) and Rice–Ramsperger–Kassel–Marcus (RRKM) theory provide tools for predicting rates based upon simple assumptions built upon the notion of reaction on the PES following the reaction coordinate.[3,4]

An often-overlooked aspect of standard reaction mechanistic thought is that it really addresses only *half* of the picture. We talk about the positions of the atoms during the course of the reaction and the relative energies of points along the reaction path, but no mention is made of the time evolution of this process. In classical mechanics, description of a reactive system requires not just the particle positions but their momenta as well. The same is true for a quantum mechanical description, though one must keep in mind the limits imposed by the Heisenberg Uncertainty Principle. A complete description of a molecular reaction requires knowledge of both the position and the momentum of every atom for the entire time it takes for reactants to convert into products. This kind of description falls under the term *molecular dynamics* (MD).

A few simple examples should suffice to demonstrate the critical role that MD plays in chemical reactions. Conservation of momentum must be enforced. Atoms and molecular fragments in motion must maintain their motion unless some barrier or force is applied to them. Consider the reaction of *cis*-2-butene with

---

Computational Organic Chemistry, Second Edition. Steven M. Bachrach.
© 2014 John Wiley & Sons, Inc. Published 2014 by John Wiley & Sons, Inc.

chloromethylenecarbene to give 1-chloro-2,3-dimethylcyclopropane. Suppose the carbene approaches in a direction perpendicular to butylene's molecular plane, coming in from the bottom, as shown in Figure 8.1a. The carbene moves upward with no twisting motion, the approach is purely translational. As the carbene carbon continues to move upwards, it starts to form the first C–C bond (Figure 8.1b). In Figure 8.1c, the C–C bond is fully formed and the C–C–C angle begins to shrink below 90°. At this point, one might consider this structure to be a diradical. The chloromethylene carbon continues to move upwards, closing the C–C angle (Figure 8.1d) until the new C–C bond is formed, creating the three-member ring product (Figure 8.1e). Since there is no twisting motion, the chlorine atom comes in on the left side (cis to the methyl groups) and remains on that side throughout the reaction, creating the all-*cis* product. Certainly, the product with chlorine on the other face (anti) is lower in energy. If there was some twisting motion, the chlorine could rotate to the other face and produce the more thermodynamically stable isomer. However, in the absence of some force to kick-start the rotation, conservation of momentum demands that chlorine remain on the same side it started from. The stereospecificity that results here is due solely to the dynamics, particularly the initial conditions of the reacting species and conservation of momentum. One need not invoke any notion of orbital control or any other quantum mechanical phenomenon.

A second example is to consider what happens to a system as it moves off of a TS. At the TS, the molecule has energy E_{TS}, which is the maximum potential energy on the PES. As it moves away from the TS, the MEP drops off in potential energy from E_{TS}. In order for the molecule to stay on the MEP, this difference in potential energy must be instantaneously removed from the molecule. In a solution, one might imagine that collisions with the bath could allow for some potential energy to be transferred to the solvent molecules, but it is highly unlikely that this transfer would be completely efficient. In the gas phase, no neighboring molecules are available for collisions. The loss in potential energy must be converted into some internal energy, as internal vibrational energy, rotational energy or kinetic energy of the whole molecule. In other words, the loss of potential energy as the molecule moves away from the TS will result in the molecule vibrating or rotating or translating more rapidly (or some combination of these). If the molecule gains sufficient translational kinetic energy, it may be able to simply climb over

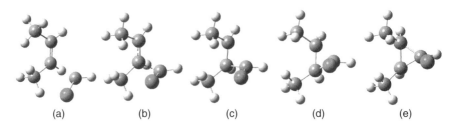

Figure 8.1 Snapshots along the reaction of ethylene with chloromethylenecarbene.

subsequent smaller TSs, passing right through intermediate structures, as indicated by the arrow in Figure 8.2.

It is unlikely that a molecular system will be able to distribute all of the lost potential energy instantaneously and continuously as it moves downhill away from the TS. Rather, the trajectory will take the molecule off of the MEP, as shown by the dashed line in Figure 8.2. The molecule will lose potential energy on this path, but not as quickly as on the MEP. As the molecule follows this non-MEP trajectory, it may sample regions of the PES far removed from the critical points on the PES, perhaps avoiding local minima or other TSs entirely.

The last issue is the possibility of nonstatistical dynamics. TST and RRKM are statistical theories. If this assumption is true, then both theories can predict reaction rates, leading to what is called *statistical dynamics*. Of relevance to this chapter is that for statistical dynamics to occur, any intermediates along the reaction pathway must live long enough for energy to be statistically distributed over all of the vibrational modes. If this redistribution cannot occur, then TST and RRKM will fail to predict reaction rates. A principle characteristic of nonstatistical dynamics is a reaction rate much faster than that predicted by statistical theories.

All three of these situations have been demonstrated in seemingly ordinary organic reactions using MD trajectory studies built on high level quantum mechanical computations. In order to understand these studies, we next present an overview of MD, with particular emphasis on techniques employed in the cases studies that follow. Readers interested in more details on MD are referred to a number of review articles.[5–9] These case studies cover a range of organic reactions, indicating that dynamic effects are not restricted to some niche category. Rather, these studies point toward a general need for physical organic chemists to rethink their models. This chapter should amply demonstrate that mechanistic thought will continue to undergo dramatic change as dynamic effects become an essential component to the way reaction mechanism are explored and discussed.

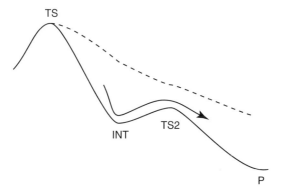

Figure 8.2 Generic PES with a shallow intermediate. The arrow indicates motion down through the intermediate and then directly over TS2 to products. The dash line indicates motion off of TS following a non-MEP path.

8.1 A BRIEF INTRODUCTION TO MOLECULAR DYNAMICS TRAJECTORY COMPUTATIONS

Classical MD is performed by determining the positions of the atoms over time using the classical equations of motion. These can be expressed in two equivalent forms: Newton's equations of motion

$$m_i \frac{d^2 q_i}{dt^2} = -\frac{\partial V(\mathbf{q})}{\partial q_i} \tag{8.1}$$

or Hamilton's equations of motion

$$\frac{\partial H}{\partial q_i} = -\frac{dp_i}{dt} \quad \text{and} \quad \frac{\partial H}{\partial p_i} = \frac{dq_i}{dt} \tag{8.2}$$

where q_i is the coordinate of atom i, p_i is its momentum, and H is the Hamiltonian, the sum of the kinetic energy T and the potential energy V. A trajectory, the time-elapsed position and momentum of the atoms in a molecular system, is computed by integrating the equation of motion starting from some initial condition. The trajectory is like a movie, each frame j contains the position and momentum of every atom at time t_j. The sequence of frames then shows how molecules evolve over time, changing from, say, reactant into product.

Solution of the equations of motion (either Eq. 8.1 or 8.2) requires three choices: (1) How will the numerical integration be performed? (2) How is the potential energy V computed? (3) What are the initial conditions for each trajectory? These three topics are the next subjects of discussion.

8.1.1 Integrating the Equations of Motion

Whether one uses Newton's or Hamilton's equations of motion, obtaining the atomic positions over time requires numerical integration. Integration of ordinary differential equations (ODE) is a well-traveled territory in numerical analysis. A number of different techniques are routinely used in MD.

The purpose of this section is not to present all the details of numerical integration methods, their benefits, and failures but rather to provide enough general sense of the methods and how this impacts choices pertaining to other aspects of the MD treatment. Suppose one is interested in a trajectory that runs for 100 fs. Given the initial position and momentum of every atom, we now want to know where the atoms are at various times up to and including 100 fs later. The computational demands of this problem are dependent on how many intermediate points are necessary in order to track the trajectory out to its conclusion. The shorter is the time interval between points, the more accurate is the trajectory. Computing the necessary information at each point takes computational effort: at least the first derivative of the energy with respect to coordinates is needed at every point along the trajectory. Therefore, one will have to make a trade-off between the number of step along each trajectory, the total time span of each trajectory, and the total number of trajectories to be computed.

The simplest numerical technique for solving ODE is Euler's method:

$$q_{n+1} = q_n + \Delta t \frac{dq}{dt} \tag{8.3}$$

where n indicates the time and Δt is the time step. While this is conceptually the simplest approach, the error is large ($O(\Delta t^2)$), requiring very small time steps to be accurate. A straightforward extension of Euler's method is to compute a number of intermediate trial steps to help correct for this error. The best known of these procedures is the Runge–Kutta method.[10] In its fourth-order implementation, the first derivatives must be computed for four different positions for each trajectory point.

The two most widely implemented numerical integration techniques within MD are the Verlet algorithm and the use of instantaneous normal mode coordinates.[6] The Verlet algorithm begins by writing the Taylor expansion for a coordinate at time $t + \Delta t$ and $t - \Delta t$:

$$q_i(t + \Delta t) = q_i(t) + \frac{dq_i}{dt}\Delta t - \frac{1}{2m_i}\frac{\partial V(q_i)}{\partial q_i}\Delta t^2 + \cdots \tag{8.4a}$$

$$q_i(t - \Delta t) = q_i(t) - \frac{dq_i}{dt}\Delta t - \frac{1}{2m_i}\frac{\partial V(q_i)}{\partial q_i}\Delta t^2 + \cdots \tag{8.4b}$$

Adding these two expressions gives the Verlet algorithm

$$q_i(t + \Delta t) = 2q_i(t) - \frac{1}{m_i}\frac{\partial V(q_i)}{\partial q_i}\Delta t^2 - q_i(t - \Delta t) \tag{8.5}$$

This algorithm provides very good computational performance.

The second method was proposed by Helgaker et al.[11] The potential about point q_0 can be approximated by the harmonic function

$$V_{\text{model}} = V_0 + \mathbf{G}\Delta\mathbf{q} + \frac{1}{2}\Delta\mathbf{q}\mathbf{H}\Delta\mathbf{q} \tag{8.6}$$

where V_0 is the potential energy at q_0, and $\mathbf{G}$ and $\mathbf{H}$ are its gradient and Hessian, respectively. Normal mode coordinates are then obtained by diagonalizing the $3N$-dimensional set of equations corresponding to the equations of motion:

$$m\frac{d^2\mathbf{q}}{dt^2} = -\mathbf{G} - \mathbf{H}\Delta\mathbf{q} \tag{8.7}$$

These equations can be analytically solved. The only concern is that the harmonic approximation is valid within some trust radius. Performance can be improved by including a corrector step, allowing for larger step sizes.[12] Also of note is that this method requires computation of the second derivatives, a time-consuming step. A variant in this method allows for numerical updating of the Hessian for a number of steps, requiring fewer explicit computations of the full Hessian matrix.[13]

These integration techniques require computation of the first derivatives of the potential energy, and some methods require computation of the second derivatives. The question now posed is how to evaluate the potential energy.

8.1.2 Selecting the PES

Until the 1990s, MD trajectories were computed using analytic PESs. Such a potential is defined by an analytic function whose major advantage is that an analytic expression of the derivative (and second derivative) is readily evaluated, as needed for the integration of the equations of motion. Analytic potential energy functions are generally fit to experimental or ab initio geometric and energetic data. The number of ab initio points needed to derive such a potential grows rapidly with the size of the molecule, and so analytic potentials were restricted to small molecules or systems with arbitrary constraints.[6] Analytic potentials may suffer from unreasonable behavior in regions of space not properly sampled by points used in the fitting procedure.

As computers became faster, especially with the development of parallel and cluster computing, MD using *direct dynamics* became feasible. In direct dynamics, an analytic potential is never employed. Rather, the energy and its derivatives are computed as needed for each trajectory point. No error is introduced in a fitting procedure—the derivatives are accurately computed given the quantum mechanical method employed.

Though direct dynamics removes the errors associated with a fitted potential, it can be enormously time consuming. A trajectory that has, say, 100 points along the path will require 100 energy and derivative computations (and perhaps 100 second-derivative computations depending on the integration technique). The computational demand of each of these individual derivatives computations depends on the size of the molecule and the quantum mechanics (QM) method employed. When one considers that a large number of trajectories will need to be computed, it is readily apparent that severe limitations may be placed on the type of systems that can reasonably be treated with direct dynamics. It is important to recognize, however, that trajectory calculations are inherently parallelizable; for example, each trajectory can be run independently of the others, making MD trajectory studies especially well suited for large computer cluster installations.

While many of the examples we discuss below involve direct dynamics with the PES computed using high level QM methods, for large molecules, this technique may be out of reach. Recognizing that semiempirical QM methods make use of parameters fit to experimental data, Truhlar[14] suggested that these parameters can be fit to ab initio data for the specific reaction of interest. Fitting a semiempirical method, as was first done with NDDO but is now more popularly carried out with AM1, to replicate ab initio data creates what is called *specific reaction parameters* or SRP. The energies and derivatives for a selection of molecular geometries are computed using some high level QM method. Selected AM1 parameters (along with perhaps additional empirical functions) are then fit to mimic the ab initio PES. This new AM1-SRP method captures the features and characteristics of the full ab initio PES. Direct dynamics are then run, using the AM1-SRP method for the energy and derivatives evaluation. The distinct advantage of the SRP approach is that it uses significantly less computer time to evaluate the needed energies and derivatives than is used with the ab initio QM treatment.

8.1.3 Initial Conditions

Trajectories can in principle be initiated anywhere on the PES. The initial geometry needs to be defined (the x-, y-, and z-coordinates of every atom) along with the initial velocity of every atom. If one uses normal modes, assigning the velocities is a matter of deciding how much energy (or how many quanta) to place into each mode, along with the phase of that vibration.

Taking the most literal approach, one might start the trajectories with reactant(s) and follow them over TSs, through intermediates and finishing at product(s). Undoubtedly, some trajectories started at reactants will proceed to products, but most of them will bounce around the neighborhood of the reactant for quite some time. It's only when the trajectory heads toward the TS that it may eventually cross over, and most trajectories that begin at reactants will not be headed in that direction.

In order to obtain trajectories that are reactive, most MD studies initiate the trajectories in the neighborhood of the TS. A trajectory is first run in one direction, say, toward product. Next, all of the velocities are reversed, and a second trajectory is again started at the TS and followed toward reactant. The union of these two trajectories defines a full trajectory from reactant to product.

A single trajectory indicates one path, one way for a molecular system to evolve from reactant to product. This one trajectory has limited value, but a sampling of trajectories, if done in a statistically meaningful way, can provide insight toward real molecular systems. Typically, an MD simulation will have hundreds to thousands of trajectories, though fewer might be run if limited interpretation is acceptable.

There are a number of different techniques for generating statistical initial conditions for the trajectories. We discuss here only one technique to give a sense of how this sampling is created. Interested readers are directed toward more extensive discussions in Refs. 5 and 6.

We demonstrate here how to create a microcanonical sampling and concern ourselves with only the distribution of energy over the vibrational modes.[15] In a microcanonical ensemble, all molecules have the same total energy, but different position in phase space, meaning different coordinates and/or momentum. The total vibrational energy E_v is given as

$$E_v = \sum_i^{3N-6} E_i = \sum_i^{3N-6} \frac{p_i^2 + \omega_i^2 q_i^2}{2} \qquad (8.8)$$

where p_i and q_i are the momentum and position within the ith vibrational mode. A uniform distribution of the vibrational energy states can be created as

$$E_i = \left[E_v - \sum_j^{i-1} E_j\right](1 - R^{1/n-1}) \tag{8.9}$$

where n is the number of vibrational modes ($3N - 6$) and R is the random number between 0 and 1. A second random number R' is chosen to assign the phase of each vibrational mode. The position and momentum for each vibrational mode is assigned as

$$q_i = \left[\frac{\sqrt{2E_i}}{\omega_i}\right]\cos(2\pi R'_i) \tag{8.10}$$

$$p_i = -\sqrt{2E_i}\sin(2\pi R'_i) \tag{8.11}$$

These coordinates and momenta can then be transformed back into Cartesian coordinates for the numerical integration. One should keep in mind that this type of sampling is exact for harmonic oscillators but will be an approximate ensemble for anharmonic oscillators.

Combining all of these ideas of this section gives rise to Born–Oppenheimer MD, sometimes also referred to as *semiclassical dynamics*. Separation of electronic and nuclear motion is assumed, namely, the Born–Oppenheimer approximation. Atoms move on the electronic PES, computed using some QM method, following the classical equations of motion. In summary, the steps for computing trajectories using direct dynamics are the following.

(1) Select a method for performing the numerical integration of the equations of motion.
(2) Decide on an appropriate quantum mechanical method to compute the energy and derivatives. Since the energy and derivatives (perhaps also the second derivatives) calculation will be repeated thousands of times, it is important to balance off the computational time against the rigor of the method.
(3) Initialize the positions and momenta for each atom in every trajectory using some sampling procedure.
(4) Compute each trajectory for a given length of time or until product or reactant is reached.

8.2 STATISTICAL KINETIC THEORIES

The reaction rate can be obtained by trajectory analysis by examining a very large number of trajectories. This can be prohibitively expensive for most reactions. An alternative approach is to use statistical theories that deal with a large ensemble of molecules. Familiar to most chemists are two major theories of this type: TST and RRKM theory. We present an overview of these two theories largely because the results of the specific dynamics studies discussed in this chapter are at odds with these theories.

Standard representation of the TS in organic chemistry textbooks is the point of maximum energy on the reaction coordinate.[16,17] More precise is the definition provided in Section 1.6: the TS is the col, a point where all the gradients vanish, and all of the eigenvalues of the Hessian matrix are positive except one, which corresponds to the reaction coordinate. In statistical kinetic theories, a slightly different definition of the TS is required.

Phase space is a $6N - 12$ dimensional representation of the atomic $(3n - 6)$ coordinates and their associated $(3N - 6)$ momenta. Reactive trajectories in phase space move from reactant to product. The TS is the hyperplane such that all trajectories that cross this plane do so only once. In other words, trajectories that cross this plane from the reactant side will go on to products without ever turning back and recrossing the plane toward reactant. Given this definition, the rate of reaction is the number of trajectories that cross the plane per unit time.[7]

In order to remove the need for explicit trajectory analysis, one makes the statistical approximation. This approximation can be formulated in a number of equivalent ways. In the microcanonical ensemble, all states are equally probable. Another formulation is that the lifetime of reactant (or intermediate) is random and follows an exponential decay rate. But perhaps the simplest statement is that intramolecular vibrational energy redistribution (IVR) is faster than the reaction rate. IVR implies that if a reactant is prepared with some excited vibrational mode or modes, this excess energy will randomize into all of the vibrational modes prior to converting to product.

RRKM[18–21] theory assumes both the statistical approximation and the existence of the TS. It assumes a microcanonical ensemble, where all the molecules have equivalent energy E^*. This energy exceeds the energy of the TS (E_0), thanks to vibration, rotation, and/or translation energy. Invoking an equilibrium between the TS (the activated complex) and reactant, the rate of reaction is

$$k(E^*) = \frac{W(E^* - E_0)}{hN^*(E^*)} \quad (8.12)$$

where $W(E^* - E_0)$ is the number of energy levels of the TS between E^* and E_0, h is the Planck constant, and $N^*(E^*)$ is the density of states of the reactant at E^*.

TST[22,23] also makes the statistical approximation and invokes an equilibrium between reactant and TS. TST invokes constant temperature instead of a microcanonical ensemble as in RRKM theory. Using statistical mechanics, the reaction rate is given by the familiar equation

$$k = \frac{k_B T}{h} e^{-\Delta G_0^\ddagger / RT} \quad (8.13)$$

where k_B is the Boltzmann constant and $\Delta G_0^\ddagger$ is the free energy of activation. TST invokes the TS hyperplane perpendicular to the reaction coordinate through the maximum on the PES. Variational transition state theory (VTST)[24] corrects for the possibility that trajectories might recross a plane defined in this manner. VTST locates the hyperplane that minimizes recrossing, which turns out to be at the maximum on the free energy surface.

With these theories in mind, we now turn to a number of examples of organic reactions studied using direct dynamics. In all of these cases, some aspect of the application of the statistical approximation is found to be in error. At minimum, the collected weight of these trajectory studies demonstrates the caution that need be used when applying TST or RRKM. More compelling though is that these studies question the very nature of the meaning of reaction mechanism.

8.3 EXAMPLES OF ORGANIC REACTIONS WITH NON-STATISTICAL DYNAMICS

8.3.1 [1,3]-Sigmatropic Rearrangement of Bicyclo[3.2.0]hex-2-ene

We begin the discussion of dynamic effects with an analysis of the rearrangement of bicyclo[3.2.0]hex-2-ene **1** to bicyclo[2.2.1]hex-2-ene **2** (Reaction 8.1). Though Carpenter's MD studies of this rearrangement involved computations on semiempirical PESs,[25,26] this study was the first to demonstrate the importance of dynamic effect in organic reactions. It also provides a fine example of some of the major symptoms of nonstatistical outcomes and evidence for dynamic effects.

(Reaction 8.1)

The rearrangement of **1** is one of the classic examples of the power and limitations of the Woodward–Hoffmann rules.[17] Thermolysis of the analogs shown in Reactions 8.2[27] and 8.3[28,29] proceeds with inversion about the migrating carbon, as predicted by orbital symmetry arguments. Interestingly, the preference for the inversion path increases with increasing temperature. *Exo*-methyl substitution, as shown in Reaction 8.4, still leads to inversion, but the *endo*-methyl analog (Reaction 8.5) rearranges predominantly with retention.[30] The explanation for this latter result is that the allowed pathway requires the methyl group to rotate downward, toward the double bond. This is sterically congested and so the retention path becomes competitive. The lack of stereorandomization in any of these cases was taken as strong evidence against a diradical intermediate.

Both the AM1 and PM3 PESs for Reaction 8.1 are characterized by a diradical intermediate sandwiched by an entrance and exit TS. Their major difference is that PM3 predicts the diradical to reside in a deeper well than does AM1 (Scheme 8.1). Semiempirical methods tend to overestimate the stability of diradicals, and so these surfaces may be in error. Carpenter argues the question as "*What if* these surfaces are correct?"[31] If the diradical is long lived, then IVR should occur, leading to

EXAMPLES OF ORGANIC REACTIONS WITH NON-STATISTICAL DYNAMICS 515

(Reaction 8.2)

307 °C, >95% / <5%

(Reaction 8.3)

276 °C 76% 24%
312 °C 89% 11%

(Reaction 8.4)

290 °C, >90% / <10%

(Reaction 8.5)

290 °C, >88% / <12%

stereorandomized products. Since the experiments indicate definite stereoselection, can a surface with a diradical intermediate lead to stereoselectivity?

To probe this question, Carpenter performed direct dynamics using both the AM1 and PM3 surfaces. An ensemble of initial states at 300 °C was prepared about the higher energy TS (**5**) and trajectories were computed in the reverse direction. The principle of microscopic reversibility implies that these same trajectories that run backwards are appropriate trajectories for the forward reaction. One hundred trajectories were run on each surface for 500 fs. The trajectories fell into two

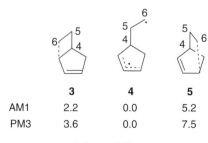

	3	4	5
AM1	2.2	0.0	5.2
PM3	3.6	0.0	7.5

Scheme 8.1

categories: (1) a set of short-lived trajectories (<250 fs) and (2) a set of long-lived trajectories.

There were four short trajectories on the PM1 surface and 10 such trajectories on the AM1 surface. All of these short-lived direct trajectories proceed with counterclockwise motion about the C_5-C_6 bond leading exclusively to inversion about C_6. No short-lived trajectory gives retention product nor do they reside in the vicinity of the diradical **4** for any appreciable time. One may consider these trajectories as following the Woodward–Hoffman allowed, concerted path, except that it occurs over a PES containing a diradical!

These direct trajectories all begin with energy in C_5-C_6 rotational vibration mode, with its initial motion in the clockwise direction. The trajectory simply conserves this angular momentum, continuing the clockwise rotation through 180° as C_6 migrates to make the new C–C bond.

How might a retention product be created? Counterclockwise rotation of 360° (or multiples thereof) while C_6 migrates would do the job. Such large angular rotations require considerably more energy in this mode than just zero-point energy. This rapidly rotating bond would then have to suddenly come to a quick halt when the new C–C bond is formed, but conservation of angular momentum prohibits this unless the momentum can be transferred to some other region of the molecule. Since it cannot form the new C–C bond, the C_5-C_6 torsional rotation continues on, giving rise to a long-lived diradical.

The other option is for clockwise rotation about the C_5-C_6 bond. This process turns out to be slower in getting started, since it leads to more rapid rupture of the C_1-C_6 bond than does counterclockwise motion. But once the clockwise C_5-C_6 does begin, it accelerates rapidly to minimize the steric interaction of the downward-pointing group on C_6 with the five-member ring below. Once again, this path generates too much angular momentum in the C_5-C_6 torsion, and it cannot be quickly stopped in order to form the new C–C bond. Clockwise C_5-C_6 trajectories are, therefore, long lived in the neighborhood of the diradical.

The short-lived trajectories indicate an effective rate constant much larger than that predicted by RRKM theory. These constitute nonstatistical dynamic paths. The longer-lived trajectories appear to follow traditional statistical dynamics. One way to reconcile these within a single statistical kinetics model is that there are two underlying reaction mechanisms—the "fast" trajectories corresponding to a concerted reaction and the "slow" trajectories corresponding to a stepwise mechanism with a long-lived diradical intermediate. Standard physical organic chemical interpretation of this idea necessitates that there be two different PESs, that is, a concerted path with one TS and a second path with two TSs and an intermediate (as in **3–5**). All the MD trajectories, however, are generated here using *one* surface, that for the stepwise mechanism! This is a fundamental break with classical organic mechanistic thought.

Carpenter has coined the term "dynamic matching" to describe these direct nonstatistical trajectories. The motion and momentum of a system coming out of one TS region directs the system straight into a second TS, neglecting any nearby local minima. The exit channel out of the first TS lines up, or matches, with the entrance channel into the second TS. For Reaction 8.1, some paths through the region about TS **4** are pointed straight at **2** and so they pass directly over this second TS and onto product. Occurrence of these directs paths, which are usually very short, is indicative of nonstatistical dynamics.

Another common symptom of reactions that display nonstatistical behavior is a bimodal distribution of trajectory lifetimes. The nonstatistical trajectories have much shorter lifetimes than predicted by TST or RRKM. Trajectories will then cluster in two (or more) groups, the very short lifetime group and those with a much longer lifetime, consistent with statistical dynamics. Both direct trajectories that seemingly avoid local minima and bimodal lifetime distributions will be seen in many of the examples to follow.

Returning to the experimental results, the dynamic model offers an explanation for the unusual observation of *greater* selectivity with higher temperature.[28,29] It is important to note that labeled norbornene represents only a very small portion (1–2 percent) of the product of Reaction 8.3; the major products are the result of a retro-Diels–Alder reaction, namely, deuterated ethylene and cyclopentadiene. The activation entropy of the retro-Diels–Alder is presumably positive, while that of the rearrangement is negative. At higher temperatures, the retro-Diels–Alder will become more favorable, so more and more of the diradical will follow this fragmentation path. On the other hand, the direct rearrangement pathway cannot lead to fragmentation and so continues to produce just the inversion product. The "bleeding off" of the diradical allows for greater stereoselectivity to result.[25]

Carpenter[25] also computed some trajectories for the rearrangements of the *exo*- and *endo*-methylbicyclo[3.2.0]hept-2-ene, **6** and **7**. The *exo* isomer **6** undergoes a direct trajectory with inversion about C_6. This pathway rotates the methyl group away from the five-member ring, very much analogous to the direct trajectories of the parent compound.

For the rearrangement of the *endo* isomer **7**, the trajectory starts off innocently enough, with a rotation about C_5-C_6 bond that brings the methyl group downward toward the five-member ring, looking just like the other direct trajectories that lead to the inversion product. However, soon along the trajectory, the methyl groups run into the five-member ring and bounce off it, reversing the torsional rotation about C_5-C_6. The net effect is now to have a *direct* trajectory that produces the retention product. This trajectory explains the preference for the retention product of Reaction 8.5.

8.3.2 Life in the Caldera: Concerted versus Diradical Mechanisms

In the limited trajectory study for Reaction 8.1, Carpenter[26] noted 10 (out of 100) direct trajectories with the AM1 surface and 4 (out of 100) direct trajectories on the PM3 surface. While the sample size is too small to say anything definitive about these percentages, he did note that the diradical is in a shallower well in AM1 than in PM3. It is reasonable that the shallower well will be more easily (i.e., directly) crossed. In particular, a trajectory is less likely to be trapped within a shallow energy well than a deep energy well.

Reactions involving intermediates in shallow energy wells may be good candidates for demonstrating unusual dynamical behavior. If this occurs in a reaction that can exhibit stereo- or regiochemistry, then this dynamic behavior can be experimentally tested. Pericyclic reactions fit this specification precisely. These reactions can show remarkable stereo- and regioselectivity, as discussed in detail in Chapter 4. Great debates have been fought over whether some pericyclic reactions are concerted or stepwise, the conflict based on the relative closeness in energy of the concerted TS and the diradical intermediate. This section presents MD trajectory studies of a variety of pericyclic reactions, all of which demonstrate nonstatistical behavior. These studies raise significant questions about the fundamental nature of pericyclic reactions. The consequence of these MD studies is in fact much larger, challenging our understanding of just what the term "mechanism" really means.

Besides being pericyclic reactions, these next examples share a characteristic PES. The diradical intermediate is energetically appreciable above the reactant and product energies. It resides in a broad shallow bowl with multiple exit channels that involve different TSs, all of which are typically of similar energy. The diradical is, therefore, poorly described as a single geometry, but rather it samples quite a range of geometries without much increase in energy. This kind of diradical has been termed *twixtyl*[32] or a *continuous diradical transition state*.[33] The term that has become more popular is *caldera*,[34] the collapsed dome of volcano,[35] which captures the shape of the surface about the diradical minimum. Figure 8.3 shows a generic PES of a caldera. The diradical intermediate (**DI**) sits in the center of the broad shallow bowl—the caldera.

To get a sense of the argument to be put forward in this chapter, imagine rolling a marble on the surface shown in Figure 8.3, starting at **TS1**, the entrance TS connecting reactant **R** with the diradical intermediate **DI**. Suppose we give the marble a slight push backwards; the marble will roll downhill toward **R** accelerating all the way as the potential energy (due to gravity) is converted into kinetic energy. This defines one trajectory.

For the second trajectory, let's again start the marble at **TS1** and give it a slight push in the forward direction, directly into the middle of the caldera, toward **DI**. The marble will again move downhill, accelerating toward **DI**. If we assume that the surface is frictionless, when the marble reaches the point **DI**, the local minimum inside the caldera, it will not stop because it has considerable momentum, gained as the potential energy is converted into kinetic energy. So, the marble continues on its straight path now climbing uphill toward **TS2**, slowing down as it gains height. If **TS2** is no greater in energy than **TS1**, the marble will roll up over **TS2** and then

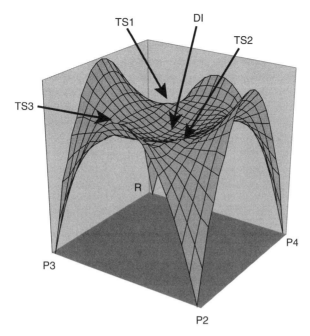

Figure 8.3 PES of a generic caldera. **TS1** connects reactant **R** with the diradical intermediate **DI**. **TS2** connects **DI** with product **P2** and **TS3** connects **DI** to **P3**.

fall downhill toward **P2**. The marble has traversed a direct path across the caldera to **P2**. It is unaware of any other features on the PES. This trajectory is indicated as the heavy line in Figure 8.4.

What is needed to get the marble to fall into one of the other products, say **P3**? Once the marble has momentum in the direction toward **TS2** it will head that way unabated, unless it is deflected by a barrier or wall. Traveling from **TS2** to **P3** requires a right-hand turn. This can be accomplished in two ways. First, the marble can be initiated with a push in the direction directly toward **TS3**, with sufficient initial momentum to carry it toward that col and not get turned inward by the caldera wall back into its interior (the long-short dashed line of Figure 8.4). A second possibility is to push the marble into the caldera but toward one of the high walls in order to rebound off that wall and toward **TS3**; the wall provides the "push" to turn the marble. This is the dashed line of Figure 8.4.

Dynamic matching is the concept that the initial motion away from the first TS (**TS1** in our model) carries the species up the entrance channel toward the second TS (**TS2**). Trajectories that start with momentum that properly matches with the second TS will directly cross that caldera and spend no time as the diradical. The trajectories will be very fast, prohibiting IVR from occurring.

Trajectories that do not match up with an exit channel will enter the caldera and bounce around the basin for a while before finally traversing up a hill and across a TS. Such trajectories do reside in the basin for an appreciable time; these

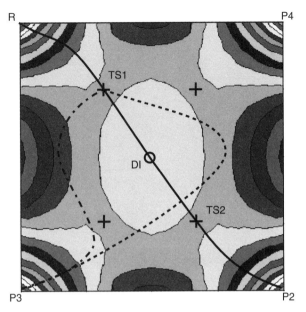

Figure 8.4 Contour plot of the PES in Figure 8.3, looking down from above. The heavy line is the direct trajectory from **TS1** across **DI** and over **TS2** to **P2**. The long-short dashed line is the semidirect trajectory from **TS1** to **P3**. The dashed line is the rebound trajectory from **TS1** to **P3**. *(See insert for color representation of this figure.)*

trajectories constitute the diradical state. During its time inside the basin, vibrational energy can be redistributed among the various modes. Statistical behavior can then be anticipated, as the diradical can now statistically exit any available channel.

As will be discussed for the specific examples below, the trajectories will tend to group in two categories: those that are direct trajectories across the caldera and those that bounce around inside the caldera for some longer time. It is the occurrence of the direct trajectories that provide the nonstatistical dynamics and call into question the traditional notions of *mechanism* and *intermediate*.

8.3.2.1 Rearrangement of Vinylcyclopropane to Cyclopentene

The thermolysis of vinylcyclopropane **8** to give cyclopentene **9** is another classic formal [1,3]-sigmatropic rearrangement. Its mechanism has been the subject of controversy. Application of the Woodward–Hoffmann orbital symmetry rules predicts that the reaction proceeds by either the *si* or *ar* routes. Here, the stereodesignations *s,a* refer to suprafacial or antarafacial participation of the allyl fragment and *r,i* indicates retention or inversion about the migrating methylene group. Baldwin[36] heated the trideuterated vinylcycloprane **10** at 300 °C and measured the ratio of the four rearrangement products (Scheme 8.2). He found that the *si:sr:ar:ai* ratio is 40 : 23 : 13 : 24. The ratio of the formally allowed (*si* + *ar*) to forbidden (*sr* + *ai*) is about 1 : 1. This result casts serious doubt on the concerted mechanism and Baldwin concluded the "diradical-mediated paths seem

Scheme 8.2

mechanistically essential." On the other hand, the secondary kinetic isotope effect (KIE) at the terminal vinyl carbon ($k_H/k_D = 1.17–1.21$) is slightly larger than that at the migrating methylene carbon ($k_H/k_D = 1.14$),[37] evidence for a concerted mechanism.

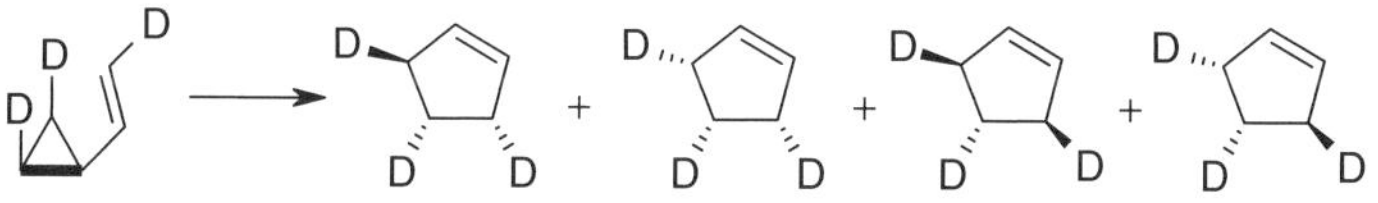

In 1997, two theoretical studies of the vinylcyclopropane to cyclopentene rearrangement were reported. Critical points at CASSCF(4,4)/6-31G* were located by Davidson and Gajewski,[38] while Houk[39] located them using (U)B3LYP/6-31G* with single-point energies computed at CASSCF(4,4)/6-31G*. Both studies agree on the fundamental characteristics of the PES for this reaction. First, no stable diradical intermediate could be located. Rather, a flat plateau region is characterized by a number of *TSs*. A TS that corresponds to the [1,3]-shift in the *si* manner is found, along with TSs with various arrangements of the terminal methylene groups. All of these TSs are within 4 kcal mol^{-1} of each other (Figure 8.5). The authors of both papers conclude that no single pathway is accountable for the rearrangement. The flatness of the PES and the lack of any diradical intermediate are suggestive of the possible role of dynamic control of this reaction.

MD trajectories have been computed for the **8 → 9** using an AM1-SRP PES fit to the MRCI/cc-pVDZ//CASSCF(4,4)/cc-pVDZ critical point energies.[40,41] Trajectories at 573 K were initiated at three TSs: TS(*si*), TS(90,0), and TS(0,0). The product distribution from these trajectories is listed in Table 8.1. The product distribution is dependent on which TS is the origin of the trajectory. Trajectories started at TS(*si*) preferentially produce the Woodward–Hoffmann allowed products, while the *ai* product is the major one produced from the TS(0,0) trajectories. Trajectories from TS(90,0) give nearly equal distribution of the four possible products. This mode selectivity strongly discounts the presence of a diradical intermediate that has some appreciable lifetime. Such a diradical would undergo intramolecular vibrational energy redistribution, and the trajectories from the three different TSs would result in identical product distributions.

The product distribution is also time dependent. Listed in Table 8.2 is the time dependence of the *s/a* and *i/r* ratios for the trajectories initiated at the three TSs. Trajectories from TS(*si*) show strong preference for *s* over *a* in the short trajectories. These are trajectories that essentially directly travel a least-motion rotation of the vinyl group to the product. Over longer times, the TS(*si*) trajectories lose specificity.

522 ORGANIC REACTION DYNAMICS

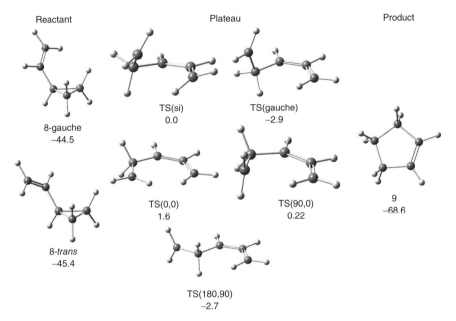

Figure 8.5 Critical points on the **8** → **9** PES. Relative energies in kilocalories per mole at CASSCF(4,4)/6-31G*.[38]

TABLE 8.1 Percent Yields from Trajectories Begun at Three TSs[a]

	TS(*si*)	TS(90,0)	TS(0,0)	Total	Experimental
si	37.7	27.5	31.0	39.6	40
sr	39.4	26.5	14.2	31.9	23
ar	12.5	17.1	13.4	12.3	13
ai	10.4	28.9	41.4	16.2	24
% React[b]	87.9	17.4	12.7		

[a] Ref. 41.
[b] Percentage of the trajectories initiated at that TS that resulted in a product.

The trajectories from TS(90,0) show strong preference for *i* over *r* at short times. This also reflects a least-motion pathway. The overall *s/a* ratio time dependence is dominated by the results of the TS(*si*) trajectories.

The time dependence suggests a bimodal trajectory distribution. The short duration trajectories are very stereoselective, mostly originating from TS(*si*) and involve direct least-motion rotation about the C_2–C_3 bond in a suprafacial motion. The longer time duration trajectories are more stereorandom and proceed through an anti conformation that requires significant geometric change to access. Even these long trajectories are not fully stereorandom; there is dependence on which TS is crossed and most trajectories maintain their angular velocity be it about C_1–C_2 or C_2–C_3.

EXAMPLES OF ORGANIC REACTIONS WITH NON-STATISTICAL DYNAMICS

TABLE 8.2 Time Dependence of the Product Ratios[a]

Time, fs	TS(si)		TS(90,0)		TS(0,0)		Total	
	s/a	i/r	s/a	i/r	s/a	i/r	s/a	i/r
100–200	463	1.2	0.5	88	1.3	7.1	23.7	1.3
200–300	11	1.7	0.8	23	0.5	9.9	3.3	2.6
300–400	2.3	0.8	1.0	9.8	0.9	0.7	1.6	1.3
400–500	0.7	0.8	1.5	9.1	1.3	0.7	0.9	1.1
500–600	0.6	0.8	0.7	10.7	1.2	1.3	0.6	1.1
600–700	0.6	0.6	1.8	2.8	1.2	0.6	0.9	0.9
700–800	0.2	0.7	0.9	2.3	0.9	1.3	0.5	1.1

[a] Ref. 41.

This nonstatistical nature for the vinylcyclopropane to cyclopentene rearrangement implies that related reactions may also occur under dynamic control. The vinylcyclobutane **11** to cyclohexene **12** rearrangement is the next higher homolog. This rearrangement proceeds with quite variable stereoselectivity, strongly dependent on the substituents. For example, the rearrangement of **13** proceeds with an allowed (*si* + *ar*) to forbidden (*sr* + *ai*) ratio of 1.1,[42] but the diastereomers **14** and **15** are much more selective: the *si/sr* ratio is greater than 9.3 for the former and less than 0.14 for the latter.[30]

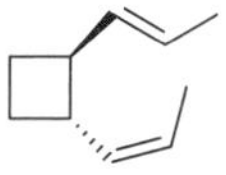

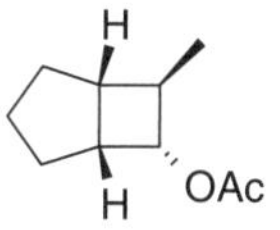

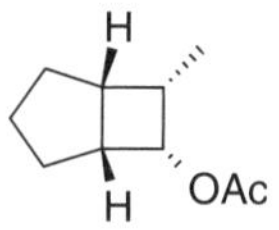

Houk[43] located the critical points along the rearrangement of **11** into **12** at (U)B3LYP/6-31G*. A diradical intermediate **16** does exist; however, three TSs (**17–19**) are found in the neighborhood of **16** (Figure 8.6). These critical points differ in energy by only 4 kcal mol^{-1} (computed at CASPT2(6,6)/6-31G//UB3LYP/6-31G*). The diradical lies in a broad flat caldera, similar to that found in the **8** → **9** rearrangement and other reactions discussed in this chapter. Though trajectory calculations were not performed, the shape of this PES strongly suggests that "the dynamic motions of diradical species moving along the flat potential surface before ring closure govern the observed stereochemical preferences."[43] In other words, dynamic effects dictate the product distribution.

Recent reviews by Baldwin[44,45] summarize the experimental and theoretical approaches to the vinylcyclopropane and vinylcyclobutane rearrangements. It is

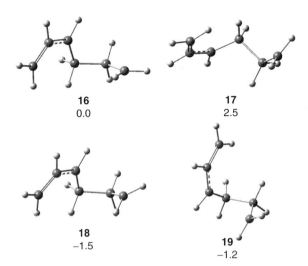

Figure 8.6 Critical point in the **11** → **12** caldera. Relative energies in kilocalories per mole at CASPT2.[43]

clear that neither orbital symmetry theory nor statistical behavior of an intermediate diradical properly accounts for the mechanisms of these rearrangement; dynamics effects are in play.

8.3.2.2 Bicyclo[3.1.0]hex-2-ene 20

The rearrangement of bicyclo[3.1.0]-hex-2-ene occurs on a flat caldera that allows for "quasi-energetic but geometrically nonequivalent pathways."[46] The rearrangement occurs through the formation of a diradical. Using the monodeuterated analog *d*-**20x** to explicate the full range of outcomes, the diradical *d*-**21** can close via (1) the [1*r*,3*s*] path to give *d*-**22x**, (2) the [1*i*,1*i*] path to give *d*-**20n**, or (3) the [1*i*,3*a*] path to give *d*-**22n** (Scheme 8.3). Baldwin[47] found that the gas-phase reaction of **20x** at 225–255 °C gave the *d*-**22x** : *d*-**20n** : *d*-**22n** product ratio of 48 : 36 : 16. Similar results were observed for the rearrangement of Δ^2-thujenes and substituted analogs.[48,49]

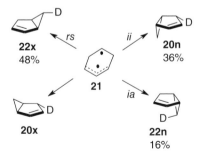

Scheme 8.3

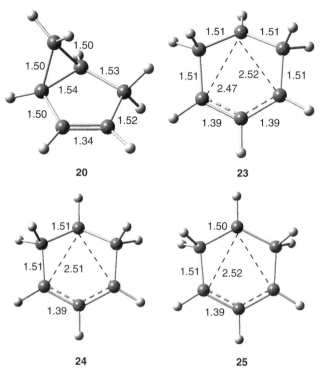

Figure 8.7 Geometries of the critical points on the PES for the rearrangement of **20**.[50]

Houk and Suhrada[50] examined the rearrangement of **20** at CASPT2(4,4)6-31G*-//CASSCF(4,4)/6-31G*. The TS **23** lies 40.0 kcal mol^{-1} above the reactant. This TSs defines the entrance into a broad flat caldera that holds three critical points: two mirror image local minima **24** and TS **25** (see Figure 8.7). These critical points lie 0.1 and 0.4 kcal mol^{-1} below **23**, respectively. Structures **23–25** possess diradical character: the distance between the secondary and the allyl radical is large, the SOMOs are of different irreducible representations, and the triplet state is lower in energy than the singlet. The caldera interior surface describes the "intermediate," but certainly no single geometry really can be ascribed to the diradical **21**. This PES is ideal for nonstatistical kinetic behavior.

Doubleday et al.[46] then performed classical trajectory simulations for the rearrangement of **20**. The trajectories were performed using direct dynamics, where the energy surface was computed using the AM1-SRP method. Here, AM1 was modified to fit to the CASPT2(4,4)/6-31G*//CASSCF(4,4)/6-31G* results discussed above. Trajectories were begun at the TS (**23**) using the TS normal mode sampling procedure to give the Boltzman distribution. The trajectories were computed at 498 and 528 K, and the percent yields of the products are reported in Table 8.3. Two important conclusions are readily drawn from this table. First, the agreement between the computed and experimental product distribution is remarkably good.

TABLE 8.3 Percent Yields from Trajectory Simulations[a]

	498 K	528 K	Experimental[b]
rs	47.0 ± 1.7	47.2 ± 1.7	48
ii	37.6 ± 1.7	37.8 ± 1.7	36
ia	15.4 ± 1.7	15.0 ± 1.7	16

[a] Ref. 46.
[b] Ref. 47.

Second, there is no difference in the product distribution at different temperatures. This is a sign of nonstatistical dynamics.

The majority of the trajectories make a single pass across the caldera and then fall into a product energy well. The time to traverse the well is less than 400 fs, insufficient time for intramolecular vibrational energy redistribution. Rather, conservation of momentum dictates that trajectories simply follow straight paths across the caldera. The product distribution should then depend principally on the initial direction into the caldera. If this is true, a trajectory that enters the caldera to the upper left of the transition vector (see Figure 8.8) will fall into the *rs* product well. Weighting this area as $(1/2) \int \cos \theta d\theta$ predicts that 42 percent of the trajectories should give the *rs* product. Those trajectories that initially head into the upper right corner will cross the caldera and fall into the *ii* well; this area corresponds to 11 percent of the trajectories. Only 6 percent of the trajectories enter the caldera heading toward the bottom right and end up in the *ia* well. These percentages are remarkably similar to the actual product distribution. The other trajectories head toward one of the "hills" on the caldera rim and will bounce about the caldera (though again generally only for a short period of time) before falling into a well.

This notion of a molecule taking a direct path suggests that product distribution should be time dependent: the path to the *rs* well is shorter than the paths across the caldera to the *ii* or *ia* wells. During the first 150 fs of trajectory time, virtually, the only product formed is *rs*. After 200 fs, the *ii* and *rs* products are formed equally. The *ia* product slowly accrues. This time dependence is another indicator of nonstatistical dynamics.

8.3.2.3 Cyclopropane Stereomutation

The stereomutation of substituted cyclopropanes remains a subject of incomplete understanding. This is particularly frustrating to physical organic chemists since the system is, at least on first look, exceedingly simple. The chiral dideutrated cyclopropane (*S,S*)-**26** undergoes thermal isomerization to its enantiomer (*R,R*)-**26** and the achiral form (*R,S*)-**26** (Scheme 8.4). One can consider these interconversions as occurring by rotation about a single carbon or rotation about two carbon centers, with relative rates of k_1 and k_{12}, respectively. Berson[51] examined the thermolysis of (*S,S*)-**26** at 695 K and concluded that the ratio of single-to-double rotation (k_1/k_2) was about 50; however, this result depends on the secondary isotope effect for the cleavage of the C–C bond. Other reasonable choices of the isotope effect can reduce k_1/k_2 to a value as low as 5. In a second experiment designed to remove the dependence on this

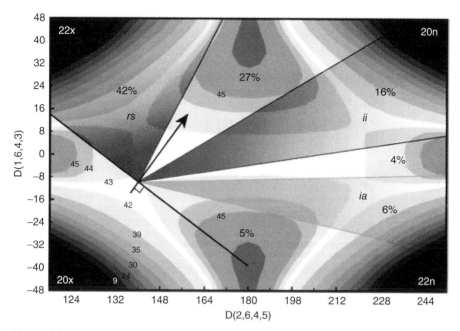

Figure 8.8 Angular regions that lead to the different products. The regions are weighted by $(1/2) \int \cos\theta d\theta$ to predict the product distributions. *Source*: Adapted with permission from Doubleday, C.; Suhrada, C. P.; Houk, K. N. *J. Am. Chem. Soc.* **2006**, *128*, 90. Copyright 2006 American Chemical Society.

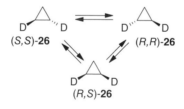

Scheme 8.4

isotope effect, Baldwin[52] examined the kinetics of the stereomutation at 470 °C of the trideuterated-^{13}C labeled cyclopropane shown in Scheme 8.5. The resultant k_1/k_2 value is 1. The results are not reconcilable by any choice of isotope effect. Furthermore, their interpretation is entirely different: Berson's result indicates little stereorandomization and predominant double rotation, while Baldwin's result indicates either stereorandomization (probably through an intermediate) or competitive single- and double-rotation rates.

Early extended Hückel calculations by Hoffmann on the stereomutation of cyclopropane invoked the intermediacy of the trimethylene radical **27**.[53] The single-rotation TS connecting **26** with **27** was estimated to be much higher than the double-rotation TS. Of the two possible double-rotation pathways,

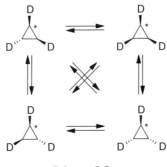

Scheme 8.5

orbital symmetry rules predict that the conrotatory pathway is preferred over the disrotatory pathway.

Given the difficulties in computing reasonable energies of diradicals, as discussed in Chapter 5, obtaining the PES of trimethylene diradical is a challenge. Multiconfiguration treatment is necessary, and the two best computations were performed at CASSCF(2,2)/VTZ(2d,p)[54] and CASPT2N/6-31G*//GVB-PP(1)/6-31G*.[55] Both computations agree on the features of the PES and structure of its critical points (Figure 8.9). The trimethylene diradical **27** has C_2 symmetry, with a conrotatory TS **28** that is essentially of equal energy. The disrotatory TS **29** about 1–1.5 kcal mol^{-1} above **28** and the TS for monorotation **30** is about 1 kcal mol^{-1} higher still. The important point is that the region describing the diradical is exceptionally flat.

In 1997, two independent MD studies of the cyclopropane stereomutation appeared back-to-back in the *Journal of the American Chemical Society*. The first, by Doubleday and Hase[56] (a follow-up full paper[57] appeared a year later), employed a PES using the AM1-SRP based upon Doubleday's earlier CASSCF(2,2)[54] study of trimethylene diradical. Borden and Carpenter,[55] the

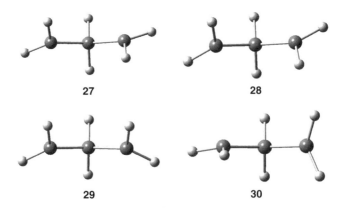

Figure 8.9 Structures of the critical points on the PES of trimethylene diradical.[55]

authors of the second paper, devised an analytical surface based on CASPT2N energies of **27–29**. Though the PESs are slightly different, the results and conclusions drawn from these two dynamics studies are in agreement.

In the Doubleday/Hase study,[56] trajectories were initiated at the conrotatory TS with 4.39 kcal mol^{-1} of energy above the zero-point energy. When all of that energy is placed into the conrotatory reaction coordinate, only conrotatory product is obtained. When 1, 3, or 7 quanta were placed into the disrotatory mode, conrotatory product remained the dominant product, but the percentages of both disrotatory and monoroation products increased with increased quanta. If the reaction proceeded by statistical intramolecular vibration energy redistribution through a trimethylene diradical, then all of the different starting conditions should have given identical product distributions. The trajectories that involve double rotations "could hardly be more direct," terminating in less than 400 fs. Borden and Carpenter also noted the generally very direct trajectories. Even the trajectories that were initiated at the disrotatory TS cross the shallow plateau and exit through a disrotatory TS, rather than through the lower conrotatory TS. This demonstrates the principle of "dynamic matching"—since entry onto the plateau is with disrotatory motion, this caries across the plateau and matches with the disrotatory exit channel.

Based on the product distribution of their computed trajectories, Doubleday and Hase found that $k_{12}/k_1 = 2.3$–3, depending on the initial state sampling. Borden and Carpenter obtained the following k_{12}/k_1 ratio for trajectories initiated at the conrotatory, disrotatory, and monorotation TSs, 5.8–6, 2.8–3.3, and 0.35–0.4, respectively, giving an overall value of 4.73. TST predicts a double-to-single rotation ratio of less than 1.5, dramatically lower than the value from two MD studies.

The very short-lived trajectories traverse the diradical plateau directly, in a dynamic matching sense. A fraction of the trajectories are long lived (>400 fs) and involve both single and double rotations and possibly multiple rotations.[57] One might expect these trajectories to result from a "true" intermediate. In fact, these long-lived trajectories give $k_{12}/k_1 = 1.4$, consistent with TST. Thus, it appears that bimodal lifetime distributions may be occurring here, namely, short-lived direct trajectories lead to nonstatistical dynamics, while the longer-lived trajectories follow statistical dynamics.

Another excellent example of dynamic matching is seen in 1,3-dipolar cycloadditions. Houk and Doubleday[58,59] examined a series of 1,3-dipoles reacting with either ethylene or acetylene. Their TSs indicate a concerted reaction. The transition vector has three major components: (1) symmetric formation/cleavage of the two new σ bonds, (2) bending of the dipolar component, or (3) symmetric bending of the hydrogens of ethylene or acetylene. In the approach of the two fragments, the dipole bend vibrates, but then after the TS, it needs to bend quickly to close the five-member ring. This means that the bending mode effectively has to "turn a corner" in phase space and, without energy in this mode, the molecules will simple bounce off of each other. Trajectory studies reveal the importance of vibrational energy in the X–Y–Z bending mode in crossing the TS: trajectories with high vibrational energy in this X–Y–Z mode produces product directly, otherwise the trajectories are long.

8.3.3 Entrance into Intermediates from Above

Another PES topology that presents an excellent opportunity for nonstatistical dynamics is characterized by a high energy entrance TS that leads into a shallow minimum. The exit TS from the intermediate is of much lower energy than the entrance TS (Figure 8.10). In such a surface, the intermediate will be prepared with excess energy for escape over the second TS. It is likely that many trajectories will be favorably aligned such that they directly traverse the intermediate and immediately exit to the final product. The following example exemplifies just such a scenario.

8.3.3.1 Deazetization of 2,3-Diazabicyclo[2.2.1]hept-2-ene 31
Thermolysis of 2,3-diazabiclco[2.2.1]hept-2-ene **31** leads to nitrogen extrusion, forming bicyclo[2.1.0]pentane **32**. The stereoselectivity of this reaction can be determined with the deuterium-labeled analog **31-d_2**: the *exo* **32x** to *endo* **32n** product ratio is $3:1$.[60] Roth proposed two possible mechanisms: the first[60] involves synchronous cleavage of the C–N bonds, with assistance provide by the backside lobes of the carbon orbitals giving rise to the *exo* isomer (Scheme 8.6), while the second[61] involves stepwise C–N cleavage and with C–C bond formation concerted with the second cleavage (Scheme 8.7). Allred and Smith offered a third mechanism (Scheme 8.8). The C–N bonds are broken synchronously, with the *exo* isomer formed from the recoil of the N_2 loss.[62] All three of these mechanisms have issues. Neither of the first two explains how the *endo* isomer is formed. The last mechanism implies the diradical intermediate is C_s but computations suggest it is C_2.

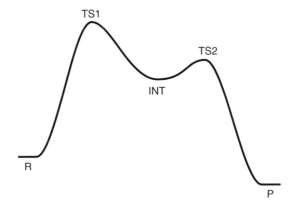

Figure 8.10 Generic PES where the entrance TS into the intermediate is much higher than the exit TS.

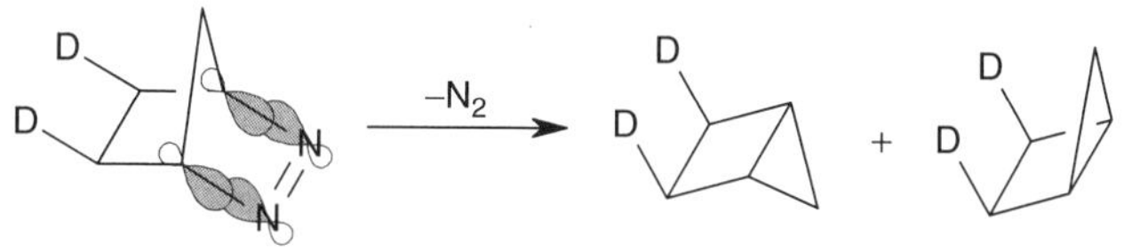

EXAMPLES OF ORGANIC REACTIONS WITH NON-STATISTICAL DYNAMICS   531

Scheme 8.6

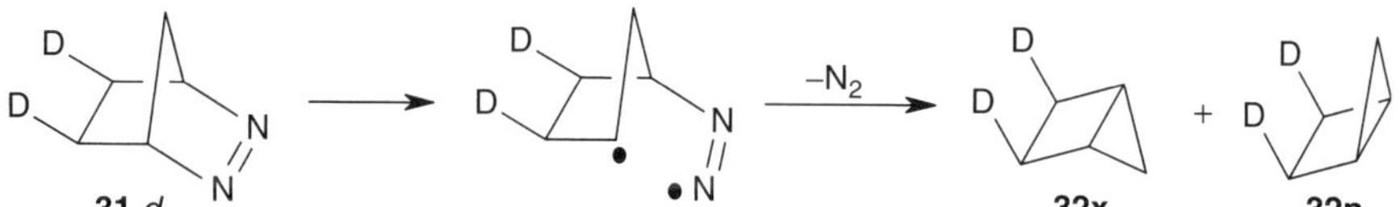

Scheme 8.7

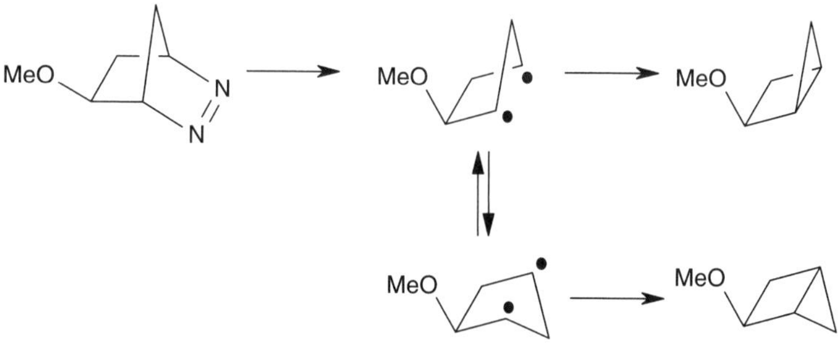

Scheme 8.8

Both the stepwise and concerted pathways have been subject to computational study.[63,64] Given the distinctly different natures of critical points along these surfaces, multiconfiguration wavefunctions with corrections for dynamic correlation is needed. Carpenter[65] examined the reaction **31→32**, optimizing the geometries using a CASSCF wavefunction and then computing single-point energies at CASPT2. The optimized geometries and their relative CASPT2 energies are shown in Figure 8.11.

The concerted pathway begins at **31** and then passes through the C_s TS **33**, with increase in energy of 34.1 kcal mol^{-1}. Next, the diradical intermediate **34** is formed, having C_2 symmetry. The barrier for formation of the new C–C bond, **35**, is only 1.2 kcal mol^{-1}, leading to **32**, with overall exothermicity of 8.7 kcal mol^{-1}. This surface mimics that shown in Figure 8.3. The stepwise path first creates the diradical **36**, which lies nearly 7 kcal mol^{-1} above the highest barrier of the concerted pathway. The TS for the stepwise path must be even greater, and so the stepwise path is not competitive with the concerted path. This discredits the mechanism shown in Scheme 8.7.

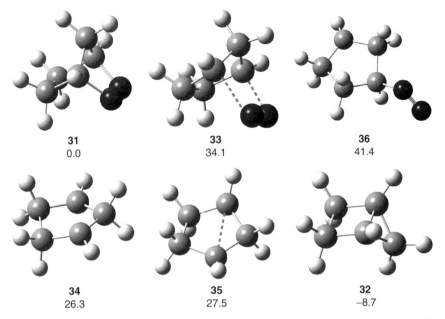

Figure 8.11 CASSCF/6-31G(d)-optimized structures of **31–36**. CASPT2 relative energies (in kcal mol^{-1}) listed below each structure.[65]

The mechanism proposed in Scheme 8.6 is also inconsistent with the CASSCF computations. This mechanism suggests that the C–C bond is formed as the C–N bonds are broken; however, this distance is longer in TS **33** (2.309 Å) than in the reactant (2.188 Å). The C_2 diradical **34**, as noted before, suggests that the mechanism of Scheme 8.8 is not correct, at least not in all of its details.

Carpenter noted that the motion associated with the reaction coordinate of **33** is inversion at both carbon atoms that break their bond to nitrogen. The reaction dynamics model for the reaction is that passage over **33** will have energy in the reaction coordinate mode that will directly lead toward the inverted product **32x**. Once momentum has been placed into this inversion mode, it will continue unless a barrier is met.[8] Carpenter pictures the mechanism as in Scheme 8.9. Molecules that cross **33** to form intermediate **34** are primarily directed onward to **32x**. Only a small portion will remain in the shallow well and get turned toward **32n**.

The nonstatistical population is the group of molecules that directly cross the diradical and produce **32x**. Carpenter[65] hypothesized that with increasing pressure, collisions will become more common such that energy will be redistributed away from the modes that lead to direct crossing of the diradical, yielding a more statistical product distribution. In other words, collisions provide the barrier so that the momentum can be redirected. The reaction of **31-d_2** was carried out in supercritical propane in order to control the pressure. The ratio of **32x** to **32n** did

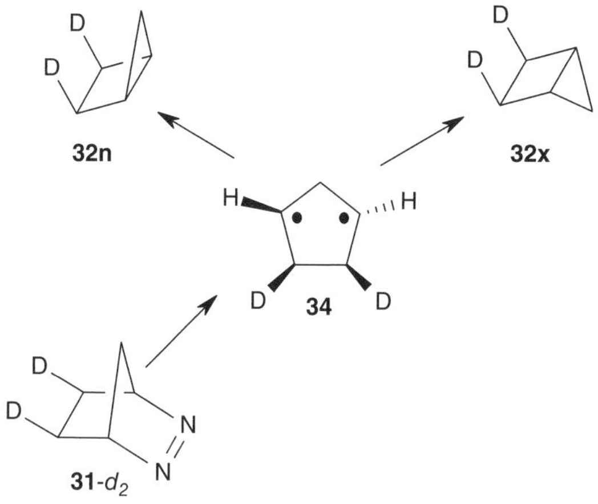

Scheme 8.9

decrease with increasing pressure, evidence that is in support of the dynamic model of Scheme 8.9.

8.3.4 Avoiding Local Minima

PESs typically display many local minima. Standard mechanistic analysis hold that local minima located between reactant and product will be sampled along the reaction pathway; we call these intervening local minima *intermediates*. The deeper the energy well associated with the intermediates, the more likely the intermediate will be sampled and the longer the lifetime of that intermediate.

The surprising result of the next examples is that local minima are available and quite energetically stable, but the reaction trajectories skirt around them. Entrance into the wells associated with these putative intermediates requires the trajectories to turn. Without some wall or barrier to deflect the trajectory, most simply pass right by deep wells and transit directly on to product, producing nonstatistical distributions and reaction times much faster than that predicted by RRKM or TST.

8.3.4.1 Methyl Loss from Acetone Radical Cation

The loss of a methyl group from acetone radical cation appears to proceed in a nonstatistical fashion. The reaction is proposed to follow the mechanism outlined in Reaction 8.6. Hydrogen migration from acetone enol radical cation **37** gives acetone radical cation **39**, which then cleaves a methyl group. Labeled reactant allows for distinguishing which methyl group is lost. If the acetone cation radical is long lived, and IVR occurs, then there should be no difference in the branching ratio: the methyl groups should be equivalent. The infrared multiple-photon (IRMP) dissociation experiments of Brauman[66] confirmed earlier studies[67–69]—the methyl group formed by the migrating hydrogen is preferentially lost. The branching ratio is greater than 1, ranging from 1.2 to 1.6.

$$\left[\underset{H_3C}{\overset{OH}{\bigtriangleup}} CH_2 \right]^{\cdot \oplus} \longrightarrow \left[\underset{H_3C}{\overset{O}{\bigtriangleup}} CH_3 \right]^{\cdot \oplus} \longrightarrow CH_3CO^+ + \cdot CH_3 \quad \text{(Reaction 8.6)}$$

$$\quad\quad\quad\quad 37 \quad\quad\quad\quad\quad\quad 39$$

The initial explanation for this nonstatistical behavior was that as the hydrogen migrates to form **39**, it deposits its energy into the antisymmetric stretch involving this newly formed methyl group.[70] One can think of this as the momentum carried by the hydrogen is then given to the methyl group, which leads to that methyl dissociation. This model is, however, at odds with the IRMP results. The branching ratio increases with additional energy. The least energetic experiment gives a small branching ratio of 1.17, even though considerable energy is available for the methyl cleavage step. It is the energy deposited into another vibrational mode that is critical. Brauman[66] argued that excitation of the C–C–O bend involving the spectator methyl group is critical because this angle must become linear in the product. Those ions with an excited C–C–O bend will be particularly inclined toward dissociation, leading to the nonstatistical products.

Carpenter[71] reported direct dynamics trajectories for Reaction 8.6. The PES was computed using AM1-SRP fit to the B3LYP/cc-pVTZ energies of the critical points. The critical points on the PES are sketched in Figure 8.12. The computed branching ratio was 1.01 ± 0.01 for the trajectories started at the acetone radical cation **39**, with a half-life of 409 fs. For trajectories initiated at the TS **38**, the branching ratio was 1.13 ± 0.01 with a half-life of about 240 fs. This branching ratio qualitatively agrees with the experimental observations, especially considering the approximate AM1-SRP PES.

The different time scale for the two sets of trajectories is interesting since all trajectories began with the same energy. The very short trajectories were those that began at **38** and never entered the region about **39**. Analysis of the C–C–O angles during the trajectories indicates that dissociation occurs when one angle is about 90° and the other is about 120°. The geometry of **38** is similar to this requirement: the angle to the forming methyl is about 98° and the other angle is about 120°. So once the TS geometry is obtained, many trajectories will simply directly dissociate. For those trajectories in the well associated with **39**, the C–C–O angles oscillate and when it nears 90°, dissociation is likely to occur. This behavior demonstrates the importance of coupling kinetic energy to the appropriate vibrational modes that lead to product formation. Also of note is that very stable intermediates can simply be ignored during the reaction.

8.3.4.2 Cope Rearrangement of 1,2,6-Heptatriene

Roth[72] examined the Cope rearrangement of 1,2,6-heptatriene **40** to 3-methylenehexa-1,5-diene **41** and was able to trap a diradical intermediate with oxygen. He proposed that about half of

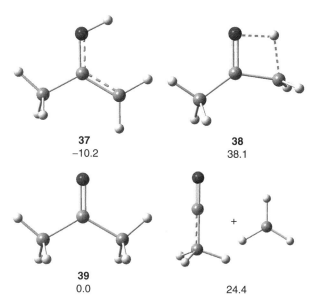

Figure 8.12 Critical points along the PES for Reaction 8.6. Relative energies in kcal mol^{-1} at B3LYP/cc-pVTZ.[71]

the reaction occurs by a concerted route and the other half occurs through a biradical intermediate **42**. This model invokes three TSs (a concerted TS, and TSs into and out of the intermediate **42**). Neither CASSCF(8,8) nor (U)B3LYP computations located a concerted TS.[73] Rather, only the entrance and exit TSs **43** and **44** could be found (Figure 8.13). A subsequent computational study[74] identified an additional conformer of the diradical.

Unlike the other pericyclic reactions discussed previously in this chapter, the diradical species here resides in a relatively deep well, greater than 12 kcal mol^{-1}, and so the PES about the diradical cannot be described as a caldera. Nonetheless, this reaction exhibits nonstatistical behavior.

Carpenter and Borden[74] first computed two trajectories using direct dynamics on the CASSCF(8,8)/6-31G* surface. Both trajectories were initiated at **43**, the TS connecting reactant to the diradical intermediate. A linear–synchronous–transit (LST) path was computed connecting **43** to the intermediate **42** and one connecting **43** to **44**. The vector corresponding to these two motions was used to assign the phases of the normal modes. The trajectory for the first case (the LST toward the intermediate) did, as expected, end at **42**. The second trajectory (the LST towards **44**) avoided the biradical and ended at product **41**.

In order to compute a ensemble of trajectorics, Carpenter and Borden created an AM1-SRP fit to the CASSCF(8,8) energies. About 83 percent of the trajectories went into the region of the biradical. Of these, about 8 percent exited the biradical and proceed to product **41** in less than 500 fs, too short a time to be trapped under Roth's experiment. In addition, 17 percent of the trajectories completely

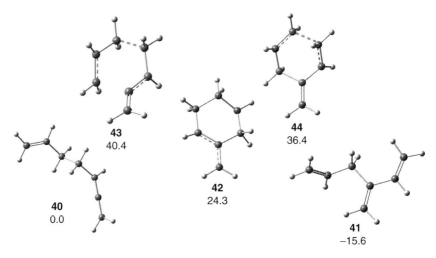

Figure 8.13 CASSCF(8,8)/6-31G* optimized geometries of the critical points for the reaction **40** → **41**. Relative energies in kilocalories per mole.[73]

bypassed the diradical and went directly to product. These trajectories, and the ones that quickly left the biradical, would constitute the "concerted" reactions that Roth noted.

Carpenter and Borden[74] made two important conclusions that raise significant concerns about the traditional physical organic notions of reaction mechanisms. First, nonstatistical dynamics can occur even when intermediates exist in relatively deep potential energy wells, not just on flat caldera-like surfaces. Second, multiple products can be formed from crossing a *single* TS. The steepest descent path from a TS can only link to a single product,[75] but reactions can follow nonsteepest descent paths that reach different products.

8.3.4.3 The S$_N$2 Reaction: HO$^-$ + CH$_3$F

The gas-phase nucleophilic substitution reaction was discussed in detail in Section 6.1.1. The generic PES for this reaction, shown in Figure 6.1, is characterized by three critical points separating reactants from products: two minima corresponding to entrance and exit ion–dipole complexes separated by a TS.

The PES for the reaction of hydroxide with fluoromethane is shown in Figure 8.14. As expected, there are two local minima separated by a single TS. The new twist on this surface is the structure of the deepest energy well. These kinds of complexes usually have the anion weakly associated with the hydrogen atoms of the methyl group, as in the case of the entrance ion–dipole **45**. There is a small plateau just after the TS where the fluoride ion is associated with the methyl hydrogens—**48** is representative of structures on this plateau. When the fluoride, however, moves away from the backside, off the O–C axis, it can form a strong hydrogen bond with the hydroxyl proton, leading to the exit ion–dipole complex **47**. One might expect this reaction to be trapped in this very deep

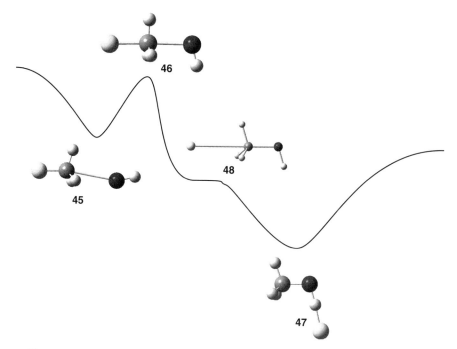

Figure 8.14 PES of the IRC for the reaction: HO$^-$ + CH$_3$F. Adapted from Hase.[76]

energy well, with the final dissociated products reflecting IVR from within this well.

Hase[76] computed direct dynamics trajectories for this reaction using MP2/6-31+G* energies and derivatives. Trajectories were initiated at the TS **46**. Just over half of these trajectories (33 out of 64) formed the entrance ion–dipole complex **45**. Of the trajectories following the exit channel, only four of them formed **47**. Of these four, one trajectory later departed this well and dissociated into products. The other 27 trajectories led directly to dissociation into methanol and fluoride, completely avoiding the deep energy well associated with **47**. Thus, most of the reactants paths followed a nonreaction coordinate pathway.

Apparently, there is weak coupling between the CH$_3$OH + F$^-$ translation mode and the O−C$\cdots$F bend mode. As F$^-$ begins to withdraw, it does not transfer energy rapidly enough into the vibrational mode, but instead continues to exit in a fairly collinear fashion to products. One might argue this in terms of conservation of momentum. There needs to be some mechanism to transfer the C–F stretching momentum into angular momentum for F$^-$ to rotate toward the hydroxyl hydrogen. Hase[76] concludes that simply examining the PES—its shape, the well depths, and barrier heights—is "insufficient for determining atomic-level mechanisms for chemical reactions." Rather, MD, determining the actual atomic motions on a reaction PES are necessary to obtain the detail needed to decipher a mechanism. This is true even in case where very stable intermediates are present. Previous standard mechanistic theory presumed that such stable intermediates would be readily accessed and internal energy completely randomized before continuing along the reaction path. This notion clearly requires serious reconsideration.

8.3.4.4 Reaction of Fluoride with Methyl Hydroperoxide

The reaction of fluoride with methyl hydroperoxide is characterized by five critical points. As characterized by Hase[77] at B3LYP/6-311+G(d,p), the reactants first form an ion–dipole complex **49** that lies 36.5 kcal mol^{-1} below separated reactants. **49** rearranges through TS **50** (with a barrier of 24.1 kcal mol^{-1}) to give the complex where fluoride is loosely associated with a methyl hydrogen (**51**). Continuing forward is a second TS (**52**) with a barrier of 4.7 kcal mol^{-1} that leads to product $CH_2(OH)_2 \cdots F^-$ (**53**), which lies in a very deep well. These critical points are shown in Figure 8.15.

What is interesting here are the experimental gas-phase results by Blanksby et al.[78] In the gas phase, the reaction produced $HF + CH_2O + OH^-$, not **53** or $HF + CH_2(OH)O^-$. Hase and coworkers[77] ran a number of trajectories simulating reaction at 300 K, the experimental condition. Reactions were started at three initial points: (1) F^- separated by 15 Å from CH_3OOH, (2) at TS **52**, or (3) at a point along the IRC of the form $HOCH_2O^- \cdots HF$. Of the 80 trajectories that start from **52**, 76 result in the formation of $HF + CH_2O + OH^-$. The majority of the trajectories that start with separated reactants produce the complex **49** (97 out of 200), reflecting its stability and high exit barriers. However, 45 of the 200 trajectories give $HF + CH_2O + OH^-$, as do all 5 trajectories that start with $HOCH_2O^- \cdots HF$. No trajectories end at **53**, the product expected from following the IRC.

Since the initial motion along the imaginary frequency of **52** is to cleave the O–O bond and the C–H bond, momentum in that direction carries the reaction over to the decomposition product rather than making a tight turn on the PES needed to lead to **53**. The computations are in complete agreement with the experimental results; the unusual decomposition products result from following a non-IRC pathway!

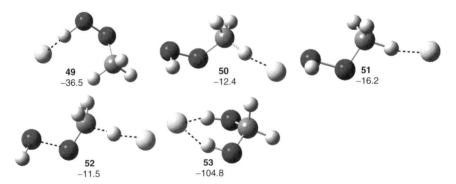

Figure 8.15 B3LYP/6-311+G(d,p) optimized geometries of the critical points on the PES for the reaction of F$^-$ with CH_3OOH.[77] Energies in kilocalories per mole relative to separated reactants.

8.3.5 Bifurcating Surfaces: One TS, Two Products

A TS is characterized by two conditions. First, all of the $3N-6$ gradients must be equal to zero. Second, the corresponding $3N-6$ eigenvalues of the Hessian matrix must all be positive and nonzero, except for one, which is negative. This implies that the TS sits at the bottom of the potential energy well with respect to motion in all $3N-7$ directions but is at the top of the hill with respect to the direction defined by the eigenvector associated with negative eigenvalues. This direction is often referred to as the *reaction coordinate*.

Motion along the reaction coordinate away from the TS will lead to lower energy structures. If one follows the reaction coordinate in the forward or reverse directions, eventually one will reach a critical point. Usually, these two critical points (the one located in the reverse direction and the one found in the forward direction) are local minima, and so we say that the TS connects reactant with product. Most importantly, there is a 1:1 correspondence between TS and product, or TS and reactant.

This implies that if a reaction gives more than one product, there must be more than one TS on the PES. Scheme 8.10 shows a number of possible ways that reactant **A** can yield products **B1** and **B2**. In Scheme 8.10a, the reaction proceeds from **A** crosses **TS1** and creates **B1** or crosses **TS2** to create **B2**. The PES of Scheme 8.10b is like Scheme 8.10a with a third TS **TS3** that allows a pathway for **B1** and **B2** to directly interconvert. Scheme 8.10c is a sequential mechanism whereby **A** converts to **B1** through **TS1**, which can then proceed further to cross **TS2** to form **B2**. Scheme 8.10d proposes an intermediate **INT** created when **A** crosses **TS1**. The intermediate can then form either product, but by crossing one of the two distinct TSs **TS2** or **TS3**. No PES exists such that **A** crosses a single TS and then directly proceeds to either **B1** or **B2** (Scheme 8.10e).

Statistical treatment requires a kinetic model like those presented in Scheme 8.10a–d. The studies presented below describe reactions that seemingly

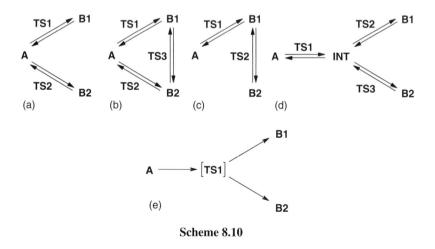

Scheme 8.10

8.3.5.1 C_2–C_6 Enyne Allene Cyclization

The chemistry and computational studies of enediynes and enyne-allenes were presented in Section 4.3, and especially in Section 4.3.4. In this section, we take on the Schmittel-type cyclization of the C_1-alkyl-substituted enyne-allenes **54**.[79,80] These compounds rearrange to give a product that can be formally considered to have undergone first a Schmittel cyclization to give the diradical **55**, which then transfers a hydrogen atom to give the product **56**. A concerted ene reaction is also possible through a single TS **57**.

Engels[81] examined the competition between these two mechanistic pathways using (U)BPW91/6-31G(d) computations for two substituted analogs of **54**: R_1=Ph, R_2=H, **58**; R_1=*t*-Bu, R_2=H, **59**. For the reaction of **58**, he was able to locate a TS for the formation of the diradical intermediate but was unable to locate a TS for the concerted path (the analog of **57**). Since the phenyl substituent will stabilize the benzyllic radical, this is not an unexpected result. For the reaction of **59**, TSs for both the concerted path and the Schmittel cyclization were located. These two TSs have nearly identical energies, and so Engels could not determine which pathway is favored. He did propose that their kinetic isotopes effects should be very different. Deuterium substitution on the methyl group of the allene will have little effect on the stepwise path. However, the computed primary isotope effect k_H/k_D for the concerted ene reaction is 1.97.

Singleton[82] carried out a combined experimental/computational/MD study of the enyne-allene cyclization problem. First, he determined that the experimental KIE for the cyclization of **60** is 1.43. This value is smaller than that usually found for concerted ene reactions, where k_H/k_D is often greater than 2.[83,84]

(U)B3LYP/6-31G(d,p) computations identified the concerted ene TS, but no TS was located for the direct formation of the diradical intermediate. Singleton speculated that there would be little geometric difference between these two potential TSs—both involve the formation of the C_2–C_6 bond and differ only in the length of a single C–H bond, which is only slightly longer in the concerted TS than in the reactant. Is it perhaps possible that these two TSs have merged into one and that a single TS leads in two directions—the radical intermediate and the ene product?

60 R = H or D

Singleton then carried out MD trajectory computations for the rearrangement of **61** into **62**. Using UB3LYP/6-31G(d,p), there is again a concerted ene TS **63**, but no TS that connects **61** to the diradical intermediate **64**. There is a TS (**65**) that takes **64** into **62**. The structures and relative energies of these critical points are shown in Figure 8.16. As discussed in Section 4.3, single-configuration methods may poorly describe the diradical character of many of the points along this type of reaction. In order to carry out direct dynamics, density functional theory (DFT) is the only computationally feasible quantum mechanical treatment, but it may fail to adequately describe the relative energies on the PES. Fortunately, the UB3LYP energies are in excellent agreement with those computed at BD(T)/6-31+G(d,p)//B3LYP/6-31G(d,p).

The general shape of the PES for this reaction is given in Figure 8.17. Note that the MEP proceeds from TS **63** to **62**, and the MEP does not bifurcate. There is, however, a ridge that some trajectories may cross, allowing for trajectories to go from **63** to **64**. This behavior is similar to that of a valley ridge inflection point, but there is no VRI on the MEP in this case.

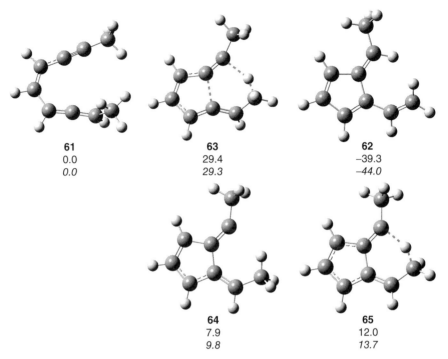

Figure 8.16 (U)B3LYP/6-31G(d,p) optimized structures of **61–65**. Relative energy at (U)B3LYP/6-31G(d,p) (normal text) and BD(T)/6-31+G(d,p)//B3LYP/6-31G(d,p) (italics).[82]

Trajectories were initiated at either the TS **63** or a random sampling of structures about this TS. Most trajectories reach a product, either **62** or **64** within 30 fs, and all finished within 85 fs. These short trajectories suggest little opportunity for IVR to occur. Most of the trajectories, 72 out of 101, end up at **62**, the product that is directly connected to the TS. The other 29 trajectories terminate at **64**. Some of these trajectories were allowed to go on for another 100 fs but none exited the area about **64**. Given sufficient time, these trajectories would cross over the barrier about **65** and make the final product **62**. The upshot is that the diradical is produced even though there is no TS that takes reactant into it. The ridge provides the opportunity for trajectories to fall into two possible energy sinks. Most follow the steepest descent downhill toward **62**, but some will divert into the neighboring minimum, the diradical **64**.

Another interesting feature of the dynamics is that 29 trajectories that end up at **62** pass through a geometry that resembles **66**. This structure is produced by a very asynchronous ene reaction where the hydrogen atom transfers to a greater degree than the formation of the C_2-C_6 bond. TS **63** also describes an asynchronous ene reaction, but one where the C–C bond formation precedes H transfer. No local energy minimum corresponding to **66** could be located. It appears that there are a number of concerted pathways that might be traversed.

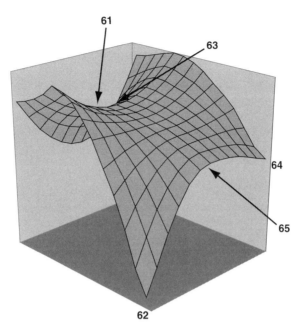

Figure 8.17 Qualitative PES for the reaction of **61** to **62**. *Source*: Adapted with permission from Bekele, T.; Christian, C. F.; Lipton, M. A.; Singleton, D. A. *J. Am. Chem. Soc.* **2005**, *127*, 9216. Copyright 2005 American Chemical Society.

66

In fact, the trajectories that take **61** into **62** fall into a broad range of possibilities. There is the route that follows the MEP, and an asynchronous ene path where H atom transfer lags behind C–C formation. There is the asynchronous ene reaction where C–C formation follows H atom transfer. And there is the stepwise path, where the C–C bond is completely formed and then the hydrogen is transferred. Trajectories intermediate these extremes also occur. Singleton concludes, "No single reaction path can adequately describe the mechanism."[82] We are confronted here by a mechanism that cannot be ascertained simply by locating the critical points on the PES and comparing their relative energies. Dynamics play a critical role in dictating just how reactant propagates into product.

8.3.5.2 Cycloadditions Involving Ketenes

We next present two examples of cycloadditions involving ketenes that invoke a bifurcating PES. The cycloaddition of ketene with cyclopentadiene had long been thought to simply give the formal [2+2] product, such as the reaction of cyclopentadiene **67** with dichloroketene **68** to give the cyclobutanone **69** (Reaction 8.7).[85] No formal [4+2] product **70** is observed. Work of Machiguchi and Yamabe[86] revealed a much more interesting reaction scenario. For the reaction of **67** with diphenylketene **71**, the initial product is the formal [4+2] adduct **72**, followed by later production of **73**, the formal [2+2] product (Reaction 8.8). The proposed mechanism, a [4+2] cycloaddition followed by a [3,3]-sigmatropic rearrangement, is shown in Reaction 8.8.

(Reaction 8.7)

(Reaction 8.8)

Singleton[87] took up this problem, using a combined experimental/computational/MD design. Careful monitoring of product formation at −20 °C showed that **73** was formed at early times even when the concentration of **72** was quite small. A kinetic model suggested that the rate constant for the formation of **73** (9.0×10^{-5} M^{-1} s^{-}) is only four times less than that for the formation of **72** (4.1×10^{-4} M^{-1} s^{-1}), while the rate constant for conversion of **72** into **73** is 2.9×10^{-5} M^{-1} s^{-1}. This is inconsistent with the proposed mechanism of Reaction 8.8.

The ^{13}C KIE for **73** is negligible at C_1 (the cyclopentadienyl carbon that joins with the carbonyl carbon), suggesting that this bond must be fully formed before the rate-limiting step. However, there is a substantial isotope effect at C_1 for **69**, implicating that bond making is occurring at this center in the rate-limiting step.

The PES for both Reactions 8.7 and 8.8 were determined using mPW1K and B3LYP. The latter method provides energies and a surface in disagreement with experimental observations. The mPW1K/6-31+G(d,p)-optimized structures and energies are shown in Figure 8.18. The TS for the [4+2] step, **74**, is 13.1 kcal mol^{-1} above reactants, much lower than that for the [2+2] step, **75**. The TS for the Claisen step, **76**, is a couple of kcal mol^{-1} lower than TS **74**. These energies are consistent with the two-step mechanism of Reaction 8.8 with the first being rate limiting but is inconsistent with the experimental observation of relatively rapid formation of **73**.

The critical points for the reaction of cyclopentadiene with dichloroketene are shown in Figure 8.19. The energetically low-lying TS **77** resembles the [4+2] cycloaddition TS **74**. The TS analogous to **75** lies 12.7 kcal mol^{-1} higher in energy. Though **77** looks like the [4+2] TS, it actually connects to **69**. Furthermore, the optimized geometry of **77** at mPW1K/6-31G* is little changed from that found with

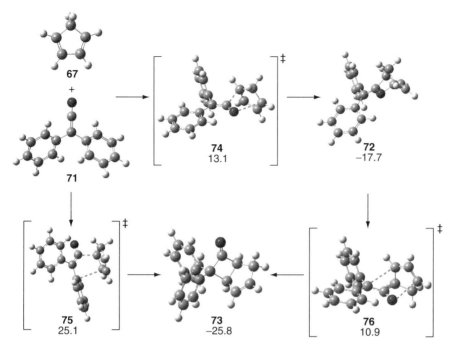

Figure 8.18 Critical points, optimized at mPW1K/6-31+G(d,p), for the reaction of **67** with **71**.[87]

the larger basis set, but it connects to **70**. This extreme sensitivity of the nature of the PES to basis set is disturbing—the predicted product differs with the basis set.

Application of TST using **77** predicts KIEs in excellent agreement with experiment. However, TST predicts a large isotope effect at C_1 if **74** is the rate-limiting step, but the agreement with experiment is much improved when using **76** as the rate-limiting step. This brings up a problem in that this suggests a rate-determining step occurs after formation of a reversible intermediate, yet conversion of **72** into **73** avoids this putative intermediate.

Direct dynamics trajectories were computed for both Reactions 8.7 and 8.8 using mPW1K/6-31G* and mPW1K/6-31+G(d,p). Trajectories were initiated at **77** for Reaction 8.7 and at **74** for Reaction 8.8. Trajectories were run until a product was reached, a recrossing back to reactant occurred, or the elapsed time reached 500 fs. These trajectory outcomes are listed in Table 8.4.

There are two critical results from these trajectory studies. First, trajectories for both reactions, initiated at a single TS on their respective surfaces, lead to both the [2+2] and the [4+2] products. This behavior is analogous to that seen in the dynamics of the C_2-C_6 enyne-allene cyclization discussed above. The PES for Reactions 8.7 and 8.8 are analogous to that shown in Figure 8.17; the majority of trajectories follow the MEP down from the TS, but the nearby ridge provides a means for traversing onto the other product as well. Second, recrossing of the TS region is a common occurrence in these simulations. These recrossing trajectories

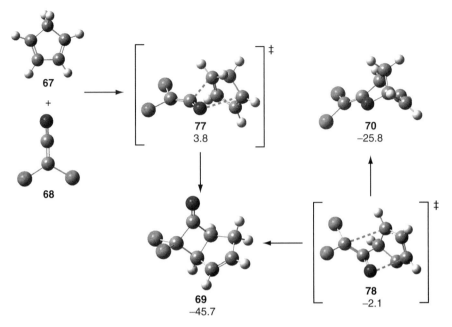

Figure 8.19 Critical points, optimized at mPW1K/6-31+G(d,p), for the reaction of **67** with **68**.[87]

TABLE 8.4 Trajectory Outcomes for Reactions 8.7 and 8.8 [a]

Starting Point	Total	[4+2]	[2+2]	Recrossing	Exceed Time Limit
		Diphenylketene **71**			
74 (mPW1K/6-31G*)	130	67	4	56	3
74 (mPW1K/6-31+G**)	24	8	1	15	0
		Dichloroketene **68**			
77 (mPW1K/6-31G*)	130	70	47	13	0
77 (mPW1K/6-31+G**)	37	8	24	5	0

[a] Ref. 87.

typically first form the C_1-C_α bond, pass through the regions associated with either **76** or **78** and then rebound off the "wall" due to a short C_1-C_α distance.

These trajectories do provide a means for reconciling all of the data into a cohesive framework. The [2+2] and [2+4] products can be formed from passage over the *same* TS. This explains why both products are observed even at short reaction times and precludes the necessity of the two-step ([4+2] followed by the Claisen rearrangement) mechanism. It obviates the requirement for climbing the very large barriers associated with the direct [2+2] pathways that were problematic when just examining the potential energy surfaces. Trajectory analysis allows for an interpretation of the KIEs. TST adequately accounts for the KIEs for Reaction 8.7.

Reaction 8.8 is problematic using TST; however, the large degree of trajectory recrossing discounts the underlying assumption of TST. In other words, TST cannot be applied to Reaction 8.8. The reaction mechanism cannot be reduced to simple models. This may be somewhat "intellectually unsatisfying"[87] and implies that understanding and predicting reaction products will be much more difficult than the application of the standard tools (characterization of critical points on the PES, TST, etc.) used over the past few decades.

Singleton[88] uncovered unusual dynamics in a second reaction involving **68**, in its reaction with ^{13}C-labeled *cis*-2-butene (Reaction 8.9). With ^{13}C at the 2-position of 2-butene, two products are observed, in a ratio of **79b** : **79a** = 0.993 ± 0.001. This is the opposite of what one might have imagined based on the carbonyl carbon acting as an electrophile.

(Reaction 8.9)

MPW1k/6-31+G**, M06, and MP2 predict a C_s transition structure. The reaction thus traverses this symmetrical TS and the falls into the symmetric energy wells associated with **79a** and **79b**. These products are also separated by a second C_s TS. Given a PES like this, one would reasonably expect to see no KIE at all.

To address this seeming paradox, Singleton employs trajectories studies on the MPW1k/6-31+G** energy surface, a task that would be impossibly long if not for the trick of making the labeled carbon super heavy—like ^{28}C, ^{44}C, ^{76}C, and ^{140}C—and then extrapolating back to ordinary ^{13}C. This predicts a product ratio of **79b** : **79a** of 0.990 in excellent agreement with the experimental value of 0.993.

Amazingly, most trajectories recross the TS, usually by reaching into the region near the second TS. However, the recrossing decreases with increasing isotopic mass. Vibrational mode 3 breaks the C_s symmetry; movement in one direction along mode 3 has no mass dependence on the recrossing but in the opposite direction, increased mass leads to decreased recrossing. Put another way, in this direction, increased mass leads more often to product, and this leads to the isotope effect.

Singleton offers a statistical interpretation as well. On the *free* energy surface, there is a variational TS for the formation of the first C–C bond, and farther on there is a set of two variational TSs, one leading to **79a** and the other leading to **79b**, associated with the formation of the second C–C bond. Application of RRKM theory to this model gives a kinetic isotope ratio of 0.992, again in agreement with experiment! As Singleton notes, it is "perplexing:" how do you reconcile the statistical view with the dynamical view?

8.3.5.3 Diels–Alder Reactions: Steps toward Predicting Dynamic Effects on Bifurcating Surfaces

Reactions 8.10 and 8.11 appear to be ordinary Diels–Alder cycloaddition reactions. One might expect to observe both products and perhaps some selectivity. In fact, the only product observed by Singleton for Reaction 8.10 is **82**, but for Reaction 8.11, the ratio of the two products **85** : **86** is 1.6 : 1.[89] Standard TST would suggest that there are two different TSs for each reaction, one corresponding to the Diels–Alder reaction where **80** is the dienophile, while the second TS has **81** as the dienophile. Thus, in Reaction 8.10, the TS leading to **82** is much lower in energy than that leading to **83**, while for Reaction 8.11, the TS state leading to **85** lies somewhat lower in energy than that leading to **86**.

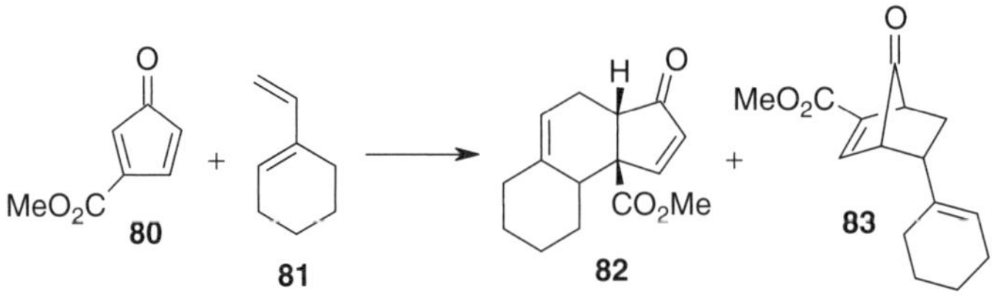

(Reaction 8.10)

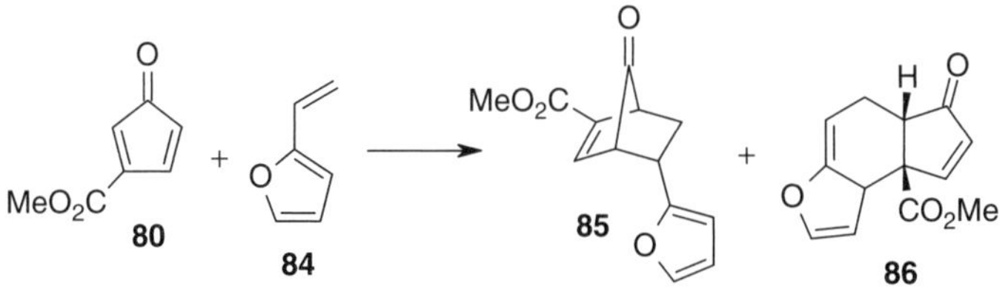

(Reaction 8.11)

However, the MPW1K/6-31+G** PES for each reaction displays only two TSs. The first is a TS for the Cope rearrangement that interconverts **82** and **83** in Reaction 8.10 and **85** and **86** in Reaction 8.11. That leaves just a single TS to connect reactants with just one of the products! These TSs are bispericyclic, as described in Section 4.4. This implies that the reactants come together, cross over a single TS, and then pass over a bifurcating surface where the lowest energy path (the IRC or reaction path) continues on to one product only. The second product, however, can be reached by passing over this same TS and then following some other nonreaction path. This sort of surface is ripe for experiencing nonstatistical behavior.

Trajectory were computed using direct dynamics and the MPW1K/6-31+G** method. Starting from the Diels-Alder TS of Reaction 8.10, 39 trajectories were followed: 28 ended with **82** and 10 ended with **83**, while one trajectory recrossed the TS. Isomerization of **83** into **82** is possible, and the predicted low barrier for this explains the sole observation of **82**. For Reaction 8.11, of the 33 trajectories, 19 led to **85** and 12 led to **86**. This distribution is consistent with the experiment which had a slight excess of **85** over **86**.

Singleton[89] provided two rules of thumb that may begin to form a basis of a predictive dynamic model. First, he noted that the geometry of the single TS that "leads" to the two products can suggest the major product. The TS geometry may be "closer" to one product over the other. For example, in the TS of Reaction 8.10 the two forming C–C bonds that differentiate the two products are 2.95 and 2.99

Å long, and the shorter distance corresponds to forming **82**, the major product. Similarly, for the TS of Reaction 8.11, the two important C–C distances are 2.83 and 3.13 Å, with the shorter distance corresponding to forming **85**. His second point has to do with the position of the second (Cope) TS connecting the two products. This TS acts to separate the PES into two product basins. The farther this second TS is from the first TS, the greater the selectivity.

Another hint toward understanding reaction dynamics on a bifurcating surface was supplied by Singleton[90] in his study of the Diels–Alder cycloaddition of acrolein with methyl vinyl ketone (Reaction 8.12). Recognizing the interconversion of **87** and **88** through a Cope rearrangement along with careful kinetic analysis led to an estimate of the ratio of the rate of formation of **87** : **88** as 2.5 : 1.

(Reaction 8.12)

The eight possible TSs for this reaction were computed with a variety of methodologies (B3LYP, MPW1K, and MP2 with different basis sets in the gas and solution phases), all providing very similar results. The lowest energy TS is **89**. A TS of type **90** (an *endo* TS having an *s-cis*-acrolein dienophile) could not be found; all attempts to optimize it collapsed to **89**. This is yet another example of a bispericyclic system.

The IRC connects TS **89** to **87**. While there is a TS that does lead to **88**, it is much higher in energy than **89**. Ordinary TST using the lowest energy TSs that lead to **87** and **88** predicts a product ratio in excess of 700 : 1, in conflict with experiment. To address the chemical selectivity on a surface like this, Singleton resorted to computing trajectories on the MP2/6-311+G** energy surface. Of the 296 trajectories that begin at **89** with motion toward product, 89 end at **87** and 33 end at **88**, an amazingly good reproduction of experimental results! Interestingly, 174 trajectories recross the TS and head back toward reactants. These recrossing trajectories result from "bouncing off" the potential energy wall of the fully formed C_4–C_5 bond, the general vicinity of the Cope TS.

Analysis of the trajectories revealed a strong correlation between the initial direction and velocity in the 98 cm^{-1} vibration, the vibration that corresponds to the closing of the second σ bond: closing the C_6–O_1 (forming **87**) in the negative direction and closing the C_3–O_8 bond (forming **88**) in the positive direction. Singleton argued that this is a type of dynamic matching. It may not just be a matching of momentum between an entrance and an exit channel on a surface, as discussed in Section 8.3.2, that are important but also matching up vibrational motion(s) that lead to specific products. This type of dynamic matching is important in [1,5]-H migration in cyclopentadiene.[91]

A broad array of organic reactions displays bifurcating PESs. An extensive review of examples was authored by Houk.[92] Tantillo has shown examples of bifurcations in terpene synthesis,[93–95] and a few examples have been found in substitution chemistry.[96,97] The importance of dynamics cannot be overstated for reactions on bifurcating surfaces and clearly additional research needs to be done, especially to aid in creating more robust predictive models.

8.3.6 Stepwise Reaction on a Concerted Surface

We have discussed reactions whose PESs have intermediates, and these intermediates are bypassed, giving rise to apparent "concerted" reactions. This section discusses a case where the PES has only one TS, but the dynamics indicate the formation of an intermediate. These trajectories are way off of the intrinsic reactions coordinate, calling into question the importance of the IRC or MEP.

8.3.6.1 Rearrangement of Protonated Pinacolyl Alcohol

Reaction 8.13 involves the loss of water and methyl migration of protonated pinacolyl alcohol **91** to produce the tertiary cation **92**. There are two proposed mechanisms for this reaction. First, the methyl group could migrate as water is lost, a concerted mechanism going through a single TS **93**. Second, a stepwise mechanism proposes that water is first lost, creating the secondary cation intermediate **94**, followed by methyl migration in a second distinct chemical step. This stepwise mechanism invokes an intermediate and two TSs, one for the water elimination and one for the migration.

(Reaction 8.13)

Ammal et al.[98] examined Reaction 8.13 by first locating all of the critical points on the PES. With the HF, MP2, and B3LYP methods and either the 6-31G* or the 6-311G** basis sets, the only TS that could be located is for the concerted process, **93** (Figure 8.20). The activation barrier is 5.5 (HF/6-31G*), 3.4 (B3LYP/6-311G**), and 8.0 kcal mol^{-1} (MP2/6-311G**). No local minimum corresponding to the secondary cation **94** was found on either the B3LYP or the MP2 PES.

MD were then performed using the HF/6-31G* PES. Fifty trajectories were initiated at the reactant position with kinetic energy equivalent to 400 K. These

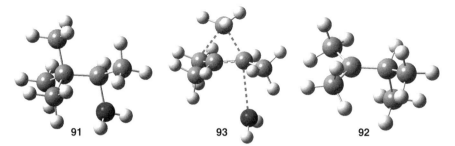

Figure 8.20 Optimized geometries of **91–93**.

trajectories were then run for 500 fs. Of these fifty, 28 remained in the reactant region. The reactive trajectories were very surprising. Only one trajectory followed a concerted pathway, leaving the reactant and directly going to rearranged product. The other 21 trajectories led to the secondary carbocation **94** created by loss of water. One of these trajectories then exited this well and led to rearranged product **92**. The 20 trajectories that stayed in the region of **94** were then run for a longer time; 7 of them rearranged to **92**, 8 of them remained in the neighborhood of **94**, and 5 returned to reactant. The secondary cation has a lifetime of as long as 4000 fs.

The vast majority of the reactive trajectories do not follow the IRC, which is the concerted mechanism. Rather, the reaction proceeds in a stepwise fashion, even though the intermediate is not a local energy minimum. The shape of the PES, and particularly the critical points on this surface, does not control the mechanism. Rather, it is the motion, the dynamics, upon this surface that defines the mechanism.

8.3.7 Roaming Mechanism

Unimolecular dissociation, often initiated by a photon, typically takes one of two courses: decomposition into two molecules over a single barrier or cleavage of a single bond resulting in two radicals. However, the photodissociation of formaldehyde into H_2 and CO reported by Moore in 1993 did not comply with these two mechanisms. When formaldehyde was excited with low energy photons, sufficient to cross the barrier leading to H_2 + CO but not enough to cleave to H + HCO, highly rotationally excited CO was observed. This is consistent with crossing over the concerted barrier. However, when formaldehyde was excited to just above the threshold for cleaving the C–H bond, a shoulder appeared on the rotational distribution curve, indicative of rotationally cold CO.[99]

Townsend et al.[100] were able to further resolve the photodissociation of formaldehyde to show that the rotationally highly excited CO was accompanied by vibrationally cold H_2, as expected for a reaction over a single-concerted TS (shown in Figure 8.21 as **95**). A smaller component, observed at higher photon energies, involves production of rotationally cold CO and vibrationally hot H_2. A PES created by fitting energies, computed at CCSD(T)/aug-cc-pVTZ, for

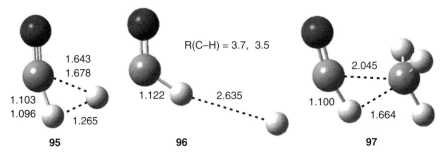

Figure 8.21 Optimized transition states for the dissociation of formaldehyde into H_2 + CO (**95** and **96**) and acetaldehyde into CH_4 + CO (**97**). Distances (in Å) computed at (U)ωB97X-D/aug-cc-pVTZ, CCSD(T)/aug-cc-pVTZ[101] (italics), and CASPT2/aug-cc-pVDZ[104] (bold).

about 80,000 different geometries of H_2CO[101] was used to follow the trajectories for dissociation. Bimodal distribution was observed in these trajectories. The major pathway crossed near the classical TS and resulted in hot CO. The smaller distribution followed a pathway whereby one hydrogen atom nearly detached and then wandered about on a very flat region of the surface until it migrated to the other side of the HCO fragment. This hydrogen then extracted the second hydrogen, forming hot H_2 and vibrational cold CO. This pathway was named the "roaming mechanism."[102,103] Subsequent reinvestigation of the PES of formaldehyde identified the roaming TS on a very flat region of the PES (**96** in Figure 8.21).[104]

Roaming is not restricted to just a small hydrogen atom. Photodissociation of acetaldehyde at 308 nm produces CO in a low vibrational state.[105] A subsequent study found that the CH_4 product is produced in a high vibrational state.[106] This dissociation was modeled in two trajectory studies, both of which gave the same results.[106,107] Both studies did two sets of trajectories. First, direct dynamics were computed at MP2/cc-pVDZ starting at the TS that connects acetaldehyde with the products CO + CH_4. This TS, **97**, is shown in Figure 8.21. The second set of trajectories were computed on a PES fit to the CCSD(T) energies a large number (>135,000) of geometries. These trajectories were initiated at the geometry of acetaldehyde. The direct dynamics trajectories (all originating at **97**) give a methane rotational distribution weighted toward lower energy (peaking at about 25 kcal mol^{-1}). This is quite different from the experimental distribution that shows high rotational energy methane, with a peak around 95 kcal mol^{-1}. Interestingly, the trajectories that originate at the acetaldehyde local minimum with added energy equivalent to a 308 nm photon results in a methane rotational energy distribution quite similar to experiment, with a peak at 85 kcal mol^{-1}. These trajectories tend to avoid the classical concerted TS **97**. Rather, first the C–C bond essentially cleaves, but the CH_3 radical never leaves the vicinity of the HCO fragment, with the maximum C···C distance of about 3.8 Å, decidedly longer than the average maximum distance (2.3 Å) in the direct trajectories that originate at **97**. The HCO fragment freely rotates and when the hydrogen is pointed toward the CH_3 radical, the two fragments approach and a hydrogen is abstracted to form methane and CO.

A MD study of 2-hydroxyethyl radical ($\cdot CH_2CH_2OH$) suggests that it too dissociates through a roaming mechanism.[108] A PES was constructed using CCSD/cc-pVDZ energies about the minima and B3LYP/cc-pVDZ energies for (>50,000) points along all exit channels. Trajectories were initiated in the vicinity of the $\cdot CH_2CH_2OH$ minimum with 44.1 kcal mol^{-1} of internal energy, corresponding to the photodissociation of 2-bromoethanol with 193 nm light. The major product is OH + ethane, as expected. However, a small amount of water and vinyl radical product is also produced. The direct fragmentation to these product requires passage over a TS that is higher than the energy provided and so cannot be the proper reaction pathway. Instead, $\cdot CH_2CH_2OH$ cleaves into ethane and OH radical which does not dissociate, but rather wanders about ethane until it eventually abstracts a hydrogen to form water. Formation of water was subsequently observed in the photodissociation of 2-bromoethanol, supporting the roaming mechanism.[109]

Photodissociation of nitrobenzene gives a bimodal distribution for the NO product.[110] There is a slow component with low J and a fast component with high J. This suggests two different operating mechanisms for dissociation. G2M(CC1)/UB3LYP/6-311+G(3df,2p) computations provide the two mechanisms. Near dissociation to phenyl radical and NO_2 allows NO_2 to roam, eventually recombining to form phenyl nitrite; phenyl nitrate then dissociates and produces the slow NO product. Rearrangement of nitrobenzene to phenylnitrite on the triplet surface eventually leads to loss of fast NO, with high rotational excitation.

8.3.8 A Roundabout S_N2 reaction

The PES of the nucleophilic substitution reaction between chloride ion and methyl iodide is characterized by an entrance and an exit ion–dipole complexes sandwiching a backside attack TS, conforming to the generic PES shown in Figure 6.1.[111] The gas-phase reaction was examined using ion–molecule crossed beam imaging at different energies. At low energies, the reaction proceeds statistically from the ion–dipole complex. At higher energies, a direct reaction dominates, whereby the ion–dipole complex is not sampled for any appreciable time but rather the nucleophile Cl$^-$ directly displaces the leaving group I$^-$. At still higher levels, a second reaction pathway is observed.

Unraveling this new pathway was accomplished through MD studies performed on the MP2/ECP/aug-cc-pVDZ surface.[111] When simulating the higher energy experiments, most of the trajectories follow the direct displacement route. However, an alternative pathway is also observed. As the nucleophile (chloride) approaches methyl iodide, the methyl group rotates toward the nucleophile. The methyl group then collides with the nucleophile, which sends the methyl group spinning about the iodine atom in the opposite direction. The methyl group rotates all the way around the iodine atom and when it approaches the chloride for a second time, the displacement reaction occurs and product is formed. Hase and

Wester term this process a "roundabout mechanism," and they have some evidence for the occurrence of the double roundabout (two rotations of the methyl group about the iodine) as well!

8.3.9 Hydroboration: Dynamical or Statistical?

Hydroboration is an essential reaction in the synthetic chemist's arsenal as it allows access to anti-Markovnikov addition. This reaction is treated in standard textbooks in a very traditional way. However, recent experiments and computations suggest that the reaction is really much more subtle.

Reaction 8.14 was carried out by Singleton[112] in an attempt to minimize the contamination of the contribution of second and third additions of alkene to borane. The observed ratio of anti-Markovnikov to Markovinkov product is 90 : 10. Assuming that this ratio derives from the difference in the TS energies leading to the two products, TST gives an estimate of the energy difference of the two activation barriers of 1.1–1.3 kcal mol^{-1}.

(Reaction 8.14)

CCSD(T)/aug-cc-pvqz//CCSD(T)/aug-cc-pvdz computations predict that the difference in the energies between the TS leading to the anti-Markovnikov and Markovnikov product is 2.4 kcal mol^{-1}, much greater than that estimated from TST. Inclusion of enthalpic corrections and entropy still leaves a free energy difference of 2.5 kcal mol^{-1}, which would indicate a product ratio of 99 : 1.

In the gas phase, formation of a precomplex is exothermic and enthalpically barrierless. (A free energy barrier for forming the complex exists in the gas phase.) When a single tetrahydrofuran (THF) molecule is included in the computations, the precomplex is formed after passing through a barrier much higher than the energy difference between the precomplex and either of the two TSs leading to the addition products. Singleton speculated that there would be little residence time within the precomplex basin, and the reaction might express nonstatistical behavior.

Classical trajectories were computed on the B3LYP/6-31G* surface. When trajectories were started at the precomplex, only 1 percent led to the Markovnikov product, consistent with TST, but *inconsistent with experiment*. When trajectories were initiated at the free energy TS for formation of the complex, 10 percent of the trajectories ended up at the Markovnikov product. Singleton argued that there are three different time domains for the reaction: short trajectories that directly proceed to product with minimal selectivity, somewhat longer trajectories which follow RRKM models, and long trajectories which linger in the precomplex and lead to very high selectivity. This picture is in fact much more consistent with other experimental observations, like little change in selectivity with varying alkene substitution and the very small H/D isotope effect. Singleton adds another interesting experimental fact that does not jibe with the classical mechanism: the selectivity is little affected by temperature, showing 10 percent Markovnikov product at 21 °C and 11.2 percent Markovnikov product at 70 °C.

Glowacki and coworkers[113] offer an alternative explanation based solely on RRKM theory, using the stochastic-energy-grained master equation. They suggest that there are hot intermediates, formed of a loose association of BH_3 with propene, that react nonselectively before cooling. The cooled intermediates react very selectively (around 99 percent) to give the anti-Markovnikov product. The upshot is that hydroboration—and by implication a whole lot of other seemingly ordinary chemistry—may in fact be much more complicated than we had previously thought. Standard TST may not always apply, and trajectory analysis may provide an incomplete perspective.

8.3.10 A Look at the Wolff Rearrangement

The last example we discuss is based on a paper from the Carpenter laboratory.[114] The paper reads like a detective story, chronicling hypotheses, tests, wrong steps, the fortuitous guidance offered by a referee, corrections, and a final summary. The discussion here holds to the storytelling in the original article.

Carpenter[114] was intrigued by the possibility of dynamical effects in carbene chemistry. He chose to explore the Wolff rearrangement taking **100** into **101**. Outside of a small isotope effect, one should observe equal amounts of **101a** and **101b** if the reaction is under statistical control (Scheme 8.11). In fact, the ratio of products is not unity, but rather **101a** : **101b** = 1 : 4.5, perhaps indicative of nonstatistical dynamics. However, the excess of **101b** could be the result of a second parallel rearrangement, **99** to **102** to **101b** (Scheme 8.12).

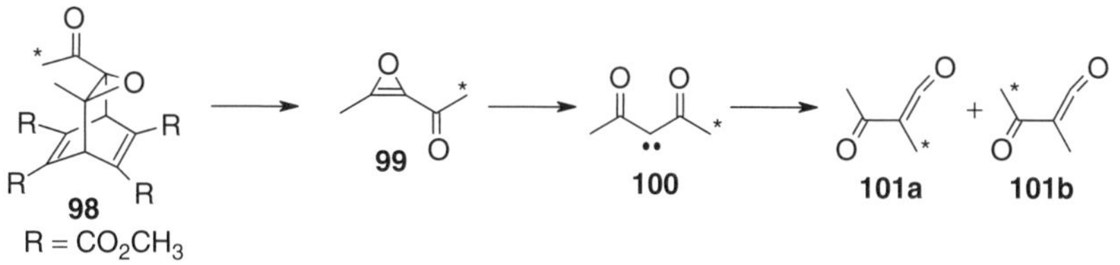

Scheme 8.11

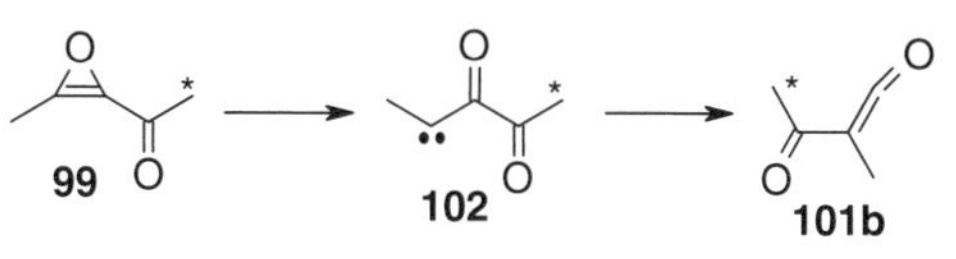

Scheme 8.12

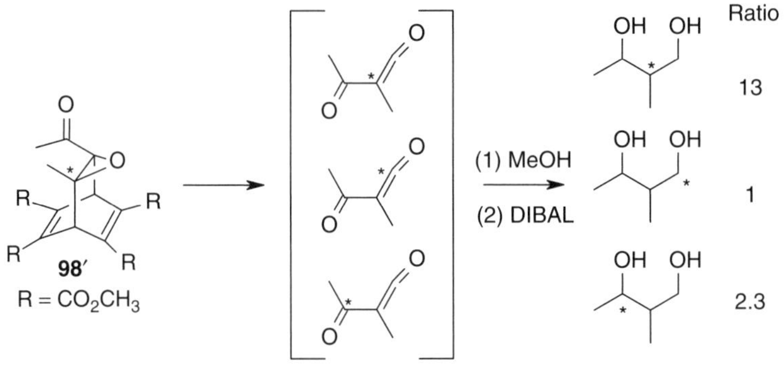

Scheme 8.13

To try to distinguish whether **102** is intervening, Carpenter carried out the photolysis of a different labeled version of **98** (namely, **98′**). The resulting product distribution is shown in Scheme 8.13. It appears that the reaction predominantly passes through **102**, but the ratio of products that come from **100** nonetheless shows nonstatistical behavior.

However, CCSD(T)/cc-pVTZ//CCSD/cc-pVDZ computations place **102** 8.7 kcal mol^{-1} above **100**, which is difficult to reconcile with the experiments that suggest predominant reaction through **102** rather than through **100**. At this point, Carpenter opted to write up the work as a communication, with the main point that nonstatistical dynamics were occurring, though without a satisfactory complete explanation.

A referee of this communication, later identified as Dan Singleton, proffered an alternative mechanism (Scheme 8.14) for the creation of **102**. This is a *pseudopericyclic reaction* (see Section 4.5 for a detailed discussion of this type of reaction) that leads from **98** directly to **102**. In fact, the TS for this pseudopericyclic reaction is 19.0 kcal mol^{-1} (MPWB1K/ 6-31+G(d,p)) lower in energy than the TS for the retro-Diels–Alder reaction of Scheme 8.11.

The mechanism was then revised to include the additional complication of the formation of **103** and is shown in Scheme 8.15, along with their relative CCSD(T) free energies. Any nonstatistical effect would occur in the transition from **102** to **100**. A direct dynamics trajectory analysis was performed starting in the neighborhood of the TS for this step using three different functionals to generate the PES.

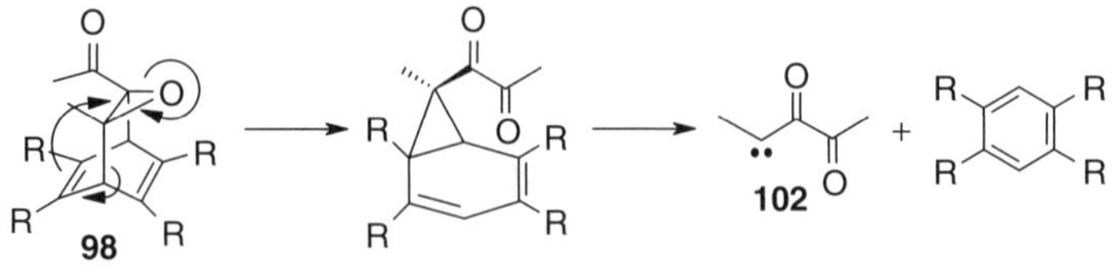

Scheme 8.14

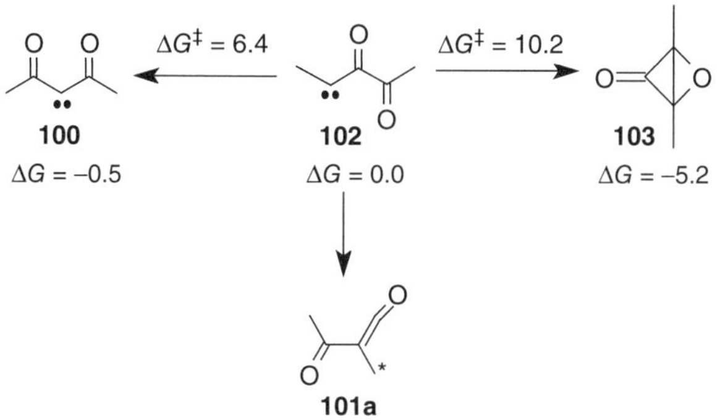

Scheme 8.15

Though only 100 trajectories were computed, the results with all three functionals are similar. About two-third of these trajectories led to **100** followed by the shift of the C_5 methyl group. Another 15 percent led to **100** followed by the shift of the C_1 methyl group. This distribution, consistent with the experiment that shows a preference for C_5 migration, implicates a nonstatistical Wolff rearrangement.

8.4 CONCLUSIONS

Reaction mechanisms are principally presented in introductory organic texts as reaction coordinate diagrams. These diagrams present a slice through the PES, where the x-axis follows the reaction path, usually interpreted as the MEP, and the y-axis is the potential energy. Students learn that a reaction follows this path. For example, an S_N1 mechanism means that from reactant R–X, the C–X bond is stretched, the energy rises up until the TS is reached, and then the energy goes down until the carbocation intermediate is formed. In the second step, the energy increases with the approach of the nucleophile until a second TS is crossed, followed by energy decrease as the C–Nuc bond is completely formed. The illustrated reaction path is just the MEP and should not be understood as the *actual path traversed*. Proper interpretation of the reaction coordinate diagram is that it identifies the critical points that *may* be sampled in the course of the reaction. Molecules will not follow a trajectory right on the MEP but will stray about it. Generally accepted is that the actual trajectories will be in the neighborhood of the MEP. Most importantly, our classical conception of mechanism is that *all* of the critical points on the reaction path will be traversed in that particular order.

Furthermore, intermediates that lie in relatively deep wells will have a lifetime that allows for complete redistribution of vibrational energy. This allows for statistical interpretation of rates for reaction coming out of these intermediates. Generally, this would imply that intermediates would scramble whatever stereo- or regiochemical information that preceded its formation. In fact, this is one of the major distinguishing features among related classes of reactions. For example, the S_N2 reaction proceeds without an intermediate and so stereochemistry is inverted, while the S_N1

reaction, which goes through an intermediate, yields racemic product. Similarly, concerted pericyclic reactions have very specific stereochemical outcomes, while radical pathways involve loss of stereochemistry.

As demonstrated by the examples in this chapter, these mechanistic concepts are simplistic if not simply wrong. Seemingly stable intermediates are avoided. Stereochemical outcome can be found for reactions with intermediates. Multiple products can be formed from crossing a single TS.

Clearly, dynamic effects can be dramatic. Most crucial is that dynamic effects rear up in a broad range of situations, including very ordinary-looking reactions. That dynamics effects are large for cases involving reactive intermediates that exist on a broad flat plateau or caldera may not be too surprising, though the number of cases where this is true is probably larger than one might have anticipated. But the fact that reactions that have stable intermediates tucked into deep energy wells also do not exhibit statistical behavior is most distressing.

The implication of these dynamics studies is compelling. Simply examining an organic reaction by locating the critical points on the PES, or even more rigorous characterization of large swathes on the PES, may provide insufficient, even misleading, information concerning the nature of the reaction mechanism.

In the first edition of this book, I stressed that more trajectory studies were needed to determine how broad the scope of nonstatistical behavior in organic reactions really is. In the meantime, many new examples of nonstatistical dynamics have been reported. Dynamic effects are real, critical to understanding experimental results, and probably more pervasive than once thought.

What is needed now are guiding principles and models for understanding and predicting when and how dynamic effects will occur. We need to identify topological features on the PES that signal for nonstatistical behavior. We need to identify patterns within the trajectories and ways to group them. A dynamics-based mechanistic model can then potentially be built, replacing our old mechanistic notions that rely solely on the PES. This pursuit will continue to be a major challenge to physical and computational organic chemists for the upcoming decade. It has the potential to be as important as the heyday of physical organic chemistry of the late 1950s through 1960s. Computational chemists will be full partners with experimentalists in developing this new more mature model of organic reactions.

8.5 INTERVIEW: PROFESSOR DANIEL SINGLETON

Interviewed June 29, 2006

Professor Daniel Singleton is a professor of chemistry at the Texas A&M University, where he has been for his entire academic career. Singleton performed his graduate studies at the University of Minnesota under Professor Paul Gassman working on synthetic methodology, but with a strong bent toward physical organic chemistry. Nearing the end of his stay at Minnesota, Singleton recalls a visit by Professor George Whitesides to the chemistry department.

Whitesides met with a group of graduate students and, according to Singleton, told them all "to get out of physical organic chemistry". Following this advice, he did a postdoctoral stint with Professor Barry Trost.

His first calculations were done as an assistant professor and they concerned the Diels–Alder reaction of vinyl boranes. "This got me a *JACS* communication in my fifth year as an assistant professor," Singleton recounts. "This was very critical in establishing a successful start of my academic career, especially in my earning tenure." Singleton basically learned how to run quantum chemical calculations on his own, with some assistance along the way. "The Hall group at A&M really helped me out," he recalls. "Preston MacDougal, who was a post-doc with Professor Hall, taught me a bunch of very fundamental things, including the basics of UNIX. One thing he taught me was how to use the *vi* editor, which shows how little I knew at the time!" A "synthetic point of view" continues to inform all of his calculations and their interpretation.

Singleton's main use of quantum chemistry is to compute KIEs, largely to aid in the interpretation of his experiments. "Most of the time our work starts with experiments and the calculations follow," Singleton explains. "We always compute the isotope effects in conjunction with the experiments, and we do them early on. I hate to admit it, but computed isotope effects are more than equal partners with experiments. That is, we really can't do without them in most of our cases."

Computations occasionally precede the experiments, but this is rare. "The experiments may fail," he points out. "Particularly now we are using trajectories in an exploratory way to look for something weird to happen and then go to the lab to try to verify it."

Singleton notes that "there is nothing special about doing the computations of isotope effects. You get the isotope effects from the frequency calculation, and people worry a great deal about the low frequency modes. If you think of it in a classical thermodynamic sense, they're going to have a tremendous influence on the rate, and these modes are not computed particularly well. So, you'd think the isotope effect calculations would be terrible. It doesn't work that way. The errors in these low modes cancel out. The major effect is in the *high* frequency modes." In computing reasonable isotope effects, Singleton worries about the quality of the optimized structure. "For a relatively synchronous Diels–Alder reaction I would bet that we would compute the isotope effects within experimental error. S_N2 reactions are much tougher. What that says is that dynamics might change things. We need more examples of this kind of thing."

Singleton computes isotope effects using the *Quiver* program. He characterized the program as "extremely user unfriendly." His offer on the Computational Chemistry List to assist people in running *Quiver* led to over two dozen requests. He has "put a lot of effort into automating the process, writing macros and scripts so that his group can now calculate thousands of isotope effects. My students used to hate me when I told them to do a *Quiver* calculation; it would take them hours to set it up. Now it's less than 10 minutes." He notes that a *Gaussian* keyword that would allow for the direct computation of the isotope effects would greatly increase the usage of computed isotope effects calculations. "Kinetic isotope effects are underutilized because there is no simple way to do this computation," he notes.

The interpretation of isotope effects in the ene reaction of singlet oxygen was a great challenge to Singleton. "We had experimental isotope effects and we tried to predict them. B3LYP couldn't even make a prediction. CASSCF allowed one to make a prediction but it was nonsense. Using a grid of single-point CCSD(T) energies gave us a TS that did well with the intermolecular isotope effect but allowed for no prediction of the intramolecular effect." The answer turned out to be a two-step no-intermediate mechanism where reaction dynamics were critical. This sparked his current interest in seeking out systems where nonstatistical dynamics are important.

Singleton feels that nonstatistical dynamics are not important in most reactions, but that they are important in a great many; "Maybe 1 out of 5 cases will have something interesting," he speculates. "If you go through our cases, what we find is mostly by accident. But it's not complete random chance that a reaction has nonstatistical behavior. Now we are designing systems to force dynamic effect on it."

Commenting on the meaning of "mechanism" in a dynamic world, Singleton says "Mechanism use to be a complete sequence of TSs and intermediates—everything on how reactants moved on to products. And the thing we really want to know about is the TS. Now we have to have an understanding of the complete energy surface and all its details. For most reactions, we can focus on just a small part of the energy surface, but for complicated mechanisms we will require knowledge of a broad portion of the PES. We will need to run trajectories for any interesting looking TS, but a single trajectory is no more costly than an IRC calculation." Though he has begun to introduce the consequences of dynamics in his graduate classes, he is "not too optimistic about its inclusion any time soon in a textbook."

Reading through Singleton's papers, one sees a growing computational component. "I love laboratory work," Singleton explains. "As a post-doc I could do it all the time. I just can't do that now—student questions just interrupt that process too much. But with calculations I can set those up and get back to them whenever is convenient, including at home. This makes it easy for me to keep up a computational front. It's also what makes it easier to teach students how to do computations than how to do the experiments. They do something wrong with a calculation, it's easy to open up their files directly on my computer and check out exactly what they did. In the lab, that's just much more difficult to do."

Though computations play such a large role in his current work, Singleton emphasizes the combination of computation and lab work when recruiting new students into his group. "I say to the student you're going to do the lab work and that's going to get you a job. But you're actually going to understand what's going on because of the computational work. That's what will get you the physical organic understanding of chemistry."

When asked to name his most important scientific contribution, Singleton quickly gave the pithy reply "I haven't done it yet!" He then acknowledged that "people are going to give me the most credit for the natural abundance isotope effect, which of course I didn't invent. We just perfected it for carbon." Singleton's recent joint computational experimental studies of dynamic effects in organic reactions is a strong indication that his first tongue-in-cheek answer is likely to be true—Singleton's most influential work appears to be in current development.

REFERENCES

1. Carey, F. A. *Organic Chemistry*; 5th ed.; McGraw-Hill: Boston, **2003**.
2. Solomons, T. W. G.; Fryhle, C. B. *Organic Chemistry*; 10th ed.; John Wiley & Sons, Inc.: Hoboken, NJ, **2011**.
3. Houston, P. L. *Chemical Kinetics and Reaction Dynamics*; McGraw-Hill: Boston, MA, **2001**.
4. Steinfeld, J. I.; Francisco, J. S.; Hase, W. L. *Chemical Kinetics and Dynamics*; Prentice Hall: UpperSaddle River, NJ, **1999**.
5. Peslherbe, G. H.; Wang, H.; Hase, W. L. "Monte Carlo sampling for classical trajectory simulations," *Adv. Chem. Phys.* **1999**, *105*, 171–201.
6. Sun, L.; Hase, W. L. "Born-Oppenheimer direct dynamics classical trajectory simulations," *Rev. Comput. Chem.* **2003**, *19*, 79–146.
7. Carpenter, B. K. In *Reactive Intermediate Chemistry*; Moss, R. A., Platz, M. S., Jones, M., Jr., Eds.; John Wiley & Sons, Inc.: Hoboken, NJ, **2004**, p 925–960.
8. Carpenter, B. K. "Nonstatistical dynamics in thermal reactions of polyatomic molecules," *Ann. Rev. Phys. Chem.* **2005**, *46*, 57–89.
9. Rehbein, J.; Carpenter, B. K. "Do we fully understand what controls chemical selectivity?," *Phys. Chem. Chem. Phys.* **2011**, *13*, 20906–20922.
10. Press, W. H.; Teukolsky, S. A.; Vetterling, W. T.; Flannery, B. P. *Numerical Recipes: The Art of Scientific Computing*; 3rd ed.; Cambridge University Press: Cambridge, UK, **2007**.
11. Helgaker, T.; Uggerud, E.; Aa. Jensen, H. J. "Integration of the classical equations of motion on ab initio molecular potential energy surfaces using gradients and Hessians: Application to translational energy release upon fragmentation," *Chem. Phys. Lett.* **1990**, *173*, 145–150.
12. Millam, J. M.; Bakken, V.; Chen, W.; Hase, W. L.; Schlegel, H. B. "Ab initio classical trajectories on the Born–Oppenheimer surface: Hessian-based integrators using fifth-order polynomial and rational function fits," *J. Chem. Phys.* **1999**, *111*, 3800–3805.
13. Bakken, V.; Millam, J. M.; Schlegel, H. B. "Ab initio classical trajectories on the Born–Oppenheimer surface: Updating methods for Hessian-based integrators," *J. Chem. Phys.* **1999**, *111*, 8773–8777.

14. Gonzalez-Lafont, A.; Truong, T. N.; Truhlar, D. G. "Direct dynamics calculations with neglect of diatomic differential overlap molecular orbital theory with specific reaction parameters," *J. Phys. Chem.* **1991**, *95*, 4618–4627.
15. Hase, W. L.; Buckowski, D. G. "Monte Carlo sampling of a microcanonical ensemble of classical harmonic oscillators," *Chem. Phys. Lett.* **1980**, *74*, 284–287.
16. Carroll, F. A. *Perspectives on Structure and Mechanism in Organic Chemistry*; 2nd ed.; John Wiley & Sons, Inc.: Hoboken, NJ, **2010**.
17. Lowry, T. H.; Richardson, K. S. *Mechanism and Theory in Organic Chemistry*; 3rd ed.; Harper and Row: New York, **1987**.
18. Kassel, L. S. "The dynamics of unimolecular reactions," *Chem. Rev.* **1932**, *10*, 11–25.
19. Marcus, R. A. "Lifetimes of active molecules. I," *J. Chem. Phys.* **1952**, *20*, 352–354.
20. Marcus, R. A. "Lifetimes of active molecules. II," *J. Chem. Phys.* **1952**, *20*, 355–359
21. Rice, O. K.; Ramsperger, H. C. "Theories of unimolecular gas reactions at low pressures," *J. Am. Chem. Soc.* **1927**, *49*, 1617–1629.
22. Evans, M. G.; Polanyi, M. "Some applications of the transition state method to the calculation of reaction velocities, especially in solution," *Trans. Faraday Soc.* **1935**, *31*, 875–894.
23. Eyring, H. "The activated complex in chemical reactions," *J. Chem. Phys.* **1935**, *3*, 107–115.
24. Truhlar, D. G.; Garrett, B. C. "Variational transition state theory," *Annu. Rev. Phys. Chem.* **1984**, *35*, 159–189.
25. Carpenter, B. K. "Dynamic matching: The cause of inversion of configuration in the [1,3] sigmatropic migration?," *J. Am. Chem. Soc.* **1995**, *117*, 6336–6344.
26. Carpenter, B. K. "Bimodal distribution of lifetimes for an intermediate from a quasi-classical dynamics simulation," *J. Am. Chem. Soc.* **1996**, *118*, 10329–10330.
27. Berson, J. A.; Nelson, G. L. "Inversion of configuration in the migrating group of a thermal 1,3-sigmatropic rearrangement," *J. Am. Chem. Soc.* **1967**, *89*, 5503–5504.
28. Baldwin, J. E.; Belfield, K. D. "Stereochemistry of the thermal isomerization of bicyclo[3.2.0]hept-2-ene to bicyclo[2.2.1]hept-2-ene," *J. Am. Chem. Soc.* **1988**, *110*, 296–297.
29. Klärner, F. G.; Drewes, R.; Hasselmann, D. "Stereochemistry of the thermal rearrangement of bicyclo[3.2.0]hept-2-ene to bicyclo[2.2.1]hept-2-ene (Norbornene). [1,3] Carbon migration with predominant inversion," *J. Am. Chem. Soc.* **1988**, *110*, 297–298.
30. Berson, J. A.; Nelson, G. L. "Steric prohibition of the inversion pathway. Test of the orbital symmetry prediction of the sense of rotation in thermal suprafacial 1,3-sigmatropic rearrangements," *J. Am. Chem. Soc.* **1970**, *92*, 1096–1097.
31. Carpenter, B. K. "Dynamic behavior of organic reactive intermediates," *Angew. Chem. Int. Ed.* **1998**, *37*, 3340–3350.
32. Hoffmann, R.; Swaminathan, S.; Odell, B. G.; Gleiter, R. "Potential surface for a non-concerted reaction. tetramethylene," *J. Am. Chem. Soc.* **1970**, *92*, 7091–7097.
33. Doering, W. v. E.; Sachdev, K. "Continuous diradical as transition state. Internal rotational preference in the thermal enantiomerization and diastereoisomerization of *cis*- and *trans*-1-cyano-2-isopropenylcyclopropane," *J. Am. Chem. Soc.* **1974**, *96*, 1168–1187.
34. Doering, W. v. E.; Cheng, X.; Lee, K.; Lin, Z. "Fate of the intermediate diradicals in the caldera: Stereochemistry of thermal stereomutations, (2 + 2)

cycloreversions, and (2 + 4) ring-enlargements of *cis-* and *trans-*1-cyano-2-(*E* and *Z*)-propenyl-*cis*-3,4-dideuteriocyclobutanes," *J. Am. Chem. Soc.* **2002**, *124*, 11642–11652.

35. *Home Ground, Language for an American Landscape*; Lopez, B., Ed.; Trinity University Press: San Antonio, TX, **2006**.

36. Baldwin, J. E.; Villarica, K. A.; Freedberg, D. I.; Anet, F. A. L. "Stereochemistry of the thermal isomerization of vinylcyclopropane to cyclopentene," *J. Am. Chem. Soc* **1994**, *116*, 10845–10846.

37. Gajewski, J. J.; Olson, L. P.; Willcott, M. R. "Evidence for concert in the thermal unimolecular vinylcyclopropane to cyclopentene sigmatropic 1,3-shift," *J. Am. Chem. Soc.* **1996**, *118*, 299–306.

38. Davidson, E. R.; Gajewski, J. J. "Calculational evidence for lack of intermediates in the thermal unimolecular vinylcyclopropane to cyclopentene 1,3-sigmatropic shift," *J. Am. Chem. Soc.* **1997**, *119*, 10543–10544.

39. Houk, K. N.; Nendel, M.; Wiest, O.; Storer, J. W. "The vinylcyclopropane-cyclopentene rearrangement: A prototype thermal rearrangement involving competing diradical concerted and stepwise mechanisms," *J. Am. Chem. Soc.* **1997**, *119*, 10545–10546.

40. Doubleday, C.; Nendel, M.; Houk, K. N.; Thweatt, D.; Page, M. "Direct dynamics quasiclassical trajectory study of the stereochemistry of the vinylcyclopropane-cyclopentene rearrangement," *J. Am. Chem. Soc.* **1999**, *121*, 4720–4721.

41. Doubleday, C. "Mechanism of the vinylcyclopropane-cyclopentene rearrangement studied by quasiclassical direct dynamics," *J. Phys. Chem. A* **2001**, *105*, 6333–6341.

42. Berson, J. A.; Dervan, P. B. "Mechanistic analysis of the four pathways in the 1,3-sigmatropic rearrangements of *trans-*1,2-*trans*, *trans-* and *trans-*1,2-*cis*,*trans*-dipropenylcyclobutane," *J. Am. Chem. Soc.* **1973**, *95*, 269–270.

43. Northrop, B. H.; Houk, K. N. "Vinylcyclobutane-cyclohexene rearrangement: Theoretical exploration of mechanism and relationship to the Diels–Alder potential surface," *J. Org. Chem.* **2006**, *71*, 3–13.

44. Baldwin, J. E.; Leber, P. A. "Molecular rearrangements through thermal [1,3] carbon shifts," *Org. Biomol. Chem.* **2008**, *6*, 36–47.

45. Baldwin, J. E.; Kostikov, A. P. "On the stereochemical characteristic of the thermal reactions of vinylcyclobutane," *J. Org. Chem.* **2010**, *75*, 2767–2775.

46. Doubleday, C.; Suhrada, C. P.; Houk, K. N. "Dynamics of the degenerate rearrangement of bicyclo[3.1.0]hex-2-ene," *J. Am. Chem. Soc.* **2006**, *128*, 90–94.

47. Baldwin, J. E.; Keliher, E. J. "Activation parameters for three reactions interconverting isomeric 4- and 6-deuteriobicyclo[3.1.0]hex-2-enes," *J. Am. Chem. Soc.* **2002**, *124*, 380–381.

48. Doering, W. v. E.; Zhang, T.-h.; Schmidt, E. K. G. "Kinetics of thermal rearrangements in the δ2-thujene system: A full quadrisection of a perturbed bicyclo[3.1.0]hex-2-ene," *J. Org. Chem.* **2006**, *71*, 5688–5693.

49. Doering, W. v.; Zhao, X. "Steric control in the thermal rearrangement of a bicyclo[3.1.0]hex-2-ene substituted at a radical-nonstabilizing position," *J. Am. Chem. Soc.* **2008**, *130*, 6430–6437.

50. Suhrada, C. P.; Houk, K. N. "Potential Surface for the quadruply degenerate rearrangement of bicyclo[3.1.0]hex-2-ene," *J. Am. Chem. Soc.* **2002**, *124*, 8796–8797.

51. Berson, J. A.; Pedersen, L. D. "Thermal stereomutation of optically active *trans*-cyclopropane-1,2-d_2," *J. Am. Chem. Soc.* **1975**, *97*, 238–240.
52. Cianciosi, S. J.; Ragunathan, N.; Freedman, T. B.; Nafie, L. A.; Lewis, D. K.; Glenar, D. A.; Baldwin, J. E. "Racemization and geometrical isomerization of (2*s*,3*s*)-cyclopropane-1-^{13}C-1,2,3-d_3 at 407 °C: Kinetically competitive one-center and two-center thermal epimerizations in an isotopically substituted cyclopropane," *J. Am. Chem. Soc.* **1991**, *113*, 1864–1866.
53. Hoffmann, R. "Trimethylene and the addition of methylene to ethylene," *J. Am. Chem. Soc.* **1968**, *90*, 1475–1485.
54. Doubleday, C. "Lifetime of trimethylene calculated by variational unimolecular rate theory," *J. Phys. Chem.* **1996**, *100*, 3520–3526.
55. Hrovat, D. A.; Fang, S.; Borden, W. T.; Carpenter, B. K. "Investigation of cyclopropane stereomutation by quasiclassical trajectories on an analytical potential energy surface," *J. Am. Chem. Soc.* **1997**, *119*, 5253–5254.
56. Doubleday, C., Jr.; Bolton, K.; Hase, W. L. "Direct dynamics study of the stereomutation of cyclopropane," *J. Am. Chem. Soc.* **1997**, *119*, 5251–5252.
57. Doubleday, C.; Bolton, K.; Hase, W. L. "Direct dynamics quasiclassical trajectory study of the thermal stereomutations of cyclopropane," *J. Phys. Chem. A* **1998**, *102*, 3648–3658.
58. Xu, L.; Doubleday, C. E.; Houk, K. N. "Dynamics of 1,3-dipolar cycloaddition reactions of diazonium betaines to acetylene and ethylene: Bending vibrations facilitate reaction," *Angew. Chem. Int. Ed.* **2009**, *48*, 2746–2748.
59. Xu, L.; Doubleday, C. E.; Houk, K. N. "Dynamics of 1,3-dipolar cycloadditions: Energy partitioning of reactants and quantitation of synchronicity," *J. Am. Chem. Soc.* **2010**, *132*, 3029–3037.
60. Roth, W. R.; Martin, M. "Stereochemistry of the thermal and photochemical decomposition of 2,3-diazabicyclo[2.2.1]hept-2-ene," *Justus Liebigs Ann. Chem.* **1967**, *702*, 1–7.
61. Roth, W. R.; Martin, M. "Zur Stereochemie der 1.2-Cycloaddition an das Bicyclo[2.1.0]system," *Tetrahedron Lett.* **1967**, *8*, 4695–4698.
62. Allred, E. L.; Smith, R. L. "Thermolysis of *exo*- and *endo*-5-methoxy-2,3-diazabicyclo [2.2.1]-2-heptene," *J. Am. Chem. Soc.* **1967**, *89*, 7133–7134.
63. Sorescu, D. C.; Thompson, D. L.; Raff, L. M. "Molecular dynamics studies of the thermal decomposition of 2,3-diazabicyclo(2.2.1)hept-2-ene," *J. Chem. Phys.* **1995**, *102*, 7911–7924.
64. Yamamoto, N.; Olivucci, M.; Celani, P.; Bernardi, F.; Robb, M. A. "An MC-SCF/MP2 study of the photochemistry of 2,3-diazabicyclo[2.2.1]hept-2-ene: Production and fate of diazenyl and hydrazonyl biradicals," *J. Am. Chem. Soc.* **1998**, *120*, 2391–2407.
65. Reyes, M. B.; Carpenter, B. K. "Mechanism of thermal deazetization of 2,3-diazabicyclo[2.2.1]hept-2-ene and its reaction dynamics in supercritical fluids," *J. Am. Chem. Soc.* **2000**, *122*, 10163–10176.
66. Osterheld, T. H.; Brauman, J. I. "Infrared multiple-photon dissociation of the acetone enol radical cation. Dependence of nonstatistical dissociation on internal energy," *J. Am. Chem. Soc.* **1993**, *115*, 10311–10316.
67. McLafferty, F. W.; McAdoo, D. J.; Smith, J. S.; Kornfeld, R. "Metastable ions characteristics. XVIII. Enolic $C_3H_6O^+$ ion formed from aliphatic ketones," *J. Am. Chem. Soc.* **1971**, *93*, 3720–3730.

68. Depke, G.; Lifshitz, C.; Schwarz, H.; Tzidony, E. "Non-ergodic behavior of excited radical cations in the gas phase," *Angew. Chem. Int. Ed. Engl.* **1981**, *20*, 792–793.
69. Turecek, F.; McLafferty, F. W. "Non-ergodic behavior in acetone-enol ion dissociations," *J. Am. Chem. Soc.* **1984**, *106*, 2525–2528.
70. Lifshitz, C. "Intramolecular energy redistribution in polyatomic ions," *J. Phys. Chem.* **1983**, *87*, 2304–2313.
71. Nummela, J. A.; Carpenter, B. K. "Nonstatistical dynamics in deep potential wells: A quasiclassical trajectory study of methyl loss from the acetone radical cation," *J. Am. Chem. Soc.* **2002**, *124*, 8512–8513.
72. Roth, W. R.; Wollweber, D.; Offerhaus, R.; Rekowski, V.; Lennartz, H. W.; Sustmann, R.; Müller, W. "The energy well of diradicals. IV. 2-Methylene-1,4-cyclohexanediyl," *Chem. Ber.* **1993**, *126*, 2701–2715.
73. Hrovat, D. A.; Duncan, J. A.; Borden, W. T. "Ab initio and DFT calculations on the cope rearrangement of 1,2,6-heptatriene," *J. Am. Chem. Soc.* **1999**, *121*, 169–175.
74. Debbert, S. L.; Carpenter, B. K.; Hrovat, D. A.; Borden, W. T. "The iconoclastic dynamics of the 1,2,6-heptatriene rearrangement," *J. Am. Chem. Soc.* **2002**, *124*, 7896–7897.
75. McIver, J. W., Jr.; Stanton, R. E. "Symmetry selection rules for transition states," *J. Am. Chem. Soc.* **1972**, *94*, 8618–8620.
76. Sun, L.; Song, K.; Hase, W. L. "A S_N2 reaction that avoids its deep potential energy minimum," *Science* **2002**, *296*, 875–878.
77. Lopez, J. G.; Vayner, G.; Lourderaj, U.; Addepalli, S. V.; Kato, S.; deJong, W. A.; Windus, T. L.; Hase, W. L. "A direct dynamics trajectory study of F^- + CH_3OOH reactive collisions reveals a major non-IRC reaction path," *J. Am. Chem. Soc.* **2007**, *129*, 9976–9985.
78. Blanksby, S. J.; Ellison, G. B.; Bierbaum, V. M.; Kato, S. "Direct evidence for base-mediated decomposition of alkyl hydroperoxides (ROOH) in the gas phase," *J. Am. Chem. Soc.* **2002**, *124*, 3196–3197.
79. Schmittel, M.; Strittmatter, M.; Kiau, S. "Switching from the Myers reaction to a new thermal cyclization mode in enyne-allenes," *Tetrahedron Lett.* **1995**, *36*, 4975–4978.
80. Schmittel, M.; Keller, M.; Kiau, S.; Strittmatter, M. "A suprising switch from the Myers-Saito cyclization toa novel biradical cyclization in enyne-allenes: Formal Diels–Alder and ene reactions with high synthetic potetnial," *Chem. Eur. J.* **1997**, *3*, 807–816.
81. Musch, P. W.; Engels, B. "The importance of the ene reaction for the C^2-C^6 cyclization of enyne-allenes," *J. Am. Chem. Soc.* **2001**, *123*, 5557–5562.
82. Bekele, T.; Christian, C. F.; Lipton, M. A.; Singleton, D. A. ""Concerted" transition state, stepwise mechanism. Dynamics effects in C^2-C^6 enyne allene cyclizations," *J. Am. Chem. Soc.* **2005**, *127*, 9216–9223.
83. Achmatowicz, O., Jr.; Szymoniak, J. "Mechanism of the dimethyl mesoxalate-alkene ene reaction. Deuterium kinetic isotope effects," *J. Org. Chem.* **1980**, *45*, 4774–4776.
84. Song, Z.; Beak, P. "Investigation of the mechanisms of ene reactions of carbonyl enophiles by intermolecular and intramolecular hydrogen–deuterium isotope effects: Partitioning of reaction intermediates," *J. Am. Chem. Soc.* **1990**, *112*, 8126–8134.
85. Ghosez, L.; Montaigne, R.; Roussel, A.; Vanlierde, H.; Mollet, P. "Cycloadditions of dichloroketene to olefins and dienes," *Tetrahedron Lett.* **1971**, *27*, 615–633.

86. Machiguchi, T.; Hasegawa, T.; Ishiwata, A.; Terashima, S.; Yamabe, S.; Minato, T. "Ketene recognizes 1,3-dienes in their *s-cis* forms through [4 + 2] (Diels–Alder) and [2 + 2] (Staudinger) reactions. An innovation of ketene chemistry," *J. Am. Chem. Soc.* **1999**, *121*, 4771–4786.
87. Ussing, B. R.; Hang, C.; Singleton, D. A. "Dynamic effects on the periselectivity, rate, isotope effects, and mechanism of cycloadditions of ketenes with cyclopentadiene," *J. Am. Chem. Soc.* **2006**, *128*, 7594–7607.
88. Gonzalez-James, O. M.; Kwan, E. E.; Singleton, D. A. "Entropic intermediates and hidden rate-limiting steps in seemingly concerted cycloadditions. Observation, prediction, and origin of an isotope effect on recrossing," *J. Am. Chem. Soc.* **2011**, *134*, 1914–1917.
89. Thomas, J. B.; Waas, J. R.; Harmata, M.; Singleton, D. A. "Control elements in dynamically determined selectivity on a bifurcating surface," *J. Am. Chem. Soc.* **2008**, *130*, 14544–14555.
90. Wang, Z.; Hirschi, J. S.; Singleton, D. A. "Recrossing and dynamic matching effects on selectivity in a Diels–Alder reaction," *Angew. Chem. Int. Ed.* **2009**, *48*, 9156–9159.
91. Goldman, L. M.; Glowacki, D. R.; Carpenter, B. K. "Nonstatistical dynamics in unlikely places: [1,5] Hydrogen migration in chemically activated cyclopentadiene," *J. Am. Chem. Soc.* **2011**, *133*, 5312–5318.
92. Ess, D. H.; Wheeler, S. E.; Iafe, R. G.; Xu, L.; Çelebi-Ölçüm, N.; Houk, K. N. "Bifurcations on potential energy surfaces of organic reactions," *Angew. Chem. Int. Ed.* **2008**, *47*, 7592–7601.
93. Hong, Y. J.; Tantillo, D. J. "A potential energy surface bifurcation in terpene biosynthesis," *Nat. Chem.* **2009**, *1*, 384–389
94. Siebert, M. R.; Zhang, J.; Addepalli, S. V.; Tantillo, D. J.; Hase, W. L. "The need for enzymatic steering in abietic acid biosynthesis: Gas-phase chemical dynamics simulations of carbocation rearrangements on a bifurcating potential energy surface," *J. Am. Chem. Soc.* **2011**, *133*, 8335–8343.
95. Siebert, M. R.; Manikandan, P.; Sun, R.; Tantillo, D. J.; Hase, W. L. "Gas-phase chemical dynamics simulations on the bifurcating pathway of the pimaradienyl cation rearrangement: Role of enzymatic steering in abietic acid biosynthesis," *J. Chem. Theor. Comput.* **2012**, *8*, 1212–1222.
96. Bogle, X. S.; Singleton, D. A. "Dynamic origin of the stereoselectivity of a nucleophilic substitution reaction," *Org. Lett.* **2012**, *14*, 2528–2531.
97. Itoh, S.; Yoshimura, N.; Sato, M.; Yamataka, H. "Computational study on the reaction pathway of α-bromoacetophenones with hydroxide ion: Possible path bifurcation in the addition/substitution mechanism," *J. Org. Chem.* **2011**, *76*, 8294–8299.
98. Ammal, S. C.; Yamataka, H.; Aida, M.; Dupuis, M. "Dynamics-driven reaction pathway in an intramolecular rearrangement," *Science* **2003**, *299*, 1555–1557.
99. van Zee, R. D.; Foltz, M. F.; Moore, C. B. "Evidence for a second molecular channel in the fragmentation of formaldehyde," *J. Chem. Phys.* **1993**, *99*, 1664–1673.
100. Townsend, D.; Lahankar, S. A.; Lee, S. K.; Chambreau, S. D.; Suits, A. G.; Zhang, X.; Rheinecker, J.; Harding, L. B.; Bowman, J. M. "The roaming atom: Straying from the reaction path in formaldehyde decomposition," *Science* **2004**, *306*, 1158–1161.
101. Zhang, X.; Zou, S.; Harding, L. B.; Bowman, J. M. "A global ab initio potential energy surface for formaldehyde," *J. Phys. Chem. A* **2004**, *108*, 8980–8986.

102. Suits, A. G. "Roaming atoms and radicals: A new mechanism in molecular dissociation," *Acc. Chem. Res.* **2008**, *41*, 873–881.
103. Herath, N.; Suits, A. G. "Roaming radical reactions," *J. Phys. Chem. Lett.* **2011**, *2*, 642–647.
104. Harding, L. B.; Klippenstein, S. J.; Jasper, A. W. "Ab initio methods for reactive potential surfaces," *Phys. Chem. Chem. Phys.* **2007**, *9*, 4055–4070.
105. Houston, P. L.; Kable, S. H. "Photodissociation of acetaldehyde as a second example of the roaming mechanism," *Proc. Nat. Acad. Sci. USA* **2006**, *103*, 16079–16082.
106. Heazlewood, B. R.; Jordan, M. J. T.; Kable, S. H.; Selby, T. M.; Osborn, D. L.; Shepler, B. C.; Braams, B. J.; Bowman, J. M. "Roaming is the dominant mechanism for molecular products in acetaldehyde photodissociation," *Proc. Nat. Acad. Sci. USA* **2008**, *105*, 12719–12724.
107. Shepler, B. C.; Braams, B. J.; Bowman, J. M. ""Roaming" dynamics in CH_3CHO photodissociation revealed on a global potential energy surface," *J. Phys. Chem. A* **2008**, *112*, 9344–9351.
108. Kamarchik, E.; Koziol, L.; Reisler, H.; Bowman, J. M.; Krylov, A. I. "Roaming pathway leading to unexpected water + vinyl products in C_2H_4OH dissociation," *J. Phys. Chem. Lett.* **2010**, *1*, 3058–3065.
109. Ratliff, B. J.; Alligood, B. W.; Butler, L. J.; Lee, S.-H.; Lin, J. J.-M. "Product branching from the CH_2CH_2OH radical intermediate of the oh + ethene reaction," *J. Phys. Chem. A* **2011**, *115*, 9097–9110.
110. Hause, M. L.; Herath, N.; Zhu, R.; Lin, M. C.; Suits, A. G. "Roaming-mediated isomerization in the photodissociation of nitrobenzene," *Nat. Chem.* **2011**, *3*, 932–937.
111. Mikosch, J.; Trippel, S.; Eichhorn, C.; Otto, R.; Lourderaj, U.; Zhang, J. X.; Hase, W. L.; Weidemüller, M.; Wester, R. "Imaging nucleophilic substitution dynamics," *Science* **2008**, *319*, 183–186.
112. Oyola, Y.; Singleton, D. A. "Dynamics and the failure of transition state theory in alkene hydroboration," *J. Am. Chem. Soc.* **2009**, *131*, 3130–3131.
113. Glowacki, D. R.; Liang, C. H.; Marsden, S. P.; Harvey, J. N.; Pilling, M. J. "Alkene hydroboration: Hot intermediates that react while they are cooling," *J. Am. Chem. Soc.* **2010**, *132*, 13621–13623.
114. Litovitz, A. E.; Keresztes, I.; Carpenter, B. K. "Evidence for nonstatistical dynamics in the Wolff rearrangement of a carbene," *J. Am. Chem. Soc.* **2008**, *130*, 12085–12094.

CHAPTER 9

Computational Approaches to Understanding Enzymes

Biochemistry examines the chemistry of molecules involved in life. These range from small molecules such as carbohydrates to large molecules such as DNA. Much of the chemistry of biological molecules is organic reactions, and so its place within this book is logical.

As might be expected, computational biochemists have applied the tools and methodologies of quantum chemistry to biological molecules. Many of the same techniques discussed in the previous chapters have been utilized to explicate reaction mechanisms, properties, and structures of a broad range of biomolecules, including proteins, DNA, and RNA. However, the scope of computational biochemistry is enormous, and a reasonable comprehensive coverage of this topic is beyond the scope of this book.

To provide a flavor of how computational chemistry has been applied to biochemical problems, this chapter focuses on a small subset of computational biochemistry, namely, computational enzymology. Presented here are some examples of how quantum chemical computations have been used to understand the mechanism of catalysis provided by enzymes. The chapter ends with a look at one of the true "holy grails" of biochemistry: the ability to design an enzyme for a specific purpose, to catalyze a particular reaction where nature provides no such option.

9.1 MODELS FOR ENZYMATIC ACTIVITY

In 1948, Pauling proposed a model for understanding enzymc activity.[1] Pauling stated

> I think that enzymes are molecules that are complementary in structure to the activated complexes of the reactions they catalyze. The attraction of the enzyme molecule for the activated complex would thus lead to a decrease in its energy and hence to a decrease in the energy of activation of the reaction, and an increase in the rate of the reaction.

Computational Organic Chemistry, Second Edition. Steven M. Bachrach
© 2014 John Wiley & Sons, Inc. Published 2014 by John Wiley & Sons, Inc.

This argument goes by the name *transition state stabilization* and remains the dominant model in the enzymatic community today! Amyes and Richard[2] recently reviewed a selection of enzymes whose behaviors are well understood within this Pauling paradigm.

Careful consideration of definitions is important in understanding the scope of *transition state stabilization* and alternatives. Warshel[3] accurately defined *transition state stabilization* to mean that the transition state (TS) in the enzyme will have a lower energy than that in water, with the reference defined as the energy of free enzyme and free substrate. This situation is often described in the literature by the energy diagram given in Figure 9.1a.[4]

The energy diagram in Figure 9.1a is misleading; it neglects to account for the formation of the Michaelis complex (MC), the binding of the substrate into the enzyme. The MC will typically lie lower in free energy than the separated enzyme and substrate (Figure 9.1b). The proper energy comparison is the free energy of the uncatalyzed reaction $\Delta G^{\ddagger}_{uncat}$ versus the free energy of the catalyzed reaction $\Delta G^{\ddagger}_{cat}$. This latter quantity is the difference in free energy between the MC and the TS. For all enzymes, $\Delta G^{\ddagger}_{cat}$ is less than $\Delta G^{\ddagger}_{uncat}$, oftentimes very much so.

Pauling implied that the complementary interaction between enzyme and substrate would be noncovalent. Zhang and Houk[5] have argued that the very large rate accelerations often observed for enzyme catalysis cannot be due solely to noncovalent stabilization of the TS. They argue that the rate enhancements (($k_{cat}/K_M)/k_{uncat}$) greater than 10^{11} exceed the energetic stabilization afforded by noncovalent interactions. Rather, they have argued that enzyme and substrate are covalently bound and that this bonding alters the reaction mechanism from that occurring in the absence of enzyme. When this takes place, direct comparison of the solution and enzyme reactions are complicated by their mechanistic differences.

Wolfenden[6] noted that "any catalyst can enhance the rate of a reaction only to the extent that it binds the substrate more tightly in the TS than in the ground state." This might be referred to as *differential stabilization*. The importance of this concept was recently reiterated in the study of the oxyanion hole (OAH) by Simón and Goodman.[7] An OAH is a region of an enzyme that possesses two or three hydrogen

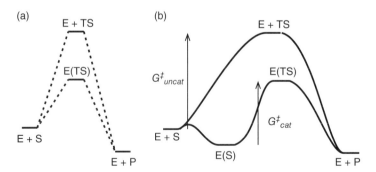

Figure 9.1 Schematic energy diagram for enzyme activation showing (a) transition state stabilization and (b) the role of the Michaelis complex (E(S)).

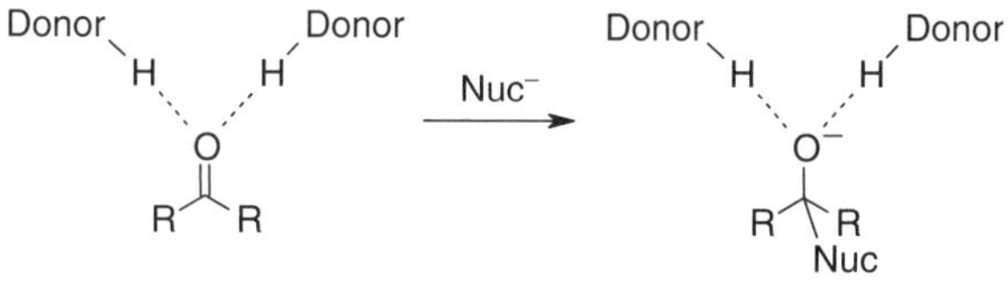

MODELS FOR ENZYMATIC ACTIVITY 571

Figure 9.2 Schematic of an oxyanion hole.

bond donors. These are typically employed for activating carbonyls toward acylation (Figure 9.2). The carbonyl oxygen is hydrogen bonded to two sites, and these sites are often backbone NH groups. As the nucleophile attacks, negative charge is built up on the oxygen atom, which should then make stronger hydrogen bonds, resulting in greater stabilization of the TS (and product) than that of the reactant.

Simón and Goodman first compared crystal structures of 249 enzymes that contain two or three hydrogen bond donors near the carbonyl of a bound substrate (obtained from the Protein Data Bank) with the crystal structures of 4618 small molecules containing two hydrogen bond donors about a carbonyl (obtained from the Cambridge Structural Database). The distributions of these two sets for a number of geometric parameters are quite similar, including the distance between the two donors r_{X-X}, the r_{X-O} distance, and the X–O–X angle (see Figure 9.3).

The only significant difference in the distribution is for the X–O=C–R dihedral angle Θ. For the small molecule set, the dihedral angle is clustered around 0°. This is in accord with the notion of the oxygen lone pairs lying in the carbonyl plane and so the H-bonds would be coplanar with the carbonyl plane. In decided contrast, the distribution of the dihedral angle for the enzyme OAH is centered at 90°. Since quite often one or both of the H-donors are the amide proton of the backbone, it is unlikely that the enzyme conformation is capable of changing during the reaction so as to twist the dihedral angle to the seemingly more logical and tighter binding 0° orientation in either the MC or the TS.

To understand this substantive difference in orientation, they looked at a model enzyme reaction, Reaction 9.1, which examines the nucleophilic attack step of esterase, peptidase, thioesterase, and (theoretical) hemiacetalase. This model invokes analogs of the triad of serine, histidine, and aspartic acid. A number of geometric constraints were employed to mimic an enzyme structure. The reactants and TSs were optimized at B3LYP/6-31G(d,p), and the single-point energies were obtained at MPWB1K/6-311++G**. For the optimized ground and TSs for all these reactions, the preferred arrangement of the water molecules comprising the

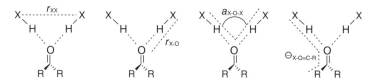

Figure 9.3 Definition of geometric parameters of the oxyanion hole.

OAH is for Θ to be $0°$, consistent with the notion of maximal hydrogen bonding to the in-plane oxygen lone pairs. This implies that the enzyme OAH structure, with its dihedral typically near $90°$, is *not optimal for stabilizing the TS*!

X = O, NH, S, CH$_2$

(Reaction 9.1)

This does not mean that the TS is not stabilized by the oxyanion when the hydrogen donors are positioned with $\Theta = 90°$; the oxyanion provides stabilization of the TS regardless of the value of Θ, but stabilization is greatest when the dihedral angle is $0°$. The same is true for the reactant state: the oxyanion stabilizes the reactant regardless of the value of Θ, but maximal stabilization is when $\Theta = 0°$. The key element here is finding the dihedral angle that minimizes the energy difference *between* the reactant and the TS and that occurs when $\Theta = 90°$. The oxyanion in enzymes is constructed for *that* purpose, not to maximally stabilize the TS but rather to *minimize the energy difference between the reactant and the TS*.

This is not an argument suggesting that enzymes catalyze a reaction by destabilizing the reactant. Rather, Simón and Goodman pointed out that one needs to look at the barrier, not just what is happening to the TS alone or the reactant alone.

Warshel[8] raised some concerns about the computations performed by Simón and Goodman, principally that model systems stripped of their protein and solution environments can be error prone. Furthermore, the reliance on crystal structures does not allow for the relaxation and conformational changes that an enzyme might make in solution, especially as it transits from reactant to TS to product.

To test for protein structure relaxation outside of the crystal, Simón and Goodman[9] performed a molecular dynamics (MD) study of the serine protease subtilisin. This enzyme uses a backbone NH and the amide NH$_2$ of asparagine in its OAH, and this should be more flexible than the typical OAH that employs both donors from the backbone. An MD run, using the OPLSAA force field, was initiated from the crystal structure, and after 10 ns, the orientation of the oxyanion donors did not change, with Θ remaining approximately equal to $90°$. Further runs were carried out with constraints that changed the force constants to twist the donor groups toward having $\Theta = 0°$, but significant rotation away from $\Theta = 90°$ occurred only when very large force constants were used, and the hydrogen bonds were broken as well. Therefore, they conclude that relaxation of the structure from the crystal is small and will not alter the structure of the OAH.

To address the concern of arbitrary constraints used on a small model of an enzyme reactive site, Simón and Goodman[9] model a large fragment of the active site of the 4-chlorobenzoyl-CoA dehalogenase enzyme using ONIOM, with the hydrogen bond donors and substrate and the nucleophilic residue treated with B3LYP/6-31G**, while the remaining residues are treated with the AMBER force field. Because of the structure of the protein, only one constraint had to be made, the distance between the two water molecules used to model the OAH. When the waters are free to move to their lowest energy position, they arrange such that Θ are near $0°$ for both the reactant and the TS. However, in the actual enzyme, the oxyanion donors are in position such that Θ is around $110°$. This arrangement has a higher energy for both the reactant and the TS, but the reactant is destabilized more than the TS. With this more precise model, the same conclusion is arrived at as before; namely, the enzyme is structured not to maximally stabilize the TS but rather to minimize the activation barrier.

Despite the need to interpret the Pauling paradigm as leading to a lowering of the activation barrier, rather than simply stabilizing the TS, it remains the dominant way to understand enzyme catalysis. Warshel[3,10] demonstrated the importance of electrostatics in stabilizing the TS within the enzyme active site, with particular emphasis on the concept of *preorganization*. Preorganization is the argument that in solution, the water molecules need to move from their orientations about the reactant to achieve proper orientations to interact with the TS; however, the enzyme is constructed in a way such that stabilizing groups are prepositioned to accommodate the TS—the enzyme is preorganized to stabilize the TS. Other proposals have been put forth to explain enzyme activity, including the role of tunneling,[11–13] dynamic effects,[14–16] and low energy barrier hydrogen bonds[17] to name a few. The remainder of this chapter presents first a general description of the approaches to computational enzymology that adopt methodologies discussed in the previous chapters of this book. This is followed by examples of the application of *ab initio* (both wavefunction and density functional theory (DFT)) methods toward explicating the mechanism of catalysis of a few enzymes. The Chapter concludes with the presentation of the use of *ab initio* methods in designing enzymes from scratch.

9.2 STRATEGY FOR COMPUTATIONAL ENZYMOLOGY

The basic approach for computations involving enzyme activity is analogous to what one does for computations involving small organic molecules. One needs to identify a suitable quantum mechanical method, obtain an initial geometry, and then optimize the structure. The situation is a bit more challenging when dealing with an enzymatic system due primarily to its size; choices are made largely to insure that the computations are tractable, while maintaining as much rigor as possible.

Prediction of the three-dimensional structure of proteins remains a major goal in computational biology and chemistry.[18,19] Typical proteins are simply way too large to be treated with any *ab initio* or DFT method. Even with a force field, the conformational space is often too large to be fully sampled. Furthermore, the inherent errors in an empirically derived force field make energy ordering of conformers suspect as best. Nonetheless, the attempt to predict protein structure in a first-principles manner, often called *free modeling*, remains an active pursuit. The approach is driven by the notion that the native form of a protein corresponds to its lowest energy conformation.[20] Most current methods that take this first-principle approach do in fact incorporate aspects of the alternative approaches, described next.

These alternatives approaches to protein structure prediction make various uses of known protein structures. *Homology* or *comparative modeling* builds a protein structure by comparison to evolutionarily-related proteins. This tactic can be quite successful when the homology is large, although interestingly even with a match of only 30 percent good structures can often be built. A related approach, called *threading*, attempts to identify sequence alignments, which can lead to similar folding, even when there is no evolutionary relationship between the target protein and the protein with a known structure. An annual contest called the *critical assessment of methods of protein structure prediction* (CASP) pits different methods against each other in trying to predict protein structures. Perusal of the results of recent contests provides an excellent view of state-of-the-art methods.[21]

Since optimization of the structure of the protein from scratch is so difficult, most computational enzymology studies begin[22] with a known protein structure, typically from an X-ray crystallography experiment. In an ideal situation, one might hope to obtain the structure of the protein bound with its substrate, even more ideally, obtain the structure of the TS within the enzyme active site. However, since enzymes are efficient catalysts, it is impossible to isolate either of these states. Rather, the next best thing is to obtain the X-ray structure of the enzyme bound with an inhibitor. This structure can then be used as the starting point for a computational examination.

Before one can start a quantum mechanics (QM) computation the X-ray structure will typically need to be preprocessed. Most X-ray structures do not locate hydrogen atoms, and so these will need to be added. The protonation state of amino acids such as aspartic acid, glutamic acid, and histidine will need to be set. Often one will also place the enzyme and substrate within a water environment, and so these water molecules need to be situated. All three of these items are typically addressed through an energy minimization. Since minimization will give just a single conformation (and a single configuration of the water arrangements), an MD trajectory is often run to provide a selection of semirandom initial geometries to begin the full QM/molecular mechanics (MM) study.

The main decisions depend on the specific choices for the QM/MM methodology. One must first decide how to partition the system into the QM and MM regions. The substrate will be included in the QM region, as should any amino acid residues that are covalently linked to the substrate at any point during the reaction sequence. Even some residues that simply provide some stabilization, through hydrogen bonding or some weaker intramolecular attraction, might be included in the QM region. The remainder of the enzyme is treated using MM, and any surrounding water molecules should be treated with an appropriate water model.

With modern computational techniques and computers, the QM region is typically evaluated with some DFT method. As with any decision to study an organic reaction, care must be taken while choosing a functional appropriate for the reaction at hand. Similarly, one needs to carefully select a basis set that provides sufficient flexibility while keeping an eye on the basis set size in order to keep the computations from becoming too long.

There are a variety of force fields that have been designed for protein modeling. The most popular are the *Amber*[23] and *Charmm*[24,25] variants. Work is ever ongoing in the development of more accurate force fields. A discussion of force fields is outside the scope of this work, and the interested reader is directed toward other sources for further information.[26–30]

Many computations also include water, often as a solvent but sometimes also as explicit molecules involved in the active site. Molecules of the later type are often included within the QM portion of the computation, especially if they are involved in proton shuttling. Solvent waters are often handled using a force field designed explicitly for this purpose. A popular solution is to model water as a rigid structure that interacts with other water molecules or with an enzyme through an electrostatic term and Lenard–Jones (L–J) terms. The TIP3P model[31] places fractional charges on each atom, with the L–J terms operative at oxygen, The TIP4P[31] model places the fractional charges on the hydrogen atoms and at a point along the bisector of the H–O–H angle. These types of water models allow for rapid computation of forces on every molecule, an important consideration within an MD or a Monte Carlo computation.

Computation of the reaction potential energy surface requires identification of reactant, TS, and product. Techniques that were discussed in Chapter 1 (and throughout the book) can be applied to enzyme/substrate systems too. Given the large number of atoms in proteins and, therefore, the large number of coordinates and gradients that must be minimized, other approaches are often more suitable.

One approach is *adiabatic mapping* whereby one or two coordinates are identified as dominant within the true reaction coordinate.[32] This might be a single bond that is made or broken in the reaction, or perhaps two such bonds. One then systematically increases or decreases the value of this coordinate(s), while optimizing the energy of the remainder of the system keeping this coordinate frozen.

A second approach addresses the notion that there are multiple reaction paths in enzymatic systems. Given the large conformational space of the enzyme–substrate complex, there will be a collection of "reactants," a number of conformational and configurational isomers that have very similar energy. Reactions can then originate from any and all of these reactant states, leading to multiple TSs (often simply conformers) and on to multiple products (again often just conformers). Consideration of enzyme reactions is, therefore, fundamentally different from that for simple organic reactions, where a single TS is often all that one needs to consider.[33] A common method for computing these multiple reaction paths is called *umbrella sampling*.[34] With multiple reaction paths in hand, one can obtain an averaged activation barrier.

9.2.1 High Level QM/MM Computations of Enzymes

When contemplating performing a computation on an enzyme–substrate system, it is natural perhaps to shy away for fear of the huge size of the molecules and the impossibly long computational times. For this reason, most QM/MM computations of enzymes have employed a semiempirical method or more recently a density functional method for the QM region simply to keep the computational cost under control. Given the limitations of these methods, many of which have been described in previous chapters, there is rightful concern over the accuracy of computations performed with a "practical" implementation of the QM/MM procedure.

In order to gauge the quality of a computational method, benchmarks are frequently performed whereby new methods are tested against experiments (where available) and, more often, against some very high level method, the so-called "gold standard." For small molecules, the gold standard is a CCSD(T) computation with some large basis set, often including an extrapolation to the complete basis set limit. To demonstrate the ability of QM/MM at this upper limit, we discuss QM/MM computations of two different enzyme–substrate systems.

The *para*-hydroxybenzoate hydrolase (PBHB) enzyme catalyzes the hydroxylation of *para*-hydroxybenzoate using flavin hydroperoxide (FAD-HOOH), Reaction 9.2. Werner and Thiel[35] examined this reaction in the following manner. The enzyme, cofactor, and substrate were subject to an MD study using AM1 for the QM region and the GROMOS force field for the MM region. They selected 10 arbitrary points along the trajectory as initial geometries for obtaining the reaction path using B3LYP/TZVP. Lastly, the energy of the reactant and the TS for each of the 10 paths were recomputed using a variety of high level *ab initio* methods. Given the large size of the QM region, this necessitated using the (aug)-cc-pVTZ basis set with a localized quantum method,[36–38] namely, LMP2, SCS-LMP2, LCCSD, and LCCSD(T0).

(Reaction 9.2)

The computed activation barriers for Reaction 9.2, averaged over the 10 different reaction paths, are listed in Table 9.1. Taking the LCCSD(T0) as the benchmark method, the barrier of 14.6 kcal mol^{-1} when corrected for zero-point vibrational energy and thermal effects gives an activation enthalpy of 13.3 kcal mol^{-1}, in very reasonable agreement with the experimental estimate. The SCS-LMP2 method, which is much more computationally affordable than LCCSD(T0), provides an excellent estimate of the barrier. The estimated barrier at B3LYP is too low by 5 kcal mol^{-1}, and so one might worry about its use in QM/MM studies of enzymes.

TABLE 9.1 Activation Barrier (in kcal mol^{-1}) for Reaction 9.2^a

Method	Barrier	rms
B3LYP	9.7	1.3
MP2	12.3	1.3
LMP2	12.0	1.2
SCS-LMP2	14.6	1.3
LCCSD	21.5	2.0
LCCSD(T0)	14.6	1.6
Experimental	12^b	
	14^c	

^{a} Computed with the (aug)-cc-pVTZ basis set and the GROMOS force field.
^{b} Activation enthalpy.[39]
^{c} Activation free energy.[40]

A second high level study of an enzymatic system was reported by Mulholland.[41] His group looked at citrase synthase and its action in catalyzing the deprotonation shown in Reaction 9.3. They obtained five different reaction paths using umbrella sampling to obtain different starting structures and the TSs were found by adiabatic mapping. These five different reactants and TSs were then reoptimized at B3LYP/6-31+G(d)/CHARMM27. Lastly, single-point energies were obtained using a variety of quantum methods. The computed activation barriers are listed in Table 9.2.

(Reaction 9.3)

TABLE 9.2 Computed Activation Energy (in kcal mol^{-1}) for Reaction 9.3 within the Citrate Synthase Enzyme ^{a,b}

Method	$\Delta E^{\ddagger}$
B3LYP/6-31+G(d)	12.7
MP2/aug-cc-pVDZ	10.5
SCS-MP2/aug-cc-pVDZ	13.0
SCS-LMP2/aug-cc-pVDZ	13.5
LCCSD(T0)/aug-cc-pVDZ	13.2

^{a} Ref. 41.
^{b} Using geometries obtained at B3LYP/6-31+G(d)/CHARMM27.

Inspection of Table 9.2 shows that all of the localized methods give similar barrier heights. Since SCS-LMP2 is the most computationally efficient of these methods, Mulholland suggests that this is the method of choice for obtaining accurate energies while not incurring an extreme computational cost. It should be noted that for this enzyme, B3LYP with the smaller basis set does a quite admirable job in predicting the barrier.

9.2.2 Chorismate Mutase

Pericyclic reactions played a central role in the development of the theory of organic reaction mechanism (see Chapter 4) and are among the most widely utilized reactions in organic synthesis. Perusal of any organic textbook, from entry level through graduate level, provides ample evidence for the centrality of this class of reaction. It may, therefore, be somewhat surprising that there appears to be only a single example of an enzyme that catalyzes a pericyclic reaction; chorismate mutase (CM) catalyzes the Claisen reaction of chorismate **1** to prephenate **2**, (Reaction 9.4).

(Reaction 9.4)

The conversion of chorismate into prephenate occurs at a critical point in the shikimic acid pathway: the biosynthesis of a variety of aromatics branch off from here. Since CM appears in lower organisms (such as fungi and bacteria) and not in mammals, it is an excellent target for the development of antibacterial and antifungal agents.

In addition to its biochemical importance, CM has drawn attention from chemists looking to study enzyme activity for three primary reasons. First, the substrate **1** binds to the enzyme CM without forming any covalent linkages, so that major electronic reorganizations do not need to be considered. Second, unlike many enzyme-catalyzed reactions, the reaction mechanism for the conversion of **1** into **2** is the same in both the enzyme environment and in solution in the absence of the enzyme. Last, the kinetics of the enzyme activity are well known, with CM increasing the rate of the reaction by 10^6 over the rate in aqueous solution.[42] This corresponds to a reduction in the activation enthalpy of 20.7 ± 0.4 kcal mol^{-1} in solution to 15.9 kcal mol^{-1} in the enzymatic environment. The activation entropy is -12.9 ± 0.4 eu in solution but is reduced to essentially nil in the enzyme. In other words, $\Delta G^{\ddagger} = 24.5$ kcal mol^{-1} in solution but only 15.4 kcal mol^{-1} in CM.[43]

Wiest and Houk[44] examined the Claisen rearrangement that takes **1** into **2** in the gas phase at B3LYP/6-31G*. The geometries of the diequatorial and diaxial form of **1**, the TS **3**, and the product **2** are shown in Figure 9.4. The diequatorial form is more stable than the diaxial form, by 12 kcal mol^{-1} at B3LYP/6-31G*,

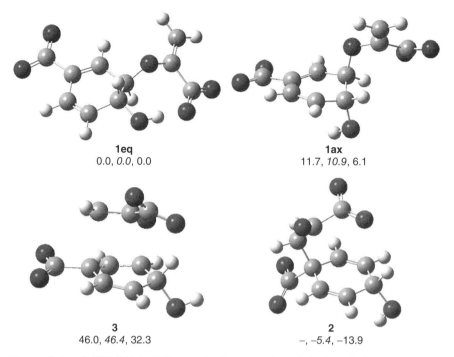

Figure 9.4 ωB97X-D/6-31G(d)-optimized geometries of **1–3**. Relative energies (in kcal mol⁻¹) ateB3LYP/6-31G*,⁴⁴ ωB97X-D/6-31G(d) (in italics), and CPCM(water)/ωB97X-D/6-31G(d).

but this higher energy conformation must be attained in order for the Claisen rearrangement to occur. The TS lies 35.3 kcal mol^{-1} above diaxial chorismate. The relative energies are little affected by using a more modern functional (ωB97X-D). However, when modeling the solvent (water) using CPCM, the diaxial conformer is much closer in energy to the diequatorial form (differing by only 6 kcal mol^{-1}) and the barrier is reduced to 26 kcal mol^{-1}, overestimating the experimental barrier by 6 kcal mol^{-1} (a not unreasonable result given the small basis set and the continuum approximation for water employed in the method). The overall reaction is exothermic, especially so in water (-14 kcal mol^{-1}).

A number of research groups have employed QM/MM methods toward understanding the activity of CM, engendering a small controversy. Bruice has argued that CM exhibits a classic example of the involvement of near attack conformers (NAC). Bruice's contention is that CM aids in the formation of the diaxial conformation, thus increasing the population of the NAC and thereby accelerating the reaction.

The NAC is the arrangement (or a group of arrangements) of reactants in an appropriate geometry to facilitate passing through the TS.[45] So, in the case of Reaction 9.4, the NAC is the collection of molecules in a diaxial conformation *that can then proceed through the TS* and on the prephenate. Some have interpreted this as simply the collection of stable conformations where the distance between the reactive atoms (in this case, the distance between the carbons coming together to form the new σ-bond) is below some threshold value. However, Bruice's concept of the NAC is more specific, requiring the atoms to be at angles similar to that in the TS. For Reaction 9.4, the NAC should be geometries where the atoms forming σ-bond are within a certain distance, and their p-orbitals oriented toward each other to form the new bond.[46] With the concept of NAC, the free energy of a activation can be partitioned as in Eq. (9.1), where ΔG_{NAC} is the free energy of formation of the NAC from the ground state and ΔG_{TS} is the energy difference between the TS and the NAC.

$$\Delta G^{\ddagger} = \Delta G_{NAC} + \Delta G_{TS} \qquad (9.1)$$

Hur and Bruice[47,48] performed MD simulations of Reaction 9.4 in water and within CM from different organisms. The MD was performed solely with a force field: CHARMM for all atoms other than water (which were modeled using TIP3P) and chorismate and the TS for which a force field was fit to B3LYP/6-31+G(d,p) energies and structures.[47] From the MD simulation in water, the mole fraction of NAC is 0.0007 percent, or ΔG_{NAC} = 8.4 kcal mol^{-1}. The mole fraction of NAC in the simulation including CM from *Escherichia coli* is 34 percent, a substantial increase over the population in water, with ΔG_{NAC} = 0.6 kcal mol^{-1}. Since $\Delta G^{\ddagger}$ = 24.5 in water, ΔG_{TS} is 16.1 kcal mol^{-1} in water. However, in the enzyme, $\Delta G^{\ddagger}$ = 15.4 kcal mol^{-1}, so that ΔG_{TS} is 14.8 kcal mol^{-1}. Bruice argues that since the activation free energies are so close in water and the enzyme, it is the formation of the NAC that accounts for the catalytic behavior of CM. A similar situation is also found using CM from *Bacillus subtilis*.[48] Essentially, the same results[46] were obtained by a one-dimensional thermodynamic integration technique.[49]

Bruice's contention that there is very little stabilization of the TS by CM, rather it is a favorable binding of the NAC that accounts for its catalytic activity, has been met with much skepticism. An early study by Wiest and Houk[44] looked at chorismate models with coordinated waters and aminidium cations in positions that matched interacting groups found in the crystal structure of CM with a coordinated substrate. These model computations suggested that the neighboring amino acid residues could be stabilizing the TS.

Far more complete computations have been carried out in the meantime that utilize the entire protein along with the full chorismate molecule. This requires use of a QM/MM strategy, where typically chorismate and the TS for the rearrangement are treated with some quantum mechanical method and the protein is treated with molecular mechanics. Jorgensen[50] examined this system using the semiempirical method AM1 to treat chorismate and its TS and used the OPLS-AA force field[51] to treat the protein—in this case, the CM from *Bacillus subtilis*—while treating all water molecules with the TIP4P[31] model. They defined the NAC as the set of all conformations having a distance of less than 3.7 Å for the forming σ-bond. This is a far more generous definition than the one imposed by Bruice. With this definition, Jorgensen finds that the population of NACs in water is nearly 82 percent, or ΔG_{NAC} = −0.9 kcal mol^{-1}. The population of the NACs inside the enzyme is nearly total, with $\Delta G_{NAC} = -9$ kcal mol^{-1}. Despite the enhanced population of NACs in the enzyme, this has no effect on the relative activation barrier since ΔG_{NAC} for both the aqueous and the protein environments are negative—the overall barrier is defined from ground-state free substrate. This means that the reaction inside the enzyme is faster than that in aqueous solution because of stabilization of the TS by CM.

Analysis of the structures of the enzyme bound to a NAC compared with the enzyme bound to the TS reveals a number of important interactions between amino acids side chains and the carboxylate groups of the substrate. This leads to an additional hydrogen bond in the TS than with the NAC. Furthermore, the TS is compressed. Both of these changes, Jorgensen claims, account for the enhanced stabilization of the TS over the NAC and the reduction in the activation barrier.

Bruice[46] countered that the very loose definition of the NAC meant that a far too large number of conformations of chorismate in water satisfied the criteria. With his more restrictive definition and using the AM1 method, a similar population was found as in his original B3LYP study.

Nonetheless, this raises the question of what is an appropriate definition of the NAC. Mulholland[52] suggests a definition that is much less arbitrary: the NAC is the conformation of the substrate bound to the enzyme, and what is critical then is the free energy needed to form this conformation in aqueous solution. Mulholland estimates this energy as 4–5 kcal mol^{-1} through a free energy perturbation calculation and MD using AM1 and CHARMM. This is half the estimate of Bruice and suggests that formation of the NAC is only partly responsible for the catalytic effect afforded by CM.

The Mulholland group employed MD to examine Reaction 9.4 within CM.[53] Chorismate was treated at B3LYP/6-31G(d), while the protein was modeled with CHARMM. They computed 16 reaction pathways. The average computed activation enthalpy was 12.7 kcal mol^{-1}, in close agreement with the experimental value. They then compared the energy of points along the reaction pathway in the enzyme with the energy in the gas phase and found that all structures are stabilized in the enzyme. Most importantly, they find that the TS is more stabilized than the reactant, by on an average 4.2 kcal mol^{-1}. It should be noted that a better comparison would be between the reaction path in the enzyme with the path in water.

Later, the Mulholland group extended this study to include the reaction in water.[54] Again using a QM/MM approach (B3LYP/6-31G(d) for the substrate and TS and CHARMM for the protein and TIP3P for water), they computed 16 reaction pathways (via MD) in the enzyme and 24 in water. They estimated that the activation enthalpy is 11.3 kcal mol^{-1} in the enzyme (in good accord with the experimental value of 12.7 kcal mol^{-1}), while the barrier in solution is about 20 kcal mol^{-1} when considering the diequatorial conformer **1eq** as the true starting point. They noted that all of the starting geometries of the reactions in both aqueous and enzyme environments satisfy any of the proposed definition of NACs, although their starting point in water is **1ax**. Since the energy difference between **1eq** and **1ax** is just a few kcal mol^{-1}, and the computed difference in activation energy for the enzyme and aqueous environments are very different, Mulholland concludes that the majority of the catalytic effect is due to TS stabilization. This is supported by the fact that all points along the reaction are stabilized by the enzyme relative to the energies in solution, and the degree of stabilization varies along the pathway. While the TS in the enzyme is compressed relative to the TS in water, Mulholland ascribes a small amount of stabilization energy to this compression.

Rather, Mulholland argues that the majority of the catalytic effect results from TS stabilization, with specific residues responsible for this stabilization. This is nicely corroborated by the QM/MM study of Ishida,[55] along with a variety of point mutations experiments. Ishida performed MD computations whereby the substrate was computed at MP4/6-31+G**//MP2/ 6-31+G**, the protein was described with the AMBER force field, and water was described with TIP3P. With this approach, the activation free energy barrier is 16.5 kcal mol^{-1}, about 5 kcal mol^{-1} greater than the barrier in water, which must be corrected for using **1ax** as the starting point. He then computed the barrier for a number of point mutations. For example, the Arg90Cit mutant has been prepared and has a barrier that is about 6 kcal mol^{-1} greater than that for the wild type.[56] The computed barrier is estimated to be 3.4 kcal mol^{-1} higher than the wild type. This mutation removes the positive-charged group that can interact with one of the carboxylate groups of **1**. Mutation of Glu78, a residue that hydrogen bonds to Asp90 and to the hydroxyl group of **1**, also shows diminished activity,[57] and the computation predict a higher barrier for these mutants. These mutants show a loss of the hydrogen bond to Asp90.

An overall consensus can be drawn, whereby CM catalyzes Reaction 9.4 primarily through stabilization of the TS through specific interactions of a small number of amino acid residues. While compression plays a very minor role in the catalysis, some 40 percent or so of the activation energy reduction by the enzyme can be attributed to coordination of the diaxial conformer **1ax**. So while there are some significant problems in uniquely defining NAC, and their role is not as dominant as first claimed by Bruice, the concept has some utility in understanding the catalytic role of enzymes.

9.2.3 Catechol-*O*-Methyltransferase (COMT)

The enzyme catechol-*O*-methyltransferase (COMT) is involved in pain regulation and regulation of neurotransmitters.[58] COMT acts to catalyze the transfer of a methyl group from the cofactor *S*-adenosyl-L-methionine (SAM, **5**) to the O$^-$ of catecholate **4** (Reaction 9.5). The catechol structure is found in neurotransmitters such as dopamine, noradrenaline, and adrenaline. A related set of enzymes are the DNA methyltransferases that appear to have similar structure, especially in their active sites, to the structure of COMT.[59]

(Reaction 9.5)

A simple model of this reaction was explored by Zheng and Bruice[60] using gas-phase and solution-phase computations. The reaction of **4** with Me_3S^+ was examined at Hartree–Fock (HF)/6-31+G(d) and B3LYP/6-31+G(d) in the gas phase. A single TS state was located, corresponding to an S_N2 mechanism. Since the reactants are of opposite charge, there is reaction complex that is much lower in energy than isolated reactants, and the TS also lies well below the energy of reactants (-76 kcal mol^{-1}). The shape of the potential energy surface (PES) alters when solvent (water) is included using the polarized continuum method (PCM). The complex remains, but now the TS lies above isolated reactants; the TS lies 17 and 9 kcal mol^{-1} above the reactants at HF and B3LYP, respectively. The significantly higher barrier in water than in the gas phase provides a hint as to how the enzyme may be catalyzing this reaction.

Zheng and Bruice estimated the kinetic isotope effect using the expression

$$\frac{k_L}{k_H} = e^{((\Delta G_L^\ddagger - \Delta G_H^\ddagger)/RT)} \tag{9.2}$$

They obtain values of $k_H/k_D = 0.80$ and $k_{12C}/k_{13C} = 1.06$. These values are in very good agreement with the experimental kinetic isotope effects obtained for the COMT-catalyzed transmethylation of 3,4-dihydroxyacetophenone and **5**: 0.83 ± 0.05 and 1.09 ± 0.05, respectively.[61] Experiments with labeled SAM demonstrated that COMT facilitates the transmethylation with stereoinversion,[62] adding further weight to operation of the S_N2 mechanism within the catalyst environment.

COMT is, for many of the same reasons as with chorismate mutase, well suited for the study with computational techniques. The reaction mechanism it catalyzes is the same mechanism that operates in the absence of the enzyme, specifically, the S_N2 mechanism, facilitating comparison of the bare solution-phase reaction with the catalyzed reaction. The substrate and cofactor do not covalently bind to the enzyme, so that defining the QM region and the MM region should be relatively uncomplicated. Lastly, the X-ray crystal structure of COMT bound with the inhibitor 3,5-dinitrocatechol has been determined with a resolution of 2 Å.[63] An interesting twist to this enzyme is that the active site includes a metal cation, Mg^{2+}. This crystal structure allows for a natural starting point for computational exploration of the means of the catalytic action of COMT. The rate acceleration provided by COMT is substantial: the reaction is 10^{16} times faster within the enzyme than in solution.[61,64]

The first computational examination of COMT was the MD study by Lau and Bruice.[65] They employed the CHARMM force field to the entire system embedded within a bath of TIP3P waters and examined the coordinates every 200 fs for a duration of 1 ns. They looked for the formation of NACs, defined here as separation of less than 3.2 Å between the nucleophilic oxyanion and methyl carbon of **5**, along with an O–C–S angle greater than 165°. Formation of the NACs occurred about 7.6 percent of the time, and Tyr68 aids in forcing the cofactor into a proper position for attack. Lau and Bruice conclude that desolvation and the positioning of the substrate and cofactor into proximity are the major roles played by the enzyme in catalyzing the reaction. This study, however, does not address the energetics of the catalyzed substitution reaction.

The first QM/MM study of COMT was performed by Kuhn and Kollman.[66] Their approach was to compute the QM region with constraints imposed by the MM region and then add on the free energy of the MM region with constraints imposed by the QM region. The QM region is modeled as the catecholate anion with Me_3S^+ using B3LYP/6-31+G* and MP2/aug-cpVDZ//HF/6-31+G(d). Adjacent atoms are used as "anchor points" that restrict the motion of the QM atoms during optimizations. For the MM energy computations, the AMBER force field was used with partial charges on the atoms in the QM region obtained from the wavefunction computed for the QM region. For the solution computations, the reaction is computed with AMBER using a periodic box of water.

To get the structure of the enzyme about the QM region, they ran an MD trajectory of the full system starting with the X-ray structure with catecholate substituting in for the inhibitor, with both oxygen atoms of catecholate bound to Mg^{2+}. However, during the trajectory run, the hydroxyl group dissociated off of the metal, replace by a water molecule. This monodentate catecholate structure was then used throughout the remaining computations.

The energy barrier for the methyl transfer computed within the QM region is about 10 kcal mol^{-1} at MP2 but only about 4 kcal mol^{-1} using B3LYP. The barrier is lower than that computed by Bruice primarily because of the constraints imposed by the anchor atoms. A more significant difference is that Kollman found a looser TS than Bruice. Since the computed kinetic isotope effects with the tighter TS are in fine agreement with experiment, Kollman reduced the computed barrier by a couple of kcal mol^{-1}. Therefore, when the QM and MM energies are summed together, the barrier is 24.5 kcal mol^{-1}, which he reduced to 21.7 kcal mol^{-1} to correct for the loose TS. This is in reasonable agreement with the experimental estimate of the barrier height of 18 kcal mol^{-1}.[67] The computed barrier in water is 29.5 kcal mol^{-1}, for a 5 kcal mol^{-1} reduction in the barrier afforded by the enzyme.

This is not a large enough barrier reduction to explain the experimental result. Kollman argues that the computation in water does not properly account for the *cratic* free energy loss, the free energy lost in bringing the reagents together. The estimate of this energy loss is 9–12 kcal mol^{-1}, which added onto the difference in barrier heights favors the enzyme by 14–16 kcal mol^{-1}, which is consistent with experiments. Kollman[68] concludes that the enzyme catalyzes the reaction in two ways. First, there is a rate enhancement brought on by bringing the reagents together, similar to Bruice's NAC argument, and there is a stabilization of the TS within the enzyme by about 5 kcal mol^{-1}. Kollman points to the Met40 residue, positioned near the sulfur cation, as affording an electrostatic stabilization.

The "cratic" correction has been subject of some concern. Part of this problem is a matter of definition. Kollman[33] uses "cratic" to mean "preorganization" and this amounts to computing the free energy cost for placing reactants into their positions in the MC in an aqueous solution, again similar to the NAC concept. Warshel[69] instead proposes to model the MC in water as simply a loose association of reactants. Further, the "cratic" energy, more widely used in the sense of the change in translational free energy when reactants come close together, has been placed on firm theoretical grounds by McCammon.[70] Kollman[33] argues that the enzyme compensates for the free energy loss of forming the MC through binding interactions that are not available in solvent.

Roca et al.[71] performed a set of QM/MM computations of COMT with AM1 as the quantum method (for catecholate **4** and **5**) and CHARMM for the MM part (the enzyme, Mg^{2+}, and 4614 water molecules). Optimization of the TS within the enzyme identified a structure where the Mg^{2+} is coordinated to the hydroxyl group of catecholate, as might be expected. However, tracing the intrinsic reaction coordinate (IRC) in the reverse direction from this TS led to a reactant structure maintaining this specific coordination. This is different from the reactant located by Kollman, which had the cation coordinated to the oxyanion. Both structures were stable within short MD runs, and so both are reasonable reactants, although Roca's has a more nucleophilic catecholate. This provides an estimate of the activation energy of 10.4 kcal mol^{-1}, but when corrected by using the difference in the MP2 and AM1 barriers for the gas-phase reaction, the barrier was estimated to be 20.7 kcal mol^{-1}, in better agreement with experiment. The AM1/TIP3P estimate for the solution-phase reaction is 25.4 kcal mol^{-1}, and even with the MP2 correction, the value of 26.9 kcal mol^{-1} underestimates the experimental estimate of 30.8 kcal mol^{-1}.

Roca and coworkers[72] next examined the potential coupling between the reactants and the COMT enzyme. Employing the same computational methods as in the previous paper, they carried out a 600-ps MD trajectory with the system constrained to remain near the TS. Using the geometry at every 50 ps, they optimized true TS structures and determined their transition vectors. An average of these vectors showed a near complete dominance of motion that one associates with the $S_N 2$ reaction and little motion in the rest of the QM or MM regions.

Next, the IRCs were traced back to reactants from each of the 12 TS structures. Most of the structural differences between the TSs and the reactants involve the positions of catecholate **4** and **5**. Changes to the rest of the environment amount to less than 1 percent of the total change, and of this, the most important is the movement of the Mg^{2+} that moves closer to the hydroxyl oxygen of the catecholate while forming the TS. In water, the hydrogen bonds to catecholate on average get longer in moving from reactant to TS. Part of this change is obvious as the charge on the oxyanion is diminishing as the methyl is transferred, but the hydrogen bonds to the hydroxyl oxygen get longer too. Thus, the nonintuitive movement of Mg^{2+} during the reaction results in stabilization of the TS. In following the projected electric field onto the S–O vector, in solution, the field acts to resist the methyl transfer, again an obvious result, since the reaction leads to a reduction in overall charge separation. In the enzyme, however, the electric field does not change much during the course of the reaction, and near the TS, the field helps to support the methyl transfer. Thus, the stabilization afforded the TS by the enzyme is electrostatic in nature.

The kinetic isotope effect for the transmethylation catalyzed by COMT was examined in a more rigorous manner by Williams and coworkers.[73] A collection of

10,000 reactant and TS structures were obtained from constrained MD trajectory QM/MM computations. The QM region (containing catecholate and **5**) was treated with AM1, the enzyme was treated with the OPLS force field and the waters were considered using TIP3P. Averaging over these reactant and TS energies gives a calculated α-d_3 kinetic isotope effect of 0.82 ± 0.5 in excellent agreement with the experimental value of 0.83 ± 0.5.[61] The ^{13}C isotope effect is slightly underestimated. Using a similar computational approach, the estimate of the α-d_3 kinetic isotope effect (KIE) is 0.99 ± 0.16, which is somewhat greater than the experimental value of 0.974 ± 0.016.[74] Although the larger inverse isotope effect in the enzyme than in solution had been suggested to arise from compression due to the enzyme, these computations show very little difference between the sums of the CO and CS distances between the enzyme and the aqueous phase TSs. Better understanding of the origin of the KIE remains to future study.

9.3 *DE NOVO* DESIGN OF ENZYMES

Computational biochemistry holds many grand challenges, and one of the more intriguing is the ability to design an enzyme to catalyze a specific reaction of interest. Nature exquisitely develops enzymes, with incredible selectivity and catalytic power. But one of nature's great advantages over human design is that she has been at the task for billions of years, compared to perhaps a decade or two of human efforts. In this section, we discuss developments in *de novo* enzyme design pioneered by the Baker and Houk laboratories. Together they have designed and built enzymes to catalyze four different organic reactions, and they have written an excellent review on their design strategy,[75] which they term the "inside-out" approach.

We begin by discussing the general strategy for computation design of an enzyme. We describe the process applied specifically to the design of an enzyme to catalyze a bimolecular Diels–Alder reaction[76] and follow that with a discussion of the other designed enzymes.

The first step in the design process is the development of a *theozyme*.[77] A theozyme is a model of a TS for a reaction in an enzyme whereby a small number of functional groups are arranged about the TS of the reaction in order to stabilize it. These functional groups are typically chosen as those available as side chains of the amino acids. As an example, serine hydrolases catalyze the hydrolysis of esters. The theozyme shown in Scheme 9.1 places an imidazole capable of accepting the proton from the alcohol nucleophile on one side of the TS. The imidazole is stabilized by an adjacent carboxylate group. On the other side of the TS are placed two water molecules to help stabilize the incipient negative charge. This entire TS, including the adjacent groups, was optimized at B3LYP/6-31+G* to obtain the theozyme structure.[78]

Baker and Houk[76] set out to design an enzyme to catalyze the intermolecular Diels–Alder reaction shown in Reaction 9.6. The substituents on both the diene and dienophile allow for the possibility of functional groups, provided by the amino acid side chains, to potentially stabilize the TS of the Diels–Alder reaction. Their

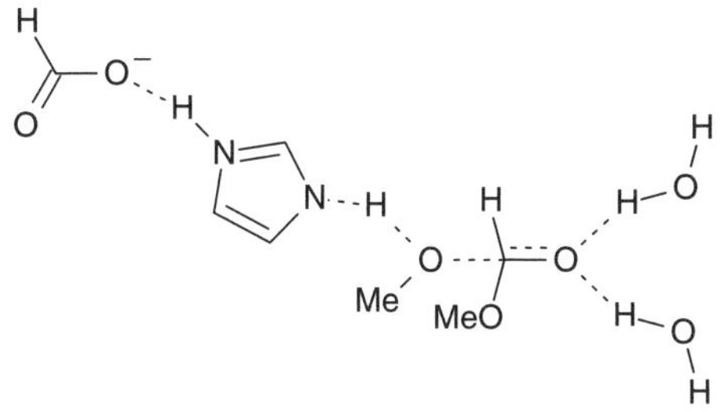

Scheme 9.1

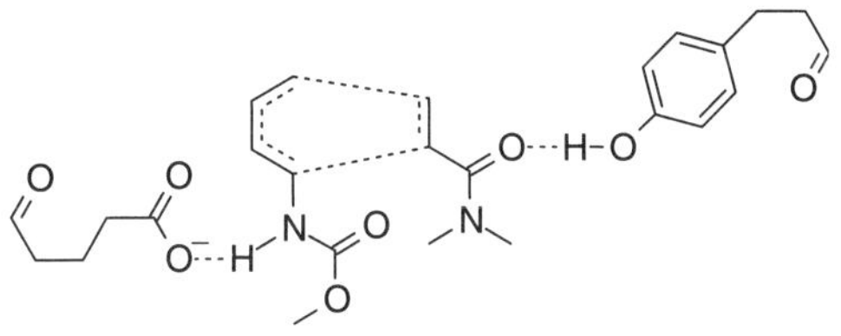

Scheme 9.2

theozyme places a hydrogen bond donor near the carbonyl of the dienophile along with a hydrogen bond acceptor near the NH bond of the diene. The actual theozyme they used is shown in Scheme 9.2, where the hydrogen bond donor is a tyrosine analog and the hydrogen bond acceptor is a glutamate analog. This theozyme was optimized at B3LYP/6-31+G(d,p) and this barrier is 2.7 kcal mol^{-1} smaller than the barrier for the unactivated parent Diels–Alder reaction.

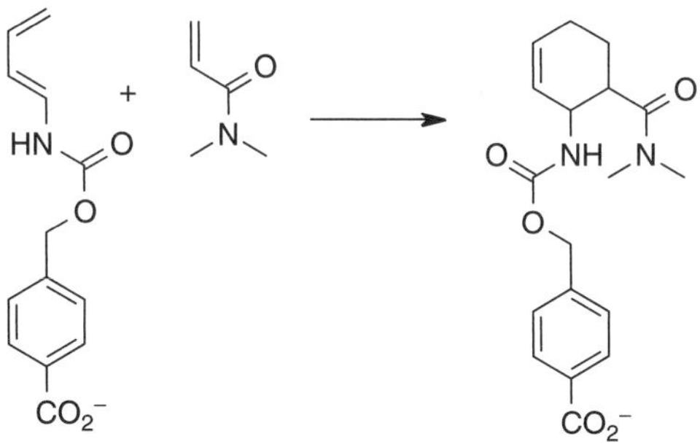

(Reaction 9.6)

Next, this theozyme geometry, with a tyrosine as the acceptor and either gulatamate or glutamine as the acceptor, becomes the template for a protein backbone. This backbone must accommodate the entire theozyme, with the two amino acids as part of the backbone. The RosettaMatch[79,80] algorithm searches through known protein structures and attempts to fit the theozyme amino acids into the backbone. The key is to identify when this can be done without incurring substantial steric interactions between any of the theozyme components and the protein chain. This is accomplished by first identifying a library of scaffold proteins and their binding pockets followed by mutations of neighboring residues to increase the binding affinity. Lastly, all feasible configurations are then fully optimized and ranked. In the case of the Diels–Alderase design, 207 known proteins were searched and all configurations that could accommodate the theozyme were then optimized using the RosettaDesign[81] algorithm. The optimization scheme here utilizes a scoring function, where the energies are obtained largely through an empirical approach. This scoring approach requires many approximations and is a primary source of error.

Ultimately, 84 designs met the energy criteria and were selected for actual preparation, carried out by encoding the appropriate gene sequences for expression in *E. coli*. Of the 84 design, only 50 proteins were soluble, and of these 50, only 2 of them showed any Diels–Alderase activity. Further mutations of each of the two designed proteins results in small improvements in performance, with the best performer having k_{cat}/k_{uncat} = 89 M, a modest catalytic effect.

A number of tests were performed to determine how well the designed enzymes actually compare to the design criteria. Of particular note is that mutation that removed one or both of the residues designed to hydrogen bond to the substrates resulted in complete loss of activity. A crystal structure of the protein (with one mutation) shows a root-mean-square deviation of the atomic positions of 0.5 Å compared to the predicted structure. To test specificity, the designed enzyme was exposed to the diene along with five slightly modified dienophiles, shown in Scheme 9.3. The rate of formation of Diels–Alder products with the alternative dienophiles was significantly less than that for the designed dimethylacrylamide. Lastly, the TS of the theozyme leads to the (3R,4S) *endo* product, which is only 47 percent of the product of the uncatalyzed reaction. However, with the designed enzyme, this stereoisomer accounts for greater than 97 percent of the product. All of these results strongly support the notion that the design strategy is actually implemented within the synthesized protein!

The turnover rate is high and suggests that these designed enzymes might have real application in chemical synthesis. The disappointing aspect of the study is the poor ratio of predicted enzymes (84) to ones that actually had activity (2). This low yield of active enzymes is also seen in the other designed proteins, which we discuss next, and remains a serious challenge to protein engineers.

For the design of a retroaldol enzyme, Baker and Houk[82] chose to build an enzyme to catalyze Reaction 9.7. Their design strategy presupposes a multistep reaction sequence whereby in the first step a lysine amine reacts to form an imine, followed by cleavage of the C–C bond, and then hydrolysis of the imine to give the

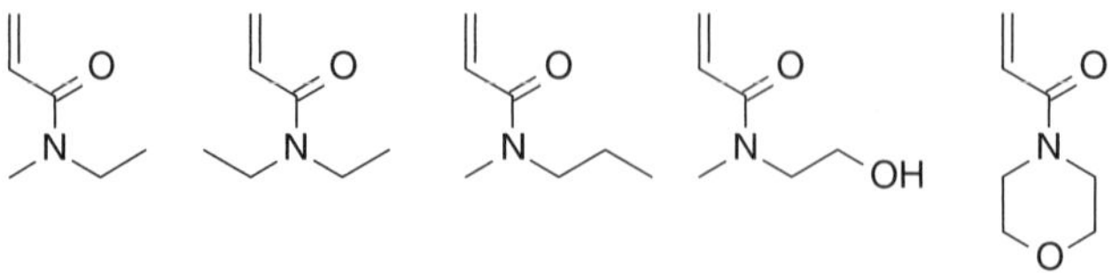

Scheme 9.3

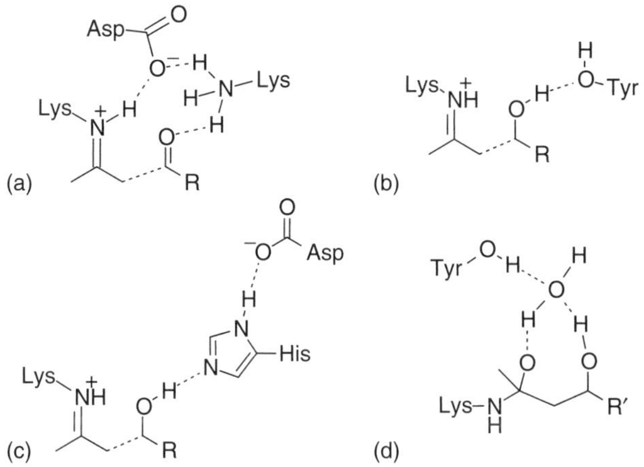

Scheme 9.4 Theozymes of the retroaldol reaction.

product carbonyl. These many steps lead to potentially many different theozymes, schematically represented in Scheme 9.4.

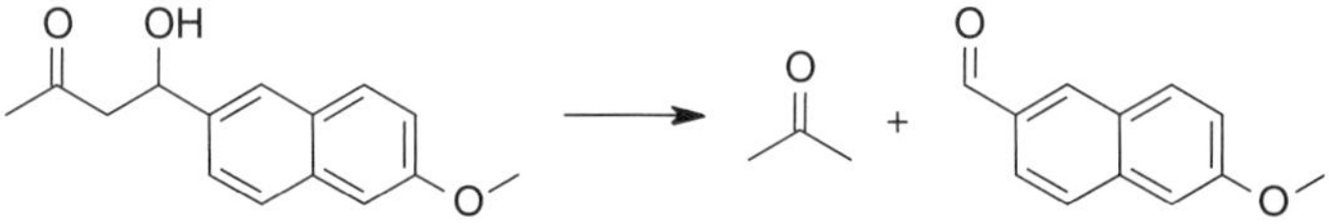

(Reaction 9.7)

The first theozyme (Scheme 9.4a) invokes a proton shuttle using lysine and aspartate. The second theozyme (Scheme 9.4b) has a tyrosine group to accept the proton from the alcohol, whereas the third theozyme (Scheme 9.4c) uses the His-Asp dyad to accept this proton. In the last theozyme (Scheme 9.4d), an explicit water molecule assists in the proton shuttling.

These four theozymes were then used as templates for fitting into 10 different protein scaffolds using the RosettaMatch algorithm. This search resulted in 72 potential enzymes, each with 8–20 residue changes from the parent scaffolds. These genes encoding these 72 enzymes were incorporated into E. coli and 70 of them led to successfully expressed protein.

There were 12 designed proteins based off of the first theozyme, but none showed any retroaldolase activity. The same was true for the 10 designed proteins based on the second theozyme. Better success was found with the third and fourth theozymes; of the 14 proteins designed for the third theozyme, 8 had activity, and of the 36 designs for the last theozyme, 23 had enzymatic activity, with rate enhancements up to 10,000 times that of the uncatalyzed reaction.

Confirmation of the design was provided in a number of ways. First, replacement of the catalytic lysine with a methionine in the active proteins led to diminishment if not outright abolishment of the activity. The X-ray structure of two of the designed proteins was obtained and they agree well with the theoretical structures. The major discrepancies occurred in backbone loops about the active sites, which points toward a path for potential improvement by allowing for greater flexibility in these regions.

In a follow-up study, Baker[83] utilized the theozyme of Scheme 9.4d but without the tyrosine residue. The modeling of the potential proteins to host this theozyme allowed for finer grained side-chain rotamer examination than in the previous study. This led to 42 designs, all of which were expressed in *E. coli* and all were soluble. Of these, 33 had activities of 10 times or more than the rate of the uncatalyzed retroaldol reaction. The higher success rate here was attributed to a design that better accommodated the TS and a polar environment to host the later steps in the reaction sequence.

Nonetheless, overall catalytic activity is poor compared to naturally occurring enzymes. Single-residue mutations led to a few new enzymes with catalytic efficiency improved by 7- to 88-fold. While this improvement points the way toward the creation of better enzymes, it is discouraging because single-point mutations are involved in the RosettaMatch process and should have been identified. Another stumbling block is that examination of the active versus the inactive protein structures did not lead to any insight regarding future design strategies.

A less successful attempt at enzyme design was reported by Baker and Houk[84] in attempting to design ester hydrolases. They developed three different theozymes, shown in Scheme 9.5, that differ in the groups making up the OAH. In the first theozyme, a backbone NH group is used to hydrogen bond to the incipient oxyanion, while in the other two theozymes, an explicit water molecule or a side-chain NH serve this purpose. Using RosettaMatch, 214 different proteins were searched to see if any of the three theozymes could be accommodated. Ultimately, 31 designs for theozyme 1, 11 designs for theozyme 2, and 12 designs for theozyme 3 were encoded and expressed. Of these 55 initial designs, 32 resulted in soluble proteins. Disappointingly, only four of the proteins designed for theozyme 1 showed any hydrolase activity, and none of the enzymes designed for theozyme 2 or 3 showed any activity at all. An identified problem is that the crystal structures of the designed proteins do not display the desired His-Cys dyad.

The last design we present here is for a catalyst of the Kemp elimination, shown in Reaction 9.8. A hapten for this reaction, utilizing acetate as the base and acetic

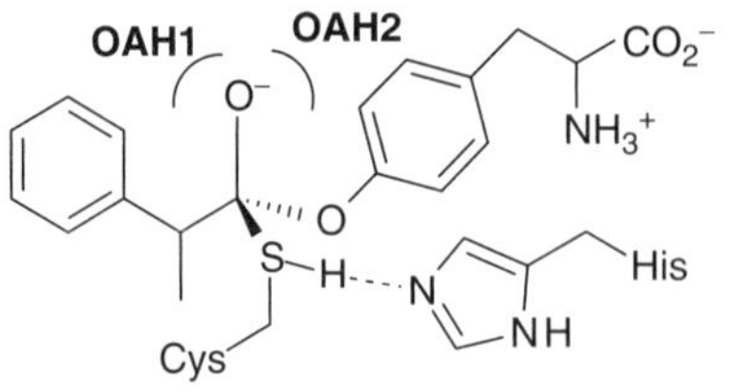

Theozyme 1: OAH1 = Backbone-NH
OAH2 = None

Theozyme 2: OAH1, OAH2 = Side chain H-Bond donor

Theozyme 3: OAH1 = Explicit water
OAH2 = Explicit water or side chain H-bond donor

Scheme 9.5

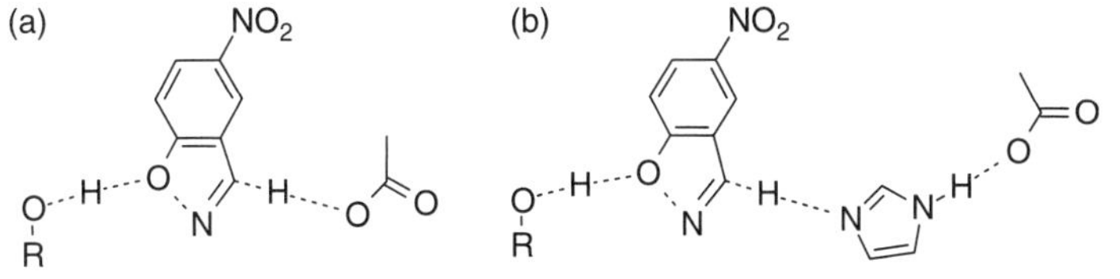

(a) and (b)

Scheme 9.6

acid as the proton source, was first developed by Houk and Hilvert[85] to design a catalytic antibody. For the theozyme component of the design of an enzyme to catalyze the Kemp elimination, Baker and Houk used the model shown in Scheme 9.6. They choose two different bases, either the carboxyl of the aspartate or glutamate chains (modeled here as acetate, Scheme 9.6a), or imidazole with a proximate aspartate (the His-Asp dyad, Scheme 9.6b). In addition, an aromatic side chain was placed above the phenyl ring in an idealized π-stacking arrangement.

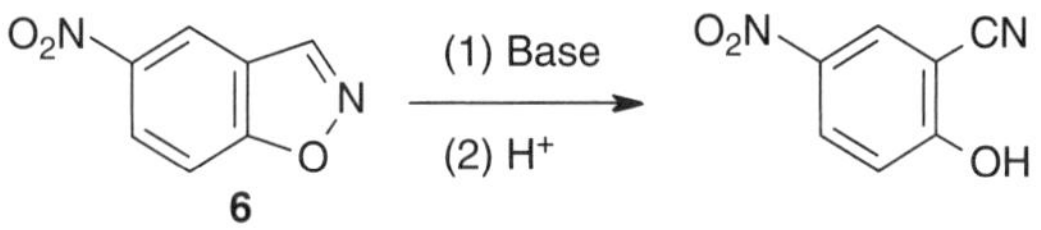

$$\text{6} \xrightarrow{\text{(1) Base}}_{\text{(2) H}^+} \text{product} \qquad \text{(Reaction 9.8)}$$

RosettaMatch was utilized to fit these two theozymes into known protein scaffolds. The hydrogen bond donor group could be Lys, Arg, Ser, Tyr, His, water, or not present at all. The π-stacking was provided by one of the three aromatic amino acids: Phe, Tyr, or Trp. After fitting into a protein scaffold and optimizing the structure, a total of 59 designs were obtained, 39 using model 9.6a and 20 using model 9.6b. Expression of all of these proteins and testing identified only eight designed proteins with catalytic activity. The two best showed a rate enhancement of 10^5, and replacement of the catalytic base with Ala or Gln/Ala essentially eliminated the activity. As with the other examples, the X-ray structure of the designed proteins matched very well with the computed structure. Nonetheless, as before, activity lagged that of natural enzymes. Directed evolution was able to improve the performance somewhat.

In a follow-up study, Baker and Jorgensen[86] performed QM/MM computations on some of the designed Kemp elimination enzymes. The QM region was computed using the PDDG/PM3 semiempirical method and the MM region was handled using the OPSL-AA force field. Water was modeled as TIP4P. As an example, for one of the original designs (KE07), only a single TS was located, indicating a concerted mechanism for the elimination. The computed free energy barrier of 8.1 kcal mol^{-1} is dramatically less than the computed barrier using hydroxide in water of 19.8 kcal mol^{-1}. The computations overestimate the stability of the TS but properly indicate some catalytic activity. Similar overestimation of the catalytic activity was found for three other designed proteins.

This type of strategy for enzyme design is not without its detractors. Warshel[87] reported a study of the KE07 enzyme designed by Baker and Houk for the Kemp elimination. He notes that the rate of the reaction of **6** with acetate extrapolated to 55 M gives an estimated free energy of activation of 21.2 kcal mol^{-1}. This estimate models the reaction of **6** with an acetate base and an acetic acid donor within a cage. Warshel notes that this is similar to the activation energy barrier with the designed enzyme (20.1 kcal mol^{-1}), suggesting that the main role that the designed enzyme is playing is simply to bring together the substrate and the base. Warshel also notes that the extended valence bond (EVB) method [88] provides a much better estimate of the activation barrier (20.2 kcal mol^{-1}) than does the QM/MM results described earlier (8.1 kcal mol^{-1}). Lastly, Warshel[89] argues that what is needed is a much better treatment of the electrostatic interactions and the reorganization effects than what is included within the Rosetta protocols.

A further concern is the role of dynamics in enzyme activity. For example, an MD study of the designed enzyme for the retroaldol (Reaction 9.2) observed that the active site as specified in the theozyme is only present in a small percentage of the protein conformations, potentially explaining its poor catalytic performance.[90] An MD study of the Diels–Alderase for Reaction 9.6 uncovered many trajectories with nonspecific binding in the active site.[91] Linder and coworkers[92] have developed an enzyme design strategy that incorporated MD and have proposed a Diels–Alderase that awaits experimental confirmation. On the other hand, Mulholland has suggested that the role of dynamics in enzyme catalysis has, in general, been overstated.[16]

Baker[93] notes that there are three key acts in designing an enzyme: creating the proper theozyme, properly fitting this theozyme into a protein structure, and having the remaining protein environment favorable for catalytic behavior. The failure of the designed proteins probably lies in all three steps to some extent. Better computational models, better force fields, and better algorithms for matching theozyme to protein are needed and remain a challenge for the future.

REFERENCES

1. Pauling, L. "Nature of forces between large molecules of biological interest," *Nature* **1948**, *161*, 707–709.
2. Amyes, T. L.; Richard, J. P. "Specificity in transition state binding: The Pauling model revisited," *Biochemistry* **2013**, *52*, 2021–2035.
3. Warshel, A.; Sharma, P. K.; Kato, M.; Xiang, Y.; Liu, H.; Olsson, M. H. M. "Electrostatic basis for enzyme catalysis," *Chem. Rev.* **2006**, *106*, 3210–3235.
4. Mobashery, S.; Kotra, L. P. In *Encylopedia of Life Sciences*; John Wiley & Sons, Ltd: Chichester, **2002**.
5. Zhang, X.; Houk, K. N. "Why enzymes are proficient catalysts: Beyond the Pauling paradigm," *Acc. Chem. Res.* **2005**, *38*, 379–385.
6. Borman, S. "Much ado about enzyme mechanisms," *Chem. Eng. News* **2004**, *82*, 35–39.
7. Simón, L.; Goodman, J. M. "Enzyme catalysis by hydrogen bonds: The balance between transition state binding and substrate binding in oxyanion holes," *J. Org. Chem.* **2009**, *75*, 1831–1840.

8. Kamerlin, S. C. L.; Chu, Z. T.; Warshel, A. "On catalytic preorganization in oxyanion holes: Highlighting the problems with the gas-phase modeling of oxyanion holes and illustrating the need for complete enzyme models," *J. Org. Chem.* **2010**, *75*, 6391–6401.
9. Simon, L.; Goodman, J. M. "Hydrogen-bond stabilization in oxyanion holes: Grand jete to three dimensions," *Org. Biomol. Chem.* **2012**, *10*, 1905–1913.
10. Warshel, A. "Electrostatic origin of the catalytic power of enzymes and the role of preorganized active sites," *J. Biol. Chem.* **1998**, *273*, 27035–27038.
11. Kohen, A.; Klinman, J. P. "Enzyme catalysis: Beyond classical paradigms," *Acc. Chem. Res.* **1998**, *31*, 397–404.
12. Benkovic, S. J.; Hammes-Schiffer, S. "A perspective on enzyme catalysis," *Science* **2003**, *301*, 1196–1202.
13. Klinman, J. P. "Linking protein structure and dynamics to catalysis: The role of hydrogen tunnelling," *Philos. Trans. R. Soc. London Ser. B* **2006**, *361*, 1323–1331.
14. Hammes-Schiffer, S.; Watney, J. B. "Hydride transfer catalysed by *Escherichia coli* and *Bacillus subtilis* dihydrofolate reductase: coupled motions and distal mutations," *Philos. Trans. R. Soc. London Ser. B* **2006**, *361*, 1365–1373.
15. Hammes-Schiffer, S. "Hydrogen tunneling and protein motion in enzyme reactions," *Acc. Chem. Res.* **2005**, *39*, 93–100.
16. Glowacki, D. R.; Harvey, J. N.; Mulholland, A. J. "Taking Ockham's razor to enzyme dynamics and catalysis," *Nat. Chem.* **2012**, *4*, 169–176.
17. Cleland, W. W.; Frey, P. A.; Gerlt, J. A. "The low barrier hydrogen bond in enzymatic catalysis," *J. Biol. Chem.* **1998**, *273*, 25529–25532.
18. Roy, A.; Zhang, Y. In *eLS*; John Wiley & Sons, Ltd: Chichester, **2001**.
19. Zhang, Y. "Progress and challenges in protein structure prediction," *Curr. Opin. Struct. Biol.* **2008**, *18*, 342–348.
20. Anfinsen, C. B. "Principles that govern the folding of protein chains," *Science* **1973**, *181*, 223–230.
21. Moult, J.; Fidelis, K.; Kryshtafovych, A.; Tramontano, A. "Critical assessment of methods of protein structure prediction (CASP)—round IX," *Proteins* **2011**, *79*, 1–5.
22. Lonsdale, R.; Harvey, J. N.; Mulholland, A. J. "A practical guide to modelling enzyme-catalysed reactions," *Chem. Soc. Rev.* **2012**, *41*, 3025–3038.
23. Salomon-Ferrer, R.; Case, D. A.; Walker, R. C. "An overview of the Amber biomolecular simulation package," *WIREs: Comput. Mol. Sci.* **2013**, *3*, 198–210.
24. MacKerell, A. D.; Bashford, D.; Bellott ; Dunbrack, R. L.; Evanseck, J. D.; Field, M. J.; Fischer, S.; Gao, J.; Guo, H.; Ha, S.; Joseph-McCarthy, D.; Kuchnir, L.; Kuczera, K.; Lau, F. T. K.; Mattos, C.; Michnick, S.; Ngo, T.; Nguyen, D. T.; Prodhom, B.; Reiher, W. E.; Roux, B.; Schlenkrich, M.; Smith, J. C.; Stote, R.; Straub, J.; Watanabe, M.; Wiórkiewicz-Kuczera, J.; Yin, D.; Karplus, M. "All-atom empirical potential for molecular modeling and dynamics studies of proteins," *J. Phys. Chem. B* **1998**, *102*, 3586–3616.
25. Brooks, B. R.; Brooks, C. L., III; Mackerell, A. D., Jr.; Nilsson, L.; Petrella, R. J.; Roux, B.; Won, Y.; Archontis, G.; Bartels, C.; Boresch, S.; Caflisch, A.; Caves, L.; Cui, Q.; Dinner, A. R.; Feig, M.; Fischer, S.; Gao, J.; Hodoscek, M.; Im, W.; Kuczera, K.; Lazaridis, T.; Ma, J.; Ovchinnikov, V.; Paci, E.; Pastor, R. W.; Post, C. B.; Pu, J. Z.; Schaefer, M.; Tidor, B.; Venable, R. M.; Woodcock, H. L.; Wu, X.; Yang, W.; York,

D. M.; Karplus, M. "CHARMM: The biomolecular simulation program," *J. Comput. Chem.* **2009**, *30*, 1545–1614.

26. Cramer, C. J. *Essentials of Computational Chemistry: Theories and Models*; John Wiley & Sons, Inc.: New York, **2002**.
27. Jensen, F. *Introduction to Computational Chemistry*; John Wiley & Sons, Ltd: Chichester, **1999**.
28. Ponder, J. W.; Case, D. A. In *Advances in Protein Chemistry*; Valerie, D., Ed.; Academic Press: New York, **2003**; Vol. *66*, p 27–85.
29. MacKerell, A. D., Jr. "Empirical force fields for biological macromolecules: Overview and issues," *J. Comput. Chem.* **2004**, *25*, 1584–1604.
30. Halgren, T. A.; Damm, W. "Polarizable force fields," *Curr. Opin. Struct. Biol.* **2001**, *11*, 236–242.
31. Jorgensen, W. L.; Chandrasekhar, J.; Madura, J. D.; Impey, R. W.; Klein, M. L. "Comparison of simple potential functions for simulating liquid water," *J. Chem. Phys.* **1983**, *79*, 926–935.
32. Eurenius, K. P.; Chatfield, D. C.; Brooks, B. R.; Hodoscek, M. "Enzyme mechanisms with hybrid quantum and molecular mechanical potentials. I. Theoretical considerations," *Int. J. Quant. Chem.* **1996**, *60*, 1189–1200.
33. Kollman, P. A.; Kuhn, B.; Peräkylä, M. "Computational studies of enzyme-catalyzed reactions: Where are we in predicting mechanisms and in understanding the nature of enzyme catalysis?," *J. Phys. Chem. B* **2002**, *106*, 1537–1542.
34. Torrie, G. M.; Valleau, J. P. "Nonphysical sampling distributions in Monte Carlo free-energy estimation: Umbrella sampling," *J. Comput. Phys.* **1977**, *23*, 187–199.
35. Mata, R. A.; Werner, H.-J.; Thiel, S.; Thiel, W. "Toward accurate barriers for enzymatic reactions: QM/MM case study on *p*-hydroxybenzoate hydroxylase," *J. Chem. Phys* **2008**, *128*, 025104–025108.
36. Schutz, M. "Low-order scaling local electron correlation methods. III. Linear scaling local perturbative triples correction (T)," *J. Chem. Phys.* **2000**, *113*, 9986–10001.
37. Werner, H.-J.; Manby, F. R.; Knowles, P. J. "Fast linear scaling second-order Moller-Plesset perturbation theory (MP2) using local and density fitting approximations," *J. Chem. Phys.* **2003**, *118*, 8149–8160.
38. Werner, H.-J.; Schuetz, M. "An efficient local coupled cluster method for accurate thermochemistry of large systems," *J. Chem. Phys.* **2011**, *135*, 144116/1–144116/15.
39. Van Berkel, W. J. H.; MÜller, F. "The temperature and pH dependence of some properties of *p*-hydroxybenzoate hydroxylase from *Pseudomonas fluorescens*," *Eur. J. Biochem.* **1989**, *179*, 307–314.
40. Senn, H. M.; Thiel, S.; Thiel, W. "Enzymatic hydroxylation in *p*-hydroxybenzoate hydroxylase: A case study for QM/MM molecular dynamics," *J. Chem. Theor. Comput.* **2005**, *1*, 494–505.
41. van der Kamp, M. W.; Żurek, J.; Manby, F. R.; Harvey, J. N.; Mulholland, A. J. "Testing high-level QM/MM methods for modeling enzyme reactions: Acetyl-CoA deprotonation in citrate synthase," *J. Chem. Phys. B* **2010**, *114*, 11303–11314.
42. Andrews, P. R.; Smith, G. D.; Young, I. G. "Transition-state stabilization and enzymic catalysis. Kinetic and molecular orbital studies of the rearrangement of chorismate to prephenate," *Biochemistry* **1973**, *12*, 3492–3498.

43. Kast, P.; Asif-Ullah, M.; Hilvert, D. "Is chorismate mutase a prototypic entropy trap? – Activation parameters for the *Bacillus subtilis* enzyme," *Tetrahedron Lett.* **1996**, *37*, 2691–2694.
44. Wiest, O.; Houk, K. N. "Stabilization of the transition state of the chorismate-prephenate rearrangement: An ab initio study of enzyme and antibody catalysis," *J. Am. Chem. Soc.* **1995**, *117*, 11628–11639.
45. Bruice, T. C.; Lightstone, F. C. "Ground state and transition state contributions to the rates of intramolecular and enzymatic reactions," *Acc. Chem. Res.* **1998**, *32*, 127–136.
46. Hur, S.; Bruice, T. C. "The near attack conformation approach to the study of the chorismate to prephenate reaction," *Proc. Nat. Acad. Sci. USA* **2003**, *100*, 12015–12020.
47. Hur, S.; Bruice, T. C. "Enzymes do what is expected (chalcone isomerase versus chorismate mutase)," *J. Am. Chem. Soc.* **2003**, *125*, 1472–1473.
48. Hur, S.; Bruice, T. C. "Just a near attack conformer for catalysis (chorismate to prephenate rearrangements in water, antibody, enzymes, and their mutants)," *J. Am. Chem. Soc.* **2003**, *125*, 10540–10542.
49. Kuczera, K. "One- and multidimensional conformational free energy simulations," *J. Comput. Chem.* **1996**, *17*, 1726–1749.
50. Guimarães, C. R. W.; Repasky, M. P.; Chandrasekhar, J.; Tirado-Rives, J.; Jorgensen, W. L. "Contributions of conformational compression and preferential transition state stabilization to the rate enhancement by chorismate mutase," *J. Am. Chem. Soc.* **2003**, *125*, 6892–6899.
51. Jorgensen, W. L.; Maxwell, D. S.; Tirado-Rives, J. "Development and testing of the OPLS all-atom force field on conformational energetics and properties of organic liquids," *J. Am. Chem. Soc.* **1996**, *118*, 11225–11236.
52. Ranaghan, K. E.; Mulholland, A. J. "Conformational effects in enzyme catalysis: QM/MM free energy calculation of the 'NAC' contribution in chorismate mutase," *Chem. Commun.* **2004**, *0*, 1238–1239.
53. Claeyssens, F.; Ranaghan, K. E.; Manby, F. R.; Harvey, J. N.; Mulholland, A. J. "Multiple high-level QM/MM reaction paths demonstrate transition-state stabilization in chorismate mutase: correlation of barrier height with transition-state stabilization," *Chem. Commun.* **2005**, *0*, 5068–5070.
54. Claeyssens, F.; Ranaghan, K. E.; Lawan, N.; Macrae, S. J.; Manby, F. R.; Harvey, J. N.; Mulholland, A. J. "Analysis of chorismate mutase catalysis by QM/MM modelling of enzyme-catalysed and uncatalysed reactions," *Org. Biomol. Chem.* **2011**, *9*, 1578–1590.
55. Ishida, T. "Effects of point mutation on enzymatic activity: Correlation between protein electronic structure and motion in chorismate mutase reaction," *J. Am. Chem. Soc.* **2010**, *132*, 7104–7118.
56. Kienhöfer, A.; Kast, P.; Hilvert, D. "Selective stabilization of the chorismate mutase transition state by a positively charged hydrogen bond donor," *J. Am. Chem. Soc.* **2003**, *125*, 3206–3207.
57. Cload, S. T.; Liu, D. R.; Pastor, R. M.; Schultz, P. G. "Mutagenesis study of active site residues in chorismate mutase from *Bacillus subtilis*," *J. Am. Chem. Soc.* **1996**, *118*, 1787–1788.
58. Männistö, P. T.; Kaakkola, S. "Catechol-O-methyltransferase (COMT): Biochemistry, molecular biology, pharmacology, and clinical efficacy of the new selective COMT inhibitors," *Pharmacol. Rev.* **1999**, *51*, 593–628.

59. Schluckebier, G.; O'Gara, M.; Saenger, W.; Cheng, X. "Universal catalytic domain structure of AdoMet-dependent methyltransferases," *J. Mol. Biol.* **1995**, *247*, 16–20.
60. Zheng, Y.-J.; Bruice, T. C. "A theoretical examination of the factors controlling the catalytic efficiency of a transmethylation enzyme: Catechol *O*-methyltransferase," *J. Am. Chem. Soc.* **1997**, *119*, 8137–8145.
61. Hegazi, M. F.; Borchardt, R. T.; Schowen, R. L. "α-Deuterium and carbon-13 isotope effects for methyl transfer catalyzed by catechol *O*-methyltransferase. S_N2-like transition state," *J. Am. Chem. Soc.* **1979**, *101*, 4359–4365.
62. Woodard, R. W.; Tsai, M. D.; Floss, H. G.; Crooks, P. A.; Coward, J. K. "Stereochemical course of the transmethylation catalyzed by catechol *O*-methyltransferase," *J. Biol. Chem.* **1980**, *255*, 9124–9127.
63. Vidgren, J.; Svensson, L. A.; Liljas, A. "Crystal structure of catechol *O*-methyltransferase," *Nature* **1994**, *368*, 354–358.
64. Mihel, I.; Knipe, J. O.; Coward, J. K.; Schowen, R. L. "α-Deuterium isotope effects and transition-state structure in an intramolecular model system for methyl-transfer enzymes," *J. Am. Chem. Soc.* **1979**, *101*, 4349–4351.
65. Lau, E. Y.; Bruice, T. C. "Importance of correlated motions in forming highly reactive near attack conformations in catechol *O*-methyltransferase," *J. Am. Chem. Soc.* **1998**, *120*, 12387–12394.
66. Kuhn, B.; Kollman, P. A. "QM–FE and molecular dynamics calculations on catechol O-methyltransferase: Free energy of activation in the enzyme and in aqueous solution and regioselectivity of the enzyme-catalyzed reaction," *J. Am. Chem. Soc.* **2000**, *122*, 2586–2596.
67. Schultz, E.; Nissinen, E. "Inhibition of rat liver and duodenum soluble catechol-O-methyltransferase by a tight-binding inhibitor OR-462," *Biochem. Pharmacol.* **1989**, *38*, 3953–3956.
68. Kollman, P. A.; Kuhn, B.; Donini, O.; Perakyla, M.; Stanton, R.; Bakowies, D. "Elucidating the nature of enzyme catalysis utilizing a new twist on an old methodology: Quantum mechanical-free energy calculations on chemical reactions in enzymes and in aqueous solution," *Acc. Chem. Res.* **2000**, *34*, 72–79.
69. Strajbl, M.; Sham, Y. Y.; Villà, J.; Chu, Z. T.; Warshel, A. "Calculations of activation entropies of chemical reactions in solution," *J. Phys. Chem. B* **2000**, *104*, 4578–4584.
70. Gilson, M. K.; Given, J. A.; Bush, B. L.; McCammon, J. A. "The statistical-thermdoynamic basis for computation of binding affinities: A critical review," *Biophysical J.* **1997**, *72*, 1047–1069.
71. Roca, M.; Martí, S.; Andrés, J.; Moliner, V.; Tuñón, I.; Bertrán, J.; Williams, I. H. "Theoretical modeling of enzyme catalytic power: Analysis of "cratic" and electrostatic factors in catechol O-methyltransferase," *J. Am. Chem. Soc.* **2003**, *125*, 7726–7737.
72. Roca, M.; Andrés, J.; Moliner, V.; Tuñón, I.; Bertrán, J. "On the nature of the transition state in catechol O-methyltransferase. A complementary study based on molecular dynamics and potential energy surface explorations," *J. Am. Chem. Soc.* **2005**, *127*, 10648–10655.
73. Kanaan, N.; Ruiz Pernia, J. J.; Williams, I. H. "QM/MM simulations for methyl transfer in solution and catalysed by COMT: Ensemble-averaging of kinetic isotope effects," *Chem. Commun.* **2008**, *0*, 6114–6116.
74. Gray, C. H.; Coward, J. K.; Schowen, K. B.; Schowen, R. L. "α-Deuterium and carbon-13 isotope effects for a simple, intermolecular sulfur-to-oxygen methyl-transfer

reaction. Transition-state structures and isotope effects in transmethylation and transalkylation," *J. Am. Chem. Soc.* **1979**, *101*, 4351–4358.
75. Kiss, G.; Çelebi-Ölçüm, N.; Moretti, R.; Baker, D.; Houk, K. N. "Computational enzyme design," *Angew. Chem. Int. Ed.* **2013**, *52*, 5700–5725.
76. Siegel, J. B.; Zanghellini, A.; Lovick, H. M.; Kiss, G.; Lambert, A. R.; St.Clair, J. L.; Gallaher, J. L.; Hilvert, D.; Gelb, M. H.; Stoddard, B. L.; Houk, K. N.; Michael, F. E.; Baker, D. "Computational design of an enzyme catalyst for a stereoselective bimolecular Diels-Alder reaction," *Science* **2010**, *329*, 309–313.
77. Tantillo, D. J.; Jiangang, C.; Houk, K. N. "Theozymes and compuzymes: Theoretical models for biological catalysis," *Curr. Opin. Chem. Biol.* **1998**, *2*, 743–750.
78. Hu, C.-H.; Brinck, T.; Hult, K. "Ab initio and density functional theory studies of the catalytic mechanism for ester hydrolysis in serine hydrolases," *Int. J. Quant. Chem.* **1998**, *69*, 89–103.
79. Zanghellini, A.; Jiang, L.; Wollacott, A. M.; Cheng, G.; Meiler, J.; Althoff, E. A.; Röthlisberger, D.; Baker, D. "New algorithms and an in silico benchmark for computational enzyme design," *Protein Sci.* **2006**, *15*, 2785–2794.
80. Richter, F.; Leaver-Fay, A.; Khare, S. D.; Bjelic, S.; Baker, D. "De novo enzyme design using Rosetta3," *PLoS ONE* **2011**, *6*, e19230.
81. Kuhlman, B.; Baker, D. "Native protein sequences are close to optimal for their structures," *Proc. Nat. Acad. Sci. USA* **2000**, *97*, 10383–10388.
82. Jiang, L.; Althoff, E. A.; Clemente, F. R.; Doyle, L.; Röthlisberger, D.; Zanghellini, A.; Gallaher, J. L.; Betker, J. L.; Tanaka, F.; Barbas, C. F.; Hilvert, D.; Houk, K. N.; Stoddard, B. L.; Baker, D. "De novo computational design of retro-aldol enzymes," *Science* **2008**, *319*, 1387–1391.
83. Althoff, E. A.; Wang, L.; Jiang, L.; Giger, L.; Lassila, J. K.; Wang, Z.; Smith, M.; Hari, S.; Kast, P.; Herschlag, D.; Hilvert, D.; Baker, D. "Robust design and optimization of retroaldol enzymes," *Protein Sci.* **2012**, *21*, 717–726.
84. Richter, F.; Blomberg, R.; Khare, S. D.; Kiss, G.; Kuzin, A. P.; Smith, A. J. T.; Gallaher, J.; Pianowski, Z.; Helgeson, R. C.; Grjasnow, A.; Xiao, R.; Seetharaman, J.; Su, M.; Vorobiev, S.; Lew, S.; Forouhar, F.; Kornhaber, G. J.; Hunt, J. F.; Montelione, G. T.; Tong, L.; Houk, K. N.; Hilvert, D.; Baker, D. "Computational design of catalytic dyads and oxyanion holes for ester hydrolysis," *J. Am. Chem. Soc.* **2012**, *134*, 16197–16206.
85. Na, J.; Houk, K. N.; Hilvert, D. "Transition state of the base-promoted ring-opening of isoxazoles. Theoretical prediction of catalytic functionalities and design of haptens for antibody production," *J. Am. Chem. Soc.* **1996**, *118*, 6462–6471.
86. Alexandrova, A. N.; Röthlisberger, D.; Baker, D.; Jorgensen, W. L. "Catalytic mechanism and performance of computationally designed enzymes for kemp elimination," *J. Am. Chem. Soc.* **2008**, *130*, 15907–15915.
87. Frushicheva, M. P.; Cao, J.; Chu, Z. T.; Warshel, A. "Exploring challenges in rational enzyme design by simulating the catalysis in artificial kemp eliminase," *Proc. Nat. Acad. Sci. USA* **2010**, *107*, 16869–16874.
88. Warshel, A.; Weiss, R. M. "An empirical valence bond approach for comparing reactions in solutions and in enzymes," *J. Am. Chem. Soc.* **1980**, *102*, 6218–6226.
89. Roca, M.; Vardi-Kilshtain, A.; Warshel, A. "Toward accurate screening in computer-aided enzyme design," *Biochemistry* **2009**, *48*, 3046–3056.

90. Ruscio, J. Z.; Kohn, J. E.; Ball, K. A.; Head-Gordon, T. "The influence of protein dynamics on the success of computational enzyme design," *J. Am. Chem. Soc.* **2009**, *131*, 14111–14115.

91. Linder, M.; Johansson, A. J.; Olsson, T. S. G.; Liebeschuetz, J.; Brinck, T. "Designing a new Diels–Alderase: A combinatorial, semirational approach including dynamic optimization," *J. Chem. Inf. Mod.* **2011**, *51*, 1906–1917.

92. Linder, M.; Johansson, A. J.; Olsson, T. S. G.; Liebeschuetz, J.; Brinck, T. "Computational design of a Diels–Alderase from a thermophilic esterase: The importance of dynamics," *J. Comput. Aid. Mol. Des.* **2012**, *26*, 1079–1095.

93. Baker, D. "An exciting but challenging road ahead for computational enzyme design," *Protein Sci.* **2010**, *19*, 1817–1819.

INDEX

(1*S*,4*S*)-norbornenone
  OR, 85
(3*Z*)-3-hexene-1,5-diyne, 235, 237, 239
  MOs, 238
[1,5]-H migration, 231
1,2,3-Propanetriol, 458–460
  Conformational energy, 459
1,2,6-heptatriene
  Cope rearrangement, 534–536
1,2-Addition to carbonyls, 392–404
  Cyclohexanone with lithium hydride, geometries, 398
  Propanal with sodium hydride, geometries, 396
1,2-Ethanediol, 453–458
  Gas-phase conformational energy, 455
  Gas-phase conformational geometry, 454
  Microsolvated geometry, 457
1,3-dipolar cycloaddition, 214
1,3-dipolar cycloadditions, 529
1-Adamantyl cation
  NMR, 81
2,2,3,3-Tetramethylbutane, 126
2,3-diazabiclco[2.2.1]hept-2-ene
2,3-hexadiene
  ORD, 88
2-Butyne
  DPE, 111
3-Ethynylpenta-1,4-diyne
  DPE, 111
4-methyl-2,6-hexadione
  Hajos-Parrish-Wiechert- Eder-Sauer reaction, 417–419
4-Triangulane
  OR, 84
5-oxo-2,4-pentadienal, 261
  MOs, 262

Absolute magnetic shielding, 151
Acetaldehyde
  Photodissociation of, 552
Acetone radical cation, 533, 534
Acidity
  Importance of solvation, 119
Activation strain. *See* Distortion energy
Adamantane
  NICS, 152
Adenine, 475–479, 483–489
  Gas-phase tautomer geometry, 477
  Microsolvated tautomer geometry, 479
  Tautomer energy, 478
Adiabatic mapping, 575, 577
Adrenaline, 582
Alanine, 121, 122, 123
  conformer relative energies, 122
Aldol reaction
  Amine-catalyzed aldol reactions, 404–429
  water catalyzed, 426–429
  Houk-List Model, 414–416
Alkanes
  Conformational isomerism, 119
Alkyl radical stability, 102
Allen, Wesley 22, 123
Allinger, Norman, 61
Amino acids, 489–492
  Acidity of, 116
  Conformations, 121
  Kinetic acidity, 117
ammonia, 45
  electron density, 46
Anh, Nguyen, 391–393, 395–397, 400–405
Annulenes, 155, 156, 157, 158, 159, 160, 161, 162, 163, 164, 165, 166
  Higher order twisted-, 165
  Relative energies, 156
  Ring current of, 160

Computational Organic Chemistry, Second Edition. Steven M. Bachrach
© 2014 John Wiley & Sons, Inc. Published 2014 by John Wiley & Sons, Inc.

Annulenes (*Continued*)
  X-ray crystal structures of, 160
  X-ray prediction, 161
Anthracene, 144, 176
  NICS, 152, 153
Antiaromatic compounds, 151, 153, 154, 155, 164, 170, 171
Aplysqualenol A
  NMR, 73
Aromatic compounds, 144, 145, 146, 147, 150, 154, 155, 171, 173
  ^{1}H NMR chemical shifts, 150
Aromatic stabilization energy (ASE), 145, 146, 147, 148, 149, 150, 153, 154, 160, 162, 166, 172
Aromaticity, 144–177
  Versus strain, 171
Artarborol
  NMR, 73
ASE see Aromatic stabilization energy (ASE)
Aspartate, 589, 591
Asymmetric induction
  1,2-addition, 392–404
  1,2-addition, Anh Model, 391, 395–397, 399–405
  1,2-addition, Cieplak Model, 395, 396, 399, 400
  1,2-addition, Cram Model, 391, 392, 402, 403
  1,2-addition, Cram chelation Model, 397
  1,2-addition, Felkin Model, 391–397, 401–404
  1,2-addition, Karabastos Model, 391–392
atomic orbitals (AO), 4
atoms-in-molecules (AIM), 47

B3LYP, 13, 25, 28
B3LYP-D, 27, 28
Bachrach, Steven, 108, 110, 111, 491
Bader, Richard 139
Baeyer strain, 139, 140, 142
Baker, David, 435, 586, 588, 590–592
Baldridge, Kim, 160, 169, 170
Baldwin, John, 520, 523, 524, 527
Barbas, Carlos, 405, 408, 425
Basis set superposition energy (BSSE) 12, 120, 121, 203–205, 240, 462
  intramolecular, 12
Basis Sets, 8–11
  Customized for NMR, 71
  Dunning correlation consistent basis sets, 11
BCP. *See* Bond critical point
BDE. *See* Bond dissociation enthalpy (BDE)
Becke, Axel, 25, 27
Benson, Sidney, 133, 135
Benzene dimer, 173–177
  Binding energy of, 173, 174
  Gas-phase experiments, 173
  substituted, 174
Benzene, 45, 143–155, 163, 166–170, 176, 177
  ASE, 145, 146
  1,3,5-Trisubstituted, 175
  bond distances, 63
  DPE, 106
  Hexasubstituted, 175
  Electron density, 46
  NICS, 153, 154
  Nonplanar structure, 145
  Stacked analogs, 176
m-Benzyne, 333, 336, 339, 341–349
  Energy, 340
  Geometry, 338
  MOs, 335
  Singlet-triplet energy gap, 345–349
  Structure, 341–345
o-Benzyne, 333, 338
  Energy, 340
  Geometry, 338
  MOs, 335
  Singlet-triplet energy gap, 345–349
p-benzyne, 235, 237–239, 243, 244, 246, 333, 336, 339, 341, 345, 346
  MOs, 238
  Energy, 340
  Geometry, 338
  MOs, 335
  Singlet-triplet energy gap, 345–347
Bergman Cyclization, 233–249
  cd Criteria, 236, 237, 244–248
  Hex-3-en-1,5-diyne to *p*-benzyne, energies, 240
  Hex-3-en-1,5-diyne to *p*-benzyne, geometry, 242
Bickelhaupt, Matthias, 383, 486
bicyclo[2.2.1]hex-2-ene, 514–517
bicyclo[3.1.0]hex-2-ene, 524–526
bicyclo[3.1.0]hexa-1,3,5-triene, 336
bicyclo[3.2.0]hex-2-ene, 514–517
Bifurcating surfaces, 539, 541, 543, 547–550
Birney, David, 261, 264, 265, 267
Bispericyclic reactions, 256–260, 548, 549
  Cyclopentadiene dimerization, 256
  Cyclopentadiene dimerization, geometry, 258
Boltzmann population, 68, 88
Bond critical point (BCP), 48, 141
Bond dissociation enthalpy (BDE), 100–103, 109, 110, 132
  C-C bonds, 102
  C-H bonds, 101
  composite methods, 101–103
  double hybrid methods, 101

R-X bonds, 102
  treatment of electron correlation, 101
Bond length
  Computed, 61–62
  R–X, 102, 103
Bond path, 139
Borden, Weston, 142, 170, 228, 229, 230, 311, 315, 318, 320, 324, 325, 328, 330, 331, 332, 337, 528, 529, 535, 536
  Interview, 278
Born–Oppenheimer Approximation, 3, 512
Brauman, John, 374, 383
Breslow, Ronald, 446, 447
Brillouin's theorem, 15, 17
Brueckner orbitals, 18, 242
Bruice, Thomas, 579–584
BSSE. *See* Basis set superposition error
Butane
  Confomational energies, 121

Caldera, 518–520, 523–526, 535, 536, 558
Calichaemicin, 233, 234, 348
  mechanism of action, 235
Canonical orbitals, 6
Carboxylic acids
  Acidity of, 113–117
Carpenter, 254
Carpenter, Barry, 254, 514–518, 528, 529, 531, 532, 534, 535, 536, 555, 556
CASPT2, 20
CASSCF. *See* Complete active space
Castro, 158, 163, 164
Catechol-O-methyltransferase (COMT), 582, 583, 584, 585
  X-ray crystal structure of, 583
CC. *See* Coupled cluster theory
Centauric model, 227
(+)-chaloxone
  OR, 85
Chameleonic model, 227
Chorismate mutase (CM), 578, 579, 580, 581, 582, 583
  from Bacillus subtilis, 580
  from Escherichia coli, 580
Chorismate, 578, 579, 580, 581, 583
CI. *See* Configuration interaction
Cieplak, Andrzej, 395, 396, 399, 400
cis-2-butene
  Reaction with dichloroketene, 547
Citrase synthase, 577
Claisen rearrangement, 578, 579
  Organocatalysis of, 429–431
CM. *See* Chorismate mutase
Complete active space (CASSCF), 20

Composite methods, 20–22
COMT. *See* Catechol-O-methyltransferase
Configuration Interaction (CI), 14–16
Conicasterol F
  NMR, 81
Continuous diradical transition state, 518
Conventional ring strain energy (CRSE), 133, 135, 136, 137, 143
Cope rearrangement, 209, 215–233, 257–260, 280, 548, 549
  1,2,6-heptatriene to 3-methylhexa-1,5-diene, 534–536
  1,5-hexadiene, energies, 220
  1,5-hexadiene, geometry, 224
  active space MOs, 218
  substituent effect, 225, 227
Corannulene
  Aromatic stabilization energy of, 149
Counterpoise, 12, 13
Coupled-cluster (CC) theory, 17, 18
  CCSD, 18
  CCSD(T), 18
  multireference coupled cluster theory (MRCC), 20
Coupling constants, 76–77, 79, 81
Cram, Donald, 391
Cramer, Christopher 2, 23, 25, 32, 33, 35, 242, 244, 254, 383, 446, 455, 457, 461, 464, 466, 468
  Interview, 493
Cremer, Dieter, 139, 140, 142, 143, 202, 241, 243, 244, 246, 336, 339, 341, 344, 347, 348
Critical assessment of methods of protein structure prediction (CASP), 574
critical points, 3, 41, 42, 44, 47, 48
CRSE. *See* Conventional ring strain energy (CRSE)
Cubane
  DPE, 108
Cyclobutadiene, 18, 155, 170
  MO diagram of, 19
  NICS, 152, 153
Cyclobutane
  Baeyer strain energies, 139
  bond energies, 141, 142
  NICS, 143
  RSE, 132, 134, 138–140, 143
Cyclobutene
  RSE, 134
Cyclohexane, 141–143
  DPE, 107
  NICS, 152, 153
Cyclohexanone
  1,2-Addition with lithium hydride, 398

Cyclohexanone (*Continued*)
  Aldol reaction, with, 413
Cyclohexene
  Rearrangement to vinylcyclobutane, 523
  trans-Cyclohexene, 202
Cyclooctatetraene
  NICS, 154
Cyclopentadiene
  Cycloaddition with ketene, 543–547
  DPE, 107
  NICS, 153
  NMR, 151
  RSE, 148
cyclopentadienyl anion
  NICS, 153
cyclopentane, 143
Cyclopentene
  Rearrangement to vinylcyclopropane, 520–523
Cyclopropane
  Baeyer strain energies, 139
  Bond energies, 141, 142
  DPE, 107, 108
  Ring current, 144
  Ring strain energy (RSE), 132, 133, 138, 139, 140, 143
  Stereomutation, 526–530
Cyclopropylhydroxycarbene
  Tunneling, 351, 354
Cyranski, Michal, 148, 172
Cysteine, 117, 118, 119, 123
  Conformational energy, 124
  conjugate base geometry, 118
Cytosine, 469–473, 476, 479, 483–489
  gas-phase tautomer geometry, 470
  Microsolvated tautomer geometry, 471
  Tautomer energy, 472

Davidson correction, 16, 329, 334
Davidson-Borden rules, 324–325
Davidson, Ernest, 324, 325, 521
De novo enzyme design, 586–592
  Bimolecular Diels–Alder reaction, 586
  Ester hydrolases, 590
  Failure of the designed proteins, 592
  Intermolecular Diels–Alder reaction, 586
  Kemp elimination, 590, 591
  Retroaldol, 588, 589
Density functional theory (DFT), 22–28, 101, 102, 106, 119–132
  Customized for NMR, 71
  Errors, 123–131
  Exchange-correlation functional, 23–26
Density matrix, 5, 6, 47

Deoxyribonucleic acids (DNA), 468, 469, 479, 486, 488
Deprotonation enthalpy (DPE), 100, 101, 104–113, 115–119
  Acetic acid, 105
  Acetone, 105
  2-Butyne, 111
  Amino acids, 117
  Cyclopentadiene, 107
  Ethane, 108
  Ethene, 106
  Ethyne, 107
  Methanol 113
  Pent-1,4-diyne, 111
  Propane, 106
  Propene, 105
  Propyne, 111
  Toluene, 107
  Tyrosine, 118
Dewar resonance energy (DRE), 147
Dewar, Michael, 143, 147, 209, 219, 279
Diamagnetic susceptibility exaltation, 150, 166
Diatropic ring currents, 144, 151, 154
Dichloroketene
  Cycloaddition with cyclopentadiene, 544
  reaction with *cis*-2-butene, 547
Dichloromethylene
  singlet-triplet energy gap, 303–304
Diels–Alder Reaction, 198–215, 232, 256, 258, 260, 261, 264, 269
  1,3-butadiene with ethylene, concerted, 199–206
  1,3-butadiene with ethylene, nonconcerted, 207–209
  Aqueous, 446–452
  Butadiene with acrolein, aqueous, 452
  Butenone with cyclopentadiene, aqueous, 448–449
  Cycloenones with cyclopentadiene, 215
  Cyclopentadiene dimerization, in solution, 446
  Fullerene with 1,3-butadiene, 206
  Kinetic isotope effect, 209–214
  Magnetic properties of TS, 205
  Reaction dynamics, 547–550
Diffuse functions, 11, 21, 106, 151
Dirac, Paul, 2, 24
Direct dynamics, 510–512, 514, 515, 525, 534, 535, 537, 541, 545, 548, 552, 556
Dispersion, 120, 129
  Chemical consequences, 131, 132
Dispersion correction, 27, 119, 130, 174
Distortion energy, 214, 215
DNA methyltransferases, 582
DNA. *See* Deoxyribonucleic acids

Doering, William, 207, 216, 217, 227, 228, 231, 279, 280
Dopamine, 582
Double hybrid functional, 26, 27, 49
 dispersion corrected, spin-component-scaled (DSD-DFT), 28
Doubleday, 525, 527–529
DP4, 75, 91
Dunitz–Shomaker strain, 140, 142
Dynamic matching, 516, 519, 529, 549, 550

ECD. See Electronic circular dichroism
Electron correlation, 8, 11, 13–22, 23
Electronic circular dichroism (ECD), 82
Electrostatic embedding, 38
Enforced hydrophobic interaction, 448
Epoxydon
 OR, 85
Esperamicin, 233, 234
Ethane
 DPE, 108
Ethanol
 Atomic charges, 114
Ethylene glycol. See 1,2-Ethanediol
Ethene
 DPE, 106
Ethyne
 DPE, 107
Exchange-correlation functional, 23–26

Felkin, Hugh, 391–397, 401–404
Flowing afterglow spectroscopy, 107, 109
Fock matrix, 5, 6, 8, 9
Formaldehyde
Formic acid
 Atomic charges, 114
 Deprotonation of, 114
Frenking, Gernot, 396–399
Furan
 ASE, 148
 NMR, 151
 RSE, 148

Gajewski, Joseph, 521
Gao, Jiali, 449
Gas-phase acidity, 104, 117
Gauge-including atomic orbitals (GIAO), 68
Gaussian functions, 9, 10
Gaussian-type orbital (GTO). See Gaussian function
Generalized Born Approximation (GB), 31
Geometry optimization, 42–45
GIAO. See Gauge-including atomic orbitals
Glucose, 452, 460–468, 493

Gas- and solution-phase population, 465
Gas-phase energy, 463
Gas-phase geometry, 462
Microsolvation geometry, 467, 468
Glycerol. See 1,2,3-Propanetriol
Glycine, 489–492
 IR frequencies, 68
 Microsolvated energy, 491
 structure of, 67
Goddard, William, 300, 301, 356
Goodman, 74, 75, 570, 571, 572, 573
 Interview, 90
Gordon, Mark, 137, 490, 491
Grimme, Stefan, 13, 17, 26, 27, 84, 125, 128, 129, 130, 131, 141, 164, 174, 176, 359
 Interview, 48
Gronert, Scott, 381
Group equivalent reaction (GER), 134, 136, 137, 138, 141, 142
Group equivalents (GE), 133, 135, 136, 137, 141
Guanine, 473–475, 483–489
 Gas-phase tautomer geometry, 474
 Microsolvated tautomer geometry, 476
 Tautomer energy, 475
GX model, 21

Hajos-Parrish-Wiechert-Eder-Sauer reaction, 405, 408, 417–420, 433
Hammond Postulate, 379
Hartree–Fock (HF), 3–7, 23
Hase, William, 528, 529, 537, 538, 553
Herges, Rainer, 162, 163, 164, 165, 166
Herzberg, Gerhard, 299, 355
Heterolytic cleavage, 99, 104
Hexacyclinol
 NMR, 78
Hexane
 Conformational energies, 119, 120, 121
Hexaphenylethane, 131
HF. See Hartree-Fock
Hilltop, 42
Hobza, Pavel, 174, 469, 470, 473–477, 479, 485
Hoffmann, Roald, 527
Hohenberg–Kohn existence theorem, 22
Homodesmotic reaction, 134, 135, 136, 137, 147, 148, 149
Homolytic cleavage, 99, 100
Houk -List model
 Aldol reaction, 414–417
Houk, Kendall, 128, 175, 176, 199, 206, 207, 209, 210, 214, 228, 229, 258, 259, 267–277, 280, 383, 394–396, 398, 399, 404, 408, 410, 412–418, 420, 421, 425, 427, 428, 521, 523,

Houk, Kendall (*Continued*)
 525, 527, 529, 550, 570, 578, 580, 586, 588, 590, 591
 Interview, 432
Hückel topology, 161, 162
Hydroboration, 554–555
Hydrophobic effect, 447–450, 452
Hydroxymethylene
 Tunneling, 349–350

Implicit Solvent Models, 29–34
Individual gauge for localized orbitals (IGLO), 68
Infrared spectroscopy, 62–67
Integrating differential equations
 Euler's method, 509
 Runge–Kutta method, 509
 Verlet algorithm, 509
Interview
 Chistopher Cramer, 493
 Daniel Singleton, 559
 Henry Schaefer, 355
 Jonathan Goodman, 90
 Kendall Houk, 432
 Paul von Rague Schleyer, 177
 Peter Schreiner, 357
 Stefan Grimme, 48
 Weston Thatcher Borden, 278
Intramolecular vibrational energy redistribution (IVR), 513, 515, 519, 533, 536, 542
Intrinsic reaction coordinate (IRC), 41, 505, 537, 538, 548, 549, 550, 551
ISE. *See* Isomerization stabilization energy (ISE)
Isodesmic reaction, 114, 133, 134, 135, 136, 137, 140, 142, 146, 147
Isomerization stabilization energy (ISE), 149, 162, 163
Isotropic shielding, 154
IVR. *See* Intramolecular vibrational energy redistribution

Jacob's Ladder, 24
Jacobsen, Eric, 429–431
Jensen, Frank, 381, 387–389
Jorgensen, William, 375, 376, 386, 448–450, 580, 581, 591

Karabatsos, Gerasimos , 391
Karney, William, 158, 163, 164, 318, 320
Kass, Steven, 106, 108, 109, 110, 116, 117, 118
ketenecycloadditon with cyclopentadiene, 543–547
Kinetic isotope effect (KIE), 420, 521, 540, 544, 545, 546, 547, 559
 Diels-Alder reaction, 209–214

Kirkwood–Onsager model, 31
Kollman, Peter 584, 585
Kraka, Elfi, 202, 241, 242, 243, 246, 336, 344, 348

laser ablation molecular beam Fourier transform microwave spectroscopy, 121, 123
Lazzeretti, Paolo, 151, 154
Linear Combination of Atomic Orbitals (LCAO) Approximation, 4–5
Linear synchronous transit (LST), 44
Lineberger, W. Carl, 301, 303, 304, 329, 330, 332, 355, 356
List, Benjamin, 405, 408, 414–417, 421, 423, 425, 433
local density approximation (LDA), 24
Lysine, 588, 589

Magnetic shielding tensor, 154
Maitotxin
 NMR, 79
Mannich reaction, 421–426
Masamune, S. 155, 156, 158, 159
MC. *See* Michaelis complex (MC)
MD. *See* Molecular dynamics
Mechanical embedding, 38
MEP. *See* Minimum energy path
Methanol
 DPE, 113
Methylene, 298–304, 305, 307, 308, 310, 326, 355, 356
 MOs, 299
 singlet-triplet energy gap, 301, 302
 triplet geometry, 299
Methylhydroxycarbene
 Tunneling, 352
 Tunneling control, 353
Methyloxirane
 OR, 86, 87
 ORD, 85
Michaelis complex (MC), 570, 571, 584, 585
Microcanonical sampling, 511, 513
Microsolvation, 28, 29, 34, 451, 455–458, 464, 465, 467, 470, 471, 474–476, 479, 482, 485, 491, 492
 1,2-Ethanediol, 457
 Adenine, 479
 Amino acids, 491
 Cytosine, 471
 Glycine, 491
 Guanine, 476
 nucleic acid base pairs, 486
 Thymine, 482
 Uracil, 482
Mills-Nixon effect, 145, 166, 167, 169, 170, 171

Minimum energy path (MEP), 41, 505–507, 541, 543, 545, 550, 557
Möbius topology, 161, 162, 164
Molecular dynamics (MD), 505, 507–512, 514, 516, 518, 521, 528, 529, 537, 540, 541, 544, 550, 553
Molecular Mechanics (MM), 36–38
Møller–Plesset (MP), 17
Momany, 461, 464, 466–468
Morokuma, Keiji 39, 331
Mulholland, Adrian, 577, 578, 581, 582, 592
Mulliken population, 46, 47
Multiconfiguration SCF (MCSCF), 18–20
Myers–Saito cyclization, 249–256
    (4Z)-1,2,4-heptatrien-6-yne, energies, 253
    (4Z)-1,2,4-heptatrien-6-yne, geometry, 252

NAC. *See* Near attack confomers
Naphthalene, 144, 160, 156, 157, 160
    NICS, 152, 153
Natural population analysis (NPA), 47, 114
Near attack confomers (NAC), 579, 580, 581, 582, 583, 584
Neocarzinostatin, 233, 254, 255
    mechanism of action, 250
Nicolaou, K. C. 360
NICS. *See* Nuclear-independent chemical shift (NICS)
NMR, 81
NMR. *See* Nuclear magnetic resonance
Nobilisitine A
    NMR, 75
Noradrenaline, 582
NPA. *See* Natural population analysis
Nuclear magnetic resonance (NMR), 66–82
Nuclear-independent chemical shift (NICS), 143, 150, 151, 152, 153, 154, 161, 162, 163, 164, 165, 166, 169, 170, 172
Nucleic acid base pairs, 479–489
    Geometry, 483
    Microsolvated geometry, 486
    Stacking energy, 484

OAH. *See* Oxyanion hole
Octane, 125, 126, 130
ONIOM, 39–40
Optical rotation (OR), 82–90
Optical rotatory dispersion (ORD), 82–90
OR. *See* Optical rotation
Orbital symmetry rules. *See* Woodward-Hoffmann rules
ORD. *See* Optical rotatory dispersion
OXA. *See* oxyallyl diradical
Oxazolidinone, 416, 417

OxetaneRSE, 134, 135
oxyallyl diradical (OXA), 330–332
    MOs, 331
    Singlet-triplet energy gap, 332
Oxyanion hole (OAH), 570–573, 590

Paddon-Row, Michael, 394, 397, 399, 400
Paracylophane, 90
Para-hydroxybenzoate hydrolase, 576
Pauling paradigm, 570, 573
Pauling, Linus, 104, 147, 166, 569, 570
PBE0, 25
PBHB. *See* para-Hydrobenzoate hydrolase
PCM. *See* Polarized continuum model
Pent-1,4-diyne
    DPE, 111
Perdew, John, 24, 25
Perturbation theory, 16–17
    MP2, 17
Petersson, George, 22
Phenanthene, 144
Phenylcarbene, 304–317, 322
    MOs, 306
    Ring expansion, 312–317
    Ring expansion, energies, 314
    singlet-triplet energy gap, 307
Phenylhydroxycarbene
    Tunneling, 350–351
Phenylnitrene, 304–324, 326
    Bond lengths, 310
    MOs, 306
    Ring expansion, 312–326
    Ring expansion, energies, 315
    Ring expansion, geometries, 316
    Ring expansion, substituent effect, 317–326
    Singlet-triplet energy gap, 310
    UHF MOs, 309
Phenyloxenium cation, 312
    MOs, 306
Photodissociation of, 551
$\pi-\pi$ stacking, 173, 176, 177
Pinacolyl alcohol, Protonated
    Rearrangement, dynamics, 550–551
Pitzer strain, 139, 140, 142
Platz, Matthew, 317–322, 365
Plumericin
    ECD, 87
Poisson equation, 30, 31, 32
Polarization functions, 11, 21
Polarized continuum model (PCM), 31, 33, 34, 583
Pople, John, 9, 11, 21, 64, 154, 355, 435
Population analysis, 45–48

Potential energy surface (PES)
  Chloride with methylchloride, 376
  Fluoride with methylchloride, 380
Preorganization, 573, 584
Prephenate, 578, 579
Prismatomerin
  OR, 87
Propane, 126, 127
  bond energies, 141
  DPE, 106
Propyne
  DPE, 111
Protein structure relaxation, 572
Pseudopericyclic reactions, 260–267
  Formylketene with ammonia, 264
  Wolff reaction, 556
PW91, 25
Pyran-2-one
  MOs, 262
Pyrenophanes, 172
Pyrrole
  NMR, 151
  Stabilization energy, 144

QM/MM, 35–40
Quadratic synchronous transit (QST), 44
quadrupole ion trap, 117

Reference configuration, 13
Restricted wavefunction, 7
Retroaldol, 590
Rice-Ramsperger-Kassel-Marcus Theory
  (RRKM), 505, 507, 513, 514, 516, 517, 533, 547, 554, 555
Ring current map, 144
Ring current, 144, 151, 153, 154, 160
Ring strain energy (RSE), 99, 112, 132, 133, 134, 135, 136, 137, 138, 139, 140, 142, 143, 146, 148
  evaluated using Group equivalent reactions, 137
  Evaluated using isodesmic reactions, 137
  Evaluated using homodesmotic reactions, 137
Roaming mechanism, 551–553
  Acetaldehyde photodissociation, 552
  Formaldehyde photodissociation, 551
  Nitrobenzene photodissociation, 553
Roothaan Method, 5–7, 23
RosettaMatch algorithm, 587, 589, 590, 591
Roundabout mechanism, 553–554
RSE. *See* Ring strain energy
Rzepa, Henry 90

Schaefer, Henry, 22, 61, 145, 151, 156, 157, 160, 161, 298–301, 303, 334

Interview, 355
Schlegel, Bernard 64
Schleyer, Paul, 106, 114, 126, 127, 141, 142, 143, 145, 147, 148, 149, 150, 151, 153, 154, 156, 160, 161, 162, 164, 357, 394
  Interview, 177
Schmittel cyclization, 249–256
  (4Z)-1,2,4-heptatrien-6-yne, energies, 253
  (4Z)-1,2,4-heptatrien-6-yne, geometry, 252
  dynamics, 540–543
Schreiner–Pascal cyclization, 256
Schreiner, Peter 125, 127, 132, 231, 242, 246–248, 254–256, 349, 350, 352–354
  Interview, 357
Schrödinger equation, 2, 3, 8, 13, 14
Secondary orbital interactions (SOIs), 258
Self-consistent field (SCF), 4
Self-consistent reaction field (SCRF), 30
Semiclassical dynamics, 512
Sherrill, C. David, 175
Shielding tensor, 154
Shikimic acid pathway, 578
SIBL. *See* Strain-induced bond localization
Siegel, Jay, 160, 169, 170
σ-aromaticity, 143, 144
Singleton, Daniel, 160, 169, 170, 540, 210, 254, 541, 543, 544, 547–549, 554–556
  Interview, 558
Size Consistency, 16
Slater determinant, 4, 14, 15
Slater-type orbitals (STO), 8
SN1 mechanism, 373, 374, 381
SN2 mechanism, 373–391
  α-Branching, 381
  α-Branching, energies, 382
  α-Branching, geometries, 383
  β-branching, 381
  Chloride with methylchloride, energy, 377
  Chloride with methylchloride, geometry, 375
  Chloride with methylchloride, PES, 376
  Chloride ion and methyliodide, dynamics 553–554
  Fluoride with methylchloride, PES, 380
  Fluoride with methylfluoride, energy, 378
  Fluoride with methylfluoride, geometry, 375
  Hydroxide with fluoromethane, dynamics, 536–537
  Microsolvation, 388
  Microsolvation, geometries, 389
  Non-identity reactions. energies, 380
  Non-identity reactions, geometries, 379
  Solvent effects, 385–391
SNF mechanism, 383, 384
  geometries, 384

Solvation model (SMx), 32–34
Specific reaction parameters (SRP), 510, 511, 521, 525, 528, 534, 535
Spin projection, 7
Spin-component scaled MP2 (SCS-MP2), 17, 28, 49
Sponer, Jiri, 479, 482, 484
Squires, Robert, 493–494
SRP. *See* Specific reaction parameters
Stanger, Amnon, 153, 167, 170, 171
STO. *See* Slater-type orbital
Strain-induced bond localization (SIBL), 166, 167, 169, 170, 171, 173, 175
Streitwieser, Andrew, 381
Subtilisin, 572
Sunoj, Raghavan, 416

Tetramethyleneethane (TME) 324–330
   MOs, 325, 327
Theozyme, 586–592
Thymine, 477–482
   Gas-phase tautomer geometry, 480
   Microsolvated tautomer geometry, 482
   Tautomer energy, 481
TME. *See* Tetramethyleneethane
TMM. *See* Trimethylenemethane
Toluene
   Aromatic stabilization energy of, 149
   DPE, 107
Tomasi, Jacopo, 446, 473
Topological electron density analysis, 47–48
Torquoselectivity, 267–278
   Electrocyclization of 1,3-cyclohexadienes, 276
   Electrocyclization of cyclobutenes, 269, 271
   Electrocyclization of cycloocta-1,3,5-trienes, 277
TPSS, 25
Transition state stabilization, 570
Transition state theory (TST), 505, 507, 513, 514, 517, 529, 533, 545–549, 554, 555
Tricyclopropabenzene, 166
Trimethylene diradical, 527–529
Trimethylenemethane (TMM), 324
   MOs, 325
Tropylium cation, 144
   NICS, 153
Truhlar, Donald, 32, 33, 35, 38, 64, 130, 383, 446, 455, 457, 461, 464, 466, 484, 492–495, 510

TST. *See* Transition state theory
Tunneling control, 353–355
Tunneling
   Carbenes, 349–355
Tyrosine, 118, 119
   DPE, 118

Umbrella sampling, 575, 577
Unrestricted wavefunction, 7
Uracil, 477–482
   Gas-phase tautomer geometry, 480
   Microsolvated tautomer geometry, 482
   Tautomer energy, 481

Valley–ridge inflection (VRI), 42, 257–259, 541
Vannusal B
   NMR, 80
Variational Principle, 8
Variational transition state theory (VTST), 514
Vibrational circular dichroism (VCD), 82–90
Vinylcyclobutane
   Rearrangement to cyclohexene, 523
Vinylcyclopropane
   Rearrangement to cyclopentene, 520–523
VRI. *See* Valley ridge inflection
VTST. *See* Variational transition state theory

Warshel, Arieh, 570, 572, 573, 584, 591, 592
Watson-Crick Base Pair, 469, 473, 474, 479, 482–487
   Geometry, 483
   Interaction energy, 484
ωB97X-D, 27
Wentzel–Kramers–Brillouin theory (WKB), 349
Wheeler, Steven, 135, 175, 176, 177
Wiberg, Kenneth 88, 106
Wiest, Olaf, 578, 580
WKB. *See* Wentzel–Kramers–Brillouin theory
Wolff rearrangement
   Dynamics, 555–557
Woodward-Hoffmann rules, 514, 516, 520, 521, 524, 528

Zero-point vibrational energy (ZPVE), 21, 45, 64, 123, 129, 149
ZPVE. *See* Zero point vibrational energy
Zwitterion, 472, 489, 490